Exploring Chemical Concepts Through Theory and Computation

Exploring Chemical Concepts Through Theory and Computation

Edited by Shubin Liu

WILEY-VCH

The Editor

Dr. Shubin Liu
University of North Carolina
211 Manning Drive
Chapel Hill, NC 27599
USA

Cover Image: © GarryKillian/ Shutterstock; Courtesy of Dr. Frédéric Guégan

All books published by **WILEY-VCH** are carefully produced. Nevertheless, authors, editors, and publisher do not warrant the information contained in these books, including this book, to be free of errors. Readers are advised to keep in mind that statements, data, illustrations, procedural details or other items may inadvertently be inaccurate.

Library of Congress Card No.: applied for

British Library Cataloguing-in-Publication Data
A catalogue record for this book is available from the British Library

Bibliographic information published by the Deutsche Nationalbibliothek The Deutsche Nationalbibliothek lists this publication in the Deutsche Nationalbibliografie; detailed bibliographic data are available on the Internet at <http://dnb.d-nb.de>.

© 2024 WILEY-VCH GmbH, Boschstraße 12, 69469 Weinheim, Germany

All rights reserved (including those of translation into other languages). No part of this book may be reproduced in any form – by photoprinting, microfilm, or any other means – nor transmitted or translated into a machine language without written permission from the publishers. Registered names, trademarks, etc. used in this book, even when not specifically marked as such, are not to be considered unprotected by law.

Print ISBN: 978-3-527-35248-7
ePDF ISBN: 978-3-527-84342-8
ePub ISBN: 978-3-527-84341-1
oBook ISBN: 978-3-527-84343-5

Typesetting   Straive, Chennai, India
Printing and Binding   CPI Group (UK) Ltd, Croydon, CR0 4YY

Contents

Preface *xv*
Foreword *xvii*
10 Questions About Exploring Chemical Concepts Through Theory and Computation *xix*

1	**Chemical Concepts from Molecular Orbital Theory** *1*	
	Feng Long Gu, Jincheng Yu, and Weitao Yang	
1.1	Introduction *1*	
1.2	Molecular Orbital Theory *2*	
1.3	Canonical Molecular Orbitals *5*	
1.4	Frontier Molecular Orbital Theory *5*	
1.5	Localized Molecular Orbitals *6*	
1.5.1	Orthogonal Localized Molecular Orbitals *7*	
1.6	Regularized Nonorthogonal Localized Molecular Orbitals *11*	
1.7	Molecular Orbitalets *15*	
	Acknowledgment *18*	
	References *18*	
2	**Chemical Concepts from *Ab Initio* Valence Bond Theory** *23*	
	Chen Zhou, Fuming Ying, and Wei Wu	
2.1	Introduction *23*	
2.2	Ab Initio Valence Bond Theory *24*	
2.2.1	Valence Bond Self-Consistent Field Method *24*	
2.2.2	Rumer Structures *26*	
2.2.3	Orbitals in VB Wave Function *29*	
2.2.4	VB Methods Involving Dynamic Correlation *30*	
2.3	Chemical Concepts in VB Theory *31*	
2.3.1	Resonance Theory *31*	
2.3.2	Conjugation, Hyperconjugation, and Aromaticity *32*	
2.3.3	Electron-Pair Bonding in Valence Bond Theory *33*	
2.3.4	Diabatic States in Valence Bond Theory *34*	
2.4	A Brief Guide to Perform VB Calculations *36*	
2.4.1	Preparing XMVB Input Files *36*

2.4.2	Reading XMVB Output Files	*38*
2.5	Concluding Remarks	*38*
	References	*38*

3   Chemical Concepts from Conceptual Density Functional Theory *43*
Frank De Proft

3.1	Introduction *43*	
3.2	The Fundamentals: Density Functional Theory (DFT) and Kohn–Sham DFT *46*	
3.3	The First Derivatives: The Electronic Chemical Potential and the Electron Density *48*	
3.4	The Second Derivatives: Chemical Hardness, Fukui Function, Linear Response Function, and Related Quantities *51*	
3.4.1	Chemical Hardness and Softness *51*	
3.4.2	The Fukui Function and the Dual Descriptor *54*	
3.4.3	Local Softness and Hardness *58*	
3.4.4	The Linear Response Function, Softness and Hardness Kernels *60*	
3.5	Perturbational Perspective of Chemical Reactivity *62*	
3.6	Conclusions *64*	
	Acknowledgment *64*	
	References *64*	

4   Chemical Concepts from Density-Based Approaches in Density Functional Theory *71*
Dongbo Zhao, Xin He, Chunying Rong, and Shubin Liu

4.1	Introduction *71*	
4.2	Four Density-Based Frameworks *72*	
4.2.1	Orbital-Free DFT (OF-DFT) *72*	
4.2.2	Conceptual DFT (CDFT) *75*	
4.2.3	Density-Associated Quantities (DAQs) *76*	
4.2.4	Information-Theoretic Approach (ITA) *77*	
4.3	Applications of Density-Based Approaches *79*	
4.3.1	Molecular Isomeric and Conformational Stability *79*	
4.3.2	Bonding and Noncovalent Interactions *79*	
4.3.3	Cooperation and Frustration *80*	
4.3.4	Homochirality and Principle of Chirality Hierarchy *82*	
4.3.5	Electrophilicity and Nucleophilicity *86*	
4.3.6	Regioselectivity and Stereoselectivity *86*	
4.3.7	Brønsted–Lowry Acidity and Basicity *90*	
4.3.8	Aromaticity and Antiaromaticity *90*	
4.3.9	Molecular Properties (Frontier Orbitals, HOMO/LUMO Gap, Oxidation States, Polarizability) *91*	

4.4	Concluding Remarks *94*
	Acknowledgments *95*
	References *95*

5	**Chemical Bonding** *101*
	Sudip Pan and Gernot Frenking
5.1	Introduction *101*
5.2	The Physical Mechanism of the Chemical Bond *103*
5.3	Bonding Models *108*
5.4	Bond Length and Bond Strength *111*
5.5	Dative and Electron-Sharing Bonds *120*
5.6	Polar Bonds *124*
5.7	Atomic Partial Charges and Atomic Electronegativity *130*
5.8	Chemical Bonding in Main-Group Compounds: N_2, CO, BF, LiF *131*
5.9	Chemical Bonding of the Heavier Main-Group Atoms *135*
5.10	Chemical Bonding in Transition Metal Complexes: $M(CO)_n$ (M = Ni, Fe, Cr, Ti, Ca; $n = 4 - 8$) *143*
5.11	Summary *146*
	Acknowledgments *147*
	References *147*

6	**Partial Charges** *161*
	Tian Lu and Qinxue Chen
6.1	Concept of Partial Charge *161*
6.1.1	What is Partial Charge? *161*
6.1.2	Theoretical Significances and Practical Applications of Partial Charge *162*
6.1.3	Limitations of Partial Charge *163*
6.1.4	What Is a Good Method of Calculating Partial Charges? *164*
6.1.5	Classification of Partial Charge Calculation Methods *165*
6.2	Methods of Calculating Partial Charges *166*
6.2.1	Partial Charges Based on Wavefunction *166*
6.2.1.1	Mulliken Method *166*
6.2.1.2	MMPA Methods *167*
6.2.1.3	Löwdin Method *168*
6.2.1.4	NPA Method *168*
6.2.2	Partial Charges Based on Real Space Partition of Electron Density *169*
6.2.2.1	AIM Method *169*
6.2.2.2	Voronoi and VDD Methods *170*
6.2.2.3	Hirshfeld Method *170*
6.2.2.4	Hirshfeld-I Method *171*
6.2.3	Partial Charges Based on Fitting Electrostatic Potential *172*

6.2.3.1	Common ESP Fitting Methods *172*	
6.2.3.2	RESP and Relevant Methods *173*	
6.2.4	Partial Charges Based on Equalization of Electronegativity *174*	
6.2.5	Partial Charges Based on Other Ideas *175*	
6.3	Partial Charges of Typical Molecules *176*	
6.4	Computer Codes for Evaluating Partial Charges *179*	
6.5	Concluding Remarks *180*	
	References *180*	

7 Atoms in Molecules *189*
Ángel Martín Pendás, Evelio Francisco, Julen Munárriz, and Aurora Costales
7.1 Introduction *189*
7.2 The Quantum Theory of Atoms in Molecules (QTAIM) *190*
7.3 QTAIM Atoms as Open Quantum Systems *194*
7.3.1 Sector Density Operators of Quantum Atoms in Molecules *195*
7.3.2 RDMs of Atoms in Molecules *197*
7.4 Interacting Quantum Atoms (IQA) *200*
References *203*

8 Effective Oxidation States Analysis *207*
Pedro Salvador
8.1 The Concept of Oxidation State *207*
8.2 Oxidation State is Not Related to the Partial Charge *208*
8.3 The Molecular Orbital Picture of the Ionic Approximation *210*
8.4 Spin-Resolved Effective Fragment Orbitals and Effective Oxidation States (EOS) Analysis *213*
8.5 EOS Analysis from Different AIM Schemes *216*
8.6 Summary *220*
References *220*

9 Aromaticity and Antiaromaticity *223*
Yago García-Rodeja and Miquel Solà
9.1 Definition of Aromaticity *223*
9.2 Physical Foundation *224*
9.3 Measures of Aromaticity *226*
9.3.1 Geometric Descriptors of Aromaticity *227*
9.3.2 Energetic Descriptors of Aromaticity *227*
9.3.3 Electronic Descriptors of Aromaticity *230*
9.3.4 Magnetic Descriptors of Aromaticity *232*
9.4 Rules of Aromaticity *233*
9.4.1 Rules for Two-Dimensional Aromaticity *234*
9.4.2 Rules for Three-Dimensional Aromaticity *237*
9.5 Metallabenzenes and Related Compounds as an Example *239*
References *243*

10 Acidity and Basicity 251
Ranita Pal, Himangshu Mondal, and Pratim K. Chattaraj

- 10.1 Introduction 251
- 10.2 Definitions and Theories 252
- 10.2.1 Arrhenius Theory 252
- 10.2.2 Brønsted–Lowry Theory 253
- 10.2.3 Lewis Theory 254
- 10.2.4 Usanovich Definition 255
- 10.2.5 Lux–Flood Definition 256
- 10.2.6 Solvent System Definition 256
- 10.3 CDFT-Based Reactivity Descriptors 257
- 10.4 CDFT-Based Electronic Structure Principles 259
- 10.4.1 Equalization Principles 259
- 10.4.2 Hard–Soft Acid–Base (HSAB) Principle 259
- 10.4.3 Maximum Hardness (MHP), Minimum Polarizability (MPP), and Minimum Electrophilicity (MEP) Principles 260
- 10.5 Systemics of Lewis Acid–Base Reactions: Drago–Wayland Equation 261
- 10.6 Strengths of Acid and Bases 262
- 10.6.1 Ionic Product 262
- 10.6.2 pH Scale 262
- 10.6.3 Ionization Constants 263
- 10.6.4 Proton Affinity 264
- 10.6.5 Electronegativity 265
- 10.6.6 Hardness 265
- 10.6.7 Electrophilicity 266
- 10.7 Effect of External Perturbation 267
- 10.7.1 Steric Effects 267
- 10.7.2 Solvent Effects 267
- 10.7.3 Periodicity 268
- 10.7.4 Inductive Effect 268
- 10.7.5 Resonance Effect 269
- 10.8 CDFT and Acidity 270
- 10.9 CDFT and ITA 272
- 10.10 Are Strong Brønsted Acids Necessarily Strong Lewis Acids? 276
- 10.11 Summary 278
- Acknowledgment 279
- Conflict of Interest 279
- References 279

11 Sigma Hole Supported Interactions: Qualitative Features, Various Incarnations, and Disputations 285
Kelling J. Donald

- 11.1 Introduction 285
- 11.1.1 What's in a Name – The Sigma Hole Terminology and Concept 285
- 11.1.2 Donor–Acceptor Interaction Continuum 286

11.2	Many Incarnations and Roles of a Single Phenomenon	*288*
11.2.1	Hydrogen Bonding	*288*
11.2.2	Halogen Bonding and Sigma Holes on Group 17 Atoms	*289*
11.2.2.1	Common Origins	*292*
11.2.2.2	Cases of Halogen Bonding	*293*
11.2.2.3	The Sigma Hole and the Whole Story	*296*
11.2.3	Chalcogens	*297*
11.2.4	Pnictogens	*298*
11.2.5	Tetrels	*299*
11.2.6	Triels	*302*
11.3	Related Interactions Elsewhere in the Main Group	*304*
11.3.1	Group 2	*304*
11.3.2	Group 1	*306*
11.3.3	Group 18	*306*
11.4	Contested Interpretations	*308*
11.5	Conclusions	*308*
	Acknowledgment	*309*
	References	*309*

12 On the Generalization of Marcus Theory for Two-State Photophysical Processes *317*
Chao-Ping Hsu and Chou-Hsun Yang

12.1	Introduction	*317*
12.2	The Golden Rule Rate Expression	*318*
12.2.1	The Marcus Theory: The Classical Treatment	*320*
12.2.2	The Marcus–Levich–Jortner Expression: A Quantum Expression for High-Frequency Modes	*320*
12.2.3	The Föster Theory: Separating Donor and Acceptor Parts in FCWD	*322*
12.3	Application	*325*
12.3.1	Electron Transfer	*325*
12.3.2	SET: Using Spectra for FCWD	*327*
12.3.3	TET and Other Energy Transfer Process with Spin Exchange	*329*
12.4	Conclusion	*330*
	Acknowledgments	*330*
	References	*330*

13 Computational Modeling of CO_2 Reduction and Conversion via Heterogeneous and Homogeneous Catalysis *335*
Yue Zhang, Lin Zhang, Denghui Ma, Xinrui Cao, and Zexing Cao

13.1	Introduction	*335*
13.2	Computational Methods	*336*
13.3	Activation and Reduction of CO_2	*338*
13.3.1	Computational Catalyst Design	*338*
13.3.1.1	Doping of Metal and Nonmetal Atoms	*338*
13.3.1.2	Structural Modification	*339*

13.3.1.3	Application of an External Electric Field *339*	
13.3.2	Electrocatalytic Reduction of CO_2 *340*	
13.3.3	Hydrogenation Reduction of CO_2 *342*	
13.4	Catalytic Coupling of CO_2 with CH_4 *345*	
13.5	Homogeneous Catalytic Conversion of CO_2 *348*	
13.5.1	Catalytic CO_2 Fixation into Cyclic Carbonates *348*	
13.5.2	CO_2 Hydrogenation Catalyzed by Metal PNP-Pincer Complexes *350*	
13.6	Conclusion and Outlook *352*	
	Acknowledgments *353*	
	References *353*	

14 Excited States in Conceptual DFT *361*
Frédéric Guégan, Guillaume Hoffmann, Henry Chermette, and Christophe Morell

14.1	Introduction *361*
14.2	Exploring Ground State Properties Thanks to Excited States *361*
14.2.1	Context and Justification *361*
14.2.2	Chemical Hardness Revisited *362*
14.2.3	State-Specific Dual Descriptors *364*
14.2.4	Polarization Interaction *367*
14.3	Exploring the Reactivity of Excited States with Excited States *371*
14.3.1	Local Chemical Potential *371*
14.3.2	Polarization in the Excited States *373*
14.4	Conclusion *375*
	References *376*

15 Modeling the Photophysical Processes of Organic Molecular Aggregates with Inclusion of Intermolecular Interactions and Vibronic Couplings *379*
WanZhen Liang, Yu-Chen Wang, Shishi Feng, and Yi Zhao

15.1	Introduction *379*
15.2	Theoretical Approaches *381*
15.2.1	Model Hamiltonian *381*
15.2.2	Parameterizing Electronic Excited-State Hamiltonian *383*
15.2.3	Calculating Electron–Phonon Couplings *387*
15.2.4	Propagating the Photophysical Dynamics *389*
15.2.5	Numerical Examples *392*
15.2.5.1	Absorption Spectra of Aggregates *392*
15.3	Concluding Remarks *397*
	Acknowledgments *398*
	References *399*

16 Duality of Conjugated π Electrons *407*
Yirong Mo

16.1	Introduction *407*

16.1.1	Conjugated Systems and the Concept of Conjugation	*407*
16.1.2	Alternative Proposal from Rogers	*409*
16.1.3	Origin of the Disparity	*411*
16.2	The New Concept of Intramolecular Multibond Strain	*412*
16.3	Theoretical Method	*413*
16.4	Computational Analysis of the Concept of Intramolecular Multibond Strain	*416*
16.4.1	π–π Repulsion in Linear Model Molecule HBBBBH (B_4H_2)	*416*
16.4.2	σ–σ Repulsion in Model Molecule B_2H_4	*417*
16.4.3	σ–π Repulsion in Model Molecule B_3H_3	*418*
16.4.4	Conjugation and Repulsion in Butadiene, Butadiyne, Cyanogen, and α-dicarbonyl	*419*
16.5	Experimental Evidence	*422*
16.5.1	Longer Carbon–Nitrogen Bond in Nitrobenzene than in Aniline	*422*
16.5.2	Abnormally Long Nitrogen–Nitrogen Bond in Dinitrogen Tetroxide (N_2O_4)	*425*
16.6	Summary	*426*
	References	*426*

17 Energy Decomposition Analysis and Its Applications *433*
Peifeng Su

17.1	Introduction	*433*
17.1.1	Single-Determinant MO-Based EDA	*434*
17.1.2	DFT-Based EDA	*435*
17.1.3	Multireference Wavefunction-Based EDA	*436*
17.1.4	Definitions of EDA Terms	*437*
17.2	Methodology	*437*
17.2.1	GKS–EDA	*437*
17.2.2	GKS–EDA(BS)	*440*
17.2.3	GKS–EDA(TD)	*441*
17.3	Applications of GKS–EDA	*442*
17.3.1	Strong Chemical Bonds	*443*
17.3.1.1	Covalent Bonds in Diatomic Molecules	*443*
17.3.1.2	Covalent Bonds (X–Y Bonds) in XH_n–YH_n Molecules (X = C or B; Y = C, N or P, n = 2 or 3)	*444*
17.3.1.3	Transition Metal–Ligand Bonds in $(CO)_5W\cdots L$ (L = CO, BF, C_2H_2, and C_2H_4) Complexes	*444*
17.3.2	Radical-Pairing Interactions	*444*
17.3.3	Noncovalent Interactions in Excited States	*445*
17.3.3.1	Hydrogen Bonds Between Aromatic Heterocycles and H_2O	*445*
17.3.3.2	Base Pair Interactions with Excited States	*446*
17.4	Conclusion	*450*
	Acknowledgments	*450*
	References	*450*

18	**Chemical Concepts in Solids** *455*	
	Peter C. Müller, David Schnieders, and Richard Dronskowski	
18.1	The Three Schisms of Solid-State Chemistry *455*	
18.2	Bloch's Theorem *457*	
18.3	Basis Sets *460*	
18.4	Interpretational Tools *462*	
18.5	Applications *470*	
18.6	Summary *477*	
	References *477*	
19	**Toward Interpretable Machine Learning Models for Predicting Spectroscopy, Catalysis, and Reactions** *481*	
	Jun Jiang and Shubin Liu	
19.1	Introduction *481*	
19.2	ML in a Nutshell *481*	
19.3	Chemistry-Based Descriptors as ML Features *485*	
19.3.1	Intrinsic Atomic Property Descriptors *486*	
19.3.2	Electronic and Structural Property Descriptors *487*	
19.3.3	Multilevel Attention Mechanisms-Identified Descriptors *488*	
19.3.4	SISSO Method-constructed Descriptors *489*	
19.3.5	Spectral Descriptors *491*	
19.4	Selected ML Applications *493*	
19.4.1	ML Prediction of IR/Raman Spectroscopy *494*	
19.4.2	ML Prediction of Surface-Enhanced Raman Spectroscopy *496*	
19.4.3	ML Prediction of Ultraviolet Absorption Spectroscopy for Proteins *497*	
19.4.4	ML Prediction of Protein Circular Dichroism Spectra *498*	
19.4.5	ML Prediction of Bond Dissociation Energy *498*	
19.4.6	ML Predictions of Catalytical Properties *499*	
19.4.7	ML Predictions for Imperfect and Small Chemistry Data *503*	
19.4.8	ML Predictions in Reactions and Retrosynthesis *505*	
19.5	Concluding Remarks *507*	
	Acknowledgments *509*	
	References *509*	
20	**Learning Design Rules for Catalysts Through Computational Chemistry and Machine Learning** *513*	
	Aditya Nandy and Heather J. Kulik	
20.1	Computational Catalysis *513*	
20.1.1	Catalyst Design with Density Functional Theory *513*	
20.1.2	Mechanistic Modeling: The Degree of Rate Control and Energetic Span Model *516*	
20.1.3	The Utility of Scaling Relationships for Catalyst Design *518*	
20.1.4	Quantifying Active Site Environments with DFT *522*	
20.1.5	Method Sensitivity in DFT: Influence on Catalysis *525*	
20.2	Machine Learning (ML) in Catalysis *529*	

	20.2.1	Utilizing ML in Catalysis for Improved Design *529*
	20.2.2	The Role of ML in the Limits of Strong and Weak Scaling Relationships *534*
	20.2.3	Uncertainty Quantification in ML and Improved Model Performance *537*
	20.2.4	The Role of Optimization Algorithms in Combination with ML Models *539*
	20.2.5	Leveraging Experimental Data for Molecular Design *542*
20.3		Summary *545*
		References *546*

Index *559*

Preface

Through abstraction, induction, and generalization in their genesis, concepts allow us to group together objects, events, and phenomena that share common characteristics. They are the foundation and building blocks of our knowledge. Chemical concepts, as an essential part of chemical understanding, are the basis on which chemical principles and empirical laws are expressed and introduced. They are mostly noumena, which, according to Immanuel Kant, are independent existences of our perception, often with no physical observables associated with them, so they are usually not measurable. Over the last few centuries, a variety of chemical concepts have been coined and widely applied in textbooks and publications. Well-known examples are atoms in molecules, chemical bonding, partial charge, acidity, aromaticity, electrophilicity, and hydrophobicity. They can be static and dynamic, related to properties like stability, function, and reactivity of molecules and other forms of matter. Chemical concepts are often fuzzy and ambiguous, making them difficult to formulate. However, theoretical and computational frameworks developed in the past century, especially the last few decades, have provided valuable insights to us about chemical concepts, such as chemical bonding from valence orbital theory, frontier orbital reactivity from molecular orbital theory, and hard and soft acids and bases (HSAB) principle from density functional theory. For this reason, it is never too much to emphasize the important role played by theoretical and computational chemistry in formulating and quantifying chemical concepts.

After finishing editing the two-volume book entitled *Conceptual Density Functional Theory: Towards a New Chemical Reactivity Theory* published by Wiley-VCH in April 2022, I realize that Conceptual DFT is only part of the big picture of our up-to-date understanding of how chemical concepts are formulated and quantified. I feel like I can do more to have a larger picture, which has been my enthusiasm for years. This is the precise reason to get this book project started. My passion can also be witnessed by the international symposium series "*Chemical Concepts from Theory and Computation*" (CCTC) that I initiated and co-organized. The first CCTC event (CCTC#1) was held on December 10–13, 2018, in Changsha, China; CCTC#2 was held (virtually due to COVID) on December 16–20, 2021; CCTC#3 was held on December 11–13, 2023, in Lyon, France; and CCTC#4 is to be held on December 15–20, 2025, in Hawaii, USA, as part of *PacifiChem2025*. I also served as the guest editor of two journals, *Acta Physico-Chimica Sinica* (2018) and *Theoretical Chemistry Accounts* (2020), for a special issue or topic collection with the same title, *Chemical Concepts from Theory and Computation*.

The 20 chapters assembled in this book are our latest effort in regard to gathering a big picture of how traditional chemical concepts can be formulated and quantified using theoretical frameworks. We started by introducing the 4 theoretical frameworks currently available in the literature, and then 16 specific topics of chemical concepts were elaborated in detail by respective experts worldwide. We also prepared 10 questions about exploring chemical concepts through theory and computation as the Foreword and asked *ChatGPT*, *Ernie Bot* (文心一言), *Google Bard*, and *Tongyi Qianwen* (通义千问), all powered by artificial intelligence to answer them. Together, in our humble view, they represent the state-of-the-art of the matter, at least for the topics that this book covers. Hopefully, this book serves as a starting point for our readers to further explore in the future.

I am most grateful to all the authors for their willingness to participate in this book initiative. I also wish to express my sincere gratitude to the publisher team for their flexibility and hard work during the publication process. This book is a collective product. Should any piece be missing, it would have failed.

April 2024

Shubin Liu
University of North Carolina, Chapel Hill, USA

Foreword

Artificial intelligence (AI) has already started becoming a part of our daily lives. Would we just like to ask questions and let computers or AI give us the answers? What kind of answers would we like to have? Should they be more than numbers? Should we keep the ideas of using chemical concepts that have been proven valuable in the past? A part of them? Would we like to keep something for our own thinking? About what? To give explanations? To formulate new questions? What might be the role of machine learning and quantum computers in chemical thinking?

Inquiries like those above motivated us to formulate a list of 10 questions below about exploring chemical concepts through theory and computation and then ask *ChatGPT, Ernie Bot* (文心一言), *Google Bard*, and *Tongyi Qianwen* (通义千问), all powered by AI, to answer them. Below is what came out.

April 2024

Andreas Savin
Sorbonne Université, Paris, France
Paul W. Ayers
McMaster University, Hamilton, Canada
Shubin Liu
University of North Carolina, Chapel Hill, USA

10 Questions About Exploring Chemical Concepts Through Theory and Computation

Answers by *ChatGPT (CG)*, *Ernie Bot (EB)*, *Google Bard (GB)*, and *Tongyi Qianwen (TQ)*

1. *What are the three advantages of exploring chemical concepts through theory and computation?*
   CG: Cost-effectiveness, enhanced understanding, and predictive power
   EB: Understanding chemical processes, predicting properties and reactions, and designing new materials
   GB: Accuracy, efficiency, and versatility
   TQ: Understanding the fundamental principles, predicting chemical behavior, and identifying optimal conditions
2. *What are the three disadvantages of exploring chemical concepts through theory and computation?*
   CG: Limited accuracy, high computational cost, and lack of experimental validations
   EB: Limited experimental data, computational errors, and computational recourses
   GB: Complexity, interpretability, and limitations of models
   TQ: Limited practical experience, dependence on computational resources, and interpretation challenges
3. *What is the best example of chemical concepts that theory and computation have helped improve its understanding?*
   CG: The concept of reaction kinetics
   EB: The concept of catalysis
   GB: The structure of DNA
   TQ: The field of quantum chemistry
4. *Is there a complete set of chemical concepts?*
   CG: No. There isn't a complete set of chemical concepts, as the field is evolving and new concepts are discovered.
   EB: Probably.
   GB: No. Chemistry is a vast and ever-evolving field, so new concepts will be discovered.
   TQ: Yes.

5. Can most, if not all, chemical concepts be formulated and quantified?
   CG: Yes, most chemical concepts can be formulated and quantified.
   EB: Yes.
   GB: Yes. Chemistry is based on the laws of physics. These laws can be expressed in mathematical formulas, and so the concepts of chemistry can also be expressed in mathematical formulas.
   TQ: Almost.
6. What are the three chemical concepts that we have understood well?
   CG: The periodic table, chemical bonding, and stoichiometry
   EB: Chemical bonding, stoichiometry, and thermodynamics
   GB: Atomic structure, chemical bonding, and chemical reactions
   TQ: Atomic structure, chemical reactions, and thermodynamics
7. What are the three chemical concepts that we don't understand yet or at all?
   CG: Dark matter, quantum gravity, and consciousness
   EB: Unified theory of chemical bonding, quantum chemistry, and complex chemical systems
   GB: The origin of life, the nature of consciousness, and the behavior of materials at the quantum level
   TQ: Quantum chemistry, electrochemistry, and non-equilibrium thermodynamics
8. What are the three areas of research that can benefit from the quantification of chemical concepts?
   CG: Materials science, environmental science, and drug discovery
   EB: Drug design, material science and engineering, and environmental science and sustainability
   GB: Drug design, sustainable chemistry, and materials science
   TQ: Materials science, energy research, and medicine
9. What will be the three roles of machine learning in quantifying chemical concepts?
   CG: Predictive modeling, data analysis, and process optimization
   EB: Data analysis and prediction, feature selection and representation, and automation and optimization
   GB: Identify patterns in chemical data, classify chemical compounds, design new materials, and predict chemical reactions
   TQ: Data analysis, model building, and concept validation
10. What will be the three roles of quantum computers in quantifying chemical concepts?
    CG: Simulation of complex chemical systems, optimization of chemical processes, and design of new materials and drugs
    EB: Accelerating quantum chemistry calculations, optimizing chemical reactions and catalyst design, and exploring quantum chemistry phenomena
    GB: Calculate the properties of molecules, model chemical reactions, and design new materials
    TQ: Improved computational methods, new computational approaches, and validation and verification

1

Chemical Concepts from Molecular Orbital Theory

Feng Long Gu[1], Jincheng Yu[2], and Weitao Yang[2]

[1]*South China Normal University, School of Environment, Ministry of Education, Key Laboratory of Theoretical Chemistry of Environment, Guangzhou 510006, People's Republic of China*
[2]*Duke University, Department of Chemistry, Box 90346, Durham, NC 27708-0346, USA*

1.1 Introduction

Quantum chemistry provides the theoretical foundations and quantitative explanations of the physical and chemical properties of atoms and molecules in terms of the physical interactions between electrons and nuclei. It is based on physics, combined with various mathematical treatments, and applies the basic principles and methods of quantum mechanics to study chemical problems [1]. Its research scope includes the microscopic study of the electronic structure properties of atoms, molecules, and bulk systems, intermolecular forces, chemical bond theory, and various spectra and chemical reactions.

The history of quantum chemistry can be traced back to 1927 just after the establishment of quantum mechanics. Till the end of the 1950s, three chemical bond theories, i.e. the valence orbital theory (VOT), the molecular orbital theory (MOT), and the coordination field theory (CFT), have been established to study molecular or crystalline systems by using quantum chemistry.

Among these three chemical bond theories, the VOT was developed by Pauling et al. [2–4] on the basis of Heitler and London's work [5] for the molecular structure of hydrogen. The result is much close to the classical atomic valence theory and generally accepted by chemists. The MOT, however, was first proposed by Mulliken and Hund [6–9] in the late 1920s to the early 1930s.

The main idea of Mulliken's work on MOT is that all electrons of atoms contribute to forming molecules, and the electrons in molecules are no longer belonging to a certain atom, but moving across the entire range of a molecular space. The state of motion of electrons in space in molecules can be described by the corresponding molecular orbitals (MOs), i.e. wave function Ψ. The main difference between MOs and atomic orbitals (AOs) is that in molecules, electrons move under the action of all nuclei potential fields. An important consequence is that the MOs can be obtained by the linear combination of atomic orbitals (LCAO) in molecules.

Exploring Chemical Concepts Through Theory and Computation, First Edition. Edited by Shubin Liu.
© 2024 WILEY-VCH GmbH. Published 2024 by WILEY-VCH GmbH.

Following Mulliken's ideas, the simplest MOT was proposed by Hückel in 1931 [10], so-called Hückel molecular orbital (HMO) method, to successfully treat conjugated molecular systems. The MOT calculation is relatively simple and now widely appreciated by chemists, and it is supported by photoelectron spectroscopy experiments, making it dominant in chemical bond theory.

From the 1960s, the main goal was to further develop the quantum chemical calculation methods, among which the *ab initio* calculation method, the semiempirical methods, and other methods have expanded the application scope of quantum chemistry and the calculation accuracy has been gradually improved. Consequently, some accurate results calculated from quantum chemistry were almost exactly the same as the experimental values. The development of computational quantum chemistry has expanded quantitative computing to large molecules and the applications of quantum chemistry into other disciplines become possible.

With the development of MOT and the upgrading of computer facilities, the systems that quantum chemistry can deal with have become larger and larger, and the calculation accuracy has been continuously improved. Quantum chemical computing programs have also become an increasingly important tool for solving chemical problems, and it is expected that more complex chemical problems can be solved in the future. At present, there are many popular program suites, such as the Gaussian series [11], GAMESS [12, 13], and others.

In the beginning of MOT, there seemed to be no direct relation between MOs and the bonds in a chemical formula, because MOs obtained from MOT normally extend over the whole molecule space and are not restricted to the region between two atoms. The difficulty was overcome by using equivalent localized molecular orbitals (LMOs) instead of the delocalized ones. The mathematical definition of equivalent MOs was given only in 1929 by Lennard-Jones [14]. The concept of localization of MOs leads to the connection between MOs and the pictures of chemical bonds. Benefits from LMOs are at least four followings (i) related to the concepts of chemical bonds, useful to isolate functional groups from different molecules; (ii) reducing the efforts for computation; (iii) transferable from one molecule to others within analogical structures; (iv) more suitable for LMOs to treat correlation.

This chapter is organized as follows, the MOT is surveyed together with the localization methods either for OLMOs and NOLMOs.

1.2 Molecular Orbital Theory

In Niels Bohr's atomic model, which is based on principles of quantum physics, electrons circle the atomic nucleus in different shells containing a fixed number of electrons. The assumption was that attractive forces between the atoms in a molecule are the result of atoms sharing electrons to fill the electron shells.

Heitler and London [5] first adopted quantum mechanics to treat hydrogen molecule in 1927, revealing the nature of the chemical bond between two hydrogen atoms, leading the typical Lewis theory to today's modern VOT. The concept of the atomic bonding created by electron sharing was introduced by Lewis in his

1916s fundamental paper [15]. It was elaborated by Langmuir [16] a few years later. Pauling et al. [2–4] introduced the concept of hybrid orbitals to greatly develop VOT and successfully applied it to the structure of diatomic molecules and polyatomic molecules.

VOT coincides with the classical concept of electron bonding familiar to chemists and has been rapidly developed as soon as it appeared. However, the calculation of VOT is more complicated, which makes the later development slow. With the increasing improvement of computing technology, there will be new developments in this theory.

VOT focuses on the contribution of unpaired electrons in the outermost orbital between bonded atoms in the formation of chemical bonds, which can successfully explain the spatial configuration of covalent molecules. However, the inner electrons of the bonding atom were not considered during the actual situation of bonding.

Meanwhile from the mid-1920s, quantum mechanics has been applied to develop sophisticated models for the movement of electrons within a molecule, so-called molecular orbitals (MOs). Under the work of Hund [6], Mulliken [9], and John Lennard-Jones [14], MOT began to arise. Thus, in the beginning, the MOT was called the Hund–Mulliken theory. The concept of the word "orbital" was first proposed by Mulliken in 1932 [9]. The first paper using MOT was published by Lennard-Jones in 1929 [14] to treat MOT in a quantitative way. The LCAO approximation was introduced for constructing MOs to study the electronic structure of oxygen molecule from quantum principles. This convinced chemists that quantum mechanics is so useful, and the success of the MOT today owes much to their great contributions.

MOT is an effective approximation method for dealing with the structure of diatomic molecules and polyatomic molecules and is an important part of chemical bond theory. It differs from VOT, which focuses on understanding chemistry by hybridizing AOs into bonds, while the former focuses on the cognition of MOs. The idea of MOT is that electrons in a molecule move around the entire molecule. MOT pays attention to the integrity of molecules, so it better illustrates the structure of polyatomic molecules. At present, MOT stands on an important position in modern covalent bond theory and is widely accepted and considered a valid and useful theory.

By the 1950s, MOs were thoroughly defined as eigenfunctions of the self-consistent field Hamiltonian operator, marking the development of MOT into a rigorous scientific theory. Hartree–Fock (HF) method is a more rigorous treatment of MOT, and MOs are expanded according to a set of basis of AOs to develop the Hartree–Fock Roothaan (HFR) equation.

$$\Psi_i = \sum_{\mu} C_{\mu i} \chi_\mu \tag{1.1}$$

Equation (1.1) is so called linear combination of atomic orbitals (LCAO), where Ψ was used in the 1930s by Hund, Mulliken [9], Hückel [10], and others to construct MOs for polyatomic molecules, also called the LCAO-MO theory.

$$\mathbf{FC} = \mathbf{SC}\epsilon \tag{1.2}$$

The Hartree–Fock Roothaan equation (Eq. (1.2)) is a method of *ab initio* calculation, and the *ab initio* method is simply to use a "correct" Hamiltonian operator, except for the most basic constants, no longer citing any experimental data, based on the Schrödinger equation, only using single-electron, nonrelativistic, and Born–Oppenheimer approximations.

On this basis, a variety of *ab initio* quantum chemical calculation methods have been developed. At the same time, MOT has also been applied to a semi-empirical calculation that uses more approximate methods, known as semi-empirical quantum chemical calculations.

MOT, based on HF approach, is a theory of chemical bonds based on single-electron approximations. The basic idea of single electron approximation is that there is a physical existence of the own behavior of a single electron, which is only constrained by the action of the nucleus and other electron average fields in the molecule, provided the Pauli exclusion principle is obeyed. The wave function that describes the behavior of a single electron is called an orbital (or orbital function), and the corresponding energy of a single electron is called an energy level. For any molecule, if its series of MOs and energy levels are found, the molecular structure can be discussed in the same way as atomic structure, and linked to a systematic interpretation of molecular properties.

MOT is widely used in modern quantum chemistry, so the HFR equation is also known as the cornerstone of modern quantum chemistry. The basic idea of the HFR equation is that the wave function of a multi-electron system is a Slater determinant constructed based on the MOs of the system. Then without changing the operator and wave function form in the equation, only changing the MO coefficients of the AOs, the system energy can reach the lowest point, this minimum energy is the approximation of the total energy of the system, and the multi-electron system wave function obtained at this point is the approximation of the system wave function.

The wave function obtained from HFR equation (1.2) is so-called the canonical molecular orbitals (CMOs), in which the Lagrangian matrix ϵ is diagonal. Any unitary transformation of CMOs does not change the properties of the system, so the total energy of the system is only related to the occupied orbitals. And some forms of wave functions obtained by linear combinations of CMOs, such as those of LMOs and hybrid orbitals, are also the solutions to the HFR equation (1.2).

The role of MOs changed dramatically with the development of Kohn–Sham density functional theory (DFT) [17–19]. In DFT, the MOs are used to represent the total electron density in principle exactly. The effects of many-electron interaction are described with the exchange-correlation density functional approximations and enter into the Kohn–Sham (KS) equations, which have a form similar to the HFR equation, but with different effective potentials. Thus the MO takes on a central role in DFT, which is now the most widely used computational approache in quantum chemistry.

There are also many other methods such as quantum chemistry composite methods, quantum Monte Carlo (QMC), configuration interaction (CI), multi-configuration self-consistent field method (MCSCF), many-body perturbation theory (MBPT), and coupled cluster (CC) theory, which are developed based on

MOT. MOT can better reflect the objective reality for dealing with polyatomic systems, explaining delocalization effects and induction effects, and can solve problems that cannot be so easily dealt by VOT.

1.3 Canonical Molecular Orbitals

CMOs obtained from the HFR equation, or the KS equations in DFT extend to the entire molecular system, that is, delocalized molecular orbitals. As an example, one of the CMOs of methane (CH_4) calculated using the SCF-LCAO-MO method can be written as:

$$\psi^{CMO} = -0.22\chi_{1s}^C + 0.63\chi_{2s}^C + 0.18 \times (\chi_{1s}^{H_1} + \chi_{1s}^{H_2} + \chi_{1s}^{H_3} + \chi_{1s}^{H_4}) \quad (1.3)$$

The CMOs normally contain the AO components of all atoms, and the electron belongs to the entire molecule, no longer limited to an atom or between two atoms, which is also called the delocalized molecular orbitals.

The CMOs cannot adopt the localized properties of electronic structure. Even though the traditional *ab initio* quantum chemical calculation method has made great achievements for small molecule systems. However, as CMOs require the calculation of the entire system, for large systems, the calculation process becomes so complicated, which makes the application of quantum chemical calculation methods based on regular CMOs to macromolecular systems a major problem. The cost of the traditional HF method increases with N^4 of the system, where N is the number of the basis functions.

Quantum chemistry calculations for large systems are normally limited by using the traditional CMO basis. Therefore, various approximation methods are developed. Kirtman [20–22] proposed the local space approximation (LSA). At the semiempirical level, LocalSCF has been proposed by Anikin et al. [23]. Yang [24] developed the Divide and Conquer method based on the approximation of DFT. Imamura et al. [25–27] developed the elongation method. The fragment molecular orbital (FMO) method has been proposed by Kitaura et al. [28] for the calculation of biological macromolecules of proteins and DNA. Li et al. constructed the fragment energy assembler (FEA) [29].

1.4 Frontier Molecular Orbital Theory

Frontier molecular orbital (FMO) theory, a kind of MOT, was developed in 1950s by Fukui et al. [30]. It has been pointed out that many properties of molecules are mainly determined by the FMOs, i.e. the highest occupied molecular orbital (HOMO) and the lowest unoccupied molecular orbital (LUMO). The FMO theory is simple, intuitive, and effective, so it has a wide range of applications in theoretical research such as chemical reactions and catalytic mechanisms. FMO plays a decisive role in the selection of the reaction route of organic synthesis. Later on, Woodward and Hoffmann applied FMO theory to study the stereochemical

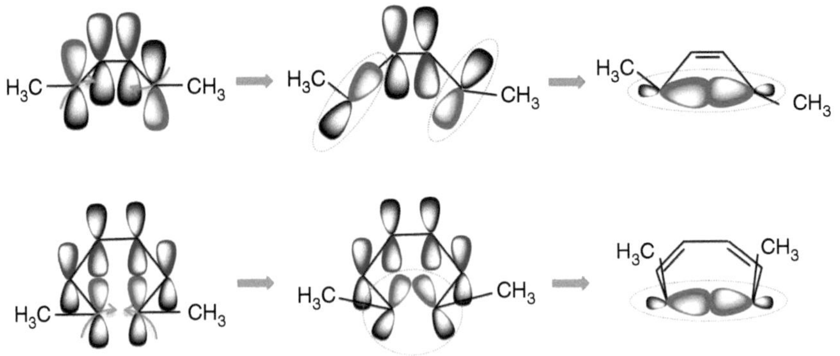

Figure 1.1 The reaction mechanisms of butadiene and hexatriene elucidated by FMO theory.

selection rule for chemical reactions, further developing it into the principle of symmetry conservation of MOs. The synthesis of vitamin B_{12} [31] is a very successful example guided by the FMO theory and the principle of conservation of MO symmetry. In 1981, Fukui and Hoffmann shared the Nobel Prize in chemistry for their intuitive explaining the occurrence of chemical reactions through the principle of conservation of MO symmetry.

For the reaction between two molecules A and B. The electrons in the HOMO of molecules A and B flow to each other's LUMO, causing the formation and breaking of chemical bonds, and a chemical reaction occurs. Electron flow is prone to occur only when the HOMO of molecule A (or B) is close to the energy of LUMO of molecule B (or A), together with the symmetry matching each other. It must be pointed out that the FMO theory applies not only to π orbits, but also to σ orbits, so it has applications in organic chemistry, inorganic chemistry, as well as surface adsorption and catalysis, quantum biology, and other fields.

As an example, in the electrocyclization reaction, conrotatory is a heteroplanar process, and disrotatory is a homoplanar process. Under the heating condition, the reaction mechanisms of butadiene and hexatriene are different. Figure 1.1 depicts how HOMOs rotate and form a σ orbital. One can see that due to the phases of the two orbital lobes at the ends are different, for butadiene, conrotatory of two π orbitals is consistent with the MO symmetry. While conversely, disrotatory of two π orbitals of hexatriene is needed for the electrocyclization reaction. This example clearly shows that the FMO theory is consistent with the experimental observations.

1.5 Localized Molecular Orbitals

In view of the computational complexity and high costs in calculating macromolecular systems based on CMOs, computational methods are highly demanded for costs increased linearly with the size of the system (linear scale). Most of the quantum chemical linear scale calculation methods developed in recent years to deal with macromolecular systems are based on LMOs, which is another form of MOs obtained by a unitary transformation of CMOs. The most important character of LMOs is the locality, that is, the properties of a spatial region are less affected by the distant spatial region. Of course, the differences in the description of LMOs and CMOs are only superficial, and they describe the same objective situation. One can think of any LMO as a "hybridization" of CMOs, just as the hybridization of AOs can form valence bond orbitals. Equation (1.4) presents the MO coefficients of one of the LMOs of CH$_4$ obtained by a unitary transformation.

$$\psi^{LMO} = 0.31\chi_{2s}^C + 0.29\chi_{2p_x}^C + 0.40\chi_{2p_y}^C + 0.54\chi_{1s}^{H_1}$$
$$+ 0.06 \times (\chi_{1s}^{H_2} + \chi_{1s}^{H_3} + \chi_{1s}^{H_4}) \quad (1.4)$$

From the comparison between Eqs. (1.3) and (1.4), one can find that LMOs are more precisely related to the AOs of directly bonded atoms, and the influence of indirectly bonded atoms on the specified molecular orbitals is reduced. After localization, the LMO, in addition to the 2s, 2p$_x$, and 2p$_y$ contributions of C atom, the 1s orbital of one hydrogen atom has the largest contribution, and the others are quite small, so that it is as if the electron motion is mainly limited between these two atoms. This confirms that the four regular CMOs of methane can be transformed by orthogonal transformations into four LMOs in the C–H bond region, corresponding to the four C–H bonds of valence bond theory.

Since the physical properties do not change under the unitary transform of MOs, the LMOs and the CMOs linked by the unitary transform are completely equivalent for the description of the molecular properties determined by the electron as a whole. If it comes to properties related to the spatial structure of molecules, the description of LMOs is more intuitive. LMOs have the advantage of being more consistent with traditional chemical concepts, and chemists are happy to use them.

From the point of view of computational quantum chemistry, one of the most fascinating aspects of the description of LMOs is its transferability. The certain molecular properties (such as bond energy and bond distance) that chemists have long determined from experience are equivalent to the "transferability" of LMOs in the single-particle approximation. If this transferability exists, LMOs obtained from small molecules are transferable to large molecules, which is meaningful to reduce the amount of computation for large molecules.

1.5.1 Orthogonal Localized Molecular Orbitals

There are two ways to generate LMOs, one is to perform a unitary transformation from CMOs to LMOs, and the other is to directly solve a certain single-electron Schrödinger equation to get LMOs. However, both methods require a localization criterion to determine the transformation matrix for the transition from a regular orbital to a localized orbital or a single-electron Schrödinger equation using the second method. Many localization methods have been proposed, and some of them artificially specify the shape and position of LMOs to be generated according to the traditional chemical concept and the symmetry nature of the molecule.

There are some popular localization methods, such as the Boys localization, the Edmiston–Ruedenberg localization, and the Pipek–Mezey localization, which are also known as orthogonal localized molecular orbital (OLMO) methods.

Any localization method requires two elements: a physically meaningful localization criterion and a computationally effective mathematical algorithm for implementing the satisfaction of this criterion. Both aspects are important for a viable localization procedure.

(1) Boys localized molecular orbitals

Foster and Boys [32, 33] propose a scheme for localized orbitals, as so called "exclusive orbitals." They are obtained by maximizing the product of the spatial distances between the centers of charge vectors R_i of all different molecular orbitals. The required consecutive iterations present convergence problems presumably because, even for moderately sized molecules, the product is of very high order in the orbitals. Inspired by Edmiston–Ruedenberg's work (see below), the Boys localization scheme now is based on minimizing the sum of the orbital's second central moment

$$\mathcal{L}_{\text{Boys}} = \sum_p^N \langle p|(\hat{r} - \langle p|\hat{r}|p\rangle)^2|p\rangle \tag{1.5}$$

where $\langle p|p \rangle = 1$ means that the AOs are normalized. This localization function is known as the Boys–Foster or Boys localization function.

The set of orbitals obtained from minimizing Eq. (1.5) is a set in which the orbitals on average are local. The Boys localization function has been widely used in chemistry and is also used in solid-state theory for the localization of Wannier functions, leading to the maximally localized Wannier functions [34].

(2) Edmiston–Ruedenberg localized molecular orbitals

Edmiston and Ruedenberg [35] introduced a localization scheme that maximizes the sum of orbital self-repulsion energies

$$\mathcal{L}_{\text{ER}} = \sum_p^N \langle pp|\frac{1}{r_{12}}|pp\rangle \tag{1.6}$$

where $\langle pp|\frac{1}{r_{12}}|pp\rangle = \iint \phi_p(r_1)\phi_p(r_2)\frac{1}{r_{12}}\phi_p(r_1)\phi_p(r_2)dr_1 dr_2$

This method has been used in different theoretical investigations, as well as in a more fundamental analysis of the origin of molecular bonding. The Edmiston and Ruedenberg method has a fifth-order computational scaling compared to a third-order scaling for the Boys and Pipek–Mezey schemes, and has, despite a reduction to asymptotically third-order scaling, been used less extensively.

(3) Pipek–Mezey localized molecular orbitals

Pipek and Mezey [36] introduced a localization function that measures the number of atomic centers over which a molecular orbital extends. The Pipek–Mezey localization function is a sum of the squared Mulliken charges for a set of N orthonormal occupied or virtual HF orbitals and is given by

$$\mathcal{L}_{PM} = \sum_p^N \sum_A \sum_{\mu \in A} \left[\langle p | \hat{P}_\mu | p \rangle \right]^2 \qquad (1.7)$$

The charge distribution operator $\hat{P}_\mu$ is defined as

$$\hat{P}_\mu = \frac{1}{2}(|^b\mu\rangle\langle\mu| + |\mu\rangle\langle^b\mu|) \qquad (1.8)$$

where $|^b\mu\rangle = \sum_v |v\rangle s^{-1}_{v\mu}$, $s^{-1}_{v\mu}$ is the inversion of the AO basis overlap. Therefore, LMOs can be obtained by taking the maximum of the functional $\mathcal{L}_{PM}$.

The Mulliken population analysis suffers from some unphysical behavior since the individual Mulliken charges for a shared electron between two atoms may have numbers greater than 1 or less than 0 [37]. This unphysical behavior is due to the fact that in a Mulliken population analysis, overlap populations occur since the AO basis is not orthogonal, and the overlap population is divided equally between the atomic centers, ignoring that different types of atoms have different electronegativity. The unphysical behavior increases when the AO basis set increases in size.

The LMOs obtained by the Boys, Edmiston and Ruedenberg (ER), and Pipek and Mezey (PM) localization schemes are the most popular orbitals used by computational chemists to compute OLMOs. Judged by their physical nature, the PM and ER orbitals are more physical than the Boys orbitals. Judged by computation speed, however, the PM and Boys schemes are much faster than that of the ER's. For large systems that can be tackled with today's computer power, the speed of orbital calculation is essential; computing LMOs should not take longer than the subsequent electron correlation calculations.

As determinantal wave functions are invariant with respect to orthogonal transformations among the orbitals, in order to cast them into localized form, intrinsic or external criteria can be used. The methods of ER and of Boys are intrinsic because only the actual molecular orbitals are used. While the method of PM is external because it is based on the overlap between the actual molecular orbitals and certain independently chosen additional linear combinations of atomic basis orbitals.

Besides the above mentioned localization methods, Löwdin [38] proposed the concept of natural orbits. A set of natural orbitals is combined into a single-electron basis function to constitute the electronic configuration of the N-particle system, so that the configuration of Ψ expansion can be achieved with fewer basis than the regular Hartree–Fock orbital basis. The work of Reed and Weinhold [39] expanded on this basis, and systematically proposed the concepts of natural spin orbital, natural bond orbital (NBO), and natural hybrid orbital, and developed into the NBO theory. Through the type of orbital and NBO analysis, one can easily find out the atomic population in a molecule, bond information, as well as intramolecular and intermolecular super-conjugate interactions.

(4) Regional localized molecular orbitals

The elongation method, proposed by Imamura et al. [25], is to theoretically synthesize a polymer chain by adding a monomer unit stepwise to a starting oligomer while keeping the degree of freedom of active space almost fixed. The elongation method works in an OLMO basis, in contrast to the conventional HF method in a CMO basis. The advantage of an LMO representation is that it allows one to freeze the region far away from the chain propagation site. This reduces the number of variational degrees of freedom in the system. Just a simple 2×2 unitary rotation was originally adopted to obtain LMOs for the elongation processing. In this 2×2 unitary rotation, pairs of CMOs are selected with an initial division of CMOs either belonging to the frozen or to the active region, and then successively rotated to form one LMO in the frozen region and another LMO in the active region. This bears that a poor selection of CMOs yields a poor localization. The slow convergence of the 2×2 localization affects the applications of the elongation method.

A different localization scheme for the elongation method has been developed based on regional localized molecular orbitals (RLMOs) [40]. This scheme is more efficient and more accurate even for covalently bonded systems with strongly delocalized π electrons. *Ab initio* test calculations have been performed even for very delocalized systems, and it was confirmed that this new scheme has big progress for the elongation method. The localization scheme is described as follows:

The AO-based density matrix is given

$$\mathbf{D}^{AO} = \mathbf{C}_{AO}^{CMO} \mathbf{d} \mathbf{C}_{AO}^{CMO\dagger} \qquad (1.9)$$

Then Löwdin's symmetric orthogonalization procedure is performed to transfer the density matrix to the orthogonal atomic orbital (OAO) basis. The transformation matrix X is obtained by diagonalizing $\mathbf{S}^{AO}$ to give

$$\mathbf{D}^{OAO} = \mathbf{X}\mathbf{D}^{AO}\mathbf{X}^\dagger \qquad (1.10)$$

After one gets OAO basis D-matrix, one can partition the D-matrix into two parts, one for the frozen region (A region) and one for the active region (B region). Region B is defined by atoms adjacent to the growing end of the chain whereas region A is at the opposite end. The purpose is to find two sets of regional LMOs (RLMOs), respectively, for both regions. The desired RLMOs can be obtained by two steps. It is something similar to the construction of natural bond orbitals (NBOs) [39] but to localized regional orbitals rather than to localized bond orbitals. A regional orbital (RO) space is generated by separately diagonalizing the subblocks of $\mathbf{D}^{OAO}$, i.e. $\mathbf{D}^{OAO}(A)$ and $\mathbf{D}^{OAO}(B)$, and the transformation from OAOs to ROs is given by the direct sum

$$\mathbf{T} = \mathbf{T}^A \oplus \mathbf{T}^B \qquad (1.11)$$

where $\mathbf{T}^A$ and $\mathbf{T}^B$ are the eigenvectors of $\mathbf{D}^{OAO}(\mathbf{A})$ and $\mathbf{D}^{OAO}(\mathbf{B})$, respectively. Then the RO-based density matrix is obtained by the following transformation,

$$\mathbf{D}^{RO} = \mathbf{T}^{\dagger}\mathbf{D}^{OAO}\mathbf{T} \tag{1.12}$$

and the transformation coefficients from ROs to CMOs may be written as

$$\mathbf{C}^{CMO}_{RO} = \mathbf{T}^{\dagger}\mathbf{X}\mathbf{C}^{CMO}_{AO} \tag{1.13}$$

At this point, one gets the non-orthogonal LMOs completely localized to either region A or region B. However, they are not completely occupied or unoccupied and of course not suitable for the elongation method. Thus, it is necessary to carry out a unitary transformation between the occupied and unoccupied blocks of $\mathbf{D}^{RO}$ in such a way as to preserve the localization as much as possible. This is done by using the Jacobi procedure. U is the transformation that diagonalizes $\mathbf{D}^{RO}$, the unitary transformation from CMO to RLMO is given as

$$\mathbf{C}^{CMO}_{RLMO} = \mathbf{U}^{\dagger}\mathbf{T}^{\dagger}\mathbf{X}\mathbf{C}^{CMO}_{AO} \tag{1.14}$$

Finally, the original AOs basis RLMOs is given by

$$\mathbf{C}^{RLMO}_{AO} = \mathbf{X}^{-1}\mathbf{T}\mathbf{U} \tag{1.15}$$

As long as sufficiently well-localized RLMOs are obtained by the scheme described above, the elongation method can proceed by using these RLMOs. It has been tested that the accuracy and the efficiency of the elongation method weaponed with this new localization scheme are greatly improved. By using RLMOs, the elongation method can be successfully applied to one-dimensional chains and any two- or three-dimensional systems no matter what type of chemical bonding is involved and regardless of whether the system is periodic or aperiodic.

1.6 Regularized Nonorthogonal Localized Molecular Orbitals

The OLMOs do reach a certain degree of locality, but due to the constraints of orthogonal normality conditions, there are long-range non-localized tails in OLMOs. This affects the accuracy and efficiency of the calculation. In order to obtain images that are more consistent with traditional chemical bonds, there was an early desire to construct non-orthogonal localized molecular orbitals (NOLMOs). NOLMOs were first applied to the calculation of electronic structure by Adams [41] and Gilbert and Lykos [42] in 1961. The nature of intramolecular interactions was investigated based on the transferability of NOLMOs and applied to mono- and difluoromethane [43]. In 2003, Sorakubo et al. proposed a partially LMO KS-DFT method based on NOLMOs [44]. Paulus et al. applied NOLMOs to the *ab initio* incremental correlation treatment at CCSD level, and their results show that the transferability of NOLMOs is much improved compared to those of OLMOs [45]. Sironi et al. applied it to valence bond theory [46]. However, these *ab initio* methods for determining NOLMOs have only been tested in small systems. In 2004, Anikin et al. [23] applied NOLMOs to semiempirical methods and completed a semiempirical program LocalSCF that can calculate the linear scale of tens of thousands of atomic systems. In 2013, Peng et al. confirmed the feasibility of the method at the Hartree–Fock level from the source, and also successfully realized the programmatic of the method [47]. Mayer et al. used extremely localized molecular orbitals (ELMO) for the transferability [48, 49].

Yang et al. developed the principle of absolute energy minimum variational to avoid the time-consuming process of diagonalizing the Fock matrix and used the conjugate gradient method to obtain the minimum value of the system energy [50, 51]. It is found that NOLMOs can obtain higher precision results than any traditional OLMOs, and the locality of NOLMOs is about 10-28% better than that of OLMOs. In 2013, Peng et al. developed an HF method based on NOLMOs [47], and the test results show that for six different systems, including $C_{18}H_{38}$ molecular system and polyglycine system, the convergence is almost comparable to the traditional method.

Relatively speaking, NOLMOs have better locality than OLMOs, because the orbitals are no longer constrained by orthogonality between NOLMOs, which also solves the problem of long-range tailing in OLMOs. However, since the NOLMO releases the constraints of orbital orthogonality, unlike the methods based on OLMOs, almost all the relevant formulas need to be re-derived.

Figure 1.2 is the comparison of the distribution in AO basis for CMOs and NOLMOs of $C_{60}H_{122}$ molecule [52]. It can be seen that NOLMOs, compared to CMOs, are much more localized in space. Due to the sparse MO coefficient matrix of NOLMOs, one can design and realize low-scaling or linear-scaling calculations of electronic structure for large systems.

The NOLMOs of a system are obtained by minimizing the following quantity

$$\Theta[\varphi_k] = \langle \varphi_k | (\mathbf{r} - \mathbf{r}_k^0)^2 | \varphi_k \rangle + \omega \langle \varphi_k | \hat{T} | \varphi_k \rangle \tag{1.16}$$

where φ_k are NOLMOs of the system, $\mathbf{r}_k^0 = (x_k^0, y_k^0, z_k^0)$ is the fixed centroid of φ_k, $\hat{T}$ is the kinetic energy operator, ω, a positive constant, is the weight of kinetic

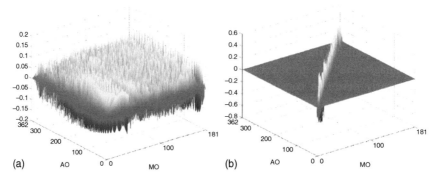

Figure 1.2 Comparison of the distribution in AO basis for CMOs (a) and NOLMOs (b) of $C_{60}H_{122}$ molecule. Source: Reproduced from Ref. [52] with permission from the Royal Society of Chemistry.

energy term. When $\omega = 0$, Eq. (1.16) is just the widely used Boys localization cost function (1.5). The added term of the kinetic energy is to regularize the solution NOLMO, or to enhance its smoothness, by reducing the kinetic energy of the orbitals. Since φ_k are occupied NOLMOs, they must be the linear combination of occupied CMOs of the system, namely

$$\varphi_k = \sum_i^{N/2} A_{ik}\psi_i \tag{1.17}$$

where N is the number of electrons in the system, ψ are CMOs obtained from conventional ways. $\Theta[\varphi_k]$ can then be rewritten as the function of coefficients A_{ik}

$$\Theta[A] = \sum_{i,j} A_{ik}^* A_{jk} \left[\langle\psi_i|(\mathbf{r}-\mathbf{r}_k^0)^2|\psi_j\rangle + \omega\langle\psi_i|\hat{T}|\psi_j\rangle\right]$$
$$= \sum_{i,j} A_{ik} A_{jk} \left[\tilde{R}_{ij} - 2x_k^0 X_{ij} - 2y_k^0 Y_{ij} - 2z_k^0 Z_{ij} + \tilde{r}_k^0 + \omega T_{ij}\right] \tag{1.18}$$

where $\tilde{R}_{ij}$ is the integration over the square of Cartesian vector $\langle\psi_i|\hat{x}^2 + \hat{y}^2 + \hat{z}^2|\psi_j\rangle$, X_{ij}, Y_{ij}, and Z_{ij} are integrals of individual Cartesian vector components $\hat{x}$, $\hat{y}$ and $\hat{z}$, respectively $\tilde{r}_k^0 = x_k^0 x_k^0 + y_k^0 y_k^0 + z_k^0 z_k^0$ is a value which equals to the square of the Cartesian vector $\mathbf{r}_k^0$.

In order to obtain the optimal value of the transformation matrix A, we need the first derivative of $\Theta[A]$ with respect to A_{ml}

$$\frac{\partial\Theta}{\partial A_{ml}} = 2\sum_j (Q_{mj}^l + \omega T_{mj})A_{jl} \tag{1.19}$$

where $Q_{mj}^l = \tilde{R}_{mj} - 2(x_l^0 X_{mj} + y_l^0 Y_{mj} + z_l^0 Z_{mj}) + \tilde{r}_l^0$. We can see that different NOLMOs are not coupled. The NOLMOs can be optimized individually. For a given centroid l, finding the best NOLMOs of centroid l is equivalent to solving the following eigenvalue equation in the system

$$\sum_j \Theta_{mj}^l A_{jl} = \theta_p^l A_{ml}^p \tag{1.20}$$

where $\Theta_{mj}^l = Q_{mj}^l + \omega T_{mj}$. The eigenvector with the lowest eigenvalue θ_0^l is the best localized NOLMO with centroid l

$$\varphi_l = \sum_m A_{ml}^0 \psi_m \tag{1.21}$$

Figure 1.3 depicts the real space representation of a π NOLMO distribution with different weights of the kinetic energy. From Figure 1.3 one can see that without the kinetic energy (i.e. $\omega = 0$) included in the cost function, the oscillation of NOLMO persists almost to the whole system, leading to poor localization. With the weight of the kinetic energy becomes larger, the orbital becomes less oscillatory and thus more localized.

However, as the weight of the kinetic energy increases, the orbitals become more diffuse, as shown in Figure 1.3. It is easy to understand the limit when the weight becomes infinity, the optimized orbital will be the lowest eigenvalue of the kinetic energy operator, and it will be delocalized in the whole space. Thus, the weight of

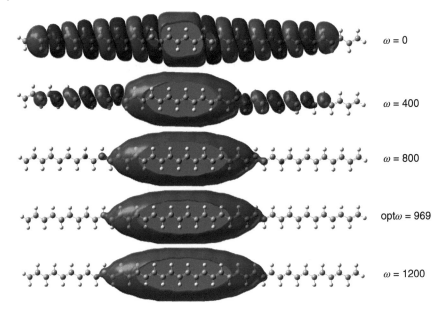

Figure 1.3 Real space representation of a π NOLMO with different weights of the kinetic energy tested on polyacetylene (H−(C=C)$_n$−H, n=20) at B3LYP/6-31G level, the buffer size is 19.0 Å, and the isosurface value is 10^{-5} e$^-$/bohr^3. ω is the weight of the kinetic energy of the cost function, while optω is the optimal value of ω. The effect of the regularization is evident – it greatly reduces the oscillation of the orbitals. Source: Reproduced with permission from Ref. [53]. Copyright 2022 American Chemical Society.

the kinetic energy term should not be too large. The weight of the kinetic energy can be determined by the minimization of the total energy of the system. It is found that with an optimized weight of the kinetic energy, one can get the most compact NOLMO.

The Divide-and-Conquer (DC) method developed by Yang [24] is to divide the entire system into fragments and then sum up the densities of the fragments for the entire system. By using the regularized NOLMOs for each fragment, NOLMO-DC is employed to obtain the electron density and the total energy. It is found that the total energy calculated by NOLMO-DC with optimal weight of the kinetic energy as regularization is more accurate than that without kinetic energy included in the cost function. Figure 1.4 depicts the accuracy in the total energy of NOLMO-DC and DC versus the buffer size with and without the kinetic energy included in the localization cost function for a conjugate system, polyacetylene (H−(C=C)$_n$−H, $n = 20$) at B3LYP/6-31G(d,p) and HF/6-31G(d,p) levels. With a small buffer size (4.0 Å) the accuracy of NOLMO-DC is higher than that of DC by one order of magnitude. The accuracy of NOLMO-DC is about 10^{-3} a.u. for a system of 40 carbon atoms. The accuracy of the original DC approaches the same level requiring a much larger buffer size (14.0 Å). With the buffer size larger than 9.0 Å the accuracy of NOLMO-DC is higher than DC by more than three orders of magnitude. While to achieve the same accuracy, the original DC will require the buffer size larger than 24.0 Å. These results

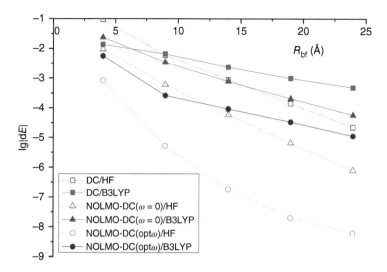

Figure 1.4 Accuracy of NOLMO-DC and DC tested on polyacetylene (H–(C=C)$_n$–H, n = 20) at B3LYP/6-31G(d,p) and HF/6-31G(d,p) levels for different values of ω and buffer sizes. ω is the weight of the kinetic energy of the cost function, while optω is the optimal value of ω determined by optimization with the analytical gradients. dE is the absolute difference between conventional energy and the DC or NOLMO-DC energy. Source: Reproduced with permission from Ref. [53]. Copyright 2022 American Chemical Society.

show that the buffer size for NOLMO-DC is only about one-third of that of the original DC method to achieve the similar accuracy level.

1.7 Molecular Orbitalets

FMOs have been successful in identifying chemically reactive groups in small systems by utilizing associated orbital energies [30, 54–59]. However, FMOs may not effectively highlight the locality of chemical reactivity in large systems due to the delocalization nature of CMOs. To address this limitation, the concept of frontier molecular orbitalets (FMOLs) was introduced to describe the reactivity of larger systems [60]. FMOLs are localized molecular orbitals that accurately reflect the frontier nature of chemical processes and offer a more comprehensive approach for identifying the localization of chemical reactivity.

Molecular orbitalets are a set of LMOs originally developed to capture the local fractional charges and spins of chemical systems for localized orbital scaling correction (LOSC) [61, 62] and fractional-spin LOSC [63] methods. These methods were designed to reduce and eliminate the delocalization error that leads to a series of systematic errors of DFT [64–68].

Orbitalets are constructed by linearly combining the CMOs, $\phi_p = \sum_q U_{pq} \varphi_q$, to minimize the following cost function,

$$F = (1 - \gamma)F_r + \gamma CF_e \qquad (1.22)$$

$$F_r = \sum_p [\langle \phi_p | \mathbf{r}^2 | \phi_p \rangle - \langle \phi_p | \mathbf{r} | \phi_p \rangle^2] \tag{1.23}$$

$$F_e = \sum_p [\langle \phi_p | \hat{h}^2 | \phi_p \rangle - \langle \phi_p | \hat{h} | \phi_p \rangle^2] \tag{1.24}$$

where $\{\phi_p\}$ are orbitalets, and $\{\varphi_p\}$ are CMOs. C is used to match magnitudes and unify units between the energy and physical spaces, and γ is the weight of the energy delocalization penalty. The cost function can be split into two parts: the physical space part F_r and the energy space part F_e. F_r, taken from Foster–Boys localization [33], shows the extent of spatial spread of the orbitalets, with higher values indicating greater spread. F_e, on the other hand, measures energy delocalization by quantifying how much the orbitalets deviate from the eigenstates of the Hamiltonian (i.e. CMOs). Lower values of F_e indicate less loss of energy information. The weights of the two parts can be adjusted accordingly by varying the value of γ. If γ is set to 1, the physical space part F_r does not contribute to the cost function, and therefore the resulting orbitalets will be the same as the CMOs. If γ is set to 0, the energy space part F_e will not contribute to the cost function, resulting in a very large deviation of the orbitalets from the eigenstates of the Hamiltonian. In this scenario, the orbitals will lose energy information and achieve maximal localization in physical space, resulting in generalized Foster–Boys localized orbitals.

We use hexadeca-1,3,5,7,9,11,13,15-octaene to illustrate the concept of orbitalets. Figure 1.5a displays a comparison between the energy structures of CMOs and LMOs obtained from Foster–Boys localization. The bar graph of CMOs clearly shows the fine energy structure. However, when γ is set to 0, the overall energy distribution is lost and there is barely any structure. Conversely, when γ is set to 0.975, although the fine structure of energy is less clear than that of CMOs, the resulting orbitalets still retain the main energy information (Figure 1.5b). Similar to the concepts of HOMO and LUMO, here we define the concepts of the highest occupied molecular orbitalet (HOMOL) and the lowest unoccupied molecular orbitalet (LUMOL). The orbitalets represent a combination of CMOs that are close in energy. As a result, a set of "frontier" orbitalets located in distinct local regions of the system can be obtained. Therefore, orbitalets can maintain the major energy information of CMOs while sacrificing some energy structure for locality information in the physical space. Moreover, when γ is close to 1, the loss of energy structure is minor, which makes FMOLs [60] a powerful tool for studying active sites, particularly in extensive chemical systems.

The use of FMOs, specifically HOMO and LUMO, has been widely employed in understanding the reactivity and regioselectivity of various chemical systems [30, 54–59]. Although FMOs have achieved great success in small systems, their delocalization nature becomes a limitation when applied to large systems. The FMOs of large systems tend to span over a significant portion of the system, leading to the loss of important locality information necessary to identify the functional groups that play a significant role in the reaction.

Explaining Diels–Alder (DA) reactions is one of the most prominent applications of FMO theory. FMOs of the diene and dienophile interact. Chemists typically focus

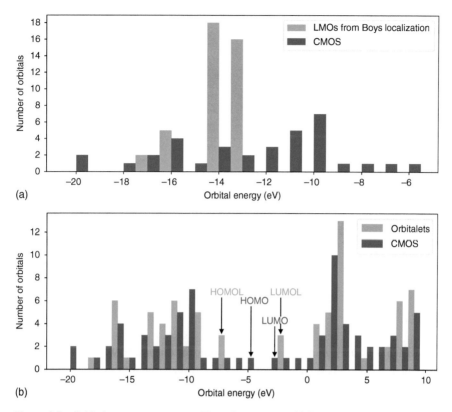

Figure 1.5 Orbital energy structures of hexadecaoctaene. (a) Energy structures shown by CMOs ($\gamma = 1$) and that shown by LMOs from Foster–Boys localization ($\gamma = 0$, only occupied orbitals are localized). (b) Energy structures shown by CMOs ($\gamma = 1$) and that shown by orbitalets ($\gamma = 0.975$). Source: Adapted from Yu et al. [60].

on "reactive functional groups" when analyzing chemical reactions. For example, in the hypothetical DA reaction between hexadeca-1,3,5,7,9,11,13,15-octaene and ethene [69], one would expect the frontier orbitals of hexadecaoctaene to be butadiene-like orbitals located at two adjacent double bonds in the middle. However, as shown in Figure 1.6a,b, the FMOs are undesirably delocalized over the entire molecule, making it challenging to use them to describe chemical reactivity. In comparison, FMOLs contain both energy and locality information, enabling HOMOLs and LUMOLs to clearly identify the most reactive site. As demonstrated in Figure 1.6c,d, the HOMOL and the LUMOL of hexadecaoctaene are mostly localized on the central adjacent double bonds and have butadiene-like shapes. This aligns with the traditional understanding of reactive functional groups. Thus, FMOLs expand the scope of FMOs beyond small systems, providing valuable energy and locality information to describe the reactivity of large systems. In addition to this, FMOLs can also generate promising results across various applications including providing fast analysis of chemical reactions based on transition states, and capturing electron transitions in excitation charge-transfer processes [60].

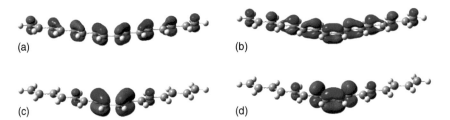

Figure 1.6 FMOs and FMOLs of hexadecaoctaene. FMOs, (a) HOMO and (b) LUMO, bury the butadiene-like model in delocalized orbitals. FMOLs, (c) HOMOL and (d) LUMOL, mainly locate on the central two double bonds, resembling the butadiene-like orbitals. Iso = 0.05, $\gamma = 0.975$.

Acknowledgment

J.Y. and W.Y. have been supported by National Science Foundation (Grant No. CHE-2154831).

References

1 Jensen, F. (2017). *Introduction to Computational Chemistry*. Wiley.
2 Pauling, L. (1932). Interatomic distances in covalent molecules and resonance between two or more Lewis electronic structures. *Proc. Natl. Acad. Sci. U.S.A.* 18: 293–297.
3 Pauling, L. (1932). The electronic structure of the normal nitrous oxide molecule. *Proc. Natl. Acad. Sci. U.S.A.* 18: 498–499.
4 Pauling, L. and Yost, D.M. (1932). The additivity of the energies of normal covalent bonds. *Proc. Natl. Acad. Sci. U.S.A.* 18: 414–416.
5 Heitler, W. and London, F. (1927). Wechselwirkung neutraler atome und homöopolare bindung nach der quantenmechanik. *Z. Angew. Phys.* 44: 455–472.
6 Hund, V.F. (1926). Zur deutung einiger erscheinungen in den molekelspektren. *Z. Angew. Phys.* 36: 657–674.
7 Hund, V.F. (1927). Zur deutung der molekelspektren. I. *Z. Phys.* 37: 742–764.
8 Hund, V.F. (1927). Zur deutung der molekelspektren. II. *Z. Phys.* 42: 93–120.
9 Mulliken, R.S. (1932). Electronic structures of polyatomic molecules and valence. *Phys. Rev.* 40: 55–71.
10 Hückel, E. (1931). Quantentheoretische beiträge zum benzolproblem, I. die elekfronenkonfigurafion des benzols und verwandfer verbindungen. *Z. Angew. Phys.* 70: 204–286.
11 Frisch, M.J., Trucks, G.W., Schlegel, H.B. et al. Wallingford, CT: Gaussian, Inc. Copyright © 1988–2017.
12 Schmidt, M.W., Baldridge, K.K., Boatz, J.A. et al. (1993). General atomic and molecular electronic structure system. *J. Comput. Chem.* 14(11): 1347–1363.
13 Gordon, M.S. and Schmidt, M.W. (2005). *Advances in Electronic Structure Theory: GAMESS a Decade Later*, 1167–1189. Elsevier.

References

14 Lennard-Jones, J.E. (1929). Electronic structures of polyatomic molecules and valence. *Trans. Faraday Soc.* 25: 668–686.

15 Lewis, G.N. (1916). The atom and the molecule. *J. Am. Chem. Soc.* 38: 762–785.

16 Langmuir, I. (1919). The arrangement of electrons in atoms and molecules. *J. Am. Chem. Soc.* 41: 868–934.

17 Hohenberg, P. and Kohn, W. (1964). Inhomogeneous electron gas. *Phys. Rev.* 136(3B): B864–B871.

18 Kohn, W. and Sham, L.J. (1965). Self-consistent equations including exchange and correlation effects. *Phys. Rev.* 140(4A): A1133–A1138.

19 Parr, R.G. and Yang, W. (1989). *Density-Functional Theory of Atoms and Molecules*. Oxford University Press.

20 Kirtman, B. (1982). Molecular electronic structure by combination of fragments. *J. Chem. Phys.* 86: 1059–1064.

21 Kirtman, B. and Demelo, C.P. (1986). Accurate local-space treatment of hydrogen bonding in large systems. *Int. J. Quantum Chem.* 29: 1209–1222.

22 Kirtman, B. and Demelo, C.P. (1987). Local space approximation for treatment of impurities in polymers. Solitons in polyacetylene. *J. Chem. Phys.* 86: 1624–1631.

23 Anikin, N.A., Anisimov, V.M., Bugaenko, V.L. et al. (2004). LocalSCF method for semiempirical quantum-chemical calculation of ultralarge biomolecules. *J. Chem. Phys.* 121(3): 1266–1270.

24 Yang, W. (1991). Direct calculation of electron density in density-functional theory. *Phys. Rev. Lett.* 66(11): 1438.

25 Imamura, A., Aoki, Y., and Maekawa, K. (1991). A theoretical synthesis of polymers by using uniform localization of molecular orbitals: proposal of an elongation method. *J. Chem. Phys.* 95(7): 5419–5431.

26 Korchowiec, J., Lewandowski, J., Makowski, M. et al. (2009). Elongation cutoff technique armed with quantum fast multipole method for linear scaling. *J. Comput. Chem.* 30(15): 2515–2525.

27 Aoki, Y. and Gu, F.L. (2012). Elongation method for delocalized nano-wires. *Prog. Chem.* 24(6): 886–909.

28 Kitaura, K., Ikeo, E., Asada, T. et al. (1999). Fragment molecular orbital method: an approximate computational method for large molecules. *Chem. Phys. Lett.* 313(3–4): 701–706.

29 Li, S., Li, W., and Fang, T. (2005). An efficient fragment-based approach for predicting the ground-state energies and structures of large molecules. *J. Am. Chem. Soc.* 127: 7215–7226.

30 Fukui, K., Yonezawa, T., and Shingu, H. (1952). A molecular orbital theory of reactivity in aromatic hydrocarbons. *J. Chem. Phys.* 20(4): 722–725.

31 Woodward, R.B. (1973). The total synthesis of vitamin B12. *Pure Appl. Chem.* 33(1): 145–178. https://doi.org/10.1351/pac197333010145.

32 Boys, S.F. (1960). Construction of some molecular orbitals to be approximately invariant for changes from one molecule to another. *Rev. Mod. Phys.* 32(2): 296–299.

33 Foster, J.M. and Boys, S.F. (1960). Canonical configurational interaction procedure. *Rev. Mod. Phys.* 32(2): 300–302.

34 Marzari, N. and Vanderbilt, D. (1997). Maximally localized generalized Wannier functions for composite energy bands. *Phys. Rev. B* 56(20): 12847–12865.

35 Edmiston, C. and Ruedenberg, K. (1963). Localized atomic and molecular orbitals. *Rev. Mod. Phys.* 35: 457–464.

36 Pipek, J. and Mezey, P.G. (1989). A fast intrinsic localization procedure applicable for ab initio and semiempirical linear combination of atomic orbital wave functions. *J. Chem. Phys.* 90: 4916–4926.

37 Pipek, J. (2000). Unique positive definite extension of Mulliken's charge populations of non-orthogonal atomic basis functions. *J. Mol. Struct. THEOCHEM* 501–502: 395–401.

38 Löwdin, P.-O. (1955). Quantum theory of many-particle systems. I. Physical interpretations by means of density matrices, natural spin-orbitals, and convergence problems in the method of configurational interaction. *Phys. Rev.* 97(6): 1474–1489.

39 Reed, A.E. and Weinhold, F. (1983). Natural bond orbital analysis of near-Hartree–Fock water dimer. *J. Chem. Phys.* 78(6): 4066–4073.

40 Gu, F.L., Aoki, Y., Korchowiec, J. et al. (2004). A new localization scheme for the elongation method. *J. Chem. Phys.* 121(1–3): 10385–10391.

41 Adams, W.H. (1961). On the solution of the Hartree-Fock equation in terms of localized orbitals. *J. Chem. Phys.* 34(1): 89–102.

42 Gilbert, T.L. and Lykos, P.G. (1961). Maximum-overlap directed-hybrid orbitals. *J. Chem. Phys.* 34(1): 2199–2200.

43 Krol, M.C. and Altona, C. (1991). Theoretical investigations of the nature of intramolecular interactions: IV. Transferability of non-orthogonal molecular orbitals and application to mono-and difluoromethane. *Mol. Phys.* 72(2): 375–393.

44 Sorakubo, K., Yanai, T., Nakayama, K. et al. (2003). A non-orthogonal Kohn-Sham method using partially fixed molecular orbitals. *Theor. Chem. Acc.* 110: 328–337.

45 Paulus, B., Rościszewski, K., Stoll, H., and Birkenheuer, U. (2003). Ab initio incremental correlation treatment with non-orthogonal localized orbitals. *Phys. Chem. Chem. Phys.* 5(24): 5523–5529.

46 Sironi, M., Famulari, A., Raimondi, M., and Chiesa, S. (2000). The transferability of extremely localized molecular orbitals. *J. Mol. Struct. THEOCHEM* 529(1–3): 47–54.

47 Peng, L., Gu, F.L., and Yang, W. (2013). Effective preconditioning for ab initio ground state energy minimization with non-orthogonal localized molecular orbitals. *Phys. Chem. Chem. Phys.* 15(37): 15518–15527.

48 Meyer, B., Guillot, B., Ruiz-Lopez, M.F., and Genoni, A. (2016). Libraries of extremely localized molecular orbitals. 1. Model molecules approximation and molecular orbitals transferability. *J. Chem. Theory Comput.* 12(3): 1052–1067.

References

49 Meyer, B., Guillot, B., Ruiz-Lopez, M.F. et al. (2016). Libraries of extremely localized molecular orbitals. 2. Comparison with the pseudoatoms transferability. *J. Chem. Theory Comput.* 12(3): 1068–1081.

50 Yang, W. (1997). Absolute-energy-minimum principles for linear-scaling electronic-structure calculations. *Phys. Rev. B* 56(15): 9294.

51 Burger, S.K. and Yang, W. (2008). Linear-scaling quantum calculations using non-orthogonal localized molecular orbitals. *J. Phys.: Condens. Matter* 20(29): 294209.

52 Cui, G., Fang, W., and Yang, W. (2010). Reformulating time-dependent density functional theory with non-orthogonal localized molecular orbitals. *Phys. Chem. Chem. Phys.* 12(2): 416–421.

53 Peng, L., Peng, D., Gu, F.L., and Yang, W. (2022). Regularized localized molecular orbitals in a divide-and-conquer approach for linear scaling calculations. *J. Chem. Theory Comput.* 18: 2975–2982.

54 Fleming, I. (1977). *Frontier Orbitals and Organic Chemical Reactions*. Wiley.

55 Fleming, I. (2011). *Molecular Orbitals and Organic Chemical Reactions*. Wiley.

56 Fujimoto, H. and Fukui, K. (1997). *Frontier Orbitals and Reaction Paths: Selected Papers of Kenichi Fukui*, vol. 7. World Scientific.

57 Nguyen, A.Q., Anh, N.T., and Nguyên, T.A. (2007). *Frontier Orbitals: A Practical Manual*. Wiley.

58 Parr, R.G. and Yang, W. (1984). Density functional approach to the frontier-electron theory of chemical reactivity. *J. Am. Chem. Soc.* 106(14): 4049–4050.

59 Woodward, R.B. and Hoffmann, R. (1969). The conservation of orbital symmetry. *Angew. Chem. Int. Ed. Engl.* 8(11): 781–932.

60 Yu, J., Su, N.Q., and Yang, W. (2022). Describing chemical reactivity with frontier molecular orbitalets. *JACS Au* 2(6): 1383–1394.

61 Li, C., Zheng, X., Su, N.Q., and Yang, W. (2017). Localized orbital scaling correction for systematic elimination of delocalization error in density functional approximations. *Natl. Sci. Rev.* 5(2): 203–215.

62 Su, N.Q., Mahler, A., and Yang, W. (2020). Preserving symmetry and degeneracy in the localized orbital scaling correction approach. *J. Phys. Chem. Lett.* 11(4): 1528–1535.

63 Su, N.Q., Li, C., and Yang, W. (2018). Describing strong correlation with fractional-spin correction in density functional theory. *Proc. Natl. Acad. Sci. U.S.A.* 115(39): 9678–9683.

64 Cohen, A.J., Mori-Sánchez, P., and Yang, W. (2008). Fractional spins and static correlation error in density functional theory. *J. Chem. Phys.* 129(12): 121104.

65 Cohen, A.J., Mori-Sánchez, P., and Yang, W. (2008). Insights into current limitations of density functional theory. *Science* 321(5890): 792–794.

66 Cohen, A.J., Mori-Sánchez, P., and Yang, W. (2011). Challenges for density functional theory. *Chem. Rev.* 112(1): 289–320.

67 Mori-Sánchez, P., Cohen, A.J., and Yang, W. (2008). Localization and delocalization errors in density functional theory and implications for band-gap prediction. *Phys. Rev. Lett.* 100(14): 146401.

68 Zheng, X., Cohen, A.J., Mori-Sánchez, P. et al. (2011). Improving band gap prediction in density functional theory from molecules to solids. *Phys. Rev. Lett.* 107(2): 026403.

69 Zhang, J.-X., Sheong, F.K., and Lin, Z. (2018). Unravelling chemical interactions with principal interacting orbital analysis. *Chem. Eur. J.* 24(38): 9639–9650.

2

Chemical Concepts from *Ab Initio* Valence Bond Theory

Chen Zhou, Fuming Ying, and Wei Wu

Xiamen University, State Key Laboratory of Physical Chemistry of Solid Surfaces, Fujian Provincial Key Laboratory of Theoretical and College of Chemistry and Chemical Engineering, No. 422, Siming South Road, Xiamen, Fujian 361005, China

2.1 Introduction

Valence bond (VB) theory, as one of the modern chemical bonding theories, provides intuitive physical pictures for classical chemical concepts and discovers the intrinsic nature of chemical bonds. The classical VB theory originated in the work of Lewis in 1916 [1] and was first developed by Heitler and London, who applied quantum mechanics to the hydrogen molecule in 1927 [2]. Shortly after, Slater extended the Heitler–London approach to polyatomic molecules [3]; Rumer proposed a rule, named Rumer's rule now, to generate independent VB structures [4]; Pauling introduced a series of chemical concepts, including covalent and ionic bonding, orbital hybridization, electron pairing, and resonance structures [5–9]. Since then, VB theory has become a popular chemical bonding theory, providing intuitive insights into various molecular structures and chemical reactions and dominating chemists' interpretations of various chemical problems until the 1950s. However, as electronic computers became popular in scientific research in the 1960s, theoretical chemists' interest gradually turned to quantitative quantum mechanical approaches. Due to the use of nonorthogonal atomic orbitals, VB theory became unhelpful for the numerical calculation of molecules. On the other hand, molecular orbital (MO) theory, developed almost at the same time as VB theory in the late 1920s, benefits from the simple form of wave function with orthogonal MOs in the Hartree–Fock method and gradually became dominant in studying molecular structure. In short, VB theory has declined since the 1950s and is considered an obsolete theory, while MO theory has become the sole wave function approach for modern quantum chemistry computation. However, thanks to the rapid development of computer science, VB theory has been enjoying a strong comeback since the 1980s, and *ab initio* quantum chemistry calculations can be performed with VB methods [10–15]. Although modern VB theory is still less popular nowadays than MO theory, it is a powerful and unique tool for understanding chemical bonding and concepts in depth.

Exploring Chemical Concepts Through Theory and Computation, First Edition. Edited by Shubin Liu.
© 2024 WILEY-VCH GmbH. Published 2024 by WILEY-VCH GmbH.

Based on the forms of many-electron wave functions and the types of orbitals used in wave function, *ab initio* VB methods are generally categorized into three classes, namely, classical, modern, and MO-based VB methods. Since we are focusing on the chemical concepts of VB theory in this chapter, we will not discuss more details of the modern and MO-based VB methods, in which the clear classical VB pictures are lost due to the use of delocalized orbitals. In the classical VB methods, the wave function is expanded in terms of many-electron VB functions built with strictly localized hybrid atomic orbitals (HAOs) or fragment-localized orbitals. The most commonly used many-electron VB functions are Heitler–London–Slater–Pauling (HLSP) functions, which are also called VB structures in classical VB theory. A VB structure exactly corresponds to a Lewis structure and is comprised of 2^n Slater determinants (n is the number of covalent bonds in a structure). Thus, VB methods are inherently multi-determinant wave function methods and are capable of intuitively describing the forming and breaking of chemical bonds and strongly correlated systems with complicated electronic structures. However, *ab initio* VB calculation still requires a higher computational cost compared with MO theory due to the use of nonorthogonal orbitals. For the last three decades, novel VB methods, algorithms, and programs have been developed that significantly reduce the cost of *ab initio* VB calculations [16–23].

The aim of this chapter is to provide some fundamental concepts and methods in VB theory and their applications to chemical concepts in the nature of chemical bonding. We will not go into VB methods and algorithms too technically in this chapter. Instead, a brief guide to perform VB calculations with Xiamen Valence Bond [24, 25] (XMVB) is presented. For the advance in *ab initio* VB methods over recent decades, interested readers can read some recently published reviews [12–14, 26]. In Section 2.2, we will introduce the fundamentals of *ab initio* VB theory. In Section 2.3, several fundamental chemical concepts derived from VB theory are presented. In Section 2.4, we provide a brief guide to perform a VB calculation with XMVB. And a concluding remark is given in Section 2.5.

2.2 Ab Initio Valence Bond Theory

2.2.1 Valence Bond Self-Consistent Field Method

In VB theory, the many-electron wave function of a molecule is expressed in terms of VB functions $\{\Phi_K\}$ as

$$\Psi = \sum_K C_K \Phi_K, \tag{2.1}$$

where Φ_K is a VB structure corresponding to a specific Lewis structure in classical VB theory, and C_K is its corresponding coefficient. As the most fundamental and significant classical VB method, valence bond self-consistent field (VBSCF) [27–29] defines VB functions by HLSP functions, and VB orbitals and structural coefficients are optimized simultaneously to minimize the total electronic energy.

2.2 Ab Initio Valence Bond Theory

In spin-free quantum chemistry, an HLSP function is a spin-adapted function, written as

$$\Phi_K = \hat{A}\Omega_K\Theta_K, \tag{2.2}$$

where $\hat{A}$ is the antisymmetrization operator, Ω_K is a direct product of occupied spatial VB orbitals $\{\phi_{K_i}\}$,

$$\Omega_K = \phi_{K_1}(1)\phi_{K_2}(2)\cdots\phi_{K_N}(N), \tag{2.3}$$

and the spin-paired eigenfunction Θ_K is

$$\Theta_K = \prod_{(ij)} 2^{-1/2}[\alpha(i)\beta(j) - \beta(i)\alpha(j)]\prod_k \alpha(k). \tag{2.4}$$

Here, (ij) runs over all n covalent bonds and k runs over all unpaired electrons, which implies that a HLSP function is expanded as a linear combination of 2^n determinants.

In order to find the electronic states of interest, one can solve the generalized eigenvalue equation

$$\mathbf{HC} = \mathbf{MC}\varepsilon, \tag{2.5}$$

where $\mathbf{H}$ and $\mathbf{M}$ are, respectively, Hamiltonian and overlap matrices constructed in the basis of HLSP functions. One of the advantages of the VBSCF method is that each HLSP function exactly corresponds to a classical VB structure, and thus VBSCF straightforwardly represents classical VB theory at the *ab initio* level, providing very clear and intuitive chemical insights into the chemical bonding nature and reactivity mechanism. As the HLSP functions are not orthogonal to each other, the importance of each VB structure is estimated by its weight W_K, instead of the square of structural coefficients used in MO theory. The definition of VB structural weights is nonunique, and the most commonly used definitions are the Coulson–Chirgwin formula [30]

$$W_K = \sum_L C_K C_L M_{KL}, \tag{2.6}$$

and the Löwdin weight formula [31]

$$W_K = \left(\sum_L C_L M_{KL}^{1/2}\right)^2 \tag{2.7}$$

to avoid negative values of the structural weights.

Here we take a VBSCF calculation (done with XMVB package) of H_2 molecule as an example. As a multiconfigurational quantum chemical method, a VBSCF calculation starts with the definition of the configuration space, i.e. active space (m,n), in which m active electrons arbitrarily distribute on n active orbitals. According to Eqs. (2.2)–(2.4), one can write totally three VB structures for hydrogen molecule by considering all singlet spin coupling patterns in the (2,2) active space. The three VB structures, namely, covalent structure Φ^{cov} and two ionic structures Φ_1^{ion} and Φ_2^{ion}, are written as

$$\Phi^{\text{cov}} = \frac{1}{\sqrt{2+2s_{ab}}}(|\phi_a\bar{\phi}_b| - |\bar{\phi}_a\phi_b|), \tag{2.8}$$

$$\Phi_1^{ion} = |\phi_a \bar{\phi}_a|, \tag{2.9}$$

$$\Phi_2^{ion} = |\phi_b \bar{\phi}_b|, \tag{2.10}$$

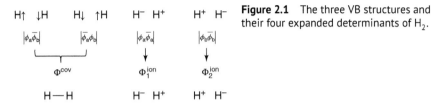

Figure 2.1 The three VB structures and their four expanded determinants of H_2.

where ϕ_a and ϕ_b are, respectively, the atomic orbitals of hydrogens a and b, s_{ab} is the orbital overlap between ϕ_a and ϕ_b, and the bar over an orbital means a β spin orbital. Since no covalent bond is formed in the two ionic structures, each of the ionic structure is described by $2^0 = 1$ determinant. The covalent structure consists of one covalent bond and is hence comprised of $2^1 = 2$ determinants, namely, $|\phi_a \bar{\phi}_b|$ and $|\bar{\phi}_a \phi_b|$. Note that neither of the determinants $|\phi_a \bar{\phi}_b|$ nor $|\bar{\phi}_a \phi_b|$ is the eigenfunction of S^2, and thus should be combined to form a spin-adapted VB structure. The three VB structures and their four expanded determinants are illustrated in Figure 2.1.

The VBSCF wave function of H_2 molecule can thus be written as

$$\Psi = C^{cov}\Phi^{cov} + C^{ion}\left(\Phi_1^{ion} + \Phi_2^{ion}\right), \tag{2.11}$$

where the coefficients of the two ionic structures are the same according to the symmetry of the H_2 molecule. With cc-pVTZ basis set, the Coulson–Chirgwin weights of the three VB structures Φ^{cov}, Φ_1^{ion}, and Φ_2^{ion} at the equilibrium bond length ($R = 0.75$ Å) are 0.82, 0.09, and 0.09, respectively, which indicates that the covalent bonding dominates the H_2 molecule, but the ionic character still plays a nonnegligible role for the homopolar bond. As the H–H bond is stretched, the weight of Φ^{cov} gradually increases while that of the ionic structures decreases. At the dissociation limit, the H–H bonding is purely covalent, which corresponds to the two neutral hydrogen atoms after bond dissociation. Figure 2.2 plots the weights of the covalent and ionic structures along the dissociation of H_2 molecule. We demonstrate again that the VB theory is naturally multi-determinantal, even when only one covalent structure is involved in the wave function for bond breaking. The multi-determinant feature enables VB theory to correctly treat bond breaking and strongly correlated systems.

One may notice that the covalent pairing patterns (ij) in Eq. (2.4) are mathematically nonunique, but not all types of covalent pairing patterns provide clear pictures that can be interpreted from a chemical perspective. One of the pairing schemes that gives clear chemical pictures is Rumer–Weyl's rule [32] (hereinafter referred to as Rumer's rule).

2.2.2 Rumer Structures

With the help of Rumer's rule, one can obtain all the linearly independent Rumer structures in a given active space. Although Rumer's rule is valid for open-shell or

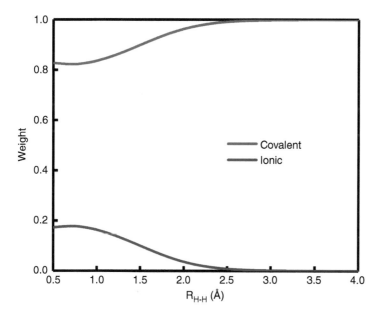

Figure 2.2 The weights of the covalent and ionic structures along the dissociation of H_2 molecule.

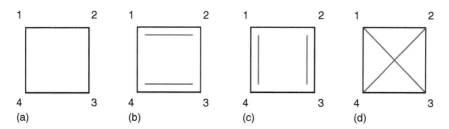

Figure 2.3 (a) Model molecule and numbering of cyclobutadiene; (b–c) two Rumer structures (Φ_1 and Φ_2) for covalent bonding in (4,4) active space; (d) the structure (Φ_3) that violates the Rumer's rule.

more complicated systems, we will focus on singlet molecules and covalent bonding in this section.

Here we present how to draw Rumer structures graphically by taking the cyclobutadiene molecule as an example. Considering the covalent bonding of electrons on the four p_π orbitals perpendicular to the cyclobutadiene plane, i.e. (4,4) active space, we first draw a regular quadrilateral as a model molecule, each vertex of the quadrilateral representing a p_π orbital, and then number the vertices clockwise or counterclockwise (shown in Figure 2.3a). We connect each of the two vertices with a straight line until each of the vertices has one and only one connection to another vertex, and note that the lines are forbidden to cross in Rumer structures. Therefore, we obtain two linearly independent Rumer structures (Φ_1 and Φ_2) for covalent bonding, shown in Figure 2.3b,c. Since there are two covalent bonds in each structure, each of the two Rumer structures can be written as a linear combination of $2^2 = 4$ determinants as

$$\Phi_1 = |(1\bar{2} - \bar{1}2)(3\bar{4} - \bar{3}4)| = -|13\bar{2}\bar{4}| - |14\bar{2}\bar{3}| - |23\bar{1}\bar{4}| - |24\bar{1}\bar{3}|, \quad (2.12)$$

$$\Phi_2 = |(1\bar{4} - \bar{1}4)(3\bar{2} - \bar{3}2)| = |13\bar{2}\bar{4}| + |12\bar{3}\bar{4}| + |34\bar{1}\bar{2}| + |24\bar{1}\bar{3}|. \quad (2.13)$$

Structure Φ_3, as shown in Figure 2.3d, has crossing lines and hence is not a Rumer structure; in fact, Φ_3 can be expressed as a linear combination of Φ_1 and Φ_2 as

$$\Phi_3 = |(1\bar{3} - \bar{1}3)(2\bar{4} - \bar{2}4)| = -|12\bar{3}\bar{4}| + |14\bar{2}\bar{3}| + |23\bar{1}\bar{4}| - |34\bar{1}\bar{2}|, \quad (2.14)$$

$$\Phi_3 = -\Phi_1 - \Phi_2. \quad (2.15)$$

The VBSCF wave function involving Φ_1 and Φ_3 (or Φ_2 and Φ_3) thus gives the same energy as that involving the two Rumer structures but gives weird structural weights due to the loss of chemical implication in non-Rumer structure Φ_3.

Another example is the benzene molecule. If the six p_π orbitals of the benzene are involved in the covalent pairing, we can write a total of five covalent Rumer structures (a)–(e), as shown in Figure 2.4. We emphasize again that one could involve, for instance, structures (a)–(d) and (f) in the VB wave function and get the equivalent energy of the benzene molecule as involving the five Rumer structures, but such a VB structure basis will lose the intuitive pictures of the VB theory.

Besides the abovementioned graphical scheme, it is also possible to generate Rumer structures by using standard Young tableaux of symmetric group, and the latter scheme is more appropriate for routinization. To generate a Rumer structure with a Young tableau, we draw a table with two rows, the first row having n_α columns, and the second row having n_β columns, where n_α and n_β are, respectively, the number of active α and β electrons. Then we fill in the table with 1, 2, …, $(n_\alpha + n_\beta)$ so that the numbers of each row increase from left to right and the numbers of each column increase from top to bottom. Figure 2.5 plots the two possible Young tableaux (YH1 and YH2) of the cyclobutadiene molecule. Each

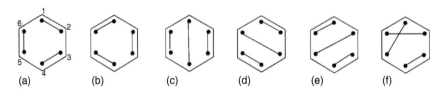

Figure 2.4 (a–e) The five covalent Rumer structures of the benzene molecule; (f) the "crossing" structure, which violates Rumer's rule.

YH1

1	2
3	4

YH2

1	3
2	4

Figure 2.5 Young tableaux of cyclobutadiene molecule.

YH1			YH2			YH3			YH4			YH5		
1	2	3	1	2	4	1	2	5	1	3	4	1	3	5
4	5	6	3	5	6	3	4	6	2	5	6	2	4	6

Figure 2.6 Young tableaux of benzene molecule.

Young tableau can be mapped to a Rumer structure by the following procedure: (1) find the largest number (x) in the first row of a Young tableau, and then pair it with the smallest number (y) in the second row, which is larger than x; (2) remove x and y from the Young tableau, and repeat step (1) until all the numbers are finally paired. We then find the pairing in YH1 is (2,3) and (1,4), and YH1 thus corresponds to the Rumer structure shown in Eq. (2.13), while the pairing in YH2 is (3,4) and (1,2), and therefore YH2 corresponds to the Rumer structure shown in Eq. (2.12).

Figure 2.6 plots the five standard Young tableaux (YH1 to YH5) for the covalent pairing of six p_π orbitals in a benzene molecule. According to the pairing procedure shown above, readers can verify that YH1 to YH5, respectively, correspond to Rumer structures (e), (b), (c), (d), and (a) shown in Figure 2.3. Overall, compared with the graphical scheme that generates Rumer structures, the Young tableau scheme provides a mathematically identical strategy that can be easily comprehended by the computer.

2.2.3 Orbitals in VB Wave Function

It is known that orbitals used in many-electron wave functions are one-electron functions that cannot be observed in reality and thus are not uniquely defined. The most commonly used orbitals are canonical MOs, which are obtained by solving the Hartree–Fock–Roothaan equation and delocalized over the whole molecule. One of the advantages of orthogonal MOs is their higher computational efficiency in *ab initio* calculations. However, the delocalization of MOs obscures the intuitive chemical picture to some extent, particularly for the nature of chemical bonding. On the contrary, in VB theory, much more localized orbitals are used for building the many-electron wave function, providing the most faithful way to retain the chemical intuition in classical VB theory. The VB orbitals may be defined as strictly localized or semi-localized in VB calculations, depending on the purpose of the study.

A strictly localized orbital, or also known as HAO, is constructed by atomic orbitals of one atom under the energy level approximation principle, but orthogonality is not necessary between hybrid orbitals located on different atoms. The hybrid orbital of an atom has a spatial orientation along the chemical bonds in the molecule, so that the bonding ability between hybrid orbitals is greatly enhanced compared with non-HAOs. Thus, the VB structures constructed by hybrid orbitals will give more intuitive bonding pictures that are intimately known to chemists. The sp, sp^2, and sp^3 hybridization schemes are the most common in text books. In VBSCF calculations, however, the hybrid coefficients of HAOs are optimized to minimize the total electronic energy of the molecule. One HAO type of p_π orbital of a benzene molecule is plotted in Figure 2.7a.

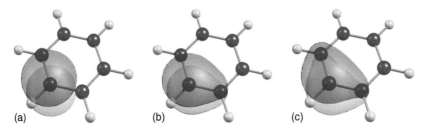

Figure 2.7 Illustration of various orbitals in VB theory. (a) HAO, (b) BDO, and (c) OEO.

Except for the HAOs, the VB orbitals are sometimes allowed to delocalize to some extent for specific purposes. If the atomic orbital is allowed to delocalize to another adjacent atom that is linked with the orbital as a covalent bond, such a VB orbital is called a bond-distorted orbital (BDO). If the atomic orbital is allowed to delocalize to a small extent over the whole molecule, such a VB orbital is called an overlap-enhanced orbital (OEO). Note that although both the OEOs and MOs are delocalized over the whole molecule, an OEO is mainly localized on one atom and has a small delocalization "tail" on other atoms. The BDO and OEO types of a p_π orbital of a benzene molecule are respectively plotted in Figure 2.7b,c. Despite the fact that the use of delocalized VB orbitals slightly loses the chemical pictures, it helps to quantitatively improve the accuracy while basically retaining the chemical intuition.

2.2.4 VB Methods Involving Dynamic Correlation

The multiconfiguration nature of VBSCF method enables VB theory to successfully describe the complicated electronic structure in strongly correlated systems. The wave function of a strongly correlated system is dominated by more than one configuration state function (CSF) or VB structure. The electronic correlation due to this degeneracy is known as static correlation. On the other hand, a more accurate description of strongly correlated systems also requires the inclusion of dynamic correlation. The dynamic correlation could be introduced by using post-SCF methods, such as perturbation theory (PT), configuration interaction (CI), coupled-cluster (CC), or density functional theory (DFT). Therefore, more precise and advanced VB methods have been proposed to involve both static and dynamic correlations. The commonly used post-VBSCF methods include breathing-orbital valence bond (BOVB) [33–35], valence bond configuration interaction (VBCI) [36, 37], valence bond perturbation theory (VBPT2) [20, 38], and density functional valence bond (DFVB) [39–43].

BOVB is a classical VB method that mimics a "breathing" effect by allowing different orbitals in different structures. For the H_2 molecule, the VBSCF wave function is shown in Eqs. (2.8)–(2.11), while the BOVB wave function is written as

$$\Psi^{BOVB} = C_1(|\phi_a \bar{\phi}_b| - |\bar{\phi}_a \phi_b|) + C_2 |\phi_a' \bar{\phi}_a'| + C_3 |\phi_b' \bar{\phi}_b'|. \tag{2.16}$$

Therefore, BOVB method gives lower energies compared with VBSCF by introducing partial dynamic correlation with extra degrees of freedom of the orbital coefficients. There are various levels of BOVB, including L-BOVB, D-BOVB, SL-BOVB, and SD-BOVB. In L-BOVB, localized HAOs are used for all orbitals, while delocalized OEOs are used for inactive orbitals in D-BOVB. "S" in SL-BOVB and SD-BOVB denotes splitting the doubly occupied orbitals for ionic bonds.

The VBCI is the most precise (though also the most expensive) classical VB method. In VBCI method, all the occupied orbitals are required to be HAOs so that the strictly localized virtual orbitals are constructed by using a projector $\mathbf{P}_A$ localized on an atom or fragment A:

$$\mathbf{P}_A = \mathbf{T}_A(\tilde{\mathbf{T}}_A \mathbf{M}_A \mathbf{T}_A)^{-1} \tilde{\mathbf{T}}_A \mathbf{S}_A, \tag{2.17}$$

where $\mathbf{T}_A$ is the coefficient of occupied orbitals localized on A, and $\mathbf{M}_A$ and $\mathbf{S}_A$ are, respectively, the overlap matrices of occupied orbitals and basis functions localized on A. The virtual orbitals localized on A are defined as the eigenvectors of $\mathbf{P}_A$ with 0 eigenvalue. The VBCI function (Φ_K^{CI}) associated with a VBSCF structure (Φ_K) is thus defined as

$$\Phi_K^{CI} = \sum_i C_K^i \Phi_K^i, \tag{2.18}$$

where $\{\Phi_K^i\}$ includes Φ_K and its excited VB structures generated by local excitations. The VBCI wave function is then expressed as a linear combination of Φ_K^{CI} as

$$\Psi^{VBCI} = \sum_K C_K^{CI} \Phi_K^{CI} = \sum_{Ki} C_{Ki} \Phi_K^i, \tag{2.19}$$

where $C_{Ki} = C_K^{CI} C_K^i$, and are optimized to minimize the VBCI energy.

The BOVB and VBCI are classical post-VBSCF methods, and the wave functions are constructed by HAOs. In VBPT2, however, the virtual orbitals are orthogonal, allowing delocalization over the whole molecule. Therefore, the excited VB structures in VBPT2 are not limited to local excitations. More details of VBPT2 can be found in Refs [20, 38].

Kohn–Sham DFT is widely applied in electronic structure calculations due to its cheap computational cost for dynamic correlation. Various DFVB methods that combine KS-DFT and VBSCF have been proposed to efficiently include dynamic correlation in VBSCF wave function. In the dynamically correlated DFVB (dc-DFVB) method [39], the DFT correlation is straightforwardly added to the VBSCF energy. In the Hamiltonian-corrected DFVB (hc-DFVB) method [40], the DFT correlation is added to each Hamiltonian matrix element. In a series of hybrid DFVB (λ-DFVB) methods [41–43], the electron–electron interaction is divided into the wave function and DFT parts with a hybrid parameter λ, which is related to the multiconfiguration character of the studied molecule. For more detailed methodology of the DFVB methods, see Refs. [39–43].

2.3 Chemical Concepts in VB Theory

2.3.1 Resonance Theory

The concept of resonance was first proposed by Pauling, and the resonance theory is still nowadays a powerful and efficient approach to understand the chemical

Table 2.1 VREs of F_2 molecule computed with VBSCF, BOVB, VBCI, and hc-DFVB methods.

Method	RE (kcal/mol)
VBSCF	41.7
L-BOVB	62.3
VBCI	54.0
hc-DFVB	49.4

The basis set is 6-31G*, and B3LYP functional is applied in hc-DFVB calculation.

bonding, molecular stability, and electron delocalization in conjugated molecules. Although resonance energy (RE) may be estimated with MO theory by using various orbital localization schemes, it can only be strictly defined with VB theory as

$$RE = E(\Psi) - E(\Phi_K), \tag{2.20}$$

where $E(\Psi)$ is the energy of the total VB wave function, while $E(\Phi_K)$ is the energy of the most stable VB structure Φ_K in Ψ.

The RE presents stabilization arising from the mixing of VB structures in classical VB theory. There are two approaches to compute the RE, namely, nonvariational and quasi-variational. For the nonvariational RE, $E(\Phi_K)$ in Eq. (2.20) is directly taken from the Hamiltonian matrix element H_{KK} in the VBSCF calculation. That is to say, the energy of structure K, $E(\Phi_K)$, is computed with the optimal orbitals of the total VBSCF wave function. For the quasi-variational RE, on the other hand, the computing of $E(\Phi_K)$ requires the orbital optimization of the specific VB structure Φ_K. The quasi-variational approach is considered to be more accurate to describe the resonance effect, while the nonvariational approach provides an approximate evaluation without orbital optimization of the upper bound of the RE.

Two kinds of RE, vertical RE (VRE) and adiabatic RE (ARE), can be defined according to whether the geometry of Φ_K is relaxed. The VRE is obtained if $E(\Phi_K)$ is computed at the same geometry as $E(\Psi)$, while the ARE is obtained if $E(\Phi_K)$ is computed at its own optimized geometry.

The RE can be evaluated at various levels of VB theory. Table 2.1 lists the VREs of F_2 molecule computed with VBSCF, BOVB, VBCI, and hc-DFVB methods. The post-VBSCF methods give larger RE compared with VBSCF with the inclusion of dynamic correlation. The VBCI method is the most precise among the four VB methods, and BOVB gives even larger RE than VBCI. This is because no orbital breathing effect is included in the calculation of $E(\Phi_K)$, and thus RE is overestimated in BOVB.

2.3.2 Conjugation, Hyperconjugation, and Aromaticity

The conjugation effect, also known as delocalization effect, mostly occurs in molecules in which the π electrons are considered to move between the parallel π

orbitals on multiple adjacent atoms. Such interaction among π electrons stabilizes the conjugation systems with smaller internal energy, and the bond lengths tend to be averaged compared with nonconjugated systems. The stabilization of conjugated systems can be evaluated as the RE between the energy of the total VB wave function and that of one VB structure with separated π bonding.

The hyperconjugation effect, on the other hand, refers to the electrons in a σ type of bond (usually C−H or C−C bond) interacting with nearby σ^* or π^* orbitals and thus stabilizing the whole system. A typical example of hyperconjugation effect is the interaction of C−H bonds in ethane molecules [44]. The stability of hyperconjugation in ethane can be directly evaluated with VB theory, and the steric hindrance effect is shown to dominate the rotational barrier in ethane.

Aromaticity is a significant and widely used chemical concept that describes the special stability of cyclic compounds. Large numbers of experimental and theoretical criteria have been proposed to diagnose aromaticity. Although it is recommended that aromaticity be integrally determined via various criteria, VB theory provides a direct approach to evaluate the extra stability of conjugated rings compared with their reference nonaromatic systems as

$$\text{ECRE} = \text{ARE(CM)} - \text{ARE(Ref)}, \tag{2.21}$$

where ECRE is the extra cyclic RE and denotes extra stability due to aromaticity; CM and Ref, respectively, denote conjugate molecule and reference molecule. Here we take the benzene molecule as an example [45]. The experimental evaluation of the ECRE is around 28.8 kcal/mol [46]. The ARE of the CM is the energy difference between the benzene molecule and the most stable VB structure ((a) or (b) in Figure 2.4). The reference molecule can be either *trans*-1,3,5-hexatriene (with the same number of double bonds as benzene) or *trans*-1,3,5,7-octatetraene (with the same number of diene conjugations); the former yields ECRE1, while the latter yields ECRE2. The VB calculation gives 36.7 and 25.7 kcal/mol for ECRE1 and ECRE2, respectively, indicating that ECRE2 is superior to ECRE1.

2.3.3 Electron-Pair Bonding in Valence Bond Theory

Electron-pair bonding is a central chemical paradigm to describe and interpret the interaction between atoms in a molecule. The concepts of covalent and ionic bonds can be dated back to the ingenious hypothesis of Lewis in 1916 and are still widely accepted and used nowadays. The covalent bonding comes from the electron spin pairing of adjacent atoms and thus strongly stabilizes the molecule, while the ionic bonding comes from the electrostatic interaction of two fragments with opposite charges. Typical covalent bonds are usually found between H, B, C, N, and O atoms, while typical ionic bonds are formed with one alkali or alkaline earth metal atom and one halogen atom. However, despite great successes of traditional electron-pair bonding in explaining molecular stability and chemical reaction mechanisms, the chemical bonding even in some simple molecules, like F_2 for instance, cannot be unambiguously classified as covalent or ionic.

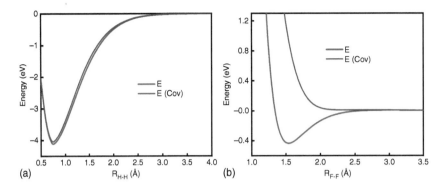

Figure 2.8 The dissociation curves of the total wave function and covalent structure for (a) H_2 and (b) F_2 molecules.

Figure 2.8 shows the dissociation curves of the total wave function and covalent structure for H_2 and F_2 molecules. For H_2, the potential energy curve (PEC) of covalent structure is a little higher than that of the total wave function, but the topology of both PECs is quite close. One thus concludes that the bonding in H_2 is dominated by the covalent interaction between the two H atoms. For F_2, however, the PEC of covalent structure is repulsive and much higher than that of the total wave function, and the topology of the two PECs is totally different, i.e. the two F atoms tend to dissociate due to the covalent interaction. Since both the covalent and ionic interactions are nonbonding in F_2, the bond dissociation energy (BDE) of F_2, which is around 38.2 kcal/mol, is totally due to the resonance of the VB structures. Based on the abovementioned novel bonding, Shaik and Hiberty et al. proposed a new type of chemical bond, charge-shift (CS) bond [47, 48], in which the RE contributes over 50% to the BDE.

The CS bond can also be characterized with the atoms in molecule (AIM) theory and electron localization function (ELF) by measuring the gradients and Laplacian of the electronic density. Overall, the concept of CS bond is now accepted to a certain extent and will find more applications in explaining chemical phenomena.

2.3.4 Diabatic States in Valence Bond Theory

Molecular dynamical process involving excited states is usually simulated under the Born–Oppenheimer adiabatic approximation, in which the adiabatic electronic states are used to construct the potential energy surfaces (PESs). Adiabatic states are eigenstates of electronic Hamiltonian and are thus unique for a given quantum chemical method. For instance, the eigenstates obtained by solving Eq. (2.5) are VBSCF adiabatic states. In nonadiabatic dynamics, to solve the quantum equation of motion requires the evaluation of nuclear-momentum couplings (NMCs) $\mathbf{F}_{KL}$

$$\mathbf{F}_{KL} = \langle \Psi_K(\mathbf{r}; \mathbf{R}) | \nabla_\mathbf{R} | \Psi_L(\mathbf{r}; \mathbf{R}) \rangle, \tag{2.22}$$

where $\Psi_K(\mathbf{r};\mathbf{R})$ is the electronic wave function of state K with parametrical dependence on the nuclear coordinates $\mathbf{R}$. Especially, $\mathbf{F}_{KL}$ in adiabatic state basis is called nonadiabatic couplings (NACs). The complexity arises near the conical intersections where the NACs have singularities due to the rapid change in the character of the adiabatic wave function. The sudden change of the adiabatic wave function can be understood as a conical intersection between two electronic stats, according to Figure 2.9. Suppose that the adiabatic state Ψ_1 is mainly covalent and Ψ_2 is mainly ionic before the avoided crossing point (R^c). Then singularity of $\mathbf{F}_{12}$ occurs at R^c when Ψ_1 rapidly changes to ionic and Ψ_2 rapidly changes to covalent. It is possible to remove such singularities by constructing the PESs with diabatic states. In diabatic state basis, the electronic Hamiltonian is not diagonal, and the effect of the NMCs should be negligible compared with the off-diagonal elements of the electronic Hamiltonian. Therefore, the definition of adiabatic states is not unique.

Various schemes based on MO theory have been proposed to construct diabatic states [49–54]. On the contrary, the construction of diabatic states with VB theory seems more straightforward, since each VB structure exactly corresponds to a Lewis structure, and thus never changes its character along the PES. In VB state correlation diagram (VBSCD) [55] proposed by Shaik and coworkers, the adiabatic states are constructed with the total VBSCF wave function, while each diabatic state is constructed with several selected VB structures that represent a distinct character (covalent/ionic, reactant/product, charge transfer states, etc.). Figure 2.10 is an illustration of two-state reaction PECs in adiabatic and diabatic states. The energy gaps of the two diabatic states at the equilibrium geometry of the reactant (R^r) and product (R^p) are respectively denoted as G^r and G^p; the energy gap between the ground adiabatic state and the crossing point of the two diabatic states is denoted as B; $\Delta E^{\neq}$ is the reaction barrier; and ΔE^{rp} is the energy gap between the reactant and product.

Figure 2.9 Illustration of adiabatic states (solid lines) and diabatic states (dashed lines).

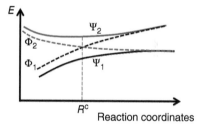

Figure 2.10 The reaction PECs of adiabatic and diabatic states.

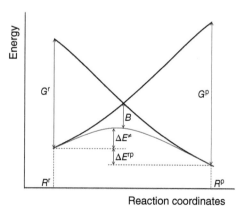

Then the reaction barrier can be approximately evaluated by the abovementioned quantities based on diabatic states. More details of VBSCD can be found in Ref. [55].

Similar to the VBSCD scheme, the block-diagonalization approach (VBBDA) [55] provides another scheme to directly construct the diabatic states with VB theory. In VBBDA, the VB structures are grouped into several subsets $\{\Phi_{Kp}\}$ according to their charge-localized characters p, q, etc., and then the rth diabatic state of the pth charge-localized character, Ψ_{rp}^{d}, is written as

$$\Psi_{rp}^{d} = \sum_{K=1}^{N_p} D_{Krp} \Phi_{Kp} \qquad (2.23)$$

where N_p is the number of VB structures in subset $\{\Phi_{Kp}\}$, and D_{Krp} is the structural coefficients. For each subspace p, the Hamiltonian and overlap matrices are constructed, and the diabatic states in this subspace and corresponding energies are solved. Finally, the diabatic states in different subspaces are orthogonalized, and the electronic couplings are shown to be very close to those obtained by diabatization schemes based on MO theory.

The abovementioned two schemes to construct diabatic states are performed according to chemical intuition. The VB-based compression approach for diabatization (VBCAD) [56], however, is a black-box-like diabatization scheme that generates diabatic states via a transformation of the adiabatic states. In VBCAD, the electronic Hamiltonian is first compressed into a low-dimensional matrix $\mathbf{H}^{\text{pre-dia}}$ containing only the interested electronic states, and then the diabatic Hamiltonian ($\mathbf{H}^{\text{dia}}$) is obtained by a transformation to $\mathbf{H}^{\text{pre-dia}}$. The transformation is determined by maximizing the separation of VB structures in different electronic states. More details of VBCAD can be found in Ref. [56]. Overall, various diabatization schemes based on VB theory are available for constructing diabatic states, either by chemical intuition or in a black-box way, and VB theory is shown to have great potential in constructing diabatic states.

2.4 A Brief Guide to Perform VB Calculations

In this section, we present a brief guide to perform a VB calculation with XMVB. Over the years, the XMVB program has been enriched continuously with more and more newly developed methods and algorithms, though we will focus on the VBSCF calculation in this guide. XMVB is distributed as a pre-compiled package free of charge on the official website (https://XACScloud.com). A more recommended access to XMVB is the Xiamen Atomistic Computing Suite (XACS) cloud computing platform (http://XACScloud.com). The manual and example calculations of XMVB can be found on the XACS official website.

2.4.1 Preparing XMVB Input Files

Running an XMVB job requires an input file with extension name "xmi," which is composed of several sections. Figure 2.11 is an xmi input file of H_2 calculation. The

Figure 2.11 The xmi input file of H_2 calculation.

```
H2 cc-pvdz
$ctrl
nstr=3 iscf=5 nae=2 nao=2 int=libcint basis=cc-pvdz
$end
$str
1 2
1 1
2 2
$end
$orb
5 5
1-5
6-10
$end
$geo
H   1.0   0.0   0.0   0.0
H   1.0   0.0   0.0   0.74
$end
```

keywords in XMVB are case-insensitive. The first line is the title of the job, followed by several sections starting with $ctrl (or $str, $orb, $geo, etc.) and ending with $end. The keywords within each section are separated by blanks and are not limited to a single line.

In an xmi file, the $ctrl and $geo sections, which correspond to global control and molecular geometry, must be provided by the user, and other sections might be default in some cases. In the $ctrl section, "nstr = 3" means 3 VB structures are included (the detailed description of these structures will be provided in $str section); "iscf" specifies the orbital optimization algorithm during the SCF; "nae" and "nao" are respectively the numbers of active electrons and active orbitals, and (2,2) active space is used in this case; "int = libcint" means the electronic integrals are obtained from Libcint library; "basis = cc-pvdz" means the cc-pVDZ basis set is used.

The $orb section defines the VB orbitals. The first line specifies the number of basis functions to expand each orbital, each number separated by blanks. In this case ("nao = 2"), "5 5" means the two orbitals are respectively expanded with 5 and 5 basis functions. With cc-pVDZ basis set, the basis functions of the first H atom (shown in the first line in $geo section) are labeled as 1–5, while those of the second H atom (the second line in $geo section) are labeled as 6–10. Thus, the second line ("1–5") in $orb section means the first orbital is expanded by the basis functions of the first H atom, while the third line "6–10" means the second orbital is expanded by the basis functions of the second H atom.

Since the number of VB structures ("nstr = 3" in $ctrl section) is specified, the user is required to provide the $str section. In the $str section, each line describes a VB structure, and "1 2," "1 1," and "2 2," respectively, denote the structures shown in Eqs. (2.8)–(2.10).

Figure 2.11 shows a fundamental framework of an xmi file, and more detailed description of the keywords and more examples are referred to the manual or XACS website.

2.4.2 Reading XMVB Output Files

The main XMVB output file has an "xmo" extension. The output file starts with the XMVB banner, followed by the description of VB structures used in calculation, nuclear repulsion energy, initial orbital guess, and a repetition of the input file. For a VBSCF calculation, the information of orbital optimization (the energy and gradients) at each iteration is printed. After the end of the SCF procedure, the VBSCF wave function is printed, including the Hamiltonian and overlap matrices, VB structure and determinant coefficients, weights of VB structures, and the optimized orbitals. Finally, the charge and spin population analysis, bond order, and dipole moments are printed. A detailed introduction can be found on XACS website.

2.5 Concluding Remarks

In this chapter, an introduction to *ab initio* VB theory with related chemical concepts is presented. In the theory and methodology section, the fundamental method of VBSCF and the definitions of VB structures and orbitals in VB theory are presented. Meanwhile, a brief introduction to the post-VBSCF methods, which further include dynamic correlation in VB computation, is also provided. Then, some chemical concepts and applications based on VB theory are presented, including resonance, conjugation, aromaticity, chemical bonds, and diabatic states in VB theory. Finally, we briefly introduce the XMVB package and show how to prepare the input files and read the output files of XMVB.

Users can now perform XMVB computations through the XMVB cloud computing service (http://XACScloud.com). More introductions, manuals, examples, and tutorials are also available on the XACS official website.

References

1 Lewis, G.N. (1916). The atom and the molecule. *J. Am. Chem. Soc.* 38 (4): 762–785.
2 Heitler, W. and London, F. (1927). Wechselwirkung neutraler Atome und homöopolare Bindung nach der Quantenmechanik. *Z. Phys.* 44 (6): 455–472.
3 Slater, J.C. (1931). Molecular energy levels and valence bonds. *Phys. Rev.* 38 (6): 1109–1144.
4 Rumer, G. (1932). Zur Theorie der Spinvalenz. *Nachr. Ges. Wiss. Gottingen, Math.-Phys. Kl.* 337–341.
5 Pauling, L. (1931). The nature of the chemical bond. Application of results obtained from the quantum mechanics and from a theory of paramagnetic susceptibility to the structure of molecules. *J. Am. Chem. Soc.* 53 (4): 1367–1400.
6 Pauling, L. (1931). The nature of the chemical bond. II. The one-electron bond and the three-electron bond. *J. Am. Chem. Soc.* 53 (9): 3225–3237.
7 Pauling, L. (1932). The nature of the chemical bond. III. The transition from one extreme bond type to another. *J. Am. Chem. Soc.* 54 (3): 988–1003.

8 Pauling, L. (1932). The nature of the chemical bond. IV. The energy of single bonds and the relative electronegativity of atoms. *J. Am. Chem. Soc.* 54 (9): 3570–3582.

9 Pauling, L. and Wheland, G.W. (1933). The nature of the chemical bond. V. The quantum-mechanical calculation of the resonance energy of benzene and naphthalene and the hydrocarbon free radicals. *J. Chem. Phys.* 1 (6): 362–374.

10 Shaik, S.S. and Hiberty, P.C. (2007). Basic valence bond theory. In: *A Chemist's Guide to Valence Bond Theory*, 40–80. Wiley.

11 Shaik, S.S. and Hiberty, P.C. (2007). Currently available ab initio valence bond computational methods and their principles. In: *A Chemist's Guide to Valence Bond Theory*, 238–270. Wiley.

12 Wu, W., Su, P., Shaik, S., and Hiberty, P.C. (2011). Classical valence bond approach by modern methods. *Chem. Rev.* 111 (11): 7557–7593.

13 Su, P. and Wu, W. (2013). Ab initio nonorthogonal valence bond methods. *WIREs Comput. Mol. Sci.* 3 (1): 56–68.

14 Chen, Z. and Wu, W. (2020). Ab initio valence bond theory: a brief history, recent developments, and near future. *J. Chem. Phys.* 153 (9): 090902.

15 Shaik, S., Danovich, D., and Hiberty, P.C. (2021). Valence bond theory—its birth, struggles with molecular orbital theory, its present state and future prospects. *Molecules* 26 (6): 1624.

16 Löwdin, P.-O. (1955). Quantum theory of many-particle systems. I. Physical interpretations by means of density matrices, natural spin-orbitals, and convergence problems in the method of configurational interaction. *Phys. Rev.* 97 (6): 1474–1489.

17 Chen, Z., Zhang, Q., and Wu, W. (2009). A new algorithm for inactive orbital optimization in valence bond theory. *Sci. China, Ser. B: Chem.* 52 (11): 1879–1884.

18 Chen, Z., Chen, X., and Wu, W. (2013). Nonorthogonal orbital based N-body reduced density matrices and their applications to valence bond theory. I. Hamiltonian matrix elements between internally contracted excited valence bond wave functions. *J. Chem. Phys.* 138 (16): 164119.

19 Chen, Z., Chen, X., and Wu, W. (2013). Nonorthogonal orbital based N-body reduced density matrices and their applications to valence bond theory. II. An efficient algorithm for matrix elements and analytical energy gradients in VBSCF method. *J. Chem. Phys.* 138 (16): 164120.

20 Chen, Z., Chen, X., Ying, F. et al. (2014). Nonorthogonal orbital based n-body reduced density matrices and their applications to valence bond theory. III. Second-order perturbation theory using valence bond self-consistent field function as reference. *J. Chem. Phys.* 141 (13): 134118.

21 Zhou, C., Chen, Z., and Wu, W. (2018). Reciprocal transformation of seniority number restricted wave function. *J. Chem. Phys.* 149 (4): 044111.

22 Zhou, C., Zeng, C., Ma, B. et al. (2019). Novel implementation of seniority number truncated valence bond methods with applications to H_{22} chain. *J. Chem. Phys.* 151 (19): 194107.

23 Ji, C., Ying, F., Su, P. et al. (2023). Implementation of molecular symmetry in valence bond calculation. *J. Chin. Chem. Soc.* 70 (3): 341–348.

24 Song, L., Mo, Y., Zhang, Q., and Wu, W. (2005). XMVB: a program for ab initio nonorthogonal valence bond computations. *J. Comput. Chem.* 26 (5): 514–521.

25 Chen, Z., Ying, F., Chen, X. et al. (2015). XMVB 2.0: a new version of Xiamen valence bond program. *Int. J. Quantum Chem.* 115 (11): 731–737.

26 Chen, Z., Song, J., Chen, X. et al. (2021). N-body reduced density matrix-based valence bond theory and its applications in diabatic electronic-structure computations. *Acc. Chem. Res.* 54 (20): 3895–3905.

27 Lenthe, J.H.V. and Balint-Kurti, G.G. (1980). The valence-bond scf (VB SCF) method. *Chem. Phys. Lett.* 76 (1): 138–142.

28 Lenthe, J.H.V. and Balint-Kurti, G.G. (1983). The valence-bond self-consistent field method (VB–SCF): theory and test calculations. *J. Chem. Phys.* 78 (9): 5699–5713.

29 Verbeek, J. and Lenthe, J.H.V. (1991). On the evaluation of non-orthogonal matrix elements. *J. Mol. Struct.: Theochem* 229: 115–137.

30 Chirgwin, B.H. and Coulson, C.A. (1950). The electronic structure of conjugated systems. VI. *Proc. R. Soc. London, Ser. A* 201 (1065): 196–209.

31 Löwdin, P.O. (1947). Model of alkali haledes. *Ark. Mat. Astron. Fys. A* 35: 30.

32 Li, X. and Paldus, J. (1992). Valence bond approach exploiting Clifford algebra realization of Rumer–Weyl basis. *Int. J. Quantum Chem.* 41 (1): 117–146.

33 Hiberty, P.C., Flament, J.P., and Noizet, E. (1992). Compact and accurate valence bond functions with different orbitals for different configurations: application to the two-configuration description of F_2. *Chem. Phys. Lett.* 189 (3): 259–265.

34 Hiberty, P.C., Humbel, S., Byrman, C.P., and van Lenthe, J.H. (1994). Compact valence-bond functions with breathing orbitals – application to the bond-dissociation energies of F_2 and FH. *J. Chem. Phys.* 101 (7): 5969–5976.

35 Hiberty, P.C. and Shaik, S. (2002). Breathing-orbital valence bond method – a modern valence bond method that includes dynamic correlation. *Theor. Chem. Acc.* 108 (5): 255–272.

36 Wu, W., Song, L.C., Cao, Z. et al. (2002). Valence bond configuration interaction: a practical ab initio valence bond method that incorporates dynamic correlation. *J. Phys. Chem. A* 106 (11): 2721–2726.

37 Song, L., Wu, W., Zhang, Q., and Shaik, S. (2004). A practical valence bond method: a configuration interaction method approach with perturbation theoretic facility. *J. Comput. Chem.* 25 (4): 472–478.

38 Chen, Z., Song, J., Shaik, S. et al. (2009). Valence bond perturbation theory. A valence bond method that incorporates perturbation theory. *J. Phys. Chem. A* 113 (43): 11560–11569.

39 Ying, F., Su, P., Chen, Z. et al. (2012). DFVB: a density-functional-based valence bond method. *J. Chem. Theory Comput.* 8 (5): 1608–1615.

40 Zhou, C., Zhang, Y., Gong, X. et al. (2017). Hamiltonian matrix correction based density functional valence bond method. *J. Chem. Theory Comput.* 13 (2): 627–634.

41 Ying, F., Zhou, C., Zheng, P. et al. (2019). Lambda-density functional valence bond: a valence bond-based multiconfigurational density functional theory with a single variable hybrid parameter. *Front. Chem.* 7: 225.

42 Zheng, P., Ji, C., Ying, F. et al. (2021). A valence-bond-based multiconfigurational density functional theory: the λ-DFVB method revisited. *Molecules* 26 (3): 521.

43 Zheng, P., Gan, Z., Zhou, C. et al. (2022). λ-DFVB(U): a hybrid density functional valence bond method based on unpaired electron density. *J. Chem. Phys.* 156 (20): 204103.

44 Mo, Y. and Gao, J. (2007). Theoretical analysis of the rotational barrier of ethane. *Acc. Chem. Res.* 40 (2): 113–119.

45 Mo, Y. and Schleyer, P.v.R. (2006). An energetic measure of aromaticity and antiaromaticity based on the Pauling–Wheland resonance energies. *Chem. Eur. J.* 12 (7): 2009–2020.

46 Schleyer, P.v.R. and Pühlhofer, F. (2002). Recommendations for the evaluation of aromatic stabilization energies. *Org. Lett.* 4 (17): 2873–2876.

47 Shaik, S., Maitre, P., Sini, G., and Hiberty, P.C. (1992). The charge-shift bonding concept – electron-pair bonds with very large ionic covalent resonance energies. *J. Am. Chem. Soc.* 114 (20): 7861–7866.

48 Braida, B., Danovich, D., Galbraith, J.M. et al. (2019). Charge-shift bonding: a new and unique form of bonding. *Angew. Chem. Int. Ed.* 132 (3): 996–1013.

49 Hoyer, C.E., Xu, X., Ma, D. et al. (2014). Diabatization based on the dipole and quadrupole: the DQ method. *J. Chem. Phys.* 141 (11): 114104.

50 Pacher, T., Cederbaum, L.S., and Köppel, H. (1988). Approximately diabatic states from block diagonalization of the electronic Hamiltonian. *J. Chem. Phys.* 89 (12): 7367–7381.

51 Ruedenberg, K. and Atchity, G.J. (1993). A quantum chemical determination of diabatic states. *J. Chem. Phys.* 99 (5): 3799–3803.

52 Shu, Y. and Truhlar, D.G. (2020). Diabatization by machine intelligence. *J. Chem. Theory Comput.* 16 (10): 6456–6464.

53 Abrol, R. and Kuppermann, A. (2002). An optimal adiabatic-to-diabatic transformation of the 1 (2)A' and 2 (2)A' states of H-3. *J. Chem. Phys.* 116 (3): 1035–1062.

54 Cave, R.J. and Newton, M.D. (1996). Generalization of the Mulliken-Hush treatment for the calculation of electron transfer matrix elements. *Chem. Phys. Lett.* 249 (1): 15–19.

55 Shaik, S. and Shurki, A. (1999). Valence bond diagrams and chemical reactivity. *Angew. Chem. Int. Ed.* 38 (5): 586–625.

56 Zhang, Y., Su, P., Lasorne, B. et al. (2020). A novel valence-bond-based automatic diabatization method by compression. *J. Phys. Chem. Lett.* 11 (13): 5295–5301.

3

Chemical Concepts from Conceptual Density Functional Theory

Frank De Proft

Vrije Universiteit Brussel (VUB), Research Group of General Chemistry (ALGC), Pleinlaan 2, Brussels B-1050, Belgium

3.1 Introduction

According to the definition by Linus Pauling, chemistry is the science of substances, i.e. their structures, properties, and the reactions that change them into other substances [1]. As immediately admitted by Pauling, however, this is a very broad definition, since many other sciences, such as e.g. physics, biology, and geology can be (partially) included within it. Practically, one thus usually defines chemistry as the science of reactions between substances and their properties that are relevant to describe these processes. In order to achieve its goals, the chemical sciences have traditionally employed a set of very powerful *concepts* [2]. These have found widespread use not only to rationalise and interpret but also to predict chemical properties and reactivity, e.g. used in the synthesis of new chemical compounds. They succeed in bringing order into the vast amount of chemical data and their use is widespread in chemistry, ranging from traditional organic and inorganic chemistry to biochemistry, catalysis, and nanosciences. A famous example is the concept of electronegativity, originally introduced by Pauling himself as the power of an atom in a molecule to attract electrons to itself [3, 4]. This concept and other ones have been essential in the description of structure, stability, and reactivity of various chemical compounds. Many of these types of concepts however can be considered as conceptual constructs, which are not quantum mechanical observables, i.e. they cannot be associated with a quantum mechanical operator enabling their, in principle, evaluation and observation. Frenking et al. compared the appearance of these heuristically developed chemical concepts with the phenomenon of unicorns from the mythical world; their appearance is known to everybody although nobody has ever seen one Additionally, the unicorn brings "law and order, health and good fortune, fame and satisfaction in an otherwise chaotic and disordered world" [2b]. A similar analogy was made by Parr et al., connecting these concepts to the notion of the "noumenon," "An object knowable by the mind or intellect, not by

Exploring Chemical Concepts Through Theory and Computation, First Edition. Edited by Shubin Liu.
© 2024 WILEY-VCH GmbH. Published 2024 by WILEY-VCH GmbH.

the senses; specifically (in Kantian philosophy) an object of purely intellectual intuition" [5]. From the early days of quantum chemistry, much attention has been devoted to the computation of these concepts and they are still very widely used and developed [2]. In the majority of cases, these concepts are applied to interpret various molecular properties and reactivity of compounds, obtained through other computational quantum chemistry methods and/or experiments. In the last 40 years, however, various accurate and powerful quantum chemical methods to compute the electronic structure of chemical compounds have been developed and implemented. Combined with the ever-increasing computer power, this has allowed (quantum) chemists to compute various (observable) atomic, molecular, and material properties to very high accuracy. In addition, the potential energy surface, i.e. the hypersurface giving the energy of the system of interest as a function of the nuclear arrangement, can now be studied very accurately, which is also the case for the relevant critical points on these surfaces, called minima and transition states. These developments have revolutionized chemistry in that the electronic structure and potential energy surfaces are central in chemistry and so it would indeed seem that all of chemistry could up to some point be computed. To interpret quantum chemistry calculations and to deal with the massive amount of data generated when solving the Schrödinger equation, the above-mentioned concepts come into play a second time, in line with very early statements of, among others, renowned quantum chemists such as Coulson ("Give us insight, not numbers !") [6] and Parr ("Accurate calculation is not synonymous to useful interpretation. To calculate a molecule is not to understand it.") [7].

When looking back, it can be stated that most of the concepts have a basis in wavefunction theory and made their appearance in either molecular orbital or valence bond approach [8]. The early 1980s saw a growing interest in, the acceptance and application of, density functional theory (DFT) [9–20] as a computational workhorse for chemists going from a model to a formally established theory with the 1964 formulation of the theorems of Hohenberg and Kohn [9]. In the late 1970s and early 1980s, Robert G. Parr and coworkers laid the foundations of the so-called conceptual density functional theory (CDFT) (for reviews see references [11, 13, 21–34]). A historical overview of the start and evolution of the CDFT fields has been given by Geerlings [34]. In addition, recently, and in retrospect, the aim of CDFT has been defined as [32] *to develop a nonempirical, mathematically and physically sound, density-based, quantum-mechanical theory for interpreting and predicting chemical phenomena, especially chemical reactions.* Implicit in the early years, the underlying fundamental precepts of (the philosophy of) CDFT, were defined as: (i) *observability* (understanding should be based on quantum-mechanical observables, which for CDFT are the energy, density, and their derivatives); (ii) *mathematical rigor* (the different tools should fit in a well-defined mathematical framework); and (iii) *universality* (the tools are not dependent on the type of calculation that is performed) [32].

A fundamental aspect of CDFT is that the propensity of a molecule to react with a given partner is governed by the change in its energy functional, $E = E[N, v(\mathbf{r})]$, due to the perturbation in its number of electrons, N, and its external potential, $v(\mathbf{r})$, induced by the partner. In the absence of external fields, the external potential is the potential felt by an electron due to the nuclei. These perturbations can then be represented in a Taylor series expansion. Using the onset of the reaction as the reference, the expansion coefficients or *response functions* are the partial and functional derivatives of the energy with respect to N and $v(\mathbf{r})$, respectively, i.e. $\dfrac{\partial^m \delta^{m'} E}{\partial^m N \delta v(\mathbf{r}_1) \ldots \delta v(\mathbf{r}_{m'})}$. They measure the sensitivity of the system toward the perturbation or the intrinsic part (i.e. without consideration of the magnitude of the perturbation) of the response of the system [25, 35]. As will be seen in Sections 3.2 and 3.3 of this chapter, the lowest (i.e. first-order) response functions are directly related to fundamental quantities of DFT, one being the electron density itself, the other being the electronic chemical potential, the Lagrangian multiplier in the variational DFT equation associated to the constraint of fixed electron-number. The identification of this electronic chemical potential as the opposite of the electronegativity was a breakthrough and provided a clear link between DFT, CDFT, and the aforementioned long-standing chemical concepts [36]. By progressing to higher N and $v(\mathbf{r})$ derivatives, a *response function tree* appears in a natural way whose components have been identified with a variety of chemical concepts such as Pearson's chemical hardness [37–40] and Fukui's frontier molecular orbital concept [41].

The CDFT concepts have been used in many applications in diverse fields of chemistry, in some cases together with principles, such as Sanderson's Electronegativity Equalization Principle [42] and Pearson's hard and soft acid and bases (HSAB) principle [37–40, 43] and maximum hardness principle (MHP) [44]. CDFT has also provided the mathematical framework to justify the validity of these principles.

In this chapter, based on the energy response function tree up to second order (Figure 3.1), an overview is given of the most commonly used CDFT chemical concepts and principles. The chapter starts by considering the fundamental aspects of DFT, introducing the DFT variational equation, followed by a brief introduction

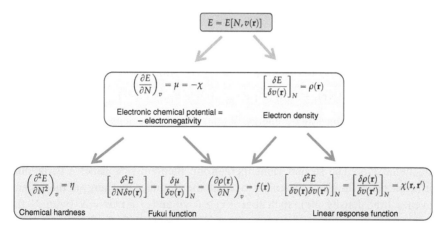

Figure 3.1 CDFT energy response function tree depicting E versus changes in the number of electrons N and the external potential $v(\mathbf{r})$.

of Kohn–Sham DFT (Section 3.2). Next, it is shown that two fundamental quantities from DFT, the electron density and the Lagrange multiplier associated with the constant electron density constraint in the DFT variational problem are in fact first-order response functions of the energy with respect to changes in the external potential and number of electrons, respectively (Section 3.3). In Section 3.4, we discuss the most important second-order response functions and introduce some functions derived from these. In a final part, all response functions are brought together in a Taylor series expansion providing a perturbative perspective on chemical reactivity. This chapter only provides a general overview of (an important selection of) chemical concepts from CDFT, neglecting e.g. many aspects such as spin-polarized reactivity concepts, influence of external forces, or temperature. For a recent detailed overview of the many different recent developments and extensions within CDFT, we refer to [34].

3.2 The Fundamentals: Density Functional Theory (DFT) and Kohn–Sham DFT

DFT was formally established by the two Hohenberg–Kohn theorems [9], which justify the use of the electron density $\rho(\mathbf{r})$ of the system as the basic variable instead of the wavefunction Ψ. The first theorem states that the external potential of the system $v(\mathbf{r})$ (for an isolated system with no external fields, this is the potential due to the nuclei) is determined, within an trivial, additive constant, by the electron density $\rho(\mathbf{r})$. The electron density of the system can be obtained from the many-electron wave function $\Psi(\mathbf{x}_1, \mathbf{x}_2, \ldots, \mathbf{x}_N)$ as

$$\rho(\mathbf{r}) = N \int \cdots \int \Psi^*(\mathbf{x}_1, \mathbf{x}_2, \ldots, \mathbf{x}_N) \Psi(\mathbf{x}_1, \mathbf{x}_2, \ldots, \mathbf{x}_N) d\sigma_1 d\mathbf{x}_2 \ldots d\mathbf{x}_N \quad (3.1)$$

where $\mathbf{x}_i$ is the combined spin and spatial coordinate of electron i. As can be seen, the integration is done over all spin and spatial coordinates of all but one electron; for this electron, only the spin is integrated out.

Based on the first Hohenberg–Kohn theorem, one can write the energy E of the atomic, molecular, or solid state system as a so-called functional of its density, i.e. $E = E[\rho(\mathbf{r})]$. More specifically, the ground state energy can be written as

$$E[\rho(\mathbf{r})] = T[\rho(\mathbf{r})] + V_{ne}[\rho(\mathbf{r})] + V_{ee}[\rho(\mathbf{r})] = V_{ne}[\rho(\mathbf{r})] + F_{HK}[\rho(\mathbf{r})] \quad (3.2)$$

where $T[\rho(\mathbf{r})]$ is the kinetic energy, $V_{ne}[\rho(\mathbf{r})]$ is the nucleus–electron attraction energy, and $V_{ee}[\rho(\mathbf{r})]$ is the electron–electron repulsion. In the second part of this equation, T and V_{ee} have been collected in the so-called Hohenberg–Kohn functional F_{HK}. The second Hohenberg–Kohn theorem establishes a variational principle using the electron density to obtain the ground state energy of the system; given a trial density $\tilde{\rho}(\mathbf{r})$, such that $\tilde{\rho}(\mathbf{r}) \geq 0, \forall \mathbf{r}$ and $\int \tilde{\rho}(\mathbf{r}) d\mathbf{r} = N$ (with N the number of electrons of the system), then $E_0 \leq E[\tilde{\rho}(\mathbf{r})]$, with E_0 the exact ground state energy of the system. Application of this variational principle, i.e. minimizing the energy of the system with respect to changes in the electron density under the constraint that the density during this process should always integrate to the number of electrons N of the system yields

$$\frac{\delta}{\delta\rho(\mathbf{r})}\left[E - \mu\left(\int \rho(\mathbf{r})d\mathbf{r} - N\right)\right] = 0 \quad (3.3)$$

where μ is the Lagrange multiplier attached to the constraint in this minimization. This yields the following analog of the Schrödinger equation

$$\mu = v(\mathbf{r}) + \frac{\delta F_{HK}}{\delta\rho(\mathbf{r})} \quad (3.4)$$

The exact functional form of F_{HK} is unfortunately unknown. Due to its magnitude, small errors in its approximation usually lead to unacceptable errors when using this equation for determining the energy of atomic and molecular systems. In order to circumvent this problem, Kohn and Sham decided to re-introduce orbitals by invoking a noninteracting reference system with an electron density exactly equal to the density of the true system [10]. For such a noninteracting system, the kinetic energy, denoted as T_s can be computed exactly as

$$T_s = \sum_{i=1}^{N}\langle\psi_i| - \frac{1}{2}\nabla^2|\psi_i\rangle \quad (3.5)$$

In addition, the electron–electron repulsion of the system is written as the classical Coulomb self-repulsion and a correction term. As a result, F_{HK} is expressed as

$$F_{HK}\rho(\mathbf{r}) = T[\rho(\mathbf{r})] + V_{ee}[\rho(\mathbf{r})] = T_s[\rho(\mathbf{r})] + J[\rho(\mathbf{r})] + E_{XC}[\rho(\mathbf{r})] \quad (3.6)$$

with

$$E_{XC}[\rho(\mathbf{r})] = \left(T[\rho(\mathbf{r})] - T_s[\rho(\mathbf{r})]\right) + \left(V_{ee}[\rho(\mathbf{r})] - J[\rho(\mathbf{r})]\right) \quad (3.7)$$

Collecting all terms, the energy functional within Kohn–Sham is given as

$$E_{KS}[\rho(\mathbf{r})] = T_s[\rho(\mathbf{r})] + J[\rho(\mathbf{r})] + \int \rho(\mathbf{r})v(\mathbf{r})d\mathbf{r} + E_{XC}[\rho(\mathbf{r})] \quad (3.8)$$

Again minimizing this energy with respect to the electron density with the constraint that the density should at all times integrate to the number of electrons, yields

$$\mu = v(\mathbf{r}) + \int \frac{\rho(\mathbf{r}')}{|\mathbf{r} - \mathbf{r}'|}d\mathbf{r}' + \frac{\delta E_{XC}}{\delta\rho(\mathbf{r})} + \frac{\delta T_s}{\delta\rho(\mathbf{r})} \quad (3.9)$$

Introducing the exchange-correlation potential $v_{XC}(\mathbf{r})$

$$v_{XC}(\mathbf{r}) = \frac{\delta E_{XC}}{\delta\rho(\mathbf{r})} \quad (3.10)$$

and

$$v_J(\mathbf{r}) = \int \frac{\rho(\mathbf{r}')}{|\mathbf{r} - \mathbf{r}'|}d\mathbf{r}' \quad (3.11)$$

yields

$$\mu = v(\mathbf{r}) + v_J(\mathbf{r}) + v_{XC}(\mathbf{r}) + \frac{\delta E_{XC}}{\delta\rho(\mathbf{r})} + \frac{\delta T_s}{\delta\rho(\mathbf{r})} = v_{KS}(\mathbf{r}) + \frac{\delta T_s}{\delta\rho(\mathbf{r})} \quad (3.12)$$

with $v_{KS}(\mathbf{r})$ the effective Kohn–Sham potential. This equation, when compared with Eq. (3.4), beautifully confirms the Kohn–Sham picture of a non-interacting system ($V_{ee} = 0$) where the electrons move in an effective potential $v_{KS}(\mathbf{r})$.

Minimization of the Kohn–Sham energy with respect to the occupied orbitals gives rise to the famous Kohn–Sham equations,

$$\left(-\frac{1}{2}\nabla^2 + v_{KS}(\mathbf{r})\right)\psi_i = \epsilon_i\psi_i \tag{3.13}$$

where ϵ_i are the Kohn–Sham orbital energies. A recent discussion concerning these energies was given in [45].

The electron density of the noninteracting system can be obtained exactly as the sum of the squares of the occupied Kohn–Sham orbitals

$$\rho(\mathbf{r}) = \sum_{i=1}^{N} |\psi_i|^2 \tag{3.14}$$

3.3 The First Derivatives: The Electronic Chemical Potential and the Electron Density

Both the number of electrons N and the external potential $v(\mathbf{r})$ of the system fix its Hamiltonian; as a result, the energy can be written as a functional of these two variables: $E = E(N, v)$. Both of these variables are essential in describing chemical reactivity; when a chemical reaction occurs, the reaction partners will perturb one another resulting in a change of the number of electrons and/or the external potential. As a result, the amount of change of these quantities during a reaction and the contribution of these changes to the total energy of the system will provide important information on the reaction process. The infinitesimal change in energy from one ground state to another resulting from the changes dN and dv can be written as

$$dE = \left(\frac{\partial E}{\partial N}\right)_v dN + \int \left[\frac{\delta E}{\delta v(\mathbf{r})}\right]_N \delta v(\mathbf{r}) d\mathbf{r} \tag{3.15}$$

This energy change at a constant external potential dE_v equals

$$dE_v = \left(\frac{\partial E}{\partial N}\right)_v dN \tag{3.16}$$

In view of the fact that

$$\int \rho(\mathbf{r}) d\mathbf{r} = N \tag{3.17}$$

the change dN results from an infinitesimal change of electron density $\delta\rho(\mathbf{r})$, so that, again at constant external potential, one can write

$$dE_v = \int \left[\frac{\delta E}{\delta \rho(\mathbf{r})}\right]_v \delta\rho(\mathbf{r}) d\mathbf{r} = \mu \int \delta\rho(\mathbf{r}) d\mathbf{r} = \mu dN \tag{3.18}$$

Comparing (3.16) with (3.18) now yields

$$\mu = \left(\frac{\partial E}{\partial N}\right)_v \tag{3.19}$$

Parr et al. identified μ with the negative of the electronegativity [36]

$$\mu = -\chi \tag{3.20}$$

This identification marks the birth of "Conceptual DFT." This electronegativity expression is in line with earlier work of Ickowski and Margrave [46]. As can be seen, the evaluation of the chemical potential or the electronegativity requires the derivative of the energy with respect to the number of electrons. In a seminal paper, Parr and coworkers showed, using an ensemble approach, that the exact E versus N curve comprises a series of straight lines [47] with derivative discontinuities at the integer N [47]. This linearity condition was also proven using a pure state approach [48]. Using a finite difference approximation, however, the μ and χ can be obtained as

$$\mu = -\chi = -\frac{I+A}{2} \tag{3.21}$$

where I and A are the vertical ionization energy and electron affinity of the system. This expression equals the Mulliken definition for the electronegativity [49]. This expression is also retrieved when using a quadratic model for the energy versus the number of electrons. Within a Koopmans type of approximation [50], the electronegativity can be obtained as the average of the highest occupied molecular orbital (HOMO) and lowest unoccupied molecular orbital (LUMO) orbital energies [37]

$$\chi = -\frac{\epsilon_{HOMO} + \epsilon_{LUMO}}{2} \tag{3.22}$$

Cárdenas et al. provided benchmark values for the chemical potential and chemical hardness of the elements from Hydrogen to Berkelium using both experimental data and an extrapolation procedure along isoelectronic series [51]. The negative of their obtained chemical potentials for these elements (i.e. the electronegativity) is plotted in Figure 3.2. As can be seen, the well-known trends for the electronegativity in the periodic table can be appreciated: the electronegativity increases from left to right in the table and decreases when going down in a given group. There is also a small but steady increase for the transition metals when going from left to right, whereas there is only little change when going through the lanthanide period.

Within conceptual DFT, theoretical justification was also provided for the electronegativity equalization principle. This principle was originally postulated by Sanderson and states that upon molecule formation, the electronegativities of the constituent atoms become equal; the equalized molecular electronegativity was then postulated by Sanderson to be the geometric average of the Sanderson electronegativities of the constituent atoms [42]. This principle has served as the basis for among others approaches to obtain partial atomic charges in molecules through e.g. Mortier's electronegativity equalization methods electronegativity equalization method (EEM) [52].

We now return to Eq. (3.15) where there is a second so-called response function present, i.e. $\left[\frac{\delta E}{\delta v(\mathbf{r})}\right]_N$.

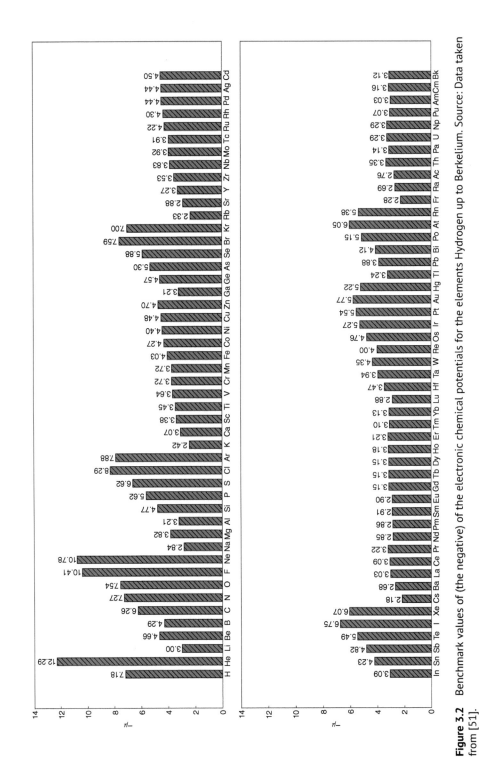

Figure 3.2 Benchmark values of (the negative) of the electronic chemical potentials for the elements Hydrogen up to Berkelium. Source: Data taken from [51].

Using Rayleigh–Schrödinger perturbation theory, this derivative can be shown to be equal to the electron density of the system, i.e.

$$\left[\frac{\delta E}{\delta v(\mathbf{r})}\right]_N = \rho(\mathbf{r}) \tag{3.23}$$

One can thus conclude that the two first-order response functions of the energy with respect to N and v are two fundamental quantities from ground state DFT.

Parr and Bartolotti have introduced the shape function $\sigma(\mathbf{r})$ (or the density per particle) as the ratio of the electron density and the number of electrons, i.e. [53]

$$\sigma(\mathbf{r}) \equiv \frac{\rho(\mathbf{r})}{N} \tag{3.24}$$

Ayers showed that also this quantity, just like the electron density, determines any observable quantity of a finite Coulombic system [54].

3.4 The Second Derivatives: Chemical Hardness, Fukui Function, Linear Response Function, and Related Quantities

3.4.1 Chemical Hardness and Softness

In line with Eq. (3.15), the infinitesimal change of the chemical potential of the system when gone from one ground state to another can be written as

$$d\mu = \left(\frac{\partial \mu}{\partial N}\right)_v dN + \int \left[\frac{\delta \mu}{\delta v(\mathbf{r})}\right]_N \delta v(\mathbf{r}) d\mathbf{r} \tag{3.25}$$

In this section, we will now focus on the first global response function in this equation. Parr and Pearson introduced this quantity as the chemical hardness η as [39]

$$\eta = \left(\frac{\partial^2 E}{\partial N^2}\right)_v = \left(\frac{\partial \mu}{\partial N}\right)_v \tag{3.26}$$

This quantity measures the system's resistance toward a charge transfer and, again within a finite difference approximation, it can be expressed as

$$\eta = I - A \tag{3.27}$$

which can also be obtained when again using a quadratic E versus N model. Adopting a Koopmans' type of approximation, one obtains

$$\eta = \epsilon_{LUMO} - \epsilon_{HOMO} \tag{3.28}$$

i.e. the HOMO–LUMO gap of the system.

The inverse of the chemical hardness is called the (global) softness S [55].

$$S = \frac{1}{\eta} = \left(\frac{\partial N}{\partial \mu}\right)_v \tag{3.29}$$

This quantity has been shown to be related to the polarizability and its cube root [56–63].

In Figure 3.3, we have depicted the evolution of the chemical hardness of the elements Hydrogen up to Berkelium, again using the benchmark values of reference [51]. The well-known trends are apparent: increase of the chemical hardness when going from left to right in the periodic table and decrease when going down in a group. The variation of this quantity for the transition metals and the lanthanides is somewhat more modest.

When studying generalized Lewis acid–base reactions of the type

$$A + :B \rightleftharpoons A-B$$

The chemical hardness was introduced in the 1960s by Pearson who proposed, based on experimental thermochemical data, to classify Lewis acids and bases as hard or soft and formulated the HSAB principle: Hard acids prefer to bond to Hard bases and Soft acids prefer to bond to Soft bases [38, 40].

Another important principle concerning the hardness is the MHP, which states that "There seems to be a rule of nature that molecules arrange themselves so as to be as hard as possible." This principle was proven but under the special circumstances of both constant chemical and external potential [44]. Numerical evidence for its validity has been provided at various instances.

Now that the first- and second-order global response functions have been introduced as the negative of the electronegativity and the chemical hardness, we can write an approximate expression for the change of the energy for the formation of a molecule between the atoms A and B neglecting changes in the external potential. The electronegativities of the atoms A and B when forming the molecule AB can be written as [11, 64]

$$\chi_A = \chi_A^0 + \eta_A^0 \Delta N_A \tag{3.30}$$

and

$$\chi_B = \chi_B^0 + \eta_B^0 \Delta N_B \tag{3.31}$$

where χ_A^0 and χ_B^0 are the electronegativities and η_A^0 and η_B^0 are the hardnesses of the isolated atoms A and B, respectively, and ΔN_A and ΔN_B are the charge transfer to A/B upon molecule formation. Since the electronegativities of both atoms will equalize when the molecule is formed and since neutrality of the resulting diatomic molecule requires that $\Delta N_A = -\Delta N_B = \Delta N$, the charge transfer can be expressed as

$$\Delta N = \frac{\chi_A^0 - \chi_B^0}{\eta_A^0 + \eta_B^0} \tag{3.32}$$

which nicely shows that the electronegativity difference drives the charge transfer, whereas the chemical hardness of A and B prevents it. From this equation, the first-order energy change can be written as

$$\Delta E = -\frac{(\chi_A^0 - \chi_B^0)^2}{\eta_A^0 + \eta_B^0} \tag{3.33}$$

These simple and appealing equations have been used in many contexts in conceptual DFT, among others in the analysis of the HSAB principle; as can be seen,

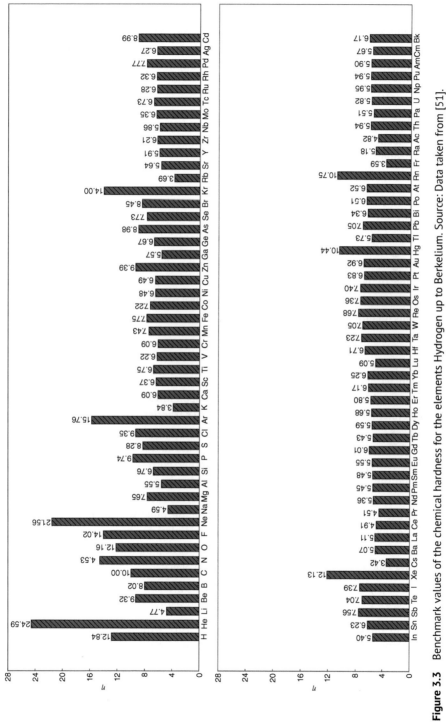

Figure 3.3 Benchmark values of the chemical hardness for the elements Hydrogen up to Berkelium. Source: Data taken from [51].

Eq. (3.33), applied to the combination of a Lewis acid/Lewis base A/B implies that soft–soft combinations of A and B will minimize the interaction energy [11]. No explanation why hard–hard combinations could be favorable however is provided.

Using Eq. (3.32), Parr et al. considered the case where an electrophile would be immersed in a free electron sea of chemical potential and derived the consequent maximum charge transfer ΔN_{max} to this electrophile as [65]

$$\Delta N_{max} = -\frac{\mu}{\eta} \qquad (3.34)$$

The negative of the accompanying energy lowering (to first order), ΔE_{max} of the electrophile, computed using Eq. (3.33), was introduced by Parr et al. as the electrophilicity index ω [65, 66]

$$-\Delta E_{max} \equiv \omega = \frac{\mu^2}{2\eta} \qquad (3.35)$$

3.4.2 The Fukui Function and the Dual Descriptor

We now come back to Eq. (3.25) expressing the change $d\mu$ upon the infinitesimal change of the number of electrons and the external potential. In this expression, in addition to the chemical hardness, the response function $\left[\frac{\delta \mu}{\delta v(\mathbf{r})}\right]_N$ appears.

This quantity was defined by Parr and Yang as the Fukui function $f(\mathbf{r})$ [67]:

$$f(\mathbf{r}) = \left[\frac{\delta \mu}{\delta v(\mathbf{r})}\right]_N = \left(\frac{\partial \rho(\mathbf{r})}{\partial N}\right)_v \qquad (3.36)$$

where we have immediately applied a Maxwell relation. As can be seen, this quantity expresses how the electron density changes upon inflow of outflow of electrons into the system. It provides an extension of Fukui's frontier molecular orbital reactivity indices.

When introducing the Fukui function, Parr and Yang put forward that the preferred approach of one reagent by another corresponds to the direction for which the initial $|d\mu|$ is a maximum ("$|d\mu|$ big is good!"). Since the first term in the change of μ is independent of the relative direction and orientation of the approaching reactant, one can assume that the preferred direction of approach is the one showing the largest value of the Fukui function at the reaction site; theoretical justification for this statement has been provided recently [68].

The Fukui function integrates to unity over space

$$\int \left(\frac{\partial \rho(\mathbf{r})}{\partial N}\right)_v d\mathbf{r} = \frac{\partial}{\partial N}\left(\int \rho(\mathbf{r}) d\mathbf{r}\right)_v = \left(\frac{\partial N}{\partial N}\right)_v = 1 \qquad (3.37)$$

Due to the discontinuity of the electron density with respect to the number of electrons, a Fukui function can be defined both for the subtraction of addition of a fraction of an electron. The derivative of the electron density when subtracting a fractional electron is the Fukui function for an electrophilic attack, i.e.

$$f^-(\mathbf{r}) = \left(\frac{\partial \rho(\mathbf{r})}{\partial N}\right)_v^- = \rho_N(\mathbf{r}) - \rho_{N-1}(\mathbf{r}) \qquad (3.38)$$

where $\rho_N(\mathbf{r})$ and $\rho_{N-1}(\mathbf{r})$ are the electron densities of the N and $N-1$ electron system, respectively, evaluated at the geometry of the N-electron system (cf. the requirement for a constant external potential). The Fukui function for a nucleophilic attack is given by density change upon addition of a fractional charge

$$f^+(\mathbf{r}) = \left(\frac{\partial \rho(\mathbf{r})}{\partial N}\right)^+_v = \rho_{N+1}(\mathbf{r}) - \rho_N(\mathbf{r}) \tag{3.39}$$

where $\rho_{N+1}(\mathbf{r})$ are now the electron densities of the $N+1$ electron systems. For a radical attack, i.e. the attack of a neutral species, the Fukui function f^0 is introduced, the average of f^+ and f^-

$$f^0(\mathbf{r}) = \frac{f^-(\mathbf{r}) + f^+(\mathbf{r})}{2} \tag{3.40}$$

The expressions for f^- and f^+ are in fact exact when the exact electron densities of the N, $N+1$, or $N-1$ are used.

The Fukui functions can be approximated as the densities of the Kohn–Sham frontier molecular orbitals [69], which provides a natural connection between frontier molecular orbital theory on one hand and conceptual DFT on the other hand [41]:

$$f^-(\mathbf{r}) = \rho_{HOMO}(\mathbf{r}) \tag{3.41}$$

and

$$f^+(\mathbf{r}) = \rho_{LUMO}(\mathbf{r}) \tag{3.42}$$

$\rho_{HOMO}(\mathbf{r})$ and $\rho_{LUMO}(\mathbf{r})$ are the densities of the HOMO and LUMO, respectively.

As can be seen when comparing Eq. (3.38) with (3.41) or Eq. (3.39) with (3.42), the Fukui function not only contains information about the frontier orbital relevant for the reaction process, but it includes also information on the orbital relaxation when electrons are subtracted or added to the system.

In Figure 3.4, the Fukui function for a nucleophilic attack $f^+(\mathbf{r})$ is plotted for the formaldehyde molecule; the function is indeed the largest around the carbon atom, correctly predicting the site for the addition of a nucleophile which is the carbon atom. This atom is also identified as the preferred site through the LUMO orbital density, also plotted in the same Figure 3.4. The Fukui function was obtained at the B3LYP [70]/Def2TZVP [71] level of theory.

Sometimes, it is however difficult to compare and appreciate changes of the Fukui function using local plots of this quantity. Yang and Mortier introduced the atom-condensed Fukui functions by taking the difference of the atomic Mulliken populations [72] of the atom in the N, $N-1$, and $N+1$ systems; this then assigns a numerical value to each atomic site in the system which is interesting to compare trends in this quantity. As such, the Fukui functions condensed to an atom k, for an electrophilic, nucleophilic and radical attack are then obtained as [73]

$$f_k^- = q_k(N) - q_k(N-1) \tag{3.43}$$

$$f_k^+ = q_k(N+1) - q_k(N) \tag{3.44}$$

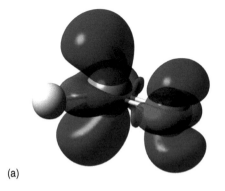

(a)

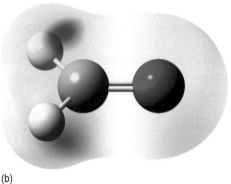

(b)

(c)

Figure 3.4 Fukui function f^+ for a nucleophilic attack for H_2CO: (a) f^+ 0.008 a.u. isosurface with positive values depicted in blue and negative in red. (b) f^+ plotted on the Van der Waals surface. (c) LUMO density (isovalue of 0.02 a.u.). The condensed Fukui function on the different atoms using Hirshfeld populations: 0.400 (C); 0.284 (O); 0.155 (H).

and

$$f_k^0 = q_k(N+1) - q_k(N-1) \tag{3.45}$$

where $q_k(N)$, $q_k(N+1)$ and $q_k(N-1)$ are the atomic populations of atom k in the N, N − 1, and N + 1 electron system, respectively. In their original contribution, Yang and Mortier used Mulliken populations to evaluate the condensed Fukui functions but also other population analysis schemes have been used for this purpose. Of course, different condensation schemes will yield different numerical values of this quantity and might even yield different predictions of the chemical reactivity based on the atom-condensed Fukui function. In addition, Bultinck et al. also pointed to

Figure 3.5 Side view of the dual descriptor for H_2CO: isosurface value of 0.01 with positive values depicted in blue and negative in red.

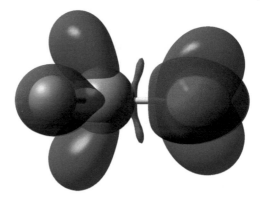

different expressions that could be used for the evaluation of the atom-condensed Fukui function, basically resulting from the order of the condensation and function evaluation [74].

In the caption of Figure 3.4, we provide that atom-condensed Fukui functions of formaldehyde based on the Hirshfeld populations [75]. It should be remarked that this atom partitioning has been derived using information theory [76]. The connection between information theory and conceptual DFT has been scrutinized on many occasions by Liu and coworkers [77].

The derivative of the Fukui function with respect to the number of electrons, highlighting the relative importance of electrophilic and nucleophilic regions in a molecule, was introduced by Morell et al. as the dual descriptor $f^2(\mathbf{r})$ [78, 79],

$$f^{(2)}(\mathbf{r}) = \left(\frac{\partial f(\mathbf{r})}{\partial N}\right)_v = \left(\frac{\delta \eta}{\delta v(\mathbf{r})}\right)_N = f^+(\mathbf{r}) - f^-(\mathbf{r}) \qquad (3.46)$$

This quantity is a third-order energy derivative and has thus not been taken up in the response function tree in Figure 3.5. In Figure 3.5, we have plotted the dual descriptor for H_2CO, clearly showing positive (blue) regions where a nucleophilic attack is preferred and negative (red) regions where an electrophile will preferably react.

A generalization of this dual descriptor to the so-called state-specific dual descriptor $\Delta f_i(\mathbf{r})$, defined for an excited state i, was established as

$$\Delta f_i(\mathbf{r}) = \rho_i(\mathbf{r}) - \rho_0(\mathbf{r}) \qquad (3.47)$$

where $\rho_0(\mathbf{r})$ and $\rho_i(\mathbf{r})$ are the electron densities of the ground state and excited state i, respectively [80]. This descriptor will describe the polarization of the molecular density resulting from an approaching reagent.

When charge is transferred to a molecule during a chemical reaction, it will induce changes in the forces acting on the nuclei in the molecule initiating the evolution to a new nuclear equilibrium constellation. The nuclear Fukui function Φ_α [81, 82] evaluated at a nucleus α probes the magnitude of the forces when the number of electrons is changing and was defined as

$$\Phi_\alpha = \left(\frac{\partial \mathbf{F}_\alpha}{\partial N}\right)_v \qquad (3.48)$$

It can be used to gain insight into the nuclear response of the system toward charge transfer and thus probes the coupling between nuclear and electronic degrees of freedom. It can be explicitly linked to changes in the nuclear position through the following Maxwell equation

$$\Phi_\alpha = \left(\frac{\partial F_\alpha}{\partial N}\right)_v = -\left(\frac{\delta \mu}{\delta \mathbf{R}_\alpha}\right)_N \tag{3.49}$$

where $\mathbf{R}_\alpha$ indicates the position of nucleus α [83].

3.4.3 Local Softness and Hardness

As can be seen from the response function tree in Figure 3.1, the Fukui function represents the first mixed derivative of the energy with respect to both the number of electrons and the external potential of the system. In the grand canonical ensemble, where the grand potential Ω is used as the state function with variables μ and $v(\mathbf{r})$ [84], this first mixed derivative has the form $\frac{\delta^2 \Omega}{\delta \mu \delta v(\mathbf{r})}$, with $\Omega = E - \mu N$. This derivative was defined as the local softness of the system $s(\mathbf{r})$ [85]

$$s(\mathbf{r}) = \left[\frac{\delta^2 \Omega}{\delta \mu \delta v(\mathbf{r})}\right] = \left(\frac{\partial \rho(\mathbf{r})}{\partial \mu}\right)_v = \left(\frac{\partial \rho(\mathbf{r})}{\partial N}\right)_v \left(\frac{\partial N}{\partial \mu}\right)_v = Sf(\mathbf{r}) \tag{3.50}$$

For a given molecule, the local softness gives the same local information as the Fukui function; it merely distributes the total softness of the molecule over the entire system. When integrating the local softness over space, one obtains the global softness of the system:

$$\int s(\mathbf{r})d\mathbf{r} = S \tag{3.51}$$

In analogy to the Fukui function, one can evaluate the local softness for a nucleophilic, electrophilic, or radical attack,

$$s^\alpha(\mathbf{r}) = Sf^\alpha(\mathbf{r}) \tag{3.52}$$

with α being either $-, +$ or 0. In addition, one can also obtain atom-condensed values for this quantity.

As can be seen in Eq. (3.50), the Fukui function is used to distribute a global quantity regionally over a molecular framework. This philosophy can also be adopted to distribute other relevant quantities, such as e.g. the electrophilicity, over the molecule, giving rise to the local electrophilicity [86, 87] and other philicity indices [86].

The Fukui function or local softness can be used to identify the soft sites in molecules and can consequently be adopted in a local application of the HSAB principle, where it will be crucial to probe soft–soft or charge transfer/orbital controlled interactions [88–91]. In order to investigate hard–hard or electrostatically controlled interactions, one thus would like to develop a counterpart of the local softness, the local hardness. When considering the expression for the local softness given in Eq. (3.50), it appears straightforward to define the local hardness as [92, 93]

$$\eta(\mathbf{r}) = \left(\frac{\delta \mu}{\delta \rho(\mathbf{r})}\right)_v \tag{3.53}$$

3.4 The Second Derivatives: Chemical Hardness, Fukui Function, Linear Response Function

When adopting this definition, the local softness and hardness integrate to one over space

$$\int s(\mathbf{r})\eta(\mathbf{r})d\mathbf{r} \tag{3.54}$$

This definition of local hardness is problematic however as it has been termed an "ambiguous constrained derivative" [92, 94–97]; for a ground state, the electron density $\rho(\mathbf{r})$ and $v(\mathbf{r})$ are dependent because of the first Hohenberg–Kohn theorem and thus taking the derivative of the chemical potential with respect to the density but meanwhile keeping v fixed yields a problem.

A possible solution could be to resort to the unconstrained derivative for the definition of local hardness as was suggested by Ayers and Parr, i.e.

$$\eta(\mathbf{r}) = \frac{\delta\mu}{\delta\rho(\mathbf{r})} \tag{3.55}$$

but problems arise also with this definition.

An alternative could be, since the local hardness should probably measure the electrostatic control in a chemical reaction, to just use the classical electrostatic potential to probe this quantity. The molecular electrostatic potential is defined as the interaction energy of a unit positive charge with the molecule neglecting geometric relaxation and polarization effects and can be expressed as [98]

$$V(\mathbf{r}) = \sum_{A=1}^{M} \frac{Z_A}{|\mathbf{r} - \mathbf{R}_A|} - \int \frac{\rho(\mathbf{r}')}{|\mathbf{r} - \mathbf{r}'|} \tag{3.56}$$

It was shown that the electronic part of the electrostatic potential divided by twice the number of electrons N of the system can be used as an approximation to the local hardness, i.e. [92, 99]

$$\eta(\mathbf{r}) = \frac{1}{2N} \int \frac{\rho(\mathbf{r}')}{|\mathbf{r} - \mathbf{r}'|} \tag{3.57}$$

The electrostatic potential can also be written as an expansion of multipoles given by (up to the quadrupole contribution)

$$V(\mathbf{r}) = \sum_{A=1}^{M} \frac{Z_A}{|\mathbf{r} - \mathbf{R}_A|}$$
$$- \sum_{A=1}^{M} \left(\frac{Q_A}{|\mathbf{r} - \mathbf{R}_A|} + \sum_i \frac{\mu_i |\mathbf{r} - \mathbf{R}_A|_i}{|\mathbf{r} - \mathbf{R}_A|^3} + \frac{1}{2} \sum_j \sum_k \frac{\Theta_{jk} |\mathbf{r} - \mathbf{R}_A|_j |\mathbf{r} - \mathbf{R}_A|_k}{|\mathbf{r} - \mathbf{R}_A|^5} + \cdots \right) \tag{3.58}$$

where μ_i is the ith component of the molecular dipole moment and Θ_{jk} the jkth element of the traceless quadrupole tensor. Despite the fact that this expression contains the atomic charges but as only one of the terms present, a possible way to probe electrostatic interactions could be using atomic charges fitted to accurately reproduce the electrostatic potential, i.e. [100]

$$V(\mathbf{r}) \approx \sum_{A=1}^{M} \frac{q_A}{|\mathbf{r} - \mathbf{R}_A|} \tag{3.59}$$

As such, even simple atomic charges can be used to probe hard–hard interactions in a chemical reaction [101].

3.4.4 The Linear Response Function, Softness and Hardness Kernels

The linear response function is defined as the second functional derivative of the energy with respect to the electron density, i.e.

$$\chi(\mathbf{r},\mathbf{r}') = \left[\frac{\delta^2 E}{\delta v(\mathbf{r})\delta v(\mathbf{r}')}\right]_N = \left[\frac{\delta \rho(\mathbf{r})}{\delta v(\mathbf{r}')}\right]_N \quad (3.60)$$

As can be seen, the linear response function (or linear response kernel) expresses how the electron density at a given point $\mathbf{r}$ changes following a change of the external potential at a point $\mathbf{r}'$. It is the static equivalent of the frequency-dependent linear response function $\chi(\mathbf{r},\mathbf{r}',\omega)$, an important quantity in time-dependent DFT [102].

Besides the study of theoretical and formal aspects connected to this quantity, it has also been a source of chemical information [103], e.g. atomic shell structure, inductive and mesomeric effects, electron delocalization, aromaticity and anti-aromaticity, electrical conductivity, and the nearsightedness of electronic matter principle of Kohn and Prodan [104, 105]. Initial studies have used the simplest coupled-perturbed Kohn–Sham (CPKS) approximation of this quantity (the independent particle approximation, IPA), but, very recently, this has been extended to its evaluation using the full CPKS [106]. The Coupled Perturbed approaches start from a single Slater determinant for the unperturbed system involving orbitals ϕ_i, solutions of the unperturbed Hartree-Fock (HF) or KS equations [107]. For a closed shell system and with the consideration of real orbitals, the first-order equations are derived as

$$\chi(\mathbf{r},\mathbf{r}') = \left(\frac{\delta \rho(\mathbf{r})}{\delta v(\mathbf{r}')}\right)_N = -4\sum_{ia}\sum_{jb}(\mathbf{M}^{-1})_{ia,jb}\phi_i^{(0)}(\mathbf{r})\phi_a^{(0)}(\mathbf{r})\phi_j^{(0)}(\mathbf{r}')\phi_b^{(0)}(\mathbf{r}') \quad (3.61)$$

This expression contains a matrix $\mathbf{M}$ of which the matrix elements in Kohn–Sham are given as

$$(\mathbf{M})_{ia,jb} = (\varepsilon_a - \varepsilon_i)\delta_{ij}\delta_{ab} + 4(ia|jb) + 4(ia|f_{XC}(\mathbf{r},\mathbf{r}')|jb) \quad (3.62)$$

ε_k denotes the Kohn–Sham orbital energies; the indices i and j refer to occupied orbitals whereas a and b refer to unoccupied orbitals. The integrals between the curly brackets are electron–electron repulsion integrals in the chemist's notation. In this expression, the exchange-correlation term is defined in terms of the operator

$$f_{XC}(\mathbf{r},\mathbf{r}') = \frac{\delta^2 E_{XC}}{\delta\rho(\mathbf{r})\delta\rho(\mathbf{r}')} \quad (3.63)$$

In the case of the IPA, often used to evaluated the linear response function, the IPA, where the $\mathbf{M}$ matrix elements are reduced to $(\varepsilon_a - \varepsilon_i)\delta_{ij}\delta_{ab}$.

We now present an illustrative example of the use of the linear response function to probe the inductive and mesomeric effect in two non-cyclic system, i.e. 1-substituted saturated and unsaturated six-membered chains, hexan-1-ol, hexan-1-amine, hexa-1,3,5-trien-1-ol, and hexa-1,3,5-trien-1-amine. In Figure 3.6,

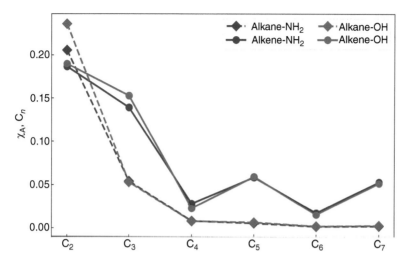

Figure 3.6 Kohn–Sham atom-condensed matrix elements χ_A, C_n of the linear response function values for six-carbon (un)conjugated chains (A = O/N and C_n where $n = 2, 3, 4, 5, 6, 7$), revealing the inductive and mesomeric effects.

the atom-condensed linear response function between the oxygen or nitrogen atom of the substituent and the six carbon atoms of both unconjugated and conjugated carbon chains evaluated using the full CPKS. The atom-condensed matrix elements of the linear response function are obtained as

$$\chi_{AB} = \int_{V_A}\int_{V_B} \chi(\mathbf{r},\mathbf{r}')d\mathbf{r}d\mathbf{r}' \tag{3.64}$$

where the linear response function (LRF) is integrated in atomic basins in regions V_A and V_B associated to atoms A and B; the atomic partitioning scheme used to obtain these values was the standard Hirshfeld approach. In the case of the saturated compounds, these matrix elements keep decreasing upon increasing distance between the heteroatom and the carbon in the chain, reflecting the decrease of the inductive effect upon increasing distance. For the unsaturated systems however, a zig-zag fluctuation with increasing distance is observed, in line with the mesomeric effect and the resonance structures that can be written for these compounds. The LRF thus clearly distinguishes between inductive and mesomeric effects along the carbon chains in these compounds.

The linear response function is the first example of a so-called kernel quantity. The softness kernel was introduced as [93]

$$s(\mathbf{r},\mathbf{r}') = \frac{\delta\rho(\mathbf{r})}{\delta u(\mathbf{r}')} = \left[\frac{\delta\rho(\mathbf{r})}{\delta v(\mathbf{r}')}\right]_\mu \tag{3.65}$$

In this expression, a modified potential $u(\mathbf{r}) = \mu - v(\mathbf{r})$ has been introduced.

This softness kernel is related to the linear response function through the Berkowitz–Parr relation [93],

$$\chi(\mathbf{r},\mathbf{r}') = -s(\mathbf{r},\mathbf{r}') + \frac{s(\mathbf{r})s(\mathbf{r}')}{S} \tag{3.66}$$

Through integration over one of the spatial coordinates of the softness kernel, one obtains the local softness

$$s(\mathbf{r}) = \int s(\mathbf{r}, \mathbf{r}')d\mathbf{r}' \tag{3.67}$$

The hardness kernel is defined as

$$\eta(\mathbf{r}, \mathbf{r}') = \frac{\delta u(\mathbf{r})}{\delta \rho(\mathbf{r}')} = \left[\frac{\delta^2 F}{\delta \rho(\mathbf{r})\delta \rho(\mathbf{r}')}\right]_\mu \tag{3.68}$$

This quantity is connected to the softness kernel through the following relation:

$$\int s(\mathbf{r}, \mathbf{r}')\eta(\mathbf{r}', \mathbf{r}'')d\mathbf{r}' = \delta(\mathbf{r} - \mathbf{r}'') \tag{3.69}$$

The hardness kernel can, in theory, be used to variationally determine Fukui function [108]. The hardness kernel was used by Ghosh and Berkowitz to introduce the local hardness as [109]

$$\eta(\mathbf{r}) = \frac{1}{N} \int \frac{\delta^2 F}{\delta \rho(\mathbf{r})\delta \rho(\mathbf{r}')} \rho(\mathbf{r}')d\mathbf{r}' \tag{3.70}$$

This quantity was termed the "total local hardness" by Ayers and Parr [96] With this definition, the local hardness can be approximated, within a Thomas–Fermi–Dirac ansatz [110], as the electronic part of the electrostatic potential divided by twice the number of electrons, as given in Eq. (3.57).

The following explicit form of the local hardness was put forward by Harbola et al. [94]

$$\eta(\mathbf{r}) = \int \frac{\delta^2 F}{\delta \rho(\mathbf{r})\delta \rho(\mathbf{r}')} \lambda(\mathbf{r}')d\mathbf{r}' \tag{3.71}$$

where $\lambda(\mathbf{r}')$ is an arbitrary function integrating to 1. Indeed, for the Ghosh/Berkowitz definition,

$$\lambda(\mathbf{r}') = \frac{\rho(\mathbf{r}')}{N} \tag{3.72}$$

Another choice for λ could be the Fukui function, yielding [94, 111]

$$\eta(\mathbf{r}) = \int \frac{\delta^2 F}{\delta \rho(\mathbf{r})\delta \rho(\mathbf{r}')} f(\mathbf{r}')d\mathbf{r}' \tag{3.73}$$

Since however

$$\int \eta(\mathbf{r}, \mathbf{r}')f(\mathbf{r}')d\mathbf{r}' = \eta \tag{3.74}$$

This definition thus gives rise to the surprising result that the local hardness is constant in space.

It is clear that the definition of the concept of local hardness remains a difficult and challenging issue.

3.5 Perturbational Perspective of Chemical Reactivity

In Sections 3.3 and 3.4, we have introduced different response functions within the framework of the "response function tree." These response functions also emerge

3.5 Perturbational Perspective of Chemical Reactivity

within a perturbation theory approach to chemical reactivity. In this approach, the change in the energy of a system upon the approach of another system is expressed in terms of the changes in its number of electrons, N, its external potential, $v(\mathbf{r})$, or both, i.e.

$$\begin{aligned}E_A[N_A^0 + \Delta N_A, v_A^0(\mathbf{r}) + \Delta v_A(\mathbf{r})] = &\; E_A[N_A^0, v_A^0(\mathbf{r})] + \mu_A \Delta N_A + \tfrac{1}{2}\eta_A(\Delta N_A)^2 \\ &+ \int \rho_A((\mathbf{r})\Delta v_A(\mathbf{r})d\mathbf{r}\Delta N_A + \int f_A(\mathbf{r})\Delta N_A \Delta v_A(\mathbf{r})d\mathbf{r} \\ &+ \tfrac{1}{2}\iint \chi_A(\mathbf{r},\mathbf{r}')\Delta v_A(\mathbf{r})\Delta v_A(\mathbf{r}')d\mathbf{r}d\mathbf{r}'\end{aligned}$$

(3.75)

This Taylor series expansion, given up to second order, expresses the energy change of a system A, with initial number of electrons N_A^0 and external potential $v_A^0(\mathbf{r})$, due to the perturbations ΔN_A and $\Delta v_A(\mathbf{r})$ resulting from the approach of another reagent. This energy change will be described most accurately when both perturbations are small, i.e. at the onset of the reaction. This approach thus allows, within certain approximations, to obtain information about the relative reaction rates of similar and competing pathways. A central assumption when adopting this methodology is that reaction profiles of similar reactions will not cross when going from the starting reagents to the transition state [112]. This implies that the CDFT descriptors are essentially to be used for kinetic considerations; their use for very late transition states and for multi-step reactions is thus less appropriate.

As can be seen, the indices in Eq. (3.75) can be used to either directly study the chemical reactivity, as probes of the system's sensitivity to chemically important perturbations, or they can be multiplied with the perturbations themselves which provides a more complete but also a more complicated analysis. An important point is the issue of the convergence of this perturbation expression. The expression is not guaranteed to converge even if all response functions would be available up to all orders; it will thus not necessarily reconstruct an accurate potential energy surface or energy profile for the reaction at state. In addition, it becomes more and more tedious to evaluate reactivity indices at higher orders, event though analytical expressions within Kohn–Sham theory are available up to the third order [113].

Considering all derivatives in the Taylor series expansion and the magnitude of their associated perturbations, although providing a complete picture of all contributing factors to chemical reactivity, will however lead to a complex analysis. One can thus ask the question if it is necessary to always compute all derivatives. The HSAB principle may provide a guiding principle in some cases. According to the local version of this principle, favorable interaction will occur between either the soft or the hard sites of the reagents. As a result, within a first approximation, all other terms in the Taylor series can be considered to be negligible compared to the terms that mediate either the *soft-soft* or *hard-hard* interactions, an approximation that might prove to be too severe however in some cases, as e.g. the case for reactions that are neither charge- or orbital controlled. In order to treat these cases, Ayers and coworkers have introduced a so-called "all purpose reactivity indicator" that enables to probe the dual reactivity behavior and to quantify the shift in the reactivity (site-selectivity) depending on the nature of the electrophile (hard or soft)

[114, 115]. Geerlings and coworkers before, within the framework of electrophilic aromatic substitution, also proposed reactivity indices consisting of a combination of local softness and a local hardness estimate to study the regioselectivity in these reactions [99].

3.6 Conclusions

In this chapter, an overview was given of chemical concepts emerging in CDFT. CDFT is connected in its essence to DFT via the first-order response functions the electron density and the electronic chemical potential. The emerging response functions can be identified with well-known concepts such as electronegativity, chemical hardness, and Fukui's frontier orbital theory. The main focus in this chapter was placed on the response functions of the $E = E[N, v(\mathbf{r})]$ functional. Additionally, we have focused on the description of these properties at zero temperature for isolated molecules and in a non-spin polarised resolution. Next to these chemical concepts, three CDFT-connected principles, Sanderson's electronegativity equalization and Pearson HSAB and MHPs were outlined. In a final part, these response functions were brought together in a Taylor series expansion affording a perturbative approach the chemical reactivity.

Acknowledgment

The author would like to acknowledge Bin Wang for preparing Figures 3.1–3.6 in this chapter.

References

1 Pauling, L. (1970). *General Chemistry*, 3e. San Francisco, CA: Freedman.
2 (a) Royal Society of Chemistry (2007). *Chemical Concepts from Quantum Mechanics, Faraday Discussion*, vol. 135. RSC Publishing. (b) Frenking, G. and Krapp, A. (2007). *J. Comput. Chem.* 28: 15. (c) Jansen, M. and Wedig, U. (2008). *Angew. Chem. Int. Ed. Engl.* 47: 10026. (d) Gonthier, J.F., Steinmann, S.N., Wodrich, M.D., and Corminboeuf, C. (2012). *Chem. Soc. Rev.* 41: 4671.
3 Pauling, L. (1932, 1932). *J. Am. Chem. Soc.* 54: 3570.
4 Pauling, L. (1960). *The Nature of the Chemical Bond*, 3e. Ithaca, NY: Cornell University Press.
5 Parr, R.G., Ayers, P.W., and Nalewajski, R.F. (2005). *J. Phys. Chem. A* 109: 3957.
6 Coulson, C.A. (1960). *Conference on Molecular Quantum Mechanics in Boulder*. Colorado, after dinner speech.
7 Parr, R.G. (1985). *Density Functional Methods in Physics* (ed. R.M. Dreizler and J. da Providencia), 141. Plenum.

8 Nye, M.J. (1994). *From Chemical Philosophy to Theoretical Chemistry: Dynamics of Matter and Dynamics of Disciplines, 1800–1950*. Berkeley, CA: University of California.
9 Hohenberg, P. and Kohn, W. (1964). *Phys. Rev. B* 136: 864.
10 Kohn, W. and Sham, L.J. (1965). *Phys. Rev. A* 140: 1133.
11 Parr, R.G. and Yang, W. (1989). *Density Functional Theory of Atoms and Molecules*. New York: Oxford University Press.
12 Dreizler, R.M. and Gross, E.K.U. (1990). *Density Functional Theory*. Berlin, Heidelberg, New York: Springer-Verlag.
13 Parr, R.G. and Yang, W. (1995). *Ann. Rev. Phys. Chem.* 46: 710.
14 Kohn, W., Becke, A.D., and Parr, R.G. (1996). *J. Phys. Chem.* 100: 978.
15 Koch, W. and Holthausen, M. (2001). *A Chemist's Guide to Density Functional Theory*, 2e. Weinheim: Wiley-VCH.
16 Cohen, A.J., Mori-Sánchez, P., and Yang, W. (2012). *Chem. Rev.* 112: 289.
17 Burke, K. (2012). *J. Chem. Phys.* 136: 150901.
18 Becke, A.D. (2014). *J. Chem. Phys.* 140: 18A301.
19 Jones, R.O. (2015). *Rev. Mod. Phys.* 87: 897.
20 Teale, A.M., Helgaker, T., Savin, A. et al. (2022). *Phys. Chem. Chem. Phys.* 24: 28700–28781.
21 Chermette, H. (1999). *J. Comput. Chem.* 20: 129.
22 Geerlings, P., De Proft, F., and Langenaeker, W. (1999). *Adv. Quant. Chem.* 33: 303.
23 De Proft, F. and Geerlings, P. (2001). *Chem. Rev.* 101: 1451.
24 Geerlings, P., De Proft, F., and Langenaeker, W. (2003). *Chem. Rev.* 103: 1793.
25 Ayers, P.W., Anderson, J.S.M., and Bartolotti, L.J. (2005). *Int. J. Quantum Chem.* 101: 520.
26 Gázquez, J.L. (2008). *J. Mex. Chem. Soc.* 52: 3.
27 Geerlings, P. and De Proft, F. (2008). *Phys. Chem. Chem. Phys.* 10: 3028.
28 Liu, S.B. (2009). *Acta Phys.-Chem. China* 25: 590.
29 Chattaraj, P.K. (ed.) (2009). *Chemical Reactivity Theory: A Density Functional View*. Boca Raton, FL: Taylor and Francis/CRC Press.
30 Geerlings, P., Ayers, P.W., Toro-Labbé, A. et al. (2012). *Acc. Chem. Res.* 55: 683.
31 Geerlings, P., Fias, S., Biosdenghien, Z., and De Proft, F. (2014). *Chem. Soc. Rev.* 43: 4989.
32 Geerlings, P., Chamorro, E., Chattaraj, P.K. et al. (2020). *Theor. Chem. Acc.* 139: 36.
33 Chakraborty, D. and Chattaraj, P.K. (2021). *Chem. Sci.* 12: 6264.
34 Liu, S. (ed.) (2022). *Conceptual Density Functional Theory: Towards a New Chemical Reactivity Theory*. Wiley-VCH GmbH.
35 De Proft, F., Geerlings, P., Heidar-Zadeh, F., and Ayers, P.W. (2022). *Reference Module in Chemistry, Molecular Sciences and Chemical Engineering, Section 3* (ed. P. Popelier). Amsterdam: Elsevier.
36 Parr, R.G., Donnelly, R.A., Levy, M., and Palke, W.A. (1978). *J. Chem. Phys.* 68: 3801.
37 Pearson, R.G. (1986). *Proc. Natl. Acad. Sci. U.S.A.* 83: 8440.

38 Pearson, R.G. (1973). *Hard and Soft Acids and Bases*. Stroudsburg, PA: Dowden, Hutchinson and Ross.
39 Parr, R.G. and Pearson, R.G. (1983). *J. Am. Chem. Soc.* 105: 7512.
40 Pearson, R.G. (1997). *Chemical Hardness*. New York: Wiley-VCH.
41 (a) Fukui, K., Yonezawa, T., and Nagata, C. (1954). *Bull. Chem. Soc. Jpn.* 27: 423. (b) Fukui, K., Yonezawa, T., and Shingu, H. (1952). *J. Chem. Phys.* 20: 722. (c) Fukui, K. (1973). *Theory of Orientation and Stereoselection*. Berlin: Springer-Verlag. (d) Fukui, K. (1982). *Science* 218: 747.
42 (a) Sanderson, R.T. (1951). *Science* 114: 670. (b) Sanderson, R.T. (1952). *Science* 116: 41. (c) Sanderson, R.T. (1952). *J. Chem. Educ.* 29: 539. (d) Sanderson, R.T. (1955). *Science* 121: 207. (e) Sanderson, R.T. (1976). *Chemical Bonds and Bond Energy*. New York: Academic Press. (f) Sanderson, R.T. (1983). *Polar Covalence*. New York: Academic Press.
43 (a) Chattaraj, P.K., Lee, H., and Parr, R.G. (1991). *J. Am. Chem. Soc.* 113: 1855–1856. (b) Ayers, P.W., Parr, R.G., and Pearson, R.G. (2006). *J. Chem. Phys.* 124: 194107.
44 (a) Pearson, R.G. (1987). *J. Chem. Educ.* 64: 561. (b) Parr, R.G. and Chattaraj, P.K. (1991). *J. Am. Chem. Soc.* 113: 1854–1855. (c) Ayers, P.W. and Parr, R.G. (2000). *J. Am. Chem. Soc.* 122: 2010–2018.
45 Baerends, E.J. (2018). *J. Chem. Phys.* 149: 054105.
46 Iczkowski, R.P. and Margrave, J.L. (1961). *J. Am. Chem. Soc.* 83: 3547.
47 Perdew, J.P., Parr, R.G., Levy, M., and Balduz, J.L. Jr., (1982). *Phys. Rev. Lett.* 49: 1691.
48 Yang, W., Zhang, Y., and Ayers, P.W. (2000). *Phys. Rev. Lett.* 84: 5172.
49 Mulliken, R.S. (1934). *J. Chem. Phys.* 2: 782.
50 Koopmans, T. (1934). *Physica* 1: 104.
51 Cárdenas, C., Heidar-Zadeh, F., and Ayers, P.W. (2016). *Phys. Chem. Chem. Phys.* 18: 25721.
52 (a) Mortier, W.J., Van Genechten, K., and Gasteiger, J. (1985). *J. Am. Chem. Soc.* 107: 829. (b) Mortier, W.J., Ghosh, S.K., and Shankar, S. (1986). *J. Am. Chem. Soc.* 108: 4315. (c) Van Genechten, K.A., Mortier, W.J., and Geerlings, P. (1987). *J. Chem. Phys.* 86: 5063. (d) Mortier, W.J. (1987). *Electronegativity, Structure and Bonding*, vol. 66 (ed. K.D. Sen and C.K. Jørgenson), 125. Berlin: Springer-Verlag.
53 Parr, R.G. and Bartolotti, L.J. (1983). *J. Phys. Chem.* 87: 2810.
54 Ayers, P.W. (2000). *Proc. Natl. Acad. Sci. U.S.A.* 97: 1959.
55 Yang, W. and Parr, R.G. (1985). *Proc. Natl. Acad. Sci. U.S.A.* 82: 6723.
56 Vela, A. and Gzquez, J.L. (1990). *J. Am. Chem. Soc.* 112: 1490.
57 (a) Ghanty, T.K. and Ghosh, S.K. (1993). *J. Phys. Chem.* 97: 4951. (b) Ghanty, T.K. and Ghosh, S.K. (1994). *J. Am. Chem. Soc.* 116: 8801.
58 Roy, R., Chandra, A.K., and Pal, S. (1994). *J. Phys. Chem.* 98: 10447.
59 Hati, S. and Datta, D. (1994). *J. Phys. Chem.* 98: 10451.
60 Simón-Manso, Y. and Fuentealba, P. (1998). *J. Phys. Chem. A* 102: 2029.
61 Ayers, P.W. (2007). *Faraday Discuss.* 135: 161.

62 Cárdenas, C., Ayers, P.W., De Proft, F. et al. (2011). *Phys. Chem. Chem. Phys.* 13: 2285.
63 Blair, S.A. and Thakkar, A.J. (2014). *J. Chem. Phys.* 141: 074306.
64 Huheey, J.E. (1965). *J. Phys. Chem.* 69: 3284.
65 Parr, R.G., Von Szentpály, L., and Liu, S. (1999). *J. Am. Chem. Soc.* 121: 1922.
66 Chattaraj, P.K., Sarkar, U., and Roy, D.R. (2006). *Chem. Rev.* 106: 2065.
67 Parr, R.G. and Yang, W. (1984). *J. Am. Chem. Soc.* 106: 4049.
68 Miranda-Quintana, R.A., Heidar-Zadeh, F., and Ayers, P.W. (2018). *J. Phys. Chem. Lett.* 9: 4344.
69 Yang, W., Parr, R.G., and Pucci, R. (1984). *J. Chem. Phys.* 81: 2862.
70 (a) Becke, A.D. (1993). *J. Chem. Phys.* 98: 5648. (b) Lee, C., Yang, W., and Parr, R.G. (1988). *Phys. Rev. B* 37: 785. (c) Becke, A.D. (1988). *Phys. Rev. A* 38: 3098. (d) Stephens, P.J., Devlin, F.J., Chabalowski, C.F., and Frisch, M.J. (1994). *J. Phys. Chem.* 98: 11623.
71 Weigend, F. and Ahlrichs, R. (2005). *Phys. Chem. Chem. Phys.* 7: 3297.
72 (a) Mulliken, R.S. (1955). *J. Chem. Phys.* 23: 1833. (b) Mulliken, R.S. (1955). *J. Chem. Phys.* 23: 1841. (c) Mulliken, R.S. (1955). *J. Chem. Phys.* 23: 2338. (d) Mulliken, R.S. (1955). *J. Chem. Phys.* 23: 2343.
73 Yang, W. and Mortier, W.J. (1986). *J. Am. Chem. Soc.* 108: 5708.
74 Bultinck, P., Van Alsenoy, C., Ayers, P.W., and Carbó-Dorca, R. (2007). *J. Chem. Phys.* 126: 144111.
75 Hirshfeld, F.L. (1977). *Theor. Chim. Acta* 44: 129.
76 Nalewajski, R.F. and Parr, R.G. (2000). *Proc. Natl. Acad. Sci. U.S.A.* 97: 8879.
77 Rong, C., Wang, B., Zhao, D., and Liu, S. (2020). *WIREs Comput. Mol. Sci.* 10: e1461.
78 Morell, C., Grand, A., and Toro-Labbé, A. (2005). *J. Phys. Chem. A* 109: 205.
79 Morell, C., Grand, A., and Toro-Labbé, A. (2006). *Chem. Phys. Lett.* 425: 342.
80 Tognetti, V., Morell, C., Ayers, P.W. et al. (2013). *Phys. Chem. Chem. Phys.* 15: 14465.
81 Cohen, M.H., Ganduglia-Pirovano, M.V., and Kudrnovsky, J. (1994). *J. Chem. Phys.* 101: 8988.
82 Cohen, M.H., Ganduglia-Pirovano, M.V., and Kudrnovsky, J. (1995). *J. Chem. Phys.* 103: 3543.
83 Baekelandt, B.G. (1996). *J. Chem. Phys.* 105: 4664.
84 Nalewajski, R.F. and Parr, R.G. (1982). *J. Chem. Phys.* 77: 399.
85 Lee, C., Yang, W., and Parr, R.G. (1988). *J. Mol. Struct. THEOCHEM* 163: 305.
86 Chattaraj, P.K., Maiti, B., and Sarkar, U. (2003). *J. Phys. Chem. A* 107: 4973.
87 Domingo, L.R., Aurell, M.J., Pérez, P., and Contreras, R. (2002). *J. Phys. Chem. A* 106: 6871.
88 Berkowitz, M. (1987). *J. Am. Chem. Soc.* 109: 4823.
89 Chattaraj, P.K. (2001). *J. Phys. Chem. A* 105: 511.
90 Melin, J., Aparicio, F., Subramanian, V. et al. (2004). *J. Phys. Chem. A* 108: 2487.
91 Stuyver, T. and Shaik, S. (2020). *J. Am. Chem. Soc.* 142 (47): 20002.
92 Berkowitz, M., Ghosh, S.K., and Parr, R.G. (1985). *J. Am. Chem. Soc.* 107: 6811.

93 Berkowitz, M. and Parr, R.G. (1988). *J. Chem. Phys.* 88: 2554.
94 Harbola, M.K., Chattaraj, P.K., and Parr, R.G. (1991). *Isr. J. Chem.* 31: 395.
95 Gal, T. (2007). *J. Math. Chem.* 42: 661.
96 Ayers, P.W. and Parr, R.G. (2008). *J. Chem. Phys.* 128: 184108.
97 Gál, T. (2012). *Theor. Chem. Acc.* 131: 1223.
98 (a) Bonaccorsi, R., Scrocco, E., and Tomasi, J. (1970). *J. Chem. Phys.* 52: 5270. (b) Scrocco, E. and Tomasi, J. (1973). *Top. Curr. Chem.* 42: 95. (c) Politzer, P. and Truhlar, D.G. (ed.) (1981). *Chemical Applications of Atomic and Molecular Electrostatic Potentials*. New York: Plenum Press. (d) Naray-Szabo, G. and Ferenczy, G.G. (1995). *Chem. Rev.* 95: 829. (e) Murray, J.S. and Sen, K.D. (ed.) (1996). *Molecular Electrostatic Potentials - Concepts and Applications, Theoretical and Computational Chemistry*, vol. 3. Amsterdam: Elsevier. (f) Suresh, C.H. and Gadre, S.R. (1998). *J. Am. Chem. Soc.* 120: 7049. (g) Murray, J.S. and Politzer, P. (2011). *WIREs Comput. Mol. Sci.* 1: 153.
99 Langenaeker, W., De Proft, F., and Geerlings, P. (1995). *J. Phys. Chem.* 99: 6424.
100 (a) Chirlian, L.E. and Francl, M.M. (1987). *J. Comput. Chem.* 8: 894. (b) Breneman, C.M. and Wiberg, K.B. (1990). *J. Comput. Chem.* 11: 361. (c) Besler, B.H., Merz, K.M., and Kollman, P.A. (1990). *J. Comput. Chem.* 11: 431. (d) Hu, H., Lu, Z., and Yang, W.J. (2007). *Chem. Theor. Comput.* 3: 1004.
101 (a) Chattaraj, P.K. (2000). *J. Phys. Chem. A* 105: 511. (b) Melin, J., Aparicio, F., Subramanian, V. et al. (2004). *J. Phys. Chem. A* 108: 2487.
102 Ullrich, C.A. (2019). *Time Dependent Density Functional Theory: Concepts and Applications*. Oxford: Oxford University Press.
103 (a) Sablon, N., De Proft, F., Ayers, P.W., and Geerlings, P. (2010). *J. Chem. Theory Comput.* 6: 3671. (b) Sablon, N., De Proft, F., and Geerlings, P. (2010). *J. Phys. Chem. Lett.* 1: 1228. (c) Sablon, N., De Proft, F., and Geerlings, P. (2010). *Chem. Phys. Lett.* 498: 192. (d) Sablon, N., De Proft, F., Solá, M., and Geerlings, P. (2012). *Phys. Chem. Chem. Phys.* 14: 3960. (e) Fias, S., Geerlings, P., Ayers, P.W., and De Proft, F. (2013). *Phys. Chem. Chem. Phys.* 15: 2882. (f) Boisdenghien, Z., Van Alsenoy, C., De Proft, F., and Geerlings, P. (2013). *J. Chem. Theory Comput.* 9: 1007. (g) Boisdenghien, Z., Fias, S., Van Alsenoy, C. et al. (2014). *Phys. Chem. Chem. Phys.* 16: 14614. (h) Fias, S., Boisdenghien, Z., De Proft, F., and Geerlings, P. (2014). *J. Chem. Phys.* 141: 184107. (i) Geerlings, P., Fias, S., Boisdenghien, Z., and De Proft, F. (2014). *Chem. Soc. Rev.* 43: 4989. (j) Stuyver, T., Fias, S., De Proft, F. et al. (2015). *J. Chem. Phys.* 142: 094103.
104 (a) Prodan, E. and Kohn, W. (2005). *Proc. Natl. Acad. Sci. U.S.A.* 102: 11635. (b) Kohn, W. (1996). *Phys. Rev. Lett.* 76: 3168.
105 Fias, S., Heidar-Zadeh, F., Geerlings, P., and Ayers, P.W. (2017). *Proc. Natl. Acad. Sci. U.S.A.* 114: 11633.
106 Wang, B., Geerlings, P., Van Alsenoy, C. et al. (2023). *J. Chem. Theory Comput.* 19: 3223.
107 Yang, W., Cohen, A.J., De Proft, F., and Geerlings, P. (2012). *J. Chem. Phys.* 136: 144110.
108 Chattaraj, P.K., Cedillo, A., and Parr, R.G. (1995). *J. Chem. Phys.* 103: 7645.

References

109 Ghosh, S.K. and Berkowitz, M. (1985). *J. Chem. Phys.* 83: 2976.

110 (a) Thomas, L.H. (1927). *Proc. Cambridge Philos. Soc.* 23: 542. (b) Fermi, E. (1927). *Rend. Accad. Naz. Lincei* 6: 602. (c) Dirac, P.A.M. (1930). *Proc. Cambridge Philos. Soc.* 26: 376.

111 Ghosh, S.K. (1990). *Chem. Phys. Lett.* 172: 77.

112 (a) Klopman, G. (1968). *J. Am. Chem. Soc.* 90: 223. (b) Klopman, G. (1974). *Chemical Reactivity and Reaction Paths* (ed. G. Klopman). Wiley; Chapter Chemical reactivity and reaction paths: general introduction.

113 Peng, D. and Yang, W. (2013). *J. Chem. Phys.* 138: 184108.

114 Anderson, J.S.M., Melin, J., and Ayers, P.W. (2007). *J. Chem. Theory Comput.* 3: 358.

115 Anderson, J.S.M., Melin, J., and Ayers, P.W. (2007). *J. Chem. Theory Comput.* 3: 375.

4

Chemical Concepts from Density-Based Approaches in Density Functional Theory

Dongbo Zhao[1], Xin He[2], Chunying Rong[3], and Shubin Liu[4,5]

[1] *Yunnan University, Institute of Biomedical Research, Kunming 650500, Yunnan, People's Republic of China*
[2] *Shandong University, Qingdao Institute for Theoretical and Computational Sciences, Qingdao 266237, Shandong, People's Republic of China*
[3] *Hunan Normal University, Key Laboratory of Chemical Biology and Traditional Chinese Medicine Research (Ministry of Education of China), Changsha, Hunan 410081, People's Republic of China*
[4] *University of North Carolina, Research Computing Center, Chapel Hill, NC 27599-3420, USA*
[5] *University of North Carolina, Department of Chemistry, Chapel Hill, NC 27599-3290, USA*

4.1 Introduction

It is commonly accepted that density functional theory (DFT) [1, 2] is now a mature and the most widely used quantum chemistry method in modeling the electronic structures of molecules and condensed-phase systems. This is because of its unprecedented simplicity and elegance in theory and decent accuracy and efficiency in computation. Lying at the heart of DFT is the electron density, which circumvents the so-called "dimensionality curse." However, less accepted is its conceptual side, which can be employed to provide new insights into numerous traditional concepts and understandings. As briefly reviewed in the previous chapter by Frank De Proft, one of the most prominent experts in the world on this topic, conceptual density functional theory (CDFT) [3–7] is such an excellent example, with which electronegativity, hardness, electrophilicity, etc. can be formulated with density-based ideas. There are other ways available in the literature [8], nevertheless, to employ density-based ideas for the use of facilitating chemical understanding, as outlined in our recent perspective article on this subject (Figure 4.1).

In this chapter, we will present our current understanding about this matter by introducing four density-based frameworks: orbital-free density functional theory (OF-DFT) [9], CDFT [3–7], density-associated quantities (DAQs), and the information-theoretic approach (ITA) [10–12], and then showcasing a few of their applications for the understanding of chemical concepts such as covalent bonding and NCIs [13, 14], cooperation [15, 16], frustration [17], homochirality [18, 19], electrophilicity [20–23], nucleophilicity [22, 23], regioselectivity [22, 23], and stereoselectivity [22, 23]. A few perspectives on possible future directions are provided at the end.

Exploring Chemical Concepts Through Theory and Computation, First Edition. Edited by Shubin Liu.
© 2024 WILEY-VCH GmbH. Published 2024 by WILEY-VCH GmbH.

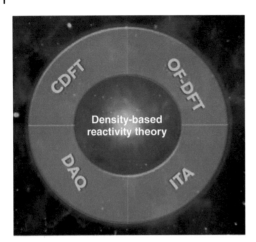

Figure 4.1 The four density-based frameworks to appreciate chemical understanding in DFT. Source: Reproduce with permission from Rong et al. [8]. Copyright 2022 American Chemical Society.

4.2 Four Density-Based Frameworks

According to the two basic DFT theorems of Hohenberg and Kohn [1, 2], the electron density of a molecular or condensed-phase system in the ground state should be sufficient to determine all electronic properties of the system, including bonding, stability, function, and reactivity. To solve the associated Euler–Lagrange equation of DFT, especially the Kohn–Sham equation, only the total energy density functional, its components, and their functional derivatives are needed. Indeed, the history of DFT in the last few decades was the history of searching for better-behaved, and more accurate forms of one of the energy components, the exchange-correlation energy density functional. For that, the well-known Jacob ladder was proposed and accomplished. However, to establish a chemical reactivity theory in DFT, which may deal with not only energetics but also many other functionalities such as stability, reactivity, and selectivity besides the energetic forms, more other forms of density functions and/or density functionals are required. These additional forms of electron density can be, but are not limited to, the ones containing density-associated functions such as density gradient and Laplacian, or density functionals and functional derivatives of different formulas. In the present literature, the following four avenues to derive these forms are available:

4.2.1 Orbital-Free DFT (OF-DFT)

The first Hohenberg and Kohn theorem [1, 2] shows that there is a one-to-one correspondence between the electron density $\rho(\mathbf{r})$ and external potential $v_{\text{ext}}(\mathbf{r})$ (composed by the atomic nuclei of a molecule) in the ground state, so the total energy E of the electronic system is a function of the electron density only,

$$E \equiv E[\rho] = F[\rho] + \int \rho(\mathbf{r})v_{\text{ext}}(\mathbf{r})d\mathbf{r}. \tag{4.1}$$

where $F[\rho]$ is the universal energy density functional. In DFT, this total energy $E[\rho]$ can be partitioned into a few components according to their physical origins, including noninteracting kinetic energy $T_s[\rho]$, nuclear–electron attraction $V_{ne}[\rho]$, classical electron–electron Coulombic repulsion $J[\rho]$, nuclear–nuclear repulsion V_{nn}, and exchange–correlation $E_{xc}[\rho]$ energy density functionals.

$$E[\rho] = T_s[\rho] + V_{ne}[\rho] + J[\rho] + E_{xc}[\rho] + V_{nn}. \tag{4.2}$$

Since three terms above, V_{ne}, V_{nn}, and J, are electrostatic in nature, they can be bundled together to form a new term called electrostatic energy E_e,

$$E_e[\rho] = V_{ne}[\rho] + J[\rho] + V_{nn}, \tag{4.3}$$

leading to the three-term sum of the total energy functional,

$$E[\rho] = T_s[\rho] + E_e[\rho] + E_{xc}[\rho]. \tag{4.4}$$

There are two unknowns in the above formula, $T_s[\rho]$ and $E_{xc}[\rho]$, whose exact analytical forms in terms of $\rho(\mathbf{r})$ are never known. The Kohn–Sham scheme introduces Kohn–Sham orbitals to bypass the need to know $T_s[\rho]$. Instead, it is expressed in terms of Kohn-Sham orbitals. It should be clear that (i) DFT does not have to use orbitals and (ii) the Kohn–Sham scheme is only a workaround. The DFT method without using orbitals is called the OF-DFT [9]. Equations (4.1)–(4.4) are in the spirit of OF-DFT.

The second Hohenberg–Kohn theorem provides the variational principle for DFT,

$$L[\rho] = E[\rho] - \mu \left(\int \rho(\mathbf{r}) d\mathbf{r} - N \right), \tag{4.5}$$

where $L[\rho]$ stands for the Lagrangian, μ denotes the chemical potential, and N represents the total number of electrons of a system. The variational process is carried out by minimizing the total energy functional $E[\rho]$ subject to the constraint that $\rho(\mathbf{r})$ is normalized to N. At the solution point, we obtain the Euler–Lagrange equation

$$\mu = \frac{\delta E[\rho]}{\delta \rho(\mathbf{r})} = \frac{\delta F[\rho]}{\delta \rho(\mathbf{r})} + v_{ext} = \frac{\delta T_s}{\delta \rho(\mathbf{r})} + v_J + v_{ex} + v_{ext} \tag{4.6}$$

where $\frac{\delta T_s}{\delta \rho(\mathbf{r})}$ is the functional derivative of the kinetic energy functional with respect to the electron density, v_{ex} and v_J are the functional derivative of the exchange–correlation potential $E_{xc}[\rho]$ and classical Coulombic repulsion $J[\rho]$, respectively, with

$$v_{xc} = \frac{\delta E_{xc}[\varrho]}{\delta \rho(\mathbf{r})} \tag{4.7}$$

and

$$v_J = \int \frac{\rho(\mathbf{r})}{|\mathbf{r} - \mathbf{r}'|} d\mathbf{r}. \tag{4.8}$$

In Eq. (4.6), both T_S and E_{xc} are unknown. It is also well known that T_S is much more difficult to approximate than E_{xc}. While there are multiple versions and implementations of OF-DFT in the literature, in what follows, we present the scheme originally proposed by March et al. In this case, the total kinetic energy T_S is partitioned into two contributions, Weizsäcker kinetic energy T_W [24] and Pauli energy T_P [25–27], with

$$T_W = \frac{1}{8} \int \frac{|\nabla \rho(\mathbf{r})|^2}{\rho(\mathbf{r})} d\mathbf{r}, \tag{4.9}$$

and

$$T_P[\rho] = T_S[\rho] - T_W[\rho] = \int [\tau(\mathbf{r}) - \tau_W(\mathbf{r})] d\mathbf{r}, \tag{4.10}$$

with $\tau(\mathbf{r})$ and $\tau_W(\mathbf{r})$ being the local density of the total and Weizsäcker kinetic energies, respectively. There are good reasons why this scheme is attractive. At first, for systems with one or two electrons, T_W equals T_S and T_P vanishes. More importantly, T_P represents the contribution of the Pauli Exclusion Principle to the total kinetic energy. That is the precise reason why this portion of the total kinetic energy is called the Pauli energy. Becke and Edgecombe [28] employed the Pauli energy to identify lone pairs in a molecule by introducing the ELF (electron localization function) index. With this OF-DFT scheme, the Euler–Lagrange equation becomes,

$$\left\{ -\frac{1}{2} \nabla^2 + v_{\text{ext}} + v_J + v_P + v_{\text{ex}} \right\} \rho^{1/2}(\mathbf{r}) = \mu \rho^{1/2}(\mathbf{r}), \tag{4.11}$$

where v_P is the Pauli potential defined by

$$v_P = \frac{\delta T_P[\rho]}{\delta \rho(\mathbf{r})} = \frac{\delta T_S[\rho]}{\delta \rho(\mathbf{r})} - \frac{\delta T_W[\rho]}{\delta \rho(\mathbf{r})}. \tag{4.12}$$

Equation (4.11) is similar to the Kohn–Sham equation. The difference is that Eq. (4.11) is just one equation involving the square root of the electron density, whereas Kohn–Sham equations are multiple ones, with each for one Kohn–Sham orbital.

Not long ago, we proposed another partition strategy [29] for the total energy functional $E[\rho]$,

$$E[\rho] = E_s[\rho] + E_e[\rho] + E_q[\rho], \tag{4.13}$$

where $E[\rho]$ is assumed to come from three independent physiochemical effects: steric effect $E_s[\rho]$, electrostatic effect $E_e[\rho]$, and Fermionic quantum effect due to the exchange–correlation effect $E_q[\rho]$. With simple algebra, we have proved that the steric effect contribution $E_s[\rho]$ can be quantified by the Weizsäcker kinetic energy $T_W[\rho]$ [24],

$$E_s[\rho] \equiv T_W[\rho] = \int \tau_W(\mathbf{r}) d\mathbf{r} = \frac{1}{8} \int \frac{|\nabla \rho(\mathbf{r})|^2}{\rho(\mathbf{r})} d\mathbf{r}, \tag{4.14}$$

where $\tau_W(\mathbf{r})$ is the Weizsäcker kinetic energy density and $\nabla \rho(\mathbf{r})$ is the density gradient. As a result, the contribution from the quantum effect, $E_q[\rho]$, is the sum of the exchange–correlation energy $E_{xc}[\rho]$ and Pauli energy $E_P[\rho]$ [25–27],

$$E_q[\rho] = E_P[\rho] + E_{xc}[\rho] = T_s[\rho] - T_W[\rho] + E_{xc}[\rho]. \tag{4.15}$$

Two total energy partitions have been applied to examine numerous chemical processes such as bond rotations [30], anomeric effect [31], and S_N2 reactions [32], and new physiochemical insights were yielded. For example, a unified view has been obtained for the bond rotation barrier height. Recent studies include understanding the internal methyl rotation [33], which is implicated in biological processes (methylation/demethylation of DNA/RNA), and employing Pauli energy to quantify chemical bonding and NCIs [13, 14]. We will talk about it more in the application section of this chapter.

Recently, we proposed a reactivity descriptor [34], expressed in terms of the kinetic energy density functional, to quantify the local temperature of a system,

$$T(\mathbf{r}) = \frac{2\tau(\mathbf{r})}{3k_B \rho(\mathbf{r})}, \qquad (4.16)$$

where k_B is Boltzmann constant, respectively. This quantity can be used as an effective descriptor of molecular reactivity, and it belongs to the territory of OF-DFT in our opinion.

4.2.2 Conceptual DFT (CDFT)

Pioneered by the late Robert G. Parr and coworkers [3–7], CDFT was introduced to quantify chemical reactivity of molecules in a unified manner using DFT language. For a system with the total number of electrons N and the external potential $v(\mathbf{r})$, the total energy changes ΔE of a chemical process due to the change of these variables can be expressed by a Taylor expansion. Up to the second order, this series with respect to these two variables is as follows,

$$\Delta E \equiv E[N + \Delta N, v(\mathbf{r}) + \Delta v(\mathbf{r})] - E[N, v(\mathbf{r})]$$
$$= \left\{ \left(\frac{\partial E}{\partial N}\right)_v \Delta N + \int \left(\frac{\delta E}{\delta v(\mathbf{r})}\right)_N \delta v(\mathbf{r}) d\mathbf{r} \right\} +$$
$$+ \frac{1}{2!} \left\{ \left(\frac{\partial^2 E}{\partial N^2}\right)_v \Delta N^2 + 2 \int \left(\frac{\partial}{\partial N}\left(\frac{\delta E}{\delta v(\mathbf{r})}\right)_N\right)_v \Delta N \delta v(\mathbf{r}) d\mathbf{r} +$$
$$+ \int \left(\frac{\delta^2 E}{\delta v^2(\mathbf{r})}\right)_N \delta v(\mathbf{r}) \delta v(\mathbf{r}') d\mathbf{r} d\mathbf{r}' \right\} \qquad (4.17)$$

where first- and second-order derivatives are assumed to be existent and well-behaved. What CDFT has accomplished in the past few decades is to make sense of these derivatives in terms of chemical insights. For example, $\left(\frac{\partial E}{\partial N}\right)_v$ is chemical potential μ and $\left(\frac{\partial^2 E}{\partial N^2}\right)_v$ is hardness η, $\left(\frac{\partial}{\partial N}\left(\frac{\delta E}{\delta v(\mathbf{r})}\right)_N\right)_v = \left(\frac{\partial \rho(\mathbf{r})}{\partial N}\right)_v$ is the Fukui function $f(\mathbf{r})$, and $\left(\frac{\delta^2 E}{\delta v^2(\mathbf{r})}\right)_N$ is the response function. In addition, electrophilicity index ω [35, 36] was defined through chemical potential and hardness as $\omega = \frac{\mu^2}{2\eta}$, and the dual descriptor $f^2(\mathbf{r})$ [37] was expressed as the response of the Fukui function, $f^2(\mathbf{r}) = \left(\frac{\partial f(\mathbf{r})}{\partial N}\right)_v$. In CDFT, the hard and soft acid and base (HSAB) principle [3–7] and MHP (maximum hardness principle) [3–7], etc., can be explained. Molecular acidity and basicity [38, 39], metal specificity [40, 41], and the proton-coupled electron transfer (PCET) mechanism [42, 43] can be appreciated using ideas from CDFT. The previous chapter was dedicated to this method, as were a few recent reviews, two status reports, and one new book. For this reason, we will skip introducing this method in this chapter.

4.2.3 Density-Associated Quantities (DAQs)

Using electron density and its associated quantities, i.e. DAQs, to appreciate bonding, function, reactivity, and other physiochemical properties has a long history in the literature. The late Richard F. W. Bader [44] was the first to employ density gradient, $\nabla\rho(\mathbf{r})$, to partition a molecule into different atomic basins using the zero-flux criterion, $\mathbf{r}\cdot\nabla\rho(\mathbf{r}) = 0$. He pioneered the topological analysis with electron density, yielding different critical points to appraise different bonding types, such as bond critical points and ring critical points. He also performed similar analyses for density Laplacian, $\nabla^2\rho(\mathbf{r})$. These analyses, together with other studies, were assembled as the AIM (atoms-in-molecules) theory or quantum theory of atoms in molecules (QTAIM).

According to the basic theorems of DFT [1, 2], the ground-state electron density alone contains all the information necessary to describe any property of a molecular system, including bonding, stability, and reactivity, so it is theoretically rigorous for the validity of QTAIM. No question about that. Plus, results from QTAIM demonstrate again and again in the literature that it is an effective and useful approach for the analysis of bonding, partial charge, and many other properties of molecular systems. Notice that, in QTAIM, behaviors of electron density $\rho(\mathbf{r})$, gradient $\nabla\rho(\mathbf{r})$, and Laplacian $\nabla^2\rho(\mathbf{r})$ were separately examined, and they were never combined in any manner.

Indeed, it is not necessary for these DAQs to behave separately. We can extend the territory of QTAIM by combining these three quantities and then examining the behavior of the combined quantity. We call this effort the DAQ approach in the density-based theory of chemical reactivity (Figure 4.1). The first such example was proposed by Johnson et al., who defined the following noncovalent interaction (NCI) index [45, 46],

$$\text{NCI} = \frac{|\nabla\rho(\mathbf{r})|^2}{\rho^{4/3}(\mathbf{r})}. \tag{4.18}$$

This index is a combination of $\rho(\mathbf{r})$ and $\nabla\rho(\mathbf{r})$. Different exponential powers of the density and gradient are necessary to make this index dimensionless with respect to the coordinate scaling. This NCI index can effectively identify NCIs in a molecular system and has found lots of applications in the literature. Very recently, we proposed another combination [14], the second example of DAQ,

$$\text{USI} = \frac{\nabla^2\rho(\mathbf{r})}{\rho^{5/3}(\mathbf{r})}. \tag{4.19}$$

We call this USI index, standing for ultra-strong interaction, because it can identify exceedingly strong covalent interactions in molecules. This USI index is a combination of $\rho(\mathbf{r})$ and $\nabla^2\rho(\mathbf{r})$. Again, their different exponential powers are required to make USI dimensionless. Is it possible to design other combinations? The answer is apparently yes, because $\nabla\rho(\mathbf{r})$ and $\nabla^2\rho(\mathbf{r})$ are yet to be combined, so are the three quantities all together. Plus, combinations of different exponential powers are possible.

4.2.4 Information-Theoretic Approach (ITA)

We invested considerable efforts in the past decade to develop this density-based ITA framework [10–12]. Three key features differentiate ITA from the other three approaches. First, this approach originated from information theory, a discipline of engineering long before DFT was invented, so all quantities in ITA were not DFT per se, because they were borrowed from elsewhere. We call the application of information theory to DFT ITA. Secondly, ever since the very early days of DFT, ITA has attracted the attention of DFT people. This was because information theory deals with probability functions with a continuous and well-defined distribution, and electron density can be thought of precisely as such a probability function, so all the tools developed in information theory can be applied to DFT. Different forms of entropy and information stemming from the electron density are its main concerns. Finally, all ITA quantities are explicit density functionals with known physical meanings. This feature is important because we have analytical formulas to deal with and process, which can be applied to correlate with physiochemical properties. This feature contrasts with energy density functionals such as $T_S[\rho]$ and $E_{XC}[\rho]$, whose exact formulas are still unknown, even though tremendous efforts were invested looking for them. Commonly used ITA quantities include Shannon entropy [47],

$$S_S = \int s_S(\mathbf{r})d\mathbf{r} = -\int \rho(\mathbf{r})\ln\rho(\mathbf{r})d\mathbf{r}, \tag{4.20}$$

Fisher information [48],

$$I_F = \int i_F(\mathbf{r})d\mathbf{r} = \int \frac{|\nabla\rho(\mathbf{r})|^2}{\rho(\mathbf{r})}d\mathbf{r}, \tag{4.21}$$

and alternative Fisher information [49],

$$I'_F = \int i'_F(\mathbf{r})d\mathbf{r} = -\int \nabla^2\rho(\mathbf{r})\ln\rho(\mathbf{r})d\mathbf{r}, \tag{4.22}$$

where $s_S(\mathbf{r})$, $i_F(\mathbf{r})$, and $i'_F(\mathbf{r})$ are the local densities of three ITA quantities, respectively. These quantities are all simple density functionals, with their analytical forms explicitly given. Shannon entropy measures the homogeneity of the density distribution, whereas Fisher information is opposite, which gauges the heterogeneity of the density. Even though their physical meanings are completely different, they are not independent, as we have proved earlier. For these three quantities, the following identity associated with their local densities must always be satisfied,

$$s_S(\mathbf{r}) = -\rho(\mathbf{r}) + \frac{1}{4\pi}\int \frac{i_F(\mathbf{r}')}{|\mathbf{r}-\mathbf{r}'|}d\mathbf{r}' - \frac{1}{4\pi}\int \frac{i'_F(\mathbf{r}')}{|\mathbf{r}-\mathbf{r}'|}d\mathbf{r}'. \tag{4.23}$$

The reason behind this is due to the basic theorems of DFT. Since electron density alone has all the information included, ITA quantities must have contained redundant information, leading to their interdependence and thus validity of Eq. (4.23).

Equations (4.20)–(4.22) are absolute entropy or information. There is a relative entropy version using reference density. Equations (4.24)–(4.26) [50, 51] are the respective relative forms of Eqs. (4.20)–(4.22),

$$I_G = \int \rho(\mathbf{r})\ln\frac{\rho(\mathbf{r})}{\rho_0(\mathbf{r})}d\mathbf{r}, \tag{4.24}$$

$$I_F^r = \int \rho(\mathbf{r}) \left[\nabla \ln \frac{\rho(\mathbf{r})}{\rho_0(\mathbf{r})} \right]^2 d\mathbf{r}, \qquad (4.25)$$

$$I_F^{r'} = \int \nabla^2 \rho(\mathbf{r}) \ln \frac{\rho(\mathbf{r})}{\rho_0(\mathbf{r})} d\mathbf{r}. \qquad (4.26)$$

where $\rho_0(\mathbf{r})$ is the reference density satisfying the same normalization condition as $\rho(\mathbf{r})$. These quantities are interdependent of each other as well, as recently proved by us from theory.

There are other forms of ITA quantities, including Ghosh–Berkowitz–Parr entropy [52], Rényi entropy of order n [53], Onicescu information energy [53], and relative Rényi entropy of order n [53]. Extensive reviews about these ITA quantities are available in the literature [10–12]. To save space, readers are referred to those reviews.

Applying ITA quantities to deepen our understanding about physiochemical properties started with the seminal work by Nalewajski and Parr [54] in 2000, who employed the AIM representation of information gain in Eq. (4.24),

$$I_G = \sum_A \int \rho_A \ln \frac{\rho_A}{\rho_A^0} d\mathbf{r} \qquad (4.27)$$

where ρ_A is the atomic density of Atom A, and ρ_A^0 is its reference density. When Eq. (4.27) is minimized subject to the condition that both densities are normalized to the total electron N,

$$\delta \left\{ I_G - \lambda \left[\sum_A \int \rho_A(\mathbf{r}) d\mathbf{r} - N \right] \right\} = 0, \qquad (4.28)$$

where λ is the Lagrange multiplier, they obtained the following density partition result,

$$\rho_A = \frac{\rho_A^0}{\sum_A \rho_A^0} \rho, \qquad (4.29)$$

which is the well-known "stockholder partition" of the electron density first proposed by Hirshfeld [55, 56]. This work provided the first example of quantifying a molecular reactivity-related property from the ITA viewpoint. Moreover, since the concept of atomic partial charges is not associated with any physical observable, it has no unique definition in principle. This work renders the first instance of deriving the atomic charge from a physiochemical principle.

Since ρ_A and ρ_A^0 are similar in identity, we introduce a new variable, $x = (\rho_A - \rho_A^0)/\rho_A$ and rewrite Eq. (4.27) as

$$I_G = \sum_A \int \rho_A \ln \frac{1}{1-x} d\mathbf{r}. \qquad (4.30)$$

Since x is small, using $\ln \frac{1}{1-x} \approx x$ as the first-order approximation, we obtain

$$I_G \approx \sum_A \int \left(\rho_A - \rho_A^0 \right) d\mathbf{r} = -\sum_A q_A \qquad (4.31)$$

where q_A is the Hirshfeld charge on atom A. This result shows that under the first-order approximation, information gain gives rise to the Hirshfeld charge distribution. Meanwhile, ρ_A and ρ_A^0 satisfy the same normalization condition, the total information gain in Eq. (4.31) must vanish, yielding

$$I_G \approx -\sum_A q_A \equiv 0. \tag{4.32}$$

Equation (4.32) suggests that under the first-order approximation, the information before and after a system is formed should be conserved. We call this result the *information conservation principle* [57].

4.3 Applications of Density-Based Approaches

In this section, we highlight a few applications of the density-based frameworks in recent literature by us in understanding chemical problems, including isomeric and conformational stability, strong covalent and noncovalent interactions [13, 14], cooperation and frustration [15–17], acidity and basicity [38, 39, 58], aromaticity and antiaromaticity [59, 60], homochirality and principle of chirality hierarchy [18, 19], electrophilicity and nucleophilicity [20–23], and regioselectivity and stereoselectivity [22, 23], among others.

4.3.1 Molecular Isomeric and Conformational Stability

The nature and origin of isomeric and conformation stability of molecules are still controversial in the literature, even for as simple systems as ethane, for which it is still unclear if steric or quantum or else effect plays the dominant role. For six simple compounds with only one rotatable bond, a unified picture emerged when we applied Eqs. (4.4) and (4.13) to appreciate which factor or factors dictate their conformational stability. The electrostatic interaction emerged as the dominant contributor to the rotation barrier height, whereas both steric repulsion and quantum effects play minor but indispensable roles, as demonstrated by Figure 4.2.

These partition schemes have also been applied to other systems to appreciate anomic effect [31, 61, 62], *cis* effect [63], fullerenes [64], *γ-gauche* effect [65], and $S_N 2$ reactions [32, 66].

4.3.2 Bonding and Noncovalent Interactions

Conventionally, one must use orbitals to understand chemical bonding. On the other hand, for NCIs, such as ionic bonds and van der Waals interactions, the picture of orbitals is inadequate. Density-based descriptors can be employed to identify and understand both covalent bonding and NCIs without resorting to orbitals. Using the Pauli energy, which is the measure of electron repulsion due to the *Pauli Exclusion Principle*, a dimensionless quantity $\zeta(\mathbf{r}) = \frac{\tau(\mathbf{r}) - \tau_W(\mathbf{r})}{\tau_{TF}(\mathbf{r})}$ is introduced, with τ_{TF} as the Thomas-Fermi kinetic energy density. With it, we introduced the strong covalent interaction (SCI) index [13],

$$\text{SCI} = \frac{1}{\zeta(\mathbf{r})}. \tag{4.33}$$

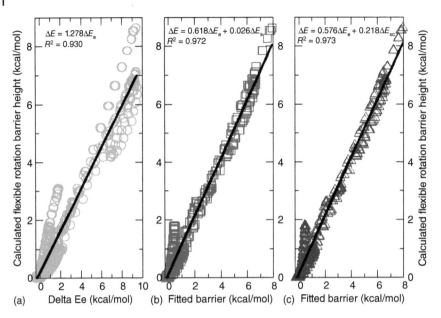

Figure 4.2 Least-square fittings with one and two energy components for all six molecules. Source: Reproduce with permission from Liu [30]. Copyright 2013 American Chemical Society.

It is closely related to the ELF index [28, 67–69], which was defined to identify electron pairs in molecules,

$$\text{ELF} = \frac{1}{1 + \zeta^2(\mathbf{r})}. \tag{4.34}$$

These indexes can be applied to identify strong covalent interactions. The basic assumption is that in areas where there exist strong covalent interactions, strong repulsion originated from the Pauli Exclusion Principle should be dominant, resulting in a signature isosurface from the SCI index for the same type of bonding multiplicity. Shown in Figure 4.3 is the comparison of the ELF and SCI isosurfaces of diatomic species with well-established double, triple, and quadruple covalent bonds. Both ELF and SCI demonstrated the same bonding pattern. Double, triple, and quadruple covalent bonds showed the signature isosurface like a dumbbell, a donut, and four beans, respectively. We tested a lot more systems. The signature isosurface was found to be identical for the same type of bonding multiplicity.

However, SCI and ELF indexes failed to identify single covalent bonds and NCIs. To resolve this issue, we recently proposed another two density-based indexes, BNI (bonding and noncovalent interactions) and USI (ultra-strong interactions), to identify both BNI. Their formulas as well as their relationships with other indexes are shown in Figure 4.4.

4.3.3 Cooperation and Frustration

Cooperation and frustration are a pair of concepts from the two sides of the same coin. Take a molecular cluster as an example. Cooperation evaluates the energetic

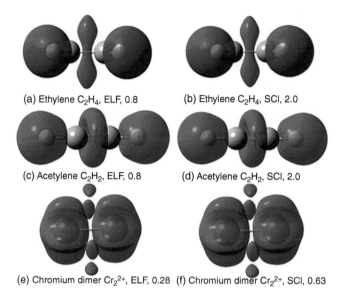

Figure 4.3 Comparison of ELF (left panel) and SCI (right panel) distributions. The C=C double bond, C≡C triple bond, and Cr::Cr quadruple covalent bonds look like a dumbbell, a donut, and four beans, respectively. Also shown in the figure are the isovalues in atomic units. Source: Reproduce with permission from Liu et al. [13]. Copyright 2018 American Chemical Society.

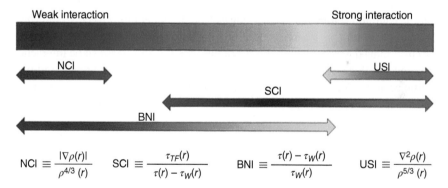

Figure 4.4 Relationship of four density-based descriptors for bonding and noncovalent interactions, NCI, BNI, SCI, and USI, in the coverage of the entire interaction spectrum. Source: Reproduce with permission from Zhong et al. [14]. Copyright 2022 American Chemical Society.

gain from cluster formation, whereas frustration quantifies the sacrifice paid by each individual molecule when such a cluster is formed. For simplicity, we assume that a molecular cluster consists of n copies of the repeating unit B, B_n, and the interaction energy per repeating unit E_n is utilized to quantify the cooperation effect κ, which is called cooperativity [15, 16],

$$\kappa = -\frac{\partial E_n}{\partial n}. \tag{4.35}$$

If $\kappa > 0$, it is positive cooperativity, indicating that adding an extra monomer into the cluster will induce stronger interactions. If $\kappa < 0$, it is negative cooperativity, meaning that adding more building blocks will lead to smaller interaction energies. If $\kappa = 0$, no cooperativity is found for the system.

With six molecular clusters [$(HF)_{20}$, $(CO_2)_{20}$, Cl_2Ar_{20}, $NH_3(H_2O)_{20}$, $F^-\cdot(H_2O)_{20}$, and $Li^+\cdot(H_2O)_{20}$] as illustrations in Figure 4.5a, it is clearly shown in Figure 4.5b that neutral complexes are positively cooperative, but charged ones are negatively cooperative. In our recent studies, we further traced the origin of these cooperativity differences using the ideas from OF-DFT and ITA.

To quantify frustration, a new reference state was introduced, where each monomer in the isolated state takes the same geometry as it is in the complex state (Figure 4.6a). The total energy difference between the monomers in their optimal state and this new reference state is defined as the energetic sacrifice paid by individual isomers. We showed that both positive and negative frustration (Figure 4.6b) are possible, same as cooperativity. Yet, its magnitude is less than that of cooperativity. Also, for small molecular clusters, no general validity for the existence of the minimum frustration principle was observed (Figure 4.6c).

4.3.4 Homochirality and Principle of Chirality Hierarchy

Homochirality, which is the uniformity of chirality for a system with multiple chiral centers, poses an intriguing challenge in biology. For example, why does nature favor L-amino acids (except Gly) in proteins and D-carbohydrates in DNA? Does this striking and common feature have something to do with cooperation? If cooperation does exist, what are the driving forces? Recently, we have showcased that the cooperativity effect indeed exists in homochiral systems, and it is the electrostatic interaction that governs this strong cooperativity effect [18]. We found that homochirality of amino acids/nucleotide sugars is related to the handedness of their corresponding helices. Specifically, the right-handed helix favors the L-chiral form of amino acids, while for deoxyribose sugars, it is the right-handed DNA helix that dictates carbohydrates assuming the D-chiral form. In addition, our results can help explain the relative scarcity of the 3_{10}-helix relative to the α-helix because 3_{10}-helixes have negative cooperativity (Figure 4.7). For two single-stranded DNA models that we built, positive cooperativity is always maintained. Based on these results, we proposed the *Principle of Chirality Hierarchy*, which states that chirality of a lower hierarchy is governed by that of higher ones. In protein helixes, for example, chirality at C_α position is zero-dimensional (0D) and chirality of the right-handed helix is one-dimensional (1D), so the chirality of C_α (0D) is determined by that of the helix (1D). In this regard, the homochirality feature of proteins and DNA is a consequence of this principle. There are 2D (planar) and 3D (supramolecular) chirality systems in the literature, and the same principle should be applicable. In a recent work, we examined and confirmed the validity of the principle for three-blade propeller systems [19].

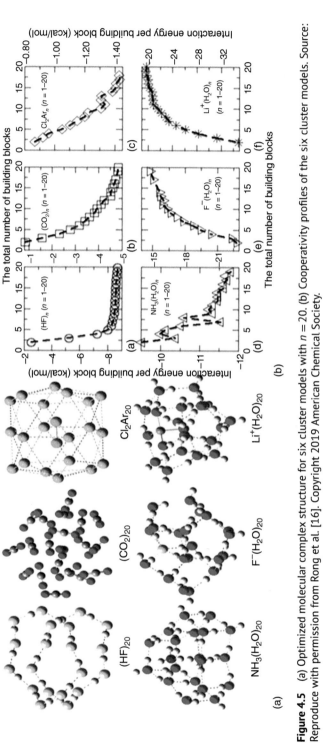

Figure 4.5 (a) Optimized molecular complex structure for six cluster models with $n = 20$. (b) Cooperativity profiles of the six cluster models. Source: Reproduce with permission from Rong et al. [16]. Copyright 2019 American Chemical Society.

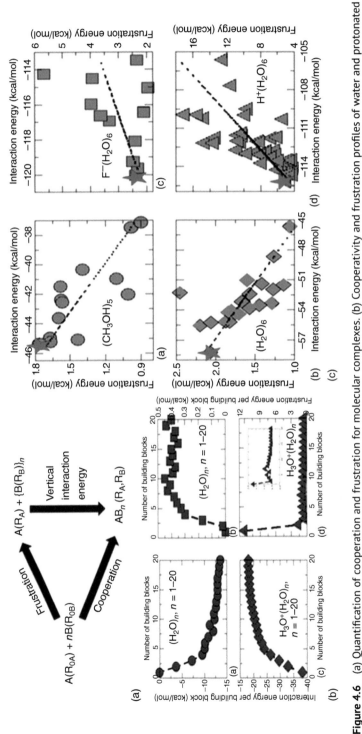

Figure 4.6 (a) Quantification of cooperation and frustration for molecular complexes. (b) Cooperativity and frustration profiles of water and protonated water clusters. (c) Relationship between stability and frustration for 4 clusters. Red stars represent the global minimum of the clusters. Source: Reproduce with permission from Liu and Rong [17]. Copyright 2021 American Chemical Society.

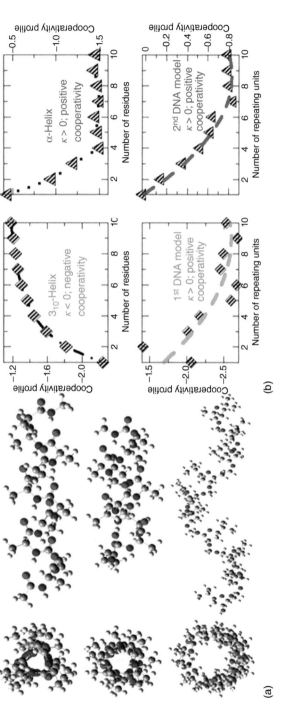

Figure 4.7 (a) Optimized structure models of 3_{10}–helix, α–helix, and single-stranded DNA helix. (b) Cooperativity profiles of secondary structure models. Source: Reproduce with permission from Liu [18]. Copyright 2020 American Chemical Society.

4.3.5 Electrophilicity and Nucleophilicity

Electrophilicity and nucleophilicity are the capabilities to accept and donate electrons by a nucleophile and electrophile, respectively. One can use Fukui functions from CDFT to qualitatively determine these properties [2–7]. In ITA, we can use information gain [50], Eq. (4.24), and its first-order approximation, Hirshfeld charge, Eq. (4.31), to quantify these properties. The reason why Hirshfeld charge can be employed to quantify electrophilicity and nucleophilicity is the following: According to *the information conservation principle* [57], up to the first-order approximation, when a new system is formed, the identity of its components will be preserved. If a component is electrophilic in nature before the new system is formed, it will still be so afterward. Moreover, to preserve the identity as much as possible, the new system adjusts its components in such a way that each component becomes charged according to the stockholder partition scheme. Therefore, the amount of the Hirshfeld charge becomes the identity indicator of each component in the new system.

Shown in Figures 4.8 and 4.9 are the strong linear relationships between the experimental scales and the information gain and Hirshfeld charge. In Figure 4.8a, we examined 21 electrophilic systems with known electrophilicity scale and, in Figure 4.8b, compared our theoretical results with experimental values. In Figure 4.9a, we considered 22 nucleophilic molecules with known experimental nucleophilicity scale and, in Figure 4.9b, compared our information gain and Hirshfeld charge results with the experimental values. In both cases, strong linear correlations between theoretical and experimental results have been obtained, suggesting that information gain and Hirshfeld charge can *simultaneously* and *quantitatively* be applied as robust descriptors of electrophilicity and nucleophilicity. We recently compared the performance of these descriptors from ITA with Fukui functions from CDFT.

4.3.6 Regioselectivity and Stereoselectivity

Regioselectivity is the susceptibility of a bond making and breaking process to preferentially take place at one site/atom of a molecule over all other possible places. A prominent example is the group-directing effect for electrophilic aromatic substitution reactions, where the position, *ortho/para* or *meta*, of the substitution depends on the nature of the group $-R$ already attached to the benzene ring. For groups like $R = -Me$, $-NH_2$, and $-OH$, the preferred substitution site is at the *ortho/para* position, but for groups such as $R = -NO_2$, $-CN$, and $-SO_3H$, the *meta* position is favored. Using information gain or Hirshfeld charge, such regioselectivity propensity can be satisfactorily rationalized. We predicted ortho/para and meta group-directing behaviors for a list of groups whose regioselectivity was previously unknown in the literature. In addition, we examined the activation energy for 18 mono-substituted-benzene molecules, nine of which are ortho/para directing and the other nine groups *meta* directing, reacting with hydrogen fluoride using BF_3 as the catalyst, and found that the barrier height of these reactions strongly

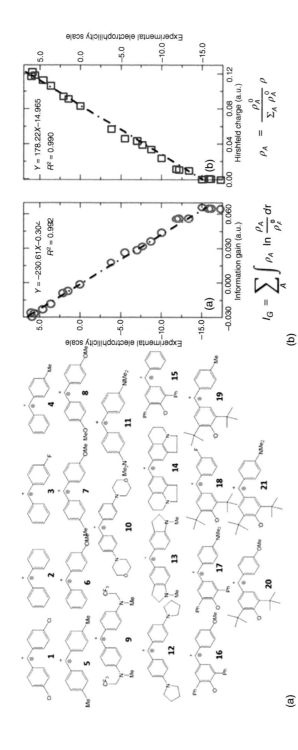

Figure 4.8 (a) Molecular systems with known experimental electrophilicity scale. (b) Strong linear relationships between the experimental electrophilicity results and computed information gain (red circle) and Hirshfeld charge (blue square). Source: Reproduce with permission from Liu et al. [57]. Copyright 2014 American Chemical Society.

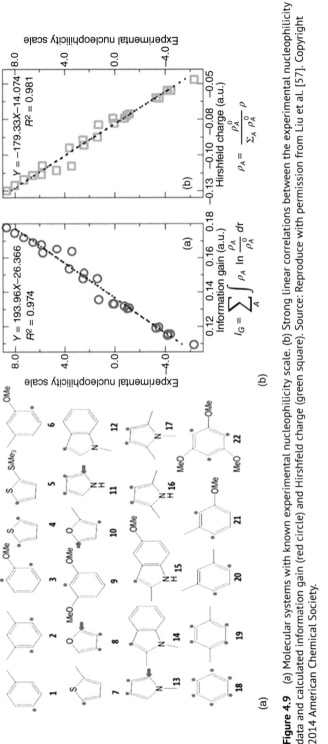

Figure 4.9 (a) Molecular systems with known experimental nucleophilicity scale. (b) Strong linear correlations between the experimental nucleophilicity data and calculated information gain (red circle) and Hirshfeld charge (green square). Source: Reproduce with permission from Liu et al. [57]. Copyright 2014 American Chemical Society.

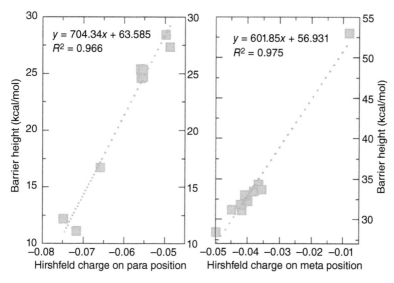

Figure 4.10 Strong correlations of the electrophilic substitution barrier height for mono-substituted benzene derivatives with para- and meta-directing groups with Hirshfeld charge. Source: Reproduce with permission from Liu [21]. Copyright 2015 American Chemical Society.

correlates with the Hirshfeld charge on the regioselective site for both ortho/para and meta directing groups, with the correlation coefficient R^2 both better than 0.96 (Figure 4.10).

On the other hand, stereoselectivity is the tendency of the preferential formation in a reaction of one stereoisomer over the other. This property can be rationalized by the quantification of the steric effect from OF-DFT. This is done through the steric potential $v_s(\mathbf{r})$ [29],

$$v_s(\mathbf{r}) = \frac{\delta E_S[\rho]}{\delta\rho(\mathbf{r})} = \frac{1}{8}\frac{|\nabla\rho(\mathbf{r})|^2}{\rho(\mathbf{r})} - \frac{1}{4}\frac{\nabla^2\rho(\mathbf{r})}{\rho(\mathbf{r})}. \tag{4.36}$$

Taking the partial derivative and Laplacian of Eq. (4.36), we obtain steric force $F_s(\mathbf{r})$ [22, 29]

$$F_s(\mathbf{r}) = -\nabla v_s(\mathbf{r}), \tag{4.37}$$

and steric charge $q_s(\mathbf{r})$ [70]

$$q_s(\mathbf{r}) = -\frac{1}{4\pi}\nabla^2 v_s(\mathbf{r}) = -\frac{1}{4\pi}\nabla^2\left(\frac{\delta E_S[\rho]}{\delta\rho(\mathbf{r})}\right). \tag{4.38}$$

With these two quantities, we can adequately explain stereoselectivity propensities for different reactions. An illustrative example, shown in Figure 4.11, is the nucleophilic addition to the carbonyl group, whose stereoselectivity depends on the size of the groups at the Cα atom. The incoming electrophile prefers the side with a smaller group on the α-carbon. This preference is unambiguously seen by the magnitude of the steric force shown for the two molecules in the figure. We also applied the same

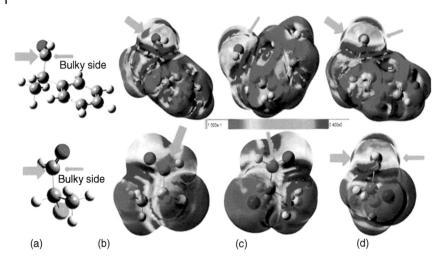

Figure 4.11 Magnitude of the steric force mapped onto the van der Waals surface for PhMeHC−CHO (top) and ClMeHC−CHO (bottom) molecules, whose nucleophilic addition product on the carbonyl group is known to be Felkin–Anh like. Their stereoselectivity is featured by smaller steric forces on the preferred attacking side (more red area coverage) of the carbonyl carbon atom. (a) Carbonyl. (b) Preferred side of attack. (c) Unfavorable side. (d) Side view. Source: Liu et al. [70]. Royal Chemical Society.

idea to $S_N 2$ and other reactions and justified the experimental finding that bulky groups on the central carbon substantially increase the reaction barrier height.

4.3.7 Brønsted–Lowry Acidity and Basicity

The pair of molecular acidity and basicity is an extremely important chemical concept and is widely used in many different disciplines of chemistry and biology. Nevertheless, quantitative determination of these intrinsic physiochemical properties from the perspective of theory and computation is still challenging. We previously proposed [71] to use two equivalent descriptors, MEP (molecular electrostatic potential) and NAO (natural atomic orbital), from the CDFT framework for this purpose. Our more recent studies using ITA quantities indicated that better results could be obtained [58]. In our study, five different categories of acidic series were employed, including singly and doubly substituted benzoic acids, singly substituted benzenesulfinic acids, singly substituted benzeneseleninic acids, phenols, and alkyl carboxylic acids. Accurate descriptions to simultaneously simulate all data sets, a total of 95 points, have been accomplished, with the correlation coefficient better than 0.96 (Figure 4.12). We recently satisfactorily applied the idea, in combination with descriptors from CDFT, to three categories of amines with a total of 195 systems.

4.3.8 Aromaticity and Antiaromaticity

Originated from the cyclic delocalization of electrons leading to extra stability and instability, aromaticity and antiaromaticity are another set of important

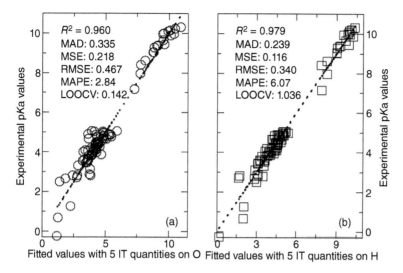

Figure 4.12 Comparison of the experimental pK_a data of all 5 series of compound studied in this work with the fitted pK_a data using 5 information-theoretic quantities from (a) the acidic atom and (b) the leaving proton. The five ITA quantities are Shannon entropy, GBP entropy, information gain, and relative Rényi entropies of second and third orders. Also shown in the plots are the mean absolute deviation (MAD), the mean square error (MSE), the root mean square error (RMSE), the mean absolute percentage error (MAPE), and leave-one-out cross-validation (LOOCV) estimate of prediction error. Source: Reproduce with permission from Cao et al. [58]. Copyright 2017 John Wiley & Sons.

chemical concepts whose appreciation and quantification are still of much recent interest in the literature. There are many different categories of aromaticity and antiaromaticity that have been reported. There are many different descriptors in the literature to characterize different properties of aromaticity and antiaromaticity as well. Using ITA quantities, we obtained new understanding about them [59, 60]. For example, using seven series of substituted fulvene derivatives, we found that cross-correlations between ITA quantities (e.g. S_{GBP}) and aromaticity indexes (e.g. ASE, aromatic stabilization energy) yielded two completely opposite patterns for different sizes of the systems with different aromaticity and antiaromaticity propensities, as shown in Figure 4.13. These ring-size-dependent correlations are in good agreement with Hückel's $4n + 2$ rule of aromaticity and $4n$ rule of antiaromaticity. These opposite cross-correlations have also been observed in other systems with other types of aromaticity and antiaromaticity.

4.3.9 Molecular Properties (Frontier Orbitals, HOMO/LUMO Gap, Oxidation States, Polarizability)

Can density-based frameworks be able to determine orbital-based properties? Using ITA quantities such as Shannon entropy and Fisher information, we evaluated and then correlated these quantities with frontier orbital levels and HOMO/LUMO gap for five polymeric systems in Figure 4.14. We demonstrated that these orbital-related properties can be simulated accurately by ITA quantities

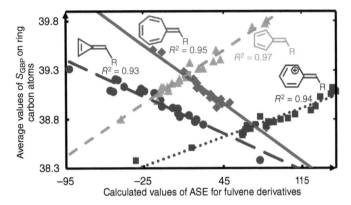

Figure 4.13 Illustrative examples of strong linear correlations between the aromaticity index ASE and the information-theoretic quantities S_{GBP} for substituted fulvene derivatives 3MR, 5MR, 6MR+, and 7MR. Source: Reproduce with permission from Yu et al. [60]. Copyright 2017 Royal Society of Chemistry.

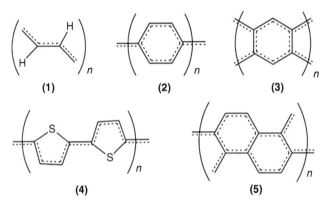

Figure 4.14 (1) Trans-ethylene chain; (2) benzene chain; (3) naphthalene chain; (4) di-thiophene chain; and (5) phenanthrene chain. Source: Reproduce with permission from Huang et al. [72]. Springer Nature.

(Figure 4.15). We also recently confirmed the case for oxidation states of different elements [73], confirming that wave-function-based theory and density-based theory such as ITA are complementary to each other.

Very recently, we applied ITA quantities to predict molecular polarizabilities (α_{iso}) for peptides and proteins (Figure 4.16), based on the strong linear correlations for 20 α-amino acids, 400 dipeptides, and 8000 tripeptides (Figure 4.17). These results confirmed the validity of applying ITA quantities as reliable descriptors of physiochemical properties. Similar results have also been observed for excited-state polarizabilities [75]. More work along this line has been extended for condensed-phase systems and the results will be presented elsewhere.

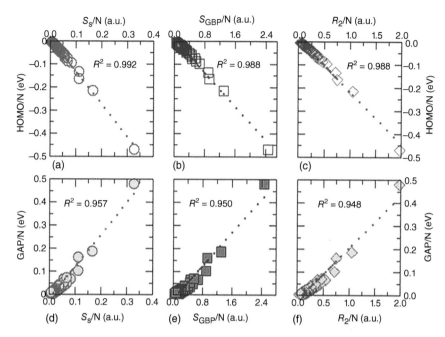

Figure 4.15 Strong linear correlations obtained at the AIM level between frontier orbital properties and ITA quantities with all data points from five polymeric systems in Figure 4.14. Source: Reproduce with permission from Huang et al. [72]. Springer Nature.

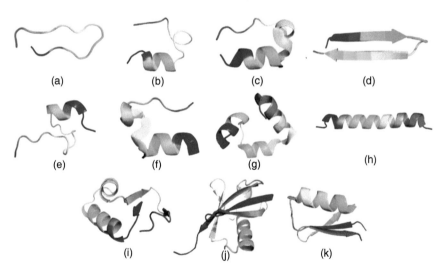

Figure 4.16 (a) 4TTK, sunflower trypsin inhibitor-1, (b) 2LDJ, Trp-cage miniprotein with D-amino acid, (c) 2LL5, cyclo-TC1 Trp-cage, (d) 2LYE, the cyclic cystine ladder motif of θ-defensins, (e) 1SP7, the cystine-rich C-terminal domain of Hydra minicollagen, (f) 1L2Y, Trp-cage miniprotein, (g) 2F4K, chicken villin subdomain, (h) 6PHM, glucagon analog fully composed of D-amino acids, (i) 1CBN, a hydrophobic protein, (j) 1UBQ, human erythrocytic ubiquitin, and (k) 5JI4. Source: Reproduce with permission from Zhao et al. [74]. Copyright 2022 Royal Chemical Society.

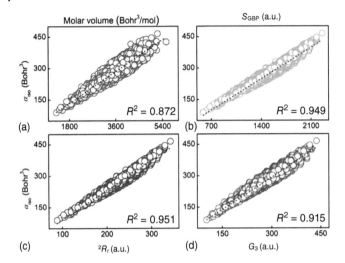

Figure 4.17 Strong correlations between molecular polarizability and (a) molecular volume, (b) S_{GBP}, (c) ^2R_r, and (d) G_3 for a total of 20 amino acids, 400 dipeptides, and 8000 tripeptides. Source: Reproduce with permission from Zhao et al. [74]. Copyright 2022 Royal Chemical Society.

4.4 Concluding Remarks

In this chapter, we overviewed the four density-based frameworks in DFT available in the literature to appreciate physiochemical properties, including OF-DFT, CDFT, DAQs, and ITA, and then highlighted a few of their applications in appreciating isomeric and conformational stability, strong covalent and noncovalent interactions, cooperation and frustration, acidity and basicity, aromaticity and antiaromaticity, homochirality and chirality hierarchy, electrophilicity and nucleophilicity, regioselectivity and stereoselectivity, etc. It is without question that these density-based frameworks can provide a lot of new insights and novel understandings for lots of chemical processes and phenomena. In other words, orbitals are not required to explain and rationalize our experimental findings. This does not mean, however, that density-based and orbital-based descriptions and understandings are mutually exclusive. Indeed, they are two different methodologies to solve the same Schrödinger equation, so these orbital-based and density-based pictures should be complementary to each other.

In the near future, following four directions to further develop the density-based theory of chemical reactivity is possible. (i) Extensions to excited states and condensed phases are needed; (ii) making interconnections among the four frameworks in Figure 4.1 should be considered; (iii) topological analyses of the quantities in the four density-based frameworks could be informative; and (iv) applying these tools to systems in biological and material sciences, especially solar energy research, should be proceeded with.

To wrap up, we wish to emphasize that any electronic structure theory for atoms and molecules should provide new understandings and insights about traditional chemical concepts. This is true for valence bond theory, and the same is true for molecular orbital theory. DFT has been well accepted as the most successful development of electronic structure theory in the last few decades. Nevertheless, for most people, it is only a computational approach. What we have shown in this chapter, together with Chapter 3, clearly shows that it is possible to establish a chemical reactivity theory using the DFT language, which can be applied to garner new insights about traditional chemical concepts such as electronegativity and hardness as well as novel understandings for phenomena from the recent literature, e.g. cooperativity and homochirality.

Acknowledgments

D.B.Z. is supported by the startup funding of Yunnan University, the Yunnan Fundamental Research Projects (Grant No. 202101AU070012), the National Natural Science Foundation of China (Grant No. 22203071) and the High Level Talents Special Support Plan. C.R. acknowledges support from the Hunan Provincial Natural Science Foundation of China (Grant No. 2022JJ30373) and the Scientific Research Fund of Hunan Provincial Education Department (Grant No. 22B0063).

References

1 Parr, R.G. and Yang, W.T. (1989). *Density Functional Theory for Atoms and Molecules*. London: Oxford University.
2 Teale, A.M., Helgaker, T., Savin, A. et al. (2022). DFT exchange: sharing perspectives on the workhorse of quantum chemistry and material science. *Phys. Chem. Chem. Phys.* 24: 28700–28781.
3 Geerlings, P., De Proft, F., and Langenaeker, W. (2003). Conceptual density functional theory. *Chem. Rev.* 103: 1793–1873.
4 Liu, S.B. (2009). Conceptual density functional theory and some recent developments. *Acta Phys. Chim. Sin.* 25: 590–600.
5 Johnson, P.A., Bartolotti, L.J., Ayers, P.W. et al. (2012). Charge density and chemical reactivity: a unified view from conceptual DFT. In: *Modern Charge Density Analysis* (ed. C. Gatti and P. Macchi). New York: Springer.
6 Geerlings, P., Chamorro, E., Chattaraj, P.K. et al. (2020). Conceptual density functional theory: status, prospects, issues. *Theor. Chem. Acc.* 139: 36.
7 Liu, S.B. (ed.) (2022). *Conceptual Density Functional Theory: Towards a New Chemistry Reactivity Theory*. Weinheim, Germany: Wiley-VCH.
8 Rong, C.Y., Zhao, D.B., He, X., and Liu, S.B. (2022). Development and applications of the density-based theory of chemical reactivity. *J. Phys. Chem. Lett.* 13: 11191–11200.
9 Wesolowski, T.A. and Wang, Y.A. (ed.) (2013). *Recent Progress in Orbital-Free Density Functional Theory*. Singapore: World Scientific.
10 Liu, S.B. (2016). Information-theoretic approach in density functional reactivity theory. *Acta Phys. Chim. Sin.* 32: 98–118.

11 Rong, C.Y., Wang, B., Zhao, D.B., and Liu, S.B. (2020). Information-theoretic approach in density functional theory and its recent applications to chemical problems. *WIREs Comp. Mol. Sci.* 10: e1461.

12 Rong, C.Y., Yu, D.H., and Liu, S.B. (2022). Information-theoretic approach. In: *Conceptual Density Functional Theory: Towards a New Chemistry Reactivity Theory* (ed. S.B. Liu). Weinheim, Germany: Wiley-VCH Chapter 15.

13 Liu, S.B., Rong, C.Y., Lu, T., and Hu, H. (2018). Identify strong covalent interactions with Pauli energy. *J. Phys. Chem. A* 122: 3087–3095.

14 Zhong, S.J., He, X., Liu, S.Y. et al. (2022). Towards density-based and simultaneous descriptions of chemical bonding and noncovalent interactions with Pauli energy. *J. Phys. Chem. A* 126: 2437–2444.

15 Rong, C.Y., Zhao, D.B., Yu, D.H., and Liu, S.B. (2018). Quantification and origin of cooperativity: insights from density functional reactivity theory. *Phys. Chem. Chem. Phys.* 20: 17990–17998.

16 Rong, C.Y., Zhao, D.B., Zhou, T.J. et al. (2019). Homogeneous molecular systems are positively cooperative but charged molecular systems are negatively cooperative. *J. Phys. Chem. Lett.* 10: 1716–1721.

17 Liu, S.B. and Rong, C.Y. (2021). Quantifying frustrations for molecular complexes with noncovalent interactions. *J. Phys. Chem. A* 125: 4910–1947.

18 Liu, S.B. (2020). Homochirality originates from the handedness of helices. *J. Phys. Chem. Lett.* 11: 8690–8696.

19 Liu, S.B. (2021). Principle of chirality hierarchy in three-blade propeller systems. *J. Phys. Chem. Lett.* 12: 8720–8725.

20 Liu, S.B. (2014). Where does the electron go? The nature of *ortho/para* and *meta* group directing in electrophilic aromatic substitution. *J. Chem. Phys.* 141: 194109.

21 Liu, S.B. (2015). Quantifying reactivity for electrophilic aromatic substitution reactions with Hirshfeld charge. *J. Phys. Chem. A* 119: 3107–3111.

22 Liu, S.B., Rong, C.Y., and Lu, T. (2017). Electronic forces as descriptors of nucleophilic and electrophilic regioselectivity and stereoselectivity. *Phys. Chem. Chem. Phys.* 19: 1496–1503.

23 Wang, B., Rong, C.Y., Chattaraj, P.K., and Liu, S.B. (2019). A comparative study to predict regioselectivity, electrophilicity and nucleophilicity with Fukui function and Hirshfeld charge. *Theor. Chem. Acc.* 138: 124.

24 von Weizsäcker, C.F. (1935). Zur Theorie der Kernmassen. *Eur. Phys. J.* 96: 431–458.

25 March, N.H. (1987). The density amplitude $\rho^{1/2}$ and the potential which generates it. *J. Comput. Chem.* 8: 375–379.

26 Levy, M. and Hui, O.-Y. (1988). Exact properties of the Pauli potential for the square root of the electron density and the kinetic energy functional. *Phys. Rev. A* 38: 625.

27 Holas, A. and March, N.H. (1991). Construction of the Pauli potential, Pauli energy, and effective potential from the electron density. *Phys. Rev. A* 44: 5521.

28 Becke, A.D. and Edgecombe, K.E. (1990). A simple measure of electron localization in atomic and molecular systems. *J. Chem. Phys.* 92: 5397–5403.

29 Liu, S.B. (2007). Steric effect: a quantitative description from density functional theory. *J. Chem. Phys.* 126: 244103.
30 Liu, S.B. (2013). Origin and nature of bond rotation barriers: a unified view. *J. Phys. Chem. A* 117: 962–965.
31 Huang, Y., Zhong, A.G., Yang, Q.S., and Liu, S.B. (2011). Origin of anomeric effect: a density functional steric analysis. *J. Chem. Phys.* 134: 084103.
32 Liu, S.B., Hu, H., and Pedersen, L.G. (2010). Steric, quantum, and electrostatic effects on S_N2 reaction barriers in gas phase. *J. Phys. Chem. A* 114: 5913–5918.
33 Wang, K.D., He, X., Rong, C.Y. et al. (2022). On the origin and nature of internal methyl rotation barriers: an information-theoretic approach study. *Theor. Chem. Acc.* 141: 68.
34 Guo, C.N., He, X., Rong, C.Y. et al. (2021). Local temperature as a chemical reactivity descriptor. *J. Phys. Chem. Lett.* 12: 5623–5630.
35 Parr, R.G., Szentpály, L.V., and Liu, S.B. (1999). Electrophilicity index. *JACS* 121: 1922–1924.
36 Chattaraj, P.K., Sarkar, U., and Roy, D.R. (2006). Electrophilicity index. *Chem. Rev.* 106: 2065–2091.
37 Morell, C., Grand, A., and Toro-Labbe, A. (2005). New dual descriptor for chemical reactivity. *J. Phys. Chem. A* 109: 205–212.
38 Liu, S.B., Schauer, C.K., and Pedersen, L.G. (2009). Molecular acidity: a quantitative conceptual density functional theory description. *J. Chem. Phys.* 131: 164107.
39 Xiao, X.Z., Cao, X.F., Zhao, D.B. et al. (2020). Quantification of molecular basicity for amines: a combined conceptual density functional theory and information-theoretic approach study. *Acta Phys. Chim. Sin.* 36: 1906034.
40 Feng, X.T., Yu, J.G., Lei, M. et al. (2009). Toward understanding metal-binding specificity of porphyrin: a conceptual density functional theory study. *J. Phys. Chem. B* 113: 13381–13389.
41 Feng, X.T., Yu, J.G., Liu, R.Z. et al. (2010). Why iron? A spin-polarized conceptual density functional theory study on metal-binding specificity of porphyrin. *J. Phys. Chem. A* 114: 6342–6349.
42 Liu, S.B., Ess, D.H., and Schauer, C.K. (2011). Density functional reactivity theory characterizing charge separation propensity in proton-coupled electron transfer reactions. *J. Phys. Chem. A* 115: 4738–4742.
43 Kumar, N., Liu, S.B., and Kozlowski, P.M. (2012). Charge separation propensity of the cozenyme B_{12}-tyrosine complex in adenosylcobalamin-dependent ethulmalonyl-CoA mutase enzyme. *J. Phys. Chem. Lett.* 3: 1035–1038.
44 Bader, R.F.W. (1990). *Atoms in Molecules: A Quantum Theory*. Oxford, England: Oxford University Press.
45 Johnson, E.R., Keinan, S., Mori-Sánchez, P. et al. (2010). Revealing noncovalent interactions. *JACS* 132: 6498–6506.
46 Contreras-Garcia, J., Johnson, E.R., Keinan, S. et al. (2011). NCIPLOT: a program for plotting noncovalent interaction regions. *J. Chem. Theory Comput.* 7: 625–632.
47 Shannon, C.E. (1948). A mathematical theory of communication. *Bell Syst. Technol. J.* 27: 379–423.

48 Fisher, R.A. (1925). Theory of statistical estimation. *Math. Proc. Cambridge Philos. Soc.* 22: 700–725.

49 Liu, S.B. (2007). On the relationship between densities of Shannon entropy and Fisher information for atoms and molecules. *J. Chem. Phys.* 126: 191107.

50 Kullback, S. (1997). *Information Theory and Statistics*. Mineola, NY: Dover Publications.

51 Liu, S.B. (2019). Identity for Kullback-Leibler divergence in density functional reactivity theory. *J. Chem. Phys.* 151: 141103.

52 Ghosh, S.K., Berkowitz, M., and Parr, R.G. (1984). Transcription of ground-state density-functional theory into a local thermodynamics. *PNAS* 81: 8028–8031.

53 Liu, S.B., Rong, C.Y., Wu, Z.M., and Lu, T. (2015). Rényi entropy, Tsallis entropy, and Onicescu information energy in density functional reactivity theory. *Acta Phys. Chim. Sin.* 31: 2057–2063.

54 Nalewajski, R.F. and Parr, R.G. (2000). Information theory, atoms in molecules, and molecular similarity. *PNAS* 97: 8879–8882.

55 Hirshfeld, F. (1977). Bonded-atom fragments for describing molecular charge densities. *Theor. Chim. Acc.* 44: 129–138.

56 Heidar-Zadeh, F., Ayers, P.W., Verstraelen, T. et al. (2018). Information-theoretic approaches to atoms-in-molecules: Hirshfeld family of partitioning schemes. *J. Phys. Chem. A* 122: 4219–4245.

57 Liu, S.B., Rong, C.Y., and Lu, T. (2014). Information conservation principle determines electrophilicity, nucleophilicity, and regioselectivity. *J. Phys. Chem. A* 118: 3698–3704.

58 Cao, X.F., Rong, C.Y., Zhong, A.G. et al. (2018). Molecular acidity: an accurate description with information-theoretic approach in density functional reactivity theory. *J. Comput. Chem.* 39: 117–129.

59 Yu, D.H., Stuyver, T., Rong, C.Y. et al. (2019). Global and local aromaticity of acenes from the information-theoretic approach in density functional reactivity theory. *Phys. Chem. Chem. Phys.* 21: 18195–18210.

60 Yu, D.H., Rong, C.Y., Lu, T. et al. (2017). Aromaticity and antiaromaticity of substituted fulvene derivatives: perspectives from the information-theoretic approach in density functional reactivity theory. *Phys. Chem. Chem. Phys.* 19: 18635–18645.

61 Zhou, X.Y., Yu, D.H., Rong, C.Y. et al. (2017). Anomeric effect revisited: perspective from information-theoretic approach in density functional reactivity theory. *Chem. Phys. Lett.* 684: 97–102.

62 Cao, X.F., Liu, S.Q., Rong, C.Y. et al. (2017). Is there a generalized anomeric effect? Analyses from energy components and information-theoretic quantities from density functional reactivity theory. *Chem. Phys. Lett.* 687: 131–137.

63 Zhao, D.B., Rong, C.Y., Jerkins, S. et al. (2013). Origin of the cis-effect: a density functional theory study of doubly substituted ethylenes. *Acta Phys. Chim. Sin.* 29: 43–54.

64 Zhao, D.B., Liu, S.Y., Rong, C.Y. et al. (2018). Towards understanding isomeric stability of fullerenes with density functional theory and information-theoretic approach. *ACS Omega* 3: 17986–17990.

65 Wang, B., Yu, D.H., Zhao, D.B. et al. (2019). Nature and origin of γ-gauche effect in sulfoxides: a density functional theory and information-theoretic approach study. *Chem. Phys. Lett.* 730: 451–459.

66 Wu, Z.M., Rong, C.Y., Lu, T. et al. (2015). Density functional reactivity theory study of S_N2 reactions from the information-theoretic perspective. *Phys. Chem. Chem. Phys.* 17: 27052–27061.

67 Silvi, B. and Savin, A. (1994). Classification of chemical bonds based on topological analysis of electron localization functions. *Nature* 371: 683–686.

68 Klein, J., Fleurat-Lessard, P., and Pilmè, J. (2021). The topological analysis of the ELF_X localization function: quantitative prediction of hydrogen bonds in the guanine-cytosine pair. *Molecules* 26: 3336.

69 Savin, A., Nesper, R., Wengert, S., and Fässler, T.F. (1997). ELF: the electron localization function. *Angew. Chem. Int. Ed.* 36: 1808–1832.

70 Liu, S.B., Liu, L.H., Yu, D.H. et al. (2018). Steric charge. *Phys. Chem. Chem. Phys.* 20: 1408–1420.

71 Liu, S.B. and Pedersen, L.G. (2009). Estimation of molecular acidity via electrostatic potential at the nucleus and valence natural atomic orbitals. *J. Phys. Chem. A* 113: 3648–3655.

72 Huang, Y., Rong, C.Y., Zhang, R.Q., and Liu, S.B. (2017). Evaluating frontier orbital energy and HOMO/LUMO gap with descriptors from density functional reactivity theory. *J. Mol. Model.* 23: 3.

73 Wu, J.Y., Yu, D.H., Liu, S.Y. et al. (2019). Is it possible to determine oxidation states for atoms in molecules using density-based quantities? An information-theoretic approach and conceptual density functional theory study. *J. Phys. Chem. A* 123: 6751–6760.

74 Zhao, D.B., Zhao, Y.L., He, X. et al. (2023). Efficient and accurate density-based prediction of macromolecular polarizabilities. *Phys. Chem. Chem. Phys.* 25: 2131–2141.

75 Zhao, D.B., He, X., Ayers, P.W., and Liu, S.B. (2023). Excited-state polarizabilities: a combined density functional theory and information-theoretic approach study. *Molecules* 18: 2576.

5

Chemical Bonding

Sudip Pan[1] and Gernot Frenking[2]

[1]*Nanjing Tech University, Institute of Advanced Synthesis, School of Chemistry and Molecular Engineering, 211816, Nanjing, China*
[2]*Philipps-Universität Marburg, Fachbereich Chemie, Hans-Meerwein-Strasse 4, D-35043 Marburg, Germany*

5.1 Introduction

There are many different perspectives in the arts and sciences that can be utilized to describe the world and make it accessible to human understanding. From a chemical point of view, the material world essentially consists of atoms as well as molecules and solids formed by chemical bonds between them, with the exception of elementary particles and so-called black matter, which are objects of physics. The scientific fundament of chemistry is atoms and the chemical bond. It is worth noting that there is no clear difference between a chemical bond as in H_2 and a weak interatomic attraction as in He_2, which is most commonly termed as van der Waals interaction. This comes clearly to the fore by the rather vague definition of the chemical bond given by the IUPAC, which goes back to a suggestion made by Linus Pauling: *"There is a chemical bond between two atoms or groups of atoms in the case that the forces acting between them are such as to lead to the formation of an aggregate with sufficient stability to make it convenient for the chemist to consider it as an independent 'molecular species'."* [1] It is a peculiarity of chemistry that fuzzy concepts and heuristic models that are not clearly defined mathematically are highly successful in explaining chemical findings. Chemistry may rightly be called the science of fuzzy concepts, which have been termed "unicorns." [2]

The use of heuristic models instead of exact mathematical expressions in chemical research stems from the fact that the physical nature of the interatomic interactions that lead to chemical bonds was a mystery to physics until 1927. It was clear that the chemical bond could only be based on Coulomb forces, but the laws of classical physics offered no mathematical framework for the strong interatomic attraction between neutral atoms that constitutes chemical bonding. It was not until the quantum theory introduced by Heisenberg and Schrödinger in 1925 and 1926 that the theoretical basis for explaining the physical nature of the chemical bond in terms of Coulomb interactions was provided. This was done in 1927 in the seminal work by Heitler and London, two young postdocs in Schrödinger's research group, where

Exploring Chemical Concepts Through Theory and Computation, First Edition. Edited by Shubin Liu.
© 2024 WILEY-VCH GmbH. Published 2024 by WILEY-VCH GmbH.

they described the interatomic interactions in the hydrogen molecule H_2 with the help of the newly developed quantum theory [3]. This is discussed below.

But chemistry was already a very successful scientific discipline in 1927, and its results also formed the foundation for an important branch of industry named after it. The Nobel Prize in chemistry was awarded to Heinrich Wieland in 1927 for his studies on the structure of bile acids and related substances, which are quite complex molecules even by current standards, without the nature of the chemical bond being known [4]. In a relatively short time, in the nineteenth and twentieth centuries, the chemical industry became a cornerstone of the economic development of many countries and contributed significantly to the prosperity of mankind. One can speculate whether chemistry would have achieved its scientific status if its research products were not so commercially utilizable. It can be said that chemistry is an example of a scientific discipline that can be pursued very successfully without understanding its basis – chemical bonding.

The historical development of chemical concepts before 1927 is an integral part of the human attempt to understand the fundamentals of physical matter. It is therefore part of human culture insofar as it attempts to grasp the essence of matter. Chemistry and chemical bonding are based on the existence of atoms as the basis of matter, and they are therefore rooted in a corpuscular view of the world. This was not the prevailing view of the physical universe for a long time. One of the first advocates of an atomic hypothesis of matter was the "laughing philosopher" Democritus, who lived from 460–370 BC and whose teaching is sometimes described with the remarkable statement "through the sensual impressions we cannot judge truth." [5] But for nearly 2000 years, the dominant view of the physical world was that of continuous matter, which cannot be divided into elementary particles. This viewpoint was challenged in the seventeenth century by Boyle and Newton, but it was John Dalton in the eighteenth century who pioneered the atomic theory of matter as the foundation of chemistry, although it was still challenged by many scientists for a long time. One of the latest prominent opponents of the atomistic view was Walter Ostwald, who acknowledged the existence of atoms only reluctantly in his later years. The historical development of chemical concepts from their very beginning is a fascinating topic of human attempts to understand nature. It is nicely described in a monograph by Mierzecki [6].

A crucial difference for theoretical approaches in chemistry is the difference between the physical mechanism of the chemical bond and a model to describe experimental observation in terms of human imagination. A model is an attempt to transfer the unlimited complexity of chemical knowledge into the limited mind of the human mind. A model is not right or wrong but is more or less useful. This was aptly formulated by Michael Dewar in 1984: *"The only criterion of a model is usefulness, not its truth."* [7] Bonding models have been suggested since the early days of chemistry. The most important is Gilbert Lewis' electron-pair model, introduced in 1916 [8], which is briefly explained further below. A source of constant controversy is the erroneous identification of the model of electron-pair bond suggested in 1916 with the physical mechanism of the chemical bond, which was not known until 1927. On the other hand, the description of molecular structures with the help of the electron-pair model, which has proven to be an unsurpassed tool for explaining chemical reactions to this day, indicates that the model has a quantum theoretical core hidden in the electron pair. It is not easy to connect the simple electron-pair model with the complicated physical mechanism of chemical bonding, which is still the subject of theoretical research today. The most relevant aspects are described in the next section.

5.2 The Physical Mechanism of the Chemical Bond

The explanation for the strong interatomic attraction between some atoms, such as hydrogen in H_2, which is curiously absent between helium atoms in He_2, has been a great mystery to physics. According to conventional physics, there are four fundamental forces that can be distinguished to describe interactions in matter. The strong and weak forces describe interactions within atomic nuclei and for processes such as free neutron decay. Gravitational forces are far too weak to explain the strong attraction in chemical bonds. This only leaves the electrical (Coulomb) forces as an explanation for chemical bonding. But according to classical physics, strong electric attraction occurs only between particles with opposite charges, not between neutral atoms like in H_2. And why should there be a strong electric attraction in H_2 and not in He_2?

The explanation for chemical bonding due to Coulombic interactions presented by Heitler and London was given within the framework of the wave mechanical description of elementary particles introduced by Schrödinger. According to quantum theory, elementary particles have a dual nature, namely as a wave function and as a particle, which are equivalent descriptions of the same species. The elementary particles for the chemical bond in a molecule are the electrons, which can be described by the charge density $\rho(\mathbf{r})$ in position space $\mathbf{r}$ or by the quantum theoretical wave function $\Psi(\mathbf{r}_1, ..., \mathbf{r}_N)$, which depends on the 3N coordinates of the number of electrons N. But while the two descriptions are equivalent, it is only the wave function $\Psi(\mathbf{r}_1, ..., \mathbf{r}_N)$ that explains the physical mechanism of the process of bond formation. The interference of the wave functions of the electrons obeying the postulates of quantum theory leads to in-phase and out-of-phase combinations that are energetically stabilizing (bonding) or destabilizing (anti-bonding) with respect to the separated atoms. This was a completely new aspect: that the interactions between two electrons can lead to two different discrete results, which depend on the quantum-theoretical state of the wave functions.

Scheme 5.1 illustrates the difference between the description of the chemical bond in dihydrogen H_2 using the classical physics approach with the electronic charge density ρ as a variable and the quantum theoretical description in the form of the wave function Ψ. The presentation of the latter interactions is simplified, where some mathematical details are ignored for the sake of clarity. A more detailed discussion is given in the literature [9]. The crucial difference between the two approaches is the occurrence of the interference term given in red in Eq. (5.9), which shows

Chemical bonding: *classical approach*

$$\rho$$

$$\rho(H_2) = \rho(H_a) + \rho(H_b) \quad (5.1)$$

$$E(H_2) = E[\rho(H_2)] \quad (5.2)$$

$$= E_{elstat} \quad (5.3)$$

Chemical bonding: *quantum chemical approach*

$$\rho(H_2) = [\Psi]^2 \quad (5.4)$$

$$\Psi(H_2) = N[\Psi(H_a) \pm \Psi(H_b)] \quad (5.5)$$

$$[\Psi(H_2)]^2 = N^2[\Psi(H_a) \pm \Psi(H_b)]^2 \quad (5.6)$$

$$= N^2\{[\Psi(H_a)]^2 + [\Psi(H_a)]^2 \pm 2[\Psi(H_a)\Psi(H_b)]\} \quad (5.7)$$

$$\quad q(H_a) \qquad q(H_b) \qquad \text{Interference}$$

$$E(H_2)_{\alpha,\beta} = E[\rho(H_2)] \pm E[\Psi(H_a)\Psi(H_b)] \quad (5.8a)$$

$$= E_{elstat} \pm E[\Psi(H_a)\Psi(H_b)] \quad (5.8b)$$

Scheme 5.1 Schematic description of the interatomic interactions between two hydrogen atoms using (a) a classical approach and (b) a quantum theoretical approach.

up only when the wave function Ψ is used as a basic entity. The key step of the quantum-theoretical approach has a mathematically simple form. It is the solution of the binomial in Eq. (5.8) that leads to Eq. (5.9). It follows that the chemical bond is a quantum theoretical effect that is due to the interference of the wave functions.

The calculation of the energy of H_2 using the classical approach yields only a shallow energy minimum with a bond energy of ~10% of the correct value. Figure 5.1 shows a re-drawing of the original curve of Heitler and London, where the classical energy is denoted as E_{11}. The energy curves E_α and E_β come from the quantum theoretical approach, which gives two solutions for Eq. (5.9) because of the $\pm$ sign for the interference term. They are related to the bonding and antibonding combinations of the wave functions. This indicates another very important aspect of the quantum theoretical description of the chemical bond, which is the symmetry of the wave functions. A stabilizing electronic interaction between two atoms or molecular fragments requires not only the opposite spin of the electrons but also proper spatial symmetry of the orbitals. The symmetry aspect is a fundament for the frontier orbital

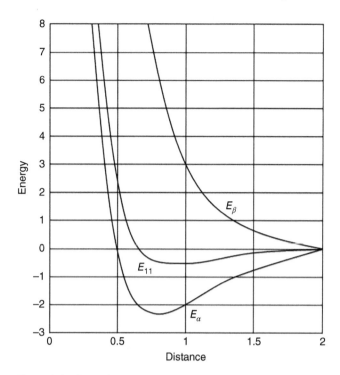

Figure 5.1 Potential energy curves of dihydrogen calculated classically (E_{11}) and using the wave function Ψ (E_α and E_β) by Heitler and London. Energy is in eV and distance in Å.

model (FOM) developed by Kenichi Fukui [10] and the orbital symmetry rules by Woodward and Hoffmann [11], which are described below.

After the wave function Ψ of a molecule is constructed from the wave functions of the atoms or fragments, the associated charge density ρ can be constructed. There is an unambiguous definition in the direction $\Psi \rightarrow \rho$, but there is no equally unambiguous definition in the opposite direction. A given wave function yields a unique charge density, but a given charge density can be constructed from an infinite number of wave functions. This means that although the bond finally formed can alternatively be fully described by the wave function Ψ or the charge density ρ, only the wave function provides a physical description of the process of bond formation.

There is one more important aspect of the physical nature of the chemical bond, which concerns the role of kinetic and potential energy in the stabilizing interaction of the chemical bond. The virial theorem for systems in an equilibrium state, such as a molecule in its equilibrium state, seems to give a straightforward answer because it states that the change in the total energy ΔE has the same sign as the change of the potential energy ΔV and the opposite sign as the kinetic energy change ΔT (Eq. (5.9)):

$$\Delta E = \Delta V + \Delta T = 0.5 \Delta V = -\Delta T \qquad (5.9)$$

But the role of kinetic and potential energy for the bonding energy is somewhat paradoxical, which was suggested for the first time in 1937 by Hellman, who suggested that the driving force for the chemical bond is actually the lowering of the kinetic energy [12]. The role of kinetic and potential energy in bonding energy has been the topic of controversial discussions for some time [13]. The controversy was finally resolved in favor of kinetic energy lowering as the driving force for bond formation by the work of Ruedenberg, published in 1962 [14], who analyzed the behavior of kinetic and potential energy not only at equilibrium geometry but also during the process of bond formation. It is important to know that the virial theorem also applies to nonequilibrium geometries, but the expression now includes an additional term for the derivative of the energy with respect to the geometric variables R (Eq. (5.10)). The numerical value for the derivative becomes zero only for equilibrium structures, and Eq. (5.9) is a special form of the virial theorem:

$$2T + V + R(dE/dR) = 0 \tag{5.10}$$

The behavior of the kinetic and potential energy is shown in Figure 5.2 using dihydrogen as a classical example.

When the hydrogen atoms approach each other, the kinetic energy along with the total energy are at first lowered whereas the potential energy rises. This is because the electronic charge experiences a larger volume in the overlapping valence space. The potential energy rises because the electron distance to the nuclei is longer than in the atoms. This is the region where the actual bond formation mainly takes place.

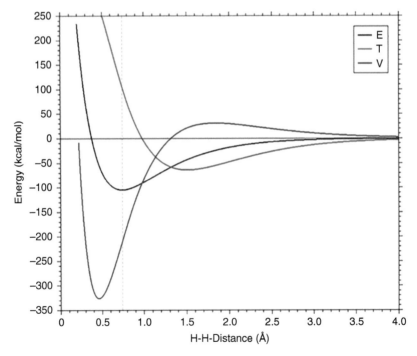

Figure 5.2 Energy curves of the total energy E, potential energy V, and kinetic energy T as a function of the H–H distance.

At shorter distances, the effect of electronic charge depletion at the nuclei dominates leading to shrinkage of the effective atomic radii. As a result, the depleted charge in the core region experiences a stronger attraction by the nucleus, which lowers V and increases T. The latter process eventually overcompensates the reverse change of T and V in the bonding region and yields an overall increase of kinetic energy and a lowering of potential energy at the equilibrium distance. The conclusion is that the driving force for the accumulation of electronic charge in the bonding region associated with the formation of the chemical bond is the lowering of the kinetic energy.

The conclusion about kinetic energy as the driving force of the chemical bond is not restricted to H_2. Ruedenberg has shown in recent work that the same mechanism is operative in heavier diatomic systems E_2 (E = Li – F) [15]. The primacy of kinetic energy for chemical bond formation has recently been challenged by Levine and Head-Gordon, who analyzed the bonding in some chemical bonds of first-row atoms. They reported that bonds between heavier elements, such as H_3C–CH_3, F–F, H_3C–OH, H_3C–SiH_3, and F–SiF_3 behave in the opposite way to H_2^+ and H_2, with kinetic energy often increasing on bringing the radical fragments together [16]. They used a particular energy decomposition analysis (EDA) termed as absolute localized molecular orbital EDA (ALMO-EDA) [17] method, which differs from other EDA methods by the mathematical determination of the energy terms, which can influence the size and even the sign of the terms. This was pointed out by Bacskay [18], who reinvestigated the systems and found an increase of the intrafragment kinetic energy but a decrease of the interfragment kinetic energy, in agreement with the original findings by Ruedenberg [14, 15].

A final aspect on the physical nature of chemical bonding concerns the role of the other three elementary forces besides Coulombic interactions on the chemical bond. As noted before, gravitational forces are much weaker and the strong and weak forces describe interactions within atomic nuclei and physical processes such as free neutron decay. But there is an observation in nature that poses a question to chemical bonding which cannot be answered when only electric forces are considered. This is the finding that many objects in nature are found preferentially with only one of the mirror images, which appear otherwise "the same and not the same." [19] Chiral compounds have two enantiomers which were for a long time believed to have the same energy, but the natural abundance of one enantiomer over the other is striking. Naturally occurring amino acids are mainly found as L enantiomers while D enantiomers are rare. The parity violation in nature is a topic of fundamental relevance for the understanding of matter and life. The explanation for molecular parity violation in terms of laws of physics can only be given when weak interactions are considered, which are the only physical force that accounts for the violation of space inversion symmetry. Since weak forces and electric forces can be unified into electroweak interactions, it follows that they are the physical origin of chirality. However, the formation of chiral compounds and the observation that only one enantiomer of chiral molecules is found in nature is a matter of discussion where several models have been suggested. The energy difference between the electronic ground states of the enantiomers is very small, it was calculated theoretically and is in the order of only 10^{-10}–10^{11} J/mol, which could not yet be experimentally detected. Its detection remains one of the great challenges of current physical–chemical stereochemistry. A recent perspective on the theoretical and experimental aspects of parity violation in chiral molecules has been presented by Martin Quack [20].

5.3 Bonding Models

The most important model for the chemical bond was introduced by Gilbert Lewis in 1916, when he suggested that *"The chemical bond is at all times and in all molecules merely a pair of electrons held jointly by two atoms."* The representation of molecular structures in terms of electron-pair bonds has become the standard in chemistry for describing the geometry and chemical bonding of molecules. More than 100 years after its first publication, it is still used with some modifications in current chemistry curricula and modern chemical research. It was derived from the observation that the stability and geometry of the majority of compounds that were known at that time could be nicely correlated with the number of electrons. It was a very bold suggestion to identify in 1916 the chemical bond with a pair of electrons because the physical laws known at that time state that electrons repel each other rather than forming an energetically stabilizing pair. Lewis was aware of the conflict between his model and the physical description of the interatomic interactions provided by classical physics. He speculated that *"Electric forces between particles which are very close together do not obey the simple law of inverse squares which holds at greater distances."* This was prior to the advent of quantum theory, which solved the puzzle of why electrons in molecules can occur in energetically stabilizing pairs. The intuitively derived electron-pair model has a hidden quantum theoretical content, the very identity of which is still the subject of theoretical work today.

The 1916 publication by Lewis did not attract much attention before Irving Langmuir recognized the importance of the electron-pair model, expanded it in subsequent publications, and established some rules that are still in use today [21]. He coined the term "covalent bond" and in 1921 postulated the octet rule and the 18-electron rule, which are still widely used concepts and prove very helpful in explaining the stability of molecules [22]. This caught the interest of the chemical community, which started to discuss the "Lewis–Langmuir" model of chemical bonding. Unfortunately, this resulted in a lifelong enmity between the two great scientists that persisted until a few hours before Lewis' mysterious death in the laboratory [23]. In 1923, Lewis expanded his concept of the electron-pair bond in a book in which he set out his ideas about the structure of molecules and solids and the nature of interatomic interactions [24].

The electron-pair model suggested by Lewis has three drawbacks. One is the limitation to localized bonds, which makes it difficult to explain the formation and stability of electron-deficient molecules such as diborane B_2H_6, where the number of bonds is higher than the number of available electron pairs. This problem was tackled by Pauling, who proposed alternating resonance forms for the description of chemical bonding. But it can lead to very clumsy expressions for strongly delocalized systems and for transition metal (TM) complexes, where MO (molecular orbital)-derived models are clearly superior. The second drawback is the failure to explain the stability of some molecules with unpaired electrons, such as O_2, which has a triplet ($^3\Sigma^-_g$) ground state. Pauling presented in 1931 the concept of one-electron and three-electron bonds [25] and he proposed writing O_2 with the structure **A**:

$$:\overset{..}{\underset{..}{O}}\vdots\overset{..}{\underset{..}{O}}:$$

A

But this suggestion is just an a-posteriori formula instead of a genuine explanation of why O_2 has a triplet ground state whereas isoelectronic C_2H_4 is a singlet, which had been a puzzle for Gilbert Lewis.

The greatest weakness of the electron-pair model is the complete absence of the symmetry of the electronic structure, which is crucially important for understanding the structure and stability of molecules. This is the third and major drawback of the Lewis model. The relevance of symmetry in the quantum theoretical description of molecular electronic structure was already recognized by Lennard-Jones in 1929, who showed that the triplet ground state of O_2 can easily be explained by the symmetry of the MOs [26]. This was based on the pioneering work of Hund, who provided between 1927 and 1930 a simple explanation for electronic states in the ground and excited states of diatomic molecules based on MO theory [27–31].

Although the weakness of the description of chemical bonding by localized electron pairs was obvious, it nevertheless became the dominant model in chemical bonding theory for quite a while. This is mainly due to the work and influence of Linus Pauling. His book "*The Nature of the Chemical Bond*," first published in 1939 with the third edition in 1960, became the most influential work in the field and greatly inspired and shaped the understanding of chemical bonding for generations of chemists [32]. Pauling had an enormous knowledge of general chemistry, and he also learned the newly developed quantum theory when he traveled to Europe in 1926 and visited the laboratories of Nils Bohr in Copenhagen, Erwin Schrödinger in Zürich, and Arnold Sommerfeld in Munich, where he also met with Werner Heisenberg. After returning to the United States of America, he began working on the application of quantum theory to chemical bonding, culminating in the book cited. Pauling realized that the heuristic electron-pair model of Lewis could easily be combined with the quantum theoretical method of Heitler and London, who used the valence bond (VB) approach for the calculation of H_2. He extended the concept of localized bonds by using the model of resonance structures to account for delocalization and polarization. He rejected the alternative MO method because the depiction of delocalized electrons did not correspond to the intuitively more reasonable description of a molecular structure in terms of the Lewis model.

The breakthrough of MO-based models came with the full recognition of the great relevance of symmetry for the electronic structure of molecules. A major impact is due to the pioneering work of Kenichi Fukui, who introduced the FMO, and to Roald Hoffmann and Robert Woodward, who suggested the orbital symmetry rules for chemical reactions. They are now standard methodologies for the explanation of chemical reactions and molecular structures. Other more recent developments comprise different versions of the EDA approach, where the energy change associated with the bond formation process is partitioned into different terms that can be interpreted in a physically inspired way. Popular variants were introduced by Keichi Morokuma [33], Tom Ziegler [34], Martin Head-Gordon [35], and Peifeng Su [36]. They are introduced and discussed in Chapter 17 of this book. The methods mentioned above focus on the bond formation process, while other approaches analyze the finally formed bond. To avoid misunderstandings, it is important to recognize the difference between the two issues, i.e. between the formation of a bond and the bond that is finally formed. Currently, popular examples of the latter type are the natural bond orbital (NBO) method by Frank Weinhold [37], the quantum theory of atoms in molecules (QTAIM) procedure developed by Bader [38], the interacting quantum atoms (IQA) method introduced by Angel Martin-Pendas et al. [39] the ELF (Electron Localization Function) method suggested by Becke and Edgecombe [40] and the CDFT (Conceptual Density Functional Theory) approach pioneered by Robert Parr [41]. The CDFT topic is covered in Chapter 3, the NBO method is described in Chapter 6, and the QTAIM and IQA topics are covered in Chapter 7 of this book.

There are currently three different types of quantum chemical methods that are used as a basis for a chemical bonding model. The historically oldest method is the VB method, whose concepts are still widely used despite its neglect of the symmetry of the electronic structure due to the influential work of Pauling and the more recent contributions of Shaik and Hiberty. Most quantum chemical calculations are nowadays based on MO theory or DFT, which in its current versions can be regarded as an efficiently parameterized Hartree–Fock method that incorporates the correlation effect through its additional parameters. The derived approaches are thus variants of MO-based methods with Kohn–Sham orbitals instead of Hartree–Fock orbitals. Chemical concepts from the three different approaches are introduced and discussed in Chapters 1–4 of this book.

The various models of bonding and the diverse descriptions of chemical bonding show the difference between the approaches of chemistry and physics to make the objects of nature accessible to the understanding of the human mind, with the aim of satisfying his curiosity for knowledge. Chemistry can be seen as the house built on the foundations of physics, which provides measurable quantities as a starting point for chemical research. Chemical models are the many different houses that have windows, doors, rooms, and other characteristics that make them habitable for chemists. They are not measurable quantities, but they serve as an organizing principle for experimental observations and as a guide for future experiments. Chemistry is thus the science of fuzzy concepts, while physics is the science of fundamental forces in nature. The infinite number of molecules and solids would be just a huge collection of data without the organizing power of models. It is the task of theorists to develop mathematically unambiguously defined models rooted in quantum theory that can serve as a guide for experimentalists to understand chemical observations and design new experiments.

5.4 Bond Length and Bond Strength

The length and the strength, together with the multiplicity (single or multiple bond), are among the most frequently mentioned characteristics of a chemical bond that are used for classification and comparison with other bonds. While length is clearly defined as the interatomic distance between the nuclei, the strength of a chemical bond is less clear. A common quantity for estimating the strength of a bond A–B is the bond dissociation energy (BDE), which gives the energy difference between the molecule AB and the fragments A and B. But there are a large number of molecules in which bond breaking results in fragments that have lower energy than the intact molecule. High-energy materials store energy that is released when a bond is broken, and multiply charged species are often metastable compared to singly charged fragments. For this reason, BDE is not suitable for a general definition of bond strength, as it is sometimes suggested [42]. In this section, the characteristics of bond length and bond strength are discussed in detail. The first section is devoted to *bond length*.

Whereas bond length is well defined, the forces that determine the equilibrium distance of a chemical bond are less known, and the physical explanation for the observed bond length given in many textbooks is incorrect. A common reasoning uses the overlap of the bonding σ orbitals $S(\sigma)$ as the most important factor for the length of a bond. This is schematically shown in Figure 5.3. The model states that the equilibrium bond length of a σ-bond coincides with the distance at which $S(\sigma)$ reaches its maximum value. Multiple bonds with σ- and π-components become slightly shorter than σ-bonds because the additional π-bond compensates for the weaker σ-interactions. This is the standard model for covalent bonds, which suggests that bond strength correlates directly with orbital overlap.

This qualitative model, which was born in the early days of MO theory, appears reasonable at first sight but, it does not agree with the explicit calculation of the orbital overlap. Figure 5.4a shows the calculated overlap integrals of the nitrogen AOs in N_2 as a function of the interatomic distance r_{NN}. It becomes obvious that the overlap of the σ orbitals (sum σ, blue values) has not yet reached its maximum value at the equilibrium value r_e. If the principle of maximum overlap is valid, the equilibrium distance should be ~0.4 Å shorter because there are also two π bonds. There is a physical force that prevents further shortening of the N–N distance below the equilibrium value. Figure 5.4b shows the curves of the total interaction energy ΔE_{int} and its three components ΔE_{orb}, ΔE_{elstat}, and ΔE_{Pauli}, which are provided by an EDA [43, 44]. ΔE_{orb} gives the orbital (covalent) contribution, which comes from the interference of the wave functions; ΔE_{elstat} gives the electrostatic (Coulomb) interactions; and ΔE_{Pauli} comes from the quantum theoretical postulate that two electrons with the same spin may not occupy the same space [45]. Figure 5.4b shows that

Figure 5.3 Schematic representation of bonding overlap of a σ bond between two sp^x hybrid orbitals and a π bond between two p_π AOs.

the covalent interactions ΔE_{orb} and the electrostatic interactions ΔE_{elstat} are both attractive components whose strength further increases when the bond becomes shorter. This holds also for the Coulomb term ΔE_{elstat}, which becomes repulsive only at a rather short distance when the nuclei are close to each other. The repulsive factor that is responsible for the bond length is the Pauli repulsion ΔE_{Pauli}, which prevents the overlap of orbitals that are occupied by electrons with the same spin.

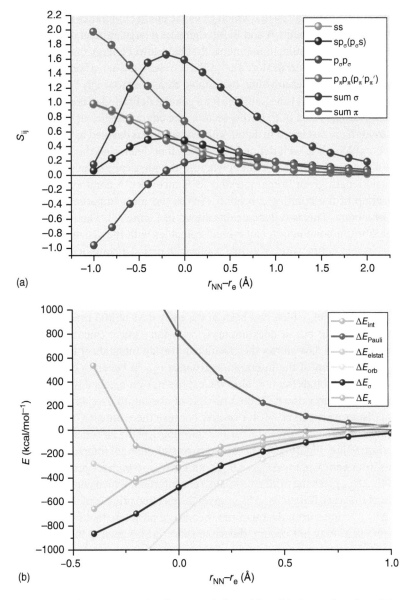

Figure 5.4 Overlap integrals of the atomic 2s and 2p orbitals as a function of the interatomic interaction relative to the equilibrium value of (a) N_2 and (c) O_2^{2+}. Energy components of EDA calculations at different interatomic distances relative to the equilibrium values of (b) N_2 and (d) O_2^{2+}.

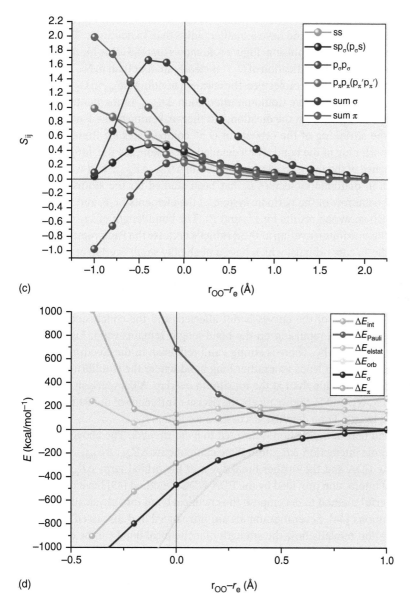

Figure 5.4 (Continued)

The decisive influence of Pauli repulsion on the chemical bond becomes clear when comparing the bond length of N_2 with the isoelectronic O_2^{2+}, which surprisingly has a shorter bond (1.061 Å) than N_2 (1.103 Å) despite the inherent Coulomb repulsion [46]. The dication O_2^{2+} is thermodynamically unstable for dissociation into $2O^+$, but it lies in a deep potential well [47], which makes it easily available for gas-phase studies [48]. Figure 5.4c shows the orbital overlap of O_2^{2+}, which looks very similar to N_2 (Figure 5.4a). Note that the absolute values of the overlaps S_{ij} for the σ and π AOs in O_2^{2+} are smaller than in N_2, although the bond length in the former is shorter than in the latter. This is because the atomic orbitals in positively charged O$^+$ are contracted and have a smaller radius than in neutral N. This also has an influence on the Pauli repulsion. Figure 5.4c shows that the value for ΔE_{Pauli} at the equilibrium distance of the dication of O$_2^{2+}$ is clearly smaller than in N$_2$. At the equilibrium distances of the two molecules, the covalent bonding ΔE_{orb} in O$_2^{2+}$ is weaker than in N$_2$, and the attractive Coulomb interaction ΔE_{elstat} in the neutral diatomic system becomes repulsive in the dication, but the weakening of the Pauli repulsion caused by the shrinking of the orbitals in O$_2^{2+}$ compensates for these forces and leads to a shortening of the bond. For a detailed discussion, see Ref. [46].

The influence of the three physical components ΔE_{orb}, ΔE_{elstat}, and ΔE_{Pauli} on the bond length in diatomic molecules E$_2$ has been studied for the atoms of the first and second octal row of the periodic system of the elements Li–F, and Na–Cl [43]. Figures 5.5a,b show the results for Li$_2$ and F$_2$. The bond length of Li$_2$ nearly coincides with the maximum overlap of the σ orbitals because the Pauli repulsion comes only from the 1s core orbitals, which have a negligibly small overlap at the equilibrium distance. The calculated values for ΔE_{Pauli} become negative for Li$_2$, which is a physically invalid result due to the self-interaction error (SIE) of the functionals used for the DFT calculations. Without the SIE, the ΔE_{Pauli} values would be slightly larger, but the shape of the curves is not affected, and the conclusion about the relevance of the Pauli repulsion on the bond length remains valid. Figures 5.5c,d exhibit the results for F$_2$. There is strong Pauli repulsion in the electron-rich difluorine molecule, which leads to a rather long bond where the equilibrium distance has a much higher value than at the maximum overlap. A comparison of the EDA results for Li$_2$, N$_2$, and F$_2$ shows nicely the decisive influence of the Pauli repulsion on the bond length. For a detailed discussion, see Ref. [43].

The second part of the section is devoted to *bond strength*. The above division of the interatomic interaction ΔE_{int} into three components ΔE_{orb}, ΔE_{elstat}, and ΔE_{Pauli} given by the EDA and the further breakdown of the orbital term ΔE_{orb} into pairwise orbital interaction provided by the EDA-NOCV method [45] have proven to be a very powerful method to decompose the chemical bond into physically meaningful contributions [44]. Several examples are introduced and discussed below. The crucial question remains how the strength of a chemical bond can be defined in a way that the bond strength of different bonds can be compared. As noted above, the BDE, which is required to break a bond, is not very useful as a generally applicable criterion for bond strength because many molecules dissociate into fragments that have a lower energy than the initial species. An example is the dication O$_2^{2+}$ that was discussed above. Also, the activation energy for breaking a bond is not a reliable indicator of bond strength because the electronic structure of the fragments may strongly change during the rupture of the bond, which obscures the bond strength at equilibrium. A striking example is the carbon–carbon bond in F$_2$C=CF$_2$, which dissociates without activation barrier into 2 CF$_2$ fragments with a BDE of $D_e = 75.5$ kcal/mol, which is even less than the standard value for a C–C single bond. The electronic state of the CF$_2$ moieties changes at longer distances from a triplet ($^3B_1$) reference state in the molecule to a singlet ($^1A_1$) ground state, which is 54 kcal/mol lower in energy [49]. This is schematically shown in Figure 5.6.

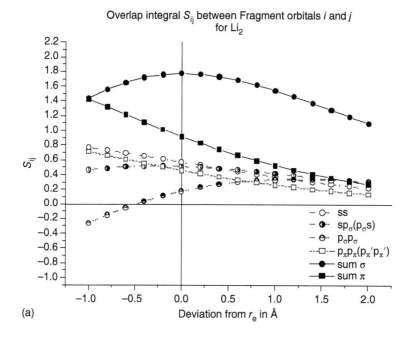

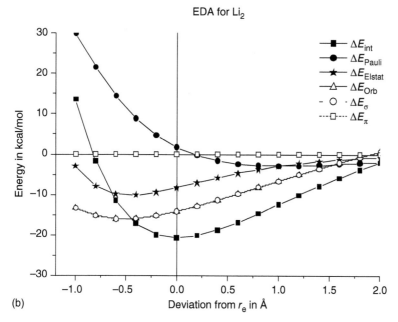

Figure 5.5 Overlap integrals of the atomic 2s and 2p orbitals as a function of the interatomic interaction relative to the equilibrium value of (a) Li_2 and (c) F_2. Energy components of EDA calculations at different interatomic distances relative to the equilibrium values of (b) Li_2 and (d) F_2.

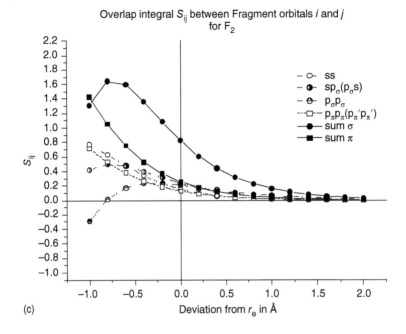

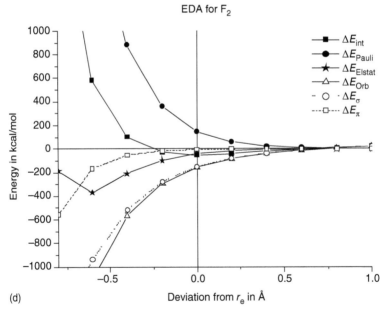

Figure 5.5 (Continued)

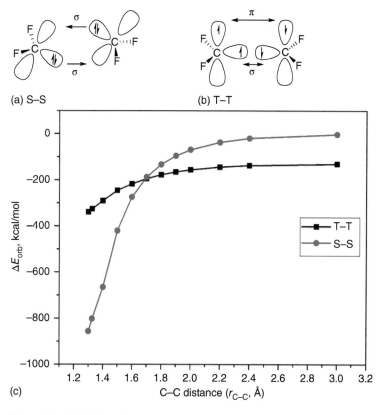

Figure 5.6 (a) Schematic representation of the orbital interactions between two CF_2 carbenes in the singlet states (S–S). (b) Schematic representation of the orbital interactions between two CF_2 carbenes in the triplet states (T–T). (c) Reaction course of the dissociation reaction $C_2F_4 \rightarrow 2CF_2$ showing the energetically more favorable orbital interaction ΔE_{orb} between the CF_2 moieties in the singlet state (red curve) or the triplet state (black curve). The best description of the bonding interaction comes from the fragments, which give the smallest ΔE_{orb} value. Source: Andrada et al. [49], John Wiley & Sons.

It becomes clear that the BDE of $F_2C=CF_2$, which relates the excited electronic triplet state of the CF_2 fragments in the molecule with the singlet ground state in free CF_2, is not a faithful criterion for the bond strength in $F_2C=CF_2$. The calculated values for the orbital interactions ΔE_{orb} suggest that the CF_2 fragments in C_2F_4 at the equilibrium distance and at somewhat longer C–C distances are best described in terms of an electron-sharing doublet bond between the triplet state $F_2C=CF_2$ (T–T, black curve in Figure 5.6c), but at longer distances they engage in dative bonds between the species in the singlet state $F_2C\rightleftarrows CF_2$ (S–S, red curve in Figure 5.6c). A better estimate of the intrinsic bond strength in $F_2C=CF_2$ would be the BDE into the triplet state of CF_2, which is much higher (183.5 kcal/mol) than the BDE into the singlet state [50]. However, the electronic reference state of the fragments is not always clear, as has been demonstrated for C_2, where the bond strength and the electronic reference state of carbon have been the topic of controversial discussions [51].

The principal problem of the BDE is the change of the electronic configuration of the fragments during bond rupture, which affects information about the bond strength of a bond at the equilibrium geometry.

The above arguments clearly show that neither the BDE nor the activation energy may be used as a general measure for the strength of a chemical bond. A property that is directly associated with the strength of a bond A–B is the vibrational stretching force constant k_e, which indicates the strength of the interatomic forces between the fragments A and B at the equilibrium distance. The advantage of using the stretching force constants k_e instead of the BDE is that they are independent of the energies and electronic structures of the free species A and B. The principal problem of using k_e as a measure for bond strength in larger molecules is the occurrence of mode coupling between different vibrational modes. This problem was solved by the work by Konkoli and Cremer [52], where it was shown that the delocalized normal vibrational modes can be transformed to local vibrational modes via a mass-decoupled version of the Wilson equation for vibration spectra [53]. The associated Konkoli–Cremer force constants k_e^{KC} are free from mode coupling, which makes it possible to derive expressions that can directly be used as a measure for the bond strength in molecules larger than diatomic species.

Table 5.1 Calculated bond lengths R_e [Å], bond dissociation energies D_e, and stretching force constants k_e (mdyn/Å).

Molecule	Bond	R_e	D_e	k_e
H_2	H–H	0.743	108.4	5.8
Li_2	Li–Li	2.725	22.3	0.23
Li_2^+	Li–Li	3.196	28.0	0.13
C_2	C–C	1.247	146.9	11.9
C_2H_2	C–C	1.207	236.6	15.8
C_2H_4	C–C	1.333	178.2	8.8
C_2H_6	C–C	1.532	93.1	3.8
N_2	N–N	1.099	236.1	23.0
$[HNNH]^{2+}$	N–N	1.080	118.7	27.9
$[FNNF]^{2+}$	N–N	1.087	−101.6	25.5
O_2	O–O	1.212	113.8	11.8
$[O_2]^{2+}$	O–O	1.047	−61.1	23.3
F_2	F–F	1.417	34.7	4.7
CO	C–O	1.128	257.4	19.1
$[CO]^+$	C–O	1.111	192.1	21.5
$[HCO]^+$	C–O	1.105	275.9	22.6
$[COH]^+$	C–O	1.154	202.6	16.1
$[HCOH]^{2+}$	C–O	1.115	86.7	22.2

Source: Zhao et al. [54], John Wiley & Jons.

Table 5.1 shows for a series of molecules the calculated stretching force constant k_e, which is for diatomic species identical to the local stretching constant k_e^{KC}, for some bonds, along with the values for the BDE and the bond lengths R_e [54]. It becomes obvious that the stretching force constants sometimes give a different ordering for the bond strength, as suggested by the BDE. For example, CO has a higher BDE but a smaller k_e value than N_2. The one-electron bond in $[Li_2]^+$ has likewise a higher BDE but a smaller k_e value than the two-electron bond in Li_2. This is because the force constants are a measure of the curvature of the bond dissociation pathway, whereas the BDE indicates the well depth. Figure 5.7 shows that the bond dissociation curve of $[Li_2]^+$, which has a very soft bond that comes largely from induction forces, is deeper but wider than for Li_2. A detailed discussion of the use of stretching force constants as a measure of bond strength is provided in Refs [54, 55].

It is important to recognize the limitations of the local force constant as a general indicator of the strength of a chemical bond. One such limitation is when the physical interactions between a pair of atoms cannot be separated from the forces between other parts of the molecule, which contribute to the same vibrational mode. Pertinent examples are substituted alkanes R_3CCR_3 with long C–C bonds and very bulky substituents R. The repulsive interaction due to Pauli forces between groups R is outweighed by attractive dispersion forces when the bond becomes longer because the Pauli repulsion depends on the overlap of the orbitals, whereas dispersion has a $1/R$ dependence. Dispersion forces are often considered to be unimportant for chemical bonding, and they are usually neglected. But it has been shown that dispersion forces between large groups can be very strong and relevant for the stability of compounds with bulky groups [56]. The local mode vibration of the stretching mode in

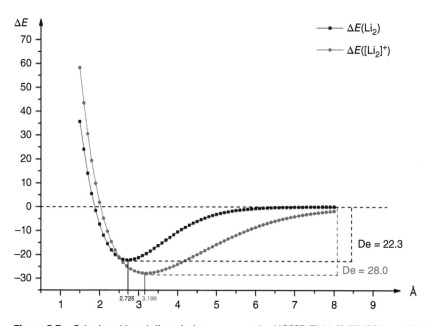

Figure 5.7 Calculated bond dissociation curves at the UCCSD(T)/def2-TZVPP level of Li_2 and Li_2^+. Source: [54], John Wiley & Sons.

R_3CCR_3 encompasses not only the direct C—C bond but also the attraction between the substituents. However, this result does not only indicate a limitation of the local force constant as an indicator of the strength of a chemical bond. It also shows that the representation and understanding of the electronic structure and bonding interactions using the Lewis bonding model do not fully cover the binding interactions between atoms.

5.5 Dative and Electron-Sharing Bonds

After Gilbert Lewis introduced the electron electron-pair bonding model in 1916, he elaborated on his understanding of chemical bonding in more detail in the monograph *"Valence and The Structure of Atoms and Molecules,"* published in 1923 [24]. Lewis provided in the book among others a general definition of acids and bases, which bear now his name Lewis acids and Lewis bases [57]. He wrote that *"a basic substance is one which has a lone pair of electrons which may be used to complete the stable group of other atoms and…an acid substance is one which can employ a lone pair from another."* [58] The idea of molecules that possess either an electron pair or a suitable vacancy was generalized by Sidgwick, who introduced in 1929 the terms donor and acceptor molecules, which are now familiar to chemical vocabulary [59]. Sidgwick also suggested the notation of an arrow for donor–acceptor bonds A→B to distinguish them from electron-sharing bonds A–B (Figure 5.8) [60]. He showed that the direction of the dative bond (as it is called now) is very helpful to understand the relationship between dipole moment and molecular structure, which sometimes deviates from the expected behavior. For example, he wrote carbon monoxide with the formula C≡O and suggested that the direction of the dative bond from oxygen to carbon is responsible for the unusual dipole moment of CO, which has a negative end at carbon [61]. Sidgwick's suggestion was later supported by quantum chemical calculations [62].

Although the distinction between electron-pair bonding into electron-sharing bonds A–B and dative bonds A→B was recognized very early, it did not play a significant role in the description of the molecular structure for a long time. This can be traced back to the influence of Linus Pauling on the description of chemical bonding. In his seminal book *"The Nature of the Chemical Bond,"* [32] which was the leading text book in the field for decades, Pauling shortly mentions the model of dative bonding but then he dismissed it, because it did not fit into his scheme of convenience: *"We shall not find it convenient to make use of these names or of these symbols."* [63] Pauling preferred to write structural formulas in terms of resonance structures, where he placed formal charges at the atomic symbols. The disadvantage is that electron-sharing bonds, which are also often written with formal charges

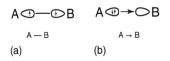

Figure 5.8 Schematic representation of (a) an electron-sharing electron-pair bond and (b) a dative (donor–acceptor) electron-pair bond.

5.5 Dative and Electron-Sharing Bonds

at the atoms, and dative bonds are not distinguishable anymore. But the physical mechanism of an electron-sharing bond A–B, where atoms A and B contribute one electron each to the bond, differs significantly from a dative bond A→B, where both electrons come from atom A. For example, the strength of a dative bond A→B is always less than that of an electron-sharing bond between the same atoms A–B. The direction of the charge flow is always from the donor to the acceptor atom, which has important consequences for the electronic charge distribution in a molecule that may be opposite to what one would expect from the electronegativities of the atoms [64].

It is interesting that the model of dative bonding became the dominant notion for chemical bonding in TM complexes after Michael Dewar explained the structure of Zeises salt $[Cl_3Pt(C_2H_4)]^-$ in terms of σ donation and π backdonation between Pt and ethylene [65]. The significance of the dative model was recognized by Chatt and Duncanson, who generalized it for other TM compounds [66]. It became known as the DCD (Dewar–Chatt–Duncanson) model for metal–ligand interactions, which is the prevailing description for chemical bonding in TM chemistry [67]. The model of dative bonding in main-group chemistry played only a minor role for a long time, although Haaland drew attention to its relevance in a 1989 review article [68], which had little effect on the way how chemists described molecular structures, however.

The situation has significantly changed since 2006, when it was shown that the bonding situation in carbodiphosphorane $C(PPh_3)_2$, which was experimentally known since 1961 [69], is best described in terms of donor–acceptor bonds $Ph_3P{\rightarrow}C{\leftarrow}PPh_3$, where a carbon(0) atom in the excited ^1D singlet state with the electron configuration $1s^22s^22p_x^0 2p_y^0 2p_z^2$ is the acceptor and the phosphane ligands PPh_3 are donors [70]. Carbodiphosphorane $C(PPh_3)_2$ can be considered as a parent system for a new class of compounds termed carbones CL_2, which became a very active field of experimental and theoretical research [71]. Figure 5.9a shows the principal features of the orbital interactions in carbones.

The breakthrough in favor of the dative bonding model came with the prediction and subsequent verification that carbodicarbenes $C(NHC)_2$ (NHC = N-heterocyclic carbene) are stable compounds that possess a similar bending angle of ~135° like $C(PPh_3)_2$ [72]. Shortly after the theoretical study appeared in 2007, two groups reported the synthesis of the first CDCs (Carbodicarbene) [73]. The experimental research on CDCs became very active after it was found that CDCs may be used as catalysts in a variety of reactions [74]. The π lone pair at carbon atom in CL_2

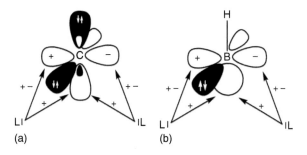

Figure 5.9 Schematic view of the orbital interactions in CL_2 and $(BH)L_2$.

(a)   (b)

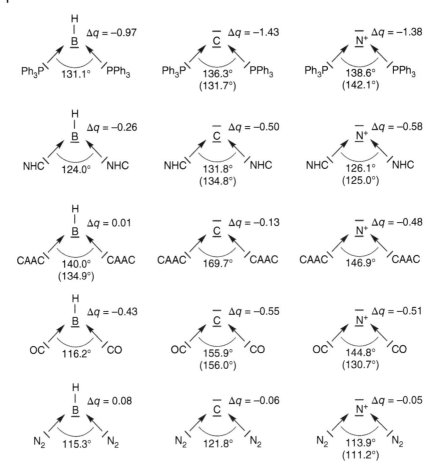

Figure 5.10 Calculated bond angles of isoelectronic molecules EL$_2$ and partial charges Δq of the central fragments E = BH, C, and N$^+$. Experimental values are given in parentheses.

induces some π backdonation, which has a strong influence on the bending angle in the carbones. Strong π acceptor ligands lead to significant π backdonation L←C→L, which yields wider bending angles. This explains the equilibrium geometry of carbon suboxide C$_3$O$_2$, which has a bending angle at the central carbon atom of 156.0° [75] that can straightforwardly be explained with the model of dative bonding for carbones C(CO)$_2$ [64b, 76]. The bending potential is very shallow and carbon suboxide has a linear structure in the solid state, where it is described with electron-sharing double bonds OC=C=CO [77]. A more appropriate description is given in terms of complementary σ and π dative interactions OC⇌C⇌CO. The weaker π donor N$^+$ in isoelectronic N$^+$(CO)$_2$, which has also been prepared under ambient conditions, has a more acute bending angle of 130.7° due to weaker π backdonation OC⇌N$^+$⇌CO [78]. Figure 5.10 shows a series of carbones and isoelectronic species EL$_2$ where E = C, BH, N$^+$ and the bending angles, where the trend nicely follows the π-donor/π-acceptor strength of E and L. A more detailed

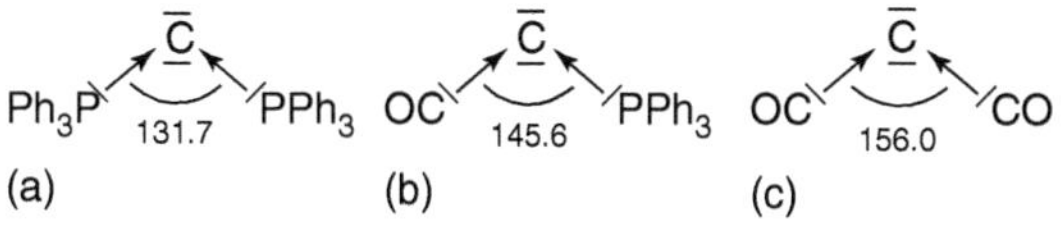

Ph₃P⟶C⟵PPh₃   OC⟶C⟵PPh₃   OC⟶C⟵CO
     131.7   145.6   156.0
(a)   (b)   (c)

Figure 5.11 Experimental bond angles at the central carbon atom of the compounds $C(PPh_3)_2$, $C(PPh_3)(CO)$, and $C(CO)_2$.

discussion is given in the literature [64]. The relevance of π backdonation for the bending angle is also obvious from the reported angles of the complexes where the phosphane ligands in $C(PPh_3)_2$ are replaced by one or two CO ligands (Figure 5.11).

The examples shown in Figure 5.10 include the complex $BH(cAAC)_2$ (cAAC = cyclic alkylaminocarbene), which was synthesized in 2011 following the discovery of carbones as donor–acceptor complexes [79]. Figure 5.9b shows the principal orbital interaction in boron complexes $(BH)L_2$ where the three-coordinated boron atom becomes a Lewis base. More recent examples of three-coordinated B(I) compounds where boron is a Lewis base are the complexes $(RB)L_2$ with L = CO and various heterocyclic carbene ligands L [80]. The model of dative bonding is very useful for isoelectronic complexes EL_2, where E = BH, C, and N^+. A theoretical study published in 2012 predicted numerous compounds that could later be synthesized [64a].

The model of dative bonding proved to be very valuable not only for compounds EL_2, where E is an atom of the first octal row of the periodic system, but also for heavier analogous. The aluminum homologue $AlH(cAAC)_2$ was recently isolated [81], and the whole series of group-14 complexes EL_2 with E = Si – Pb could be prepared, which were previously predicted by quantum chemical calculations [82]. The generic name tetrylones was suggested for the EL_2 (E = C – Pb) species [71a] with individual names for the different elements (Table 5.2). The first genuine silylones SiL_2 and germylones GeL_2 were reported in 2013 [83] followed by the first stannylone SnL_2 [84]. Plumbylones PbL_2 are very unstable, and they could only be prepared very recently [85]. The chemistry of nitrogen homologs N^+L_2, for which the name nitreone was coined, has been explored by Bharatam and coworkers [86]. Further examples of ylidone complexes EL_2 with atoms E of groups 3, 14, 15, and 16 are now experimentally known. An overview is given in Table 5.2.

There are not only single-center ylidones EL_2 but also two-center adducts E_2L_2 and cyclic three-center complexes $(cyc-E_3)L_3$ known, whose structures and reactivities can nicely be explained with dative bonds [71b]. Several recent review articles summarize theoretical and experimental studies in the field [71, 87]. After initial resistance against the description of chemical bonds in terms of donor–acceptor interactions using arrows, which sparked a controversial but constructive and very helpful discussion in the literature [88], the model of dative bonding, as originally introduced already by Lewis [57] and Sidgwick [59–61], has become a widely used tool for describing molecular structures and reactivity.

We want to point out that during the formation of a chemical bond between two atoms, the initial dative interaction may change so that the final bond becomes an electron-sharing bond. One example has already been discussed in the section about bond strength, where the carbon–carbon bond in C_2F_4 at the equilibrium

Table 5.2 Overview of isoelectronic ylidone compounds EL_2 and their proposed names.

Experimentally known examples	Ylidone	Name
Yes	$[BL_2]^-$	Borylone
Yes	$[AlL_2]^-$	Aluminone
No	$[GaL_2]^-$	Gallylone
No	$[InL_2]^-$	Indylone
Yes	CL_2	Carbone
Yes	SiL_2	Silylone
Yes	GeL_2	Germylone
Yes	SnL_2	Stannylone
Yes	PbL_2	Plumbylone
Yes	$[NL_2]^+$	Nitreone
Yes	$[PL_2]^+$	Phosphorone
Yes	$[AsL_2]^+$	Arseone
Yes	$[SbL_2]^+$	Stibione
Yes	$[BiL_2]^+$	Bismutone
Yes	$[OL_2]^{2+}$	Oxygeone
Yes	$[SL_2]^{2+}$	Sulfurone
Yes	$[SeL_2]^{2+}$	Selenone
Yes	$[TeL_2]^{2+}$	Telurone

Source: Petz and Frenking [71d].

geometry has a classical double bond $F_2C=CF_2$ built by the interactions between CF_2 in the triplet state, but at longer distances the CF_2 fragments engage in dative bonds between the singlet state $F_2C\rightleftarrows CF_2$ (Figure 5.6) [49]. The same situation was found in the sodium–boron bond in $NaBH_3^-$, which was recently reported as the first example of a Lewis adduct with an alkalide as Lewis base that has a $Na^-\rightarrow BH_3$ dative bond [89]. This assignment was challenged because the Na–B interactions are only at longer distances well described in terms of a dative bond $Na^-\rightarrow BH_3$, but they change at shorter distances when they are finally better described as electron-sharing bond $Na-BH_3^-$ [90].

5.6 Polar Bonds

The introduction to chemical bonding commonly used simple molecules with nonpolar bonds such as H_2, N_2, H_3C-CH_3, $H_2C=CH_2$, and $HC\equiv CH$ as archetypical examples. Polar bonds receive less attention and are often misleadingly described. For example, LiF is usually introduced to have an ionic bond between Li^+ and

5.6 Polar Bonds

F$^-$ with little covalent bonding. This holds to a good approximation for solid LiF and generally for ionic solids, but not for diatomic LiF nor for any other molecule. Chemical bonding in any molecule comes from the interference of the wave functions (Section 5.2), which leads to a charge accumulation in the bonding region. The charge accumulation in unpolar bonds A–A is at the middle of the bond, but in polar bonds A–B, it is polarized toward the more electronegative atom. The charge accumulation at the interatomic region also provides some electrostatic (Coulomb) attraction to the chemical bond, and the electrostatic contribution increases when the bond becomes polar, but it remains a covalent electron-pair bond just as in unpolar bonds. This has been discussed in a recent paper about the nature of the polar bond [91].

The frequent assignment of ionic character to polar bonds comes from the description of the bonding interactions using the VB approach, which does not have an explicit term for polar bonds. The basic VB expansion for interatomic interactions in a bond A–B has the electron-sharing covalent term $(\lambda_a - \lambda_b)$ ("Heitler–London (HL)" term) and the two ionic terms $(\lambda_a|^- \lambda_b|^+)$ and $(\lambda_a|^+ \lambda_b|^-)$ besides the mixing term Mix [92]:

$$\Psi_0^{VB} = \Sigma c_1 (\lambda_a - \lambda_b) + \Sigma c_2 (\lambda_a|^- \lambda_b|^+) + \Sigma c_3 (\lambda_a|^+ \lambda_b|^-) + \text{Mix} \qquad (5.11)$$

The coefficients c_2 and c_3 of the pair functions $\lambda_a \lambda_b$ that are occupied by two electrons give the contributions of the ionic terms, which are considered as "ionic character" of the bonding. But this is a physically misleading expression because the polarity of the bond is due to the different electronegativity of the atoms, which pushes the electron density toward the more electronegative atom, while the physical mechanism is the same as in nonpolar bonds. Genuine ionic bonds exist between atoms with negligible overlap of the valence orbitals, and they require the concomitant attraction of an atomic ion by several ions with opposite charge. *There is an electrostatic contribution to the covalent bond but there is no ionic bonding in molecules!*

Pauling favored VB theory over MO theory, and his very biased viewpoint in the influential book "*The Nature of the Chemical Bond*" [32] led to the still ubiquitous description of polar bonds in terms of covalent character and ionic character given in Eq. (5.11), although by now most people use MO or DFT methods for quantum chemical calculations. The "covalent" and "ionic" terms in Eq. (2.13) led to a seemingly easy correlation to the Lewis structures with covalent or ionic bonds, which is the reason why Pauling and still many chemists are using the ill-fated notation [93]. The situation became even more confused recently when it was found that VB calculations of the covalent bond in F_2 give neither of the three terms $c_1 - c_3$ as leading contributions to the chemical bond. Instead, the mixing term MIX between the covalent forms is the dominant term in the VB calculation. It is well known from MO calculations that electron-rich species like F_2 and H_2O_2 require higher orders of the theoretical method to accurately describe the electronic structure. Thus, F_2 is predicted to be thermodynamically unstable for dissociation and it requires correlation terms to give the correct results. This holds, not surprisingly, also for VB calculations, but the interpretation of the nature of the chemical bond in terms of "covalent" and "ionic" contributions becomes difficult. Amazingly, the authors suggested that F_2

and related molecules do not have a covalent bond. A new type of bond was created termed "charge-shift bond" (CSB), without any physical origin [94]. Meanwhile, it has become fashionable to determine the "CSB" contribution in chemical bonds, the value of which is as redundant as a crop [95]. *New bonding models, and in particular the introduction of new types of chemical bonds, should not be based simply on the appearance within mathematical terms of a quantum chemical method, but should have a physical meaning.*

It is instructive to consider the representation of polar bonds in the light of frequently used indicators of chemical bonding in the literature. It is common to present calculated bond orders as a fingerprint for comparing different bonds. Table 5.1 shows the bond orders for diatomic molecules with increasing polarity N_2, CO, BF, and LiF using two popular methods, i.e. the Wiberg and Mayer approaches. It is important to be aware of the theoretical fundament of the methods in order to estimate the relevance of the numerical results. The Wiberg bond orders were originally defined for semiempirical methods [96] and they are calculated from orthogonalized basis functions, whereas the Mayer bond orders consider the overlap matrix [97]. This is particularly important for polar bonds, where the Mayer bond order should therefore give a more faithful representation of the bond multiplicity.

The Wiberg bond orders P_{Wiberg} show a gradually decreasing order: N_2 > CO > BF > LiF, where the value for N_2 (3.03) agrees with the assignment of a triple bond. A naïve interpretation of the values for BF (0.85) and LiF (0.12) would suggest less than a single bond and a nearly negligible covalent character of the Li–F bond. BF and LiF, however, result from the interactions between neutral boron or lithium and neutral fluorine. There are covalent (electron-sharing) σ bonds between the unpaired electrons of B or Li and F, which are further complemented by π donation from the fluorine $p(\pi)$ electrons into vacant $p(\pi)$ AOs of B or F. There is therefore more than a single electron-pair bond in BF and LiF, which is not represented by the bond orders. The same trend, N_2 > CO > BF > LiF, is calculated by the Mayer bond orders P_{Mayer}, which show a slightly smaller value for N_2 but larger values for CO, BF, and LiF. The value for BF (1.50) suggests some multiple bond character, but the value for LiF (0.50) indicates less than a single bond. But there is definitely a single bond that comes from the pairing of the unpaired electrons of Li and F. The bond order must not be taken as a measure for the number of electron-pair bonds. They give a rough estimate for the polarity of the bond; but the degree of the polarity is not obvious from the bond order. Bond orders do not provide much insight into the nature and multiplicity of the bond, they are numerical indicators with questionable information. This becomes obvious when the values of B–F are compared with those of H_2B–F, which are also given in Table 5.3. There are two orthogonal π donor bonds B⇌F besides the electron-sharing σ bond in the diatomic molecule, but there is only one dative bond in the latter species H_2B←F. Table 5.3 shows that the BF bond in H_2BF is significantly longer (1.331 Å) than in diatomic BF (1.278 Å), which comes from the loss [98] of the in-plane $\pi_\parallel$ donation H_2B←F. The bond shortening of the BF bond from H_2BF to BF indicates the nonnegligible effect of the π bonding in BF. But the Wiberg bond order increases

5.6 Polar Bonds

Table 5.3 Calculated Wiberg and Mayer bond orders, partial charges q, and bond lengths r_{E-X} (Å) at the BP86/def2-TZVP level.

Molecule	Bond order			Partial charges				Bond length
	P_{Wiberg}	P_{Mayer}	q_{NBO}	$q_{Hirshfeld}$	$q_{Voronoi}$	q_{QTAIM}		r_{E-X}
N_2	3.03	2.78	0.00	0.00	0.00	0.00		1.103
CO	2.30	2.44	0.46 (C)	0.07 (C)	0.07 (C)	1.15 (C)		1.137
			−0.46 (O)	−0.07 (O)	−0.07 (O)	−1.15 (O)		
BF	0.85	1.50	0.54 (B)	0.06 (B)	0.04 (B)	0.87 (B)		1.278
			−0.54 (F)	−0.06 (F)	−0.04 (F)	−0.87 (F)		
LiF	0.12	0.50	0.94 (Li)	0.57 (Li)	0.51 (Li)	0.90 (Li)		1.577
			−0.94 (F)	−0.57 (F)	−0.51 (F)	−0.90 (F)		
H_2BF	0.92	1.27	0.70 (B)	0.22 (B)	0.22 (B)	2.01 (B)		1.331 (B−F)
			−0.45 (F)	−0.12 (F),	−0.10 (F)	−0.81 (F)		
			−0.13 (H)	−0.05 (H)	−0.06 (H)	−0.60 (H)		

from BF (0.85) to H_2BF (0.92) while the Mayer bond order decreases from BF (1.50) to H_2BF (1.27). The trend of the latter values is physically more reasonable than that of the former, which suffers from the neglect of the orbital overlap, but the nature of the polar bonds is not obvious from either method.

A physically very helpful insight into the polarity of the bonds is given by the QTAIM method developed by Bader [38]. Figure 5.12 shows the Laplacian $\nabla^2 \rho(\mathbf{r})$ of the five molecules N_2, CO, BF, LiF, and H_2BF with the bond paths and bond critical points and the zero-flux surfaces in the chosen molecular planes. The areas of charge accumulation ($\nabla^2 \rho(\mathbf{r}) < 0$, red dotted lines) exhibits nicely the feature of charge migration due to the bond formation. It is important to realize that the areas of charge depletion ($\nabla^2 \rho(\mathbf{r}) > 0$, blue solid lines) do not indicate the absence of charge but the curvature of the charge distribution. The droplet-like appendages in CO and BF clearly show the formation of local charge concentrations on carbon and boron atoms due to the occurrence of lone-pair orbitals, which are absent in LiF and H_2BF. The center of the local negative charge accumulation at C and O in CO and BF is located away from the positively charged nuclei, whereas the charges at O and particularly at F are more symmetrical around the nuclei. This explains the somewhat paradoxical dipole moments of CO (0.12 D) and BF (0.82 D), where the negative end is at the more electropositive atoms C and B, respectively [62, 91, 99]. Dipole moments are vector properties, and they must not be confused with scalar values such as atomic partial charges, which do not provide any information about the orientation of the charge. LiF has in contrast to CO and BF, a dipole moment of 6.28 D with the negative end at the more electronegative fluorine atom because it does not have a lone-pair orbital at the electropositive atom [100].

There is still more information that is gained by the QTAIM method about the polar bonds in the five molecules. The location of the bond critical points (bcp) r_c

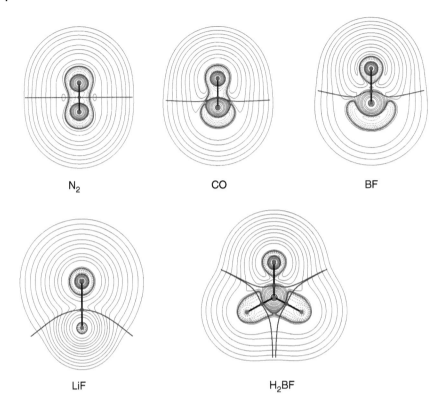

Figure 5.12 Contour line plots of the Laplacian $\nabla^2 \rho(r)$ of the electron density for (a) N_2, (b) CO, (c) BF, (d) LiF, and (e) H_2BF in the electronic ground state. The blue solid lines and violet dotted lines show the area of $\nabla^2 \rho(r) > 0$ and $\nabla^2 \rho(r) < 0$, respectively. The black lines show the zero-flux surfaces in the molecular plane. The blue dots indicate the bond critical point.

indicates the polarity of the bonds. Table 5.4. gives the distance of r_c from the nucleus of the more electronegative atoms and the shift from the midpoint of a hypothetical nonpolar bond r_{NP} of the atoms. The latter were calculated from the sum of the bonding radii of homoatomic species E_2 and X_2 and scaled by the actual bond length EX [101]. Table 5.4 shows a clear shift $\Delta r_{c\text{-NP}}$ for the more electronegative atom, which indicates the greater polarity of the bond EX where X becomes more electronegative. Note that the longer B–F bond in H_2BF than in BF also has a higher polarity because the π backdonation B←F in the former species is weaker than in the latter. It is a very useful feature of the QTAIM analysis that it directly indicates the polarity of a chemical bond with respect to a nonpolar reference value. It is possible on this way to give the effective electronegativity of an atom in a given bond, which may be quite different from the global value for the electronegativity of an atom. This is discussed in Section 2.7.

Table 5.4 also gives some numerical QTAIM values at the bcp of the five molecules. It has been suggested that the charge density $\rho(r_c)$, the value of the Laplacian $\nabla^2 \rho(r_c)$,

Table 5.4 Numerical results of the QTAIM calculations.

Bond E–X	Δr_c Å	Δr_{c-NP}	$\rho(r_c)$	$\nabla^2 \rho(r_c)$	$H(r_c)$
N–N	0.552	0.0	0.693	−2.616	−1.252
C–O	0.755	0.195	0.494	0.611	−0.921
B–F	0.843	0.242	0.228	1.486	−0.154
Li–F	0.968	0.423	0.076	0.662	0.013
H_2B-F	0.882	0.256	0.198	1.206	−0.137

Distance Δr_c of the bcp r_c from atom X and distance Δr_{c-NP} from a hypothetical nonpolar point r_{NP} of the bond E–X. Calculated values of the density $\rho(r_c)$, Laplacian $\nabla^2 \rho(r_c)$, and energy $H(r_c)$ at the bcp. All values were calculated at the BP86/def2-TZVP level.

and particularly the energy density $H(r_c)$ may be used as probes for the nature of the chemical bond [38, 102]. The charge density $\rho(r_c)$ has decreasing values in the order $N_2 > CO > BF > H_2BF > LiF$, which nicely agrees with the trend of the polarity. The values of the Laplacian $\nabla^2 \rho(r_c)$ change the sign from N_2 to the other bonds, but it can also change the sign for different unpolar bonds, as noted before. It is not a reliable indicator for the nature of the bond, which has been discussed and explained before. The most useful information comes from the values for the energy density $H(r_c)$, which do not only give a correct trend of the polarity but also show that the bond in LiF is best described by closed-shell interactions between Li^+ and F^- rather than by electron-sharing bonding. It was shown by Cremer and Kraka that electron-sharing bonds possess negative values for the energy density $H(r_c)$, whereas dative bonds with closed-shell interactions or van der Waals complexes typically have value of $H(r_c) \geq 0$ [102]. A detailed analysis of the bonds in N_2, CO, BF, and LiF is given below in Section 5.8.

The polarity of a bond A–B is often discussed in terms of the atomic partial charges of atoms A and B, which are subject to charge partitioning methods. Table 5.3 shows the calculated values for N_2, CO, BF, LiF, and H_2BF using four popular approaches [103]. It becomes obvious that the methods give very different values, which indicates that a quantitative interpretation of the polarity of the bonds in terms of atomic partial charges is dubious. This holds particularly for CO, where the partial charge for the oxygen atom varies from −1.15e given by QTAIM [38] to −0.07e suggested by the Voronoi [104] and Hirshfeld [105] methods. Unfortunately, partial charges are often presented in the literature using only one method, and they are then taken as evidence for a particular interpretation of the bond and as support for the chemical intuition of the author, which results in an arbitrarily flexible statement with doubtful value. Note that the values of partial charges are not only subject to different methods; they are also global values, which do not provide any information about the topology of the electronic charge, which is crucially important to the physical and chemical properties of the molecule.

5.7 Atomic Partial Charges and Atomic Electronegativity

The notions of partial atomic charges in molecules and electronegativity as a measure of the attractive force of an atom in a covalent bond are among the most important concepts in chemistry. Like other chemical entities, they are not observable, and their numerical value depends on their definition, where numerous different variants have been suggested in the past. The topic of partial charges is explicitly covered in Chapter 6 of this book, and the less controversial concept of electronegativity has been used in the sections above and can be found in nearly all chapters of this book. This short section is dedicated to an aspect of partial charges and electronegativity that is often overlooked, although it is of crucial importance for molecular structures and reactivity.

The numerical values for the partial charge and the electronegativity of an atom in a molecule refer to the entire atomic species, regardless of the method used to determine them. But the electronic structure of an atomic species in most molecules is not spherically symmetrical, but it exhibits distinct areas of electronic charge depletion and charge accumulation. This comes clearly to the fore when the Laplacian of the electron density distribution $\nabla^2 \rho(\mathbf{r})$ is considered, which nicely visualizes the degree of anisotropic charge distribution of an atom in a molecule. This becomes visible in the examples shown in Figure 5.12. The areas of charge concentration ($\nabla^2 \rho(\mathbf{r}) < 0$, red dotted lines) around Li and F in LiF are nearly perfectly roundly shaped, but the other atoms in N_2, CO, BF, and H_2BF show some distortion of the density distribution, which is particularly strong for C in CO and B in BF and H_2BF. This induces important information about the chemical and physical properties of the molecules, which is not obvious when only the atomic partial charge is considered. As mentioned above, this holds, for example, for the dipole moments of CO (0.12 D) and BF (0.82 D) having the negative end at the more electropositive atoms C and B. But the anisotropic charge distribution also has an effect on the electronegativity of the atoms, whose value depends on the local orientation of the bond.

Figures 5.9 and 5.10 show schematically the areas of vacant orbitals in the isoelectronic species BH, C, and N^+, where donor ligands with suitable occupied orbitals may bind. Such donor–acceptor interaction may give rise to chemical bonds A→B where the globally less electronegative acceptor atom B has a higher negative charge than the more electronegative donor atom A. Examples are the borane complexes (BH)(cAAC)$_2$ and (BR)(CO)$_2$, which have been mentioned above [64a, 79, 80]. The B–C bonds in these complexes have a charge distribution that is opposite to the atomic electronegativity.

It may be argued that the Laplacian distribution gives the total area of charge distribution and that the structures and reactivities of the molecules are mainly determined by the valence electrons. But the deformation of the core electrons is negligibly small, and the Laplacian distribution can therefore be identified with the distortion of the valence shell. The important message of this chapter is the emphasis on the anisotropy of the valence electrons, which has a profound influence on the chemical behavior of the molecules. This information is neglected when the global values of atomic partial charges and electronegativity are considered [106].

5.8 Chemical Bonding in Main-Group Compounds: N$_2$, CO, BF, LiF

This book chapter focuses on fundamental aspects of chemical bonding rather than on specific classes of compounds. We present the results of a bonding analysis only for a small number of selected molecules that are often covered in chemistry textbooks to show what kind of information is available with modern quantum chemistry methods. The molecules are partly covered already in Section 5.6 on polar bonds, and the information given here complements the bonding analysis with different methods. We use an EDA approach that has been shown to provide deep insights into interatomic interactions, combining modern quantum chemical models with the Lewis electron-pair model and the physical analysis of chemical bonding proposed by Ruedenberg. The EDA-NOCV method has the further advantage that the results cannot only be given in terms of numbers, but that the change in the electronic structure associated with the bond formation can be nicely visualized using deformation densities. More details of the EDA-NOCV approach are described in recent review articles [44]. A comprehensive presentation of chemical bonding in main-group compounds is available in a recent review article [44c].

A crucial and important part of any EDA method is the choice of the interacting fragments and their spatial arrangement. This gives the method the flexibility to compare the results using different moieties, which is not always trivial. In the present case, we used the atoms in the electronic reference state, which is associated with bond formation. This is not always the electronic ground state, but the atomic species that we used for the bonding analysis of N$_2$, CO, BF, and LiF are the ground state of the atoms or ions in the arrangements, which are shown in Figure 5.13.

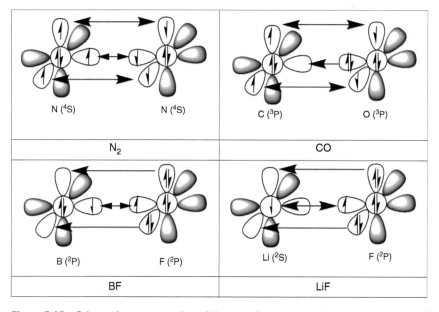

Figure 5.13 Schematic representation of the most important orbital interaction in N$_2$, CO, BF, and LiF, which lead to electron-sharing bonds A−B or dative bonds A→B.

Table 5.5 shows the numerical results of the EDA-NOCV calculations of the four molecules. The calculated values indicate that the attractive interactions in N_2 originate mainly (70%) from the (covalent) orbital term ΔE_{orb} due to the lowering of the kinetic energy and 30% from the electrostatic (Coulomb) term ΔE_{elstat}. The σ-orbitals contribute two-thirds to the covalent bond, and the degenerate π-orbitals contribute one-third to ΔE_{orb}. Note that the strongest interaction is due to the Pauli repulsion ΔE_{Pauli}, which is often neglected in the model of orbital interactions, although it has a profound effect on the equilibrium bond length, as shown above. The deformation densities $\Delta\rho$ shown in Figure 5.14 nicely illustrate the charge flow from the atoms toward the interatomic bonding region. We want to point out that the sum of the energy terms in the EDA approach gives a total interaction energy ΔE_{int}, which directly correlates with the experimentally available BDE.

The EDA-NOCV values for CO suggest a stronger total interaction energy ΔE_{int} as in N_2 in agreement with experiment. Inspection of the energy terms shows that the attractive forces ΔE_{orb} and ΔE_{elstat} are actually weaker in CO than in N_2. The bigger interaction energy comes from the Pauli repulsion ΔE_{Pauli}, which is significantly weaker in CO than in N_2. Pauling's original assumption that the stronger binding in CO than in N_2 is due to an additional ionic (electrostatic) attraction [32] is not confirmed by the calculated values for ΔE_{elstat}, which is weaker in CO than in N_2. Another interesting feature concerns the relative strength of the σ and π bonds in CO. The total π bonding in CO is nearly as strong as the σ bond, which is in striking contrast to the result for N_2 (Table 5.5). This can be explained by the nature of the σ and π orbital interaction in the two molecules. Figure 5.13 shows that the σ bond in N_2 comes from electron-sharing interactions N–N, whereas the σ bond in CO is a dative bond C←O. Dative bonds are always weaker (but not necessarily weak) [107] than electron-sharing bonds.

Table 5.5 gives the numerical EDA-NOCV results for BF and LiF using neutral atoms and ions as interacting fragments. The results nicely demonstrate the large differences between the interactions, which address two different questions that are often confused. The choice of neutral atoms and the associated data deal with the change in electronic structure when the neutral atoms approach each other and form a chemical bond. Both molecules undergo a significant charge flow B→F and Li→F along with bond formation and the electronic structures of the eventually formed molecules. BF and LiF are best described in terms of ionic fragments B^+–F^- and Li^+–F^-. This comes to the fore by calculating values for the orbital term ΔE_{orb}, which indicates the magnitude of charge deformation between the fragments. Those fragments that change least are best suited to describe the finally formed bond, which is not the same as the change in the electronic structure during the formation of the bond. It is important to recognize the difference between the two issues.

The data in Table 5.5 indicate that the bond formation of BF and LiF when the two atoms approach each other comes mainly from the orbital interaction, which provides two-thirds of the attraction in BF and even 91% in LiF. It comes mainly from the formation of the σ bond, whereas the contribution of the π orbitals in BF is very small (11%) and it is negligible for LiF (2%). The contribution of the π interaction $\Delta E_{orb(2)}$ in the finally formed bond to the covalent bonding between the charged

Table 5.5 The results of EDA-NOCV calculations of N_2, CO, BF, and LiF at the BP86/TZ2P//BP86-/def2-TZVPP level.

		N_2	CO	BF		LiF	
Energy	Orbital interaction	2 N (4S)	C (3P) + O (3P)	B (2P) + F (2P)	B^+ (1S) + F^- (1S)	Li (2S) + F (2P)	Li^+ (1S) + F^- (1S)
ΔE_{int}		−240.5	−267.4	−186.1	−309.3	−140.4	−190.6
ΔE_{Pauli}		801.4	582.5	483.1	355.0	55.2	41.2
ΔE_{elstat} [a]		−312.3 (30.0%)	−240.0 (28.2%)	−213.7 (31.9%)	−402.2 (60.5%)	−17.3 (8.8%)	−207.9 (89.5%)
ΔE_{orb} [a]		−729.6 (70.0%)	−610.0 (71.8%)	−455.5 (68.1%)	−262.1 (39.5%)	−178.4 (91.0%)	−24.0 (10.3%)
$\Delta E_{orb(1)}$ [b]	A−B σ-interaction	−478.0 (65.5%)	−290.6 (47.6%)	−392.9 (86.3%)	−186.5 (71.2%)	−176.0 (98.7%)	−10.6 (44.2%)
$\Delta E_{orb(2)}$ [b]	A−B π-interaction	−124.8 (17.1%)	−149.2 (24.5%)	−25.6 (5.6%)	−36.5 (13.9%)	−1.6 (0.9%)	−6.3 (26.3%)
$\Delta E_{orb(3)}$ [b]	A−B π′-interaction	−124.8 (17.1%)	−149.2 (24.5%)	−25.6 (5.6%)	−36.5 (13.9%)	−1.6 (0.9%)	−6.3 (26.3%)
$\Delta E_{orb(rest)}$ [b]		−2.0 (0.3%)	−21.0 (3.4%)	−11.4 (2.5%)	−2.6 (1.0%)	0.8 (−0.4%)	−0.8 (3.3%)
D_e ($= -\Delta E_{int}$)		240.5	267.4	186.1	309.3	140.4	190.6

a) The values in parentheses give the percentage contribution to the total attractive interactions $\Delta E_{elstat} + \Delta E_{orb}$.
b) The values in parentheses give the percentage contribution to the total orbital interactions ΔE_{orb}.

fragments is much larger; it amounts to 28% in BF, and it is even larger (52%) than the σ bond in LiF. However, the total attraction between the charged fragments comes mainly from the electrostatic term ΔE_{elstat}, which contributes 61% to the attraction in BF and 90% in LiF. The nonnegligible contribution of the π orbital interaction in BF becomes clear by the bond lengthening bond in H_2BF (1.331 Å) compared with BF (1.278 Å), which can be attributed to the loss of one π bond in the former molecule.

Figure 5.14 illustrates the charge flow, which is associated with the σ and π orbital interaction of the neutral and charge fragments of BF and LiF. The figure also gives

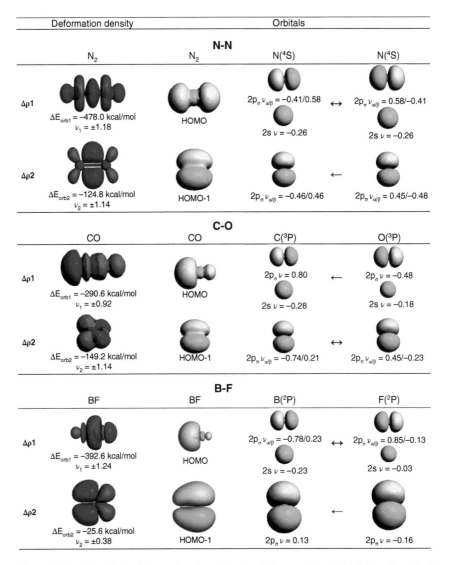

Figure 5.14 Plot of the deformation densities $\Delta\rho$ of the σ and π orbital interactions in (N_2, CO, BF, and LiF, which indicate the direction of the charge flow red→blue. Shape of the most important AOs of the atomic fragments and the diatomic molecules. The eigenvalues υ give the relative size of the charge transfer.

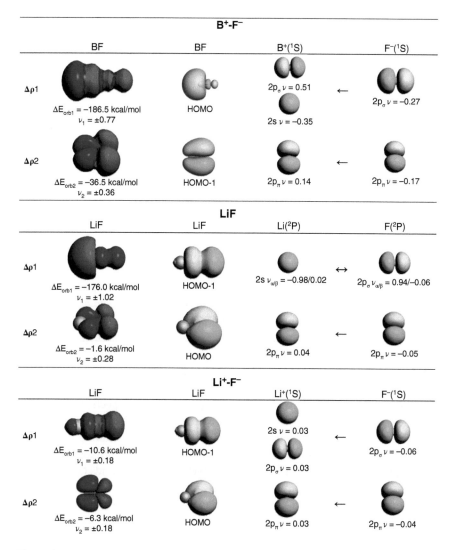

Figure 5.14 (Continued)

the eigenvalues of the deformation densities, which indicate the size of the charge migration.

5.9 Chemical Bonding of the Heavier Main-Group Atoms

A major puzzle in the early days of quantum chemistry was the considerable differences between the structures and stabilities of molecules of atoms from the first octal series of the periodic table Li–F and the heavier homologues. Lewis' heuristic bonding model was mainly based on experimental findings on compounds of lighter atoms, as these were much more extensively studied and better known than heavier

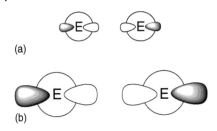

Figure 5.15 Schematic representation of the relative size of the (n)s and (n)p AOs in atoms (a) of the first octal row and (b) in heavier main-group atoms.

systems due to their generally higher stability. The structures and reactivities of the lighter systems were considered as reference systems that were regarded as normal, while the geometries and stabilities of the heavier molecules were viewed as abnormal. Quantum chemical studies later showed that the atoms of the first octal series are chemically abnormal, while the heavier atoms show more regular and normal behavior.

Figure 5.15 shows the principal difference between the valence orbitals of atoms E of the first octal series of the periodic table and heavier atoms. It concerns the ratio between the radii of the ns and np valence AOs of the atoms, which significantly changes from the first to the higher octal-row atoms. The 2p orbitals of the first octal-row atoms may penetrate rather deeply into the core because there are no lower lying occupied p functions, whereas the 2s orbitals are constraint to be orthogonal to the 1s AO. The radii of 2s and 2p orbitals are thus very similar, which leads to effective sp hybridization in chemical bonds of the lighter atoms. In contrast, the occupied np AOs ($n > 2$) of the heavier atoms are antisymmetric to the p core orbitals, which causes larger radii for the np than for the respective ns AOs. The spatial regions of ns and np AOs ($n > 2$) are more separate than those of 2s and 2p AOs, which induces a less effective hybridization of the heavier main-group atoms. The model of s/p hybridization, which is commonly used to account for the σ bonds of carbon in single bonds (sp^3), double bonds (sp^2), and triple bonds (sp), does not hold for heavier group-14 atoms. Whereas CH_4 has nearly perfect sp^3 hybridization at carbon, the bonding orbitals of SiH_4 have a much higher p character due to the larger radius of 3p than for 3s [108]. The different s/p ratio of heavier main-group atoms was already pointed out by Pyykko [109] and the particular feature of their chemical bonds has been analyzed and discussed in detail by Kutzelnigg [110].

Two pertinent examples for the differences between molecules of first octa-row atoms and heavier homologues shall shortly be discussed. A more detailed analysis is found in the literature [44c, 111, 112]. The first example concerns the structure of acetylene HCCH and the heavier homologues HEEH (E = Si – Pb). Acetylene has a linear geometry with a carbon–carbon triple bond HC≡CH but the heavier group-14 species possess unusual equilibrium geometries. Figure 5.16 shows the calculated energy minimum structures of the heavier E_2H_2 species. The nonplanar, double hydrogen-bridged structure **A** is the overall energy minimum for all E_2H_2 molecules. The energetically stable isomer next is the single-bridged form **B**, which has a curious inward bend of the terminal hydrogen atom. In the case of species **B**, all three atoms are on the same side of the triply coordinated atom E. This is a highly unusual bonding situation for a group 14 atom. Both E_2H_2 isomers **A** and **B**

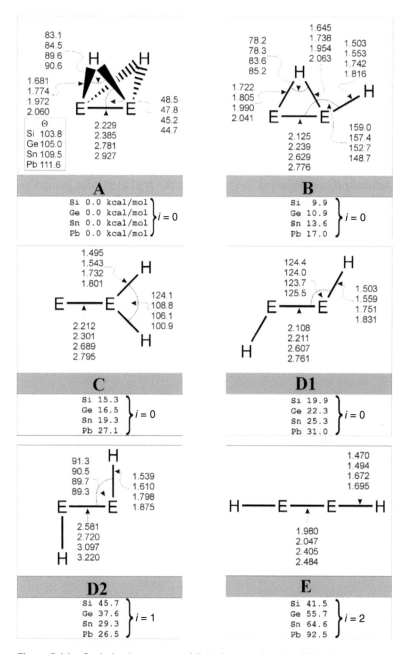

Figure 5.16 Optimized structures of E_2H_2 isomers **A – E** at BP86/QZ4P. Bond lengths are given in Å, angles in degree. The values for Θ give for structure **A** the dihedral angle between the E_2H and E_2H' planes. The relative energies with respect to **A** are given at the bottom of each entry in kcal/mol together with the number of imaginary frequencies i.
Source: Reproduced from Ref. [111].

have been observed for all elements E = Si – Pb in low-temperature matrix studies [113]. The vinylidene species **C** is the following energetically higher-lying isomer. It is the only energy minimum structure that is common for carbon and the heavier group-14 atoms Si–Pb. Finally, there are two *trans*-bent structures **D1** and **D2**, which possess significantly different E–E bond lengths and E–E–H angles. Substituted homologues E_2R_2 with bulky groups R featuring **D1** have been synthesized and structurally characterized for E = Si, Ge, Sn [114]. The lead analogue Pb_2R_2 exhibits structure **D2** [115]. The linear form **E**, which is the global energy minim for acetylene, is an energetically high-lying second-order saddle point for all heavy group-14 species E_2H_2.

The peculiar structure of the heavier systems with the atomic connectivity HEEH can straightforwardly be explained by considering the $X^2\Pi$ electronic ground and the $a^4\Sigma^-$ first excited states of the EH fragments, which are shown in Figure 5.17. The linear structure HE≡EH requests the $a^4\Sigma^-$ excited state of EH for the formation of an electron-sharing triple bond. The excitation energy $X^2\Pi \to a^4\Sigma^-$ of the heavier EH species is significantly larger than for CH. Table 5.6 shows the calculated bond dissociation energies D_e of the linear structures HE≡EH yielding $(a^4\Sigma^-)$ EH. Subtracting the excitation energies for two EH fragments to the $a^4\Sigma^-$ state ΔE_{exc} from the D_e values gives the net stabilization energy of the electron-sharing triple bond. The value of 240.0 kcal/mol for CH is sufficiently large to surpass the stabilization of an electron-sharing single bond, which would be provided by coupling the unpaired

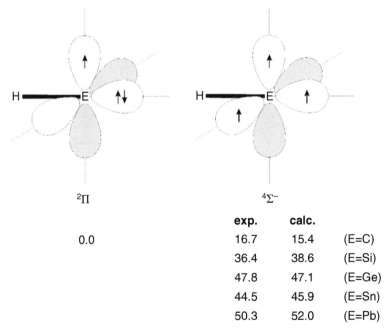

Figure 5.17 Schematic representation of the electron configuration of the $^2\Pi$ electronic ground state and the $a^4\Sigma^-$ excited state of EH (E = C – Pb). The experimental [101] and calculated (BP86/QZ4P) excitation energies are given in kcal/mol.

Table 5.6 Calculated bond dissociation energies D_e (kcal/mol) of linear HE≡EH → 2 EH ($a\Sigma^-$) and $X\Pi \rightarrow a\Sigma^-$ excitation energies ΔE_{exc} (kcal/mol) of EH at BP86/QZ4P.

E	D_e	ΔE_{exc}	$D_e - 2 \times \Delta E_{exc}$
C	270.9	15.45	240.0
Si	121.6	38.6	44.4
Ge	113.3	47.1	19.1
Sn	89.4	45.9	−2.4
Pb	69.0	52.0	−35.0

electrons of the CH fragments in the $X^2\Pi$ ground state. The net stabilization energy $D_e - 2\Delta E_{exc}$ for Si and Ge is much smaller (Table 5.6). It does not compensate for the energy gain of a E–E single bond. The excitation energies of two SnH and PbH fragments are even higher than the BDE of linear HE≡EH. It follows that the bonding interactions between two EH species for E = Si – Pb take place from the $X^2\Pi$ ground state, which induces nonlinear geometries.

Figure 5.18 shows three possible arrangements of the EH fragments in their $X^2\Pi$ ground state where the unpaired electrons are coupled to an electron-sharing σ bond. The assignments (a) and (c) are unfavorable, because they leave the unoccupied $p(\pi)$ orbitals in the resulting species vacant. Consequently, the structures are not energy minima but transition states ($i = 1$). The arrangement (c) leads to structure **D2**, which becomes an energy minimum for E = Pb when bulky substituents prevent the formation of the electronically more favored isomers [116]. The arrangement (b) in Figure 5.32 places the EH bonds in perfect position to donate the associated electron pairs into the respectively vacant $p(\pi)$ orbitals. This leads to the global energy minimum form **A** of the heavier E_2H_2 molecules. It becomes clear that the structure **A** has a triple bond, which consists of one electron-sharing σ component and a degenerate EH donor component.

Figure 5.19 shows three different placements of the ($X^2\Pi$) EH fragments where the unpaired electrons yield a π bond. Placement (a) shows that the electron-sharing π bond is supported by a lone-pair donor bond and an EH donor bond. The optimal EH donation requires a tilting of the vacant $p(\pi)$ orbital of the right EH fragment, which leads to the inward bending of the associated E–H bond. The formation of the EH donor bond nicely explains the unusual equilibrium structure of the isomer **B**, which has a triple bond that includes an electron-sharing π bond, a lone-pair donor bond, and an EH donor bond. Rotation of one EH fragment leads to structure **G**, which is not an energy minimum, although it is lower in energy than **B**. Structure **G** possesses two EH donor bonds, which are stronger than one EH and one lone-pair donor bond in **B**. But **G** is the transition state for the degenerate flip-flap rearrangement of the global energy minimum **A**. Finally, there is the energetically highest lying energy minimum **D1**, which has a triple bond consisting of an electron-sharing π bond and two lone-pair donor bonds. Substitution of hydrogen by

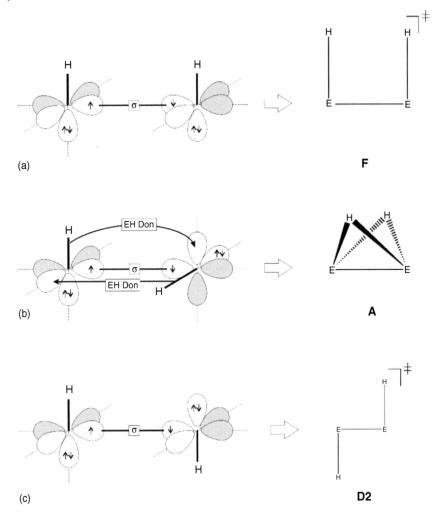

Figure 5.18 Qualitative model for the orbital interactions between two EH molecules in different orientations where the unpaired electrons yield a σ bond. Source: Reproduced from Ref. [111].

bulky groups in E_2H_2 stabilizes **D1** so much that it can be isolated. When E = Pb, the isomeric form **D2** becomes lower in energy than **D1**, because the electron-sharing π bond is replaced by a stronger σ bond and the lone-pair donation of lead is rather weak.

The model presented above easily explains the experimentally observed structures of E_2H_2, which possess equilibrium geometries that are not directly accessible by the Lewis bonding model. They can be straightforwardly explained with an extension of the model by Trinquier and Malrieu [117] and by Carter and Goddard [118], that was suggested for heavier homologues of ethylene. It impressively demonstrates the importance of orbital symmetry for understanding the structures of molecules,

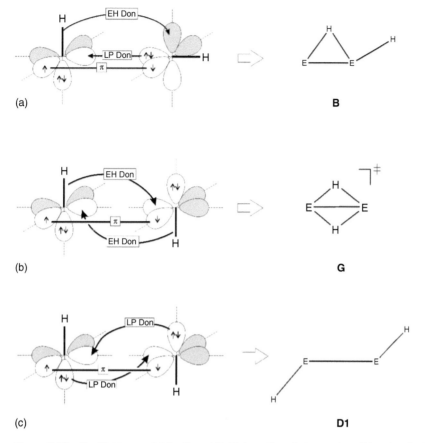

Figure 5.19 Qualitative model for the orbital interactions between two EH molecules in different orientations where the unpaired electrons yield a π bond. Source: Reproduced from Ref. [111].

which is not considered in Lewis' pre-quantum electron-pair model. The crucial importance of orbital symmetry also comes to the fore in so-called hypervalent compounds of heavier main group atoms, which will be discussed next.

Figure 5.20a shows schematically the orbital correlation diagrams of octahedral EX_6 which possess six E–X σ bonds which are formed from the (n)s and (n)p valence orbitals of atom E and symmetry adapted σ orbitals of the ligand cage X_6. Figure 5.20b displays a related correlation diagram for trigonal bipyramidal EX_5. The orbital correlation diagram holds for so-called hyperconjugated molecules like SF_6 and PF_5, which are very stable compounds while their lighter homologues OF_6 and NF_5 are unknown. The central question is how six σ bonds can be built in EF_6 and EF_5 although the central atom E has only four valence orbitals. The appearance of additional π bonding is irrelevant for the topic. Figure 5.20a shows that the (n)s and (n)p atomic orbitals of atom E split into an a_{1g} and a triply degenerate set of t_{1u} AOs, whereas the six σ orbitals of the ligand cage X_6 are divided into a_{1g}, t_{1u}, and e_g

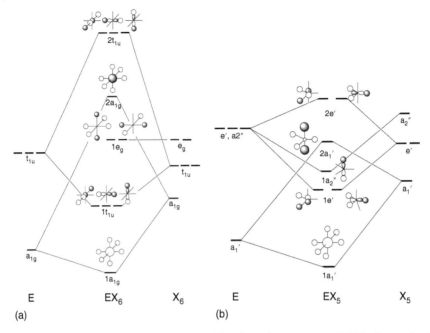

Figure 5.20 Splitting of the $(n)s$ and $(n)p$ AOs of a main-group atom E (a) in the octahedral field of six ligands X in EX_6 and (b) in the trigonal pyramidal field of five ligands X in EX_5.

MOs. The molecule SF_6 has twelve valence electrons that occupy six σ MOs. There are only four MOs in EX_6 (a_{1g} and t_{1u}) that have the right symmetry for interference between the AOs of atom E and the ligand cage. The degenerate e_g MO does not enable covalent E–X bonding due to lack of the right symmetry. This does not mean that the electrons in the e_g MO do not contribute to the stabilization of the molecule. The electronic charge of the ligand cage X_6 may be strongly stabilized through Coulomb interactions with the nucleus of atom E [119]. The crucial finding that comes from the orbital correlation diagram is that six E–X σ bonds can be formed in octahedral EX_6, although the central atom E has only four valence AOs.

A similar reasoning holds for the five E–X σ bonds in EX_5. Figure 5.20b shows that the AOs of E and σ MOs of X_5 split in the D_{3h} symmetric model compound EX_5 and give rise to σ orbitals possessing a_1', a_2'', and e' symmetry, which are occupied in PF_5 by ten electrons. The $1a_1'$, $1a_2''$, and the degenerate $1e'$ MOs are bonding orbitals that give four delocalized electron-pair bonds, which result from the interference of the wave functions. The $2a_1'$ orbital is the HOMO in EX_5, but the coefficient of the $(n)s$ valence AO is very small, which makes it effectively a nonbonding MO. There is negligible interference between the atomic orbitals of phosphorus and fluorine in the $2a_1'$ MO [120]. Covalent bonding refers to the number of bonds, which are due to the interference of the wave functions. There are four such doubly occupied MOs in the molecules. The bonding situation in EF_6 and EF_5 agrees with the octet rule [121]. This is possible because the symmetry of the orbitals is considered.

5.10 Chemical Bonding in Transition Metal Complexes: M(CO)$_n$ (M = Ni, Fe, Cr, Ti, Ca; $n = 4 - 8$)

Chemical bonding of TM compounds is traditionally divided from main-group compounds by the nature of the valence orbitals of the atoms. TMs use their $(n-1)$ d orbitals besides the (n)s and (n)p orbitals, and they obey the 18-electron rule for the valence shell, which was already suggested by Langmuir in 1921 [22]. In contrast, main-group atoms use their (n)s and (n)p AOs, and they obey the octet rule for filling the valence shell. Another difference between the present standard description of the chemical bond is that TM compounds are commonly described with dative bonds in the framework of the DCD model [67] whereas main-group molecules are usually introduced with electron-sharing bonds [44c]. But the dichotomy of the two types of covalent bonds has been recognized as useful model for understanding different variants of TM compounds such as the carbene/carbyne (Fischer) complexes versus alkylidene/alkylidyne (Schrock) compounds [122] or the TM alkene/alkyne complexes versus metallacyclic species [123]. The chemical bonding in TM compounds has been described in previous articles [67, 124]. Here we briefly present two examples that illustrate the current state of chemical bond description in TM compounds and that also show the crucial importance of orbital symmetry and the not-so-strict separation of the main group atoms of TM compounds using d-orbital participation as a criterion.

The first example concerns coordinatively saturated homoleptic carbonyl complexes TM(CO)$_n$ which may be considered as parent compounds of TM atoms. The group 10–6 complexes Ni(CO)$_4$, Fe(CO)$_5$ and Cr(CO)$_6$ and their heavier homologues are classical examples of the 18-electron rule. The extension to the group 4 atoms Ti, Zr, Hf in the direction of hypothetical heptacarbonyl complexes TM(CO)$_7$, which could be expected by electron counting, remained puzzlingly unsuccessful for a long time, which was aptly expressed in a theoretical study of 2008 *"No evidence for M(CO)$_7$ for any of the three metals has yet been suggested from these experiments despite the fact that our theoretical studies predict M(CO)$_7$ to be reasonable molecules for all three group 4 metals."* [125] Experimental evidence for coordinatively saturated carbonyl complexes of the group-4 atoms was finally provided in 2020, when a gas-phase studied showed that these are actually octacarbonyls TM(CO)$_8$ rather than heptacarbonyls TM(CO)$_7$ [126]. Theoretical studies of the octa-coordinated species, which have cubic (O_h) symmetry, showed that there is one occupied valence MO with a$_{2u}$ symmetry which does not mix with the valence (n)s, (n)p, or $(n-1d)$ AOs of the metal but it is a pure ligand MO. The complexes TM(CO)$_8$ have 20 valence electrons, but still fulfil the 18-electron rule, since only the electrons occupying mixed orbitals between metal and ligand need to be considered. The situation is comparable to the hyperconjugated main-group compounds EX$_6$ and EX$_5$ mentioned above, which fulfil the octet rule although the compounds have 12 and 10 electrons occupying valence σ orbitals.

Figure 5.21 shows the orbital correlation diagrams for coordinatively saturated carbonyl complexes TM(CO)$_n$ (n = 4–8) which surprisingly also include the group-2 metals Ca, Sr, and Ba (Figure 5.21e). The experimental observation of the

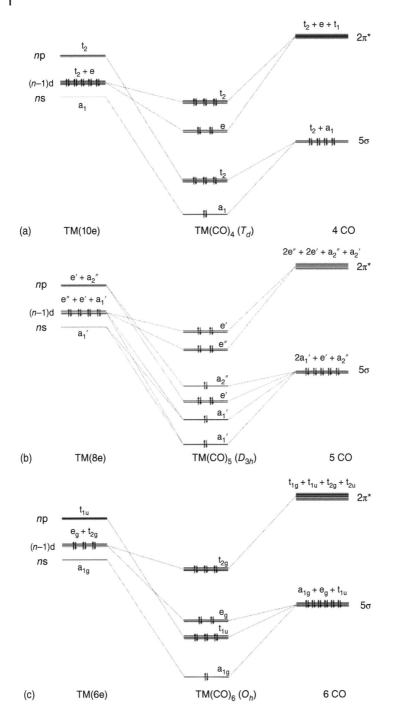

Figure 5.21 Correlation diagram of the splitting of the valence orbitals of the metals TM and the ligands $(CO)_n$ in the carbonyl complexes $TM(CO)_n$ and occupation of the σ and π orbitals. (a) Tetrahedral $TM(CO)_4$ (T_d). (b) Trigonal bipyramidal $TM(CO)_5$ (D_{3h}). (c) Octahedral $TM(CO)_6$ (O_h). (d) Cubic $TM(CO)_8$ (O_h), where TM has four valence electrons. (e) Cubic $TM(CO)_8$ (O_h), where TM has two valence electrons.

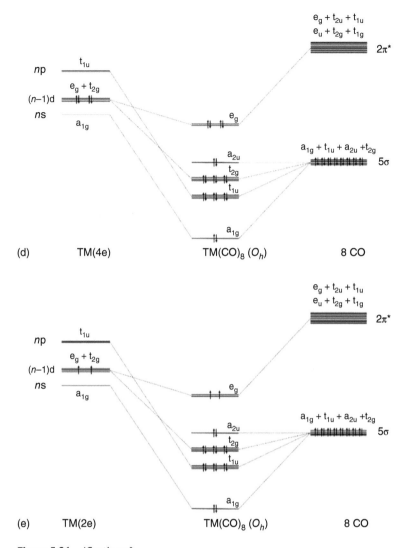

Figure 5.21 (Continued)

octa-coordinated carbonyl complexes of the alkaline earth (Ae) atoms came as a great surprise, and they drastically changed the understanding of the chemical bonds of Ca, Sr, and Ba [127]. The two valence electrons of the metal atoms occupy $(n-1)d$ AOs, and they engage in strong Ae→CO π backdonation, which is the main driving force for the metal–CO interaction. The complexes Ae(CO)$_8$ (Ae = Ca, Sr, Ba) have 18 valence electrons, but the cubic (O_h) structure has one occupied ligand-only a_{2u} MO, which makes the complexes 16-electron species. Since the HOMO is doubly occupied, the molecules have a triplet ($^3A_{1g}$) ground state. Further experimental studies showed that the dinitrogen complexes Ae(N$_2$)$_8$ [128] and the tribenzene adducts Ae(Bz)$_3$ (Bz = benzene) [129] can also be observed in the gas phase. A systematic study of the valence orbitals of alkaline earth atoms showed that Ca, Sr, and Ba use their $(n-1)$d orbitals for chemical bonding [130]. If the nature of the valence orbitals that are used in chemical bonds is used as a criterion for the division between main-group atoms and TMs, the Ae atoms Ca, Sr, and Ba should be considered as TMs [131].

5.11 Summary

The most important aspects of chemical bonding can be summarized as follows:

- The physical mechanism of covalent bonding has a quantum theoretical origin. It is due to the interference of the wave function. Chemical bonds do not arise from the coupling of two electrons, as is often claimed. Covalent bonds are already formed when an electron occupies a bonding orbital. The observation that only two electrons can be found in a stabilizing orbital results from the Pauli principle (exclusion principle), which states that a maximum of two electrons may occupy the same spatial region, and only if they have opposite spin. If the electrons have the same spin or if another degenerate orbital with the same energy is present, the second electron occupies a different orbital.
- The energy stabilization by the covalent bond comes from the decrease of the kinetic energy of the electrons in the bonding region. The overall stabilization of the bond by the lowering of the potential energy given by the virial theorem comes from the remaining electronic charge near the nuclei encountering a stronger attraction by the nuclei, which increases their kinetic energy. This explains the paradoxical finding that the kinetic energy of the electrons in the bonding region decreases but the total kinetic energy of the electronic charge increases.
- Over time, numerous bonding models have been proposed. The most important is Gilbert Lewis' heuristic electron-pair model, which was introduced before the advent of modern quantum theory. Its unsurpassed success shows that it mysteriously contains quantum theoretical information. In the early days of quantum chemistry, many new bonding models were derived, the most famous being those proposed by Linus Pauling and based on VB theory. Later studies have shown that some of the plausible arguments suggested by Pauling could not be justified in the light of advanced quantum chemical studies.
- The distinction between electron-sharing bonds A–B and dative bonds A→B as two types of covalent electron-pair bonds is a very useful tool for understanding the structures and reactivities of main group and TM complexes. It was already proposed by Gilbert Lewis and further developed by Nevil Sidgwick, but discarded by Pauling. Recent developments have led to a re-evaluation of the importance of dative bonding in main group chemistry, resulting in new classes of complexes with unusual bonds.
- It is very important to distinguish between models that describe the change in the electronic structures of fragments A and B during the bond formation A–B and the analysis of the finally formed bond. The former considers the change in the electronic structure of the structures when the bond is formed, while the latter takes

the deformed electronic structures of A and B in the molecule AB as a starting point.
- Another important point concerns the difference between the spatial distribution of the electronic charge at the atoms in a molecule and global properties such as atomic partial charge. The spatial distribution is pivotal for the chemical behavior such as the reactivity of a molecule, and the physical property such as the dipole moment. Global descriptors do not provide any information about local features, which can lead to misleading conclusions.
- The most important finding of quantum chemical analysis of chemical bonding in molecules and their electronic structure concerns the symmetry of the wave function and the associated MOs, which explains experimental observations and is a very useful tool for predicting new molecules and chemical reactions. Pioneering works are the FOM of Fukui and the orbital symmetry rules of Woodward and Hoffmann.
- Recent developments have shown that Pauli repulsion between electrons with the same spin, which is a quantum theoretical theorem, is a very important part of the interatomic interaction, along with orbital (covalent) attraction and Coulomb (electrostatic) forces. Pauli repulsion leads to equilibrium bond lengths that are significantly larger than the distance resulting from the maximum overlap of the bond orbitals.

Acknowledgments

One of the authors (GF) acknowledges very helpful discussions in the course of time about the chemical bond and insightful comments by W. H. Eugen Schwarz (Siegen), Klaus Ruedenberg (Ames), Ralf Tonner (Leipzig), Roald Hoffmann (Ithaca), Angel Pendas (Oviedo), and Alfred Paulus (Köln). This work was financially supported by the Institute of Advanced Synthesis, the School of Chemistry and Molecular Engineering, Nanjing Tech University.

References

1 Muller, P. (1994). Glossary of terms used in physical organic chemistry (IUPAC Recommendations 1994). *Afr. J. Pure Appl. Chem.* 66: 1077.
2 Frenking, G. and Krapp, A. (2007). Unicorns in the world of chemical bonding models. *J. Comput. Chem.* 28: 15.
3 Heitler, W. and London, F. (1927). Wechselwirkung neutraler Atome und homöopolare Bindung nach der Quantenmechanik. *Z. Angew. Phys.* 44: 455.
4 https://www.nobelprize.org/prizes/chemistry/1927/summary/.
5 https://www.geniuses.club/genius/democritus.
6 Mierzecki, R. (1991). The historical development of chemical concepts. *J. Chem. Edu.* 69: 0021.
7 Dewar, M.J.S. (1984). Chemical implications of sigma conjugation. *JACS* 106: 669.

8 Lewis, G.N. (1916). The atom and the molecule. *JACS* 38: 762.
9 Zhao, L., Schwarz, W.H.E., and Frenking, G. (2019). The Lewis electron-pair bonding model: the physical background, one century later. *Nat. Chem. Rev.* 3: 35.
10 (a)Fukui, K. (1971). Recognition of stereochemical paths by orbital interaction. *Acc. Chem. Res.* 4: 57. (b) Fukui, K. (1975). Chemical reactivity theory. In: *Theory of Orientation and Stereoselection*, 8. Berlin: Springer Verlag.
11 Woodward, R.B. and Hoffmann, R. (1970). *The Conservation of Orbital Symmetry*, 250. Weinheim: Verlag Chemie.
12 Hellmann, H. (2015). *Einführung in die Quantenchemie*. Deuticke, Leipzig and Wien, 1937. The book has recently been republished with biographical notes of the son Hans Hellmann jr. by D. Andrae (Ed.). Heidelberg: Springer Spektrum.
13 Kutzelnigg, W. (1973). The physical mechanism of the chemical bond. *Angew. Chem. Int. Ed.* 12: 546.
14 Ruedenberg, K. (1962). The physical nature of the chemical bond. *Rev. Mod. Phys.* 34: 326.
15 (a) Schmidt, M.W., Ivanic, J., and Ruedenberg, K. (2014). Covalent bonds are created by the drive of electron waves to lower their kinetic energy through expansion. *J. Chem. Phys.* 140: 0021. (b) Schmidt, M.W., Ivanic, J., and Ruedenberg, K. The physical origin of the chemical bond. In: *The Chemical Bond. 1. Fundamental Aspects of Chemical Bonding* (ed. G. Frenking and S. Shaik). Weinheim: Wiley-VCH *The Chemical Bond*. 2014, *1*, 1.
16 Levine, D.S. and Head-Gordon, M. (2020). Clarifying the quantum mechanical origin of the covalent chemical bond. *Nat. Commun.* 11: 4893.
17 (a) Levine, D.S., Horn, P.R., Mao, Y., and Head-Gordon, M. (2016). Variational energy decomposition analysis of chemical bonding. 1. Spin-pure analysis of single bonds. *J. Chem. Theory Comput.* 12: 4812. (b) Levine, D.S. and Head-Gordon, M. (1967). Quantifying the role of orbital contraction in chemical bonding. *J. Phys. Chem. Lett.* 2017: 8. (c) Levine, D.S. and Head-Gordon, M. (2017). Proc. Energy decomposition analysis of single bonds within Kohn–Sham density functional theory. *Proc. Natl. Acad. Sci. U.S.A.* 114: 12649.
18 Bacskay, G.B. (2022). Orbital contraction and covalent bonding. *J. Chem. Phys.* 156: 20.
19 Hoffmann, R. (1995). *The Same and Not the Same*. New York, NY: Columbia University Press.
20 Quack, M., Seyfang, G., and Wichmann, G. (2022). Perspectives on parity violation in chiral molecules: theory, spectroscopic experiment and biomolecular homochirality. *Chem. Sci.* 13: 10598.
21 (a) Langmuir, I. (1919). The arrangement of electrons in atoms and molecules. *JACS* 41: 868. (b) Langmuir, I. (1919). Isomorphism, isosterism and covalence. *JACS* 41: 1543. (c) Langmuir, I. (1920). The octet theory of valence and its applications with special reference to organic nitrogen compounds. *JACS* 42: 274.
22 Langmuir, I. (1921). Types of valence. *Science* 54: 59.
23 Coffey, P. (2008). *Cathedrals of Sciences – The Personalities and Rivalries That Made Modern Chemistry*. Oxford: Oxford University Press.

24 Lewis, G.N. (1923). *Valence and the Structure of Atoms and Molecules*. New York: American Chemical Society Monograph Series The book is out of print but an online version is found at: http://babel.hathitrust.org/cgi/pt?id=uc1.$b35072;view=1up;seq=1.

25 Pauling, L. (1931). The nature of the chemical bond. II. The one-electron bond and the three-electron bond. *JACS* 53: 3225.

26 Lennard-Jones, J.E. (1929). The electronic structure of some diatomic molecules. *Trans. Faraday Soc.* 25: 668.

27 Hund, F. (1927). Zur Deutung der Molekelspektren, Part I. *Z. Angew. Phys.* 40: 742.

28 Hund, F. (1927). Zur Deutung der Molekelspektren, Part II. *Z. Angew. Phys.* 42: 93.

29 Hund, F. (1927). Zur Deutung der Molekelspektren, Part III. *Z. Angew. Phys.* 43: 805.

30 Hund, F. (1928). Zur Deutung der Molekelspektren, Part IV. *Z. Angew. Phys.* 51: 759.

31 Hund, F. (1930). Zur Deutung der Molekelspektren, Part V. *Z. Angew. Phys.* 63: 719.

32 Pauling, L. *The Nature of the Chemical Bond*. Ithaca, New York: Cornell University Press. first edition 1939, second edition 1948, third edition 1960.

33 Kitaura, K. and Morokuma, K. (1976). A new energy decomposition scheme for molecular interactions within the Hartree-Fock approximation. *Int. J. Quantum Chem.* 10: 325.

34 Ziegler, T. and Rauk, A. (1977). On the calculation of bonding energies by the Hartree Fock Slater method. *Theor. Chim. Acta* 46: 1.

35 (a) Levine, D.S., Horn, P.R., Mao, Y., and Head-Gordon, M. (2016). Variational energy decomposition analysis of chemical bonding. 1. Spin-pure analysis of single bonds. *J. Chem. Theory Comput.* 12: 4812. (b)Levine, D.S. and Head-Gordon, M. (2017). Inaugural article by a recently elected academy member: energy decomposition analysis of single bonds within Kohn–Sham density functional theory. *Proc. Natl. Acad. Sci. U.S.A.* 114: 12649.

36 (a) Su, P. and Li, H. (2009). Energy decomposition analysis of covalent bonds and intermolecular interactions. *J. Chem. Phys.* 131: 0021. (b) Tang, Z., Song, Y., Zhang, S. et al. (2021). XEDA, a fast and multipurpose energy decomposition analysis program. *J. Comput. Chem.* 42: 2341.

37 (a) Reed, A.E., Curtiss, L.A., and Weinhold, F. (1988). Intermolecular interactions from a natural bond orbital, donor-acceptor viewpoint. *Chem. Rev.* 88: 899. (b) Landis, C.R. and Weinhold, F. (2014). The NBO view of chemical bonding. In: *The Chemical Bond: Fundamental Aspects of Chemical Bonding* (ed. G. Frenking and S. Shaik), 9. Weinheim: Wiley-VCH. (c) Weinhold, F. and Landis, C.R. (2012). *Discovering Chemistry With Natural Bond Orbitals*. New Jersey: Wiley. (d) Landis, C.R. and Weinhold, F. (2005). *Valency and Bonding: A Natural Bond Orbital Donor-Acceptor Perspective*. Cambridge: Cambridge University Press.

38 Bader, R.F.W. (1990). *Atoms in Molecules: A Quantum Theory*. Oxford: Oxford University Press.

39 (a) Pendás, A.M., Francisco, E., and Blanco, M.A. (2006). Binding energies of first row diatomics in the light of the interacting quantum atoms approach. *J. Phys. Chem. A* 110: 12864. (b) Blanco, M.A., Martín Pendás, A., and Francisco, E. (2005). Interacting quantum atoms: a correlated energy decomposition scheme based on the quantum theory of atoms in molecules. *J. Chem. Theory Comput.* 1: 1096. (c) Francisco, E., Pendás, A.M., and Blanco, M.A. (2006). A molecular energy decomposition scheme for atoms in molecules. *J. Chem. Theory Comput.* 2: 90. (d) Menéndez, M., Boto, R.Á., Francicso, E., and Pendás, A.M. (2015). One-electron images in real space: natural adaptive orbitals. *J. Comput. Chem.* 36: 833.

40 Becke, A.D. and Edgecombe, K.E. (1990). A simple measure of electron localization in atomic and molecular systems. *J. Chem. Phys.* 92: 5397.

41 Parr, R.G., Donnelly, R.A., Levy, M., and Palke, W.E. (1978). Electronegativity: the density functional viewpoint. *J. Chem. Phys.* 68: 3801.

42 Kaupp, M., Danovich, D., and Shaik, S. (2017). Chemistry is about energy and its changes: a critique of bond-length/bond-strength correlations. *Coord. Chem. Rev.* 344: 355.

43 Krapp, A., Bickelhaupt, F.M., and Frenking, G. (2006). Orbital overlap and chemical bonding. *Chem. Eur. J.* 12: 9196.

44 (a) Zhao, L., von Hopffgarten, M., Andrada, D.M., and Frenking, G. (2018). Energy decomposition analysis. *WIREs Comput. Mol. Sci.* 8: 1345. (b) Zhao, L., Hermann, M., Schwarz, W.H.E., and Frenking, G. (2019). The Lewis electron-pair bonding model: modern energy decomposition analysis. *Nat. Rev. Chem.* 3: 48. (c) Zhao, L., Pan, S., Holzmann, N. et al. (2019). Chemical bonding and bonding models of main-group compounds. *Chem. Rev.* 119: 8781. (d) Zhao, L., Pan, S., and Frenking, G. (2022). Energy decomposition analysis of the chemical bond: scope and limitation. In: *Reference Module in Chemistry, Molecular Sciences and Chemical Engineering*. Elsevier https://doi.org/10.1016/B978-0-12-821978-2.00021-0.

45 (a) Michalak, A., Mitoraj, M., and Ziegler, T. (2008, 1933). Bond orbitals from chemical valence theory. *J. Phys. Chem. A* 112. (b) Mitoraj, M.P., Michalak, A., and Ziegler, T. (2009). A combined charge and energy decomposition scheme for bond analysis. *J. Chem. Theory Comput.* 5: 962.

46 Fu, M., Pan, S., Zhao, L., and Frenking, G. (2020). Bonding analysis of the shortest bond between two atoms heavier than hydrogen and helium: O_2^{2+}. *J. Phys. Chem. A* 124: 1087.

47 Wong, M.W., Nobes, R.H., Bouma, W.J., and Radom, L. (1989). Isoelectronic analogs of molecular nitrogen: tightly bound multiply charged species. *J. Chem. Phys.* 91: 2971.

48 (a) Parkes, M.A., Lockyear, J.F., and Price, S.D. (2013). Reactions of O_2^{2+} with CO_2, OCS and CS_2. *Int. J. Mass Spectrom.* 39: 354. (b) Sigaud, L., Ferreira, N., and Montenegro, E.C. (2013). Absolute cross sections for O_2 dication production by electron impact. *J. Chem. Phys.* 139: 02.

49 Andrada, D.M., Casalz-Sainz, J.L., Pendas, A.M., and Frenking, G. (2018). Dative and electron-sharing bonding in C_2F_4. *Chem. Eur. J.* 24: 9083.

50 The value is calculated from the BDE into the singlet ground state of CF_2 and the S-T excitation energy which are given in Ref. [49].

51 (a) Shaik, S., Rzepa, H.S., and Hoffmann, R. (2013). One molecule, two atoms, three views, four bonds? *Angew. Chem. Int. Ed.* 52: 3020. (b) Frenking, G. and Hermann, M. (2013). Critical comments on "One molecule, two atoms, three views, four bonds?". *Angew. Chem. Int. Ed.* 52: 5922. (c) Danovich, D., Shaik, S., Rzepa, H.S., and Hoffmann, R. (2013). A response to the critical comments on "One molecule, two atoms, three views, four bonds?". *Angew. Chem. Int. Ed.* 52: 5926. (d) Danovich, D., Hiberty, P.C., Wu, W. et al. (2014). The nature of the fourth bond in the ground state of C_2: the quadruple bond conundrum. *Chem. Eur. J.* 20: 6220. (e) Hermann, M. and Frenking, G. (2016). The chemical bond in C_2. *Chem. Eur. J.* 22: 4100. (f) Shaik, S., Danovich, D., Braida, B., and Hiberty, P.C. (2016). The quadruple bonding in C_2 reproduces the properties of the molecule. *Chem. Eur. J.* 22: 4116. (g) Frenking, G. and Hermann, M. (2016). Comment on "the quadruple bonding in C_2 reproduces the properties of the molecule". *Chem. Eur. J.* 22: 18975. (h) Shaik, S., Danovich, D., Braida, B., and Hiberty, P.C. (2016). A response to a comment by G. Frenking and M. Hermann on:"The quadruple bonding in C_2 reproduces the properties of the molecule". *Chem. Eur. J.* 22: 18977. (i) Piris, M., Lopez, X., and Ugalde, J.M. (2016). The bond order of C_2 from a strictly N-representable natural orbital energy functional perspective. *Chem. Eur. J.* 22: 4109. (j) Cooper, D.L., Ponec, R., and Kohout, M. (2016). New insights from domain-averaged Fermi holes and bond order analysis into the bonding conundrum in C2. *Mol. Phys.* 114: 1270. (k) Zou, W. and Cremer, D. (2016). C_2 in a box: determining its intrinsic bond strength for the X1Σg+ ground state. *Chem. Eur. J.* 22: 4087. (l) de Sousa, D.W.O. and Nascimento, M.A.C. (2016). Is there a quadruple bond in C_2? *J. Chem. Theory Comput.* 12: 2234. (m) Xu, L.T. and Dunning, T.H. Jr., (2014). Insights into the perplexing nature of the bonding in C_2 from generalized valence bond calculations. *J. Chem. Theor. Comput.* 10: 195.

52 (a) Konkoli, Z. and Cremer, D. (1998). A new way of analyzing vibrational spectra. IV. Application and testing of adiabatic modes within the concept of the characterization of normal modes. *Int. J. Quantum Chem.* 67: 1. (b) Konkoli, Z., Larsson, J.A., and Cremer, D. (1998). A new way of analyzing vibrational spectra. IV. Application and testing of adiabatic modes within the concept of the characterization of normal modes. *Int. J. Quantum Chem.* 67: 11. (c) Konkoli, Z. and Cremer, D. (1998). A new way of analyzing vibrational spectra. IV. Application and testing of adiabatic modes within the concept of the characterization of normal modes. *Int. J. Quantum Chem.* 67: 29. (d) Konkoli, Z., Larsson, J.A., and Cremer, D. (1998). A new way of analyzing vibrational spectra. IV. Application and testing of adiabatic modes within the concept of the characterization of normal modes. *Int. J. Quantum Chem.* 67: 41.

53 Wilson, B., Decius, J.C., and Cross, P.C. (1955). *Molecular Vibrations*. New York: McGraw-Hill.

54 Zhao, L., Zhi, M., and Frenking, G. (2022). The strength of a chemical bond. *Int. J. Quantum Chem.* 122: e26773.

55 (a) Cremer, D. and Kraka, E. (2017). Generalization of the Tolman electronic parameter: the metal–ligand electronic parameter and the intrinsic strength of the metal–ligand bond. *Dalton Trans.* 46: 8323. (b) Kalescky, R., Kraka, E., and Cremer, D. (2013). Identification of the strongest bonds in chemistry. *J. Phys. Chem. A* 117: 8981.

56 (a) Wagner, J.P. and Schreiner, P.R. (2015). London dispersion in molecular chemistry—reconsidering steric effects. *Angew. Chem. Int. Ed.* 54: 12274. (b) Goerigk, L. and Grimme, S. (2011). A thorough benchmark of density functional methods for general main group thermochemistry, kinetics, and noncovalent interactions. *Phys. Chem. Chem. Phys.* 13: 6670. (c) Zhao, Y. and Truhlar, D.G. (2006). A density functional that accounts for medium-range correlation energies in organic chemistry. *Org. Lett.* 8: 5753. (d) Zhao, Y. and Truhlar, D.G. (1967). Attractive noncovalent interactions in the mechanism of Grubbs second-generation Ru catalysts for olefin metathesis. *Org. Lett.* 2007: 9. (e) Truhlar, D.G. (2019). Dispersion forces: neither fluctuating nor dispersing. *J. Chem. Ed.* 96: 1671.

57 Lewis, G.N. (1938). Acids and bases. *J. Franklin Inst.* 226: 293.

58 Ref. 24, p. 142.

59 Sidgwick, N.V. (1929). *The Electronic Theory of Valency*. Oxford: *Clarendon Press*.

60 Sidgwick, N.V. and Plant, S.G.P. (1925). XXXIV. Some co-ordinated compounds of the alkali metals. *J. chem. Soc. Trans.* 127: 209.

61 Sidgwick, N.V. (1931). Structure of divalent carbon compounds. *Chem. Rev.* 9: 77.

62 Frenking, G., Loschen, C., Krapp, A. et al. (2007). Electronic structure of CO – an exercise in modern chemical bonding theory. *J. Comput. Chem.* 28: 117.

63 Ref. [32], p. 9 and 10.

64 For pertinent examples see: (a)Celik, M.A., Sure, R., Klein, S. et al. (2012). Borylene complexes (BH)L$_2$ and nitrogen cation complexes (N+) L$_2$: isoelectronic homologues of carbones CL$_2$. *Chem. Eur. J.* 18: 5676. (b) Tonner, R. and Frenking, G. (2008). Divalent carbon (0) chemistry, part 1: parent compounds. *Chem. Eur. J.* 14: 3260. (c) Uddin, J. and Frenking, G. (2001). Energy analysis of metal-ligand bonding in transition metal complexes with terminal group-13 diyl ligands (CO)$_4$Fe-ER, Fe(EMe)$_5$ and Ni(EMe)$_4$ (E = B – Tl; R= Cp, N(SiH$_3$)$_2$, Ph, Me) reveals significant π bonding in homoleptical molecules. *JACS* 123: 1683. (d) Boehme, C., Uddin, J., and Frenking, G. (2000). Chemical bonding in mononuclear transition metal complexes with Group 13 diyl ligands ER (E = B Tl): Part X: theoretical studies of inorganic compounds. *Coord. Chem. Rev.* 197: 249.

65 Dewar, M.J.S. (1951). A review of π complex theory. *Bull. Soc. Chim. Fr.* 18: C79.

66 Chatt, J. and Ducanson, L.A. (1953). Olefin co-ordination compounds. Part III. Infra-red spectra and structure: attempted preparation of acetylene complexes. *J. Chem. Soc.* 2939.

67 (a) Frenking, G. and Fröhlich, N. (2000). The nature of the bonding in transition-metal compounds. *Chem. Rev.* 100: 717. (b) Frenking, G. (2001). Understanding the nature of the bonding in transition metal complexes: from Dewar's molecular orbital model to an energy partitioning analysis of the metal–ligand bond. *J. Organomet. Chem.* 635: 9. (c) Leigh, G.J. and Winterton, N. (ed.) (2002). *The Legacy of Joseph Chatt*, 111. London: The Royal Society.

68 Haaland, A. (1989). Covalent and dative bonds to main group metals-a useful distinction. *Angew. Chem.* 101: 1017; Haaland, A. (1989). Covalent versus dative bonds to main group metals, a useful distinction. *Angew. Chem., Int. Ed.* 28: 992.

69 Ramirez, F., Desai, N.B., Hansen, B., and McKelvie, N. (1961). Hexaphenylcarbodiphosphorane, $(C_6H_5)_3PCP(C_6H_5)_3$. *JACS* 83: 3539.

70 Tonner, R., Öxler, F., Neumüller, B. et al. (2006). Carbodiphosphoranes: the chemistry of divalent carbon (0). *Angew. Chem. Int. Ed.* 45: 8038.

71 (a) Frenking, G., Tonner, R., Klein, S. et al. (2014). New bonding modes of carbon and heavier group 14 atoms Si–Pb. *Chem. Soc. Rev.* 43: 5106. (b) Frenking, G., Hermann, M., Andrada, D.M., and Holzmann, N. (2016). Donor-acceptor bonding in novel low-coordinated compounds of boron and group-14 atoms C-Sn. *Chem. Soc. Rev.* 45: 1129. (c) Zhao, L., Chai, C., Petz, W., and Frenking, G. (2020). Carbones and carbon atom as ligands in transition metal complexes. *Molecules* 25: 4943. (d) Petz, W. and Frenking, G. (2022). Neutral and Charged Group 13 – 16 Homologues of Carbones EL_2 (E = B^- – In^-; Si – Pb; N^+ - Bi^+, O^{2+} - Te^{2+}). *Prog. Inorg. Chem.* 79: 243.

72 Tonner, R. and Frenking, G. (2007). $C(NHC)_2$: divalent carbon (0) compounds with N-heterocyclic carbene ligands—theoretical evidence for a class of molecules with promising chemical properties. *Angew. Chem. Int. Ed.* 46: 8695.

73 (a) Dyker, C.A., Lavallo, V., Donnadieu, B., and Bertrand, G. (2008). Synthesis of an extremely bent acyclic allene (a "carbodicarbene"): a strong donor ligand. *Angew. Chem. Int. Ed.* 47: 3206. (b) Fernandez, I., Dyker, C.A., DeHope, A. et al. (2009). Exocyclic delocalization at the expense of aromaticity in 3, 5-bis (π-donor) substituted pyrazolium ions and corresponding cyclic bent allenes. *JACS* 131: 11875. (c) Fürstner, A., Alcarazo, M., Goddard, R., and Lehmann, C.W. (2008). Coordination chemistry of ene-1, 1-diamines and a prototype "carbodicarbene". *Angew. Chem. Int. Ed.* 47: 3210. (d) Alcarazo, M., Lehmann, C.W., Anoop, A. et al. (2009). Coordination chemistry at carbon. *Nat. Chem.* 1: 295.

74 (a) Pranckevicius, C., Fan, L., and Stephan, D.W. (2015). Cyclic bent allene hydrido-carbonyl complexes of ruthenium: highly active catalysts for hydrogenation of olefins. *JACS* 137: 5582. (b) Hsu, Y.-C., Shen, J.-S., Lin, B.-C. et al. (2015). Synthesis and isolation of an acyclic tridentate bis (pyridine) carbodicarbene and studies on its structural implications and reactivities. *Angew. Chem. Int. Ed.* 54: 2420. (c) Goldfogel, M.J., Roberts, C.C., and Meek, S.J. (2014). Intermolecular hydroamination of 1, 3-dienes catalyzed by bis (phosphine) carbodicarbene–rhodium complexes. *JACS* 136: 6227. (d) Roberts, C.C., Matías, D.M., Goldfogel, M.J., and Meek, S.J. (2015). Lewis acid activation of carbodicarbene catalysts for Rh-catalyzed hydroarylation of dienes. *JACS* 137: 6488. (e) Chen, W.-C., Shen, J.-S., Jurca, T. et al. (2015). Expanding the ligand framework diversity of carbodicarbenes and direct detection of boron activation in the methylation of amines with CO_2. *Angew. Chem. Int. Ed.* 54: 15207. (f) Chen, W.-C., Shih, W.-C., Jurca, T. et al. (2017). Carbodicarbenes: unexpected π-accepting ability during reactivity with small molecules. *JACS* 139: 12830.

75 (a) Koput, J. (2000). An ab initio study on the equilibrium structure and CCC bending energy levels of carbon suboxide. *Chem. Phys. Lett.* 320: 237. (b) Jensen, P. and Johns, J.W.C. (1986). The infrared spectrum of carbon suboxide in the v6 fundamental region: experimental observation and semirigid bender analysis. *J. Mol. Spectrosc.* 118: 248.

76 Tonner, R. and Frenking, G. (2009). Divalent carbon(0) compounds. *Pure Appl. Chem.* 81: 597.

77 Ellern, A., Drews, T., and Seppelt, K. (2001). The structure of carbon suboxide, C_3O_2, in the solid state. *ZAAC* 627: 73.

78 Bernhardi, I., Drews, T., and Seppelt, K. (1999). Isolation and structure of the $OCNCO^+$ ion. *Angew. Chem. Int. Ed.* 38: 2232.

79 Kinjo, R., Donnadieu, B., Celik, M.A. et al. (2011). Synthesis and characterization of a neutral tricoordinate organoboron isoelectronic with amines. *Science* 333: 610.

80 (a) Kong, L., Li, Y., Ganguly, R. et al. (2014). Isolation of a bis(oxazol-2-ylidene)– phenylborylene adduct and its reactivity as a boron-centered nucleophile. *Angew. Chem. Int. Ed.* 126: 9434. (b) Braunschweig, H., Dewhurst, R.D., Hupp, F. et al. (2015). Multiple complexation of CO and related ligands to a main-group element. *Nature* 522: 327.

81 Mellerup, S.K., Cui, Y., Fantuzzi, F. et al. (2019). Lewis-base stabilization of the parent Al (I) hydride under ambient conditions. *JACS* 141: 16954.

82 (a) Takagi, N., Shimizu, T., and Frenking, G. (2009). Divalent silicon (0) compounds. *Chem. Eur. J.* 15: 3448. (b) Takagi, N., Shimizu, T., and Frenking, G. (2009). Divalent E(0) compounds (E = Si–Sn). *Chem. Eur. J.* 15: 8593.

83 (a) Xiong, Y., Yao, S., Inoue, S. et al. (2013). A cyclic silylone ("siladicarbene") with an electron-rich silicon (0) atom. *Angew. Chem. Int. Ed.* 52: 7147. (b) Xiong, Y., Yao, S., Tan, G. et al. (2013). A cyclic germadicarbene ("germylone") from germyliumylidene. *JACS* 135: 5004.

84 (a) Kuwabara, T., Nakada, M., Hamada, J. et al. (2016). (η^4-Butadiene) Sn (0) complexes: a new approach for zero-valent p-block elements utilizing a butadiene as a 4π-electron donor. *JACS* 138: 11378. (b) Xu, J., Dai, C., Yao, S. et al. (2022). A genuine stannylone with a monoatomic two-coordinate tin (0) atom supported by a bis(silylene) ligand. *Angew. Chem. Int. Ed.* 61: e202114073.

85 (a) Xu, J., Pan, S., Yao, S. et al. (2022). The heaviest bottleable metallylone: synthesis of a monatomic, zero-valent lead complex ("plumbylone"). *Angew. Chem.* 134: e202209442. *Angew. Chem. Int. Ed.* 2022, 61, e202209442;. (b) Chen, M., Zhang, Z., Qiao, Z. et al. (2023). An isolable bis(germylene)-stabilized plumbylone. *Angew. Chem. Int. Ed.* 62: e202215146.

86 (a) Patel, N., Sood, R., and Bharatam, P.V. (2018). NL_2^+ systems as new-generation phase-transfer catalysts. *Chem. Rev.* 118: 8770. (b) Patel, N., Arfeen, M., Singh, T. et al. (2020). Divalent N(I) compounds: identifying new carbocyclic carbenes to design nitreones using quantum chemical methods. *J. Comput. Chem.* 61: 2624.

87 (a) Nesterov, V., Reiter, D., Bag, P. et al. (2018). NHCs in main group chemistry. *Chem. Rev.* 118: 9678. (b) Légaré, M.-A., Pranckevicius, C., and Braunschweig, H. (2019). Metallomimetic chemistry of boron. *Chem. Rev.* 119: 8231. (c) Wang, Y. and Robinson, G.H. (2023). Counterintuitive chemistry: carbene stabilization of zero-oxidation state main group species. *JACS* 145: 5592. (d) Wang, Y. and Robinson, G.H. (2009). Unique homonuclear multiple bonding in main group compounds. *Chem. Commun.* 5201. (e) Yao, S., Xiong, Y., and Driess, M. (2026). A new area in main-group chemistry: zerovalent monoatomic silicon compounds and their analogues. *Acc. Chem. Res.* 2017: 50. (f) Hadlington, T.J., Driess, M., and Jones, C. (2018). Low-valent group 14 element hydride chemistry: towards catalysis. *Chem. Soc. Rev.* 47: 4176. (g) Power, P.P. (2020). An update on multiple bonding between heavier main group elements: the importance of Pauli repulsion, charge-shift character, and London dispersion force effects. *Organometallics* 39: 4127.

88 (a) Himmel, D., Krossing, I., and Schnepf, A. (2014). Dative bonds in main-group compounds: a case for fewer arrows! *Angew. Chem. Int. Ed.* 53: 370. (b) Frenking, G. (2014). Dative bonds in main-group compounds: a case for more arrows! *Angew. Chem. Int. Ed.* 53: 6040. (c) Himmel, D., Krossing, I., and Schnepf, A. (2014). Dative or not dative? *Angew. Chem. Int. Ed.* 53: 6047. (d) Schmidbaur, H. (2007). Réplique: a new concept for bonding in carbodiphosphoranes? *Angew. Chem. Int. Ed.* 46: 2984. (e) Frenking, G., Neumüller, B., Petz, W. et al. (2007). Reply to réplique: a new concept for bonding in carbodiphosphoranes? *Angew. Chem. Int. Ed.* 46: 2986.

89 Liu, G., Fedik, N., Martinez-Martinez, C. et al. (2019). Realization of Lewis basic sodium anion in the $NaBH_3^-$ cluster. *Angew. Chem. Int. Ed.* 58: 13789.

90 (a) Pan, S. and Frenking, G. (2020). Comment on "Realization of lewis basic sodium anion in the $NaBH_3^-$ cluster". *Angew. Chem. Int. Ed.* 59: 8756. (b) Liu, G., Fedik, N., Martinez-Martinez, C. et al. (2020). Reply to the comment on "Realization of Lewis basic sodium anion in the $NaBH_3^-$ cluster". *Angew. Chem. Int. Ed.* 59: 8760.

91 Zhao, L., Pan, S., Wang, G., and Frenking, G. (2022). The nature of the polar covalent bond. *J. Chem. Phys.* 157: 034105.

92 Shaik, S.S. and Hiberty, P.C. (2007). *A Chemist's Guide to Valence Bond Theory*. Wiley.

93 Pan, S. and Frenking, G. (2021). A critical look at Linus Pauling's influence on the understanding of chemical bonding. *Molecules* 26: 4695.

94 (a) Shaik, S., Maitre, P., Sini, G., and Hiberty, P.C. (1992). The charge-shift bonding concept. Electron-pair bonds with very large ionic-covalent resonance energies. *JACS* 114: 7861; (b) Shaik, S., Danovich, D., Wu, W., and Hiberty, P.C. (2009). *Nat. Chem.* 1: 443.

95 Frenking, G. (2017). Covalent bonding and charge shift bonds: Comment on "The Carbon–Nitrogen Bonds in Ammonium Compounds Are Charge Shift Bonds". *Chem. Eur. J.* 23: 18320.

96 Wiberg, K. (1968). Application of the Pople-Santry-Segal CNDO method to the cyclopropylcarbinyl and cyclobutyl cation and to bicyclobutane. *Tetrahedron* 24: 1083.

97 (a) Mayer, I. (1984). Bond order and valence: relations to Mulliken's population analysis. *Int. J. Quantum Chem.* 26: 151. (b) Mayer, I. (1983). Charge, bond order and valence in the AB initio SCF theory. *Chem. Phys. Lett.* 97: 270.

98 There remains a weaker in-plane $\pi_\parallel$ donation $H_2B \leftarrow F$ into the $\pi_\parallel^*$ orbital of the BH2 fragment which is not relevant for the topic.

99 (a) Muenter, J.S. (1975). Electric dipole moment of carbon monoxide. *J. Mol. Spectrosc.* 55: 490. (b) Zhang, K.Q., Guo, B., Braun, V. et al. (1995). Infrared emission spectroscopy of BF and AlF. *J. Mol. Spectrosc.* 170: 82. (c) Fantuzzi, F., Cardozo, T.M., and Nascimento, M.A.C. (2015). Nature of the chemical bond and origin of the inverted dipole moment in boron fluoride: a generalized valence bond approach. *J. Phys. Chem. A.* 119: 5335. (d) Al Shawa, S., El-Kork, N., Younes, G., and Korek, M. (2016). Theoretical study with dipole moment calculation of new electronic states of the molecule BF. *Nanotechnol. Rev.* 5: 363.

100 Bittner, D.M. and Bernath, P.F. (2018). Line lists for LiF and LiCl in the $X1\Sigma+$ ground state. *Astrophys. J. Suppl.* 235: 8.

101 Huber, K.P. (1979). Constants of diatomic molecules. *Mol. Spectra Mol. Struct.* 4: 146. The bonding radii of homoatomic species E_2 and X_2 are taken as one half of the experimental values of the bond lengths in the electronic ground states. They are taken from: Huber, K.P. and Herzberg, G. (1979). *Constants of Diatomic Molecules*, Van Nostrand Reinhold, New York.

102 Cremer, D. and Kraka, E. (1984). Chemical bonds without bonding electron density—does the difference electron-density analysis suffice for a description of the chemical bond? *Angew. Chem., Int. Ed.* 23: 627.

103 S. Pan, G. Frenking, to be published.

104 (a) Fonseca Guerra, C., Handgraaf, J.-W., Baerends, E.J., and Bickelhaupt, F.M. (2004). Voronoi deformation density (VDD) charges: assessment of the Mulliken, Bader, Hirshfeld, Weinhold, and VDD methods for charge analysis. *J. Comput. Chem.* 25: 189. (b) Nieuwland, C., Vermeeren, P., Bickelhaupt, F.M., and Fonseca Guerra, C. (2023). Understanding chemistry with the symmetry-decomposed Voronoi deformation density charge analysis. *J. Comput. Chem.* 44: 2108.

105 Hirshfeld, F.L. (1977). Bonded-atom fragments for describing molecular charge densities. *Theor. Chim. Acta* 44: 129.

106 There have been early studies where different values for the electronegativity of the valence orbitals in an atom were proposed, but they are hardly ever used and it seems that this approach is forgotten: (a)Hinze, J. (1999). The concept of electronegativity of atoms in molecules. In: *Theoretical and Computational*

Chemistry, vol. 6 (ed. Z.B. Maksic and W.J. Orville-Thomas), 189. (b) Bergmann, D. and Hinze, J. (1996). Electronegativity and molecular properties. *Angew. Chem. Int. Ed.* 35: 150.

107 Liu, R., Qin, L., Zhang, Z. et al. (2023). Genuine quadruple bonds between two main-group atoms. Chemical bonding in AeF$^-$ (Ae = Be – Ba) and isoelectronic EF (E = B – Tl) and the particular role of d orbitals in covalent interactions of heavier alkaline-earth atoms. *Chem. Sci.* 14: 4872.

108 The NBO method gives wrongly sp^3 hybridization of all bonding orbitals in EH$_4$ (E = C – Pb) because it neglects the off-diagonal matrix elements of the overlap matrix.

109 Pyykkö, P. (1979). Interpretation of secondary periodicity in the periodic system. *J. Chem. Res., Synop.* 380.

110 Kutzelnigg, W. (1984). Chemical bonding in higher main group elements. *Angew. Chem. Int. Ed.* 23: 272.

111 Lein, M., Krapp, A., and Frenking, G. (2005). Why do the heavy-atom analogues of acetylene E$_2$H$_2$ (E = Si–Pb) exhibit unusual structures? *JACS* 127: 6290.

112 Lein, M. and Frenking, G. (2004). Chemical bonding in octahedral XeF$_6$ and SF$_6$. *Aust. J. Chem.* 57: 1191.

113 (a) Bogey, M., Bolvin, H., Demuyneck, C., and Destombes, J.-L. (1991). Nonclassical double-bridged structure in silicon-containing molecules: experimental evidence in Si$_2$H$_2$ from its submillimeter-wave spectrum. *Phys. Rev. Lett.* 66: 413. (b) Cordonnier, M., Bogey, M., Demuynck, C., and Destombes, J.-L. (1992). Nonclassical structures in silicon-containing molecules: the monobridged isomer of Si$_2$H$_2$. *J. Chem. Phys.* 97: 7984. (c) Wang, X., Andrews, L., and Kushto, G. (2002). Infrared spectra of the novel Ge$_2$H$_2$ and Ge$_2$H$_4$ species and the reactive GeH$_{1,2,3}$ intermediates in solid neon, deuterium and argon. *J. Phys. Chem. A* 106: 5809. (d) Wang, X., Andrews, L., Chertihin, G.V., and Souer, P.F. (2002). Infrared spectra of the novel Sn$_2$H$_2$ species and the reactive SnH$_{1,2,3}$ and PbH$_{1,2,3}$ intermediates in solid neon, deuterium, and argon. *J. Phys. Chem. A* 106: 6302. (e) Wang, X. and Andrews, L. (2003). Infrared spectra of group 14 hydrides in solid hydrogen: experimental observation of PbH$_4$, Pb$_2$H$_2$ and Pb$_2$H$_4$. *JACS* 125: 6581.

114 (a) Sekiguchi, A., Kinjo, R., and Ichinohe, M. (2004). A stable compound containing a silicon-silicon triple bond. *Science* 305: 1755. (b) Stender, M., Phillips, A.D., Wright, R.J., and Power, P.P. (2002). Synthesis and characterization of a digermanium analogue of an alkyne. *Angew. Chem. Int. Ed.* 41: 1785–1787. (c) Phillips, A.D., Wright, R.J., Olmstead, M.M., and Power, P.P. (2002). Synthesis and characterization of 2,6-dipp$_2$-H$_3$C$_6$SnSnC$_6$H$_3$-2,6-dipp$_2$ (dipp = C$_6$H$_3$-2,6-Pr$^i_2$): a tin analogue of an alkyne. *JACS* 124: 5930.

115 Pu, L., Twamley, B., and Power, P.P. (2000). Synthesis and characterization of 2,6-trip$_2$H$_3$C$_6$PbPbC$_6$H$_3$-2,6-trip$_2$ (trip = C$_6$H$_2$-2,4,6-i-Pr$_3$): a stable heavier group 14 element analogue of an alkyne. *JACS* 122: 3524.

116 Chen, Y., Hartmann, M., Diedenhofen, M., and Frenking, G. (2001). Turning a transition state into a minimum – the nature of the bonding in diplumbylene compounds RPbPbR (R=H, Ar). *Angew. Chem. Int. Ed.* 40: 2051.

117 (a) Trinquier, G. and Malrieu, J.P. (1987). Nonclassical distortions at multiple bonds. *JACS* 109: 5303. (b) Malrieu, J.P. and Trinquier, G. (1989). Trans-bending at double bonds. Occurrence and extent. *JACS* 111: 5916.

118 Carter, E.A. and Goddard, W.A. (1986). Relation between singlet-triplet gaps and bond energies. *J. Phys. Chem.* 90: 998.

119 The Coulombic attraction between neutral sulfur atom and F_6 in SF_6 using a spherically symmetric charge distribution at S has been estimated to provide 22% of the total attraction, see Ref. [112].

120 For an early discussion of the bonding in pentacoordinated phosphorous compounds see:Hoffmann, R., Howell, J.M., and Muetterties, E.L. (1972). Molecular orbital theory of pentacoordinate phosphorus. *JACS* 94: 3047.

121 An excellent presentation of orbital correlation diagrams and molecular structures for a variety of main-group compounds and transition metal complexes is given by:Albright, T.A., Burdett, J.K., and Whangbo, M.-H. (2013). *Orbital Interactions in Chemistry*, 2e. New York: Wiley.

122 Vyboishchikov, S.F. and Frenking, G. (1998). Structure and bonding of low-valent (Fischer type) and high-valent (Schrock type) transition metal carbene complexes. *Chem. Eur. J.* 4: 1428.

123 (a) Pidun, U. and Frenking, G. (1996). The bonding of acetylene and ethylene in high-valent and low-valent transition metal compounds. *J. Organomet. Chem.* 525: 269. (b) Jerabek, P., Schwerdtfeger, P., and Frenking, G. (2019). Dative and electron-sharing bonding in transition metal compounds. *J. Comput. Chem.* 40: 247.

124 (a) Frenking, G. and Pidun, U. (1997). *Ab Initio* studies of transition-metal compounds: the nature of the chemical bond to a transition metal. *J. Chem. Soc., Dalton Trans.* 1653. (b) Boehme, C., Uddin, J., and Frenking, G. (2000). Chemical bonding in mononuclear transition metal complexes with group 13 diyl ligands ER (E = B – Tl). *Coord. Chem. Rev.* 197: 249.

125 Luo, Q., Li, Q.S., Yu, Z.H. et al. (2008). Bonding of seven carbonyl groups to a single metal atom: theoretical study of M(CO)n (M = Ti, Zr, Hf; n = 7, 6, 5, 4). *JACS* 130: 7756–7765.

126 Wang, Q., Pan, S., Wu, Y. et al. (2020). Filling a gap: the coordinatively saturated group 4 carbonyl complexes $TM(CO)_8$ (TM = Zr, Hf) and $Ti(CO)_7$. *Chem. Eur. J.* 26: 10487. The octacarbonyl $Ti(CO)_8$ is a minimum on the potential energy surface, but it is thermodynamically unstable upon loss of one CO due to steric repulsion.

127 Wu, X., Zhao, L., Jin, J. et al. (2018). Observation of alkaline earth complexes $M(CO)_8$ (M = Ca, Sr, Ba) that mimic transition metals. *Science* 361: 912.

128 Wang, Q., Pan, S., Lei, S.J. et al. (2019). Octa-coordinated alkaline earth metal–dinitrogen complexes $M(N_2)_8$ (M = Ca, Sr, Ba). *Nat. Commun.* 10: 3375.

129 Wang, Q., Pan, S., Wu, Y.B. et al. (2019). Transition-metal chemistry of alkaline-earth elements: the trisbenzene complexes M(Bz)$_3$ (M = Sr, Ba). *Angew. Chem. Int. Ed.* 58: 17365.

130 Fernández, I., Holzmann, N., and Frenking, G. (2020). The valence orbitals of the alkaline-earth atoms. *Chem. Eur. J.* 26: 14194.

131 Frenking, G. and Zhou, M. (2021). The transition metal chemistry of the heavier alkaline earth atoms Ca, Sr, Ba. *Acc. Chem. Res.* 54: 3071.

6

Partial Charges

Tian Lu and Qinxue Chen

Beijing Kein Research Center for Natural Sciences, Beijing 100024, People's Republic of China

6.1 Concept of Partial Charge

6.1.1 What is Partial Charge?

In a chemical system, due to the formation of ionic bonds or polar covalent bonds, the action of an external electric field, electron ionization and attachment, and so on, atoms can have a non-integer net charge, which is known as partial atomic charge and can also simply be referred to as partial charge or atomic charge. Partial charge is generally represented by a point charge located at nuclear position, which is one of the simplest and most intuitive ways to describe charge distribution in chemical systems.

Partial charge is never a physically observable quantity, and it is therefore impossible to have a unique definition. Partial charges may be determined indirectly through some experimental data [1], such as molecular multipole moment, infrared spectrum intensity and frequency, ligand field splitting energy, NMR shift, and so on. However, the data acquisition in this way is relatively inconvenient, the correspondence between the data and partial charges is often highly empirical, and unstable systems or transient electronic states cannot be studied. The development of computational chemistry has made it possible to obtain partial charges conveniently, quickly, and reasonably [2–7]. Since the very famous Mulliken charge [8–10] was proposed in 1955, at least 50 methods of evaluating partial charges have been proposed so far. Many researchers are still trying to improve the calculation methods in recent years. These methods have different characteristics and emphases, and researchers need to choose the most appropriate one according to their actual needs.

Partial charge uses *e* as the unit, which is the abbreviation of elementary charge and corresponds to the amount of charge carried by a single proton. Elementary charge is the charge unit in the atomic unit (a.u.) system; therefore, a unit of partial charge can also be written as a.u. In the literature, the unit of partial charge is often omitted since this will not cause any ambiguity.

Notice that partial charge is quite different from the oxidation state (OS); the latter is always an integer and corresponds to the hypothetically charged state if all

Exploring Chemical Concepts Through Theory and Computation, First Edition. Edited by Shubin Liu.
© 2024 WILEY-VCH GmbH. Published 2024 by WILEY-VCH GmbH.

heteronuclear bonds formed by the atom are assumed to be fully ionic. Usually, the magnitude of OS is by far larger than the partial charge. For example, the partial charge of oxygen in water molecules calculated by most methods is in the range of −0.3 to −1.0, while its OS is −2. The partial charge is also in sharp contrast to formal charge, which is defined based on Lewis structure and calculated as N (valence electrons in a free atom) − N (nonbonding electrons) − N (electrons in related bonds)/2, where N indicates the number of electrons of the corresponding kind. Like OS, formal charge is also an integer, but their values are often different. In the case of water molecules, both hydrogen and oxygen have a formal charge of zero. It is obvious that the partial charges calculated in a meaningful way could reflect actual charge distribution in chemical systems significantly better than OSs and formal charges. The main practical uses of OSs and formal charges are just bookkeeping and classification.

6.1.2 Theoretical Significances and Practical Applications of Partial Charge

The theoretical significance of calculating and investigating partial charges is quite evident. First, partial charges can help chemists easily understand the charged state of atoms and can be conveniently and quantitatively compared between different geometries, electronic states, and external environments. Moreover, there are wave-function analysis methods utilizing partial charges. For example, the condensed Fukui function [11] and condensed dual descriptor [12], which are quite popular in predicting preferential reactive sites, can be evaluated based on partial charges corresponding to different charged states of a molecule. Note that the partial charge itself is also important in identifying reaction sites; namely, the more negative (positive) the partial charge is, the more likely the atom is a favorable electrophilic (nucleophilic) reaction site [13–15]. In addition, the partial charges of the atoms in a fragment can be summed up to form a fragment charge, which is also quite valuable. For instance, fragment charge of a functional group can be used to compare and discuss its electron-withdrawing or electron-donating capabilities, and fragment charge can be used to evaluate the percentage of charge transfer character (CT%) of electron excitations.

Partial charge also has numerous practical applications in the field of computational chemistry, some of which are as follows:

1) Partial charge is an important atomic descriptor that can be used to predict or explain many properties related to atoms. For example, the partial charge of the site containing a dissociable proton is closely related to the corresponding pK_a [16, 17]. Partial charges show a strong correlation with NMR chemical shifts and core electron binding energies [18–20]. C−O bond dissociation energy (BDE) of alkoxy roaming reactions is found to be correlated with the partial charge of carbon in the reactant molecules [21].
2) Partial charge at the site where a chemical reaction occurs is closely related to the reaction barrier. It is frequently found that reaction barriers of analogous systems can be successfully explained and accurately predicted based

on the partial charges. For example, when Liu et al. studied reactions of monosubstituted-benzene molecules reacting with hydrogen fluoride, they found that there is a nice linear correlation between the Hirshfeld charge on the regioselective carbon atoms and reaction barriers. Furthermore, they found that the Hirshfeld charge of the reacting atom in more than twenty organic molecules has an excellent linear relationship with the experimental nucleophilicity and electrophilicity scales of Mayr [22]. In addition, a prominent correlation between barrier heights of keto–enol tautomer reactions and Mulliken charges of the keto/aldehyde carbons was detected by Heufer and coworkers [21].

3) Partial charge plays a crucial role in the molecular force field. Most molecular force fields have a fairly simple potential function, which is generally based on partial charges to rapidly calculate electrostatic interactions between atoms. Therefore, molecular dynamics (MD) and Monte Carlo simulations based on molecular force fields are inseparable from partial charges, and the choice of calculation method for partial charges directly affects the quality of the simulation result. In the field of molecular docking, partial charges are also extensively adopted in the design of forcefield-based scoring functions [23].

4) Partial charge is employed in a wide variety of computational chemistry methods. For example, In GFN-xTB [24] and SCC-DFTB [25] theories, partial charges are used to very quickly estimate electronic energy; partial charge is involved in DFT-D4 dispersion correction to estimate atomic polarizabilities and C_6 coefficients [26]; many implicit solvation models such as SM12 [27], uESE [28], generalized Born [29] and Poisson–Boltzmann [30] calculate electrostatic part of solvation energy by means of partial charges; electrostatic potential (ESP) of huge systems such as protein can be generated by partial charges with minimal cost; comparative molecular field analysis (CoMFA) often employs partial charges to rapidly yield ESP on evenly distributed grids for large amount of ligands [31]; molecular surface can be divided into polar and nonpolar parts according to magnitude of partial charge of exposed atoms; energy decomposition analysis based on forcefield (EDA-FF) employs partial charges to estimate electrostatic interaction contribution to overall interaction energy [32].

5) In quantum mechanics/molecular mechanics (QM/MM) [33] and embedded cluster [34] calculations, it is the standard way of employing partial charges to represent the electrostatic field due to MM atoms or environmental atoms on the subsystem explicitly described by quantum chemistry method.

6) Partial charges are very valuable in studying weak interactions. According to the signs of partial charges, it can be easily judged which atoms have electrostatic attraction and electrostatic mutual repulsion effects. Furthermore, according to the magnitude of partial charges, the strength of electrostatic-dominated interactions, such as hydrogen bonding [35], can be estimated.

6.1.3 Limitations of Partial Charge

It is important to emphasize that there are evident limitations in describing the charge distribution of chemical systems in terms of partial charges. As the point charge at the position of the nucleus, partial charge essentially represents atoms having a spherically symmetrical charge distribution. Therefore, the anisotropy character of charge distribution within atomic space cannot be captured by partial charge at all. Consideration of anisotropy can sometimes be important in the investigation of non-covalent interaction. For example, a halogen bond is formed by a covalently bonded halogen atom via its σ-hole region as Lewis acid and an electronegative atom mainly through electrostatic interaction. If the charge distribution of the halogen atom is simply represented as a partial charge, then the σ-hole will not exist and the halogen bond cannot be discussed at all [36]. Another example is provided by Lu et al., who pointed out that the electrostatic interaction between two N_2 or two H_2 [35], and between two cyclo [18] carbon molecules [37], has a significant control effect on the dimer configuration. If only partial charges are employed to describe the molecular charge distributions, it is obviously impossible to reveal the role of electrostatic interaction on the configuration and binding strength of the dimer because all atoms in these molecules have exactly zero partial charge due to molecular symmetry.

It is noteworthy that if charge distribution of an atom is represented in terms of electric multipole expansion, then partial charge corresponds to monopole moment, and the description of charge distribution anisotropy needs to simultaneously consider atomic dipole, quadrupole, or even higher order of moments [38, 39]. Alternatively, additional point charges away from the atom center may be introduced to describe the anisotropy effect. For example, it was found that simply adding a point charge in the direction toward σ-hole from the nucleus for a halogen atom is adequate to describe its halogen bond [40]. The various limitations of partial charges in describing charge distribution were extensively discussed by Kramer et al. [41]

6.1.4 What Is a Good Method of Calculating Partial Charges?

Due to the experimental unobservability of partial charge, there is no absolute right or wrong way of calculating it. But there are three requirements that must be met. At least the deviation should be very small; otherwise, the method will be physically meaningless. The requirements are: (i) Partial charges should have good rotation invariance. In the absence of an external field, if the calculated partial charges differ evidently before and after rotating the system, then the method must be unreliable because the orientation of the system is essentially arbitrary. (ii) Sum of partial charges should be equal to the overall net charge of the system. (iii) Distribution of partial charges should be consistent with the symmetry of the studied structure.

In addition, an ideal partial charge calculation method should also meet the following requirements:

1) For calculation methods based on the quantities derived by quantum chemistry, the partial charges should converge to constant values smoothly when the basis set is gradually approaching completeness limit. A method well satisfying this condition is known to have a good basis set stability or low basis set sensitivity.
2) Calculation method should have a concise and clear physical meaning, and introducing excessive empirical parameters should not be avoided.

3) Partial charges calculated for typical chemical systems should conform to conventional chemical concepts, such as the rule of electronegativity. However, notice that the electronegativity rule is not valid for some special systems. As an example, dipole moment of carbon monoxide is almost zero, implying the two atoms should have comparable net charges, though the electronegativity of oxygen is by far greater than that of carbon.
4) Calculated partial charges should be able to well reproduce the observable properties related to charge distribution, such as electric dipole and multipole moments, ESP.
5) Applicability of calculation method should be as broad as possible. Ideally, the method should be applicable to atoms involving any form of chemical bonds, any element, any molecular shape (linear, planar, cluster, etc.), equilibrium and largely distorted structures, both isolated and periodic systems, both ground and excited states, and wavefunctions represented by both all-electron and pseudopotential basis sets, with and without external field perturbation.
6) Partial charges in functional groups should have transferability to some extent due to the independent nature of functional groups, and partial charges should not vary greatly because of a minor change in the chemical environment, such as a slight change in molecular conformation during MD simulation.
7) Algorithm should be easy to implement, less computationally intensive, have low memory requirement, and have good numerical stability.

6.1.5 Classification of Partial Charge Calculation Methods

The partial charge calculation methods that have been proposed so far can be divided into the following categories according to their main ideas:

1) Partial Charges Based on Wavefunction: In this category, the number of electrons carried by each atom (atomic population) is calculated directly based on electronic wavefunction or density matrix derived from quantum chemistry calculations. Then, the partial charge of an atom is simply the difference between its nuclear charge and atomic population. Representative methods in this category include Mulliken [8–10], MMPA [42–44], Löwdin [45], and NPA [46].
2) Partial Charges Based on Real Space Partition of Electron Density: The commonality of this kind of method is that the entire three-dimensional space is divided into subspaces corresponding to different atoms, and electron density is integrated in each subspace to obtain atomic population. Different ways of partitioning atomic spaces correspond to different methods of calculating partial charges. Representative methods in this category include atom-in-molecules (AIM) [47, 48], Hirshfeld [49], Hirshfeld-I (HI) [50], Voronoi deformation density (VDD) [51], minimal basis iterative stockholder (MBIS) [52], density-derived electrostatic and chemical (DDEC) [53], and so on.
3) Partial Charges Based on Fitting Electrostatic Potential (ESP): These methods determine partial charges by making them maximally reproduce the ESP calculated by quantum chemistry methods in the region close to molecular van der

Waals (vdW) surface. Representative realizations of this idea include MK [54], CHELPG [55], RESP [56], RESP2 [57], REPEAT [58], and AM1-BCC [59, 60].
4) Partial Charges Based on Equalization of Electronegativity: The methods in this category generate partial charges mainly based on the principle of Sanderson's equalization of electronegativity. Representative methods include EEM [61, 62], QEq [63], PEOE (Gasteiger) [18, 64], PEPE [65], MPEOE [66].
5) Partial Charges Based on Other Ideas: There are many other ways to determine partial charges using different ideas from the above; well-known ones include ADCH [38], CM5 [67], MMFF94 [68], and GAPT [69].

In the next section, we will review the known partial charge calculation methods listed above. Then, in Section 6.3, some molecules will be taken to compare the results of some popular calculation methods. Section 6.4 will briefly mention computer programs that can calculate partial charges. The final section will conclude this chapter and provide our suggestions for the choice of partial charges.

6.2 Methods of Calculating Partial Charges

6.2.1 Partial Charges Based on Wavefunction

6.2.1.1 Mulliken Method

The Mulliken method [8–10] is the oldest method of population analysis and deriving partial charges. Its algorithm is quite simple and the calculation cost is negligible compared to single-point energy calculation; therefore, almost all quantum chemistry programs support calculating Mulliken charges and even print them by default.

The idea of Mulliken method is fairly easy to understand. First, consider the normalization condition of molecular orbitals or natural orbitals as follows (orbitals are assumed to be real for simplicity, similarly hereinafter).

$$\int [\varphi_i(\mathbf{r})]^2 d\mathbf{r} = 1 \tag{6.1}$$

where φ_i denotes the wavefunction of orbital i and $\mathbf{r}$ is the coordinate vector in three-dimensional space. Orbital can be expressed as a linear combination of basis functions, χ

$$\varphi_i = \sum_\mu C_{\mu,i} \chi_\mu \tag{6.2}$$

where $C_{\mu,i}$ is the coefficient of the basis function μ in orbital i. By substituting Eq. (6.2) into Eq. (6.1), we have

$$\sum_\mu C_{\mu,i}^2 + \sum_\mu \sum_{\nu \neq \mu} C_{\mu,i} C_{\nu,i} S_{\mu,\nu} = 1 \tag{6.3}$$

in which the first and second terms on the left-hand side are known as local term and cross term, respectively. $S_{\mu,\nu} = \int \chi_\mu(\mathbf{r}) \chi_\nu(\mathbf{r}) d\mathbf{r}$ is an element of overlap matrix. The Mulliken method defines the composition of basis function μ in orbital i as

$$\Theta_{\mu,i} = C_{\mu,i}^2 + \sum_{\nu \neq \mu} C_{\mu,i} C_{\nu,i} S_{\mu,\nu} \tag{6.4}$$

That is, the local term is fully assigned to the corresponding basis function, while each cross term is divided equally between the corresponding two basis functions. Then, the population of an atom can be straightforwardly obtained as

$$p_A = \sum_i \eta_i \sum_{\mu \in A} \Theta_{\mu,i} \tag{6.5}$$

where η stands for orbital occupation number. Finally, the Mulliken charge of an atom is simply obtained as $q_A = Z_A - p_A$, with Z_A being nuclear charge. It is noteworthy that the atomic Mulliken population can also be equivalently evaluated based on single-particle reduced density matrix $\mathbf{P}$:

$$p_A = \sum_{\mu \in A} (\mathbf{PS})_{\mu,\mu} \tag{6.6}$$

There are some known serious problems of the Mulliken population and its partial charge: (i) The bisection of the cross terms is somewhat arbitrary. The difference between atoms is not taken into account, which is the main reason why Mulliken charges sometimes underestimate the ionicity of bonds. (ii) Composition of a basis function in an orbital, that is the Θ term defined by Eq. (6.4), may be negative or larger than 1, which obviously lacks physical meaning and shows inherent shortcoming of the theory of the Mulliken population analysis. (iii) Basis set sensitivity is extremely large. The Mulliken charge does not smoothly converge with increase size of basis set, and the result is often quite unreasonable when diffuse functions are employed. For example, with B3LYP [70]/aug-cc-pVTZ [71] wavefunction of ethanol, the Mulliken charge of the carbon in methyl group (−0.67) is even much more negative than that of the oxygen (−0.46), which obviously violates the fact that oxygen has much larger electronegativity than carbon. The charges of the carbon and oxygen at the B3LYP/cc-pVTZ level, namely −0.28 and −0.34, respectively, look much more reasonable. Due to the above defects, the use of Mulliken charge is generally deprecated.

The reason why the Mulliken charge is incompatible with diffuse functions is easy to understand. Assuming that two atoms A and B are bonded and a basis set containing diffuse functions is adopted, since the diffuse functions of atom A conspicuously cover the atomic space of atom B, it will cause a certain number of electrons that should belong to atom B to be wrongly assigned to atom A, this clearly undermines the reasonableness of Mulliken charges. At the same time, it should also be noted that when calculating Mulliken charges, using a larger basis set never necessarily leads to a better result. For example, Mulliken charges calculated with a high-quality 4-zeta basis set such as def2-QZVP [72] may even be worse than those with a 2-zeta basis set such as 6-31G* [73], because some basis functions of def2-QZVP exhibit a semidiffuse feature[74].

6.2.1.2 MMPA Methods
In view of the unreasonableness of Mulliken population in dividing cross terms, some researchers have proposed different ways to improve it; they are collectively known as modified Mulliken population analysis (MMPA). For example, Ros and Schuit defined the composition of a basis function in an orbital as

$$\Theta_{\mu,i} = C_{\mu,i}^2 / \sum_v C_{v,i}^2 \tag{6.7}$$

This method is commonly referred to as C-squared population analysis (SCPA). Ostensibly, SCPA avoids explicit partition of the cross terms; however, we have proved that it is equivalent to dividing each cross term according to the ratio of square of basis function coefficient to the sum of squares of all coefficients [75]. An advantage of SCPA compared to Mulliken method is that Θ is always in the range of [0,1], and in the meantime, overlap matrix is not explicitly needed. Stout and Politzer proposed to divide each cross term according to the ratio of the squares of coefficients of the corresponding two basis functions [44], but the partial charge calculated by this method does not have unitary transformation invariance of basis functions and degenerate MOs [76]. Bickelhaupt et al. proposed to divide each cross term according to the ratio of the sums of the local terms of the corresponding two basis functions in all occupied MOs [43].

Although the MMPA methods yield more reasonable results than the Mulliken method sometimes, the serious basis set sensitivity problem was not resolved, which keeps them from becoming popular.

6.2.1.3 Löwdin Method

If Löwdin orthogonalization for basis functions is performed prior to Mulliken population analysis, the resulting partial charges are known as Löwdin charges [45]. Löwdin charges also suffer from excessive basis set dependency problems like Mulliken charges, and the Löwdin orthogonalization does not have a clear physical meaning. There is no obvious advantage to Löwdin charges over Mulliken charges.

Redistributed Löwdin population analysis (RLPA) aims to alleviate sensitivity of Löwdin charges to inclusion of diffuse basis functions. It redistributes the population of the diffuse functions to different atoms, so that the result is approximately consistent with that obtained when the diffuse functions are not added. RLPA never became popular.

6.2.1.4 NPA Method

The natural population analysis (NPA) proposed by Weinhold and coworkers [46] is a key ingredient of the famous natural bond orbital (NBO) theory framework [77]. The partial charge derived by NPA is known as NPA charge or natural charge and is sometimes incorrectly referred to as NBO charge in literature. The key idea of NPA is to elegantly transform the wavefunction described by the original basis set, which only has mathematical meaning, to that mainly described by a set of orthogonal minimal bases, which has strong physical meaning. This treatment greatly suppressed the aforementioned basis set sensitivity issue of the Mulliken method; diffuse functions can be safely used, and in the meantime, the difficulty of dividing cross terms is implicitly bypassed. The practical result of NPA is usually more reasonable than the Mulliken method, especially for systems containing ionic bonds. Due to these advantages, NPA has become one of the most popular methods of deriving partial charges nowadays. Note that NPA charges need to be used with caution for systems containing transition metal, lanthanide, and actinide atoms [78–80].

Although the partial charge defined based on quantum chemistry introduced in the next sections must also have a certain degree of basis set dependence, after all basis set quality directly determines the quality of electronic structure, their basis set sensitivity is much smaller than that of Mulliken method, generally similar to or lower than NPA [7]. This comes from the fact that these partial charges are calculated based on physically observable properties, such as electron density and ESP, which converge smoothly as the basis set gradually approaches the complete limit.

6.2.2 Partial Charges Based on Real Space Partition of Electron Density

6.2.2.1 AIM Method

The AIM theory framework proposed by Bader defines a way of calculating partial charge, which is known as AIM charge and occasionally referred to as the Bader charge. The AIM theory defines the zero-flux surface of electron density as the interface between atoms. This means that every point $\mathbf{r}'$ in the surface satisfies the condition $\nabla \rho(\mathbf{r}') \cdot \mathbf{n}(\mathbf{r}') = 0$, where ρ is the electron density and $\mathbf{n}$ is the unit normal vector of the surface. The independent space of each atom divided by the interfaces is called atomic basin, which corresponds to the atomic space defined by AIM theory. The AIM charge is calculated as the difference between nuclear charge and integral of ρ in the AIM atomic basin (Ω):

$$q_A = Z_A - \int_{\Omega_A} \rho(\mathbf{r}) d\mathbf{r} \tag{6.8}$$

The definition of AIM atomic basins has a clear mathematical meaning and also has a certain physical meaning; that is, each atomic basin satisfies the Virial theorem. However, there is no evident chemical meaning of the basins, and in rare cases, the basins cannot be obtained. For example, there are so-called pseudoatoms between lithium atoms in the lithium crystal [81]. Due to the existence of basins corresponding to the pseudoatoms, the atomic basins of lithium atoms no longer fill the whole space, and certainly the AIM charge of the atoms cannot be meaningfully evaluated. In addition, an atom does not necessarily have a corresponding basin. In the KrH$^+$ system, since the widely distributed electrons of the krypton heavily submerge the hydrogen, there is no basin individually corresponding to each atom, and thus AIM charges cannot be calculated. Furthermore, as will be illustrated in Section 6.3, the AIM charge often has poor chemical meaning, and its magnitude is usually significantly larger than any other partial charges, and sometimes it violates the concept of electronegativity. Because the interatomic boundaries defined by the AIM method are irregular, the algorithm for integrating atomic basins is complicated. This is another disadvantage of the AIM method.

Due to the aforementioned shortcomings of the AIM charge, it has very limited practical significance and is rarely used in molecular systems. However, for periodic systems, AIM charge is widely used in literature. We believe the main reason is that most first-principle programs are based on plane waves, and many methods such as Mulliken cannot be used directly in this case. On the other hand, very few partial charge calculation methods are supported in mainstream first-principles programs, while these programs can usually generate grid data files of electron density, by which AIM charges can be quickly and easily obtained using additional codes like Bader and Multiwfn via grid-based algorithms.

6.2.2.2 Voronoi and VDD Methods

The Voronoi method sets up a vertical plane at the midpoint between each pair of adjacent atoms. These vertical planes divide the entire space into Voronoi polyhedra, and each one corresponds to an atom. Voronoi polyhedra also correspond to Wigner–Seitz unit cells in periodic systems. Rousseau et al. integrate ρ in the Voronoi polyhedra to obtain atomic populations and corresponding partial charges [82]. Note that they adjusted the position of the intersection point between the vertical plane and the line linking adjacent atoms, so that the distances between the intersection point and the two atoms are proportional to the vdW radii of the two atoms. Another closely related method is Voronoi deformation density (VDD) [51], which directly integrates deformation density in Voronoi polyhedra to obtain partial charges. These partial charge calculation methods did not receive much attention because of their ambiguous physical meaning and no better performance than the popular Hirshfeld method described in the next section.

6.2.2.3 Hirshfeld Method

The AIM and Voronoi methods described above partition the space discretely; there is another class of partition methods known as fuzzy partition. Their commonality is that an atomic weighting function is defined to describe the space belonging to each atom; the function varies smoothly everywhere with a value range of [0,1], and the sum of the values of all atomic weighting functions at any position is exactly 1.0. Obviously, the atomic space defined in this way does not have a sharp boundary.

The Hirshfeld method is the most representative and oldest fuzzy partition method; its atomic weighting function is defined as

$$w_A^{\text{Hirsh}}(\mathbf{r}) = \frac{\rho_A^{\text{free}}(\mathbf{r})}{\sum_B \rho_B^{\text{free}}(\mathbf{r})} \tag{6.9}$$

where ρ_A^{free} denotes the spherically averaged electron density of atom A in its free state. The denominator in Eq. (6.9) corresponds to the promolecular density, which corresponds to the ρ of the system prior to the presence of electron polarization and transfer due to formation of interatomic interactions. The Hirshfeld partition has a clear idea, and its meaning was also interpreted from perspective of information theory [83]. Hirshfeld charge is defined as

$$q_A^{\text{Hirsh}} = Z_A - \int w_A^{\text{Hirsh}}(\mathbf{r})\rho(\mathbf{r})d\mathbf{r} \tag{6.10}$$

The presence of atomic weighting function in the integrand constrained the integration performed in the corresponding atomic space.

Hirshfeld charge is easy to implement and robust; the result is generally meaningful and fairly insensitive to the choice of basis set, so it is widely used in literature. In particular, it is worth mentioning that Hirshfeld charge is more suitable for predicting reaction sites than any other atomic charges [13–15]. The main disadvantage of Hirshfeld charge is that its overall magnitude is evidently too small, and ESP and dipole moment are poorly reproduced [7].

It is worth mentioning that the Hirshfeld charges given by different programs are different to a certain extent, mainly because the free-state atomic densities employed by them to calculate the Hirshfeld weighting functions are somewhat different.

6.2.2.4 Hirshfeld-I Method

The choice of the free state of the atoms used in calculating Hirshfeld weighting functions notably influences the resulting Hirshfeld charges; however, the choice is somewhat arbitrary. For example, in calculation of Hirshfeld charges for NaCl, the free states can be chosen as the neutral Na and Cl, or the ionic Na$^+$ and Cl$^-$; the corresponding results are significantly different. The HI method introduces an iterative process to the Hirshfeld method to eliminate the dependency on the initial choice of the free states [50]. In HI method, the weighting function of atom A at iteration n is defined as

$$w_A^{(n)}(\mathbf{r}) = \frac{\rho_A^{\text{free},(n-1)}(\mathbf{r})}{\sum_B \rho_B^{\text{free},(n-1)}(\mathbf{r})} \tag{6.11}$$

The free-state atomic density of the current iteration is obtained by linear interpolation between adjacent charged states according to the partial charge at last iteration:

$$\rho_A^{\text{free},(n)}(\mathbf{r}) = \left(q_{\text{up}} - q_A^{(n-1)}\right) \rho_{A,\text{low}}^{\text{free}}(\mathbf{r}) + \left(q_A^{(n-1)} - q_{\text{low}}\right) \rho_{A,\text{up}}^{\text{free}}(\mathbf{r}) \tag{6.12}$$

where q_{up} and q_{low} are upper and lower integers of partial charge of atom A at iteration $n-1$, while $\rho_{A,\text{up}}^{\text{free}}$ and $\rho_{A,\text{low}}^{\text{free}}$ are spherically averaged free-state density of atom A at the corresponding two charged states, respectively. The iteration is performed until convergence of partial charges is sufficiently reached.

HI is more physically sound than the Hirshfeld method, as it considers the response of atomic spaces to actual chemical environment. The value of the HI charges is significantly larger than that of the Hirshfeld charge and more in line with common chemical intuition. HI charges have conspicuously better capability of reproducing ESP than Hirshfeld charges [84]. A key disadvantage of HI is that it generally takes 20–30 iterations to obtain converged result; hence, the calculation is significantly more time-consuming than Hirshfeld for large systems. Another critical problem with HI is that nonexistent anion, such as O^{2-}, may be involved during calculation. Although the density of the anion state can be produced normally by a calculation using a finite atomic center basis set, it is essentially unphysical and shows very strong diffuse character. As a result, the atomic weighting function of the corresponding atom will excessively extend to other atomic spaces, possibly leading to an unreasonable partial charge.

There are some variants of HI, such as Hirshfeld-E [85], Hirshfeld-Iλ [86], and the fractional occupation Hirshfeld-I (FOHI) [87]. Besides, there are also other partial charge calculation methods related to the idea of iterative refinement of atomic spaces, such as iterated stockholder atoms (ISA) [88], minimal basis iterative stockholder (MBIS) [52], and DDEC [53]. The DDEC method is constantly evolving and the latest version is DDEC6 [89]. Among them, MBIS and DDEC have received increasing attention in recent years due to their satisfactory performance in reproducing ESP. Limited to the length of this chapter, these methods will not be introduced in detail. Interested readers are referred to the dedicated review of these methods by Ayer et al. [90] A known problem with MBIS is that it does not perform well for molecular anions [90].

6.2.3 Partial Charges Based on Fitting Electrostatic Potential

6.2.3.1 Common ESP Fitting Methods

In chemical systems, ESP is defined as follows: it measures the interaction energy of present system and a unit point charge placed at point **r** without consideration of charge polarization effect

$$V(\mathbf{r}) = \sum_A \frac{Z_A}{|\mathbf{r} - \mathbf{R}_A|} - \int \frac{\rho(\mathbf{r}')}{|\mathbf{r} - \mathbf{r}'|} d\mathbf{r}' \tag{6.13}$$

where **R** is the nuclear position. It can be seen that ESP comes from both nuclear and electronic contributions. In the region close to nuclei, ESP is always positive because nuclear charges play a dominant role. However, the ESP in the region outside vdW surface can be either positive or negative depending on electronic structure of nearby regions, and thus has unique chemical significance [91].

The ESP produced by partial charges is simply expressed as

$$V^q(\mathbf{r}) = \sum_A \frac{q_A}{|\mathbf{r} - \mathbf{R}_A|} \tag{6.14}$$

It is the consensus that the ESP derived from partial charges should be able to well reproduce the exact ESP (usually derived quantum chemically according to Eq. (6.13) outside vdW surface). This condition is particularly important if the partial charges will be used for molecular force field to represent electrostatic interaction.

ESP fitting refers to a class of methods of deriving partial charges by minimizing the difference between V^q and V in the regions of chemical interest. Most ESP fitting methods realize this purpose by least square fitting to minimize the following error function with a constraint to keep the sum of all partial charges equal to the net charge of the system.

$$F(q_1, q_2 \cdots q_N) = \sum_i [V^q(\mathbf{r}_i) - V(\mathbf{r}_i)]^2 \tag{6.15}$$

where the summation loops over all fitting points and N is the number of atoms. A unique advantage of deriving partial charges by ESP fitting is that additional constraints may be imposed to make net charge of a fragment correspond to an expected value and make equivalent atoms share exactly identical charges. These constraints can be employed for special purposes, for example, making the net charge of a repeating unit in a polymer an integer so that the partial charges of the unit have transferability. However, these constraints must reduce the reproducibility of ESP to some extent.

Cox–Williams [92], Merz–Kollman (MK) [54], charges from electrostatic potentials (CHELP) [93], and CHELP using a grid-based method (CHELPG) [55] are four practical realizations of the philosophy of ESP fitting. They only differ by the distribution of fitting points and their results are generally comparable. CHELPG is the most popular among them. In this method, the fitting points are uniformly distributed in a rectangular box surrounding the whole molecule. The nearest distance between the box and any atom is 2.8 Å. The fitting points within vdW surface and those farther than 2.8 Å away from vdW surface are removed. The smaller the spacing between the fitting points, the more accurate the result and the better the rotational invariance, but the more expensive the calculation. Usually, a spacing of 0.3 Å is adequate to produce satisfactory result.

The partial charges obtained in this way certainly have better ESP reproducibility than any other kind of method. Since ESP and electric multipole moments are both reflections of charge distribution, the ESP fitting charges can usually reproduce molecular dipole moment and quadrupole moment very well. Some fitting methods deliberately take accurate reproduction of dipole moment as an additional constraint in the fitting [94], but in practice it is never necessary. Note that, as mentioned in Section 6.1.3, partial charge is just a very simple model of representing charge distribution; one should not expect that ESP fitting charges can reproduce ESP nicely anywhere outside vdW surface. For example, the featured ESP distribution due to presence of lone pair, σ-hole, and π electrons cannot be well represented by V^q [41].

For periodic systems, Eq. (6.15) cannot be used for ESP fitting because only relative ESP value between different positions is meaningful, while absolute ESP value is ill-defined due to the arbitrariness of the reference. Repeating Electrostatic Potential Extracted ATomic (REPEAT) focuses on deriving partial charges from ESP for surface and porous systems [58]. Its key difference to other ESP fitting methods is that it essentially fits variation rather than value of ESP, and meantime periodicity is also properly taken into account. The fitting points of REPEAT are distributed on vacuum region over surface systems or within cavities of bulk systems.

6.2.3.2 RESP and Relevant Methods

In principle, ESP fitting charges are most suitable for MD simulation based on molecular force field because of their superior reproducibility of ESP. However, there are three obvious problems with common ESP fitting charges, such as CHELPG, that hinder their application in force fields, especially for the flexible molecules whose conformation frequently changes during MD process: (i) ESP fitting charges have large dependence on molecular conformation. (ii) The charges fitted for a single geometry are often not in line with the equivalence of chemically equivalent atoms (such as the three hydrogens of methyl group). (iii) Charges of heavily buried atoms show evident numerical instability in the ESP fitting procedure since they are far from the fitting points. See ref. [95] for a detailed discussion about these aspects.

Restrained electrostatic potential (RESP), proposed by Kollman and coworkers, aims to circumvent the above three problems [56]. A hyperbolic penalty function is introduced with adjustable parameters for each non-hydrogen atom to suppress the magnitude of its fitted charges. In addition, equivalent constraints are imposed to guarantee that chemically equivalent hydrogens have identical fitted charges. With appropriately selected parameters and an elaborately designed two-stage fitting process of RESP, the aforementioned three problems are largely resolved. Here we do not introduce the implementation details of the RESP method; we refer interested readers to the original paper and the detailed introduction in Section 3.9.16 of Multiwfn program manual. RESP has become one of the most popular ways of evaluating partial charges in the field of forcefield-based MD simulation, and the very famous AMBER [96] and GAFF [97] forcefields employ RESP as the standard way of deriving partial charges.

Evaluation of RESP charges for macromolecules or huge number of small molecules is computationally demanding. AM1-BCC is a method to approximately produce RESP charges of HF/6-31G* level with significantly lower cost. In this method, Mulliken charges are first obtained based on the very cheap AM1 semi-empirical method, and then empirical correction is applied according to bonding relationship between atoms [59]. TPACM4 is another inexpensive approximation to RESP [98].

RESP2 charge [57] is expressed as $q_A^{RESP2} = (1-\delta)q_A^{gas} + \delta q_A^{water}$, where q^{gas} and q^{water} are RESP charges evaluated under vacuum and the water environment represented by implicit solvation model, respectively. It is found that the RESP2 charge with δ of 0.5–0.6 based on DFT density is ideal for MD simulation in water and more recommended than RESP.

6.2.4 Partial Charges Based on Equalization of Electronegativity

The methods in this section are all based on the electronegativity equilibrium principle proposed by Sanderson [99], which argues that the greater the electronegativity of an atom, the stronger its ability to receive electrons; when atoms form bonds, the electrons of the atoms with lower electronegativity will flow to the atoms with greater electronegativity, and in this process, electronegativity of the former and the latter will increase and decrease, respectively. When electronegativities of all atoms become equal, the charge distribution corresponds to actual equilibrium state.

Electronegativity equalization method (EEM) is a very rapid way of determining partial charges based on the above idea, which is independent of wavefunction and only requires geometry information and empirical parameters to generate the charges. In this method, electronegativity of an atom is defined as

$$\chi_A = \left(\chi_A^0 + \Delta\chi_A\right) + 2\left(\eta_A^0 + \Delta\eta_A\right)q_A + \sum_{B \neq A} \frac{q_B}{|\mathbf{R}_B - \mathbf{R}_A|} \tag{6.16}$$

where χ^0 and η^0 are Sanderson electronegativity and Parr–Pearson hardness of the corresponding element, respectively, while $\Delta\chi$ and $\Delta\eta$ are fitted parameters. EEM charges are easily determined by simultaneously solving linear equations corresponding to the following conditions:

$$\chi_1 = \chi_2 = \cdots = \chi_N$$
$$\sum_A q_A = Q \tag{6.17}$$

where Q is the net charge of the whole system and N is the number of atoms. The result of EEM is largely determined by the parameters used. Different researchers have fitted different EEM parameters [100, 101]. For example, if the parameters proposed for reproducing MK charges of B3LYP/6-31G* level are used [102], then the resulting EEM charges will also be close to those charges. Although EEM provides a very fast, cheap, and convenient method to calculate partial charges, its scope of application is heavily limited, namely that the applicable systems must be highly analogous to the training set of the parameters, and it can hardly be used for ionized states, excited states, transition states, and so on.

The popular charge equilibration method (QEq) [63] can be viewed as a variant of EEM; they only differ by the definition of atomic electronegativity. There is no very evident advantage of QEq over EEM. QTPIE modified the form of QEq, so that the partial charges can exhibit correct asymptotic behavior during dissociation process of chemical bond [103]. E-QEq [104] and I-QEq [105] extended QEq specifically to derive partial charges of metal–organic frameworks and perform significantly better than QEq for this type of system.

Partial equalization of orbital electronegativity (PEOE) charge is also known as Gasteiger and Gasteiger–Marsili charges [18, 64]. PEOE defines atomic electronegativity in a different way to EEM, and interatomic connectivity rather than 3D structure is needed to derive PEOE charges. PEOE charges are calculated via an iterative process. In every iteration, a certain amount of electron is transferred between each pair of bonded atoms. In contrast to the EEM, PEOE iteration does not finally meet but partially meets the electronegativity equalization condition due to the damping factor in the equation for evaluating interatomic electron transfer. Because the computational cost of PEOE is negligible even for fairly large systems and merely 2D structure information is needed as input, PEOE has been supported by many molecular design, molecular docking, and cheminformatics softwares. Originally, PEOE could only be used on organic molecules without π conjugation; the partial equalization of π-electronegativity (PEPE) method eliminates this limitation [65]. PEOE has very poor capability in reproducing molecular dipole moment and ESP [7]. To improve PEOE in this aspect, PEOE was modified by various researchers, which can be collectively referred to as modified PEOE (MPEOE) [66, 106, 107].

In general, EEM, PEOE, and similar methods are highly empirical, with limited applicability and reliability. They are only suitable for use when partial charges based on quantum chemistry cannot be readily calculated due to excessive computational cost and complexity.

6.2.5 Partial Charges Based on Other Ideas

It was found that the main reason why Hirshfeld charges have a too small magnitude and have poor reproducibility of ESP is that atomic dipole moments are completely neglected. In the atomic dipole moment-corrected Hirshfeld (ADCH) charges [38], atomic dipole moment of each atom is expanded to correction charges placed at neighboring atoms, and the ADCH charge is just the sum of Hirshfeld charge and the correction charge. ADCH charge is found to be reasonable in chemical sense; its magnitude is notably larger than the Hirshfeld charge and in better agreement with chemical intuition. Molecular dipole moment produced by ADCH charges is proved to be exactly identical to that calculated quantum chemically, and the ESP reproducibility of ADCH charges is significantly improved over that of Hirshfeld charges [7]. ADCH does not bring detectable computational overhead over the Hirshfeld method, and its basis set stability is as good as Hirshfeld. These advantages have led to the increasing use of ADCH charges in the literature.

CM5 charge [67] is somewhat akin to the ADCH charge; both of them were defined as a post-correction to the Hirshfeld charge. Unlike ADCH, which is free of empirically fitted parameters, calculation of correction charge of CM5 involves global parameters as well as parameters for individual elements, and it is dependent on interatomic distances. The parameters were optimized for the best reproduction of highly accurate experimental or theoretical molecular dipole moment. It is noteworthy that the CM5 charge calculated under vacuum enhanced by a factor of 1.2, namely 1.20*CM5 charge, is well-suited to be used with OPLS-AA forcefield to perform MD simulation in water phase [108]. In our own viewpoint, the correction charge of ADCH is somewhat more elegant than that of CM5, as no element-dependent fitted parameters are involved, and it is yielded fully based on real electronic structure.

MMFF94 is a popular molecular forcefield for organic systems; it also defines an easy way of generating partial charges [68]. The initial charge is determined directly by atom type, then post-corrected to yield final charge to take into account polarity of bonds formed by the atom. The involved empirical parameters were derived from fitting molecular dipole moments and interaction energies of HF/6-31G* level. Like PEOE, MMFF94 charges only depend on 2D structure, and the time consumption is extremely low. Although the reproducibility of ESP is obviously not as good as that of ESP fitting charges, at least it is much better than PEOE [7].

Generalized atomic polar tensor (GAPT) is defined as the average of the diagonal elements of the atomic polarization tensor [69]. GAPT is computationally demanding; its cost is equivalent to performing a harmonic frequency analysis, while it does not show obvious advantage in terms of reasonableness and reproducibility of ESP, so few literatures employ GAPT charges nowadays.

In addition to the methods described above, there are also many other partial charge calculation methods. But since they have never been popular and some methods are only suitable for certain special types of systems and applications, they are not covered in this chapter.

6.3 Partial Charges of Typical Molecules

In this section, some typical molecules are taken to illustrate the results of some partial charge calculation methods introduced earlier, so that readers will have an intuitive understanding of the basic characters of different methods. The data are collectively presented in Table 6.1. All partial charges except for NPA were calculated by the Multiwfn 3.8 program [109] developed by us based on the high-quality B3LYP/def2-TZVP wavefunction generated by Gaussian 16 [110]. The geometries were optimized at the same level, and there is no imaginary frequency. NPA was calculated using the same condition via NBO 7.0.7 code [111]. Note that a comprehensive comparison between various methods in different aspects, including reproducibility of dipole moment and ESP, basis set dependency, relationship with electronegativity, and so on, has been made by us [7]. There are also comparison and correlation analysis articles about partial charges by other researchers [52, 112–114].

First, from the statistical data in the last row of Table 6.1, it can be seen that there is a general relationship between the magnitude of partial charges: AIM > NPA ≥ HI ≥ MBIS > CHELPG > Mulliken ≈ ADCH > CM5 > Hirshfeld. In most cases, AIM and Hirshfeld methods severely overestimate and underestimate partial charges, respectively.

Although there is no strict reference to measure the reasonability of partial charges, it can still be seen from the data in Table 6.1 that some partial charges are clearly unreasonable for specific molecules. The charge of N atom in HCN calculated by the Mulliken method is nearly zero, which is unlikely to be reasonable compared to the charges calculated by other methods. The Mulliken charge of Mg atom in MgO is only about 0.7, which largely underestimates the ionicity of the Mg–O bond. The Hirshfeld charge is not only small as a whole but also ridiculously small for certain systems. For example, the Hirshfeld charge of O atom in water is merely −0.3, that of S atom in SO_4^{2-} is only 0.27, and that of C atom in CLi_4 is only −0.81, whose magnitudes are even lower than half of the corresponding charges calculated by other methods. The post-correction on Hirshfeld charges introduced by ADCH method significantly improves the representation of charge distribution for these systems. ADCH shows chemically meaningful result for all atoms in Table 6.1, and at the same time, the values of ADCH charges are within a reasonable range, neither generally high nor generally low. HI also greatly improves the Hirshfeld method; however, from Table 6.1 it can be found that the overall magnitude of HI charges is somewhat too large, and ionicity is overestimated to a certain extent. The partial charges of MBIS are close to those of HI, but ionicity of individual systems is exaggerated even more, contrary to common sense. For example, the MBIS charge of Li atom in CLi_4 even reaches 1.07, which is obviously false as each Li atom cannot transfer more than one electron to the C atom. CM5 charges in Table 6.1 show some obvious problems. For example, the partial charge of N atom in CH_3NO_2 calculated by CM5 is only 0.07, and that of N atom in NO_3^{-} is merely 0.09. For these atoms, the post-correction introduced by CM5 not only fails to improve the issue of the Hirshfeld charges being generally too small, but it further worsens the problem. In contrast, the ADCH method performs well in these cases, implying that the post-correction scheme of ADCH is often more reliable. Compared with other partial charges, it can be found that NPA charge is somewhat too large for atoms in many systems, such as SO_2 and ClF_3. The dipole moment of carbon monoxide is almost zero. From the results of ADCH method, which can fully accurately reproduce molecular dipole moment, it can be recognized that the partial charges of C and O atoms in this system should also be very close to 0. However, the partial charge of C atom calculated by NPA method is as high as 0.49, which clearly cannot be considered as an acceptable result. AIM charge has

6 Partial Charges

Table 6.1 Partial charges of typical molecules calculated by some popular methods. Clearly unreasonable values are highlighted.

Molecule	Atom	Mulliken	Hirshfeld	ADCH	HI	MBIS	CM5	CHELPG	NPA	AIM
H_2O	H	0.316	**0.153**	0.368	0.457	0.442	0.319	0.376	0.460	0.562
	O	−0.632	**−0.307**	−0.735	−0.915	−0.885	−0.639	−0.751	−0.919	−1.125
HCCH	C	−0.190	−0.093	−0.247	−0.210	−0.260	−0.146	−0.230	−0.232	−0.168
	H	0.190	0.092	0.247	0.210	0.260	0.146	0.230	0.232	0.168
HCN	H	0.189	0.127	0.283	0.207	0.227	0.187	0.183	0.227	0.215
	C	−0.180	0.055	0.003	0.083	0.091	0.127	0.185	0.074	0.947
	N	**−0.009**	−0.182	−0.287	−0.290	−0.318	−0.314	−0.367	−0.301	**−1.161**
CH_3NO_2	C	−0.244	−0.020	−0.219	−0.516	−0.431	−0.108	−0.322	−0.478	**0.233**
	H	0.152	0.059	0.136	0.175	0.181	0.125	0.133	0.227	0.080
	H	0.150	0.065	0.137	0.178	0.181	0.129	0.140	0.232	0.077
	N	0.379	0.252	0.401	0.831	0.709	**0.067**	0.792	0.512	0.483
	O	−0.295	−0.208	−0.295	−0.421	−0.411	−0.169	−0.438	−0.360	−0.476
CO	C	−0.010	0.084	−0.020	0.165	0.135	0.130	0.008	**0.487**	**1.201**
SO_2	S	0.656	0.454	0.500	1.071	0.893	0.542	0.561	**1.586**	2.417
	O	−0.328	−0.227	−0.250	−0.535	−0.446	−0.271	−0.281	−0.793	−1.208
ClF_3	Cl	0.720	0.482	0.551	0.915	0.691	0.528	0.604	**1.250**	1.301
	F(axial)[b]	−0.304	−0.214	−0.240	−0.375	−0.291	−0.228	−0.270	−0.470	−0.474
	F(equat.)[b]	−0.111	−0.054	−0.070	−0.165	−0.110	−0.072	−0.065	−0.311	−0.354
$FeCl_3$	Fe	0.871	0.496	0.363	1.549	/[a]	0.667	0.933	1.357	1.500
	Cl	−0.290	−0.165	−0.121	−0.516	/[a]	−0.222	−0.311	−0.452	−0.500
$Ni(CO)_4$	Ni	0.248	−0.122	−0.244	−0.166	/[a]	0.114	0.047	0.279	0.546
	C	−0.054	0.125	0.124	0.245	/[a]	0.116	0.121	0.384	1.036
	O	−0.008	−0.094	−0.063	−0.203	/[a]	−0.144	−0.135	−0.454	−1.173
CLi_4	C	−1.853	**−0.806**	−1.765	−3.219	**−4.275**	**−1.095**	−1.848	−3.317	−3.275
	Li	0.463	**0.202**	0.441	0.805	**1.069**	**0.274**	0.458	0.829	0.819
NaCl	Na	0.648	**0.579**	0.757	0.898	**0.946**	0.694	0.753	0.910	0.883
MgO	Mg	**0.697**	**0.573**	0.879	1.029	0.933	0.744	0.885	1.067	1.242
SO_4^{2-}	S	0.875	**0.268**	0.691	2.565	1.824	0.417	1.604	2.551	**3.862**
	O	−0.719	**−0.567**	−0.673	−1.141	−0.956	−0.604	−0.901	−1.138	−1.466
NO_3^-	N	0.425	0.220	0.492	1.109	1.038	**0.088**	1.049	0.679	0.899
	O	−0.475	−0.407	−0.497	−0.703	−0.679	**−0.363**	−0.683	−0.560	−0.629
Sum of |value|[c]		11.2	6.8	11.2	19.2	18.7	8.5	14.1	20.2	25.7

a) Cannot be calculated because current implementation of MBIS in Multiwfn does not support elements heavier than Ar.
b) ClF_3 has two axial and one equatorial fluorine atoms.
c) Sum of absolute values of all listed partial charges except for $FeCl_3$ and $Ni(CO)_4$.

overall largest magnitude, and charges of many atoms are severely overestimated, as highlighted by the bolded values in Table 6.1. Furthermore, sometimes AIM charge fully lacks chemical meaning. For example, in CH_3NO_2 molecule, the AIM charge of the C atom is larger than that of H by 0.15, which is fully contrary to the principle of electronegativity. Such a misleading result may lead researchers to qualitatively misjudge actual charge distribution of a chemical system.

6.4 Computer Codes for Evaluating Partial Charges

In this section, we provide an incomplete list of computer codes for evaluating some popular partial charges to facilitate readers to utilize them in practical studies.

- Mulliken and Löwdin: Multiwfn [109], almost all mainstream quantum chemistry programs including Gaussian [110] and ORCA [115], CP2K [116]
- MMPA: Multiwfn
- NPA: NBO 3.1 module in Gaussian, NBO [111], JANPA [117]
- AIM: Multiwfn, AIMALL [118], AIM2000 [119], Bader [120]
- Hirshfeld: Multiwfn, Gaussian, ORCA, CP2K, VASP [121], Horton [122]
- Hirshfeld-I: Multiwfn, CP2K, VASP, Horton
- ADCH: Multiwfn
- CM5: Multiwfn, Gaussian
- MBIS: Multiwfn, Horton
- CHELPG: Multiwfn, Gaussian, ORCA
- RESP: Multiwfn, AmberTools [123], CP2K (note that atomic equivalences, charge constraints, and restraints can be set in CP2K, but standard two-stage fitting of RESP is not supported).
- REPEAT: CP2K
- MMFF94: OpenBabel, Avogadro
- EEM: Multiwfn, EEM SOLVER [124], VCharge [125], NEEMP [101]
- Gasteiger: Multiwfn, AmberTools, OpenBabel [126]
- QEq: Gaussian, OpenBabel
- AM1-BCC: AmberTools
- GAPT: Gaussian

In the above list, Gaussian and ORCA are two of the most popular quantum chemistry programs, which mainly calculate isolated systems. CP2K and VASP are two of the most popular first-principle programs, which mostly aim at the calculation of periodic systems. Other softwares in the list are designed for different purposes. Multiwfn is a versatile wavefunction analysis code; evaluation of partial charges is one of the functions it is very good at. From the list, it can be seen that Multiwfn supports very rich methods, most of which not only support molecules and clusters but are also applicable to periodic systems. Multiwfn is freely accessible at http://sobereva.com/multiwfn.

6.5 Concluding Remarks

In this chapter, we introduced concept, significance, calculation methods, and related computer programs of partial charge. We hope this chapter can help readers better understand and apply partial charge in studying practical problems. In the end, we provide some suggestions for the selection of partial charge calculation methods:

1) For studying charge distribution characteristics of molecular systems, ADCH is a good choice; HI, MBIS, and NPA can also be considered simultaneously. Although there are some known issues in Mulliken method, it is still useful for rough discussions and comparisons as it is the cheapest and most widely supported method by calculation programs; however, the use of diffuse functions must be avoided in this case.
2) In the case of employing partial charges in MD simulations based on classical force fields, for rigid molecules, CHELPG is generally satisfactory; for flexible molecules, RESP and RESP2 methods are more suitable; for solid surfaces and porous systems, REPEAT is our most recommended method; and for dense solids, MBIS and DDEC6 may be preferential choices. To quickly generate partial charges in a very large organic system, EEM based on suitable parameters and MMFF94 are very useful.
3) For predicting reaction sites and discussing reactivity, the Hirshfeld charge is robust and highly recommended [13–15].

References

1 Meister, J. and Schwarz, W.H.E. (1994). Principal components of ionicity. *J. Phys. Chem.* 98: 8245–8252.
2 Bachrach, S.M. (1994). *Reviews in Computational Chemistry*, vol. 5 (ed. K.B. Lipkowitz and D.B. Boyd), 171–227. New York: VCH Publishers.
3 Cioslowski, J. (1998). Electronic wavefunction analysis. In: *Encyclopedia of Computational Chemistry 2* (ed. P.V.R. Schleyer), 892–905. West Sussex: Wiley.
4 Jensen, F. (2007). *Introduction to Computational Chemistry*. West Sussex: Wiley.
5 Cramer, C.J. (2004). *Essentials of Computational Chemistry*. West Sussex: Wiley.
6 Young, D.C. (2001). *Computational Chemistry*. New York: Wiley.
7 Lu, T. and Chen, F. (2012). Comparison of computational methods for atomic charges. *Acta Phys. Chim. Sin.* 28: 1–18.
8 Mulliken, R.S. (1955). Electronic population analysis on LCAO-MO molecular wave functions. I. *J. Chem. Phys.* 23: 1833–1840.
9 Mulliken, R.S. (1955). Electronic population analysis on LCAO-MO molecular wave functions. II. Overlap populations, bond orders, and covalent bond energies. *J. Chem. Phys.* 23: 1841–1846.
10 Mulliken, R.S. (1955). Electronic population analysis on LCAO-MO molecular wave functions. III. Effects of hybridization on overlap and gross AO populations. *J. Chem. Phys.* 23: 2338–2342.

11 Yang, W. and Mortier, W.J. (1986). The use of global and local molecular parameters for the analysis of the gas-phase basicity of amines. *JACS* 108: 5708–5711.

12 Morell, C., Grand, A., and Toro-Labbé, A. (2004). New dual descriptor for chemical reactivity. *J. Phys. Chem. A* 109: 205–212.

13 Cao, J., Ren, Q., Chen, F. et al. (2015). Comparative study on the methods for predicting the reactive site of nucleophilic reaction. *Sci. China Chem.* 45: 1281–1290.

14 Fu, R., Lu, T., and Chen, F. (2014). Comparing methods for predicting the reactive site of electrophilic substitution. *Acta Phys. Chim. Sin.* 30: 628–639.

15 Wang, B., Rong, C., Chattaraj, P.K. et al. (2019). A comparative study to predict regioselectivity, electrophilicity and nucleophilicity with Fukui function and Hirshfeld charge. *Theor. Chem. Acc.* 138: 124.

16 Cao, X., Rong, C., Zhong, A. et al. (2018). Molecular acidity: an accurate description with information-theoretic approach in density functional reactivity theory. *J. Comput. Chem.* 39: 117–129.

17 Haslak, Z.P., Zareb, S., Dogan, I. et al. (2021). Using atomic charges to describe the pKa of carboxylic acids. *JCIM* 61: 2733–2743.

18 Gasteiger, J. and Marsili, M. (1980). Iterative partial equalization of orbital electronegativity--a rapid access to atomic charges. *Tetrahedron* 36: 3219–3228.

19 Mullay, J. (1988). A method for calculating atomic charges in large molecules. *J. Comput. Chem.* 9: 399–405.

20 de Oliveira, A.E., Guadagnini, P.H., Haiduke, R.L.A. et al. (1999). A simple potential model criterion for the quality of atomic charges. *J. Phys. Chem. A* 103: 4918–4924.

21 Döntgen, M., Fenard, Y., and Heufer, K.A. (2020). Atomic partial charges as descriptors for barrier heights. *JCIM* 60: 5928–5931.

22 Liu, S., Rong, C., and Lu, T. (2014). Information conservation principle determines electrophilicity, nucleophilicity, and regioselectivity. *J. Phys. Chem. A* 118: 3698–3704.

23 Pujadas, G., Vaque, M., Ardevol, A.B. et al. (2008). Protein-ligand docking: a review of recent advances and future perspectives. *Current Pharm. Anal.* 4: 1–19.

24 Grimme, S., Bannwarth, C., and Shushkov, P. (2017). A robust and accurate tight-binding quantum chemical method for structures, vibrational frequencies, and noncovalent interactions of large molecular systems parametrized for all spd-block elements ($Z = 1$–86). *J. Chem. Theory Comput.* 13: 1989–2009.

25 Elstner, M., Porezag, D., Jungnickel, G. et al. (1998). Self-consistent-charge density-functional tight-binding method for simulations of complex materials properties. *Phys. Rev. B* 58: 7260–7268.

26 Caldeweyher, E., Ehlert, S., Hansen, A. et al. (2019). A generally applicable atomic-charge dependent London dispersion correction. *J. Chem. Phys.* 150: 154122.

27 Marenich, A.V., Cramer, C.J., and Truhlar, D.G. (2013). Generalized born solvation model SM12. *J. Chem. Theory Comput.* 9: 609–620.

28 Vyboishchikov, S.F. and Voityuk, A.A. (2021). Fast non-iterative calculation of solvation energies for water and non-aqueous solvents. *J. Comput. Chem.* 42: 1184–1194.

29 Qiu, D., Shenkin, P.S., Hollinger, F.P. et al. (1997). The GB/SA continuum model for solvation. A fast analytical method for the calculation of approximate born radii. *J. Phys. Chem. A* 101: 3005–3014.

30 Baker, N.A. (2004). Poisson Boltzmann methods for biomolecular electrostatics. In: *Methods Enzymol*, vol. 383 (ed. J.N.A.M.I. Simon), 94–118. Elsevier Inc.

31 Madhavan, T., Gadhe, C.G., Kothandan, G. et al. (2012). Various atomic charge calculation schemes of CoMFA on HIF-1 inhibitors of moracin analogs. *Int. J. Quantum Chem.* 112: 995–1005.

32 Lu, T., Liu, Z., and Chen, Q. (2021). Comment on "18 and 12 – member carbon rings (cyclo[n]carbons) – a density functional study". *Mater. Sci. Eng., B* 273: 115425.

33 Clemente, C.M., Capece, L., and Martí, M.A. (2023). Best practices on QM/MM simulations of biological systems. *JCIM* 63: 2609–2627.

34 Dittmer, A., Izsák, R., Neese, F. et al. (2019). Accurate band gap predictions of semiconductors in the framework of the similarity transformed equation of motion coupled cluster theory. *Inorg. Chem.* 58: 9303–9315.

35 Lu, T. and Chen, F. (2013). Revealing the nature of intermolecular interaction and configurational preference of the nonpolar molecular dimers (H2)2, (N2)2, and (H2)(N2). *J. Mol. Model.* 19: 5387–5395.

36 Politzer, P., Murray, J.S., and Concha, M.C. (2008). σ-hole bonding between like atoms; a fallacy of atomic charges. *J. Mol. Model.* 14: 659–665.

37 Liu, Z., Lu, T., and Chen, Q. (2021). Intermolecular interaction characteristics of the all-carboatomic ring, cyclo[18]carbon: focusing on molecular adsorption and stacking. *Carbon* 171: 514–523.

38 Lu, T. and Chen, F. (2012). Atomic dipole moment corrected Hirshfeld population method. *J. Theor. Comp. Chem.* 11: 163–183.

39 Kramer, C., Gedeck, P., and Meuwly, M. (2012). Atomic multipoles: electrostatic potential fit, local reference axis systems, and conformational dependence. *J. Comput. Chem.* 33: 1673–1688.

40 Soteras Gutiérrez, I., Lin, F.-Y., Vanommeslaeghe, K. et al. (2016). Parametrization of halogen bonds in the CHARMM general force field: Improved treatment of ligand–protein interactions. *Bioorg. Med. Chem.* 24: 4812–4825.

41 Kramer, C., Spinn, A., and Liedl, K.R. (2014). Charge anisotropy: where atomic multipoles matter most. *J. Chem. Theory Comput.* 10: 4488–4496.

42 Ros, P. and Schuit, G.C.A. (1966). Molecular orbital calculations on copper chloride complexes. *Theor. Chem. Acc.* 4: 1–12.

43 Bickelhaupt, F.M., van Eikema Hommes, N.J.R., Fonseca Guerra, C. et al. (1996). The carbon-lithium electron pair bond in (CH3Li)n (n = 1, 2, 4). *Organometallics* 15: 2923–2931.

44 Stout, E.W. and Politzer, P. (1968). An investigation of definitions of the charge on an atom in a molecule. *Theor. Chem. Acc.* 12: 379–386.

45 Cusachs, L.C. and Politzer, P. (1968). On the problem of defining the charge on an atom in a molecule. *Chem. Phys. Lett.* 1: 529–531.

46 Reed, A.E., Weinstock, R.B., and Weinhold, F. (1985). Natural population analysis. *J. Chem. Phys.* 83: 735–746.

47 Bader, R.F.W. and Beddall, P.M. (1972). Virial field relationship for molecular charge distributions and the spatial partitioning of molecular properties. *J. Chem. Phys.* 56: 3320–3329.

48 Bader, F.W. (1994). *Atoms in Molecules: A Quantum Theory*. New York: Oxford University Press.

49 Hirshfeld, F.L. (1977). Bonded-atom fragments for describing molecular charge densities. *Theor. Chem. Acc.* 44: 129–138.

50 Bultinck, P., Van Alsenoy, C., Ayers, P.W. et al. (2007). Critical analysis and extension of the Hirshfeld atoms in molecules. *J. Chem. Phys.* 126: 144111–144119.

51 Guerra, C.F., Handgraaf, J.-W., Baerends, E.J. et al. (2004). Voronoi deformation density (VDD) charges: assessment of the Mulliken, Bader, Hirshfeld, Weinhold, and VDD methods for charge analysis. *J. Comput. Chem.* 25: 189–210.

52 Verstraelen, T., Vandenbrande, S., Heidar-Zadeh, F. et al. (2016). Minimal basis iterative stockholder: atoms in molecules for force-field development. *J. Chem. Theory Comput.* 12: 3894–3912.

53 Manz, T.A. and Sholl, D.S. (2010). Chemically meaningful atomic charges that reproduce the electrostatic potential in periodic and nonperiodic materials. *J. Chem. Theory Comput.* 6: 2455–2468.

54 Besler, B.H. Jr., Merz, K.M. Jr., and Kollman, P.A. (1990). Atomic charges derived from semiempirical methods. *J. Comput. Chem.* 11: 431–439.

55 Breneman, C.M. and Wiberg, K.B. (1990). Determining atom-centered monopoles from molecular electrostatic potentials. The need for high sampling density in formamide conformational analysis. *J. Comput. Chem.* 11: 361–373.

56 Bayly, C.I., Cieplak, P., Cornell, W. et al. (1993). A well-behaved electrostatic potential based method using charge restraints for deriving atomic charges: the RESP model. *J. Phys. Chem.* 97: 10269–10280.

57 Schauperl, M., Nerenberg, P.S., Jang, H. et al. (2020). Non-bonded force field model with advanced restrained electrostatic potential charges (RESP2). *Commun. Chem.* 3: 44.

58 Campañá, C., Mussard, B., and Woo, T.K. (2009). Electrostatic potential derived atomic charges for periodic systems using a modified error functional. *J. Chem. Theory Comput.* 5: 2866–2878.

59 Jakalian, A., Bush, B.L., Jack, D.B. et al. (2000). Fast, efficient generation of high-quality atomic charges. AM1-BCC model: I. Method. *J. Comput. Chem.* 21: 132–146.

60 Jakalian, A., Jack, D.B., and Bayly, C.I. (2002). Fast, efficient generation of high-quality atomic charges. AM1-BCC model: II. Parameterization and validation. *J. Comput. Chem.* 23: 1623–1641.

61 Mortier, W.J., Ghosh, S.K., and Shankar, S. (1986). Electronegativity-equalization method for the calculation of atomic charges in molecules. *JACS* 108: 4315–4320.

62 Mortier, W.J., Van Genechten, K., and Gasteiger, J. (1985). Electronegativity equalization: application and parametrization. *JACS* 107: 829–835.

63 Rappe, A.K. and Goddard, W.A. (1991). Charge equilibration for molecular dynamics simulations. *J. Phys. Chem.* 95: 3358–3363.

64 Gasteiger, J. and Marsili, M. (1978). A new model for calculating atomic charges in molecules. *Tetrahedron Lett.* 34: 3181–3184.

65 Marsili, M. and Gasteiger, J. (1980). Pi-charge distributions from molecular topology and Pi-orbital electronegativity. *Croat. Chem. Acta* 53: 601–614.

66 No, K.T., Grant, J.A., and Scheraga, H.A. (1990). Determination of net atomic charges using a modified partial equalization of orbital electronegativity method. 1. Application to neutral molecules as models for polypeptides. *J. Phys. Chem.* 94: 4732–4739.

67 Marenich, A.V., Jerome, S.V., Cramer, C.J. et al. (2012). Charge model 5: an extension of Hirshfeld population analysis for the accurate description of molecular interactions in gaseous and condensed phases. *J. Chem. Theory Comput.* 8: 527–541.

68 Halgren, T.A. (1996). Merck molecular force field. II. MMFF94 van der Waals and electrostatic parameters for intermolecular interactions. *J. Comput. Chem.* 17: 520–552.

69 Cioslowski, J. (1989). A new population analysis based on atomic polar tensors. *JACS* 111: 8333–8336.

70 Stephens, P.J., Devlin, F.J., Chabalowski, C.F. et al. (1994). Ab initio calculation of vibrational absorption and circular dichroism spectra using density functional force fields. *J. Phys. Chem.* 98: 11623–11627.

71 Kendall, R.A., Dunning, T.H., and Harrison, R.J. (1992). Electron affinities of the first-row atoms revisited. Systematic basis sets and wave functions. *J. Chem. Phys.* 96: 6796–6806.

72 Weigend, F. and Ahlrichs, R. (2005). Balanced basis sets of split valence, triple zeta valence and quadruple zeta valence quality for H to Rn: design and assessment of accuracy. *Phys. Chem. Chem. Phys.* 7: 3297–3305.

73 Hariharan, P.C. and Pople, J.A. (1973). The influence of polarization functions on molecular orbital hydrogenation energies. *Theor. Chem. Acc.* 28: 213–222.

74 Grimme, S., Antony, J., Ehrlich, S. et al. (2010). A consistent and accurate ab initio parametrization of density functional dispersion correction (DFT-D) for the 94 elements H-Pu. *J. Chem. Phys.* 132: 154104–154119.

75 Lu, T. and Chen, F. (2011). Calculation of molecular orbital composition. *Acta Chim. Sin.* 69: 2393–2406.

76 Grabenstetter, J.E. and Whitehead, M.A. (1972). Comment on a paper by Stout and Politzer. *Theor. Chem. Acc.* 26: 390–390.

77 Glendening, E.D., Landis, C.R., and Weinhold, F. (2012). Natural bond orbital methods. *WIREs Comput. Mol. Sci.* 2: 1–42.

78 Clark, A.E., Sonnenberg, J.L., Hay, P.J. et al. (2004). Density and wave function analysis of actinide complexes: What can fuzzy atom, atoms-in-molecules, Mulliken, Löwdin, and natural population analysis tell us? *J. Chem. Phys.* 121: 2563–2570.

79 Landis, C.R. and Weinhold, F. (2007). Valence and extra-valence orbitals in main group and transition metal bonding. *J. Comput. Chem.* 28: 198–203.

80 Maseras, F. and Morokuma, K. (1992). Application of the natural population analysis to transition-metal complexes. Should the empty metal p orbitals be included in the valence space? *Chem. Phys. Lett.* 195: 500–504.

81 Mei, C., Edgecombe, K.E., Smith, V.H. et al. (1993). Topological analysis of the charge density of solids: bcc sodium and lithium. *Int. J. Quantum Chem.* 48: 287–293.

82 Rousseau, B., Peeters, A., and Alsenoy, C.V. (2001). Atomic charges from modified Voronoi polyhedra. *J. Mol. Struct. THEOCHEM* 538: 235–238.

83 Nalewajski, R.F. and Parr, R.G. (2000). Information theory, atoms in molecules, and molecular similarity. *PNAS* 97: 8879–8882.

84 Van Damme, S., Bultinck, P., and Fias, S. (2009). Electrostatic potentials from self-consistent Hirshfeld atomic charges. *J. Chem. Theory Comput.* 5: 334–340.

85 Verstraelen, T., Ayers, P.W., Van Speybroeck, V. et al. (2013). Hirshfeld-E partitioning: AIM charges with an improved trade-off between robustness and accurate electrostatics. *J. Chem. Theory Comput.* 9: 2221–2225.

86 Ghillemijn, D., Bultinck, P., Van Neck, D. et al. (2011). A self-consistent Hirshfeld method for the atom in the molecule based on minimization of information loss. *J. Comput. Chem.* 32: 1561–1567.

87 Geldof, D., Krishtal, A., Blockhuys, F. et al. (2011). An extension of the hirshfeld method to open shell systems using fractional occupations. *J. Chem. Theory Comput.* 7: 1328–1335.

88 Lillestolen, T.C. and Wheatley, R.J. (2008). Redefining the atom: atomic charge densities produced by an iterative stockholder approach. *Chem. Commun.* 5909–5911.

89 Manz, T.A. and Limas, N.G. (2016). Introducing DDEC6 atomic population analysis: part 1. Charge partitioning theory and methodology. *RSC Adv.* 6: 47771–47801.

90 Heidar-Zadeh, F., Ayers, P.W., Verstraelen, T. et al. (2018). Information-theoretic approaches to atoms-in-molecules: Hirshfeld family of partitioning schemes. *J. Phys. Chem. A* 122: 4219–4245.

91 Murray, J.S. and Politzer, P. (2011). The electrostatic potential: an overview. *WIREs: Comp. Mol. Sci.* 1: 153–163.

92 Cox, S.R. and Williams, D.E. (1981). Representation of the molecular electrostatic potential by a net atomic charge model. *J. Comput. Chem.* 2: 304–323.

93 Chirlian, L.E. and Francl, M.M. (1987). Atomic charges derived from electrostatic potentials: a detailed study. *J. Comput. Chem.* 8: 894–905.

94 Woods, R.J., Khalil, M., Pell, W. et al. (1990). Derivation of net atomic charges from molecular electrostatic potentials. *J. Comput. Chem.* 11: 297–310.

95 Franc, M.M. and Chirlian, L.E. (2000). The pluses and minuses of mapping atomic charges to electrostatic potentials. In: *Reviews in Computational Chemistry*, vol. 14 (ed. K.B. Lipkowitz and D.B. Boyd), 1–31. New York: VCH Publishers.

96 Wang, J., Cieplak, P., and Kollman, P.A. (2000). How well does a restrained electrostatic potential (RESP) model perform in calculating conformational energies of organic and biological molecules? *J. Comput. Chem.* 21: 1049–1074.

97 Wang, J., Wolf, R.M., Caldwell, J.W. et al. (2004). Development and testing of a general amber force field. *J. Comput. Chem.* 25: 1157–1174.

98 Mukherjee, G., Patra, N., Barua, P. et al. (2011). A fast empirical GAFF compatible partial atomic charge assignment scheme for modeling interactions of small molecules with biomolecular targets. *J. Comput. Chem.* 32: 893–907.

99 Sanderson, R.T. (1951). An interpretation of bond lengths and a classification of bonds. *Science* 114: 670–672.

100 Bultinck, P., Langenaeker, W., Lahorte, P. et al. (2002). The electronegativity equalization method II: applicability of different atomic charge schemes. *J. Phys. Chem. A* 106: 7895–7901.

101 Raček, T., Pazúriková, J., Svobodová Vařeková, R. et al. (2016). NEEMP: software for validation, accurate calculation and fast parameterization of EEM charges. *J. Cheminf.* 8: 57.

102 Jiroušková, Z., Vařeková, R.S., Vaněk, J. et al. (2009). Software news and updates electronegativity equalization method: parameterization and validation for organic molecules using the Merz-Kollman-Singh charge distribution scheme. *J. Comput. Chem.* 30: 1174–1178.

103 Chen, J. and Martínez, T.J. (2007). QTPIE: charge transfer with polarization current equalization. a fluctuating charge model with correct asymptotics. *Chem. Phys. Lett.* 438: 315–320.

104 Wilmer, C.E., Kim, K.C., and Snurr, R.Q. (2012). An extended charge equilibration method. *J. Phys. Chem. Lett.* 3: 2506–2511.

105 Wells, B.A., De Bruin-Dickason, C., and Chaffee, A.L. (2015). Charge equilibration based on atomic ionization in metal–organic frameworks. *J. Phys. Chem. C* 119: 456–466.

106 Hammarström, L.-G., Liljefors, T., and Gasteiger, J. (1988). Electrostatic interactions in molecular mechanics (MM2) calculations via PEOE partial charges I. Haloalkanes. *J. Comput. Chem.* 9: 424–440.

107 Cho, K.-H., Kang, Y.K., No, K.T. et al. (2001). A fast method for calculating geometry-dependent net atomic charges for polypeptides. *J. Phys. Chem. B* 105: 3624–3634.

108 Dodda, L.S., Vilseck, J.Z., Tirado-Rives, J. et al. (2017). 1.14*CM1A-LBCC: localized bond-charge corrected CM1A charges for condensed-phase simulations. *J. Phys. Chem. B* 121: 3864–3870.

109 Lu, T. and Chen, F. (2012). Multiwfn: a multifunctional wavefunction analyzer. *J. Comput. Chem.* 33: 580–592.

110 Frisch, M.J., Trucks, G.W., Schlegel, H.B., et al. (2016). Gaussian 16 A.03. Wallingford, CT.

111 Glendening, E.D., Landis, C.R., and Weinhold, F. (2019). NBO 7.0: new vistas in localized and delocalized chemical bonding theory. *J. Comput. Chem.* 40: 2234–2241.

112 Zhao, J., Zhu, Z.-W., Zhao, D.-X. et al. (2023). Atomic charges in molecules defined by molecular real space partition into atomic subspaces. *Phys. Chem. Chem. Phys.* 25: 9020–9030.

113 Cho, M., Sylvetsky, N., Eshafi, S. et al. (2020). The atomic partial charges arboretum: trying to see the forest for the trees. *ChemPhysChem* 21: 688–696.

114 Manz, T.A. (2020). Seven confluence principles: a case study of standardized statistical analysis for 26 methods that assign net atomic charges in molecules. *RSC Adv.* 10: 44121–44148.

115 Neese, F., Wennmohs, F., Becker, U. et al. (2020). The ORCA quantum chemistry program package. *J. Chem. Phys.* 152: 224108.

116 Kühne, T.D., Iannuzzi, M., Del Ben, M. et al. (2020). CP2K: An electronic structure and molecular dynamics software package – quickstep: efficient and accurate electronic structure calculations. *J. Chem. Phys.* 152: 194103.

117 Nikolaienko, T.Y., Bulavin, L.A., and Hovorun, D.M. (2014). JANPA: an open source cross-platform implementation of the Natural Population Analysis on the Java platform. *Comput. Theor. Chem.* 1050: 15–22.

118 Keith, T. A. (2019) AIMALL, TK Gristmill Software, Overland Park KS, USA (aim.tkgristmill.com).

119 Biegler-König, F. and Schönbohm, J. (2002). Update of the AIM2000-Program for atoms in molecules. *J. Comput. Chem.* 23: 1489–1494.

120 Bader program. https://theory.cm.utexas.edu/henkelman/code/bader/ (accessed 7 July 2023).

121 VASP program. https://www.vasp.at (accessed 7 July 2023).

122 Verstraelen, T., Tecmer, P., Heidar-Zadeh, F., et al. (2017). HORTON 2.1.1, https://github.com/theochem/horton (accessed 7 July 2023).

123 Case, A., Aktulga, H.M., Belfon, K. et al. (2023). AmberTools. *J. Chem. Inf. Model.* 63: 6183–6191.

124 Vařeková, R.S. and Koča, J. (2006). Optimized and parallelized implementation of the electronegativity equalization method and the atom-bond electronegativity equalization method. *J. Comput. Chem.* 27: 396–405.

125 VCharge program. https://www.verachem.com/products/vcharge/ (accessed 7 July 2023).

126 O'Boyle, N.M., Banck, M., James, C.A. et al. (2011). Open babel: an open chemical toolbox. *J. Cheminf.* 3: 33.

7

Atoms in Molecules

Ángel Martín Pendás, Evelio Francisco, Julen Munárriz, and Aurora Costales

Universidad de Oviedo, Departamento de Química Física y Analítica, C/Julián Clavería, 8, Oviedo 33006, Spain

7.1 Introduction

Developing a universally accepted framework for extracting chemical insights from quantum chemical calculations has proven more challenging than enhancing the accuracy of the calculations themselves. The task of adapting our cherished chemical concepts – such as single and multiple bonds, lone pairs, and others – to the rigorous framework of quantum mechanics has been approached from various angles, but while valence bond and molecular orbital theories, for instance, may converge to the true wavefunction under well-defined limits, the same cannot be said about their associated chemical interpretations. Consequently, this discrepancy has led to considerable confusion and ongoing debates.

A viable solution to this problem necessitates a direct analysis of the wavefunction Ψ of a given system. The resulting interpretation should be independent of how Ψ is constructed, and this naturally leads to the consideration of orbital invariant entities, such as reduced density matrices (RDMs) or reduced densities (RDs), expressed in either real or momentum space. Since chemists typically perceive molecules as tangible entities evolving in real space, most approaches, albeit not all [1, 2], are based on real-space RDMs. Rejecting the notion of atom-centered functions as an indispensable part of the analysis of Ψ would result in dissolving atoms within the sea of the N-electron wavefunction, effectively eradicating the very essence of chemistry. In this regard, the statements made by Ruedenberg and Schmidt advocating for the preservation of atoms within molecules resonate with credibility [3].

The emergence of the concept of atoms in molecules finds its theoretical foundation in Kato's cusp theorem [4]. The lowest (first) order reduced density, namely, the electron density, exhibits logarithmic cusps at the positions of atomic nuclei, with slopes determined by their nuclear charges. Although these ideas were not the historical origin of utilizing ρ in the context of chemical bonding, they suitably serve our purposes by allowing us to identify the constituent atoms of a molecule based on its shape. The examination of ρ should enable the association of a specific atom with its nuclear position as well as its surroundings. Consequently, developing a

Exploring Chemical Concepts Through Theory and Computation, First Edition. Edited by Shubin Liu.
© 2024 WILEY-VCH GmbH. Published 2024 by WILEY-VCH GmbH.

program to partition real space based on the analysis of the density or, equivalently, decomposing the density into atomic contributions becomes an attainable goal. This endeavor was initiated several decades ago by Richard F. W. Bader [5], who proposed employing the topology induced by a scalar field such as the electron density to divide space into nonoverlapping regions. This approach, known as the quantum theory of atoms in molecules (QTAIM), has enjoyed considerable success and has paved the way for numerous other atomic partitions.

7.2 The Quantum Theory of Atoms in Molecules (QTAIM)

By analyzing the gradient of $\rho(r)$, we can intuitively partition molecular or periodic space into different regions associated with atoms, similar to how we naturally divide a mountain range based on its peaks.

A dynamical system (DS), corresponds to a vector field y defined over an n-dimensional manifold M. Note that, in this context, a manifold refers to a topological space where each point has a neighbor that is homeomorphic (i.e. a field that preserves the topological properties and relationships of the original one) to an n-dimensional Euclidean space. The trajectories of a differentiable field, denoted as $r(t)$, are uniquely determined with $\frac{dr}{dt} = y$. In the following lines, some features of gradient dynamical systems in $\mathbb{R}^3$ are explained by taking as a case function the electron density, ρ, a scalar function that depends on r.

Given $\rho(r)$, we can associate a DS system to it via its gradient field $\nabla \rho(r)$. The trajectories (also known as flux, force, field, or gradient lines) of the DS are then defined by a system of ordinary differential equations, $\dot{r} = \nabla \rho$, whose solution can be written in terms of the following parameterized curves in $\mathbb{R}^3$:

$$r(t) = r(t_0) + \int_{t_0}^{t} \nabla \rho(r(t)) dt \tag{7.1}$$

The trajectories of such curves exhibit several noteworthy properties, which can be summarized as follows: (i) Each point in space is uniquely associated with a trajectory, ensuring that field lines never intersect except at specific points of interest. (ii) At any given point, the vector $\nabla \rho(r)$ is tangent to the corresponding field line. (iii) The trajectories of $\nabla \rho(r)$ are perpendicular to the isoscalar lines. (iv) Field trajectories originate or terminate at points characterized by $\nabla \rho(r) = 0$, or they extend to infinity.

As mentioned previously, the analysis of the electron density and related scalar fields using topological methods provides an intuitive and orbital-invariant perspective on quantum mechanical calculations. By examining $\rho(r)$, it becomes possible to partition physical space into regions associated with atoms in the chemical realm. These regions' pairwise relationships can be correlated with chemical bonds, thereby recovering the familiar molecular model of interconnected atoms. This forms the basis of the QTAIM developed by R. F. W. Bader et al. [6], who also sought to demonstrate that the topology induced by $\rho(r)$ arises from the extension of quantum mechanics (QM) to subsystems or regions in $\mathbb{R}^3$. By partitioning space using the gradient of electron density, $\nabla\rho(\mathbf{r})$, QM principles are maintained within these open regions (see Section 7.3) [5].

It is important to note that extending QM to subsystems presents challenges, including the presence of surface terms that may not vanish within the finite boundaries of the subsystems. This is particularly significant as the Hermitian nature of QM's fundamental operations, based on the momentum operator, relies on the vanishing of these surface terms. Fortunately, the QTAIM offers a potential solution to this issue, although a detailed explanation lies beyond the scope of this discussion.

The QTAIM represents a theory of open subsystems since the number operator $\hat{N}$ does not commute with position operators. Consequently, it is not possible to assign a definite number of electrons to a region in $\mathbb{R}^3$ through any spatial partitioning. Within a subsystem, one can only speak of the average number of particles, and the boundaries of such subsystems are necessarily permeable to electrons. The QTAIM provides expectation values for traditional quantum chemically relevant operators within these subsystems, which we refer to as atomic observables. Importantly, most standard molecular observables can be derived from these atomic observables.

Within this framework, the atomic observables and properties provided by QTAIM satisfy various theorems, many of which generalize those satisfied by total molecular expectation values. Combining atoms to form functional groups, as known in chemistry, leads to group properties exhibiting the additivity commonly observed in experiments. For instance, QTAIM's group dipole moments, energies, and polarizabilities, among other properties, show excellent agreement with experimentally tabulated values. Thus, QTAIM establishes a quantum mechanical foundation for the chemical transferability of functional groups, contributing significantly to our understanding of molecular systems.

It should be noted that Kato's theorem shows that $\rho(\mathbf{r})$ is not a truly differentiable field. We can nevertheless build a homeomorphic (topologically equivalent) field that is equal to $\rho(\mathbf{r})$ at all points save small neighborhoods around the nuclei, which are substituted by differentiable local maxima.

Under these conditions, we can analyze the shape of $\rho(\mathbf{r})$ with simple mathematical function analysis tools from the field of topology. Returning to the analogy of a mountain range, we previously identified the peaks as maxima associated with nuclei. Consequently, these maxima are referred to as nuclear critical points (NCPs). It has been found that other critical points (CPs, points where $\nabla\rho = 0$), can be also associated to chemical objects. Bond critical points (BCPs), which are first-order saddle points of the electron density, emerge between bonded atoms. At these BCPs, the electron density is at its minimum along the bonding line and reaches its maximum across the perpendicular plane.

To analyze and characterize CPs, we can examine the components of the Hessian matrix of $\rho(\mathbf{r})$ with respect to the coordinates $\mathbf{r} = (x, y, z)$ at the CP as represented by (7.2):

$$H(\mathbf{r}) = \begin{bmatrix} \partial^2\rho/\partial x^2 & \partial^2\rho/\partial x\partial y & \partial^2\rho/\partial x\partial z \\ \partial^2\rho/\partial y\partial x & \partial^2\rho/\partial y^2 & \partial^2\rho/\partial y\partial z \\ \partial^2\rho/\partial z\partial x & \partial^2\rho/\partial z\partial y & \partial^2\rho/\partial z^2 \end{bmatrix} \quad (7.2)$$

By diagonalizing $H(r)$, we rotate the reference frame so that the three curvatures λ_1, λ_2, and λ_3 along its principal axes are maximal. Notice that the Laplacian of the electron density, $\nabla^2 \rho(r)$, is just the sum of these diagonal eigenvalues. In the context of critical points, it is customary to use the notation (r, g), where the rank r indicates the number of nonzero eigenvalues and the signature g is defined as the sum of the eigenvalues divided by their absolute values. Four types of rank 3 critical points (the most common, nondegenerate ones) can be found in molecules.

- $(3, -3)$ points, or NCPs, in general, correspond to local maxima of $\rho(r)$. As previously stated, these points are typically found at nuclear positions within the molecule, representing the nuclei of atoms. Nonnuclear maxima can also rarely occur, having sparked some controversy [7–9].
- $(3, -1)$ points, the so-called BCPs, exhibiting a maximum along two of the H eigenvectors and a minimum along the other orthogonal direction. Traditionally, they have been associated with chemical bonds, and they thus provide important information about their nature and strength.
- $(3, +1)$ points are generally known as ring critical points (RCPs). Contrarily to BCPs, RCPs exhibit a minimum along two H eigen-directions and a maximum along the other. In the chemical realm, they are typically found at the centers of bonded rings.
- $(3, +3)$ points correspond to cage critical points (CCPs). They are local minima of $\rho(r)$, commonly observed at the centers of 3D-bonded "cages." CCPs represent regions of relatively low electron density within a molecule and are often associated with cavities or void spaces among bonded atoms.

Figure 7.1 provides some illustrative examples of each type of critical point. The critical points for [2.2.2]propellane are illustrated in Figure 7.1a. No BCP appears between the two bridge carbon atoms, something that has again sparked significant controversy [10]. The QTAIM analysis of benzene-1,2-diol (Figure 7.1b) reveals two RCPs, at the center of the benzene-based ring, and at a point in between the C–C, C–O, O–H, and the hydrogen bond, which shows a BCP. Figure 7.1c shows a CCP at the center of the octafluorocubane structure. Notice that a CP analysis of hexahelicene (Figure 7.1d) also reveals two ring critical points (RCPs) along the axis of the helix, together with some exotic BCPs between certain sterically crowded hydrogen and carbon atoms.

Given that one can draw two ascending gradient paths of $\rho(r)$ from each BCP that generally end up at two nuclei – this is the bond path of a BCP –, the graph formed by the set of bond paths in a molecule is a nonempirical molecular graph (see again Figure 7.1). Except in some exotic topologies that break this general rule, the possibility of obtaining a molecular graph without the use of either chemical intuition or tables of bond distances or radii has to be understood as a great success of the QTAIM. As a result, a molecular picture in terms of Lewis-like entities, i.e. atoms and bonds that connect them, is retrieved solely in terms of the electron density. This facilitates the determination of bonding patterns that might not be evident at first sight, such as those taking place in periodic systems. Note that, in general, paths are not exactly coincident with the shortest internuclear lines. This is especially clear in

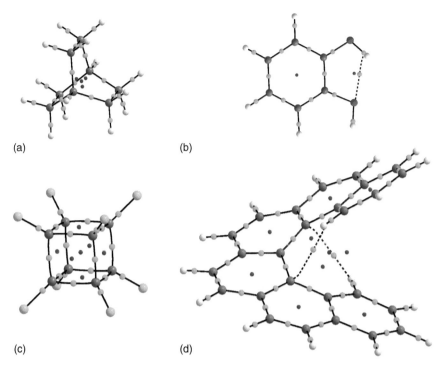

Figure 7.1 CPs of $\rho(r)$ and molecular graph for different molecules calculated at the PBE/def2-SVP level of theory: (a) [2.2.2]propellane, (b) benzene-1,2-diol, (c) octafluorocubane, and (d) hexahelicene. CPs are shown as spheres. BCPs, RCPs, and CCPs are colored green, purple, and blue, respectively.

octafluorocubane (Figure 7.1c), as a consequence of structural strain. In this regard, note that the difference between both distances can be used as an indicator of bond strain.

The Laplacian of the electron density, $\nabla^2\rho(r)$, which is related to local charge accumulation ($\nabla^2\rho(r) < 0$) or depletion ($\nabla^2\rho(r) > 0$), can also be used to obtain relevant chemical information [11]. The shell structure of atoms and molecules is revealed by $\nabla^2\rho(r)$ as alternating regions of electron charge concentration and depletion, the valence-shell charge concentration (VSCC) being the outer shell with $\nabla^2\rho(r) < 0$. Shared interactions (typically associated to covalent bonds) between two atoms are revealed by fusion of their VSCCs, while closed-shell interactions (such as ionic, van der Waals, metallic, or hydrogen bond interactions) correspond to nonoverlapping VSCCs. Some examples are provided in Figure 7.2. In this context, the presence of ionic bonding in Li_2F_2 is evident from the nonoverlapping VSCCs of the different atoms, along with a region of charge depletion (indicated by a blue isocontour) observed at the interatomic distance. It is also remarkable to note that even though the Li–Li distance is much shorter than the F–F one (2.215 and 2.606 Å, respectively), the BCP analysis unravels a F–F bonding interaction. Shared interactions in the formic acid dimer are depicted in Figure 7.2b, where the VSCCs of all covalently bonded atoms (i.e. O, C, and H) are fused. In Figure 7.2b, hydrogen bonds are

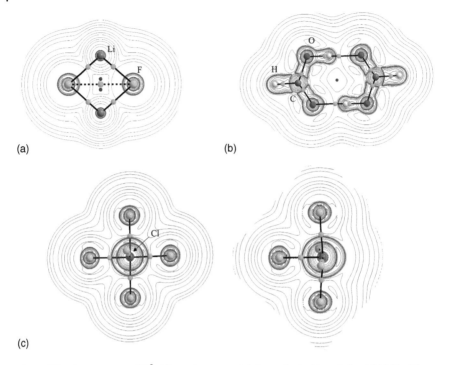

Figure 7.2 Isocontours of $\nabla^2\rho(r)$ on planes containing selected bonds for (a) Li_2F_2, (b) formic acid dimer, and (c) ClF_5. Negative Laplacian values are depicted in red, while positive ones are shown in blue. Calculations were performed at the PBE/def2-SVP level of theory, and topological analysis was made by means of the AimAll. Source: [13] Todd A. Keith.

also revealed as closed-shell interactions with nonoverlapping VSCCs. The Laplacian is also useful to reveal lone pairs, by means of nonbonded charge concentrations (NCCs), thus completing a valence shell electron pair repulsion theory (VSEPR) theory from the electron density alone [12]. This feature is exemplified for ClF_5 (see Figure 7.2c), a classical example of a VSEPR-following molecule, which exhibits a square pyramid geometry that derives from a octahedron in which a ligand position has been substituted by a lone pair. At first sight, $\nabla^2\rho(r)$ reveals a closed-shell bonding between Cl and F atoms, although with non-negligible covalent character that can be grasped from the location of the BCP at $\nabla^2\rho(r) = 0$, very close to the point where the Laplacian changes sign. In addition, the lone pairs of F and Cl atoms are revealed by distortion of the spherical atomic symmetry.

7.3 QTAIM Atoms as Open Quantum Systems

After the above short introduction to the QTAIM, we turn to some of its more modern developments. In this section, we examine [14] how the consideration of real space regions in general, or QTAIM atomic basins in particular, as open quantum systems (OQS) allows us to gain a deeper understanding of different aspects of quantum chemical topology (QCT). To do that, [14] we start by summarizing how to build closed expressions for the density operator of a quantum subsystem, showing that it can be expressed as a direct sum of electron sectors with weights that are equal to the probabilities that sector houses a given number of electrons. This approach provides a new way to understand chemical bonding through orbital invariant descriptors and establishes a deep connection between chemical concepts and quantum entanglement.

7.3.1 Sector Density Operators of Quantum Atoms in Molecules

The meaning of the term open system in this chapter is the same as that usually used in Chemical Thermodynamics: a closed region of the physical space that can exchange matter (in our case, electrons) with its environment. In short, if a molecule with a fixed number of electrons N is divided into two regions, Ω_A and its complementary, $\overline{\Omega}_A \equiv \Omega_B = R^3 - \Omega_A$, a snapshot of the system at a given instant of time, can locate in Ω_A from 0 to N electrons. This simple idea brings as a consequence that a quantum mechanical observable of Ω_A can be determined as a statistical average of their values for all possible electronic populations of Ω_A, $0 \leq n \leq N$. The statistical weight associated to a given n will be simply the probability that Ω_A has exactly n electrons. The essential minimum to easily follow the rest of the remaining of this section can be found in Ref. 14, and in the books *The Theory of Open Quantum Systems* by H.P. Breuer and F. Petruccione [15], and *Quantum Computation and Quantum Information* by Nielsen and Chuang [16]. In what follows, a region Ω_A (or simply A) will be identified either with a single QTAIM atom or with a grouping of an arbitrary number of them [17]. However, the results in what follows are of general applicability to arbitrary spatial domains.

The concept of indicator function of a domain Ω will be essential in the following. It is simply a Heaviside-like domain weight function, $\omega_\Omega(r)$, defined as $\omega_\Omega(r) = 0$ for $r \notin \Omega$ and $\omega_\Omega(r) = 1$ for $r \in \Omega$. If a spin-projection is assigned to a domain Ω, the indicator function $\omega_\Omega(x)$, where $x = r\sigma$ denotes a general spin(σ)-spatial(r) coordinate, can also be defined. In this case, $\omega_\Omega(x) = 1$ only when $r \in \Omega$ and σ coincides with the spin-projection assigned to Ω, and zero otherwise. Using the above definition, it turns out that the full density operator $\hat{\rho}$ for an N-electron system, $\hat{\rho} = \Psi(x_1 \ldots, x_N)\Psi^\star(x'_1 \ldots, x'_N)$, can be written in the form [14]:

$$\hat{\rho} = 1^N 1'^N \hat{\rho} \tag{7.3}$$

where $1^N = \prod_{i=1}^N [\omega_{\Omega_A}(x_i) + \omega_{\Omega_B}(x_i)]$ is an N-electron indicator function, with a similar definition of $1'^N$. The identity (7.3) follows from the property $\omega_{\Omega_A}(x_i) + \omega_{\Omega_B}(x_i) = 1$, since $\Omega_A \cup \Omega_B = R^3$ and $x_i \in R^3$. Equation (7.3) shows that $\hat{\rho}$ becomes a sum of 2^{2N} terms in which the primed and unprimed electrons are separated into the Ω_A and Ω_B domains. Each of the 2^N terms of 1^N and $1'^N$ has the form $\prod_i^N \omega_{\Omega_i}(x_i)(\Omega_i = \Omega_A$ or $\Omega_B)$ and $\prod_j^N \omega_{\Omega_j}(x'_j)(\Omega_j = \Omega_A$ or $\Omega_B)$, respectively. The reduced density operator of domain Ω_A, $\hat{\rho}^A$, is obtained from $\hat{\rho}$ by performing a partial spatial trace over the Ω_B region [15], with the usual $x'_i \to x_i$ identification before integration. This means that, from the 2^{2N} terms resulting from the product $1^N 1'^N$, only those with $\omega_{\Omega_i} = \omega_{\Omega_j}$ survive (i.e. the diagonal ones), leaving 2^N terms. Each of these corresponds to a given number of electrons in Ω_A (say n). The value of n coincides with the number of Ω_i's in the expansion $\prod_i^N \omega_{\Omega_i}(x_i) \times \omega_{\Omega_i}(x'_i)$ that are equal to Ω_A. No matter which of the $\binom{N}{n}$ possible ways of choosing n out of N Ω_i's as equal to Ω_A, electronic indistinguishability guarantees that its contribution to $\hat{\rho}^A$ is the same [18]. The 2^N terms can then be grouped by a common value of n, allowing us to write

$$\hat{\rho}^A = \sum_{n=0}^N \hat{\rho}_n^A \qquad (7.4)$$

where $\hat{\rho}_0^A = \int_B \Psi^\star(x_1 \ldots x_N) \Psi(x_1 \ldots x_N) dx_1 \ldots dx_N$ and, for $n \geq 1$

$$\hat{\rho}_n^A(x_{i \leq n}; x'_{i \leq n}) = 1_n^A 1_n^{'A} \binom{N}{n} \int_B \Psi^\star(x'_{i \leq n}, x_{i > n}) \Psi(x_{i \leq n}, x_{i > n}) dx_{i > n} \qquad (7.5)$$

with

$$1_n^A 1_n^{'A} = \prod_{i=1}^n \omega_{\Omega_A}(x_i) \omega_{\Omega_A}(x'_i) \qquad (7.6)$$

$x_{i \leq n} \equiv x_1 \ldots x_n$ and $x_{i > n} \equiv x_{n+1} \ldots x_N$. Each $\hat{\rho}_n^A$ is the reduced density operator of Ω_A associated to having n electrons in this domain. It is usually named the n-electron sector reduced density operator in the theory of OQSs [14]. Subsystem A is thus described by a mixed-density operator with $N+1$ possible values for its number of electrons. In the same way that $\Psi^\star \Psi$ is the N-electron density matrix of the full system, each $\hat{\rho}_n^A$ can be interpreted as an n-electron density matrix in its sector, ρ_n^A. However, in contrast to $\Psi^\star \Psi$, that meets the normalization condition $\int \Psi^\star \Psi dx_1 \ldots dx_N = 1$, ρ_n^A is not normalized. Actually

$$\text{Tr} \rho_n^A = \int_{\Omega_A} \rho_n^A(x_{i \leq n}; x'_{i \leq n})|_{x'_i \to x_i} dx_{i \leq n} = \int_{\Omega_A} \rho_n^A(x_{i \leq n}) dx_{i \leq n} = p^A(n) \qquad (7.7)$$

where $p^A(n)$ is the probability that n and only n electrons reside in spatial domain Ω_A [18–22]. This fact can be seen more easily if we combine Eqs. (7.5) and (7.7) into a single one,

$$p^A(n) = \binom{N}{n} \int_D \Psi^\star(x_1, \ldots, x_N) \Psi(x_1, \ldots, x_N) dx_1 \ldots dx_N \qquad (7.8)$$

where D is a N-dimensional domain such that electrons 1 to n are integrated in Ω_A and electrons $n+1$ to N in Ω_B. The previous result, although quite old, has been revitalized in recent years, since the statistical analysis of the probabilities $p^A(n)$ and its generalization to the case in which physical space is partitioned into more than two domains, is a very promising tool in chemical bonding theory [22]. Be that as it may, in the present context, we will only make use of the previous equation to define a normalized n-electron density matrix $\tilde{\rho}_n^A$ by

$$\tilde{\rho}_n^A = \rho_n^A / p^A(n) \qquad (7.9)$$

With this,

$$\rho^A = \sum_{n=0}^N p^A(n) \tilde{\rho}_n^A \qquad (7.10)$$

In this way, $\tilde{\rho}_n^A$ can be manipulated as a pseudo-pure system n-electron density matrix that can be manipulated on its own in each sector.

7.3.2 RDMs of Atoms in Molecules

Each sector density operator ρ_n^A describes a system of n electrons. In the same way that reduced density matrices (RDM) of Quantum Chemistry are obtained from the total density operator $\hat{\rho}$ by integrating a subset of the electron coordinates on which it depends [23], we can obtain m-electron RDMs associated to ρ_n^A as

$$\rho_n^{A,m}(\boldsymbol{x}_{i\leq m};\boldsymbol{x}'_{i\leq m}) = \frac{n!}{(n-m)!}\int \rho_n^A(\boldsymbol{x}_{i\leq n};\boldsymbol{x}'_{i\leq n})d\boldsymbol{x}_{i>m} \tag{7.11}$$

with trace $\operatorname{Tr}\rho_n^{A,m} = n!/(n-m)!$, and the spinless m-th RDM as

$$\rho_n^{A,m}(\boldsymbol{r}_{i\leq m};\boldsymbol{r}'_{i\leq m}) = \int \rho_n^{A,m}(\boldsymbol{x}_{i\leq m};\boldsymbol{x}'_{i\leq m})|_{\sigma'_i \to \sigma_i}d\sigma_{i\leq m} \tag{7.12}$$

(From now on, an integral symbol ($\int$) without a subindex on it will indicate integration to the entire space R^3.) It can be shown from Eqs. (7.5) and (7.11) that $\rho_n^{A,m}$ can also be written in the form

$$\rho_n^{A,m}(\boldsymbol{x}_{i\leq m};\boldsymbol{x}'_{i\leq m}) = \mathbf{1}_m^{'A}\mathbf{1}_m^A \Lambda_{N,n}^m \int_D \rho(\boldsymbol{x};\boldsymbol{x}')d\boldsymbol{x}_{i>m} \tag{7.13}$$

where $\Lambda_{N,n}^m = N!/[(N-n)!(n-m)!]$ is a normalization factor that takes into account electron indistinguishability. At difference with Eq. (7.8), the integration domain D in the above expression is $(N-m)$-dimensional since only electrons $m+1$ to n are integrated over Ω_A (and electrons $m+1$ to N in its complementary domain $\overline{\Omega}_A$). As a direct comparison of Eqs. (7.8) and (7.13), the particular case $m=n$ transforms $\rho_n^{A,m}(\boldsymbol{x}_{i\leq m};\boldsymbol{x}'_{i\leq m})$ into a single number equal to $p^A(n)$. It is interesting to mention at this point that, except for the $\Lambda_{N,n}^m$ factor defined above, $\rho_n^{A,m}(\boldsymbol{x}_{i\leq m};\boldsymbol{x}'_{i\leq m})$ is a particular case of the so-called coarse-grained RDMs, introduced a few years ago [24].

Another aspect that deserves to be commented is that, in the same way that we have decomposed $\hat{\rho}^A$ into a sum of the $N+1$ sector densities $\hat{\rho}_n^A$ (Eq. (7.4)), doing the reverse process with the $\hat{\rho}_n^{A,m}$'s leads to a relevant expression. Indeed, after some algebraic work, it is possible to show that the sum over n of all $\Lambda_{N,n}^m \int_D \rho(\boldsymbol{x};\boldsymbol{x}')\boldsymbol{x}_{i>m}$ reconstructs the m-th order RDM of the full system [14], i.e. $\sum_{n=0}^N \Lambda_{N,n}^m \int_D \rho(\boldsymbol{x};\boldsymbol{x}')\boldsymbol{x}_{i>m} \equiv \rho^m$, where

$$\rho^m(\boldsymbol{x}_{i\leq m};\boldsymbol{x}'_{i\leq m}) = \frac{N!}{(N-m)!}\int_{R^3} \Psi(\boldsymbol{x}_1\ldots,\boldsymbol{x}_N)\Psi^\star(\boldsymbol{x}'_1\ldots,\boldsymbol{x}'_N)d\boldsymbol{x}_{i>m} \tag{7.14}$$

As a direct consequence,

$$\rho^{A,m} = \sum_n \rho_n^{A,m} = \mathbf{1}_m^{'A}\mathbf{1}_m^A \rho^m \tag{7.15}$$

This equation has recently been used in the development of a strategy to determine the Lewis structures of arbitrary molecules under the viewpoint of OQSs alternative [25], for example, to the well-known natural bonding orbital (NBO) method or the adaptive natural density partitioning scheme developed some time ago for the same purpose by Boldyrev and coworker [26].

Sector RDMs can be manipulated in a way almost entirely analogous to that used with the usual RDMs of Quantum Chemistry. For example, the first-order n-electron sector RDM, $\rho_n^{A,1}$, integrates to n, the number electrons of the sector, $\rho_n^{A,2}$ integrates to the number of electron pairs of the sector, $n(n-1)$, etc. Moreover, each of the mRDMs of a n–electron sector can be diagonalized to obtain sector-specific natural orbitals ($m = 1$), geminals ($m = 2$), etc. The algebra is relatively simple in the case of single determinant wavefunctions (SDW). Then, it is always possible to find a one-electron basis of molecular orbitals (MO), $\overline{\phi}_i$, simultaneously orthogonal in Ω_A and $\overline{\Omega}_A$ [27]. This is achieved by diagonalizing S^A, the Hermitian matrix defined as $S_{ij}^A = \langle \varphi_i | \varphi_j \rangle_A$, where φ_i is another basis of MOs (usually the canonical MOs or the basis of Kohn–Sham MOs in a density functional theory (DFT) calculation [28]), by a unitary transformation U, $U^\dagger S^A U = \text{diag}(s_i) \equiv s$. The $\overline{\phi}_i$'s defined by $|\overline{\phi}_i\rangle = \sum_j |\varphi_j\rangle U_{ji}$ enjoy the properties $\langle \overline{\phi}_i|\overline{\phi}_j\rangle_A = \delta_{ij} s_i = \delta_{ij} - \langle \overline{\phi}_i|\overline{\phi}_j\rangle_{\bar{A}} = \delta_{ij}(1 - s_i)$. Simultaneously, the $\overline{\phi}_i$ basis is orthonormal in R^3, $\langle \overline{\phi}_i|\overline{\phi}_j\rangle = \delta_{ij}$. It is particularly relevant for chemical bonding purposes that the $\overline{\phi}_i$ basis coincides exactly with that proposed by Robert Ponec for SDWs, [14, 29, 30] and also that the s_i's have a statistical interpretation: s_i is equal to the probability of finding an electron described by $|\overline{\phi}_i\rangle$ in the domain Ω_A ($\overline{\Omega}_A$), and $1 - s_i$ that of finding it in its complementary domain $\overline{\Omega}_A$. They are usually called domain natural orbitals (DNO) and have been successfully used to extract chemical information and interpreted in statistical terms [19, 27].

If an N-electron SDW is written in the DNO basis as

$$|\Psi\rangle = (N!)^{-1/2} \det |\overline{\phi}_1(x_1) \ldots \overline{\phi}_N(x_N)| \tag{7.16}$$

it can be proven that

$$\rho_n^A(x_{i\leq n}; x'_{i\leq n}) = 1_n^{'A} 1_n^A \times \sum_k |\phi_k\rangle p_n^k \langle \phi_k| \tag{7.17}$$

where ϕ_i is an orthonormal basis in Ω_A defined as $\phi_i = s_i^{-1/2}\overline{\phi}_i$, $k = \{k_1, \ldots, k_n\}$ is a set of n ordered integers ($k_1 < \cdots < k_n$), the sum runs over the $\binom{N}{n}$ possible choices of k, $p_n^k = \prod_i^N p_i$ with $p_i = s_i$ if $i \in k$ and $p_i = 1 - s_i$ if $i \notin k$, and $|\phi_k\rangle$ is the n-electron Slater determinant made of the spin-orbitals $\phi_{k_1}, \ldots, \phi_{k_n}$, i.e.

$$|\phi_k\rangle = \frac{1}{\sqrt{n!}}|\phi_{k_1}(x_1)\cdots \phi_{k_n}(x_n)\rangle \tag{7.18}$$

We must note that, since the $|\phi_i\rangle$'s are orthonormal in Ω_A, one has $\langle \phi_k|\phi_k\rangle_A = 1$. For obvious reasons, the $|\phi_i\rangle$'s will be called normalized DNOs.

Equation (7.17) enjoys a certainly beautiful interpretation. For a SDW, the n-electron sector RDM is a statistical mixture of all possible ket, bra dyads of n-electron determinants that can be obtained by selecting from the original N-electron wavefunction, expressed in the basis ϕ_i, n of these ϕ_i in all possible ways. Each contribution has a statistical weight equal to p_n^k. This weight can be interpreted as the probability of finding the subsystem A in the state $|\phi_k\rangle$. Since

$\text{Tr}|\phi_k\rangle\langle\phi_k| = 1$, the probability of having exactly n electrons in Ω_A and $N - n$ electrons in $\overline{\Omega}_A$ will be given by $p^A(n) = \sum_k p_n^k$. For instance, for a 4-electron molecule, $p^A(3) = s_1 s_2 s_3 (1 - s_4) + s_1 s_2 (1 - s_3) s_4 + s_1 (1 - s_2) s_3 s_4 + (1 - s_1) s_2 s_3 s_4$.

The sector RDMs are relatively easy to obtain in the DNO basis: Equation (7.17) is a sum of single-determinant contributions. Hence, the standard Löwdin expressions [23] for the RDMs apply to each term. For instance, the sector 1RDM, $\rho_n^{A,1}(x;x')$, is directly diagonal, and it is given by

$$\rho_n^{A,1}(x;x') = 1_1'^A 1_1^A \times \sum_{j=1}^{N} \phi_j^\star(x') n_{n,j}^{A,1} \phi_j(x) \tag{7.19}$$

where $n_{n,j}^{A,1}$ are the sector domain natural occupations, $n_{n,j}^{A,1} = s_j \times p_j^A(n-1)$, with $p_j^A(n-1)$ representing the probability that $n-1$ electrons lie in Ω_A and $(N-n)$ electrons in $\overline{\Omega}_A$ for a hypothetical $(N-1)$-electron SDW built with all ϕ_i's except ϕ_j. Taking again the case with $N = 4$ and $n = 3$, we have $p_1^A(2) = s_2 s_3 (1 - s_4) + s_2 s_4 (1 - s_3) + s_3 s_4 (1 - s_2)$, $p_2^A(2) = s_1 s_3 (1 - s_4) + s_1 s_4 (1 - s_3) + s_3 s_4 (1 - s_1)$, etc. Given the relation between normalized DNOs (ϕ_i) and Ponec DNOs ($\overline{\phi}_i$), $\rho_n^{A,1}(x;x')$ in the basis of the latter is

$$\rho_n^{A,1}(x;x') = 1_1'^A 1_1^A \times \sum_{j=1}^{N} \overline{\phi}_j^\star(x') p_j^A(n-1) \overline{\phi}_j(x) \tag{7.20}$$

Since the ϕ_i's are normalized in Ω_A, the trace of the n-electron sector 1RDM is equal to $\sum_{j=1}^{N} n_{n,j}^{A,1}$. Performing this sum for the $N=4$, $n=3$ case, one has $\sum_{j=1}^{N} n_{n,j}^{A,1} = 3[s_1 s_2 s_3(1-s_4) + s_1 s_2(1-s_3)s_4 + s_1(1-s_2)s_3 s_4 + (1-s_1)s_2 s_3 s_4] = 3 \times p^A(3)$. Of course, this result is completely general: the trace of the n-electron sector 1RDM is equal to $n \times p^A(n)$. The average electron population in Ω_A from an electron-counting perspective is $\langle n_A \rangle = \sum_{n=0}^{N} n \times p^A(n)$, so that $n \times p^A(n)$ is simply the contribution of sector n to the average population of the domain.

At the single-determinant level that we are considering, Eqs. (7.19) and (7.20) are valid for all values of n. Therefore, DNOs are the sector natural orbitals for any sector. This is no longer true in the case that electron correlation is taken into account (for instance, when Ψ is multideterminant). The total occupation number of DNO ϕ_j is given by a sum of the $n_{n,j}^{A,1}$ for all sectors, $n_j^{A,1} = \sum_n n_{n,j}^{A,1}$, and it can be shown to be equal to the overall probability that ϕ_j is found occupied in all $|\phi_k\rangle$'s in the sector multiplied by $s_j = \langle \overline{\phi}_j | \overline{\phi}_j \rangle_A$, the probability that a ϕ_j electron be found in Ω_A. Finally, being diagonal the RDMs of all the sectors in the DNO basis, it is clear that the total domain 1RDM, $\rho^{A,1}(x;x')$, is also diagonal in this basis.

Let us show how this machinery can be put to work in a very simple example. We present in Table 7.1 a brief summary of the results obtained for the LiH molecule described at the RHF//6-311G(d,p) level. Since DNOs are the sector natural orbitals for any sector, adding for any $\overline{\phi}_j$ the $p_j^A(n-1)$ values of Eq. (7.20) for all values of n one obviously obtains 2.0, the full contribution of this orbital to the electron density of the molecule $\rho(r)$. Similarly, if we add for each n the five $p_j^A(n-1)$'s, the result is the contribution of sector n to $\int_{R^3} dr \rho(r)$. On the other hand, from Eq. (7.19) or (7.20), the contribution of the j-th DNO to the electron population of fragment A in

Table 7.1 Sector densities $\rho_n^{A,1}$ of the LiH molecule at the RHF//6-311G(d,p) level in the basis of Ponec DNOs φ_j. A is the Li atomic basin. The first and second parts of the Table are the $p_j^A(n-1)$ and $n_{n,j}^A(n-1)$ coefficients of Eqs. (7.20) and (7.19), respectively, and the s_j's are the overlaps of the DNOs in A.

n	$p_1^A(n-1)$	$p_2^A(n-1)$	$\sum_j p_j^A(n-1)$
4	0.0067	0.0000	0.0069
3	1.8103	0.0141	1.8244
2	0.1785	1.8926	2.0711
1	0.0044	0.0933	0.0978
Σ	2.0000	2.0000	4.0000
s_j	0.9963	0.04700	
n	$n_{n,1}^{A,1}$	$n_{n,2}^{A,1}$	$\sum_j n_{n,j}^{A,1}$
4	0.0067	0.0000	0.0067
3	1.8036	0.0007	1.8043
2	0.1779	0.0889	0.2668
1	0.0045	0.0044	0.0088
Σ	1.9926	0.0940	2.0866

the sector n is given by $n_{n,j}^{A,1} = s_j \times p_j^A(n-1)$, so that $n_{n,j}^{A,1}$ is obtained by multiplying each $p_j^A(n-1)$ of Table 7.1 by s_j. The average electron population of fragment A is given by $\langle n_A \rangle = \sum_{n,j} n_{n,j}^{A,1}$. However, since $\sum_n p_j^A(n-1) = 2$ for any j, $\langle n_A \rangle = 2\sum_j s_j$.

All the above equations for sector 1RDMs in SDWs can be generalized to multideterminant Ψ's. We will not discuss them in this chapter. However, as in the SDW case, their expressions are much simpler when they are expressed in a basis simultaneously orthogonal in Ω_A and $\overline{\Omega}_A$. Likewise, sector natural geminals, $\rho_n^{A,2}(x_1, x_2; x_1', x_2')$, take a simple form in SDWs when expressed in the DNO basis, although they become a little bit complicated in the case of correlated wavefunctions.

7.4 Interacting Quantum Atoms (IQA)

The interacting quantum atoms (IQA) method is an exact orbital invariant energy partitioning of a molecule based on the QTAIM [31, 32]. The total energy E of a molecule described by a Coulomb Hamiltonian can be written as

$$E = \text{Tr}[\hat{h}\rho_1(r; r_1')] + \frac{1}{2} \text{Tr}[r_{12}^{-1}\rho_2(r_1; r_2)] + V_{nn} \qquad (7.21)$$

where $\hat{h}_i = \hat{T} - \sum_\alpha Z^\alpha / r_{i\alpha}$ is its one-electron part, α runs over all the nuclei of the molecule, with nuclear charge Z^α, $\rho_1(r; r_1')$ and $\rho_2(r_1; r_2)$ are the spinless one-particle and (diagonal) two-particle RDMs, obtained from the general definition (Eq. (7.14)) by integrating the spin coordinates of all the electrons, and V_{nn} is the total inter-nuclear repulsion. An arbitrary exhaustive partition of R^3 in domains A allows to re-write Eq. (7.21) as

$$E = \sum_A \int_A d\mathbf{r}_1 \hat{h} \rho_1(\mathbf{r}_1; \mathbf{r}_1')|_{\mathbf{r}_1' \to \mathbf{r}_1}$$
$$+ \frac{1}{2} \sum_{A,B} \int_A d\mathbf{r}_1 \int_B d\mathbf{r}_2 r_{12}^{-1} \rho_2(\mathbf{r}_1, \mathbf{r}_2) + V_{nn} \quad (7.22)$$

The IQA method corresponds to the choice of region A as equal to the atomic basins of the QTAIM [31]. Separating the double sum over A and B into the contributions with $A = B$ and with $A \neq B$, E becomes

$$E = \sum_A E^A_{\text{self}} + \frac{1}{2} \sum_{A \neq B} E^{AB}_{\text{int}} \quad (7.23)$$

where E^A_{self} is the self-energy of atom A, that includes the kinetic energy of the electrons in A (T^A), the nuclear attraction between electrons in A and the nucleus of this domain (V^{AA}_{ne}), and the electron repulsion of electrons in A with themselves (V^{AA}_{ee}), i.e. $E^A_{\text{self}} = T^A + V^{AA}_{ne} + V^{AA}_{ee}$. These three terms are given, respectively, by

$$T^A = \int_A d\mathbf{r}_1 \hat{T} \rho_1(\mathbf{r}_1; \mathbf{r}_1')|_{\mathbf{r}_1' \to \mathbf{r}_1} \quad (7.24)$$

$$V^{AA}_{ne} = -Z^A \int_A d\mathbf{r}_1 \rho(\mathbf{r}_1)/r_{1A} \quad \text{and} \quad (7.25)$$

$$V^{AA}_{ee} = \int_A d\mathbf{r}_1 \int_A d\mathbf{r}_2 r_{12}^{-1} \rho_2(\mathbf{r}_1; \mathbf{r}_2) \quad (7.26)$$

E^{AB}_{int} in Eq. (7.23) is equal to $V^{AB}_{nn} + V^{AB}_{ne} + V^{BA}_{ne} + V^{AB}_{ee}$, where V^{AB}_{nn} is the repulsion between nuclei A and B, V^{AB}_{ne} and V^{BA}_{ne} the attraction between the nucleus of A and the electrons in B and vice versa, and V^{AB}_{ee} is the repulsion between the electrons in A and electrons in B. The IQA partition, despite its relatively high computational cost, is valid at any molecular geometry, making it very useful for visualizing the energy changes between different fragments of a system along a chemical reaction. This contrasts with the standard QTAIM, where the virial theorem is invoked to write the total energy of an atomic basins A as equal to minus the kinetic energy of this basin, $E^A = -T^A$, that can be exclusively applied at the equilibrium geometry.

The IQA scheme only requires $\rho_1(\mathbf{r}_1; \mathbf{r}_1')$ and $\rho_2(\mathbf{r}_1, \mathbf{r}_2)$ as input, regardless of how they are obtained. This flexibility allows the method to work with electronic structure methods that yield a wavefunction, as it is the case of the Hartree–Fock (HF), complete active space (CAS), or full Interaction of Configurations (full-CI) methods, but also with procedures where the wavefunction is typically unavailable but forms for the 1RDM and 2RDMs can be obtained, such as in the case of the Coupled Cluster (CC) method, which is highlighted for its high accuracy in medium and small systems. It is even possible to apply it in the DFT realm where a true $\rho_2(\mathbf{r}_1, \mathbf{r}_2)$ is not available [33], expanding the potential for conducting IQA calculations on larger systems that were previously unexplored. IQA can also be applied to systems in excited states [34].

The self-energy of atom A in the molecule, E^A_{self}, is the local expectation value of the atomic Hamiltonian in the molecule and contains the same energetic contributions as a single isolated atom A. The difference $E^A_{def} = E^A_{self} - E^{A,0}_{self}$, where $E^{A,0}_{self}$ is the total energy of isolated A, is usually called the deformation energy of A, and it provides the change in the total energy that this atom suffers when it passes from being isolated to being part of the molecule. The binding energy of a molecule, measured with respect to the isolated atom references, is written as

$$E_{bind} = E - \sum_A E^{A,0}_{self} = \sum_A E^A_{def} + \frac{1}{2}\sum_{A \neq B} E^{AB}_{int} \qquad (7.27)$$

E^A_{def} can be viewed as the sum of two effects, $E^A_{def} = E^A_{def}(CT) + E^A_{def}(CR)$. The first one is due to the charge transfer (CT) effects that take place when the molecule is formed from the isolated atoms, and the second one is the atomic charge re-organization (CR) energy, which is nonzero even in the absence of CT (for instance, when a homonuclear diatomic molecule A_2 is formed from the isolated atoms A).

Probably, the most valuable contributions that the IQA method has made to the understanding of chemical bonding comes from the splitting of E^{AB}_{int} in two contributions endowed with a clear physical content. If we separate the 2RDM as

$$\rho_2(r_1; r_2) = \rho(r_1)\rho(r_2) - \rho_{xc}(r_1; r_2), \qquad (7.28)$$

where $\rho(r) \equiv \rho_1(r)$ is the electron density, $\rho(r_1)\rho(r_2)$ is the Coulomb contribution to $\rho_2(r_1, r_2)$, and $\rho_{xc}(r_1, r_2)$ is the exchange-correlation (xc) density, E^{AB}_{int} can be written as

$$E^{AB}_{int} = V^{AB}_{cl} + V^{AB}_{xc} \qquad (7.29)$$

The first term or classical (cl) interaction contains the full electrostatic interaction between the nucleus and electrons of atom A with the nucleus and electrons of B,

$$V^{AB}_{cl} = V^{AB}_{nn} + V^{AB}_{ne} + V^{BA}_{ne} + V^{AB}_{Coul} \quad \text{with} \qquad (7.30)$$

$$V^{AB}_{Coul} = \int_A dr_1 \int_B dr_2 \rho_1(r_1)\rho_1(r_2) r_{12}^{-1} \qquad (7.31)$$

The second term in Eq. (7.29), V^{AB}_{xc}, is exchange-correlation interaction energy, which collects the covalent or bond energy contributions to the interaction between both atoms [35]. Its exact expression is given by

$$V^{AB}_{xc} = -\int_A dr_1 \int_B dr_2 \rho_{xc}(r_1; r_2) r_{12}^{-1} \qquad (7.32)$$

For well-separated atomic densities, the classical term V^{AB}_{cl} can be approximated by a multipolar series. The leading term in the expansion is $E^{AB}_Q = Q^A Q^B / R_{AB}$, where Q^A and Q^B are the net atomic charges of A and B. This energy is by far the most important in highly ionic systems. As we just noted V^{AB}_{xc} is related to covalent-like contributions between atoms A and B [35]. The bond order in QTAIM is quantified by the delocalization index, δ^{AB}, which counts the number of shared-electron pairs between both atoms [36]. Since

$$\delta^{AB} = 2\int_A dr_1 \int_B dr_2 \rho_{xc}(r_1, r_2) \qquad (7.33)$$

is seems clear that integrals (7.32) and (7.33) are clearly correlated.

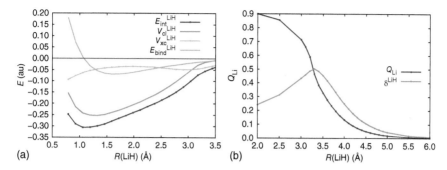

Figure 7.3 IQA results for a CAS(2,2)/6-311G(p) calculation in the ground state of LiH. (a) Shows the interaction energy components together with the total binding energy, and (b) the topological charge of the Li atom as well as total Li,H delocalization index.

The IQA method and all the energy expressions discussed so far can be generalized so that A and B represent fragments and not individual atoms. For example, the self-energy of a fragment $\mathcal{G}$ and the xc interaction between two fragments $\mathcal{G}$ and $\mathcal{H}$ are given by the expressions [17]

$$E^{\mathcal{G}}_{self} = \sum_{A \in \mathcal{G}} E^A_{self} + \sum_{A \geq B \in \mathcal{G}} E^{AB}_{int} \quad \text{and} \tag{7.34}$$

$$V^{\mathcal{G}\mathcal{H}}_{xc} = \sum_{A \in \mathcal{G}} \sum_{B \in \mathcal{H}} V^{AB}_{xc} = -\int_{\mathcal{G}} d\mathbf{r}_1 \int_{\mathcal{H}} d\mathbf{r}_2 \rho_{xc}(\mathbf{r}_1; \mathbf{r}_2) r_{12}^{-1} \tag{7.35}$$

At the equilibrium geometry, the total energy in the QTAIM is exactly given as a sum of atomic contributions, $E = \sum_A E^A = -\sum_A T^A$. However, the greater generality of IQA, in the sense of its validity for any molecular geometry, means that the second of the two previous equalities does not hold in general. However, it is still possible to define additive atomic (or group) energies, $E^{\mathcal{G}}_{add}$, such that their sum exactly recovers E, $E = \sum_A E^{\mathcal{G}}_{add}$. This definition is

$$E^{\mathcal{G}}_{add} = E^{\mathcal{G}}_{self} + \frac{1}{2} \sum_{\mathcal{G} \neq \mathcal{H}} E^{\mathcal{G}\mathcal{H}}_{int} \tag{7.36}$$

Let us show a minimal example of IQA insights. We have selected again LiH, so that the reader can compare with the previous discussion regarding the OQS approach. Figure 7.3 shows some CAS(2,2)/6-311G(p) results. Notice how the interaction energy is dominated at large distances by the exchange-correlation contribution, like in a covalent system, while this changes abruptly at about 3.3 Å. This coincides (right panel) with a peak in the delocalization index and an inflection point in the Li topological charge, signaling the avoided crossing between the neutral and the ionic states. At equilibrium, in agreement with the OQS analysis, Li is basically ionized, and the main occupied electron sector is that with 2 electrons.

References

1 Kohout, M., Pernal, K., Wagner, F.R., and Grin, Y. (2004). Electron localizability indicator for correlated wavefunctions. I. Parallel-spin pairs. *Theor. Chem. Acc.* 112 (5-6): 453–459.

2 Kohout, M., Pernal, K., Wagner, F.R., and Grin, Y. (2005). Electron localizability indicator for correlated wavefunctions. II Antiparallel-spin pairs. *Theor. Chem. Acc.* 113 (5): 287–293.

3 Ruedenberg, K. and Schmidt, M.W. (2009). Physical understanding through variational reasoning: electron sharing and covalent bonding. *J. Phys. Chem. A* 113 (10): 1954–1968.

4 Kato, W.A. (1957). On the eigenfunctions of many-particle systems in quantum mechanics. *Commun. Pure Appl. Math.* 10: 151.

5 Bader, R.F.W. (1990). *Atoms in Molecules*. Oxford: Oxford University Press.

6 Matta, C.F. and Boyd, R.J. (ed.) (2007). *The Quantum Theory of Atoms in Molecules*. Wiley.

7 Martín Pendás, A., Blanco, M.A., Costales, A. et al. (1999). Non-nuclear maxima of the electron density. *Phys. Rev. Lett.* 83 (10): 1930–1933.

8 Luaña, V., Mori-Sánchez, P., Costales, A. et al. (2003). Non-nuclear maxima of the electron density on alkaline metals. *J. Chem. Phys.* 119 (12): 6341–6350.

9 Timerghazin, Q.K. and Peslherbe, G.H. (2007). Non-nuclear attractor of electron density as a manifestation of the solvated electron. *J. Chem. Phys.* 127 (6): 064108.

10 Wu, W., Gu, J., Song, J. et al. (2009). The inverted bond in [1.1.1]propellane is a charge-shift bond. *Angew. Chem. Int. Ed.* 48 (8): 1407–1410.

11 Bader, R.F.W. and Essén, H. (1984). The characterization of atomic interactions. *J. Chem. Phys.* 80 (5): 1943–1960.

12 Bader, R.F.W., Gillespie, R.J., and MacDougall, P.J. (1988). A physical basis for the VSEPR model of molecular geometry. *J. Am. Chem. Soc.* 110 (22): 7329–7336.

13 Keith, T.A. (2019). AIMAll 19.10.12, TK Gristmill Software. Overland Park, KS, USA. https://aim.tkgristmill.com/

14 Martín Pendás, A. and Francisco, E. (2018). Quantum chemical topology as a theory of open quantum systems. *J. Chem. Theory Comput.* 15 (2): 1079–1088.

15 Breuer, H.-P. and Petruccione, F. (2002). *The Theory of Open Quantum Systems*. Oxford, New York: Oxford University Press. ISBN: 978-0199213900.

16 Nielsen, M. and Chuang, I.L. (2010). *Quantum Computation and Quantum Information*. Cambridge, New York: Cambridge University Press.

17 Martín Pendás, A., Blanco, M.A., and Francisco, E. (2006). Chemical fragments in real space: definitions, properties, and energetic decompositions. *J. Comput. Chem.* 28 (1): 161–184.

18 Francisco, E., Martín Pendás, A., and Blanco, M.A. (2007). Electron number probability distributions for correlated wave functions. *J. Chem. Phys.* 126 (9): 094102.

19 Martín Pendás, A., Francisco, E., and Blanco, M.A. (2007). Spin resolved electron number distribution functions: how spins couple in real space. *J. Chem. Phys.* 127 (14): 144103.

20 Martín Pendás, A., Francisco, E., and Blanco, M.A. (2007). An electron number distribution view of chemical bonds in real space. *Phys. Chem. Chem. Phys.* 9 (9): 1087–1092.

21 Francisco, E. and Martín Pendás, A. (2014). Electron number distribution functions from molecular wavefunctions. Version 2. *Comput. Phys. Commun.* 185 (10): 2663–2682.

22 Martín Pendás, Á. and Francisco, E. (2019). Chemical bonding from the statistics of the electron distribution. *ChemPhysChem* 20 (21): 2722–2741.

23 Löwdin, P.-O. (1955). Quantum theory of many-particle systems. I. Physical interpretations by means of density matrices, natural spin-orbitals, and convergence problems in the method of configurational interaction. *Phys. Rev.* 97 (6): 1474–1489.

24 Martín Pendás, A., Francisco, E., and Blanco, M.A. (2007). Pauling resonant structures in real space through electron number probability distributions. *J. Phys. Chem. A* 111 (6): 1084–1090.

25 Francisco, E., Costales, A., Menéndez-Herrero, M., and Martín Pendás, A. (2021). Lewis structures from open quantum systems natural orbitals: Real space adaptive natural density partitioning. *J. Phys. Chem. A* 125 (18): 4013–4025.

26 Zubarev, D.Y. and Boldyrev, A.I. (2008). Developing paradigms of chemical bonding: adaptive natural density partitioning. *Phys. Chem. Chem. Phys.* 10 (11): 5207–5217.

27 Francisco, E., Martín Pendás, A., and Blanco, M.A. (2009). A connection between domain-averaged Fermi hole orbitals and electron number distribution functions in real space. *J. Chem. Phys.* 131 (12): 124125.

28 Parr, R.G. and Yang, W. (1989). *Density-Functional Theory of Atoms and Molecules*. New York: Oxford University Press.

29 Ponec, R. (1997). Electron pairing and chemical bonds. Chemical structure, valences and structural similarities from the analysis of the Fermi holes. *J. Math. Chem.* 21: 323–333.

30 Ponec, R. (1998). Electron pairing and chemical bonds. Molecular structure from the analysis of pair densities and related quantities. *J. Math. Chem.* 23: 85–103.

31 Blanco, M.A., Martín Pendás, A., and Francisco, E. (2005). Interacting quantum atoms: a correlated energy decomposition scheme based on the quantum theory of atoms in molecules. *J. Chem. Theory Comput.* 1: 1096–1109.

32 Francisco, E., Martín Pendás, A., and Blanco, M.A. (2006). A molecular energy decomposition scheme for atoms in molecules. *J. Chem. Theory Comput.* 2: 90–102.

33 Francisco, E., Casals-Sainz, J.L., Rocha-Rinza, T., and Martín Pendás, Á. (2016). Partitioning the DFT exchange-correlation energy in line with the interacting quantum atoms approach. *Theor. Chem. Acc.* 135: 170.

34 Fernández-Alarcón, A., Casals-Sainz, J.L., Guevara-Vela, J.M. et al. (2019). Partition of electronic excitation energies: the IQA/EOM-CCSD method. *Phys. Chem. Chem. Phys.* 21 (25): 13428–13439.

35 Martín Pendás, Á., Francisco, E., Blanco, M.A., and Gatti, C. (2007). Bond paths as privileged exchange channels. *Chem. Eur. J.* 13 (33): 9362–9371.

36 Bader, R.F.W. and Stephens, M.E. (1975). Spatial localization of the electronic pair and number distributions in molecules. *J. Am. Chem. Soc.* 97 (26): 7391–7399.

8

Effective Oxidation States Analysis

Pedro Salvador

Universitat de Girona, Institut de Química Computacional i Catàlisi, Departament de Química, C/ Mª Aurèlia Capmany 69, 17003 Girona, Catalonia, Spain

8.1 The Concept of Oxidation State

Computational chemistry methods have reached an unprecedented level of accuracy, either from the most sophisticated density functional approximations [1] or by making genuine wavefunction (WF) methods affordable, like the successful domain-based local pair natural orbital coupled-cluster (DLNO-CC) theories [2]. Combined with artificial intelligence [3] and increasing computational power, these methods are leading to black-box strategies to explore complex reaction mechanisms [4] or to simulate from first principles electron ionization mass spectrometry [5] or fully spin-coupled NMR spectra [6] with minimal input or human guidance. So, as Grimme and Schreiner anticipated [7], perhaps in the near future, human-less computations will truly *reveal* new chemistry, rather than assist experimental evidence.

Contrary to Molecular Physics, heuristic concepts play a key role in chemical knowledge. According to quantum mechanics, all physical information on a system is contained in the WF, and well-defined observables can be obtained via the expectation values using the appropriate operator. Unfortunately, chemical concepts such as aromaticity, atom-in-molecule (AIM), or the mere chemical bond are not observable in a strict quantum mechanical sense, so there are few detractors claiming their inherent arbitrariness. Still, they have undoubtedly proven very useful for shedding light on chemical phenomena and, more importantly, for achieving *true* predictions (i.e. without actually performing an experiment or a computational exercise).

Often, the problem with many of these chemical concepts arises when it comes to their quantification, as recently stressed by Grunenberg [8]. There are a myriad of aromaticity indicators or different realizations to compute bond orders or partial atomic charges that sometimes may lead to different interpretations emerging from the same WF/density. There is, however, a chemical concept of utmost relevance in chemistry whose flaw is not how it is quantified but rather its very general definition, namely the oxidation state (OS).

For years, the entry of the term "oxidation state" on IUPAC's Gold Book discussed a set of "agreed upon" rules for deriving the OS, but no formal definition was given. Consequently, considerable debate can be found in the literature discussing over misconceptions, inconsistencies, or alternative OS assignment in nontrivial bonding situations over the years [9–13].

On the other hand, the concept of oxidation number was associated with the central atom in a coordination entity, being the *charge the central atom would bear if all the ligands were removed along with the electron pairs shared with the central atom*. Nowadays, both OS and oxidation number are used almost indistinctively.

Rightly so, in 2009, the IUPAC set up a task group led by Prof. Karen aimed at tackling the conundrum. Their conclusions were made public in 2014 with an extensive IUPAC Technical Report [14], gathering over a hundred examples, and in 2015 with an accompanying essay [15]. Final recommendations and a summary of the task group were later provided [16], based on which the IUPAC Gold Book entries for OS and ON have been most recently updated [17] A new generic definition for OS of an atom has been given, namely *"the atom's charge after ionic approximation of its heteronuclear bonds,"* as well as practical algorithms applicable to molecules and solids. The burden is thus on the application of the ionic approximation (IA). In the case of molecular systems, with the algorithm of moving bonds, *"Bond electrons are moved onto the more negative bond partner identified by ionic approximation, and atom charges are evaluated, giving the OS"* [14]. The algorithm proceeds by first establishing the appropriate Lewis structure for the molecule. Then, it allocates electron pairs from each bond to the more electronegative atom based on Allen's electronegativity scale [18]. By knowing the number of electrons assigned to each atom, the OS values can be determined through simple subtraction. The implementation of this new OS definition along with its corresponding algorithm signifies a substantial step forward from earlier guidelines, offering a broadly efficient approach for assigning OSs. Due to its simplicity and overall effectiveness, the IUPAC method should be the first choice when determining OSs in newly encountered compounds of interest.

8.2 Oxidation State is Not Related to the Partial Charge

Noticeably, despite OS being intrinsically related to the physical electron distribution around atoms, the IUPAC report tiptoed around the role of quantum-chemical calculations for OS assignment. The eventual success of ascertaining formal OS from first principles has been largely hindered by the tacit assumption that their role is merely to provide partial atomic charges and bond orders. It is still striking that many disputes in the literature have revolved around the association of partial atomic charges with OS. It is known that partial atomic charges do not *match* with OS, and the recurring arguments are whether or not they can (or should) be correlated with

8.2 Oxidation State is Not Related to the Partial Charge

Table 8.1 Iron partial charges (Q_{Fe}) and spin populations (p_sFe) from a series of [Fe(PyTACN)] complexes obtained using the TFVC atomic definition.

System	Q_{Fe}	OS	p_s Fe	p_s (ideal) Fe
[Fe(Pytacn)(H$_2$O)$_2$]$^{2+}$	1.28	+2	—	—
[Fe(Pytacn)(H$_2$O)(OH)]$^+$	1.24	+2	—	—
[Fe(Pytacn)(OH)$_2$]$^+$	1.68	+3	4.10	5
[Fe(Pytacn)(OH)$_2$]$^{2+}$	1.51	+4	1.88	2
[Fe(Pytacn)O(H$_2$O)]$^{2+}$	1.45	+4	1.30	2
[Fe(Pytacn)O(OH)]$^+$	1.52	+4	3.14	4
[Fe(Pytacn)O(OH)]$^{2+}$	1.50	+5	2.10	3

PyTACN = 1-(2'-pyridylmethyl)-4,7-dimethyl-1,4,7-triazacyclononane.
Source: Results are extracted from Ref. [22].

OS [10–12, 19–21] An illustrative example is the series of iron-based compounds described in Ref. [22] and compiled here in Table 8.1. The formal OS of the hexa-coordinate iron center ranges from Fe(+2) in the bis-aquo complex to up to Fe(+5) in the oxo-hydroxo species. One can immediately see from the data that the partial atomic charges on Fe do not even correlate with OS. In fact, a formal Fe(+3) case exhibits a larger partial charge (more positive) than the most oxidized species. The same occurs with the integrated atomic spin densities, or spin populations. These are nothing else but the difference of partial atomic charges computed for the two spin channels. Being a difference, the effect of choosing one or another AIM to compute the charges is minimized in the case of spin populations. Yet, in order to relate them to OS, one must know beforehand the spin state of the complex, and of course, they trivially vanish for closed-shell species.

When extracting chemical information from WF analysis, one should clearly distinguish the function that is scrutinized from the method chosen to perform the analysis. The latter, in most analyses, refers to how atoms are identified within the molecule, which is essential in the assignation of the OS. In the so-called Hilbert-space analyses, the linear combination of atomic orbital approaches to generate the molecular orbitals (LCAO-MO) framework is exploited to collect atomic contributions. When the density function is the one-electron density $\rho(r)$ and the AOs used are those of the underlying one-electron basis set, the well-known Mulliken [23] population analysis is recovered. Orthonormalized AOs such as those provided by Löwdin orthogonalization [24], Weinhold's natural atomic orbitals (NAOs) [25], or Ruedenberg's quasiatomic orbitals [26], or Knizia's intrinsic atomic orbitals (IAOs) [27] lead to more robust atomic populations (i.e. less dependent on the one-electron basis set choice). On the other hand, in real space analyses, the atoms are identified by a region of the three-dimensional physical space that may be disjoint like in Bader's quantum theory of atoms in molecules (QTAIM) [28], or overlapping, like in the different flavors of Hirshfeld-type approaches [29] or the topological fuzzy Voronoi cells (TFVC) [30]. Real-space analyses are very robust with respect to the basis set but still bear the burden of arbitrariness. We do not consider one approach conceptually better than another. In fact, a link has been found between Hilbert-space and real-space analyses by means of the effective atomic orbitals (eff-AOs) [31–33]. The recipe is simple: (i) Pick a real-space atomic definition such as Hirshfeld's approach and obtain the corresponding numerical eff-AOs in that framework; (ii) Expand the original WF in terms of these eff-AOs and perform a Mulliken-type analysis of the electron density. The original Hirshfeld's atomic populations will be *exactly* recovered. Such a numerical exercise evidences that there is nothing fundamentally flawed with Mulliken's approach; it all depends on the underlying basis set used for the analysis [32]. In our opinion, one does not need a more suitable or tailored AIM definition for population analysis when it comes to OS prediction, but rather to figure out a scheme that overcomes partial charges and fits more faithfully with the concept of OS.

8.3 The Molecular Orbital Picture of the Ionic Approximation

The authors of the IUPAC report first appealed to the MO picture shown in Figure 8.1 to apply the IA [14]. Accordingly, the values of coefficients c_A and c_B of AOs of atoms/fragments A and B should decide the fate of the electron pair associated to the A–B bond. The atom/fragment that has larger contribution to the bonding MO is the one that should keep the electron pair in a winner-takes-all fashion.

But, can one really carry out the IA and ascertain the OSs by solely examining the orbital coefficients from a Kohn–Sham density functional theory (KS-DFT) or a WF calculation? Addressing this question necessitates taking several considerations into account. Firstly, in the context of a single-determinant approach, one must decide whether to work with canonical or localized orbitals. Canonical MOs have two key attributes: (i) they conform to the molecular symmetry, and (ii) they are often delocalized across multiple centers. For instance, in the case of a relatively simple molecule like water, using the IA for its two O–H bonds would ideally require MOs associated with each specific bonding interaction N–H. However, canonical orbitals yield A_1 and B_1 orbitals, with contributions from all three centers. In more real-life applications, there are situations where a canonical MO might be relatively

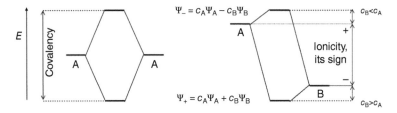

Figure 8.1 The ionic approximation in light of contributions to the bonding MO. Source: Adapted from Walter de Gruyter GmbH [14].

well-associated with a particular bonding interaction. Nevertheless, this isn't always the case, and the MO of interest often contains contributions from other ligands. Should the user go through MO after MO in search of an unmixed one, either due to symmetry or serendipity? Determining OSs demands an approach that isn't contingent on whether individual MOs are sufficiently "pure."

In this context, orbital localization emerges as the appropriate path to follow. For single-determinant WFs, one can execute unitary transformations within their occupied space to generate a different set of orbitals that exhibit greater localization compared to the canonical orbitals, as determined by specific criteria. Localized orbitals typically provide a clear distinction between core orbitals, lone pairs, and valence two-center bonding orbitals. This perspective aligns with the concept of the Lewis structure, which serves as the foundation for IUPAC algorithms. However, it's important to note that there isn't a single criterion for orbital localization, thus presenting users with a choice to make. As an example, the widely known Boys localization scheme [34] does not facilitate the separation of σ and π orbitals, leading to what are known as "banana orbitals" in the case of double bonds. On the other hand, the Pipek-Mezey scheme [35] allows for the separation of σ and π orbitals, making it a more suitable option for the current objective. Orbital localization methods may encounter challenges in symmetric systems, where the localization function may have different minima, corresponding to different resonance structures. Many of the schemes specifically designed to extract OSs from first principles that have been put forward over the last few years rely in one or another way on localized orbitals [36–40].

The next consideration pertains to the one-electron AO basis used to expand the MOs. Should one retain the original AO basis employed in the electronic structure calculation? In the LCAO-MO framework, the MOs are written as

$$\varphi_i(\vec{r}) = \sum_A \sum_{\mu \in A} c_{\mu i} \chi_\mu^A(\vec{r}). \tag{8.1}$$

The contribution of a specific AO to the MO is determined by its corresponding coefficient. But in general, the original underlying AO basis is not orthogonal, so the normalization condition for the MO reads

$$\langle \varphi_i | \varphi_i \rangle = \sum_{\substack{A,B \\ \mu \in A \\ \nu \in B}} c_{\mu i}^* \langle \chi_\mu^A | \chi_\nu^B \rangle c_{\nu i} = \sum_{\substack{A,B \\ \mu \in A \\ \nu \in B}} c_{\mu i}^* S_{\mu \nu} c_{\nu i} = 1, \tag{8.2}$$

where S is the AO overlap matrix. One can readily see that in the original AO framework, the double summation on Eq. (8.2) prevents a straightforward assessment of the contribution of a particular AO coefficient to the MO charge distribution (Mulliken solved it in the population analysis scheme by systematizing over atomic contributions for only one of the two subscripts). More importantly, the corresponding elements of the overlap matrix modulate their contributions, so that external AOs with large overlap with the neighboring atoms may have larger contributions than expected by looking merely at the AO coefficient. In this framework, the individual values of coefficients c_A and c_B in Figure 8.1 would not be sufficient to decide the IA.

Of course, if the AO basis were orthogonal, the AO coefficients would represent probability amplitudes, and their squared magnitudes could be viewed as probabilities. The distribution of individual MO charges is then discretized into contributions from all AOs. These contributions (modulus square of the orbital coefficients) can be further aggregated for all AOs associated with a particular atom or molecular fragment to gauge the atom's overall contribution to the MO charge distribution. These contributions are the values that can potentially be compared when implementing the IA for a bond between atoms A and B, provided that the MO exhibits sufficient localization on these centers (as discussed earlier). The modulus square is useful as it permits to add up contributions from the same atom (i.e. in case the bond orbital is formed by a hybrid, with two or more AOs contributing to it), as well as to properly deal with complex's AO coefficients. In this vein, Löwdin's orthogonalization [24] is just one of the infinite different ways the original AO basis can be transformed into an orthogonal basis, so, once again, the user would have the decision to take on which orthogonal basis to rely on.

A third dimension to consider is the nature of the WF employed in the electronic structure calculation. Thus far, we have primarily discussed the simplest scenario of a single-determinant approach, where a sole configuration is taken into account and the overall charge distribution results from the summation of all MO contributions, whether they are localized or canonical. However, advancements in both technical capabilities and algorithmic methodologies have now made it feasible to utilize multireference methods, such as CASSCF, or post-Hartree–Fock approaches like DPLNO-CCSD(T), where static or dynamic correlation plays a pivotal role, respectively. In CASSCF, it is common to include both bonding and corresponding antibonding MOs within the active space. The role of the antibonding orbital is to induce spin polarization in the bond effectively.

In a straightforward 2×2 scenario, the greater the weight of the configuration involving the antibonding orbital, the less classical covalent character the bond demonstrates. The extreme case where both configurations share the same weight leads to a perfect diradical, where a homolytic cleavage of the "bond" is the anticipated outcome (in such cases, one would typically expect $c_A \approx c_B$). However, intermediate situations with a substantial population of the antibonding orbitals give rise to what are known as diradicaloids, for which a homolytic cleavage might also offer the most accurate description [41]. Returning to Figure 8.1, it's evident that the values of c_A and c_B for the bonding orbital alone cannot distinguish the diradicaloid scenario, urging special considerations to assess the role of the antibonding orbital as well.

In summary, IUPAC's guidelines for implementing the IA, as depicted in Figure 8.1, should not be considered literal. For practical application, both orbital localization and AO basis transformations are essential, but neither procedure is uniquely defined. Additionally, considerations beyond the single-determinant description must be accounted for. Until these uncertainties are resolved, well-established and tested computational methods specifically designed for extracting OSs from WF analysis represent a much more robust alternative to IUPAC's rules and algorithms.

8.4 Spin-Resolved Effective Fragment Orbitals and Effective Oxidation States (EOS) Analysis

Coincidentally, almost at the same time as Karen's report, Ramos-Cordoba et al. introduced a new and general scheme to derive OS from first principles [22]. The method is formally applicable on equal footing to any molecular system and at any level of theory, even in the absence of an underlying atom-centered basis set. The so-called effective oxidation state (EOS) analysis relies on the occupation number of Mayer's eff-AOs [31] obtained for all atoms or molecular fragments defined. In the process to assign the OS, the eff-AOs are considered as either occupied or empty, leading to an effective configuration of these atoms or fragments within the molecular system, which directly determines their OS.

Let us consider a system with n orthonormalized occupied MOs $\varphi_i(r)$, $i = 1, 2, \ldots, n$ of a given spin case (alpha or beta), and a *fuzzy* division of the rea-space into N_{at} atomic domains Ω_A defined, e.g. by a continuous atomic weight function $w_A(r)$, fulfilling $w_A(r) \geq 0$ and $\sum_A w_A(r) \equiv 1$.

Let us for each atom A ($A = 1, 2, \ldots, N_{at}$) form the $n \times n$ Hermitian matrix $\mathbf{Q}^A$ with the elements

$$Q_{ij}^A = \int w_A^*(r)\phi_i^*(r)\phi_j(r)w_A(r)dr \tag{8.3}$$

The matrix $\mathbf{Q}^A$ is essentially the *net* atomic overlap matrix on the basis of the MOs $\{\varphi_i(\vec{r})\}$. Furthermore, for each atom A, we define the *intraatomic* part $\varphi_i^A(r)$ of every MO as $\varphi_i^A(\vec{r}) \equiv w_A(\vec{r})\varphi_i(\vec{r})$. Thus, $Q_{ij}^A = <\varphi_i^A | \varphi_j^A>$ i.e. $\mathbf{Q}^A$ is the overlap matrix of the orbitals $\{\varphi_i^A(r)\}$, for every atom A.

We diagonalize the Hermitian matrix $\mathbf{Q}^A$ by the unitary matrix $\mathbf{U}^A$:

$$\mathbf{U}^{A+}\mathbf{Q}^A\mathbf{U}^A = \mathbf{\Lambda}^A = \text{diag}\{\lambda_i^A\}. \tag{8.4}$$

It can be shown that every $\lambda_i^A \geq 0$, as is the case for a proper overlap matrix. For each atom A we can define n_A ($n_A \leq n$) effective atomic orbitals $\chi_\mu^A(r)$ as linear combinations of the *intraatomic* parts $\{\varphi_i^A(r)\}$ of the MOs as

$$\chi_\mu^A(r) = \frac{1}{\sqrt{\lambda_\mu^A}} \sum_{i=1}^n U_{i\mu}^A \varphi_i^A(r) \quad \mu = 1, 2, \ldots, n_A. \tag{8.5}$$

where n_A is the number of nonzero eigenvalues λ_μ^A. The occupation number of each eff-AO is given by the eigenvalues $0 \leq \lambda_\mu^A \leq 1$. The sum of the occupation number of the n_A eff-AOs is the net population of atom A for the given spin case:

$$N_{net}^A = \sum_\mu^{n_A} \lambda_\mu^A. \tag{8.6}$$

The shape and occupation number of the eff-AOs faithfully reproduce the core and valence shells of the atoms; those with occupation numbers close to 1 are associated with core orbitals or lone pairs, whereas those with smaller but significant occupations are identified with the AOs directly involved in the bonds. The remaining eff-AOs are marginally occupied and have no chemical significance. For most atoms, the number of hybrids with significant occupation numbers always coincides with the classical minimal basis set, except for those that exhibit hypervalent character.

Notice that the eff-AOs and their occupation numbers can be obtained in the framework of real-space analysis even in the absence of an underlying atom-centered basis set, i.e. for plane-wave calculations [42]. Another relevant aspect is that the eff-AOs can be easily obtained for any level of theory, provided a first-order density matrix is available (in the case of Kohn–Sham DFT the latter is approximated by the usual Hartree-Fock-like expression). As noted by Mayer [31], the eff-AOs of a given atom A can also be obtained from the diagonalization of the matrix $\mathbf{PS}^A$, where $\mathbf{P}$ is the LCAO density matrix and $\mathbf{S}^A$ is the intra-atomic overlap matrix in the actual AO basis. This permits the straightforward generalization to correlated WFs, from which the $\mathbf{P}$ matrix is usually available, and it also is the usual form when using Hilbert-space-based AIM to obtain the eff-AOs. An alternative way for correlated calculations is to rely on the natural orbital basis and write the Q^A matrix elements as

$$Q_{ij}^A = n_i^{1/2} n_j^{1/2} \int w_A^*(r) \phi_i^*(r) \phi_j(r) w_A(r) dr, \qquad (8.7)$$

where n_i and n_j are the (spin) natural orbital occupations [32].

Once all eff-AOs and their occupations have been obtained for all atoms, one might naïvely use them to establish the effective valence state of the atoms in the molecule. However, when the number of atoms of the system is large, and especially if there are covalent apolar bonds (e.g. C–C, C–H, etc...), there will be a large number of eff-AOs with very similar and close to 0.5 occupations. These are valence hybrids that participate in the bonds in which each atom is involved. Also, one is usually interested in the valence or OS of the transition metal atoms and the formal charge of their ligands as a whole. Take, for instance, cyclopentadienyl ligand, (C_5H_5), which is generally considered as formally anionic (−1) in order to complete the sexted of the π-system. Any atom-based decomposition of the charge (either real or formal) would necessarily lead to a 1/5 contribution from each C center, and hence a complete indetermination of each individual C atom's (integer) formal charge. Hence, soon we realized that in order to assign OS, one *must* define molecular fragments beforehand and generalize Mayer's eff-AOs to effective fragment orbitals (EFOs) [22].

It is worth stressing that each of the predefined molecular fragments is simply built up by a set of atomic centers. There is no need, contrary to other schemes such as energy decomposition analysis, to perform separated free-fragment reference calculations or to specify the fragment's electronic state. As we will see, the latter is in fact obtained as an output of the EOS procedure.

That is, instead of eff-AOs, we obtain EFOs by using *fragment weight functions* of the form

$$w_P(\vec{r}) = \sum_{i \in P} w_i(\vec{r}), \qquad (8.8)$$

where the sum runs for all atoms of molecular fragment P. The EOS analysis algorithm, depicted in Figure 8.2, goes as follows: (i) the alpha EFOs that are

8.4 Spin-Resolved Effective Fragment Orbitals and Effective Oxidation States (EOS) Analysis | 215

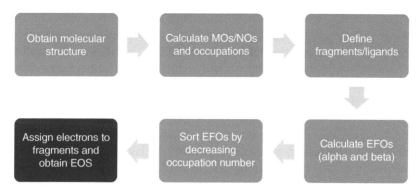

Figure 8.2 The effective oxidation states analysis scheme flowchart.

significantly populated are collected for all fragments, (ii) the EFOs are sorted according to decreasing occupation number, and (iii) integer alpha electrons are assigned to the EFOs of the fragments with higher occupation number until the number of alpha electrons is reached. Then, proceed analogously for the beta electrons. By this procedure, an effective electronic configuration is obtained for each atom/fragment. The EOS of each atom/fragment is simply given by the difference between its atomic number and the number of alpha and beta electrons that have been assigned to it. This scheme can be safely applied to basis sets, including effective core potentials, simply by readily assigning the electrons described by the atomic core potential to the given atom.

It is important to note that the occupation numbers of the EFOs are not simply rounded to the closest integer. Instead, they are sorted in decreasing order for each spin case, and the first n_σ EFOs are considered *occupied*, being n_σ the total number of electrons for spin case σ. By comparing the occupation numbers, the effects of applying one or another AIM to obtain the EFOs are minimized [22]. Such a strategy also underlines the fact that the OS depends on all atoms of the system [10] and, of course, on the total number of electrons. Hence, in the case of electron-deficient systems (e.g. involving boron atoms), it is possible to have the last EFO considered as occupied by the EOS with an occupation smaller than 0.5. Contrarily, electron-rich systems such as formal Cu(III) complexes with CF_3 ligands can exhibit unoccupied EFOs with occupation numbers larger than 0.5.

The occupation numbers of these *frontier* EFOs, namely the last occupied, λ_{LO}^σ and the first unoccupied, λ_{FU}^σ, can be used, for each spin case σ, to indicate how close the formal picture given by the EOS is to the actual electronic distribution of the system. When λ_{LO}^σ and λ_{FU}^σ differ by more than half an electron (i.e. a full electron rounding up the difference in occupation number), the assignation of EOS is considered as fully indisputable. For each spin case, the reliability index $R^\sigma(\%)$ reads

$$R^\sigma(\%) = 100 \min\left(1, \max\left(0, \lambda_{LO}^\sigma - \lambda_{FU}^\sigma + 1/2\right)\right), \quad (8.9)$$

and then $R(\%) = \min(R^\alpha(\%), R^\beta(\%))$. That is, the overall $R(\%)$ index is the minimum value obtained for either the alpha or beta electrons. The larger the $R(\%)$ value the closer the overall assignation of the EOS is to the actual electronic structure of the system. Note that $R(\%)$ can take values formally from 0 to 100, where values below 50% indicate that the assignment of the electrons has not followed *an Aufbau* principle according to the occupation numbers of the EFOs. The latter avenue can be used to measure to what extent the molecular system conforms to a given set of OSs, rather than which are the most appropriate formal OSs.

If the frontier EFOs for any spin case are degenerated (same occupation number) and belong to different fragments, a value of $R = 50\%$ would be obtained. In that case, however, one may choose to assign a half-electron to each of the two atoms/fragments involved (or, in general, a fraction of the last m electrons that must be distributed among n_d degenerated EFOs). Then, the λ_{FU}^σ value to be used to evaluate $R^\sigma(\%)$ is the one immediately below the degenerated λ_{LO}^σ value. We use such an approach only when the degeneracies are due to symmetry. Alternatively, one might define a (small) threshold to consider two or more EFOs as pseudo-degenerated when their occupation numbers are close enough. It is also important to stress that the reliability index does not measure any *probability*. A given OS assignment with $R = 65\%$ does not necessarily imply that it is 65% likely to be the "right" assignment. Instead, it should be considered as a similarity measure between the formal ionic picture and the actual electron distribution. As such, the particular expression given in Eq. (1.9) is to some extent arbitrary. What is relevant are which are the frontier EFOs, what shape they have, and how their respective occupations compare.

Another important aspect of the EOS analysis is that, unlike other approaches, individual bonds or localized MOs are not formally considered. The latter are in essence the WF representation of a Lewis structure, whereas the resonance phenomena are usually put into correspondence with multireference character or static correlation (except aromatic resonance, which is quite well described with single-reference methods). Schemes based on a Lewis structure (e.g. algorithm of assigning bonds) or on actually computing it [38–40] are restricted to systems with a single dominant resonance structure. On the contrary, the EOS method goes beyond this paradigm and focuses on atomic/fragment EFOs. The smaller the number of EFOs significantly populated for a given fragment, the greater (more positive) its OS is. While admittedly the method loses the direct connection with IUPAC's philosophy, the EOS analysis can be applied on equal footing for single- and multireference WFs. In other words, to tackle, with no additional provisions, molecular systems with more than one dominant Lewis structure [22].

8.5 EOS Analysis from Different AIM Schemes

The procedure to obtain EFOs uses an underlying AIM definition or atomic partition scheme. In Eqs. (8.3)–(8.6), we have provided formulae in the general framework of real-space fuzzy atoms, described by appropriate atomic (or fragment) weight functions. But one can obtain EFOs using any other AIM scheme. Indeed, in Mayer's original formulation, the eff-AOs were obtained by requiring stationary properties of Mulliken's net atomic populations corresponding to some localized orbitals [31]. Later on, Mayer also showed how to generalize the scheme to disjoint real-space frameworks, such as QTAIM [43], where he already anticipated the possibility of considering groups of atoms for the analysis. Salvador and Mayer also compared the effect of different fuzzy atom schemes on the nature of the EFOs [32].

Experience gained over time has shown that the number and shape of the EFOs remain relatively unaffected by the specific AIM method used to generate them. In the case of atoms, the numerical calculations show that there are always as many eff-AOs, with occupation numbers considerably differing from zero, as many orbitals are contained in the classical minimal basis of the given atom, disregarding the size of the underlying AO basis in which the molecular calculation has been performed. When setting molecular fragments (e.g. ligands), one typically recovers the MO diagram of the free fragment, but with fractional occupations. However, there is variation in the occupancies when different AIM schemes are employed. As a result, it is eventually possible to face the undesirable situation where two different AIM schemes yield different assignments for OSs. In the original study, Ramos-Cordoba et al. explored the effect of different AIM in the OS assignations for a set of rather coordinated compounds. The assignations were robust with the AIM [22]. But in more recent challenging applications to borderline systems, different OS assignations have been eventually observed [44].

While choosing one or another AIM might be to some extent a matter of taste, for the specific task of resolving the IA, some schemes appear to be more appropriate than others. In particular, it is essential that the AIM is able to recover the partial ionic character of the bonded atoms, and to do so, the same atoms should not be treated on equal footing in different chemical environments. In other words, the AIM definition should somehow depend on a computed molecular property (e.g. the electron density). Both Hilbert-space and real-space schemes can suffer from this drawback. It is widely known that Mulliken-type partitioning in the original AO basis can lead to unreliable results in combination with extended basis sets. Transformation to Löwdin basis alleviates the problem but still is not able to capture the chemical environment of the atoms, as the matrix transformation only depends on the original AO basis overlap matrix. In this vein, weighted-Löwdin schemes such as NAO or the recent meta-Löwdin scheme [45] are more appropriate as they incorporate the distinction between core, valence, and virtual spaces in the orthogonalization procedure.

Concerning real-space AIM schemes, QTAIM is the paradigmatic example where the atomic boundaries depend explicitly on the electron density. The fuzzy-atom TFVC method is based on Becke's multicenter scheme [46], but borrows elements of QTAIM (e.g. the shape of the Voronoi cells is derived from the position of the bond critical points) so that it is also density-dependent. The atomic weight functions in the original Hirshfeld approach, on the contrary, only depend on precomputed spherically averaged atomic promolecular densities. However, Hirshfeld-iterative [47] or more recent developments [48] are dependent on the molecular electron density, so they are more appropriate in this context.

The choice of AIM introduces another distinction in the occupations of the EFOs. As stated before, the original formulation was based on the stationarity of Mulliken's net atomic population. Consequently, the sum of the occupation numbers of all

EFOs of a given atom (up to min(n,m), where n is the number of occupied MOs and m is the rank of the AO basis of the atom) equals the atom's net population, not the gross population. This is also the case of real-space schemes with overlapping atoms. Since the atomic weights must be entered twice in Eq. (8.3), the occupation numbers of the EFOs do not add up to the total number of electrons, but a smaller number.

This is not the case when using disjoint domains in real-space schemes, such as QTAIM, because of the idempotency of the atomic weight functions. Indeed, in QTAIM, one has $w_A(r) = 1$ if $r \in \Omega_A$ and $w_A(r) = 0$ otherwise, where Ω_A represents the atomic basin of attractor A. One can trivially see that the Q^A matrices are in fact equivalent to the atomic overlap matrices (AOM) in the MO basis, S^A

$$Q^A_{ij} = \int w^*_A(r)\phi^*_i(r)\phi_j(r)w_A(r)dr = \int w_A(r)\phi^*_i(r)\phi_j(r)dr = \int_{\Omega_A} \phi^*_i(r)\phi_j(r)dr = S^A_{ij}$$

(8.10)

The lack of atomic overlap essentially means that there is no distinction between net and gross atomic/fragment populations. The occupation number of the EFOs (eigenvalues of S^A matrix) adds up to the population of the fragment because the trace of S^A is conserved upon diagonalization.

This desirable property can be recovered in the fuzzy atom framework by considering the untruncated EFOs of Eq. (8.5), namely

$$\chi^{un}_\mu(r) = \sum_{i=1}^{n} U^A_{i\mu}\phi_i(r)$$

(8.11)

And then performing population analysis on the corresponding atom/fragment

$$\lambda^{A,gross}_\mu = \int w_A(r)\chi^{un,*}_\mu(r)\chi^{un}_\mu(r)dr = \sum_{i=1}^{n}\sum_{j=1}^{n} U^{A,+}_{\mu i} S^A_{ij} U^A_{j\mu} = (U^{A,+}S^A U^A)_{\mu\mu}$$

(8.12)

In the original formulation of the EOS scheme and subsequent applications [22–49], the authors used the net EFO occupations to obtain the OS assignations and reliability indices. While these won't be much affected if using gross EFO occupations, the latter are probably preferred because they permit a better comparison between the occupation numbers of disjoint and overlapping AIM schemes.

An analogous situation occurs also in the Hilbert-space-based formulations of the EFOs. In this case, it is the use of an orthonormalized basis that makes any overlap population vanish, and hence, no distinction again between net and gross populations. In Hilbert-space schemes, the formulae are usually expressed in the AO basis, where the systematization according to the indices affords the decomposition. There is, however, a simple alternative approach that permits it to work on MO/NO basis and yet use an underlying Hilbert-space AIM scheme.

To illustrate it, let us consider, for simplicity, a restricted single-determinant WF. The atomic population using a general real-space AIM scheme can be written in

8.5 EOS Analysis from Different AIM Schemes

terms of the trace of the aforementioned AOM matrix as

$$N^A = \int w_A^*(\mathbf{r})\rho(\mathbf{r})d\mathbf{r} = 2\sum_i^{occ} \int w_A(\mathbf{r})\phi_i^*(\mathbf{r})\phi_i(\mathbf{r})d\mathbf{r} = 2\mathrm{tr}(\mathbf{S}^A) \tag{8.13}$$

On the other hand, Mulliken's gross atomic population expressed in terms of the elements of the MO coefficients ($\mathbf{C}$) and the AO overlap ($\mathbf{S}$) matrices reads

$$N^{A,\mathrm{Mull}} = \sum_{\nu \in A}\sum_{\mu} D_{\nu\mu}S_{\mu\nu} = 2\sum_i^{occ}\sum_{\nu \in A}\sum_{\mu} C_{i\mu}^+ S_{\mu\nu} C_{\nu i}. \tag{8.14}$$

The left-hand side of Eq. (8.14) cannot be readily expressed as a trace because of the restriction on the summation over ν. However, one can introduce the block-truncated unit matrix η^A with all elements equal to zero except $\eta_{ii}^A = 1 \forall i \in A$ and express the atomic population as

$$N^{A,\mathrm{Mull}} = 2\sum_i^{occ}\sum_{\nu\mu\sigma} C_{i\mu}^+ S_{\mu\nu} \eta_{\nu\sigma}^A C_{\sigma i} = 2\mathrm{tr}(\mathbf{C}^+\mathbf{S}\eta^A\mathbf{C}). \tag{8.15}$$

It is now easy to recognize that the AOM matrix in MO basis consistent with a Mulliken-type partitioning is given by the mapping

$$\mathbf{S}^A \Leftarrow \mathbf{C}^+\mathbf{S}^{AO}\eta^A\mathbf{C} \tag{8.16}$$

An analogous mapping can also be set when using an orthonormalized basis. In the case of the Löwdin basis, the atomic populations are expressed as

$$N^{A,\mathrm{Low}} = \sum_{\nu \in A}\sum_{\mu\sigma} S_{\nu\mu}^{1/2} D_{\nu\sigma} S_{\sigma\mu}^{1/2} = 2\sum_i^{occ}\sum_{\nu \in A}\sum_{\mu\sigma} C_{i\sigma}^+ S_{\sigma\mu}^{1/2} S_{\mu\nu}^{1/2} C_{\nu i}. \tag{8.17}$$

Making use of the atomic truncation matrix, one can readily see

$$N^{A,\mathrm{Low}} = 2\sum_i^{occ}\sum_{\nu\mu\sigma\lambda} C_{i\mu}^+ S_{\mu\nu}^{1/2} \eta_{\nu\sigma}^A S_{\sigma\lambda}^{1/2} C_{\lambda i} = 2\mathrm{tr}(\mathbf{C}^+\mathbf{S}^{1/2}\eta^A\mathbf{S}^{1/2}\mathbf{C}), \tag{8.18}$$

so the mapping now becomes

$$\mathbf{S}^A \Leftarrow \mathbf{C}^+\mathbf{S}^{1/2}\eta^A\mathbf{S}^{1/2}\mathbf{C} \tag{8.19}$$

In a more general fashion, applicable, for instance, to NAO or IAO, the mapping can be written as

$$\mathbf{S}^A \Leftarrow \mathbf{C}^+\mathbf{T}^{-1,+}\eta^A\mathbf{T}^{-1}\mathbf{C}, \tag{8.20}$$

where $\mathbf{T}$ is the matrix that transforms the original AO basis into the orthogonalized one.

Finally, it can be readily seen that the EFOs in the framework of Mulliken-type partitioning can still be obtained using Eqs. (8.3)–(8.6), but replacing the real-space-based $\mathbf{Q}^A$ matrix by

$$\mathbf{Q}^A \Leftarrow \mathbf{C}^+\eta^A\mathbf{S}^{AO}\eta^A\mathbf{C}, \tag{8.21}$$

whereas in the orthogonalized basis, because of the absence of overlap population, $\mathbf{Q}^A \equiv \mathbf{S}^A$.

8.6 Summary

We have discussed in detail the EOS algorithm to assign OSs from the first principles. Even though there are nowadays a number of alternative schemes for OS assignment, EOS is currently the only one that works on equal footing for both single-determinant and correlated WFs. It can also accommodate different AIM definitions and even be applied to solids (i.e. in the absence of an underlying atomic-centered AO basis) because of the unique properties of the effective atomic or fragment orbitals in which it is rooted.

The efforts from the IUPAC to clarify the term OS are acknowledged, and the current definition and recommended algorithms are, in general, satisfactory and useful. Yet, there remain ambiguities in the application of the IA that are critically discussed here. Until these are further clarified, we advocate for the general use of properly devised chemical bonding tools such as EOS, which properly take into account the chemical environment of the atoms within the molecule without any additional input or provision for exceptions.

References

1 Mardirossian, N. and Head-Gordon, M. (2017). *Mol. Phys.* 115: 2315.
2 Riplinger, C. and Neese, F. (2013). *J. Chem. Phys.* 034106.
3 Ramakrishnan, R. and von Lilienfeld, O.A. (2017). Machine learning, quantum chemistry, and chemical space. In: *Reviews in Computational Chemistry, Volume 30* (ed. A.L. Parrill and K.B. Lipkowitz). Hoboken, NJ, USA: Wiley.
4 Bergeler, M., Simm, G.N., Proppe, J., and Reiher, M. (2015). *J. Chem. Theory Comput.* 11: 5712–5722.
5 Grimme, S. (2013). *Angew. Chem., Int. Ed.* 52: 6306–6312.
6 Grimme, S., Bannwarth, C., Dohm, S. et al. (2017). *Angew. Chem. Int. Ed.* 56: 14763.
7 Grimme, S. and Schreiner, P.R. (2017). *Angew. Chem. Int. Ed.* 56: 2–9.
8 Grunenberg, J. (2017). *Int. J. Quantum Chem.* 117: e25359.
9 Klein, J.E.M.N., Miehlich, B., Holzwarth, M.S. et al. (2014). *Angew. Chem. Int. Ed.* 53: 1790–1794.
10 Jansen, M. and Wedig, U. (2008). *Angew. Chem. Int. Ed.* 47: 10026.
11 Resta, R. (2008). *Nature* 453: 735.
12 Raebiger, H., Lanny, S., and Zunger, A. (2008). *Nature* 453: 763.
13 Yu, H.S. and Truhlar, D.G. (2016). *Angew. Chem. Int. Ed.* 55: 9004.
14 Karen, P., McArdle, P., and Takats, J. (2014). *Pure Appl. Chem.* 86: 1017.
15 Karen, P. (2015). *Angew. Chem. Int. Ed.* 54: 2.
16 Karen, P., McArdle, P., and Takats, J. (2016). *Pure Appl. Chem.* 88: 831.
17 IUPAC (2019). Oxidation state. https://doi.org/10.1351/goldbook.O04365.
18 Mann, J.B., Meek, T.L., and Allen, L.C. (2000). *J. Am. Chem. Soc.* 122: 2780.
19 Aullón, G. and Alvarez, S. (2009). *Theo. Chem. Acc.* 123: 67–73.

20 Walsh, A.A., Sokol, J., Buckeridge, D.O. et al. (2017). *Chem. Phys. Lett.* 8: 2074–2075.
21 Koch, D. and Manzhos, S. (2017). *J. Phys. Chem. Lett.* 8: 1593–1598.
22 Ramos-Cordoba, E., Postils, V., and Salvador, P. (2015). *J. Chem. Theory Comput.* 11: 1501.
23 Mulliken, R.S. (1955). *J. Chem. Phys.* 23: 1833–1840.
24 Löwdin, P.O. (1950). *J. Chem. Phys.* 18: 365–375.
25 Reed, E., Weinstock, R.B., and Weinhold, F. (1985). *J. Chem. Phys.* 83: 735–746.
26 Lu, W.-C., Wang, C.-Z., Schmidt, M.W. et al. (2004). *J. Chem. Phys.* 120: 2629–2637.
27 Knizia, G. (2013). *J. Chem. Theory Comput.* 9: 4834–4843.
28 Bader, R.F.W. (1990). *Atoms in Molecules: A Quantum Theory.* Oxford, UK: Oxford University Press.
29 Hirshfeld, F.L. (1977). *Theor. Chim. Acta* 44: 129.
30 Salvador, P. and Ramos-Cordoba, E. (2013). *J. Chem. Phys.* 139: 071103.
31 Mayer (1996). *J. Phys. Chem.* 100: 6249.
32 Salvador, P. and Mayer, I. (2009). *J. Chem. Phys.* 130: 234106.
33 Ramos-Cordoba, E., Salvador, P., and Mayer, I. (2013). *J. Chem. Phys.* 138: 2.
34 Boys, S.F. (1960). *Rev. Mod. Phys.* 32: 296–299.
35 Pipek, J. and Mezey, P.G. (1989). *J. Chem. Phys.* 90: 4916–4926.
36 Sit, P.H.-L., Car, R., Cohen, M.R., and Selloni, A. (2011). *Inorg. Chem.* 50: 10259.
37 Sit, P.H.-L., Zipoli, F., Chen, J. et al. (2011). *Chem. Eur. J.* 17: 12136–12143.
38 Thom, J.W., Sundstrom, E.J., and Head-Gordon, M. (2009). *Phys. Chem. Chem. Phys.* 11: 11297.
39 Gimferrer, M., Van der Mynsbrugge, J., Bell, A.T. et al. (2020). *Inorg. Chem.* 59: 15410–15420.
40 Gimferrer, M., Aldossary, A., Salvador, P., and Head-Gordon, M. (2022). *J. Chem. Theory Comput.* 18: 309–322.
41 Salvador, P., Vos, E., Corral, I., and Andrada, D.M. (2021). *Angew. Chem. Int. Ed.* 60: 1498.
42 Mayer, I., Bako, I., and Stirling, A. (2011). *J. Phys. Chem. A* 115: 12733–12737.
43 Mayer, I. (1996). *Can. J. Chem.* 74: 939–942.
44 Gimferrer, M., Danés, S., Vos, E. et al. (2022). *Chem. Sci.* 13: 6583–6591.
45 Sun, Q. and Chan, G.K. (2014). *J. Chem. Theory Comput.* 10 (9): 3784–3790.
46 Becke, A.D. (1988). *J. Chem. Phys.* 88: 2547.
47 Bultinck, P., Van Alsenoy, C., Ayers, P.W., and Carbó-Dorca, R. (2007). *J. Chem. Phys.* 126: 144111.
48 Heidar-Zadeh, F., Ayers, P.W., Verstraelen, T. et al. (2018). *J. Phys. Chem. A* 122: 4219–4245.
49 Postils, V., Delgado-Alonso, C., Luis, J.M., and Salvador, P. (2018). *Angew. Chem. Int. Ed.* 57: 10525.

9

Aromaticity and Antiaromaticity

Yago García-Rodeja and Miquel Solà

Universitat de Girona, Institut de Química Computacional i Catàlisi and Departament de Química, C/Mª Aurèlia Capmany 69, 17003, Girona, Catalonia, Spain

9.1 Definition of Aromaticity

Misunderstandings and conflicts about aromaticity's definition and the aromatic character of particular systems are common in the field of aromaticity. The chemical community has produced a large number of articles, reviews, and conferences, but there is still no complete consensus on what aromaticity is. According to the IUPAC definition of 1999 [1]:

> "The concept of spatial and electronic structure of cyclic molecular systems displaying the effects of cyclic electron delocalization which provide for their enhanced thermodynamic stability (relative to acyclic structural analogues) and tendency to retain the structural type in the course of chemical transformations. A quantitative assessment of the degree of aromaticity is given by the value of the resonance energy. It may also be evaluated by the energies of relevant isodesmic and homodesmotic reactions. Along with energetic criteria of aromaticity, important and complementary are also a structural criterion (the lesser the alternation of bond lengths in the rings, the greater is the aromaticity of the molecule) and a magnetic criterion (existence of the diamagnetic ring current induced in a conjugated cyclic molecule by an external magnetic field and manifested by an exaltation and anisotropy of magnetic susceptibility). Although originally introduced for characterization of peculiar properties of cyclic conjugated hydrocarbons and their ions, the concept of aromaticity has been extended to their homoderivatives (see homoaromaticity), conjugated heterocyclic compounds (heteroaromaticity), saturated cyclic compounds (σ-aromaticity) as well as to three-dimensional organic and organometallic compounds (three-dimensional aromaticity). A common feature of the electronic structure inherent in all aromatic molecules is the close nature of their valence electron shells, i.e., double electron occupation of all bonding molecular orbitals (MOs) with all

Exploring Chemical Concepts Through Theory and Computation, First Edition. Edited by Shubin Liu.
© 2024 WILEY-VCH GmbH. Published 2024 by WILEY-VCH GmbH.

antibonding and delocalized nonbonding MOs unfilled. The notion of aromaticity is applied also to transition states."

This definition has been criticized by several authors [2] because new types of aromaticity such as excited state aromaticity, metalloaromaticity, spherical aromaticity, multiple aromaticity, and conflicting aromaticity are not mentioned in the IUPAC definition. Additionally, it ignores employing electron delocalization methods to quantify aromaticity and places too much weight on the magnetic criteria for measuring it. Finally, it overemphasizes resonance energies, which are frequently very challenging to evaluate.

In his book "The Same and Not the Same" [3], Roald Hoffmann writes:

> "There are some concepts which cannot be mathematicized, they cannot be defined unambiguously, but they are of fantastic utility to our science."

Chemists use many ill-defined concepts that lack a strict physical basis. Aromaticity is one of them, but we can also mention here the chemical bond, bond order, atomic charge, (hyper)conjugation, strain energy, etc. [4, 5]. These concepts are very useful to chemists when are used properly to understand chemical reactivity, but they cannot be directly experimentally measured, which is a consequence of the nonexistence of a quantum mechanical operator that can extract the value of these concepts from the wave function. For some of these concepts, the definition is clear though the diverse ways to measure them lead to different results. This is the case of atomic charges, which can be defined as the electronic charge assigned to an atom in a molecule. Here, the fuzziness comes from how space is partitioned into atomic domains leading to different ways to determine atomic charges. As an extreme case, the charge on Cl in endohedral Cl@B_{39} was found to range from 0.758e (Löwdin charges) to −0.618e (QTAIM charges) depending on the method used to obtain atomic charges [6].

In the case of aromaticity, the problem is even worse since there is no consensus on what aromaticity exactly is; the concept of aromaticity is itself controversial [7, 8]. In a recent perspective [2], some authors even consider that a universal definition of aromaticity is impractical or noncompatible with the general laws constituting chemical theory. Others think that although it is not possible to offer an all-inclusive general definition of this concept, it is necessary to keep updated the current definition of the aromaticity concept by the IUPAC of 1999 [1]. Probably, the most widely accepted definition nowadays was provided in 2005 by Chen et al. [9] who defined this concept as "a manifestation of electron delocalization in closed circuits, either in two or in three dimensions. This results in energy lowering, often quite substantial, and a variety of unusual chemical and physical properties. These include a tendency toward bond length equalization, unusual reactivity, and characteristic spectroscopic features."

9.2 Physical Foundation

Despite the controversy on the definition of aromaticity, what cannot be denied is the fact that the concept is deeply rooted in fundamental quantum mechanics. In a recent paper, Solà and Bickelhaupt [10] showed that the phenomenon of aromaticity already appears by comparing the solutions of the particle in a box (PIB) and the particle on a ring (POR) simple physical models. For a box and a ring of the same length, the cyclic POR model leads to more stable states than the corresponding PIB model, with the most significant factors being that the cyclic system has a significantly more stable lowest-energy eigenvalue and, in addition, it has degenerate higher-energy eigenvalues in which it can accommodate twice as many particles per "shell" before the next higher-energy eigenfunctions are populated. As a result, the POR and PIB model systems recover the existence of the aromatic stabilization energy (ASE) and provide insight into the underlying physics.

There are many examples showing the importance of aromaticity in the chemical stability of (anti)aromatic compounds and their chemical reactivity. It is commonly known that stability and aromaticity go hand in hand. In comparison to their linear counterparts, aromatic compounds have improved energetic (thermodynamic) and chemical (kinetic) stability. For instance, in comparison to its isomer anthracene, phenanthrene is more aromatic and more stable by ca. 5 kcal/mol [11]. Figure 9.1a depicts phenanthrene with its two π-sextets and anthracene with one migrating π-sextet (the Clar's π-sextet theory will be discussed in the coming section). All kinked benzenoids, not just phenanthrene, are more stable and aromatic than their linear polyacenes isomers [11]. Similarly, 1,2-benzoquinones are more stable and aromatic than their 2,3-isomers, and this trend is also true for larger benzoquinones [12]. Interestingly, double oxidation of anthracene and phenanthrene produces an anthracene dication that is more stable than the phenanthrene dication [13]. The reverse trend in stability found for the dications as compared to the neutral species can be explained by the larger aromaticity of the anthracene dication, despite the difference in aromaticity between the isomers is now smaller, together with the presence of H···H repulsion in the bay region of phenanthrene [13]. However, in general, the most stable isomer is not always the most aromatic among a group of isomers or various electronic states. One of the numerous elements that influence the relative energy of isomers is aromaticity. But other factors, such as strain energy, hyperconjugation, the presence of hydrogen bonds, long-range interactions, and so forth, may have, in many cases, a greater influence.

Aromaticity and reactivity are two ideas that are closely related. Already in 1825, Michael Faraday observed that benzene was far less reactive than *trans*-2-butene when he first produced it [14]. Since then, this lower reactivity has been regarded as an experimental property of aromatic compounds. By either stabilizing a transition state or product through aromaticity or by destabilizing a reactant through antiaromaticity, aromaticity can be utilized to make reactions faster. In particular, an antiaromatic reactant will typically boost a reaction both kinetically and thermodynamically, whereas aromaticity achieved in a transition state favors reactions kinetically and in a product does thermodynamically. As an example, we can mention the double group transfer reactions to 1,2-benzyne analyzed by Fernández and Cossío [15]. In this reaction, the benzyne intermediate (**1**) generated from an hexahydro-Diels–Alder reaction (HDDA) [16] (see Figure 9.1b) accepts two hydrogen atoms from a saturated alkane (**2**), leading to the formation of the corresponding benzenoid product (**3**) and alkene (**4**). Their investigation

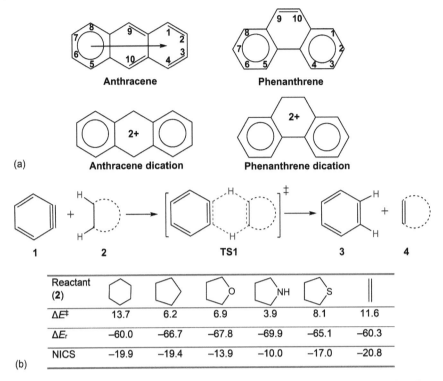

Figure 9.1 (a) Schematic representation of anthracene and phenanthrene neutral and dication with their π-sextets; (b) energy barriers and reaction energies for the reaction of benzyne with cyclic hydrocarbons ($\Delta E^{\ddagger}$ and ΔE_r, in kcal/mol) and NICS (in ppm) computed at the [3,+1] ring critical point of the ring containing the two transferred H atoms computed at the M06-2X/6-311+G(d,p) level of theory.

found a concerted deprotonation of a hydrocarbon by the intermediate benzyne. Interestingly, these reactions have relatively low reaction activation barriers via in-plane aromatic transition states (**TS1**) according to DFT calculations. The low energy barriers found, despite the aromaticity of reactant **1**, are attributed to the aromaticity of the transition states. Such aromaticity was confirmed by the negative NICS values from −10.0 to −20.8 ppm computed at the (3,+1) ring critical point of the ring formed in the transition state (see Figure 9.1b) and the presence of a diatropic ring current in the anisotropy of the induced current density (ACID) plots.

9.3 Measures of Aromaticity

Aromaticity is a quality that cannot be directly quantified by any physical or chemical experiment since it is not well-defined. The measurement of several characteristics that are typically found in aromatic compounds, such as energetic stabilization, bond length equalization, magnetic behavior linked to induced ring currents, and electron delocalization, is then used to indirectly quantify the aromaticity of a given compound. Sections 9.3.1–9.3.4 provide descriptions for the most widely used descriptors of aromaticity of these four categories. It is worth emphasizing that conclusions reached by using a set of aromaticity descriptors based on different properties are always more reliable than those obtained with a single indicator of aromaticity [17].

9.3.1 Geometric Descriptors of Aromaticity

The decreased variation of bond lengths between aromatic compounds and their unsaturated acyclic analogs is the foundation for the geometric descriptor of aromaticity. The harmonic oscillator model of aromaticity (HOMA) index is the most frequently utilized descriptor derived from molecular structures. It is defined as [18–20]:

$$\text{HOMA} = 1 - \frac{\alpha}{n}\sum_{i=1}^{n}(R_{opt} - R_i)^2 = 1 - \left[\alpha(R_{opt} - R_{av})^2 + \frac{\alpha}{n}\sum_{i=1}^{n}(R_{av} - R_i)^2\right]$$
$$= 1 - \text{EN} - \text{GEO} \qquad (9.1)$$

where n is the number of bonds considered and α is an empirical constant selected to give HOMA = 0 for a nonaromatic model system. When HOMA = 1, the system with all bonds equal to an optimal R_{opt} value is assumed to be fully aromatic. For C–C, C–N, C–O, N–N [20], C–B [21], and B–B [22] bonds, α = 257.7, 93.5, 157.4, 130.3, 104.5, and 244.1, whereas R_{opt} = 1.388, 1.334, 1.265, 1.309, 1.424, and 1.567 Å, respectively. R_{av} is the mean bond length of the ring and R_i stands for a running bond length. HOMA value can be split into energetic (EN) and geometric (GEO) contributions [20, 23], according to the relationship HOMA = 1 − EN − GEO (Eq. (9.1)). The GEO contribution measures the decrease/increase of bond length alternation (BLA), whereas the EN term takes into account the lengthening/shortening of mean bond lengths of the ring. Table 9.1 collects the strengths and weaknesses of some of the most popular indices of aromaticity. In general, HOMA cannot be applied to study the aromaticity of excited states. However, very recently Arpa and Durbeej have provided reference values for the calculation of HOMA for the lowest-lying triplet excited states of molecules having C–C, C–N, C–O, and N–N bonds [24].

9.3.2 Energetic Descriptors of Aromaticity

The energetic stability of the delocalized structure of cyclic π-conjugated molecules is one of the most significant characteristics of the aromaticity phenomena [25]. The so-called ASE is a well-established parameter for assessing the aromaticity or antiaromaticity of cyclic conjugated compounds that can be calculated theoretically or experimentally from suitable isodesmic [26, 27] or homodesmotic [28, 29] reactions. Isodesmic reactions demand an equal number of formal single and double bonds in products and reactants. Homodesmotic reactions are a particular case of isodesmic reactions in which there is the same number of bonds between given atoms in each hybridization state. The number of hydrogen atoms joined

Table 9.1 Strengths and weaknesses of some of the most widely used indices of aromaticity.

Descriptor	Strengths	Weaknesses
HOMA	Applicable to structures derived from experiments and theory. It can classify aromatic, nonaromatic, and antiaromatic species. In macrocycles or nanographenes, it can be used to find the most favorable circuit for electron delocalization.	It cannot be applied to structures far from equilibrium (e.g. transition states) or excited states. It cannot be applied to structures having bonds in the ring with no reference values. Separation into α and β components is not possible. Separation into $\pi/\sigma/\delta$... contributions is not possible.
ASE	The thermodynamic stabilization due to aromaticity is the most important primary effect of aromaticity. It can classify aromatic, nonaromatic, and antiaromatic species.	The selection of proper reference molecules is not always clear, especially in the case of metallaaromatic compounds, transition states, or excited states. Separation into α and β components is not possible. Separation into $\pi/\sigma/\delta$... contributions is not possible.
FLU	Electron delocalization is a primary property of aromaticity. Separation into α and β components is possible. There is a FLU$_\pi$ version of this index that can be applied to planar π-conjugated species that does not require reference values. In macrocycles or nanographenes, it can be used to find the most favorable circuit for electron delocalization	It cannot be applied to structures far from equilibrium (e.g. transition states). It cannot be applied to structures having bonds in the ring with no reference values. As reference values are derived for the ground states, attention must be taken when applying FLU to excited states. It cannot easily differentiate between nonaromatic and antiaromatic compounds.
MCI/I_{ring}	Electron delocalization is a primary property of aromaticity. Separation into α and β components is possible. It can be applied to metallaaromatic species. Several studies consider this index the most reliable indicator of aromaticity. Separation into $\pi/\sigma/\delta$... contributions is possible.	It cannot distinguish between nonaromatic and antiaromatic compounds. When using QTAIM partitioning, it can only be used for small to medium-sized rings (fewer than 12 atoms) due to numerical inaccuracies. The high computational cost for large rings. It is size dependent. The larger the ring, the smaller the value. However, this can be solved using a normalized version of these indices: $MCI^{1/N}/I_{ring}^{1/N}$

(continued)

Table 9.1 (Continued)

Descriptor	Strengths	Weaknesses
NICS	It is one of the easiest to calculate indicators of aromaticity. The modern versions of this index such as $NICS_{zz}$, $NICS_{\pi}$, NICS scans, or NICS-XY-scan provide reliable results in most cases. It can classify aromatic, nonaromatic, and antiaromatic species. Separation into α and β components is possible. Separation into $\pi/\sigma/\delta\ldots$ contributions is possible.	In some cases, contributions from electrons not related to aromaticity (e.g. σ-electrons in classical aromatic compounds) can be important. NICS is likely to be overestimated in excited states because of small energy differences between frontier orbitals. It is not always a reliable indicator of aromaticity for metallaaromatic species. In polycyclic aromatic hydrocarbons, the influence of the induced magnetic fields of adjacent rings can be substantial. The size of the ring can influence the NICS values. NICS can indicate the aromaticity of systems that are not aromatic such as $(HF)_3$, carborane-fused rings, or large macrocycles.
Ring currents	As a magnetic indicator of aromaticity, it is more reliable than NICS. Separation into α and β components is possible. Separation into $\pi/\sigma/\delta\ldots$ contributions is possible. It can classify aromatic, nonaromatic, and antiaromatic species. Results of NMR chemical shifts, which can be calculated from ring currents, can also be obtained experimentally.	The size of the ring can influence the ring current strengths. Ring currents can indicate the aromaticity of large macrocycles that are nonaromatic. For systems with small HOMO–LUMO gaps, the (anti)aromaticity can be overestimated due to the high intensity of the ring currents. Ring currents are likely to be overestimated in excited states because of small energy differences between frontier orbitals.

to the atoms in given hybridization states must also coincide in products and reactants. Homodesmotic reactions are preferred because they minimize errors due to compensation of strain, hyperconjugation, anomeric effects, etc. [30, 31]. In principle, the energies of substrates and products in these reactions can be taken either from the quantum chemical calculations or from thermochemical (calorimetric) measurements. In the case of benzene, several isodesmic or homodesmotic reactions can be constructed, as shown in Table 9.2. Cyrański collected fifteen ASEs calculated from isodesmodic and homodesmotic reactions. The ASE values obtained range from 66.9 to 18.4 kcal/mol [25]. When conjugation, hyperconjugation, and

Table 9.2 Five different aromatic stabilization energies (in kcal/mol) for benzene at B3LYP/6-311+G** (+ZPE).

Reactions – B3LYP/6-311+G** (+ZPE)	ASE
benzene + 6 CH$_4$ → 3 CH$_3$-CH$_3$ + 3 CH$_2$=CH$_2$	66.9[a]
benzene + 2 cyclohexane → 3 cyclohexene	37.5[b]
benzene + pentadiene → cyclohexene + hexadiene	36.5[c]
benzene + 3 butene → 3 hexadiene	19.3[d]
toluene (CH$_3$) → methylenecyclohexadiene (CH$_2$)	33.2[e]

a) Experimental values: 61.1 kcal/mol [26], 64.2 kcal/mol [32], and 64.7 kcal/mol [33].
b) Experimental values: 35.6 kcal/mol [34, 35] and 35.9 kcal/mol [26].
c) Experimental value: 33.4 kcal/mol [33].
d) Experimental value: 22.5 kcal/mol [33].
e) Experimental value: 38 and 23.1 kcal/mol [36, 37].

protobranching corrections are taken into account, the widely divergent ASE of benzene assessments that have been reported can be refined to around 29 kcal/mol [38, 39].

The isomerization stabilization energies (ISEs) were first described by Schleyer and Pühlhofer [35] as a specific method of obtaining ASEs. The energy difference between a methyl derivative of an annulene and an isomeric species with an acyclic conjugation and an exocyclic methylene group is known as ISE for a particular (anti)aromatic molecule. For benzene, for example, the value ISE can be obtained from a nonaromatic methylenecyclohexadiene and toluene (see isomerization reaction 5 in Table 9.2). In many cases, it is possible to create a number of nonaromatic methylenecyclohexadiene isomers. If so, one must average out all potential ISEs. Excited states can be challenging for the ISE approach. For instance, the state that is most akin to the lowest excited state of the (anti)aromatic isomer may not be the lowest excited state of the nonaromatic isomer.

Finally, Fernández and Frenking [40] found that it is possible to use the ΔE_π term of the orbital interaction component in an energy decomposition analysis (EDA) [41–44] of cyclic and acyclic conjugated systems to estimate the strength of π-conjugation in (anti)aromatic species.

9.3.3 Electronic Descriptors of Aromaticity

One of the primary and key characteristics of aromatic compounds is the cyclic delocalization of mobile electrons in two or three dimensions. As this electronic delocalization is not observable, there is no experimental attribute that enables direct measurement of it. The absence of an observable for electron delocalization suggests that there is not a single widely used way to quantify it from a theoretical perspective. Indeed, there are various electronic descriptors of aromaticity [8, 45, 46], and we mention here only some of the most widely used.

The aromatic fluctuation index (FLU) [47, 48] measures the uniformity of the electron delocalization along the molecular ring and its bonding difference with respect to some aromatic reference, using Eq. (9.2):

$$\text{FLU}(A) = \frac{1}{N} \sum_{i=1}^{N} \left[\left(\frac{V(A_i)}{V(A_{i-1})} \right)^\alpha \left(\frac{\delta(A_i, A_{i-1}) - \delta_{\text{ref}}(A_i, A_{i-1})}{\delta_{\text{ref}}(A_i, A_{i-1})} \right) \right]^2 \quad (9.2)$$

where the ring considered is formed by atoms in the string $\{A\} = \{A_1, A_2, \ldots A_N\}$, $A_0 \equiv A_N$ and the atomic delocalization is defined by Eq. (9.3),

$$V(A) = \sum_{A \neq B} \delta(A, B) \quad (9.3)$$

where the delocalization index between atoms A and B, $\delta(A,B)$, for monodeterminantal closed-shell wavefunctions, is obtained from Eq. (9.4),

$$\delta(A, B) = 4 \sum_{i,j}^{occ.MO} S_{ij}(A) S_{ij}(B) \quad (9.4)$$

The summation in Eq. (9.4) runs over all occupied molecular orbitals (MOs). $S_{ij}(A)$ is the overlap between MOs i and j within the basin of atom A. Delocalization indexes in Eq. (9.4) reduce to Wiberg–Mayer bond orders [49] if the integrations over atomic basins are replaced by a Mulliken-like partitioning of the corresponding integrals. In Eq. (9.2), α is a function that ensures that the ratio of atomic delocalizations is always greater or equal to 1,

$$\alpha = \begin{cases} 1 & V(A_i) > V(A_{i-1}) \\ -1 & V(A_i) \leq V(A_{i-1}) \end{cases} \quad (9.5)$$

The reference values of C—C and C—N bonds are taken from benzene and pyridine in their ground state. FLU is close to 0 in aromatic rings and increases as the rings become less and less aromatic.

The definition of delocalization index can be generalized to study multicenter bonding by defining a multicenter delocalization index in Eq. (9.4) among the N centers A_1 to A_N. According to Giambiagi and coworkers [50], the closed-shell form of this index for monodeterminantal closed-shell wavefunctions is provided by Eq. (9.6),

$$I_{\text{ring}}(A) = 2^N \sum_{i_1, i_2, i_3, \ldots, i_N}^{occ.MO} S_{i_1 i_2}(A_1) S_{i_2 i_3}(A_2) \ldots S_{i_N i_1}(A_N) \quad (9.6)$$

As an extension of the I_{ring} index, Bultinck and coworkers defined the multicenter index, MCI [51] as:

$$\text{MCI}(A) = \frac{1}{2N} \sum_{P(A)} I_{\text{ring}}(A) = \frac{1}{2N} \sum_{P(A)} \sum_{i_1, i_2, i_3, \ldots, i_N}^{occ.MO} S_{i_1 i_2}(A_1) S_{i_2 i_3}(A_2) \ldots S_{i_N i_1}(A_N) \quad (9.7)$$

where P represents all possible permutations among centers A_1 to A_N and the internal summation runs over all occupied MOs. Both I_{ring} and MCI give a measure of the electron sharing through the whole studied ring. For planar species, $S_{ij}(A_k) = 0$ for $i \in \sigma$ and $j \in \pi$ orbital symmetries, thus I_{ring} and MCI can be exactly split into σ- and π-contributions, i.e. MCI_σ and MCI_π, respectively. The more positive the I_{ring} or MCI, the more aromatic the ring is. The values of I_{ring} and MCI are dependent on the size of the ring. The larger the ring, the smaller the I_{ring} and MCI values. To avoid or reduce this problem, there is a normalized version of the I_{ring} and MCI indexes, the so-called I_{NG} and I_{NB} [52].

Finally, some authors [53, 54] proposed the use of the hardness as a measure of aromaticity. The hardness, η, is a measure of the resistance of a chemical species to change its electronic configuration and, therefore, of its stability. It is defined as the second-order partial derivative of the total electronic energy with respect to the total number of electrons. The operational definition of this quantity is the difference between the ionization potential and electron affinity of the chemical species considered [55]. It is expected that the higher the hardness, the larger the aromatic character of a molecule. The problem is that one can only provide global aromaticities but not local aromaticities in polycyclic aromatic hydrocarbons (PAHs) or nanographenes.

9.3.4 Magnetic Descriptors of Aromaticity

As aromatic molecules have delocalized electrons in a closed cyclic circuit, in the presence of an external magnetic field, they show a ring current that creates an induced magnetic field that is proportional to the external magnetic field $\mathbf{B}_0$ according to Eq. (9.8):

$$\vec{B}_{\text{ind}} = -\vec{\vec{\sigma}} \vec{B}_0 \tag{9.8}$$

where $\vec{\vec{\sigma}}$ is the magnetic shielding tensor. If the molecule is aromatic, it produces a diatropic ring current, whereas antiaromatic molecules generate paratropic ring currents. The strength of the ring currents can be used as an indicator of (anti)aromaticity. The more intense the ring current, the more (anti)aromatic the molecule. The two most widely used methods to compute ring currents are the continuous transformation of the origin of the current density (CTOCD) [56, 57] and the gauge-including magnetically induced currents (GIMIC) [58].

One of the clearest indirect experimental evidence of the presence of a ring current is given by the observed nuclear magnetic resonance (NMR) chemical shifts. Because the induced magnetic field impacts the ^{1}H NMR chemical shifts of the H atoms linked to (anti)aromatic rings, ring currents in aromatic molecules are important to NMR spectroscopy [59]. In aromatic species, the induced magnetic field in the position of the H atoms has the same direction as the external field, protons in aromatic rings undergo a deshielding action. For instance, in contrast to the 5.6 ppm vinyl proton in cyclohexene, the chemical shift of protons in benzene is around 7.3 ppm. Compared to aromatic species, ring currents in antiaromatic molecules flow in the opposite direction. Therefore, protons that are shielded in aromatic species are consequently deshielded in antiaromatic compounds. The ^{1}H NMR values can be taken as a measure of aromaticity.

The nucleus-independent chemical shift (NICS) is a simple and widely available index of aromaticity [9, 60, 61]. It is defined as a negative value of the absolute isotropic shielding computed at the center of a ring or at some other interesting point of the system. It is obtained by averaging the diagonal elements of the magnetic shielding tensor in Eq. (9.8) over all directions, as shown in Eq. (9.9):

$$\text{NICS} = -\frac{1}{3}(\sigma_{xx} + \sigma_{yy} + \sigma_{zz}) = -\sigma_{av} \tag{9.9}$$

NICS can be computed at any point of space. Aromatic rings used to have negative NICS calculated in the center of the ring (NICS(0)). The more negative the NICS value, the more aromatic the ring is. It is worth noting that the NICS at the ring center of benzene, which is −8.9 ppm, is the result of a negative NICS(π) of −20.7 ppm and a positive NICS(σ) of 13.8 ppm [9]. To remove the effect of σ-electrons that are not related to aromaticity, one can use NICS(1), i.e. the NICS value at 1 Å above or below the center of the ring plane, which is considered to better reflect the π-electron effects [62, 63], or with the corresponding out-of-plane tensor component (NICS(1)$_{zz}$). Another option is to use the π contribution to the out-of-plane magnetic shielding tensor component computed at the center of the ring or at 1 Å above or below the ring plane (NICS(0)$_{\pi zz}$ and NICS(1)$_{\pi zz}$, respectively). Among many NICS-related definitions in organic compounds, NICS(1)$_{zz}$ and NICS(0)$_{\pi zz}$ have been reported to be the best measures of aromaticity [64, 65]. Although widely used, NICS is not exempt from criticism (see weaknesses in Table 9.1) [66–70].

9.4 Rules of Aromaticity

As we have seen in Sections 9.2 and 9.3, there are several ways to detect and quantify aromaticity. In practical applications, the different descriptors of aromaticity do not speak with the same voice [17, 31, 65, 71]. For this reason, first, it is highly advisable to use several indicators of aromaticity, if possible based on different criteria (energetic, geometric, magnetic, and electronic); and, second, if one detects (anti)aromaticity, it is important to rationalize the result obtained based on the existing rules of aromaticity. In keeping with Coulson's philosophy of "give us insight, not numbers" [72], numbers that prove that our system is (anti)aromatic are necessary, but we also need an understanding of the reason(s) why our system displays (anti)aromaticity.

A frequent feature of classic aromatic compounds is symmetry. High symmetry is present in archetypical aromatic compounds like C_6H_6, Al_4^{2-}, $B_{12}H_{12}^{2-}$, C_{60}^{10+}, or $C_5H_5^+$ in its lowest-lying triplet state. From the point of view of valence bond (VB) theory, highly symmetric species with highly delocalized electrons are found in systems that require for the description of its electronic structure the use of resonance hybrid structures, i.e. a combination of resonance structures of the same or similar weight (see Figure 9.2). Therefore, in VB theory, the requirement of resonance hybrid structures in cyclic species, with two- or three-dimensional (2D or 3D) closed

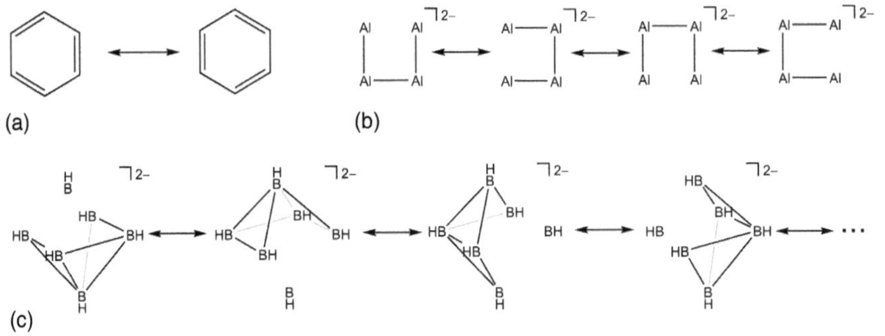

Figure 9.2 (a) Resonance structures for benzene with 6π-electrons, (b) Al_4^{2-} with 14 valence electrons (8 out of 14 electrons remain in the 2s orbitals of Al), and (c) $B_6H_6^{2-}$ with 14 cage electrons (only 4 out of the 792 possible resonance structures are depicted).

circuits, to describe its delocalized electronic structure provides a justification of its aromaticity.

From the point of view of the MO theory, understanding can be attained from the molecular distribution of our system and the counting rules of aromaticity [73]. Symmetric structures found in aromatic species result in shells of degenerate highest-occupied molecular orbitals (HOMOs), which may be completely filled to form a closed-shell structure, or may the last shell be half-filled with the same spin electrons. These two types of electronic structure stabilize aromatic molecules in a similar manner main-group elements are stabilized with eight valence electrons in their valence shell (the octet rule), transition metals with eighteen valence electrons or d^5 parallel spin occupation, or f-elements with thirty-two valence electrons [74]. This shell-closed or last shell half-filled electronic structure is the source of the several rules of aromaticity that will be discussed in the next subsections. These rules can not only be applied to highly symmetric species but also to less or nonsymmetric species related to aromatic compounds (e.g. pyridine).

9.4.1 Rules for Two-Dimensional Aromaticity

The work by Hückel [75] in 1931 was the germ of the most widely known rule of aromaticity, the Hückel's rule, finally established by Doering and Detert [76]. It states that monocyclic π-conjugated annulenes of D_{Nh} symmetry with $4n + 2$ π-electrons (n being an integer) are aromatic, whereas those with $4n$ π-electrons are antiaromatic [77]. The source of this rule is the MO distribution of π-orbitals in D_{Nh} annulenes shown in Figure 9.3a. The lowest in-energy MO is formed by the in-phase interactions of all p-atomic orbitals. The rest of the orbitals, except the last one for even N values that have all p-orbitals out of phase, come by pairs. Then, closed-shell electronic structures are obtained for the following number of π-electrons: 2, 6, 10, 14…, i.e. $4n + 2$ π-electrons.

In 1972, Colin Baird [78–80] found that monocyclic π-conjugated annulenes of D_{Nh} symmetry in their lowest-lying triplet state (T_1) are aromatic with $4n$

9.4 Rules of Aromaticity

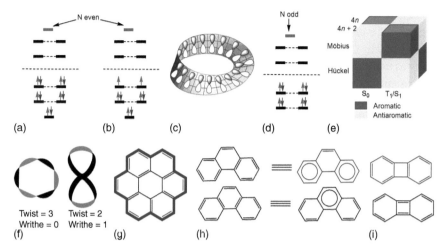

Figure 9.3 (a) Orbital distribution in the ground state of D_{Nh} annulenes with Hückel topology and $4n+2$ π-electrons, (b) orbital distribution in the T_1 state of D_{Nh} annulenes with Hückel topology and $4n$ π-electrons, (c) the lowest-lying molecular orbital for annulenes with Möbius topology, (d) orbital distribution in the ground state of annulenes with Möbius topology and $4n$ π-electrons, (e) Ottosson's cube for combined $4n$ versus $4n+2$ π-electrons, S_0 versus T_1/S_1 electronic states and Hückel versus Möbius topology, (f) two strips with linking number L_k equal to 3. The strip on the right has three twists and the one on the left has two twists and one writhe. (g) Platt's perimeter rule for coronene, (h) application of Clar's π-sextet rule in phenanthrene, (i) application of the Glidewell–Lloyd rule in biphenylene.

π-electrons (leading to a last shell half-filled with same spin electrons, Figure 9.3b) and antiaromatic with $4n+2$. Baird's and Hückel's rules were generalized to states of different spin by Soncini and Fowler [81] by stating that monocyclic π-conjugated annulenes of $4n+2$ π-electrons at the lowest-lying electronic state with even spin (singlet, quintet, …) and those of $4n$ π-electrons at the lowest-lying electronic state with odd spin (triplet, septet, …) are aromatic. Interestingly, by considering the α and β electrons separately, Mandado et al. [82] showed that the Hückel and Baird rules, as well as the Soncini–Fowler extension, can be fused into a single rule stating that D_{Nh} annulenes with an odd number $(2n+1)$ of α and β electrons are aromatic, while those with an even number $(2n)$ of α and β electrons are antiaromatic. A similar rule reported by Valiev et al. considers that molecules are aromatic (or antiaromatic) if they have an odd (or even) number of doubly and singly occupied π-conjugated valence orbitals [83], suggesting that Mandado's rule can be expanded to the singlet excited S_1 state with the same electron configuration as the ππ* T_1 state.

Möbius aromaticity is the particular electron delocalization found in cyclic conjugated species having a molecular topology that resembles that of a Möbius strip [84–86]. These species present a phase inversion in the most stable π-MO (see Figure 9.3c). The most stable MO is double degenerate. The rest of the orbitals, except the last one for odd N values, also come in pairs as shown in Figure 9.3d. Closed-shell electronic structures, which give aromatic stabilization, are then obtained with 4, 8, 12, …, i.e. $4n$ π-electrons ($n = 1, 2, 3…$). The Möbius rule is opposed to that of the Hückel rule and, consequently, Möbius-type annulenes with $4n+2$ π-electrons are antiaromatic. Möbius aromaticity was first described theoretically by Craig and Paddock [87] for organometallics systems in 1958 and in 1964 was expanded by Heilbronner for annulenes [84]. It is worth mentioning that a more recent study has shown that all metallacycles are Hückel–Möbius hybrids [88].

The "aromaticity cube" shown in Figure 9.3e was introduced by Ottosson et al. [89] to summarize in a simple scheme several combinations of (anti)aromaticity in annulenes depending on the number of π-electrons ($4n$ versus $4n+2$), electronic-state (S_0 versus T_1/S_1), and topology (Hückel versus Möbius). For instance, a $4n$ π-electron system in its T_1 state and with Möbius topology is antiaromatic as indicated by the top, right, and back, yellow cube in Figure 9.3e.

A generalization of the Hückel and Möbius rules can be achieved through the linking number (L_k) that represents the number of times that an annulene winds (taking into account both writhes and twists) [90]. As an example, two strips with $L_k = 3$ (which feature 3 distinct domains) are illustrated in Figure 9.3f; the strip on the left has three twists and the one on the right has two twists and one writhe. Cyclic conjugated annulenes that have an even L_k follow Hückel's rule of aromaticity, whereas those having an odd L_k obey Möbius' rule.

The $4n+2$ Hückel rule is strictly obeyed only for monocyclic systems like benzene and cyclooctatetraene. PAHs such as coronene (Figure 9.3g) do not obey this rule. Coronene has 24 π-electrons, and, consequently, it is antiaromatic according to Hückel's rule. However, its physicochemical properties are more similar to aromatic than antiaromatic species. A first attempt to extend Hückel's rule to PAHs was Platt's ring perimeter model [91], which divided PAHs into the perimeter and the inner core. The aromatic character of the PAH is that of the annulene of the perimeter (18 π-electrons in coronene), and the inner core represents a minor perturbation of this aromatic character. This simple adaptation of Hückel's rule, however, fails to account for the aromaticity of many non-benzenoid polycyclic conjugated hydrocarbons (PCHs) [92], such as biphenylene that has 12 π-electrons in the perimeter.

An improvement of the Platt's ring perimeter model to describe aromaticity in PAHs [93, 94] is the so-called π-sextet model. In this model, aromaticity is regarded as a local property of six-membered ring fragments of a larger molecule. An aromatic π-sextet is defined as a single benzene-like ring, with six localized π-electrons separated from adjacent rings by formal C–C single bonds. For instance, the resonance structure of phenanthrene in the top of Figure 9.3h has one more π-sextet than that of the bottom, which according to the Clar model means that it represents a more realistic description of phenanthrene. Indeed, the X-ray structure of phenanthrene shows that the C9–C10 bond (Figure 9.1) is the shortest among all C–C bonds [95]. According to this model, the two external rings of phenanthrene are the most aromatic, and the C9–C10 bond is the most reactive. Moreover, phenanthrene with two π-sextets is more stable than its isomer anthracene with a unique π-sextet [11]. Theory and computation have provided extensive support to the predictions made by Clar's π-sextet rule [94, 96].

The Clar π-sextet rule has an essential restriction that it may only be used with benzenoid species. To overcome this limitation and to extend Clar's rule to non-benzenoid PCHs, Glidewell and Lloyd formulated a more general rule [97] stating that the total population of π-electrons in conjugated polycyclic systems tends to form the smallest $4n + 2$ groups and to avoid the formation of the smallest $4n$ groups. For benzenoid systems (i.e. PCHs having only 6-MRs), Glidewell and Lloyd's rule reduces to Clar's π-sextet rule. Figure 9.3i shows an example of a non-benzenoid PCHs (biphenylene) in which application of the Glidewell and Lloyds rule leads to the conclusion that the structure in blue is more relevant than the structure in black to explain its electronic and molecular properties. The Glidewell and Lloyd rule is followed by the majority of PCHs, according to a systematic study [92] in a large series of PCHs.

9.4.2 Rules for Three-Dimensional Aromaticity

The concept of 3D aromaticity was introduced by King and Rouvray [98] in 1977 and by Aihara [99] in 1978 when they analyzed polyhedral boranes using a Hückel-like MO approach. A study by Ottosson et al. [100] identified the prerequisites for a chemical compound to be 3D-aromatic: (i) (at least) triply degenerate MOs as found in tetrahedral or higher symmetry molecules, (ii) a closed-shell electronic structure, which leads to a $6n+2$ electron count in the case of tetrahedral or octahedral molecules, (iii) extensive electron delocalization involving the complete 3D molecule leading to resonance stabilization, and (iv) similar (optoelectronic and magnetic) properties in the three xyz directions. Such conditions are obeyed by a few compounds. The first known species with 3D-aromaticity were the *closo* boranes, $B_nH_n^{-2}$ (Figure 9.4a), such as $[B_{10}H_{10}]^{2-}$ or $[B_{12}H_{12}]^{2-}$ and derivatives, synthesized at the end of the 1950s [101, 102]. These species have a structure of polyhedron with triangular faces and are highly stable [103]. They obey Wade's $2N + 2$ electron rule, where N is the number of vertexes of the polyhedron or Mingos' rule that is $4N + 2$ [104, 105]. Wade and Mingos' rules are equivalent, but Wade's rule refers to the skeletal electrons (all valence electrons except for those of the B—H bonds), whereas Mingos' rule also incorporates the exo electrons of the B—H bonds, thus the $4N + 2$ in Mingos' rule designates the total number of valence electrons. It has been demonstrated that Wade–Mingos' rule, which was first formulated for borane clusters, is also applicable to other main group clusters as well as Zintl phases and ions [106, 107].

The Wade–Mingos' rule refers to the aromaticity due to electron delocalization of electrons inside the cage. The Hirsch's $2(n + 1)^2$ rule [108] is a rule for the spherical aromaticity of the electrons on the surface of the sphere. It is based on the fact that the Schrödinger equation of a uniform electron gas surrounding the surface of a sphere is the same as that of the rigid rotor and, consequently, the wavefunctions of this model are characterized by the angular momentum quantum number l ($l = 0, 1, 2...$) similar to the situation found for atomic orbitals. Each energy level is $2l + 1$ times degenerated, and, therefore, π-shells are completely filled when we have 2, 8, 18, 32, 50, 72, ... electrons, i.e. $2(n + 1)^2$ electrons ($n = 0, 1, 2, ...,$

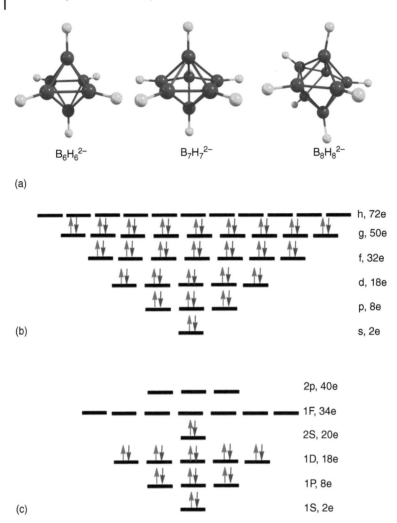

Figure 9.4 (a) *Closo* borane anions, $B_nH_n^{2-}$, $n = 6–8$, (b) with 50 π-electrons, C_{60}^{+10} follows the Hirsch rule of spherical aromaticity; (c) with its 20 valence electrons, Na_{20} is aromatic according to the jellium model.

see Figure 9.4b). Considering that the π-electron system of an icosahedral fullerene can be approximated to that of a spherical electron gas, then charge icosahedral fullerenes such as C_{20}^{2+}, C_{60}^{10+}, and C_{70}^{2-} (but not C_{60} or C_{70}, for instance) among others are aromatic fullerenes, as they have 18, 50, and 72 electrons, respectively. The spherical systems with a same-spin half-filled last energy level and the rest being double-filled should be aromatic, much as Baird's $4n$ rule reflected the extension of Hückel's $4n + 2$ rule to lowest-lying triplet states. This open-shell spherical aromaticity [109] is reached when the spherical compounds have $2n^2 + 2n + 1$ electrons and with a spin $S = n + \frac{1}{2}$, as in the case of C_{60}^{-1} with $S = 11/2$ ($n = 5$).

The spherical jellium model was able to satisfactorily explain the reported experimental abundances of alkali, alkaline earth metals, and gold clusters of 2, 8, 18, 20,

34, and 40 atoms recorded in experimental mass spectra [110]. The energy levels of valence electrons for such a model are $1S^2 1P^6 1D^{10} 2S^2 1F^{14} 2P^6 1G^{18} 2D^{10}\ldots$ and, therefore, with 2, 8, 18, 20, 34, 40… electrons one gets a closed-shell electronic structure (see Figure 9.4c), and, consequently, can be considered jellium aromatic. Similar to the Wade–Mingos' rule, the jellium rule refers to the inner aromaticity of skeletal electrons of the cage, whereas the Hirsch's rule describes the outer aromaticity of the electrons delocalized on the surface of the spherical species. An open-shell jellium aromaticity rule has been also formulated [111]. Finally, certain orbital occupations have been connected to disk [112], cylindrical [113], octahedral [114], and cubic [115] aromaticities.

9.5 Metallabenzenes and Related Compounds as an Example

The first metallabenzene was theoretically predicted in 1979 by Thorn and Hoffmann [116]. The carbon moiety can be treated as a monoanionic ($C_5H_5^-$) interacting with the metal, therefore, serving as a four electron donor to the metal (see Figure 9.5).

Only three years later, the first metallabenzene (see Figure 9.6) was synthesized and characterized by Roper and coworkers [117]. Since then, metallaaromatic chemistry has experienced a fast development due to its importance in both, theoretical and experimental areas [118–120].

Nowadays, seven major types of metallaaromatic compounds are found: metallabenzenes, metallabenzynes, heterometallaaromatics, dianion metalloles,

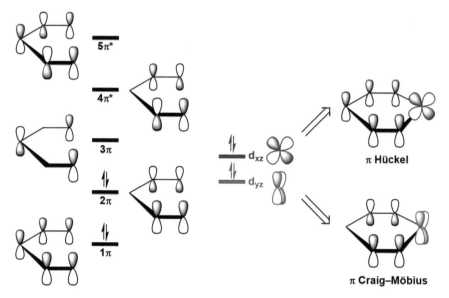

Figure 9.5 π orbitals of the $C_5H_5^-$ and d orbitals of the metal moieties involved in the formation of a metallabenzene with the two occupied π-molecular orbitals.

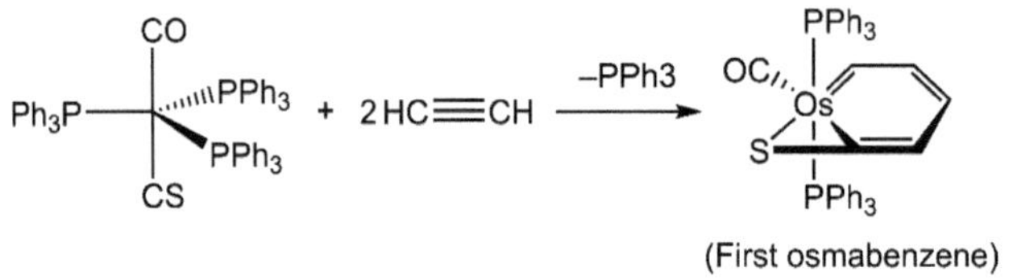

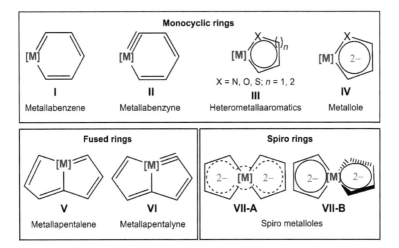

Figure 9.6 The first metallabenzene was developed by Roper and coworkers in 1982.

Figure 9.7 Metallaaromatic scaffolds. Source: Used with permission from ref. [118].

metallapentalenes, and metallapentalynes (also known as carbolongs) as well as a recent class of aromatic system, which has no analog in the carbon chemistry, the spiro metalloles (see Figure 9.7) [118–121].

A few years ago, Jia et al. summarized the reactivity and synthesis of metallabenzenes and metallabenzynes [122]. In 2015, Fernández, Merino, and Frenking analyzed from a computational point of view the aromaticity of metallabenzenes and heterometallaaromatic systems [123]. Two years later, Wright and co-workers edited a book focused on the chemistry of these metallabenzenes and metallabenzynes complexes [119]. In 2018, Saito et al. presented the progress on transition metal complexes featuring dianion metalloles of group 14 elements [124], and Xi et al. reported a review concerning dianion metalloles and spiro metalloles [125], whereas Xia et al. summarized the carbolong chemistry [126]. Afterward, in 2019, Liu et al. developed from tetra- to heptacyclic irida PAHs [127] and in July 2022, Xia et al. synthesized the first metalla-aromatic conjugated polymer through consecutive carbyne shuttling processes [128].

The presence of a transition metal atom in the molecule makes the metallaaromatic reactivity different from that observed in the organic congeners. However, as aromatic complexes, metallaaromatic compounds possess various benzene-like reactions, including electrophilic substitution, nucleophilic aromatic substitution, or π-adduct formation (Figure 9.8 left).

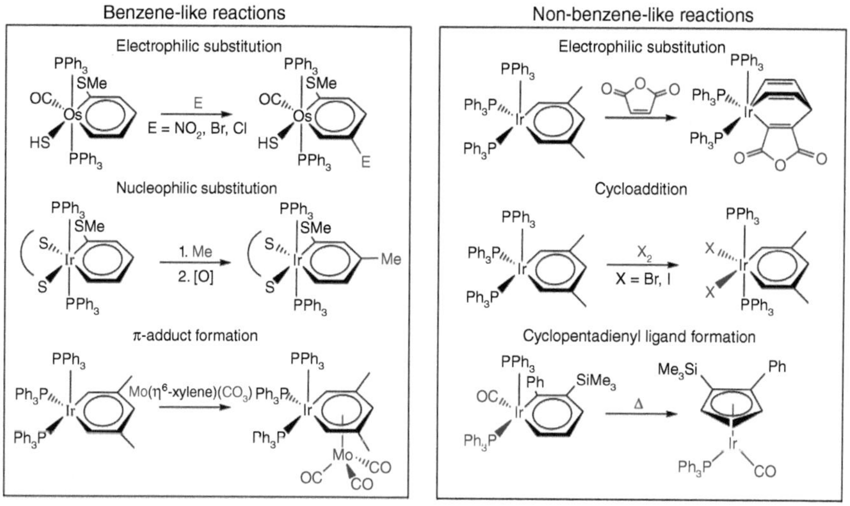

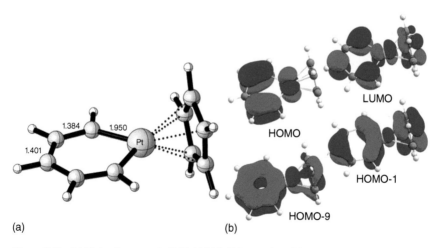

Figure 9.8 Illustrative examples of *benzene-* and *non-benzene-like* reactivity of metallabenzenes.

At the same time, these compounds present non-benzene-like chemistry, such as oxidative additions, cycloaddition reactions, or rearrangement to cyclopentadienyl ligands (Figure 9.8 right) [120]. This chemistry is not only restricted to transition metal aromatic but to heavier-main group aromatic complexes as well [129–134].

As an example, the platinabenzene Pt[C_5H_5](η^5-Cp) has been extensively studied by using different aromaticity criteria. On one hand, from a geometrical point of view, the planarity of the metallacycle and the bond-length equalization (Figure 9.9a), as well as the presence of delocalized π MOs (Figure 9.9b), denote an aromatic character.

Figure 9.9 (a) Main distances in Pt[C_5H_5](η^5-Cp) complex; (b) representative molecular orbitals (isosurface value of 0.04). Bond lengths in Å. Level of theory: BP86/TZVP and BP86/def2-TZVP//BP86/TZVP, respectively.

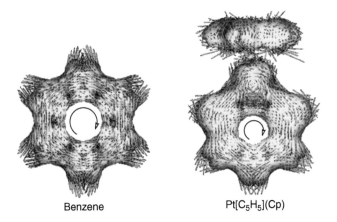

Benzene   Pt[C$_5$H$_5$](Cp)

Figure 9.10 ACID plots (isosurface value of 0.03) of benzene and Pt[C$_5$H$_5$](η^5-Cp) complexes. Source: Adapted from ref. [123].

On the other hand, within the magnetic descriptors, the NICS(0) and NICS(1) are −2.6 and −6.4 ppm, respectively [135]. Moreover, the ACID plot depicted in Figure 9.10 clearly shows the presence of diatropic ring currents (clockwise vectors) within the metallacycle moiety (as in the benzene), therefore, validating again the aromaticity of the Pt[C$_5$H$_5$](η^5-Cp) complex.

As mentioned in Section 9.3.2, Fernández and Frenking developed an energetic descriptor based on the EDA to compute the ASE (Eq. (9.10)) that has been used to predict the (anti)aromatic nature of metallaaromatic complexes, organic and heteroaromatic compounds, and the hyperconjugative aromaticity [40, 123, 136].

$$\text{ASE} = \Delta E_\pi^{\text{cyclic}} - \Delta E_\pi^{\text{linear}} \tag{9.10}$$

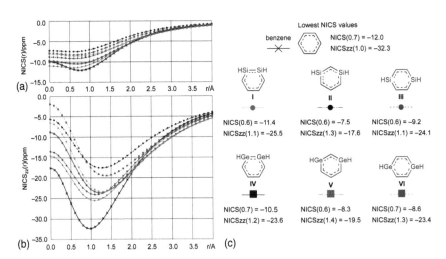

Figure 9.11 Left: (a) NICS(r) plot; (b) NICS$_{zz}$(r) plot. Right: (c) the lowest NICS values for benzene, disilabenzenes (I, II, III), and digermabenzenes (IV, V, VI), calculated at the GIAO-MP2/6-311G(3d)//MP2/6-311G(3d,p) level of theory. Source: Figure adapted from Sasamori [134].

By using this equation, the complex Pt[C$_5$H$_5$](η^5-Cp) presents a positive ASE of 37.6 kcal/mol. This result, in conjunction with his geometrical parameters, aromatic MOs, and the magnetic descriptor represented by the ACID plot, confirms the aromatic nature of the complex.

These methods have been also used, not only for metallaaromatic systems but also for heavier-main group aromatic systems. For example, the magnetic descriptors NICS and NICS$_{zz}$ have been used for heavier-main group aromatic compounds, disila- and digermabenzenes (see Figure 9.11).

The results showed that the substitution of the two carbon atoms for heavier ones keeps the aromatic character of the systems, independently of the position of the Si and Ge on their structures [134].

Something clear is that the utilization of more than one descriptor to investigate the (anti)aromatic nature of these complexes is mandatory to obtain reliable results.

References

1 Minkin, V.I. (1999). Glossary of terms used in theoretical organic chemistry. *Pure Appl. Chem.* 71: 1919–1981.
2 Merino, G., Solà, M., Fernández, I. et al. (2023). Aromaticity: Quo Vadis. *Chem. Sci.* 14: 5569–5576.
3 Hoffmann, R. (1995). *The Same and Not the Same.* Chichester: Columbia University Press.
4 Gonthier, J.F., Steinmann, S.N., Wodrich, M.D. et al. (2012). Quantification of "fuzzy" chemical concepts: a computational perspective. *Chem. Soc. Rev.* 41: 4671–4687.
5 Grunenberg, J. (2017). Ill-defined chemical concepts: the problem of quantification. *Int. J. Quantum Chem.* 117: e25359.
6 Stasyuk, A.J., Solà, M., and Voityuk, A.A. (2018). Reliable charge assessment on encapsulated fragment for endohedral systems. *Sci. Rep.* 8: 2882.
7 Fernández, I. (ed.) (2021). *Aromaticity. Modern Computational Methods and Applications.* Dordrecht: Elsevier.
8 Solà, M., Boldyrev, A.I., Cyrański, M.C. et al. (2023). *Aromaticity and Antiaromaticity: Concepts and Applications.* New York: Wiley-VCH.
9 Chen, Z., Wannere, C.S., Corminboeuf, C. et al. (2005). Nucleus-independent chemical shifts (NICS) as an aromaticity criterion. *Chem. Rev.* 105: 3842–3888.
10 Solà, M. and Bickelhaupt, F.M. (2022). Particle on a ring model for teaching the origin of the aromatic stabilization energy and the Hückel and Baird rules. *J. Chem. Edu.* 99: 3497–3501.
11 Poater, J., Visser, R., Solà, M. et al. (2007). Polycyclic benzenoids. Why kinked is more stable than straight. *J. Org. Chem.* 72: 1134–1142.
12 Szatylowicz, H., Krygowski, T.M., Solà, M. et al. (2015). Why 1,2-quinone derivatives are more stable than their 2,3-analogues? *Theor. Chem. Acc.* 134: 1–14.

13 Poater, J., Duran, M., and Solà, M. (2018). Aromaticity determines the relative stability of kinked vs. straight topologies in polycyclic aromatic hydrocarbons. *Front. Chem.* 6: 561.

14 Faraday, M. (1825). On new compounds of carbon and hydrogen and on certain other products obtained during the decomposition of oil by heat. *Philos. Trans. R. Soc. Lond.* 115: 440–446.

15 Fernández, I. and Cossío, F.P. (2016). Interplay between aromaticity and strain in double group transfer reactions to 1,2-benzyne. *J. Comput. Chem.* 37: 1265–1273.

16 Fluegel, L.L. and Hoye, T.R. (2021). Hexadehydro-Diels–Alder reaction: benzyne generation via cycloisomerization of tethered triynes. *Chem. Rev.* 121: 2413–2444.

17 Poater, J., García-Cruz, I., Illas, F. et al. (2004). Discrepancy between common local aromaticity measures in a series of carbazole derivatives. *Phys. Chem. Chem. Phys.* 6: 314–318.

18 Kruszewski, J. and Krygowski, T.M. (1972). Definition of aromaticity basing on the harmonic oscillator model. *Tetrahedron Lett.* 13: 3839–3842.

19 Krygowski, T.M. (1993). Crystallographic studies of inter- and intra-molecular interactions reflected in benzenoid hydrocarbons. Nonequivalence of indices of aromaticity. *J. Chem. Inf. Comp. Sci.* 33: 70–78.

20 Krygowski, T.M. and Cyrański, M.K. (2001). Structural aspects of aromaticity. *Chem. Rev.* 101: 1385–1419.

21 Zborowski, K.K., Alkorta, I., Elguero, J. et al. (2012). Calculation of the HOMA model parameters for the carbon–boron bond. *Struct. Chem.* 23: 595–600.

22 Zborowski, K.K., Alkorta, I., Elguero, J. et al. (2013). HOMA parameters for the boron–boron bond: how the introduction of a BB bond influences the aromaticity of selected hydrocarbons. *Struct. Chem.* 24: 543–548.

23 Krygowski, T.M. and Cyrański, M.K. (1996). Separation of the energetic and geometric contributions to the aromaticity of π-electron carbocyclics. *Tetrahedron* 52: 1713–1722.

24 Arpa, E.M. and Durbeej, B. (2023). HOMER: a reparameterization of the harmonic oscillator model of aromaticity (HOMA) for excited states. *Phys. Chem. Chem. Phys.* 25: 16763–16771.

25 Cyrański, M.K. (2005). Energetic aspects of cyclic π-electron delocalization: evaluation of the methods of estimating aromatic stabilization energies. *Chem. Rev.* 105: 3773–3811.

26 Hehre, W.J., Ditchfie, R., Radom, L. et al. (1970). Molecular orbital theory of electronic structure of organic compounds. 5. Molecular theory of bond separation. *JACS* 92: 4796–4801.

27 Pople, J.A., Radom, L., and Hehre, W.J. (1971). Molecular orbital theory of the electronic structure of organic compounds. VII. Systematic study of energies, conformations, and bond interactions. *JACS* 93: 289–300.

28 George, P., Trachtman, M., Bock, C.W. et al. (1976). Homodesmotic reactions for the assessment of stabilization energies in benzenoid and other conjugated cyclic hydrocarbons. *J. Chem. Soc. Perkin Trans.* 2: 1222–1227.

29 George, P., Trachtman, M., Brett, A.M. et al. (1977). Comparison of various isodesmic and homodesmotic reaction heats with values derived from published ab initio molecular orbital calculations. *J. Chem. Soc. Perkin Trans.* 2: 1036–1047.

30 Cyrański, M.K., Schleyer, P.V.R., Krygowski, T.M. et al. (2003). Facts and artifacts about aromatic stability estimation. *Tetrahedron* 59: 1657–1665.

31 Cyrański, M.K., Krygowski, T.M., Katritzky, A.R. et al. (2002). To what extent can aromaticity be defined uniquely? *J. Org. Chem.* 67: 1333–1338.

32 George, P., Bock, C.W., and Trachtman, M. (1984). The evaluation of empirical resonance energies as reaction enthalpies with particular reference to benzene. *J. Chem. Edu.* 61: 225.

33 Afeely, H.Y., Liebman, J.F., and E. S.S. (2021). Neutral thermochemical data. In: *NIST Chemistry WebBook, NIST Standard Reference Database No. 69*. Gaithersburg MD: National Institute of Standards and Technology.

34 Pedley, J.B., Naylor, D.R., and Kirby, S.P. (1986). *Thermodynamical Data of Organic Compounds*. London: Chapman & Hall.

35 Schleyer, P.V.R. and Pühlhofer, F. (2002). Recommendations for the evaluation of aromatic stabilization energies. *Org. Lett.* 4: 2873–2876.

36 Bally, T., Hasselmann, D., and Loosen, K. (1985). The molecular ion of 5-methylene-1,3-cyclohexadiene: electronic absorption spectrum and revised enthalpy of formation. *Helv. Chim. Acta* 68: 345–354.

37 Bartmess, J.E. (1982). Gas-phase ion chemistry of 5-methylene-1,3-cyclohexadiene (o-isotoluene) and 3-methylene-1,4-cyclohexadiene (p-isotoluene). *JACS* 104: 335–337.

38 Wodrich, M.D., Wannere, C.S., Mo, Y. et al. (2007). The concept of protobranching and its many paradigm shifting implications for energy evaluations. *Chem. Eur. J.* 13: 7731–7744.

39 Ciesielski, A., Stepień, D.K., Dobrowolski, M.A. et al. (2012). On the aromatic stabilization of benzenoid hydrocarbons. *Chem. Commun.* 48: 10129–10131.

40 Fernández, I. and Frenking, G. (2007). Direct estimate of conjugation and aromaticity in cyclic compounds with the EDA method. *Faraday Discuss.* 135: 403–421.

41 te Velde, G., Bickelhaupt, F.M., Baerends, E.J. et al. (2001). Chemistry with ADF. *J. Comput. Chem.* 22: 931–967.

42 Hopffgarten, M. and Frenking, G. (2012). Energy decomposition analysis. *WIREs Comput. Mol. Sci.* 2: 43–62.

43 Morokuma, K. (1977). Why do molecules interact? The origin of electron donor-acceptor complexes, hydrogen bonding and proton affinity. *Acc. Chem. Res.* 10: 294–300.

44 Ziegler, T. and Rauk, A. (1977). On the calculation of bonding energies by the Hartree-Fock-Slater method. *Theor. Chim. Acta* 46: 1–10.

45 Matito, E. and Solà, M. (2009). The role of electronic delocalization in transition metal complexes from the electron localization function and the quantum theory of atoms in molecules viewpoints. *Coord. Chem. Rev.* 253: 647–665.

46 Feixas, F., Matito, E., Poater, J. et al. (2015). Quantifying aromaticity with electron delocalisation measures. *Chem. Soc. Rev.* 44: 6434–6451.

47 Matito, E., Duran, M., and Solà, M. (2005). The aromatic fluctuation index (FLU): a new aromaticity index based on electron delocalization. *J. Chem. Phys.* 122: 014109.

48 Matito, E., Duran, M., and Solà, M. (2006). Erratum: "The aromatic fluctuation index (FLU): A new aromaticity index based on electron delocalization" [J. Chem Phys. 122, 014109 (2005)]. *J. Chem. Phys.* 125: 059901.

49 Mayer, I. (1986). Bond orders and valences from ab initio wave functions. *Int. J. Quantum Chem.* 29: 477–483.

50 Giambiagi, M., de Giambiagi, M.S., dos Santos, C.D. et al. (2000). Multicenter bond indices as a measure of aromaticity. *Phys. Chem. Chem. Phys.* 2: 3381–3392.

51 Bultinck, P., Ponec, R., and Van Damme, S. (2005). Multicenter bond indices as a new measure of aromaticity in polycyclic aromatic hydrocarbons. *J. Phys. Org. Chem.* 18: 706–718.

52 Cioslowski, J., Matito, E., and Solà, M. (2007). Properties of aromaticity indices based on the one-electron density matrix. *J. Phys. Chem. A* 111: 6521–6525.

53 Zhou, Z. and Parr, R.G. (1989). New measure of aromaticity: absolute hardness and relative hardness. *JACS* 111: 7371–7379.

54 De Proft, F. and Geerlings, P. (2004). Relative hardness as a measure of aromaticity. *Phys. Chem. Chem. Phys.* 6: 242–248.

55 Pearson, R.G. (1986). Absolute electronegativity and hardness correlated with molecular orbital theory. *Proc. Natl. Acad. Sci. U.S.A.* 83: 8440–8441.

56 Keith, T.A. and Bader, R.F.W. (1993). Calculation of magnetic response properties using a continuous set of gauge transformations. *Chem. Phys. Lett.* 210: 223–231.

57 Lazzeretti, P., Malagoni, M., and Zanasi, R. (1994). Computational approach to molecular magnetic properties by continuous transformation of the origin of the current density. *Chem. Phys. Lett.* 220: 299–304.

58 Jusélius, J., Sundholm, D., and Gauss, J. (2004). Calculation of current densities using gauge-including atomic orbitals. *J. Chem. Phys.* 121: 3952–3963.

59 Mitchell, R.H. (2001). Measuring aromaticity by NMR. *Chem. Rev.* 101: 1301–1315.

60 Bühl, M. and van Wüllen, C. (1995). Computational evidence for a new C_{84} isomer. *Chem. Phys. Lett.* 247: 63–68.

61 Schleyer, P.V.R., Maerker, C., Dransfeld, A. et al. (1996). Nucleus-independent chemical shifts: a simple and efficient aromaticity probe. *JACS* 118: 6317–6318.

62 Corminboeuf, C., Heine, T., Seifert, G. et al. (2004). Induced magnetic fields in aromatic [n]-annulenes – interpretation of NICS tensor components. *Phys. Chem. Chem. Phys.* 6: 273–276.

63 Schleyer, P.V.R., Manoharan, M., Jiao, H.J. et al. (2001). The acenes: is there a relationship between aromatic stabilization and reactivity? *Org. Lett.* 3: 3643–3646.

References

64 Fallah-Bagher-Shaidaei, H., Wannere, C.S., Corminboeuf, C. et al. (2006). Which NICS aromaticity index for planar rings is best? *Org. Lett.* 8: 863–866.

65 Feixas, F., Matito, E., Poater, J. et al. (2008). On the performance of some aromaticity indices: a critical assessment using a test set. *J. Comput. Chem.* 29: 1543–1554.

66 Lazzeretti, P. (2000). Ring currents. *Prog. Nucl. Magn. Reson. Spectrosc.* 36: 1–88.

67 Lazzeretti, P. (2004). Assessment of aromaticity via molecular response properties. *Phys. Chem. Chem. Phys.* 6: 217–223.

68 Aihara, J. (2002). Nucleus-independent chemical shifts and local aromaticities in large polycyclic aromatic hydrocarbons. *Chem. Phys. Lett.* 365: 34–39.

69 Poater, J., Solà, M., Viglione, R.G. et al. (2004). The local aromaticity of the six-membered rings in pyracylene. A difficult case for the NICS indicator of aromaticity. *J. Org. Chem.* 69: 7537–7542.

70 Islas, R., Martínez-Guajardo, G., Jiménez-Halla, J.O.C. et al. (2010). Not all that has a negative NICS is aromatic: the case of the H-bonded cyclic trimer of HF. *J. Chem. Theory Comput.* 6: 1131–1135.

71 Solà, M., Feixas, F., Jiménez-Halla, J.O.C. et al. (2010). A critical assessment of the performance of magnetic and electronic indices of aromaticity. *Symmetry* 2: 1156.

72 Coulson, C.A. (1960). Present state of molecular structure calculations. *Rev. Mod. Phys.* 32: 170–177.

73 Solà, M. (2022). Aromaticity rules. *Nat. Chem.* 14: 585–590.

74 Dognon, J.-P., Clavaguéra, C., and Pyykkö, P. (2007). Towards a 32-electron principle: Pu@Pb$_{12}$ and related systems. *Angew. Chem. Int. Ed.* 46: 1427–1430.

75 Hückel, E. (1931). Quantentheoretische Beiträge zum Benzolproblem I. Die Elektronenkonfiguration des Benzols und verwandter Verbindungen. *Z. Phys.* 70: 104–186.

76 Doering, W.V.E. and Detert, F.L. (1951). Cycloheptatrienylium oxide. *JACS* 73: 876–877.

77 Breslow, R. (1965). Antiaromaticity. *Chem. Eng. News* 43: 90–100.

78 Baird, N.C. (1972). Quantum organic photochemistry. II. Resonance and aromaticity in the lowest $^3\pi\pi^*$ state of cyclic hydrocarbons. *JACS* 94: 4941–4948.

79 Karas, L.J. and Wu, J.I. (2022). Baird's rules at the tipping point. *Nat. Chem.* 14: 723–725.

80 Ottosson, H. (2012). Organic photochemistry: exciting excited-state aromaticity. *Nat. Chem.* 4: 969–971.

81 Soncini, A. and Fowler, P.W. (2008). Ring-current aromaticity in open-shell systems. *Chem. Phys. Lett.* 450: 431–436.

82 Mandado, M., Graña, A.M., and Pérez-Juste, I. (2008). Aromaticity in spin-polarized systems: can rings be simultaneously alpha aromatic and beta antiaromatic? *J. Chem. Phys.* 129: 164114.

83 Valiev, R.R., Kurten, T., Valiulina, L.I. et al. (2022). Magnetically induced ring currents in metallocenothiaporphyrins. *Phys. Chem. Chem. Phys.* 24: 1666–1674.

84 Heilbronner, E. (1964). Hückel molecular orbitals of Möbius-type conformation of annulenes. *Tetrahedron Lett.* 5: 1923–1928.

85 Ajami, D., Oeckler, O., Simon, A. et al. (2003). Synthesis of a Möbius aromatic hydrocarbon. *Nature* 426: 819–821.
86 Rzepa, H.S. (2005). Möbius aromaticity and delocalization. *Chem. Rev.* 105: 3697–3715.
87 Craig, D.P. and Paddock, N.L. (1958). A novel type of aromaticity. *Nature* 181: 1052–1053.
88 Szczepanik, D.W. and Solà, M. (2019). Electron delocalization in planar metallacycles: Hückel or Möbius aromatic? *ChemistryOpen* 8: 219–227.
89 Rosenberg, M., Dahlstrand, C., Kilså, K. et al. (2014). Excited state aromaticity and antiaromaticity: opportunities for photophysical and photochemical rationalizations. *Chem. Rev.* 114: 5379–5425.
90 Rappaport, S.M. and Rzepa, H.S. (2008). Intrinsically chiral aromaticity. rules incorporating linking number, twist, and writhe for higher-twist Möbius annulenes. *JACS* 130: 7613–7619.
91 Platt, J.R. (1949). Classification of spectra of cata-condensed hydrocarbons. *J. Chem. Phys.* 17: 484–495.
92 El Bakouri, O., Poater, J., Feixas, F. et al. (2016). Exploring the validity of the Glidewell–Lloyd extension of Clar's π-sextet rule: assessment from polycyclic conjugated hydrocarbons. *Theor. Chem. Acc.* 135: 205.
93 Clar, E. (1972). *The Aromatic Sextet*. New York: Wiley.
94 Solà, M. (2013). Forty years of Clar's aromatic π-sextet rule. *Front. Chem.* 1: 22.
95 Kay, M.I., Okaya, Y., and Cox, D.E. (1971). Refinement of structure of room temperature phase of phenanthrene, $C_{14}H_{10}$, from X-ray and neutron diffraction data. *Acta Cryst. B* 27: 26–33.
96 Randić, M. (2014). Novel insight into Clar's aromatic π-sextets. *Chem. Phys. Lett.* 601: 1–5.
97 Glidewell, C. and Lloyd, D. (1984). MNDO study of bond orders in some conjugated bi- and tri-cyclic hydrocarbons. *Tetrahedron* 40: 4455–4472.
98 King, R.B. and Rouvray, D.H. (1977). Chemical applications of group theory and topology. 7. A graph-theoretical interpretation of the bonding topology in polyhedral boranes, carboranes, and metal clusters. *JACS* 99: 7834–7840.
99 Aihara, J. (1978). Three-dimensional aromaticity of polyhedral boranes. *JACS* 100: 3339–3342.
100 El Bakouri, O., Szczepanik, D.W., Jorner, K. et al. (2022). Three-dimensional fully π-conjugated macrocycles: when 3D-aromatic and when 2D-aromatic-in-3D? *JACS* 144: 8560–8575.
101 Lipscomb, W.N., Pitochelli, A.R., and Hawthorne, M.F. (1959). Probable structure of the $B_{10}H_{10}^{-2}$ ion. *JACS* 81: 5833–5834.
102 Pitochelli, A.R. and Hawthorne, F.M. (1960). The isolation of the icosahedral $B_{12}H_{12}^{-2}$ ion. *JACS* 82: 3228–3229.
103 Pitt, M.P., Paskevicius, M., Brown, D.H. et al. (2013). Thermal stability of $Li_2B_{12}H_{12}$ and its role in the decomposition of $LiBH_4$. *JACS* 135: 6930–6941.
104 Wade, K. (1971). The structural significance of the number of skeletal bonding electron-pairs in carboranes, the higher boranes and borane anions, and various

transition-metal carbonyl cluster compounds. *J. Chem. Soc. D Chem. Commun.* 792–793.

105 Mingos, D.M.P. (1972). A general theory for cluster and ring compounds of the main group and transition elements. *Nat. Phys. Sci.* 236: 99–102.

106 Hirsch, A., Chen, Z., and Jiao, H. (2001). Spherical aromaticity of inorganic cage molecules. *Angew. Chem. Int. Ed.* 40: 2834–2838.

107 Liu, C., Popov, I.A., Chen, Z. et al. (2018). Aromaticity and antiaromaticity in Zintl clusters. *Chem. Eur. J.* 24: 14583–14597.

108 Hirsch, A., Chen, Z., and Jiao, H. (2000). Spherical aromaticity in icosahedral fullerenes: the $2(N+1)^2$ rule. *Angew. Chem. Int. Ed.* 39: 3915–3917.

109 Poater, J. and Solà, M. (2011). Open-shell spherical aromaticity: the $2N^2+2N+1$ (with $S = N+1/2$) rule. *Chem. Commun.* 47: 11647–11649.

110 Cohen, M.L., Chou, M.Y., Knight, W.D. et al. (1987). Physics of metal clusters. *J. Phys. Chem.* 91: 3141–3149.

111 Poater, J. and Solà, M. (2019). Open-shell jellium aromaticity in metal clusters. *Chem. Commun.* 55: 5559–5562.

112 Tai, T.B., Havenith, R.W.A., Teunissen, J.L. et al. (2013). Particle on a boron disk: ring currents and disk aromaticity in B_{20}^{2-}. *Inorg. Chem.* 52: 10595–10600.

113 Duong, L.V., Pham, H.T., Tam, N.M. et al. (2014). A particle on a hollow cylinder: the triple ring tubular cluster B_{27}^+. *Phys. Chem. Chem. Phys.* 16: 19470–19478.

114 El Bakouri, O., Duran, M., Poater, J. et al. (2016). Octahedral aromaticity in $^{2S+1}A_{1g} X_6^q$ clusters (X = Li-C and Be-Si, S = 0-3, and q = -2 to +4). *Phys. Chem. Chem. Phys.* 18: 11700–11706.

115 Cui, P., Hu, H.-S., Zhao, B. et al. (2015). A multicentre-bonded $[Zn^I]_8$ cluster with cubic aromaticity. *Nat. Commun.* 6: 6331.

116 Thorn, D.L. and Hoffmann, R. (1979). Delocalization in metallocycles. *Nouv. J. Chim.* 3: 39–45.

117 Elliott, G.P., Roper, W.R., and Waters, J.M. (1982). Metallacyclohexatrienes or 'metallabenzenes.' Synthesis of osmabenzene derivatives and X-ray crystal structure of [Os(CSCHCHCHCH)(CO)(PPh$_3$)$_2$]. *J. Chem. Soc. Chem. Commun.* 811–813.

118 Chen, D., Hua, Y., and Xia, H. (2020). Metallaaromatic chemistry: history and development. *Chem. Rev.* 120: 12994–13086.

119 Wright, L.J. (ed.) (2017). *Metallabenzenes: An Expert View*. Chichester: Wiley 328 p.

120 Frogley, B.J. and Wright, L.J. (2018). Recent advances in metallaaromatic chemistry. *Chem. Eur. J.* 24: 2025–2038.

121 Zhang, Y., Wei, J., Chi, Y. et al. (2017). Spiro metalla-aromatics of Pd, Pt, and Rh: synthesis and characterization. *JACS* 139: 5039–5042.

122 Chen, J. and Jia, G. (2013). Recent development in the chemistry of transition metal-containing metallabenzenes and metallabenzynes. *Coord. Chem. Rev.* 257: 2491–2521.

123 Fernandez, I., Frenking, G., and Merino, G. (2015). Aromaticity of metallabenzenes and related compounds. *Chem. Soc. Rev.* 44: 6452–6463.

124 Saito, M. (2018). Transition-metal complexes featuring dianionic heavy group 14 element aromatic ligands. *Acc. Chem. Res.* 51: 160–169.
125 Wei, J., Zhang, W.-X., and Xi, Z. (2018). The aromatic dianion metalloles. *Chem. Sci.* 9: 560–568.
126 Zhu, C. and Xia, H. (2018). Carbolong chemistry: a story of carbon chain ligands and transition metals. *Acc. Chem. Res.* 51: 1691–1700.
127 Hu, Y.X., Zhang, J., Wang, X. et al. (2019). One-pot syntheses of irida-polycyclic aromatic hydrocarbons. *Chem. Sci.* 10: 10894–10899.
128 Chen, S., Peng, L., Liu, Y. et al. (2022). Conjugated polymers based on metalla-aromatic building blocks. *Proc. Natl. Acad. Sci. U.S.A.* 119: e2203701119.
129 Hong, J.-H., Boudjouk, P., and Castellino, S. (1994). Synthesis and characterization of two aromatic silicon-containing dianions: the 2,3,4,5-tetraphenylsilole dianion and the 1,1'-disila-2,2',3,3',4,4',5,5'-octaphenylfulvalene dianion. *Organometallics* 13: 3387–3389.
130 Sekiguchi, A., Tsukamoto, M., and Ichinohe, M. (1997). A free cyclotrigermenium cation with a 2π-electron system. *Science* 275: 60–61.
131 Saito, M. and Yoshioka, M. (2005). The anions and dianions of group 14 metalloles. *Coord. Chem. Rev.* 249: 765–780.
132 Ota, K. and Kinjo, R. (2020). Inorganic benzene valence isomers. *Chem. Asian J.* 15: 2558–2574.
133 Ota, K. and Kinjo, R. (2021). Heavier element-containing aromatics of [4n+2]-electron systems. *Chem. Soc. Rev.* 50: 10594–10673.
134 Sasamori, T. (2021). Disila- and digermabenzenes. *Chem. Sci.* 12: 6507–6517.
135 Iron, M.A., Lucassen, A.C.B., Cohen, H. et al. (2004). A computational foray into the formation and reactivity of metallabenzenes. *JACS* 126: 11699–11710.
136 Fernández, I., Wu, J.I., and von Ragué, S.P. (2013). Substituent effects on "hyperconjugative" aromaticity and antiaromaticity in planar cyclopolyenes. *Org. Lett.* 15: 2990–2993.

10

Acidity and Basicity

Ranita Pal[1], Himangshu Mondal[2], and Pratim K. Chattaraj[3]

[1] Indian Institute of Technology, Kharagpur, Advanced Technology Development Centre, Kharagpur, 721302, West Bengal, India
[2] Indian Institute of Technology, Kharagpur, Department of Chemistry, Kharagpur, 721302, West Bengal, India
[3] Birla Institute of Technology, Mesra, Department of Chemistry, Ranchi, 835215, Jharkhand, India

10.1 Introduction

The term "acid" derived from the Latin word "ac-", meaning the substance being "pointed" or sharp, akin to the taste of vinegar, was originated in the seventeenth century. The fundamentals of chemistry teach us to recognize acids based on their characteristic sour taste, the ability to turn blue litmus red, form H_2 gas on reaction with some metals, and react with bases to produce salt and water [1]. Bases, on the other hand, are recognized in their aqueous state by their bitter taste, a soapy feeling, turning red litmus blue, and reacting with acids to produce salts. A term synonymous with base is "alkali," which has its roots in Arabic. However, the etymology of the word can be traced back to the Latin term "kalium" or potash, which also gave rise to the chemical symbol for potassium. However, the initial conceptualization of acids in chemistry was flawed. In 1787, Antoine Lavoisier classified acids as a distinct group of "complex substances" within his comprehensive taxonomy of compounds. He postulated that their unique properties were attributable to the presence of a fundamental constituent, which he named *oxygen* (meaning "acid former" in Greek). He had assigned this term to a gaseous element identified a few years earlier by Joseph Priestley as the vital component in combustion. Given that several combustion products (oxides) yield acidic solutions, and many acids contain oxygen, Lavoisier's misconception was understandable [2]. In 1811, Humphry Davy conducted experiments that revealed muriatic acid (hydrochloric acid), previously believed to be an element by Lavoisier, did not contain oxygen. However, this finding only led to the belief among some scientists that chlorine might be an oxygen-containing compound rather than an element itself. Despite the discovery of several oxygen-free acids by the year 1830, it was not until approximately 10 years later that the hydrogen theory of acids gained general acceptance. By then, the term "oxygen" had already been well-established, and changing its name was considered impractical [3].

Exploring Chemical Concepts Through Theory and Computation, First Edition. Edited by Shubin Liu.
© 2024 WILEY-VCH GmbH. Published 2024 by WILEY-VCH GmbH.

In this chapter, we will discuss various definitions and theories of acids, bases, and salts, along with an analysis of their strength, and a conceptual density functional theory (CDFT) perspective, among others. Our treatment of the topic is mostly qualitative, with some quantitative treatment of the equilibrium systems.

10.2 Definitions and Theories

A significant breakthrough in comprehending acids, bases, and salts emerged with Michael Faraday's discovery in the mid-nineteenth century. Faraday's observation that solutions of salts, referred to as electrolytes, have the ability to conduct electricity indicated the presence of charged particles capable of migration on the application of an electric field. These particles were named ions, or "wanderers," by Faraday. Subsequent investigations into electrolytic solutions suggested that the characteristics associated with acids stem from an abundance of hydrogen ions within the solution [4]. Over the years, many definitions of acids and bases have been put forward. The significant ones are discussed briefly to understand different perspectives.

10.2.1 Arrhenius Theory

During the 1880s, Svante Arrhenius, a chemist from Sweden, had developed the first significant theory regarding acids. According to Arrhenius, an acidic substance is characterized by having at least one hydrogen atom within its molecular structure that is capable of dissociating or ionizing when dissolved in water. This process results in the formation of a hydrated hydrogen ion (H^+) and an accompanying anion. In a strict sense, an "Arrhenius acid" is defined as a substance that contains hydrogen. However, certain substances, despite lacking hydrogen themselves, can still generate H^+ ions when dissolved in water through a reaction with water molecules. Thus, for practical purposes, a more useful operational definition of an acid is a system that produces an abundance of H^+ ions on dissolution in water. Bases, on the other hand, dissociate in aqueous solutions to generate hydroxide ions (OH^-). According to this theory, an acid and a base increase the concentrations of H^+ and OH^- ions, respectively, in solutions. However, this theory can only explain aqueous solutions and does not take into consideration acid–base reactions occurring in nonaqueous solvents or gas-phase systems.

Certain crucial aspects should be understood regarding hydrogen in acids. First, while "Arrhenius acids" do contain H atoms, not all of them are capable of dissociating, e.g. the H atoms in the methyl group of acetic acid (CH_3COOH) are incapable of dissociation and are known as "non-acidic." Second, the extent of dissociation may vary among the H atoms that can dissociate, which gives an idea of the strength of the acidity. Strong acids, viz., HCl and HNO_3, are almost entirely dissociated (close to 100%) in a solution. On the other hand, most organic acids, including acetic acid, are weak acids, undergoing very little dissociation in most solutions. Examples of weak inorganic acids include HF and HCN. Third, acids containing more than one dissociable H atom are known as polyprotic acids. Sulfuric and phosphoric acids are two well-known examples. The intermediate forms that have the ability to accept and release protons are known as ampholytes (e.g. HPO_4^{2-}).

10.2.2 Brønsted–Lowry Theory

Arrhenius' original perspective regarded acids as systems that generate H^+ ions and bases that generate OH^- ions upon dissociation. While this conceptualization has been valuable, it falls short in explaining certain observations. For example, it fails to clarify why NH_3, which lacks OH^- ions, is considered a base rather than an acid, and why $FeCl_3$ solution exhibits acidity, while that of Na_2S is alkaline in nature. These limitations of the Arrhenius definition prompted the development of alternative acid–base theories that provide a more comprehensive explanation for the behavior of these substances.

In 1923, J.N. Brønsted put forth a theory that proved to be simpler as well as more universally applicable, which is that an acid and a base are proton donor and acceptor, respectively. That year itself, T.M. Lowry also set forth some similar ideas without prescribing any proper definition. However, as it reinforced, the Brønsted concept is commonly known as the Brønsted–Lowry theory. This definition has a crucial implication that an acid cannot function as such without a corresponding base present that will act as the proton acceptor, and vice versa. Therefore, an acid–base reaction involves proton exchange, i.e.,

$$AH + B \rightleftharpoons A^- + BH^+ \tag{10.1}$$

where AH and B are the acid and base, respectively. Note that the resulting product BH^+ can now release the recently acquired proton to another acceptor, making it a potential acid as well. In Equation (10.1), A^- is conjugate to AH, and BH^+ is conjugate to B. In these types of reactions involving equivalent concentrations of acids and bases, it proceeds spontaneously in the direction that leads to the formation of the weaker acid and base.

This formulation of acid–base behavior differs from other theories in several aspects, which are as follows [5]:

(1) It establishes that acid–base behavior is independent of any specific solvent and is not limited to a particular solvent type. Instead, it encompasses the actions of all protolytic solvents in a generalized manner.
(2) Ions, which were previously considered a highly exceptional subclass, are now recognized as acids and bases on an equal footing with uncharged molecules.
(3) An acid reacts with a base not to form salt and a solvent, as traditionally assumed, but rather to produce a new acid and a new base. Consequently, the conventional notion of solvent "neutrality" loses its significance in this context.
(4) Here, solvated acids like HCl (in H_2O) and NH_4Cl are not considered acids but rather are salts containing the acids H_3O^+ and NH_4^+. Similarly, KOH and KNH_2 are not classified as bases, but salts containing the bases, OH^- and NH_2^-.

(5) All other factors being equal, the strength of an acid increases with a higher positive charge on the species, while the strength of a base increases with a higher negative charge. However, it is important to note that certain anions, such as ClO_4^- and RSO_3^-, exhibit minimal basicity despite their negative charge. Similarly, many cations, including Fe^{3+}, NR_4^+, do not act as acids, although their hydrogen-containing solvates can exhibit acidic properties.

(6) Certain substances, such as SO_3, BCl_3, and CO_2, which might have been classified as acids by early chemists, are excluded from the acid category due to the absence of hydrogen. Similarly, compounds like German's "phosgenoaluminic acid" ($COAl_2Cl_5$) are also not considered acids as they do not contain hydrogen.

(7) This system enables several informative comparisons with that of oxidizing and reducing agents.

A redox system can be represented by Equation 10.2.

$$\text{Reductant} \rightleftharpoons \text{oxidant} + n \text{ (electrons)} \tag{10.2}$$

A thermodynamic approach, involving proton activity, equilibrium constants, and acidity potentials, can be applied to Brønsted acids and bases in a manner similar by replacing the electron activities with proton activities. This allows for a consistent application of thermodynamic principles to describe the behavior of Brønsted acids and bases.

Certain substances have the capability to donate as well as accept protons, exhibiting both acidic and basic behavior. Examples of such substances include $H_2PO_4^-$, HCO_3^-, NH_3, and H_2O. These versatile substances are referred to as amphiprotic, and the dissolved species are known as ampholytes.

10.2.3 Lewis Theory

The Lewis Theory, proposed in 1923, is a comprehensive and expansive theory that surpasses the earlier theories of Arrhenius and Brønsted. It is an electronic theory that does not rely on the presence of any specific solvent for its definitions of acids and bases. An acid, according to this theory, is characterized as an electron-deficient species capable of accepting a pair of electrons or tends to associate with molecular species possessing available electron pairs. Examples of such acids include H^+, NO_2^+, BF_3, and $AlCl_3$. Bases, on the other hand, are substances that possess a pair of donatable electrons. Examples of bases include Cl^-, H_2O, OH^-, RNH_2, ethers, esters, and ketones. According to the Lewis theory, an acid–base reaction involves electron pair transfer from a base to an acid, formatting a coordinate covalent bond between them. Notably, the formation of this bond is always regarded as the initial step, even though ionization may occur subsequently.

Lewis aimed to expand the scope of the acid–base definition by considering both experimental and theoretical aspects. From an experimental standpoint, he identified four criteria: neutralization, titration with indicators, displacement, and catalysis. Lewis defined acids and bases as substances that demonstrate the capability to engage in these "typical" functions. On the theoretical side, he linked these properties to the acceptance and donation of electron pairs, regardless of whether the transfer of protons was involved. This electronic theory allows for a more comprehensive definition of acidity and basicity and has a broad scope that encompasses the proton-transfer definition. Furthermore, since electron-donor molecules have the ability to associate with protons, the Lewis concept of a base includes the Brønsted–Lowry definition. In contrast, the Lewis definition of an acid includes a wide range of species that do not contain an H atom, resulting in a significant expansion of the number of acids compared to those defined by the Brønsted concept.

For instance, the reaction between aluminum chloride and pyridine follows the same principles as the conventional neutralization of pyridine by an acid that donates protons. One significant drawback of the Lewis system becomes apparent when considering its quantitative aspect. The group of protonic acids demonstrates more uniformity in comparison to the non-protonic acids defined by Lewis when evaluating acid–base strengths within a simple system. Another drawback is that certain substances, such as HCl and CO_2, exhibit acidic behavior experimentally, despite their electronic formulas, as commonly written, not indicating their ability to accept electron pairs. Lewis referred to these acids and bases as "secondary" in contrast to his "primary" acids and bases that involve the sharing of electron pairs. This introduces a cumbersome terminology for substances that are commonly encountered and raises the question of the practicality of the term "acid" in its conventional usage.

10.2.4 Usanovich Definition

In 1938, the Russian physical chemist Mikhail Usanovich formulated a comprehensive theory regarding acids and bases. Similar to Lewis, Usanovich expressed concerns regarding the limitations of the Brønsted acid–base theory, which restricts acid properties exclusively to hydrogen compounds. However, Usanovich expanded upon Lewis's ideas by associating general electropositive or electrophilic [6] characteristics with acidity, and nucleophilic [6] behavior with basicity. The definition of acids in this theory includes:

i. Formation of salts with bases and participation in neutralization reactions.
ii. Release of cations, which may include H⁺ or other positively charged species.
iii. Incorporation of anions and free electrons into the acid structure.

This viewpoint suggests that the concepts of oxidation and reduction can be seen as specific instances of the acid–base behavior within Usanovich's framework.

By incorporating the aspects of oxidation–reduction, Usanovich revisits the viewpoint of Kossel [7], which emphasizes the influence of ionic charges on molecular properties. Additionally, he emphasizes the significance of "coordination unsaturation," which refers to the ability of an atom to increase its covalence. According to Usanovich, the acid function is determined by the presence of a positively charged particle with coordination unsaturation, while the basic function is associated with a similarly unsaturated negative atom.

Under similar conditions, the acidity or basicity of a molecule is determined by the charge or polar valence of its constituent atoms. Typically, the influence of the atom with the highest valence is predominant. Thus, oxides of low-valence atoms, such as alkalis and alkaline earth metals, tend to exhibit primarily basic properties, whereas those of elements from the third group and higher tend to be primarily acidic. Similarly, the corresponding halides show lower basicity compared to the oxides, e.g. $AsCl_3$ and AlF_3 demonstrate greater acidity than their respective oxides. It should be noted that most compounds exhibit some degree of amphoteric behavior, and factors beyond ionic charge come into play. Otherwise, compounds like TiO_2 would exhibit acidity comparable to that of CO_2.

10.2.5 Lux–Flood Definition

Unlike the Bronsted–Lowry theory, which focuses on protons as the primary species in acid–base reactions, the Lux–Flood theory describes acid–base behavior in terms of oxide ions, i.e. acids accept oxide ions while bases provide them.

An example of the above definition is the reaction between calcium oxide and silicon dioxide.

$$CaO + SiO_2 \rightarrow CaSiO_3$$

where CaO is the oxide ion donor and hence the base, $CaO \rightleftharpoons Ca^{2+} + O^{2-}$ and SiO_2 is the oxide ion acceptor and hence the acid, $SiO_2 + O^{2-} \rightleftharpoons SiO_3^{2-}$

10.2.6 Solvent System Definition

The limitations of the proton-centered definitions of acids and bases, such as Arrhenius and Brønsted–Lowry, restrict their applicability to protic solvents. However, a more inclusive approach is the solvent-system acid–base definition, which allows for the description of acid–base reactions in non-protic solutions as well. According to this concept, solvents often undergo self-ionization or auto-ionization, resulting in the formation of solvent cations and solvent anions. In this context, substances that produce solvent cations when dissolved in a particular solvent are referred to as acids, while substances that yield solvent anions when dissolved in the solvent are known as bases.

Similar to the self-dissociation of water, i.e. $2H_2O \rightleftharpoons H_3O^+ + OH^-$, the concept of acidity and alkalinity could be expanded to the corresponding self-dissociations of other solvents as well, such as NH_3, SO_2, and $COCl_2$ [8] (Equations 10.3-10.5)

$$2NH_3 \rightleftharpoons NH_4^+ + NH_2^- \tag{10.2}$$

$$2SO_2 \rightleftharpoons SO^{2+} + SO_3^{2-} \tag{10.2}$$

$$COCl_2 \rightleftharpoons CO^{2+} + 2Cl^- \tag{10.5a}$$

or

$$COCl_2 \rightleftharpoons COCl^+ + Cl^- \tag{10.5b}$$

10.3 CDFT-Based Reactivity Descriptors

CDFT-based global reactivity descriptors are fundamental quantities used to analyze the reactivity of molecules based on their electronic structure. These descriptors provide a conceptual framework for analyzing the structure, stability, interaction capabilities, and reactivity of chemical systems.

Electronegativity (χ), as described by Pauling, measures the ability of an atom to attract electrons toward itself in a chemical bond [9]. Significant contributions from various scientists, such as Mulliken [10, 11], Sanderson [12], Allred–Rochow [13, 14], Iczkowski and Margrave [15], Pearson [16], and Allen [17], have led to the evolution of the concept throughout the years. Parr's [18] description is perhaps the most important in DFT where he equates χ as the negative of the chemical potential (μ) value (Equation 10.6). The latter is the Lagrange multiplier related to the normalization constraint.

$$\mu = \left(\frac{\partial E}{\partial N}\right)_{v(r)} = -\chi \tag{10.6}$$

μ measures the escaping nature of the electron cloud within a system [19] and utilizing an approximation of finite difference, it can be represented as the negative average of the electron affinity (EA) and ionization potential (IP). To simplify the computational process, Koopmans' theorem [20] can be applied which equates IP and EA to the negative values of the highest occupied (ε_{HOMO}) and the lowest unoccupied (ε_{LUMO}) molecular orbital energies, respectively (Equation 10.7).

$$\mu = -\chi \approx -\frac{1}{2}(IP + EA) \approx \frac{1}{2}(\varepsilon_{LUMO} + \varepsilon_{HOMO}) \tag{10.7}$$

The hard–soft acid–base (HSAB) principle [21–23] explains how systems achieve thermodynamic stability through hard–hard or soft–soft pairing but lacks a quantitative measure of hardness or softness. Parr and Pearson [24–26] addressed this limitation by describing hardness (η) as the variation of the μ with respect to the number of electrons (N) under a fixed external potential, $v(r)$. It can be interpreted as the curvature of the energy (E) versus N plot. Considering finite difference approximation, hardness can be written as the difference between the IP and EA (Equation 10.8).

$$\eta = \left(\frac{\partial^2 E}{\partial N^2}\right)_{v(r)} = \left(\frac{\partial \mu}{\partial N}\right)_{v(r)} \approx IP - EA \tag{10.8}$$

Since softness quantifies the extent of electron cloud spreading within a molecular system, it can be mathematically defined as [27] (Equation 10.9),

$$S = \frac{1}{2\eta} = \frac{1}{2}\left(\frac{\partial N}{\partial \mu}\right)_{v(r)} \approx \frac{1}{2(IP - EA)} \tag{10.9}$$

Chemical compounds can be classified as electrophilic or nucleophilic based on their affinity toward electron-rich or -deficient species, respectively. To quantify this behavior, Parr et al. [28] defined the electrophilicity index (ω) as follows (Equation 10.10):

$$\omega = \frac{\mu^2}{2\eta} = \frac{\chi^2}{2\eta} \tag{10.10}$$

Combining the concepts of Frontier molecular orbital theory (FMOT) and Maxwell relation from classical thermodynamics, Parr and Yang [29] defined the Fukui function (FF) as a local reactivity parameter (Equation 10.11).

$$f(r) = \left(\frac{\partial \mu}{\partial v(r)}\right)_N = \left(\frac{\partial \rho(r)}{\partial N}\right)_{v(r)} \tag{10.11}$$

where $\rho(r)$ is the electron density and can be defined as (Equation 10.12),

$$\rho(r) = \left(\frac{\delta E}{\delta v(r)}\right)_N \tag{10.12}$$

Owing to the discontinuous nature of the $\rho(r)$ versus N plot, FFs for different types of reactions can be calculated. For electrophilic, nucleophilic, and radical attacks, the corresponding condensed-to-atom FFs are calculated as follows [30, 31] (Equations 10.13-10.15):

$$f_k^- = q_k(N) - q_k(N-1); \text{ for electrophilic attack} \tag{10.13}$$

$$f_k^+ = q_k(N+1) - q_k(N); \text{ for nucleophilic attack} \tag{10.14}$$

$$f_k^0 = [q_k(N+1) - q_k(N-1)]/2; \text{ for radical attack} \tag{10.15}$$

The true local reactivity of different regions in the molecule can be better understood from the derivative of FF with respect to N, which is known as the dual descriptor, $\Delta f(r)$ and is presented as (Equation 10.16),

$$\Delta f(r) = \left(\frac{\partial f(r)}{\partial N}\right)_{v(r)} \approx f^+(r) - f^-(r) \tag{10.16}$$

Another popular local reactivity parameter, known as the multiphilic descriptor ($\Delta\omega(r)$) [32], is the difference between the philicity indices ($\omega(r)$) [33] for nucleophilic and electrophilic attacks (Equation 10.17).

$$\Delta\omega(r) = \omega^+(r) - \omega^-(r) = \omega \cdot \Delta f(r) \tag{10.17}$$

where

$$\omega^\alpha(r) = \omega \cdot f^\alpha(r) \tag{10.18}$$

and condensed-to-atom,

$$\omega_k^\alpha = \omega \cdot f_k^\alpha \tag{10.19}$$

where $\alpha = +, -$, and 0 represents nucleophilic, electrophilic, and radical attacks, respectively (Equations 10.18 and 10.19).

Group quantities can be obtained by summing up the philicity parameters for a group of atoms (Equations 10.20 and 10.21):

$$\omega_g^- \equiv \sum_{k=1}^n \omega_k^- \text{ electrophilic attack} \tag{10.20}$$

$$\omega_g^+ \equiv \sum_{k=1}^n \omega_k^+ \text{ nucleophilic attack} \tag{10.21}$$

10.4 CDFT-Based Electronic Structure Principles

10.4.1 Equalization Principles

Interaction between two chemical species with different μ and χ values at constant $v(r)$ causes a transfer of electron cloud from a region of higher μ (lower χ) to that with lower μ (higher χ) so that both regions have equal μ (and χ). This phenomenon is known as the electronegativity equalization introduced by Sanderson [34, 35]. The equalized intermediate χ is the geometric mean (GM) of those of the isolated moieties (with P is the number of atoms) (Equation 10.22).

$$\chi_{GM} \approx \left(\prod_{k=1}^{P} \chi_k \right)^{1/P} \tag{10.22}$$

Since the ratio between η and χ is more or less constant for atoms of the same group and similar molecules [36, 37] the equalization principle for the hardness can be described [38–40] in a similar fashion (Equation 10.23).

$$\eta_{GM} \approx \left(\prod_{k=1}^{P} \eta_k \right)^{1/P} \tag{10.23}$$

That of electrophilicity can also be derived (Equation 10.24) by combining the Equations 10.22 and 10.23 [41]

$$\omega_{GM} \approx \left(\prod_{k=1}^{P} \omega_k \right)^{1/P} \tag{10.24}$$

10.4.2 Hard–Soft Acid–Base (HSAB) Principle

The HSAB principle is a well-known concept extensively employed in the study of acid–base chemistry, according to which, hard acids preferably interact with hard bases, and soft acids do with soft bases. It revolves around the phenomenon of charge transfer (CT) between a donor (Lewis base, B) and an acceptor (Lewis acid, A). Parr and Pearson [24] introduced a quantitative approach to this qualitative principle by relating the total electronic CT (ΔN) and change in energy (ΔE) between A and B to their χ and η. Equations (10.25) and (10.26) demonstrate a systematic decrease in energy as CT occurs:

$$\Delta N = \frac{\chi_A^0 - \chi_B^0}{2(\eta_A + \eta_B)} \tag{10.25}$$

$$\Delta E = -\frac{(\chi_A^0 - \chi_B^0)^2}{4(\eta_A + \eta_B)} \tag{10.26}$$

To gain deeper understanding of these principles, it is essential to analyze the MOs of the interacting species. In a "hard" molecule, the HOMO–LUMO energy gap is larger compared to a "softer" molecule. A larger energy gap indicates a reduced perturbation effect on the electron cloud during a reaction, leading to lower reactivity and increased stability.

According to Parr and Yang [27, 29], the reactivity of atoms toward electrophilic or nucleophilic attacks is influenced by their softness. This observation is also supported by multiple research groups. [42–45] Furthermore, Gazquez and Méndez found that the stabilization energy between two systems reacting with each other, denoted as A and B, is higher when the FFs of their respective atoms are larger [46]. These findings corroborate earlier reports by Parr and Yang [29]. However, it is important to note that A and B do not necessarily interact exclusively through their softest atoms. Instead, it occurs through specific atoms (such as kth and lth atoms in A and B, respectively) whose FFs (f_{Ak} and f_{Bl}) are more or less similar, denoted as $f_{Ak} \approx f_{Bl}$. The derivation of the equality of FFs assumes the global HSAB principle, where the overall softness of A is equal to that of B ($S_A = S_B$). Under this assumption, the local softness of the kth atom in A and the lth atom in B is also considered to be approximately equal (denoted as $s_{Ak} \approx s_{Bl}$). According to Gazquez and Méndez, the equal local softness of atoms participating in a reaction can be directly obtained by minimizing the grand canonical potential, without the need for assuming $S_A \approx S_B$. This expanded statement provides a more general perspective of their previous description of the local HSAB principle, which identifies specific atoms through which the A and B react [47]. This principle has a broader interpretation and a more general version of the global HSAB principle.

The combination of the global HSAB principle and the frontier orbital theory, augmented by Klopman's concepts [48], can provide a comprehensive explanation for both reactivity and selectivity in chemical reactions. Frontier orbitals primarily govern soft–soft interactions in chemical reactions and are predominantly covalent. In these interactions, the preferred site for reaction is determined by the maximum value of the FF, which indicates the propensity for electron density transfer. On the other hand, hard–hard interactions are primarily influenced by charge distribution and have a predominantly ionic nature. In such cases, the preferred site for interaction is the one with the maximum net charge, which may sometimes coincide with the site associated with the minimum value of the FF [49].

10.4.3 Maximum Hardness (MHP), Minimum Polarizability (MPP), and Minimum Electrophilicity (MEP) Principles

Pearson [50, 51] investigated the relationship between hardness and stability and introduced the concept of the maximum hardness principle (MHP), which states that, "there seems to be a rule of nature that molecules arrange themselves so as to be as hard as possible." A corollary to this principle is the minimum polarizability principle (MPP) [52, 53], which states that "the natural direction of evolution of any system is toward a state of minimum polarizability." This principle has also been demonstrated to be applicable in time-independent scenarios [54]. Numerous investigations into various aspects of chemistry, including aromaticity [55], molecular vibrations [56–59], excited states [60, 61], internal rotations [62–64], and other chemical reactions, have provided substantial support for the validity of both MHP and MPP. The proof of MHP [65] is derived using DFT and statistical mechanics under the constraints of constant $v(r)$ and μ. However, the rare occurrence of these constraints simultaneously, some relaxations are allowed within the framework of the MHP. In a recent report [66] involving the isomeric transformation between ligand-stabilized metal isocyanide and its cyanide form (LMNC → LMCN), the validity of the MHP was demonstrated even though the expected variation of μ along the isomerization path did not occur as anticipated.

The minimum electrophilicity principle (MEP) [67–70] states that the isomer with the highest stability is characterized by the minimum in the electrophilicity index (ω) value. It is important to note that the relationship between minimum energy and minimum electrophilicity is contingent on the simultaneous behavior of hardness and chemical potential. In other words, the ω reaches its minimum (or maximum) value only when both η and μ achieve their maximum (or minimum) values at any given point during the reaction path.

10.5 Systemics of Lewis Acid–Base Reactions: Drago–Wayland Equation

The EC (containing the electrostatic and covalent parameters, E, and C, respectively) model is a valuable semiquantitative approach used to characterize and predict the strength of interactions between Lewis acids and Lewis bases. The EC model, along with its modified versions, provides a quantitative framework for understanding acid–base interactions complementing the qualitative HSAB theory. While Pearson hardness values are commonly employed to assess the hardness of acids and bases, they are not directly applicable for estimating the interaction energy of Lewis acid semi-quantitative base pairs. However, the EC model, pioneered by Drago and Wayland [71], offers a solution to this limitation. In an acid semi-quantitative base reaction forming an adduct (Equation 10.27), the formation enthalpy can be represented in terms of the E and C parameters indicating the tendency of the acid and base to form strong electrostatic and covalent interactions with each other (Equation 10.28). These parameters have a unit of kcal$^{1/2}$ mol$^{-1/2}$.

$$A + B \rightleftharpoons AB \tag{10.27}$$

$$-\Delta H_{AB} = E_A E_B + C_A C_B \tag{10.28}$$

The basic EC model focuses solely on electrostatic and covalent factors and does not take into account steric effects, lattice energy, and other potential contributions to the overall interaction energy. As a result, its applicability is limited in analyzing the energies of interaction of the sterically unhindered adducts where the impact of solvation energy and other factors is negligible. A few modifications to include the said factors are as follows:

(a) Steric effect: An additional parameter, D, is introduced to include the steric effects (Equation 10.29).

$$-\Delta H_{AB} = E_A E_B + C_A C_B - D_A D_B \tag{10.29}$$

(b) Charge transfer: Drago and Wong [72] expanded the EC model by introducing receptance factors, which consider the capacity of the acid to accept electron density (R_A), and the transference factors, reflecting the ability of the base to donate electron density (T_B) (Equation 10.30).

$$-\Delta H_{AB} = E_A E_B + C_A C_B + R_A T_B \tag{10.30}$$

(c) Energy required to cleave a dimer to produce the Lewis acid: Drago and Vogel [73] further developed the EC model to incorporate a constant energy term, W ($= W_A + W_B$), to form the following ECW model (Equation 10.31).

$$-\Delta H_{AB} = E_A E_B + C_A C_B + W \tag{10.31}$$

Apart from estimating enthalpies of the formation of Lewis acid–base complexes, EC and associated models can also be used to estimate the relative importance of covalent, ionic, and steric factors in the complex formation.

10.6 Strengths of Acid and Bases

10.6.1 Ionic Product

The extend of dissociation of water is very small, yet is important to utilize the H^+ and OH^- ion concentrations in pure water to represent the equilibrium constant (Equation 10.32).

$$H_2O \rightleftharpoons H^+_{(aq.)} + OH^-_{(aq.)} \tag{10.32}$$

At a temperature of 25 °C, the equilibrium concentrations of both H^+ ions and OH^- ions in pure water are approximately equal to 1.0×10^{-7} (Equation 10.33)

$$[H^+][OH^-] = 10^{-7} * 10^{-7} = K_w = 10^{-14} \tag{10.33}$$

This is the ionic product of water, which is applicable not only to pure water but to all aqueous solutions as well. This has significant implications, as it indicates that a rise in the concentration of H^+ ions will reduce that of the OH^-, and vice versa. Also, it indicates that H^+ ions are not exclusively present in acidic solutions, but in all aqueous ones. Their concentration, however, differs in various solutions, e.g. [H^+] is greater than, less than, and equal to [OH^-] in acidic, alkaline, and neutral solutions, respectively. For the neutral solution [H^+] = [OH^-] = 1.0 * 10^{-7} M at 25 °C.

10.6.2 pH Scale

The [H^+] and [OH^-] values can be represented on a compressed logarithmic scale as follows (Equations 10.34 -10.36):

$$pH = -\log_{10}[H^+] \tag{10.34}$$

$$pOH = -\log_{10}[OH^-] \tag{10.35}$$

and

$$pK_w = -\log_{10} K_w \tag{10.36}$$

From Equation (10.31), we have Equation 10.37,

$$pH + pOH = pK_w \ (= 14.0 \text{ at } 25\,°C \text{ in pure water}) \tag{10.37}$$

where pH [74] and pOH are the measures of the [H⁺] and [OH⁻] ion concentrations in a solution and are calculated as the negative logarithm (base 10) of [H⁺] and [OH⁻]. The pH scale spans between 0 and 14, with pH = 7 indicating neutrality. pH < 7 suggests acidity, while pH > 7 suggest alkalinity. In general, lower pH values indicate higher concentrations of H⁺ ions, corresponding to stronger acids. Conversely, higher pH values indicate lower concentrations of H⁺ ions and stronger bases.

10.6.3 Ionization Constants

The reaction where an acid reacts with a base forming a salt and water only is called the neutralization reaction (Equation 10.38).

$$HA + BOH \rightleftharpoons BA + H_2O \tag{10.38}$$

The acidity, alkalinity, or neutrality of a solution is determined by the interaction between a salt and water, i.e. the reverse process known as hydrolysis. When a salt BA is formed from the combination of a weak acid HA and a strong base BOH, the resulting solution in water exhibits alkaline properties. This is attributed to the limited release of H⁺ ions from the hydrolyzed acid, while the hydrolyzed base generates a significant concentration of OH⁻ ions (represented in Equation 10.39),

$$B^+ + A^- + H_2O \rightleftharpoons HA + B^+ + OH^- \tag{10.39}$$

At equilibrium, the law of mass action gives the hydrolysis constant as,

$$K_h = \frac{[\text{base}][\text{acid}]}{[\text{unhydrolyzed salt}]} = \frac{[OH^-][HA]}{[A^-]} \tag{10.40}$$

For a weak acid and a strong base in water, the relevant equilibria are represented in Equations 10.41-10.43,

$$H_2O \rightleftharpoons H^+ + OH^-; K = \frac{[H^+][OH^-]}{[H_2O]} \tag{10.41}$$

$$HA \rightleftharpoons H^+ + A^-; K_a = \frac{[H^+][A^-]}{[HA]} \tag{10.42}$$

$$A^- + H_2O \rightleftharpoons HA + OH^-; K'_h = \frac{[HA][OH^-]}{[A^-][H_2O]} \tag{10.43}$$

Since the [H₂O] is more or less constant in dilute solutions, we can introduce the ionic constant, K_w instead of K, and a new constant K_h, to replace K_h'. K_h is thus defined as,

$$K_h = \frac{[HA][OH^-]}{[A^-]} = \frac{K_w}{K_a} \tag{10.44}$$

Assuming that that only the pure aqueous solution of the salt of HA is present here, and that the $[OH^-]_{\text{ionized water}} \ll [OH^-]_{\text{salt hydrolysis}}$, we have, $[OH^-] = [HA]$ during salt hydrolysis (Equations 10.45-10.47), i.e.

$$K_h = \frac{K_w}{K_a} = \frac{[OH^-]^2}{[A^-]} \Rightarrow [OH^-] = \sqrt{\frac{[A^-] \cdot K_w}{K_a}} \quad (10.45)$$

$$[H^+] = \frac{K_w}{[OH^-]} = \sqrt{\frac{K_w \cdot K_a}{[A^-]}} \quad (10.46)$$

Hence,

$$pH = \frac{1}{2}pK_w + \frac{1}{2}pK_a + \frac{1}{2}\log c \quad (10.47)$$

Here c is the stoichiometric molar concentration of the salt (only applicable for a low extent of salt hydrolysis). Also, it is to be noted that pH is the negative logarithm of the H^+ ion concentration rather than its activity to maintain simplicity in all the calculations.

Similar to Equation (10.47), in the case of a strong acid and a weak base system, the pH can be determined in terms of the ionization constant of the base, K_b, using Equation 10.48.

$$pH = \frac{1}{2}pK_w - \frac{1}{2}pK_b - \frac{1}{2}\log c \quad (10.48)$$

10.6.4 Proton Affinity

The Brønsted–Lowry acid–base concept, which defines acidity in terms of the donation and acceptance of hydrogen ions, offers a comprehensive understanding of acidity and basicity across various media such as liquids, solids, and gases. This viewpoint becomes especially significant when examining the thermodynamics of hydrogen ion transfer, as the energies of gaseous species remain unaffected by solvation factors. The proton affinity (PA) is a theoretical concept used to evaluate the tendency of a species to accept a proton. The PA of a base, B, is the negative of the change in enthalpy when it takes up an H^+ ion in the gas phase (Equation 10.49), and hence, is easier to consider the reverse reaction, i.e. the dissociation of the corresponding conjugate acid (Equation 10.50).

$$H^+_{(g)} + B_{(g)} \rightleftharpoons BH^+_{(g)} \quad (10.49)$$

$$BH^+_{(g)} \rightleftharpoons H^+_{(g)} + B_{(g)} \quad (10.50)$$

The $\Delta H_{\text{reaction}}$ is the enthalpy corresponding to the dissociation of the B–H bond and also measures the PA of the base B. The corresponding $\Delta G_{\text{reaction}}$ is known as the absolute or gas phase basicity (GB) of B. Similarly, for an acid HA, the gas phase acidity (GA) is the $\Delta G_{\text{reaction}}$ of the following reaction (Equation 10.51).

$$HA_{(g)} \rightleftharpoons H^+_{(g)} + A^-_{(g)} \quad (10.51)$$

Higher PA and GB correspond to a stronger base and weaker conjugate acid. Among known bases, the CH_3^- and $CH_3CH_2^-$ exhibit the highest PAs (1743 and 1758 kJ/mol, respectively), which are slightly higher than the H^- ion (1675 kJ/mol) [75], resulting in CH_4 and C_2H_5 being the weakest proton acids in the gas phase, followed by H_2. On the other hand, the He atom is recognized as the weakest base (177.8 kJ/mol) [76], making the HeH^+ ion the strongest proton acid.

10.6.5 Electronegativity

It is a fundamental concept in chemistry that measures the tendency of an atom to attract bonded pair of electrons toward itself [15]. It helps in understanding the polarity and reactivity of chemical compounds. Electronegativity values are assigned to each element on a scale, such as the Pauling scale [77], which allows for comparison and prediction of chemical behavior. To elaborate on this concept, we can infer that atoms possessing higher electronegativity exhibit a reduced tendency to share their electrons with a proton. Consequently, when the electronegativity of the atom involved in electron sharing increases, its basicity decreases. Weaker bases correspondingly yield stronger conjugate acids. Therefore, it can be deduced that an increase in the electronegativity of an atom results in an elevation in the acidity of the attached proton (Equation 10.52).

$$A^- + H^+ \rightleftharpoons HA \tag{10.52}$$

In this equation, A^- represents an anion or a negatively charged species with lower electronegativity, and HA represents the conjugate acid formed when A^- accepts a proton (H^+). As electronegativity increases, the tendency of A^- for proton acceptance decreases, leading to reduced basicity and increased acidity of the conjugate acid (HA).

In the context of acid–base chemistry, electronegativity influences the behavior of acids and bases. Acids with more electronegative elements exhibit greater electron-withdrawing tendencies, making them stronger acids. Bases, on the other hand, are influenced by the electronegativity of the atom or group that donates an electron pair. Electronegative atoms or groups tend to stabilize the negative charge on the conjugate base, making the base stronger.

10.6.6 Hardness

It is a theoretical concept used to explain the stability and reactivity of molecules or ions based on their electron density distribution and energy levels. It is often associated with the strength of the chemical bonds within a system. Hardness provides insights into the resistance of a species to deformation or change in its electronic structure. Hard acids and bases are defined by their strong bonding interactions, typically involving localized electrons and a significant difference in energy between the HOMO and the LUMO.

$$HA + HB \longrightarrow \text{Complex} \tag{10.53}$$

In Equation 10.53, HA represents a species with hard characteristics (e.g. cations of alkali and alkaline earth metals), while HB represents a species with hard base properties (e.g. hydroxide and fluoride ions). The reaction between the hard acid HA and the hard base HB forms a stable complex.

For the soft acid and soft base reaction, the equation can be written as:

$$SA + SB \longrightarrow \text{Complex, Charge Transfer} \tag{10.54}$$

In Equation 10.54, SA represents a species with soft acid characteristics (e.g. transition metal ions), while SB represents a species with soft base properties (e.g. larger, polarizable anions or molecules). The interaction between the soft acid SA and the soft base SB results in the formation of a complex and may involve CT. These species typically have higher-lying frontier orbitals and a smaller HOMO–LUMO energy gap.

It is to be noted that in scenarios where electron-transfer effects play a significant role and other factors are less influential, the HSAB principle is primarily driven by the exceptional stability observed in the product formed by a soft acid and a soft base. Conversely, in cases where the reactivity is dominated by electrostatic effects, the HSAB principle is driven by the superior stability exhibited by the product resulting from a hard acid and a hard base. By examining the acid–base exchange reactions, it becomes possible to discern whether the reactivity of a given reagent is primarily governed by electron-transfer effects or electrostatic effects. Notably, as the former effect favors interactions between soft acids and soft bases, while the latter favors that between hard acids and hard bases, the HSAB principle becomes valid when the electronic chemical potentials of the acids and bases involved in the reaction are similar [78].

10.6.7 Electrophilicity

It, in the context of acids and bases, refers to the ability of a species to act as an electrophile or nucleophile in acid–base reactions [28]. It characterizes the reactivity of acids and bases toward electron-rich or electron-deficient species, providing insights into their behavior in chemical reactions.

$$A \text{ (acid)} + B \text{ (base)} \rightleftharpoons A^- \text{ (conjugate base)} + B^+ \text{ (conjugate acid)} \tag{10.55}$$

In this equation, A represents an acid species, which acts as an electrophile, and B represents a base species, which acts as a nucleophile. The acid A accepts an electron pair from the base B, resulting in the formation of a conjugate base A^- and a conjugate acid B^+. These reactions exemplify the electrophilic behavior of acids, which accept an electron pair, and the nucleophilic behavior of bases, which donate an electron pair.

The electrophilicity of acids and the nucleophilicity of bases are interconnected concepts in acid–base reactions. Acids with a higher electrophilicity have a greater tendency to accept an electron pair, making them stronger acids. Similarly, bases with higher nucleophilicity exhibit a stronger tendency to donate an electron pair, making them stronger bases. The electrophilic or nucleophilic character of acids and bases is influenced by several factors, including their electronic structure, steric effects, and the presence of functional groups. For example, an acid with a more electron-deficient center or an electron-withdrawing group (EWG) attached to it will exhibit increased electrophilicity. Likewise, a base with a more available lone pair of electrons or an electron-donating group (EDG) can enhance its nucleophilicity.

10.7 Effect of External Perturbation

10.7.1 Steric Effects

It plays a crucial role in determining the acid–base behavior of chemical species by considering the spatial arrangement and hindrance caused by bulky substituents. These effects arise from the repulsive interactions between atoms or groups in close proximity, leading to distortions in molecular geometry and affecting the accessibility of protons. In the context of acidity and basicity, steric hindrance refers to the obstruction of proton transfer due to the presence of bulky substituents in close proximity to the reacting atoms by hindering the approach of reactants or impeding the formation of acid–base complexes (or transition states). As a result, the equilibrium position and the corresponding pK_a or pK_b values can be significantly altered [79, 80]. Introducing bulky substituents near the acidic or basic center of a molecule can lead to a decrease in the acidity or basicity. The steric hindrance imposed by these substituents limits the accessibility of the proton donor or acceptor, making it more difficult for the reaction to occur. Consequently, the strength of the acid or base is diminished. Steric interactions can influence the mechanism and rate of proton transfer reactions [81]. In cases where steric hindrance is significant, alternative pathways may be favored, such as the involvement of neighboring groups or the formation of more stable intermediates. These pathways can alter the kinetics and thermodynamics of the acid–base reaction, leading to different reaction outcomes. Understanding the steric effects on acidity and basicity is crucial for rationalizing and predicting the behavior of organic molecules and functional groups.

10.7.2 Solvent Effects

It also plays a significant role by influencing the interactions between solute molecules and the surrounding solvent molecules. The nature and properties of the solvent (especially the solvent polarity) [82] can impact the stability of ions and the ability of molecules to donate or accept protons, thereby affecting the acid–base equilibria [83]. Polar solvents, such as water or polar organic solvents, have the ability to stabilize charged species through solvation. This stabilization can influence the relative strengths of acids and bases by altering the dissociation constants (pK_a and pK_b values) of the solutes. Highly polar solvents enhance the solvation of ions, thereby decreasing the availability of protons and reducing the acidity or basicity of a species [84]. Solvent effects on acidity and basicity can be categorized as specific or nonspecific. Specific solvent effects arise from direct interactions between the solvent and solute molecules, e.g. in protic solvents, the hydrogen bonding interactions can play a crucial role in determining acid–base equilibria. Nonspecific solvent effects, on the other hand, result from the bulk properties of the solvent, such as its polarity, dielectric constant, and viscosity, which can influence solute–solvent interactions and, subsequently, the acid–base behavior. Solvents can exhibit their own acid–base behavior, known as solvent basicity or solvent acidity. Basic solvents can act as proton acceptors, competing with basic solutes for protons and decreasing their basicity. Conversely, acidic solvents can act as proton donors, enhancing the acidity of acidic solutes. The solvent basicity or acidity can be quantified using solvent acidity functions or basicity functions, which provide a measure of their ability to interact with protons. Solvents can participate in donor–acceptor interactions with solutes, influencing their acidity and basicity [85]. For example, in Lewis acid–base interactions, solvents can serve as Lewis bases by donating electron pairs to Lewis acids or as Lewis acids by accepting them from Lewis bases [86]. These solvent–solute interactions can modify the electronic properties and reactivity of the solutes, ultimately impacting their acid–base behavior.

10.7.3 Periodicity

It refers to the systematic variation of properties across periods and groups in the periodic table. These variations have a significant impact on the acidity and basicity of chemical species. Understanding periodicity effects allows for the prediction and explanation of trends in acid–base behavior and provides insights into the underlying electronic structure and reactivity of compounds. Periodic trends, such as atomic size, electronegativity, and electron affinity, directly affect the acidity and basicity of compounds [87]. As one moves across a period from left to right, the atomic size decreases, leading to stronger electrostatic attractions between the positively charged nucleus and the surrounding electrons. This increased attraction makes it more difficult for atoms to accept protons, resulting in lower basicity. Moving down a group in the periodic table, the atomic size increases, resulting in weaker electrostatic attractions. As a result, atoms in lower periods have a greater tendency to accept protons, exhibiting higher basicity. The influence of periodicity can be observed in trends such as the increasing basicity of hydroxides (OH^-) from Group 1 to Group 2 elements [88]. IP and EA are closely related to acidity and basicity. Generally, elements with high IPs and low EAs tend to exhibit higher acidity, as they are less likely to donate or accept electrons. Conversely, elements with low IPs and high EAs tend to display higher basicity. Understanding periodicity effects allows chemists to predict and explain trends in acid–base reactivity. For example, the trend in acid strength across halogens (F > Cl > Br > I) can be attributed to periodic trends in electronegativity and atomic size [89].

10.7.4 Inductive Effect

It refers to the electron-withdrawing or electron-donating influence of neighboring atoms or groups on the acidity or basicity of a compound. It arises from the polarizability of chemical bonds and the ability of electronegative or EDGs to withdraw or donate electron density through sigma bonds [90]. Understanding the inductive effect is crucial for predicting and manipulating the acid–base behavior of molecules. Inductive effects influence the distribution of electron density in a molecule, thereby affecting the stability of charged species involved in acid–base equilibria. EWGs, such as halogens or carbonyl groups, exert a pull on the electrons in the sigma bonds, resulting in a decrease in electron density at the atom or group involved in the acid–base reaction. Conversely, EDGs, such as alkyl or amino groups, donate electron density, increasing the electron density at the reactive site. EWGs decrease the electron density at the atom or group involved in acid–base reactions, thereby increasing acidity or decreasing basicity. Examples of EWGs include halogens (e.g. chlorine, fluorine), nitro groups (NO_2), and carbonyl groups (C=O). Conversely, EDGs increase the electron density at the reactive site, leading to increased basicity or decreased acidity. Examples of EDGs include hydroxyl groups (OH), alkyl groups (e.g. methyl, ethyl), and amino groups (NH_2) [91]. Inductive effects are also observed in conjugated systems and aromatic compounds. In conjugated systems, the delocalization of electrons along the pi system allows the inductive effect to propagate through the molecule, influencing the acidity or basicity of multiple sites. Aromatic compounds, such as benzene rings, exhibit significant stabilization due to delocalization of electrons, leading to decreased acidity and increased basicity compared to their nonaromatic counterparts.

10.7.5 Resonance Effect

Also known as mesomeric effect, it plays a crucial role in determining the acidity and basicity of molecules by redistributing electron density through the delocalization of pi electrons. Resonance occurs when a molecule can be represented by multiple Lewis structures with varying electron distribution. Understanding the resonance effect is essential for predicting and explaining the acid–base behavior of conjugated systems and molecules containing multiple resonance structures. Resonance effects stabilize charged species by delocalizing electron density over multiple atoms [92]. In the context of acidity, the formation of a conjugate base is stabilized through resonance, leading to increased stability and decreased basicity. Similarly, in the case of acidic species, resonance can enhance the delocalization of positive charge, resulting in increased stability and decreased acidity. Resonance effects can significantly impact the acidity or basicity of specific centers within a molecule. In the case of acids, the presence of a resonance structure that distributes a negative charge away from the acidic hydrogen can increase the acidity of that proton. Conversely, for bases, the availability of resonance structures that can delocalize the electron pair can enhance basicity by stabilizing the resulting conjugate acid. Resonance is closely linked to the concept of aromaticity, which refers to the stability of cyclic compounds with conjugated pi systems [93]. Aromatic compounds, such as benzene, possess delocalized pi electrons that are spread over the entire ring, leading to exceptional stability. The presence of aromaticity can significantly influence the acid–base properties of molecules, as aromatic systems tend to exhibit decreased acidity and increased basicity compared to nonaromatic analogs. Conjugated systems, characterized by alternating single and multiple bonds, exhibit extensive delocalization of pi electrons. The presence of multiple resonance structures allows for the spread of electron density over the entire system. This delocalization leads to increased stability of the system and affects the acid–base behavior of functional groups within the conjugated system. The understanding and manipulation of resonance effects are crucial in organic synthesis and drug design. By utilizing resonance to stabilize reactive intermediates or transition states, chemists can direct and control chemical reactions. Additionally, the presence or absence of resonance structures can influence the selectivity of reactions and the regio-chemistry of bond formation, enabling the synthesis of complex molecules with desired properties.

10.8 CDFT and Acidity

CDFT focuses on understanding the electronic structure of molecules and their reactivity in terms of fundamental chemical concepts. It provides a powerful framework for understanding acid–base chemistry, which is central to many areas of chemistry. Acidity is often quantified by the acid dissociation constant, pK_a, which is a measure of the tendency of a molecule to lose a proton. CDFT provides a way to calculate pK_a based on the electronic structure of a molecule, which can be used to predict the acidity of different compounds. CDFT identifies the key factors that contribute to the acidity of a molecule, including the charge distribution, the electron density, and the frontier molecular orbitals. The charge distribution and electron density influence the ability of a molecule to donate or accept electrons, while the frontier molecular orbitals determine the reactivity of a molecule. By using CDFT, chemists can design and optimize molecular structures to enhance or reduce their acidity, which is important in many applications, including catalysis, drug design, and materials science. Overall, CDFT is a useful tool for understanding and forecasting molecular behavior in a wide range of chemical environments. The acid–base equilibrium is a fundamental type of reaction in chemistry that has been studied extensively for structure–reactivity relationships, leading to the establishment of substituent constants. The structure of a molecule impacts its acidity or basicity in various ways; however, due to the many factors at work in most molecules, distinguishing the observed changes in acidity or basicity becomes difficult. It is crucial to consider the solvent or medium where the structure-reactivity relationships are determined. DFT-based reactivity descriptors offer valuable insights into such acid–base reactions and structure-activity data.

In the study by Gupta et al. [94], a collection of 45 organic acids with aromatic properties, such as derivatives of cinnamic acid, benzoic acid, benzohydroxamic acid, phenol, and anilinium ion, were examined through quantum chemical calculations in aqueous conditions at 298 K. The IP, EA, η, μ, and ω of molecules were determined using the B3LYP/6-31G(d) level of theory, along with the calculation of some local descriptors. The basis set 6-31+G(d) was used for substituted benzoic and cinnamic acids to assess the effect of diffuse functions. The aromatic acids were then

10.8 CDFT and Acidity

reacted with a strong base (OH⁻), and various parameters such as the Gibbs free energy of deprotonation (ΔG), associated energy change (ΔE), fractional number of electrons transferred (ΔN), and electrophilicity-based CT index (ECT) were computed. These parameters, along with ω^+_g and group charges, were then correlated with the experimental pK_a values of the acids. The study observed trends in acidity for related molecules and the substituent effect on the descriptors. The parameters were also correlated with experimental Hammett substituent constants (σ). The results showed that ECT, ΔN, and group charge had a strong correlation with pK_a and σ in separate aromatic acid groups. The study also tested the minimum energy principle, along with MEP and MHP for the acid–base reactions. The pK_a values of 45 acids from five sets were experimentally measured and plotted against their estimated pK_a values using ΔN, ECT, and ΔN, ECT, ω^+_g in Figures 10.1 and 10.2, respectively. The σ values also exhibited positive correlations across all groups of aromatic acids. When the ECT and ΔN were considered as indicators of Lewis acidity, the correlation between Brønsted acidity (pK_a) and Lewis acidity was relatively weaker when acids of different structures were analyzed. However, when acids of similar structures were examined, the correlation significantly improved. The ΔE values for the reactions of the OH⁻ containing aromatic acids consistently showed negative values, while the $\Delta \eta$ values were consistently positive, adhering to the MHP. The MEP was found to be valid for all the cases when using the 6-31G(d) basis set.

Parthasarathi et al. [95] conducted a study proposing a novel method to predict pK_a values based on group philicity, which measures the tendency of specific functional groups within a molecule to act as electrophilic or nucleophilic centers. They hypothesized that certain known functional groups could serve as indicators for predicting pK_a values. They selected diverse classes of compounds, including

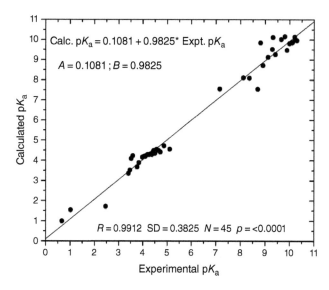

Figure 10.1 Correlation between the experimental and predicted pK_a values of aromatic acids based on the ECT and ΔN. Source: Reproduced from Ref. [94] with permission from Royal Society of Chemistry.

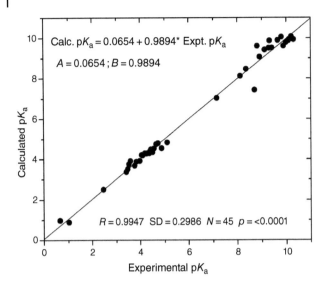

Figure 10.2 Correlation between the experimental and predicted pK_a values of aromatic acids with respect to the ΔN, ECT, and group philicity (ω^+_g). Source: Reproduced from Ref. [94] with permission from Royal Society of Chemistry.

carboxylic acids, substituted phenols, alcohols, phosphoric acids, and, anilines and calculated the group philicity values (ω^+_g) for each compound class using the relevant functional groups present in the molecules. Experimental pK_a values from previous studies were collected for comparison. Linear regression analyses were performed to establish the relationship between calculated group philicity and experimental pK_a values. The results indicated a strong inverse relationship between the ω^+_g and pK_a values within each compound class, suggesting that increasing group philicity corresponds to decreasing pK_a values and hence higher acidity. The obtained correlation coefficients confirmed the effectiveness of group philicity as a descriptor for predicting pK_a values (Figures 10.3a,b). Polynomial regression analyses and second-order relationships were also explored to gain further insights into the relationship between group philicity and pK_a values (Figure 10.3c), demonstrating its effectiveness in predicting pK_a values for diverse classes of small molecules.

10.9 CDFT and ITA

CDFT and ITA (Information Theoretical Approach) are two different theoretical frameworks used in chemistry to understand the electronic structure of molecules and their reactivity. ITA is a theoretical framework based on information theory, which is used to quantify the amount of information contained in a chemical system. ITA provides a way to calculate information-theoretic (IT) properties of molecules, such as entropy, mutual information, and channel capacity, which are used to understand the relationship between the structure and function of molecules. While both

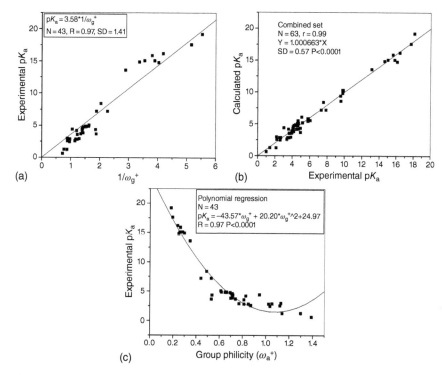

Figure 10.3 Relationship between (a) experimental pK_a values and reciprocal of group philicity index, (b) experimental pK_a values with predicted pK_a values using group philicity index. (c) Polynomial second-order regression analysis of experimental pK_a values and group philicity index for the series of molecules. Source: Reproduced from Ref. [95] with permission from American Chemical Society.

CDFT and ITA are used to understand the electronic structure of molecules, they approach the problem from different perspectives. CDFT focuses on the relationship between electronic properties and chemical reactivity, while ITA focuses on the amount of information contained in a chemical system and its relationship to function. Overall, CDFT and ITA can complement each other when used together to provide a more complete understanding of the electronic structure and reactivity of molecules.

Molecular acidity is an essential property of a molecular system, which determines its behavior and reactivity. However, accurately calculating and predicting the acidity of a molecule has been a long-standing challenge in the field of chemistry. To address this issue, a new approach based on IT quantities in density functional reactivity theory has been proposed by Cao et al. [96]. This novel approach presents a fresh perspective on molecular acidity by incorporating IT quantities such as Shannon entropy (S_S), Fisher information (I_F), information gain (I_G), Onicescu information energy, Ghosh–Berkowitz–Parr entropy (S_{GBP}), and relative Rényi entropy (R_n). These density-dependent quantities enable the simultaneous prediction of experimental pK_a values for various categories of acidic compounds. To demonstrate the efficacy of this approach, five different categories of acidic series were investigated, viz., singly and doubly substituted benzoic acids, singly substituted benzeneseleninic acids, benzenesulfinic acids, alkyl carboxylic acids, and phenols. Experimental pK_a values of these compounds were successfully predicted using the IT quantities, illustrating the universality of the approach. Compared to traditional descriptors such as molecular electrostatic potential and natural atomic orbital (NAO) energy, the IT quantities offer a more generalizable and consistent approach for predicting pK_a values. While molecular electrostatic potential and NAO have been shown to be effective in predicting pK_a values for different categories of compounds, their strong linear correlations exhibit different slopes and intercepts, making it difficult to develop a generalized predictive model. In contrast, the IT quantities proposed in this work offer a more consistent and generalizable approach to predicting pK_a values across different categories of compounds. This new approach provides a promising solution to the long-standing challenge of predicting molecular acidity accurately and should be applicable to other systems as well. The experimental pK_a data of both types of substituted benzoic acids exhibit strong correlations with six IT quantities. Figure 10.4 illustrates six fitted lines representing the experimental pK_a data, each associated with one of the IT quantities: S_S, I_G, S_{GBP}, I_F, and relative R_2 and R_3. These lines have vastly different fitting parameters, suggesting that they measure different aspects of molecular acidity. Although the I_F plot (Figure 10.4b) has two lines, one for singly substituted benzoic acids and the other for the doubly substituted ones, all other IT quantities have only one line for both types of compounds. The slope and

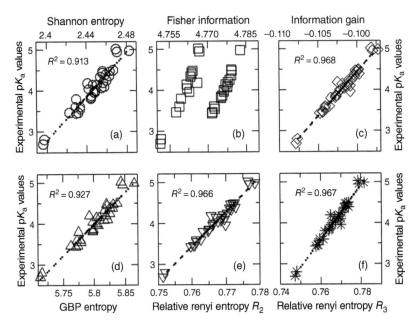

Figure 10.4 Strong linear correlations observed between the experimental pK_a values of singly and doubly substituted benzoic acid derivatives and various IT quantities: (a) S_S, (b) I_F, (c) I_G, (d) S_{GBP}, (e) R_2, and (f) R_3 for the dissociating proton of singly substituted benzoic acid derivatives. Source: Reproduced from Ref. [96] with permission from Wiley-VCH.

intercept of these lines are also markedly different. Together, these IT quantities offer a more comprehensive understanding of molecular acidity, enabling more accurate predictions of pK_a values.

CDFT introduced molecular electrostatic potential and NAO as tools for determining molecular properties. Later, the density functional reactivity theory incorporated IT quantities such as S_S, I_F, and I_G to quantify reactivity aspects such as electrophilicity, nucleophilicity, regioselectivity, and stereoselectivity. These approaches were successfully utilized to determine molecular acidity in various systems. In the study conducted by Xiao et al. [97], a novel combination of these approaches was proposed as a new set of descriptors to quantify molecular basicity. The effectiveness of this new approach was demonstrated using primary, secondary, and tertiary amines, accurately predicting their experimental pK_a values. By incorporating descriptors such as molecular electrostatic potential, natural valence atomic orbital energy, and IT quantities, the new approach allows for determining the molecular basicity of diverse molecular species, making it a robust method for understanding acidity and basicity concepts. The results showed that using either CDFT descriptors (MEP and NAO) or all eight ITA quantities alone exhibited reasonable correlations, but the best results were achieved by combining both CDFT and ITA quantities.

In Figure 10.5, the results from utilizing all ITA quantities together to simulate the experimental pK_a data for each of the three categories of amines are shown. The R^2 values obtained were 0.981 for primary amines, 0.924 for secondary amines, and 0.879 for tertiary amines, respectively. These results are comparable to what has been obtained earlier for other systems. From this figure, we can see that ITA quantities are able to quantitatively simulate the experimental pK_a values for amines. In Figure 10.6, the results from three possible ways of simulating three categories of amines together in a single least-square fitting are shown. The three methods employ either or both CDFT and ITA quantities. The results show that using the descriptor NAO alone from CDFT yields a correlation coefficient of 0.916 for the entire dataset, which is as good as what has been previously obtained for other systems. With all

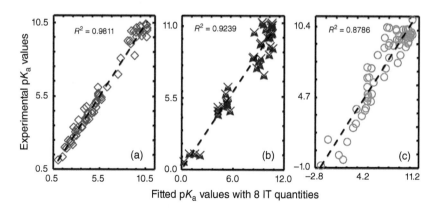

Figure 10.5 Comparison of experimental pK_a values with the calculated ones using all eight ITA quantities for (a) primary, (b) secondary, and (c) tertiary amines. Source: Reproduced from Ref. [97].

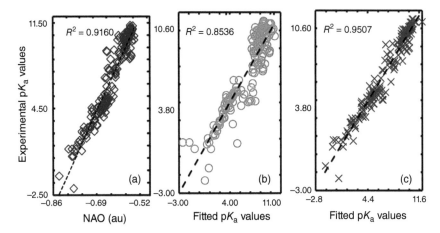

Figure 10.6 Comparison of experimental pK_a data with fitted pK_a values utilizing either or both CDFT and ITA quantities for all three categories of amines combined. Fitting using (a) only NAO (b) eight ITA quantities. (c) Both CDFT and ITA quantities. Source: Reproduced from Ref. [97].

eight ITA quantities alone, the correlation becomes worse, with $R^2 = 0.854$. However, when both CDFT and ITA quantities are employed, a significantly better fitting can be yielded, with the correlation coefficient R^2 equal to 0.951. These results demonstrate that combining CDFT and ITA quantities is an effective and robust way to quantify molecular basicity, enabling us to accurately determine the pK_a values for diversified categories of molecular systems.

10.10 Are Strong Brønsted Acids Necessarily Strong Lewis Acids?

The Brønsted–Lowry theory of acid–base focuses on the ability to donate or accept protons, while the Lewis acid–base theory is more concerned with the ability to accept or donate electron pairs. In an attempt to reconcile these two theories, Gupta et al. conducted a study [98] aiming to determine if they can be integrated. They employed DFT calculations to determine IP, EA, χ, and ω for 58 different types of organic and inorganic acids. ΔN and associated ΔE were also to establish possible descriptors for Lewis acidity by reacting the acids with trimethylamine. Figure 10.8 illustrates the correlation between experimental pK_a values and the $-(\Delta E)$ associated with the transfer of electron from trimethylamine to various organic and inorganic acids in the gas phase. A larger absolute value of $-(\Delta E)$ indicates a stronger Lewis acid and has a smaller pK_a value, indicating a stronger Brønsted acid. However, it is important to note that this trend may not hold for every pair of acids. On average, a stronger Lewis acid also tends to be a stronger Brønsted acid. The regression model for predicting pK_a values using $-(\Delta E)$ is provided in Figure 10.7a, and a reasonably good correlation between the experimental and calculated pK_a values is depicted in Figure 10.7b.

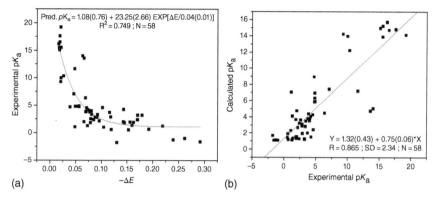

Figure 10.7 (a) Correlation between the experimental pK_a values of inorganic and organic acids and $(-\Delta E)$ in the gas phase. (b) Comparison of the experimental and calculated pK_a values of the acids when reacting with the soft base, trimethylamine. Source: Gupta et al. [98]/Elsevier.

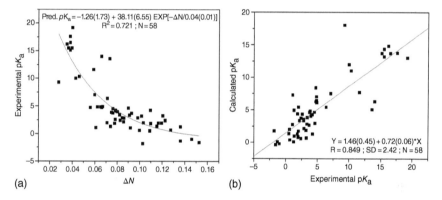

Figure 10.8 (a) Correlation between the experimental pK_a values of inorganic and organic acids and ΔN in the gas phase. (b) Comparison of the experimental and calculated pK_a values of the acids when reacting with the soft base, trimethylamine. Source: Gupta et al. [98]/Elsevier.

In Figure 10.8, the correlation between pK_a and ΔN is displayed for the gas phase. Here also, exponential decay regression models were utilized to establish pK_a as a function of ΔN in gas phase. The findings reveal that there is a corresponding exponential relationship between pK_a and ΔN, which implies that the Brønsted and Lewis definitions of acidity and basicity generally align. A greater ΔN value indicates a stronger Lewis acid–base pair with more electron transfer. As the same strong base was employed for all acid–base pairs studied, a higher ΔN value would indicate a stronger acid. Therefore, if the ΔN value is larger, the acid is stronger, which does not contradict the definition of a stronger Brønsted acid having a smaller pK_a value. In general, this relationship is valid, but it may not apply to every acid–base pair.

The acidity behavior of molecules is typically determined by the type of functional group present, such as –COOH or –OH. However, this study takes a different approach by considering both global and local factors for molecules

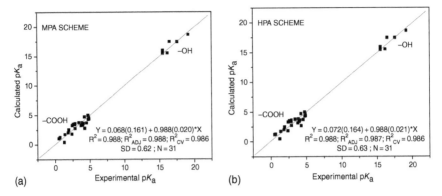

Figure 10.9 Correlation between the experimental and predicted pK_a values for carboxylic acids and alcohols, considering the effects of ω^+_g and ΔN, in (a) MPA and (b) HPA schemes. Source: Gupta et al. [98]/Elsevier.

containing carboxylic acids and alcohols. The global factor, represented by ΔN ($-\Delta E$), and the local factor, represented by ω^+_g, play a role in determining the preferred Lewis acid sites. The study demonstrates a strong correlation between the estimated pK_a values, calculated using a two-parameter linear regression model, and the actual values. The correlation coefficients (R^2, R^2_{ADJ}, and R^2_{CV}) are all above 0.98 in every case, as shown in Figure 10.9. Carboxylic acids and alcohols follow a similar trend, aligning closely to a line with a slope near unity and an intercept near zero. However, they appear in different regions of the line, indicating variations in the sets of functional groups. Stronger Brønsted acids also tend to exhibit greater Lewis acidity. Although the HSAB principle holds for acids and bases with similar strengths, it may not always apply to the preference for hard or soft acid and base. In these cases, the values of $-\Delta E$ and ΔN are relatively small. By establishing correlations between these descriptors and the acid's cation-releasing or anion-accepting ability, a comprehensive acid–base theory encompassing redox and electrophile–nucleophile reactions could be developed.

10.11 Summary

Acid–base theory represents a pivotal area of study in chemistry, underpinning a wide range of scientific disciplines and technological applications. Through the lens of historical development, fundamental concepts, measurement techniques, and contemporary advancements, this chapter provides a comprehensive overview to the acid–base theory. Various definitions of acids and bases, as they evolve over time, are analyzed. Two important aspects are the transfer of electrons and protons. Measurement of strengths of acids and bases is intimately related to those concepts. Conceptual DFT-based reactivity descriptors, different information theory-based quantities, and the associated electronic structure principles lend additional support in understanding the characteristics of acids and bases and also in the determination of their strength. Effects of possible external perturbations and electron redistribution such as in the presence of solvent, steric effect, inductive effect, resonance, periodicity on the variation in the behavior of acids and bases are highlighted. Continued research and exploration in acid–base theory will undoubtedly yield profound insights and pave the way for transformative advancements in the realm of chemistry and beyond.

Acknowledgment

PKC would like to thank Professor Shubin Liu for kindly inviting him to contribute a chapter to the book entitled, "Exploring Chemical Concepts through Theory and Computation." He thanks DST, New Delhi for the J. C. Bose National Fellowship, grant number SR/S2/JCB-09/2009. He is also thankful to all his students and collaborators whose work is discussed here. RP and HM thank CSIR for their fellowships.

Conflict of Interest

The authors declare that they have no conflict of interest, financial and/or otherwise.

References

1 Chen, L. and Neibling, H. (2014). Anaerobic digestion basics. *Univ. Idaho Extens.* 2: 6.
2 McQuilton, P., St Pierre, S.E., Thurmond, J., and FlyBase Consortium (2012). FlyBase 101--the basics of navigating FlyBase. *Nucl. Acids Res.* 40(Database issue): D706–D714.
3 Brooks, H., Lebleu, B., and Vives, E. (2005). Tat peptide-mediated cellular delivery: back to basics. *Adv. Drug Deliv. Rev.* 57: 559–577.
4 Johnson, J.W. (2003). Acid tests of N-methyl-d-aspartate receptor gating basics. *Mol. Pharmaco.* 63: 1199–1201.
5 Hall, N.F. (1940). Systems of acids and bases. *J. Chem. Educ.* 17: 124–128.
6 Ingold, C.K. (1934). Principles of an electronic theory of organic reactions. *Chem. Rev.* 15: 225–274.
7 Kossel, W. (1916). Über Molekülbildung als Frage des Atombaus. *Ann. Phys.* 354: 229–362.
8 Germann, A.F.O. (1925). A general theory of solvent systems. *J. Am. Chem. Soc.* 47: 2461–2468.
9 Pauling, L. (1960). *The Nature of the Chemical Bond*. Ithaca, NY: Cornell University Press.
10 Mulliken, R.S. (1934). A new electroaffinity scale; together with data on valence states and on valence ionization potentials and electron affinities. *J. Chem. Phys.* 2: 782–793.
11 Mulliken, R.S. (1935). Electronic structures of molecules XI. Electroaffinity, molecular orbitals and dipole moments. *J. Chem. Phys.* 3: 573–585.

12 Sanderson, R.T. (1952). Electronegativities in inorganic chemistry. *J. Chem. Educ.* 29: 539–544.

13 Allred, A.L. and Rochow, E.G. (1958). A scale of electronegativity based on electrostatic force. *J. Inorg. Nucl. Chem.* 5: 264–268.

14 Allred, A.L. (1961). Electronegativity values from thermochemical data. *J. Inorg. Nucl. Chem.* 17: 215–221.

15 Iczkowski, R.P. and Margrave, J.L. (1961). Electronegativity. *J. Am. Chem. Soc.* 83: 3547–3551.

16 Pearson, R.G. (1985). Absolute electronegativity and absolute hardness of Lewis acids and bases. *J. Am. Chem. Soc.* 107: 6801–6806.

17 Allen, L.C. (1989). Electronegativity is the average one-electron energy of the valence-shell electrons in ground-state free atoms. *J. Am. Chem. Soc.* 111: 9003–9014.

18 Parr, R.G., Donnelly, R.A., Levy, M., and Palke, W.E. (1978). Electronegativity: the density functional viewpoint. *J. Chem. Phys.* 68: 3801–3807.

19 Parr, R.G. and Yang, W. (1989). *Density Functional Theory of Atoms and Molecules*. New York: Oxford University Press.

20 Koopmans, T. (1934). Über die Zuordnung von Wellenfunktionen und Eigenwerten zu den Einzelnen Elektronen Eines Atoms. *Physica* 1: 104–113.

21 Pearson, R.G. (1963). Hard and soft acids and bases. *J. Am. Chem. Soc.* 85: 3533–3539.

22 Pearson, R.G. (1968). Hard and soft acids and bases, HSAB, part 1: fundamental principles. *J. Chem. Educ.* 45: 581.

23 Pearson, R.G. (1968). Hard and soft acids and bases, HSAB, part II: underlying theories. *J. Chem. Educ.* 45: 643.

24 Parr, R.G. and Pearson, R.G. (1983). Absolute hardness: companion parameter to absolute electronegativity. *J. Am. Chem. Soc.* 105: 7512–7516.

25 Pearson, R.G. (1990). Hard and soft acids and bases—the evolution of a chemical concept. *Coord. Chem. Rev.* 100: 403–425.

26 Pearson, R.G. (2005). Chemical hardness and density functional theory. *J. Chem. Sci.* 117: 369–377.

27 Yang, W. and Parr, R.G. (1985). Hardness, softness, and the fukui function in the electronic theory of metals and catalysis. *Proc. Nat. Acad. Sci.* 82: 6723–6726.

28 Parr, R.G., Szentpály, L.V., and Liu, S. (1999). Electrophilicity index. *J. Am. Chem. Soc.* 121: 1922–1924.

29 Parr, R.G. and Yang, W. (1984). Density functional approach to the frontier-electron theory of chemical reactivity. *J. Am. Chem. Soc.* 106: 4049–4050.

30 Yang, W., Parr, R.G., and Pucci, R. (1984). Electron density, Kohn–Sham frontier orbitals, and Fukui functions. *J. Chem. Phys.* 81: 2862–2863.

31 Yang, W. and Mortier, W.J. (1986). The use of global and local molecular parameters for the analysis of the gas-phase basicity of amines. *J. Am. Chem. Soc.* 108: 5708–5711.

32 Roy, D.R., Parthasarathi, R., Padmanabhan, J. et al. (2006). Careful scrutiny of the philicity concept. *J. Phys. Chem. A* 110: 1084–1093.

33 Chattaraj, P.K., Maiti, B., and Sarkar, U. (2003). Philicity: a unified treatment of chemical reactivity and selectivity. *J. Phys. Chem. A* 107: 4973–4975.

34 Sanderson, R.T. (1951). An interpretation of bond lengths and a classification of bonds. *Science* 114: 670–672.

35 Sanderson, R.T. (1955). Partial charges on atoms in organic compounds. *Science* 121: 207–208.

36 Yang, W., Lee, C., and Ghosh, S.K. (1985). Molecular softness as the average of atomic softnesses: companion principle to the geometric mean principle for electronegativity equalization. *J. Phys. Chem.* 89: 5412–5414.

37 Chattaraj, P.K. (1991). Atomic and molecular properties from the density-functional definition of electronegativity. *Curr. Sci.* 61: 391–395.

38 Datta, D. (1986). Geometric mean principle for hardness eualization: a corollary of Sanderson's geometric mean principle of electronegativity equalization. *J. Phys. Chem.* 90: 4216–4217.

39 Chattaraj, P.K., Nandi, P.K., and Sannigrahi, A.B. (1991). Improved hardness parameters for molecules. *Proc. Ind. Acad. Sci.* 103: 583–589.

40 Chattaraj, P.K., Ayers, P.W., and Melin, J. (2007). Further links between the maximum hardness principle and the hard/soft acid/base principle: insights from hard/soft exchange reactions. *Phys. Chem. Chem. Phys.* 9: 3853.

41 Chattaraj, P.K., Giri, S., and Duley, S. (2010). Electrophilicity Equalization Principle. *J. Phys. Chem. Lett.* 7: 1064–1067.

42 Langenaeker, W., Coussement, N., De Proft, F., and Geerlings, P. (1994). Quantum chemical study of the influence of isomorphous substitution on the catalytic activity of zeolites: an evaluation of reactivity indexes. *J. Phys. Chem.* 98: 3010–3014.

43 Damoun, S., Langenaeker, W., Van de Woude, G., and Geerlings, P. (1995). Acidity of halogenated alcohols and silanols: competition of electronegativity and softness in second and higher row atoms. *J. Phys. Chem.* 99: 12151–12157.

44 Corma, A., Llopis, F., Viruela, P., and Zicovich-Wilson, C. (1994). Acid softness and hardness in large-pore zeolites as a determinant parameter to control selectivity in orbital-controlled reactions. *J. Am. Chem. Soc.* 116: 134–142.

45 Baekelandt, B.G., Mortier, W.J., Lievens, J.L., and Schoonheydt, R.A. (1991). Probing the reactivity of different sites within a molecule or solid by direct computation of molecular sensitivities via an extension of the electronegativity equalization method. *J. Am. Chem. Soc.* 113: 6730–6734.

46 Gazquez, J.L., and Mendez, F. (1994). The hard and soft acids and bases principle: an atoms in molecules viewpoint. *J. Phys. Chem.* 98: 4591–4593.

47 Krishnamurty, S., Roy, R.K., Vetrivel, R. et al. (1997). The local hard–soft acid–base principle: a critical study. *J. Phys. Chem. A* 101: 7253–7257.

48 Klopman, G. (1968). Chemical reactivity and reaction paths. *J. Am. Chem. Soc.* 90: 223.

49 Chattaraj, P.K. (2001). Chemical reactivity and selectivity: local HSAB principle versus frontier orbital theory. *J. Phys. Chem. A* 105: 511.

50 Pearson, R.G. (1993). The principle of maximum hardness. *Acc. Chem. Res.* 26: 250–255.

51 Pearson, R.G. (1999). Maximum chemical and physical hardness. *J. Chem. Educ.* 76: 267.

52 Chattaraj, P.K. and Sengupta, S. (1996). Popular electronic structure principles in a dynamical context. *J. Phys. Chem.* 100: 16126–16130.

53 Chattaraj, P.K. and Sengupta, S. (1997). Dynamics of chemical reactivity indices for a many-electron system in its ground and excited states. *J. Phys. Chem. A* 101: 7893–7900.

54 Ghanty, T.K. and Ghosh, S.K. (1996). A density functional approach to hardness, polarizability, and valency of molecules in chemical reactions. *J. Phys. Chem.* 100: 12295–12298.

55 Zhou, Z. and Parr, R.G. (1989). New measures of aromaticity: absolute hardness and relative hardness. *J. Am. Chem. Soc.* 111: 7371–7379.

56 Pearson, R.G. and Palke, W.E. (1992). Support for a principle of maximum hardness. *J. Phys. Chem.* 96: 3283–3285.

57 Makov, G. (1995). Chemical hardness in density functional theory. *J. Phys. Chem.* 99 (23): 9337–9339.

58 Chakraborty, D., and Chattaraj, P.K. (2021). Conceptual density functional theory based electronic structure principles. *Chem. Sci.* 12: 6264–6279.

59 Chattaraj, P.K., Fuentealba, P., Jaque, P., and Toro-Labbé, A. (1999). Validity of the minimum polarizability principle in molecular vibrations and internal rotations: an ab initio SCF study. *J. Phys. Chem. A* 103: 9307–9312.

60 Chattaraj, P.K. and Poddar, A. (1999). Chemical reactivity and excited-state density functional theory. *J. Phys. Chem. A* 103: 1274–1275.

61 Fuentealba, P., Simón-Manso, Y., and Chattaraj, P.K. (2000). Molecular electronic excitations and the minimum polarizability principle. *J. Phys. Chem. A* 104: 3185–3187.

62 Gutiérrez-Oliva, S., Letelier, J.R., and Toro-Labbé, A. (1999). Energy, chemical potential and hardness profiles for the rotational isomerization of HOOH, HSOH and HSSH. *Mol. Phys.* 96: 61–70.

63 Parthasarathi, R., Padmanabhan, J., Subramanian, V. et al. (2003). Chemical reactivity profiles of two selected polychlorinated biphenyls. *J. Phys. Chem. A* 107: 10346–10352.

64 Chattaraj, P.K., Gutiérrez-Oliva, S., Jaque, P., and Toro-Labbé, A. (2003). Towards understanding the molecular internal rotations and vibrations and chemical reactions through the profiles of reactivity and selectivity indices: an ab initio SCF and DFT study. *Mol. Phys.* 101: 2841–2853.

65 Parr, R.G. and Chattaraj, P.K. (1991). Principle of maximum hardness. *J. Am. Chem. Soc.* 113: 1854–1855.

66 Pal, R., Jana, G., and Chattaraj, P.K. (2020). Ligand stabilized transient "MNC" and its influence on MNC → MCN isomerization process: a computational study (M = Cu, Ag, and Au). *Theor. Chem. Acc.* 139: 15.

67 Chamorro, E., Chattaraj, P.K., and Fuentealba, P. (2003). Variation of the electrophilicity index along the reaction path. *J. Phys. Chem. A* 107: 7068–7072.

68 Parthasarathi, R., Elango, M., Subramanian, V., and Chattaraj, P.K. (2005). Variation of electrophilicity during molecular vibrations and internal rotations. *Theor. Chem. Acc.* 113: 257–266.

69 Chattaraj, P.K., Sarkar, U., and Roy, D.R. (2006). Electrophilicity Index. *Chem. Rev.* 6: 2065–2091.

70 Pan, S., Solà, M., and Chattaraj, P.K. (2013). On the validity of the maximum hardness principle and the minimum electrophilicity principle during chemical reactions. *J. Phys. Chem. A* 117: 1843–1852.

71 Drago, R.S. and Wayland, B.B. (1965). A double-scale equation for correlating enthalpies of Lewis acid-base interactions. *J. Am. Chem. Soc.* 87: 3571–3577.

72 Drago, R.S. and Wong, N.M. (1996). The role of electron-density transfer and electronegativity in understanding chemical reactivity and bonding. *J. Chem. Educ.* 73: 123.

73 Vogel, G.C. and Drago, R.S. (1996). The ECW model. *J. Chem. Educ.* 73: 701.

74 Bates, R.G. (1948). Definitions of pH scales. *Chem. Rev.* 42: 1–61.

75 Bartmess, J.E., Scott, J.A., and McIver, R.T. Jr., (1979). Scale of acidities in the gas phase from methanol to phenol. *J. Am. Chem. Soc.* 101: 6046–6056.

76 Lias, S.G., Liebman, J.F., and Levin, R.D. (1984). Evaluated gas phase basicities and proton affinities of molecules; heats of formation of protonated molecules. *J. Phys. Chem. Ref. Data* 13: 695–808.

77 Pauling, L. (1932). The nature of the chemical bond. IV. The energy of single bonds and the relative electronegativity of atoms. *J. Am. Chem. Soc.* 54: 3570–3582.

78 Ayers, P.W., Parr, R.G., and Pearson, R.G. (2006). Elucidating the hard/soft acid/base principle: a perspective based on half-reactions. *J. Chem. Phys.* 124: 194107.

79 Rochester, C.H. and Rossall, B. (1967). Steric hindrance and acidity. Part I. The effect of 2,6-di-t-butyl substitution on the acidity of phenols in methanol. *J. Chem. Soc. B Phys. Org.* 743. https://doi.org/10.1039/J29670000743.

80 Namboori, C.G.G. and Haith, M.S. (1968). Steric effects in the basic hydrolysis of poly(ethylene terephthalate). *J. Appl. Polym. Sci.* 12: 1999–2005.

81 Markle, T.F., Rhile, I.J., and Mayer, J.M. (2011). Kinetic effects of increased proton transfer distance on proton-coupled oxidations of phenol-amines. *J. Am. Chem. Soc.* 133: 17341–17352.

82 Cox, B.G. (2013). Introduction. In: *Acids and Bases: Solvent Effects on Acid-Base Strength*, 1–9. Oxford Academic.

83 Laturski, A.E., Gaffen, J.R., Demay-Drouhard, P. et al. (2023). Probing the impact of solvent on the strength of Lewis acids via fluorescent Lewis adducts. *Precision Chem.* 1: 49–56.

84 Fawcett, W.R. (1993). Acidity and basicity scales for polar solvents. *J. Phys. Chem.* 97: 9540–9546.

85 Catalan, J., Sanchez-Cabezudo, M., de Paz, J.L.G., and Elguero, J. (1988). Acidity and basicity of azoles: solvent effects. *THEOCHEM* 166: 415–420.

86 Krygowski, T.M. and Fawcett, W.R. (1975). Complementary Lewis acid-base description of solvent effects. I. Ion-ion and ion-dipole interactions. *J. Am. Chem. Soc.* 97: 2143–2148.

87 De Proft, F., Langenaeker, W., and Geerlings, P. (1995). Acidity of first- and second-row hydrides: effects of electronegativity and hardness. *Int. J. Quantum Chem.* 55: 459–468.

88 Rich, R.L. (1985). Periodicity in the acid-base behavior of oxides and hydroxides. *J. Chem. Educ.* 62: 44.

89 Kozuch, S. and Martin, J.M.L. (2013). Halogen bonds: benchmarks and theoretical analysis. *J. Chem. Theory Comput.* 9: 1918–1931.

90 Catalan, J. (1996). Influence of inductive effects and polarizability on the acid-base properties of alkyl compounds. Inversion of the alcohol acidity scale. *J. Phys. Org. Chem.* 9: 652–660.

91 Exner, O. and Böhm, S. (2002). Inductive effects in isolated molecules: 4-substituted bicyclo[2.2.2]octane-1-carboxylic acids. *Chem. Euro. J.* 8: 5147–5152.

92 Holt, J. and Karty, J.M. (2003). Origin of the acidity enhancement of formic acid over methanol: resonance versus inductive effects. *J. Am. Chem. Soc.* 125: 2797–2803.

93 Barbour, J.B. and Karty, J.M. (2004). Resonance energies of the allyl cation and allyl anion: contribution by resonance and inductive effects toward the acidity and hydride abstraction enthalpy of propene. *J. Org. Chem.* 69: 648–654.

94 Gupta, K., Giri, S., and Chattaraj, P.K. (2008). Acidity of meta- and para-substituted aromatic acids: a conceptual DFT study. *New J. Chem.* 32: 1945–1952.

95 Parthasarathi, R., Padmanabhan, J., Elango, M. et al. (2006). pKa prediction using group philicity. *J. Phys. Chem. A* 110: 6540–6544.

96 Cao, X., Rong, C., Zhong, A. et al. (2018). Molecular acidity: an accurate description with information-theoretic approach in density functional reactivity theory. *J. Comput. Chem.* 39: 117–129.

97 Xiao, X., Xiaofang, C., Dongbo, Z. et al. (2020). Quantification of molecular basicity for amines: a combined conceptual density functional theory and information-theoretic approach study. *Acta Phys. Chim. Sin.* 36: 1906034.

98 Gupta, K., Roy, D.R., Subramanian, V., and Chattaraj, P.K. (2007). Are strong Brønsted acids necessarily strong Lewis acids? *THEOCHEM* 812: 13–24.

11

Sigma Hole Supported Interactions: Qualitative Features, Various Incarnations, and Disputations

Kelling J. Donald

University of Richmond, Gottwald Center for the Sciences, Department of Chemistry, 28 Westhampton Way, Richmond, VA 23173, USA

11.1 Introduction

11.1.1 What's in a Name – The Sigma Hole Terminology and Concept

Giving a new name to an old thing does not make it new. Moreover, renaming runs the risk of obscuring rather than elucidating, which is especially threatening for scientific inquiry where the light of the past is the nourishment of progress. Yet, in the study of chemical bonding, as in science more broadly, progress is rarely linear, and two contexts in which a new name comes to be imposed on old phenomena are (i) rediscovery (where a new observer thinks mistakenly that an observation is novel and labels it as such) [1, 2][1] and (ii) reinterpretation (where a new observer offers a new perspective or articulates, with evidence, the importance of a previously un- or under-emphasized feature of a well-documented phenomenon).

The sigma hole concept emerged from a context akin to the latter [3].[2] The polarization of the electron density of atomic centers in molecules is not a new phenomenon – it has existed since the first few molecules were formed in the universe. By the time the term sigma hole was introduced [3], the region of depleted electron density in molecules to which the term refers had already been identified 15 years earlier [4]. And weak noncovalent bonding interactions that such electron-deficient regions help to stabilize were already well known under different appellations

---

1 The claims of simultaneous discoveries and inventions may be included in this category. See Ref. [1].
2 Broadly defined, the term "sigma hole," as we will use it in this chapter, refers to an area on an atom, M, that is depleted of electron density due to the polarization of M by its sigma bonded substituent(s). That depletion leads to a maximum in the surface electrostatic potential, $V_{s,max}$, in the sigma hole region (the least negative V_s compared to the surrounding surface), which can even become positive if the polarization is substantial enough such that the influence of the nucleus rather than the electron density dominates the potential in that region of the surface.

Exploring Chemical Concepts Through Theory and Computation, First Edition. Edited by Shubin Liu.
© 2024 WILEY-VCH GmbH. Published 2024 by WILEY-VCH GmbH.

[5–7].[3] Yet the introduction of the sigma hole concept emerged out of a search for a deeper understanding of (and provided language for discussing) broad sub-classes of interactions [8–10]. The sigma hole concept gave justifiable emphasis as well to the role of electrostatics in accounting for what appeared, oddly, to be the formation of bonding interactions between two negatively charged atomic centers (as in halogen bonding, about which we will say more shortly).

How were such interactions characterized and discussed in the earlier literature? The history of chemical bonding, including many decades of discussions of Lewis-type acid–base chemistry [5], is rife with the naming and renaming (and creative imaginings) of stabilizing interactions. In his 1968 review of what the title describes as "donor–acceptor interactions," Henry Bent listed 19 other terms that had been used up to that point to describe the general 'acid-base' bonding phenomenon to which his title referred. And the list that he provided includes terms such as "charge transfer" and "filling of antibonding orbitals," and others that might elicit giggles from contemporary students of chemical bonding theory, such as, "bumps-in-hollows," "saturation of residual affinities," "adhesion by the attraction of positive and negative patches in molecules," and "electron clutching" [5].[4] Sigma hole bonding which we consider in this chapter refers to a class of attractive interactions that are related to classical donor–acceptor interactions. They have certain specific characteristics and tend to be at the weaker end of a continuum of bonding interactions that span strong coordinate covalent (dative) bonds and the weakest noncovalent interactions. The general thinking among proponents, therefore, is that – unlike some of the terms mentioned above that have fallen out of use – the term 'sigma hole bonding' emphasizes certain key and previously unacknowledged features of an important category of noncovalent interactions.

11.1.2 Donor–Acceptor Interaction Continuum

Simple donor–acceptor interactions, such as that between BH_3 and NH_3 in the dative ammonia borane complex ($H_3B \cdot NH_3$), may be represented by the interaction diagram shown in Figure 11.1. Implicit in that picture is the assumption that the thermodynamic stability of an A·Y adduct compared to the total energy of the isolated electron acceptor or acid unit, A, and isolated electron donor or base, Y, is rooted in the transfer of electron density from Y to A (along with all of the associated rearrangements of the overall electron density of the two bonding partners under the influence of each other) to form the happy complex. That focus is reasonable for strong dative bonds, but it does not address explicitly certain aspects of chemical bonding that become crucial as we move along the bonding continuum from dative to hydrogen bonding, for example, where other contributions to the interaction energy apart from charge transfer take on heightened importance.

---

3 Ref. [7] is a separately published version of the Nobel lecture cited in Ref. [6].
4 See Bent's list on pp. 609–610 of his review [5].

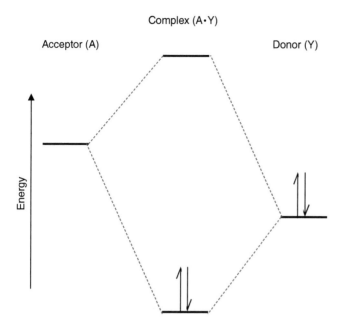

Figure 11.1 Model of a donor–acceptor interaction diagram in which A is the Lewis acid (acceptor) and Y is the Lewis base (donor).

In the $H_3B \cdot NH_3$ complex, the electron acceptor site (B in BH_3) is hypovalent. If, however – like the H center in a hydrogen bond, R-H—Y – the electron acceptor site on A in the A—Y complex is not formally electron deficient, the corresponding A—Y interaction tends to be rather weak (compared to a dative bond), and the minimum energy separation between the two atomic centers in A and Y that are involved in the A—Y bond is usually long – closer to the sum of their van der Waals radii than the sum of their covalent radii. And, of particular relevance to this discussion, the interaction energies for such systems may be dominated by contributions other than charge transfer. Indeed, even in the case of the $H_3B \leftarrow NH_3$ complex, if the B—N distance is stretched or compressed, the extent of the charge transfer between the B and N centers would change, and so too would the relative importance of other contributions to the interaction energy, such as electrostatic and dispersion interactions.

Since a range of different electronic changes and (de)stabilizing forces can contribute, therefore, in different proportions to the overall stability of a (weak or strong) chemical bond, identifying any net bonding interaction exclusively with one specific influence or effect (such as electrostatics or charge transfer) leads almost inevitably to controversy [11]. The situation can become even more fraught if the interaction itself eventually comes to be named after a specific feature (e.g. the sigma hole) of the interacting fragments, rather than the less disputable outcome (e.g. noncovalent interaction or coordinate covalent [dative] bonding). Nonetheless, we will examine the sigma hole concept briefly here to see how it manages, despite limitations, to unify what would otherwise appear to be distinct modes of weak bonding interactions.

11.2 Many Incarnations and Roles of a Single Phenomenon

11.2.1 Hydrogen Bonding

In the case of hydrogen bonding interactions (R-H—Y), in which H is polarized significantly by R, electrostatic interactions become quite important. That is so even if – as is commonly understood nowadays to be the case – charge transfer (and other energy contributions) can play decisive roles in hydrogen bonding depending on the chemical environment and identities of R and Y [12, 13].

Clear evidence of the polarization of the electron density of the H center in the H_2O molecule may be obtained from a comparison of the electrostatic potential (ESP, V) in the vicinity of the H versus the O centers on a selected surface of the molecule (Figure 11.2). Based solely on the difference in the electronegativity of the H versus O atoms, the ESPs are expected to be positive at the H and negative about the O centers, which is precisely what the computed potentials show in Figure 11.2. In that figure, the bluest regions are the most positive, and the reddest regions are the most negative. And the former – the areas of positive potential (in blue) on the H site in Figure 11.2 – is considered by some to be an elaborate example of a sigma hole.

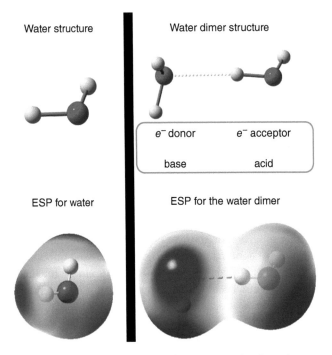

Figure 11.2 Structure of an optimized water molecule and a water dimer, with the corresponding electrostatic potential (ESP) maps immediately below them. The structures in the ESP maps are oriented in roughly the same way as the model structures above them.

Since the ESP maps tend to shroud the molecular structure, we show the relevant structures in Figure 11.2 (and in some later figures as well) along with the semitransparent overlay of the ESP maps. In all cases, the ESP color spectrum runs from blue (positive) to red (negative) for ±0.07 au on the 0.001 au isodensity surface. The molecules and complexes shown in this work are optimized minimum energy structures that have been confirmed to be minima by vibrational frequency analyses. The calculations were all carried out (using the Gaussian 16 suite of programs) [14]. Except for results taken from elsewhere in the literature, these results were obtained using the ωB97X-D method [15] and all-electron correlation-consistent triple-ζ (cc-pVTZ) basis sets [16] for H through to Kr, and small-core MDF relativistic pseudopotentials provided by the Stuttgart/Cologne group [17] combined with the associated triple zeta (valence) basis sets for heavier atoms. Molecular representations and ESP maps were generated using the Chemcraft and Gaussview 6 graphical user interfaces [18, 19], and ESP maxima were determined using Multiwfn [20, 21].

In Figure 11.2, models of the optimized structures and ESP maps are provided for an isolated water molecule (left) and for a water dimer (H_2O—H-OH; right). For ease of analysis, one of the monomer units in the dimer complex is oriented in roughly the same way as that of the isolated molecule on the far left in Figure 11.2. By comparing the ESP maps for the isolated monomer and the (H_2O—H-OH) dimer, we see readily that one of the H atoms on the 'acid' H-OH fragment in the dimer intersects in the O—H overlap region with the lone pair site on the O in the 'base' H_2O fragment. That is, the sigma hole region at H in one H_2O unit (see the blue region on the ESP of the isolated molecule) overlaps with the lone pair region on O in the other H_2O monomer unit (see the red region on the ESP of the isolated molecule) to form the bound pair (Figure 11.2). Hydrogen bonds have been claimed, therefore, as sigma hole interactions simply because of the direct involvement of the area of positive potential (the sigma hole) on H in the bonding region. To be clear, however, that is not to say that other influences such as charge transfer or other stabilizing interactions are irrelevant, insignificant, or absent.

But what of atomic centers other than H? Do other sites in molecules exhibit sigma hole type interactions? We pick up on that question presently and will begin with the elements of group 17. We will consider the other elements of group 1 briefly later on, but the group 17 atoms have been crucial historically in the evolution of the sigma hole concept. Additionally, much less work has been done on the group 1 systems in this regard, and, after all, group 17 is a natural progression from the H atom, since H itself belongs arguably to group 17! [22].

11.2.2 Halogen Bonding and Sigma Holes on Group 17 Atoms

One place where people might not expect to find a region of positive potential is on an atomic center that has a negative partial charge. And the best-known examples of such electronegative centers in the main group of the periodic table are the atoms in groups 15–17 and the halogen (X) atoms in particular.

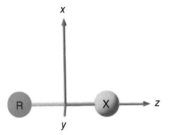

Figure 11.3 A representation of an R–X molecule with the principal axis pointing along the z-axis.

A negative point charge is associated with an accumulation of electron density on an atomic center in a molecule, so the reason for the emergence of an electron-deficient region on the very electronegative group 17 atomic center in a molecule is not immediately obvious. Yet it is now very well known that electron density is not distributed evenly over the surface of any atomic center in a molecule. The formation of directional (covalent) bonds involves the accumulation of some of the electron density of, e.g. atoms A and B, in the A–B bonding region, and that results necessarily in an uneven or polarized distribution of the electron density over the surface of both A and B.

For halogen atoms, each having both ns and np electrons in the valence shell, the valence electron configuration is quite different from that of the hydrogen atom. There are seven valence electrons in halogen atoms: two in the valance ns orbital and two each in the np_x and np_y orbitals (based on the orientation of the principal axis in Figure 11.3) plus one unpaired valence electron in the np_z orbital. That is an already substantial difference compared to the H atom, which has just one electron in its 1s orbital, even if we ignore for the moment any hybridization (of the ns and np_z orbitals) at the halogen center in the R–X bond.

The two fully occupied np orbitals of a halogen atom in the R–X bond (see Figure 11.3) are in the x–y plane perpendicular to the z-axis and form together a torus of electron density around the X center in the x–y plane. If R is sufficiently electron-withdrawing, the electron density in the valence p_z orbital will be polarized, upon bond formation, toward the R–X bonding region. Under such conditions, it has been shown that a region of depleted electron density (a sigma hole) [3] tends to arise readily on the X atom opposite the R–X bonding region around the extension of the R–X bond [4, 23]. That sigma hole region on the molecular surface is often positive (especially for X = Cl, Br, and I) [3], and the likelihood that a given R group will induce a positive potential on X increases as R becomes more electron-withdrawing and as X gets softer or more polarizable (going down group 17 from F to I) and less electronegative [24, 25].

That positive sigma hole region may exist on the pole of the X atom even if an integration over all of the electron density in the atomic basin assigned to X in the R–X bond would return a net negative point charge for that X center [25]. All that is required for a sigma hole to emerge on X in any R–X molecule is a significant shift of electron density away from the pole of X toward the σ bonding region between

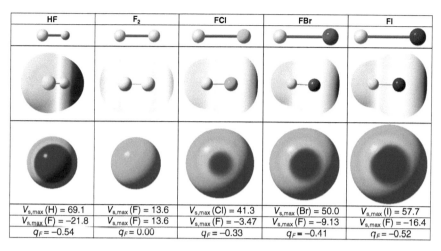

HF	F$_2$	FCl	FBr	FI
$V_{s,max}$ (H) = 69.1	$V_{s,max}$ (F) = 13.6	$V_{s,max}$ (Cl) = 41.3	$V_{s,max}$ (Br) = 50.0	$V_{s,max}$ (I) = 57.7
$V_{s,max}$ (F) = −21.8	$V_{s,max}$ (F) = 13.6	$V_{s,max}$ (F) = −3.47	$V_{s,max}$ (F) = −9.13	$V_{s,max}$ (F) = −16.4
q_F = −0.54	q_F = 0.00	q_F = −0.33	q_F = −0.41	q_F = −0.52

Figure 11.4 Structures of optimized HF and FX molecules for X = F, Cl, Br, and I, and the corresponding electrostatic potential (ESP) maps ([side view] faded to show structures beneath, and [front view] fully opaque, with shadow and spotlight, showing the sigma hole on H and X). The maximum potential in the center of the sigma hole on the cap of both F and X in the compounds ($V_{s,max}$) in kcal/mol unit are listed as well as the net charge on F, in units of electronic charge (e), that we obtained from a "natural population analysis" (NPA).

R and X; it is not necessary for the electron density to be so substantially polarized toward R that the R fragment has a net negative charge and X is positive.

Figure 11.4 shows, for example, the optimized structures and ESPs (faded side view) for F−H and for the group 17 R−X molecules for R = F and X = F, Cl, Br, and I, with positive sigma holes induced on H and X (and on F in some cases) visible in blue. The front views of the same ESPs are shown looking down the bond axis at the sigma holes on the H and X atoms.

Broadly defined, a sigma hole is a region of lowered electron density compared to the surroundings on the atomic center, so the ESP values in the sigma hole, V_s, (even the extremum of the sigma hole ESPs, $V_{s,max}$) do not have to be positive. The sigma hole only needs to be electron deficient relative to the local surroundings, but sigma holes are of particular interest for bonding analyses if $V_{s,max}$ is positive. As we see in Figure 11.4, for instance, the sigma hole on F becomes more negative (less positive) going to the right in Figure 11.4 from F$_2$ to FCl, FBr, and FI. In the FX molecules considered, both F and X have sigma holes, but F is more polarizing than Cl, Br, and I, and it ends up with a negative sigma hole in all of those cases. The sigma holes are all positive for H and for X = F (in the F$_2$ molecule), Cl, Br, and I (Figure 11.4).

In the case of the F$_2$ molecule, the two F atoms mutually polarize each other so substantially that a positive sigma hole actually emerges on both F sites. It is relatively rare for a positive sigma hole to be induced on F, and in general $V_{s,max}$ on X in a molecule R−X gets more positive and larger as X gets larger for any given R. Notice, however, that the sign of the net charge assigned to an atomic center tells us

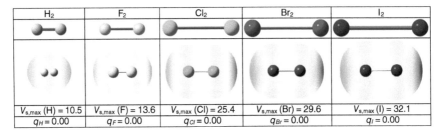

Figure 11.5 Structure of optimized H_2 and X_2 molecules for X = F, Cl, Br, and I, and corresponding electrostatic potential (ESP) maps. The maximum potential in the center of the sigma hole on the cap of both H and X in the compounds ($V_{s,max}$) in kcal/mol units is listed. The net charges are necessarily zero for homonuclear diatomic molecules.

nothing about the presence or absence of a positive sigma hole on that center. F_2 and all of the other group 17 homonuclear diatomics are extremes in which the potential is positive (Figures 11.4 and 11.5) but the point change at X is zero.

11.2.2.1 Common Origins

We started out with an overview of hydrogen bonding and pointed out that the interaction in that case, which may involve a significant degree of change transfer, is supported as well by the presence of a positive ESP induced on the H surface due to a substantial shift of electron density from H into the overlap region of the R−H bond. That is fundamentally the same phenomenon that is observed in group 17 R−X molecules, except that for the group 17 atoms (i) the R−X bond involves a hybridized X orbital with a substantial valance p contribution (rather than simply an s orbital as is the case for H in R−H), and (ii) the expansion of the sigma hole across the surface of X opposite the bonding region (in the manner observed for H in the H_2O molecule in Figure 11.2) is restricted by the electron belt or torus (made up of the valence np_x^2 and np_y^2 electrons) in the x–y plane (see Figure 11.3). So, the positive potential induced on the H surface (see Figure 11.2) and that on the X surface (Figures 11.4 and 11.5) – thus hydrogen and halogen bonding – have common roots.

The presence of the $np_x + np_y$ belt of electron density on X [which is faint in the semi-transparent ESP maps in Figures 11.5 and 11.5, but shows up as a prominent green region in the opaque ESPs in Figure 11.4] restricts the spatial extent of the positive sigma hole region on X in a way that does not arise for hydrogen in R−H. Note: in the opaque (front view) ESP in Figure 11.4, the positive potential (in bright blue) on H seems to be surrounded too by a negative belt (green), but that is not the case. The green area in that front view is the larger F atom behind the H atom. In the other opaque (front view) ESPs, the X atoms are larger than F, so F is hidden completely behind the X atoms. For the hydrogen atom in any R−H molecule, there is a single 1s electron and no other occupied orbital on the atom, so the positive potential induced on H is able to expand unchecked over the majority of the H surface as its limited electron density is pulled into the R−H bonding region (see the H sites in the water molecule in Figure 11.2, for instance).

11.2 Many Incarnations and Roles of a Single Phenomenon

The spatial extent of the sigma hole on X is limited, thus, to a neatly circumscribed area immediately opposite the R–X bonding region. And that severe restriction of the sigma hole on X (compared to the unfettered sprawl of a sigma hole on H) has remarkable consequences for the nature of sigma hole interactions to X (vs. hydrogen bonds), some of which we will highlight shortly.

11.2.2.2 Cases of Halogen Bonding

In the same way that hydrogen bonding is supported by electrostatic interactions between H and an electron-rich (nucleophilic) site on a base, Y (R-H—Y), it is possible for a polarized halogen atom center with a positive sigma hole, such as those discussed above (see Figures 11.4 and 11.5) to support analogous interactions of the form R-X—Y. Those weak interactions were classified under the general banner of halogen bonds (X-Bonds) in the mid-1970s [26–29], and we show in Figure 11.6 a local minimum energy structure that features such an interaction. The model case that we consider in Figure 11.6 employs the H–O–I Lewis acid analog of the H_2O case shown for hydrogen bonding in Figure 11.2.

The severe spatial restrictions imposed on the sigma hole of the halogen atomic centers by the negative torus (made clear in the opaque rendering in Figure 11.4) lead to one of the big practical distinctions between H-bonds and X-bonds. Like hydrogen bonds, halogen bonds are usually strongest when the R–X—Y bond angles are very close to 180° [25, 30, 31], where the lone pair of Y points directly toward the sigma

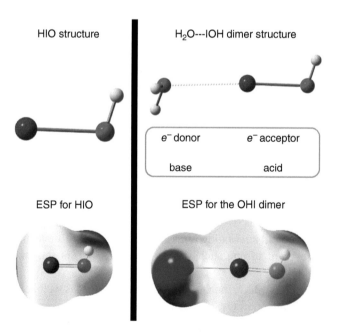

Figure 11.6 Structure of the optimized HIO molecule and an H_2O–IOH pair, with the corresponding electrostatic potential (ESP) maps immediately below them. The structures in the ESP maps are oriented in roughly the same way as the model structures above them.

hole (and the R−X σ* orbital) along the extension of the R−X bond. But X-bonding interactions tend to fall off quite rapidly as the R−X—Y bond angle decreases – more rapidly than they do for hydrogen bonds for which the polarized surface is not bound by any other electron on that atomic center (compare, e.g., H and I in F−H and F−I, respectively, in Figure 11.4).

Wilcken et al. have provided, for instance, some very telling plots (*cf.* Figures 7 and 8 in Ref. [31]) that quantify the increased stabilization of halogen bonding interactions going from X = Cl to I down group 17 and the significant orientational dependence of those halogen bonding interactions [31]. They point out, for example, that, for a Ph−X—O interaction between the PhX molecule and the carbonyl O lone pair of N-methylacetamide, "deviations of 25–30° [in the Ph-X—O bond angle from linearity, by rotating the Ph-X bond about the X center and keeping the position of X and O fixed] reduce the energy of complex formation to 50% of its maximum value." As the bond angle is reduced beyond about 140° – as the sigma hole on X begins to point decidedly away from the O atom and the electron belt on X is increasingly oriented toward the lone pair on O – the attractive (halogen bonding) interaction is lost [31].

While sigma holes can be induced on fluorine centers in molecules where R is exceptionally withdrawing (as in F_2 – see Figure 11.4 – where the positive potential is relatively small but present) [32, 33], the sigma hole potentials are usually weak (or negative) for F atoms, and any stabilizing sigma hole interaction that they manage to form will be flimsy. The significant cases of halogen bonding that have been observed experimentally in crystal structures, for example, involve typically the larger, softer (more polarizable), and less electronegative halogen atoms.

We mentioned that the terms 'sigma hole' [3] and 'halogen bonding' [26–29] are relatively new in the history of chemistry, but the interactions to which they refer are not late twentieth-century discoveries [34, 35]. Evidence of those types of interactions was apparent even in the mid-to-late-1800s [36–38]. And, by the time Odd Hassel was awarded the 1969 Nobel Prize for "contributions to the development of the concept of conformation and its application in chemistry," based in part on structural investigations of so-called *charge transfer complexes* [6, 7] (some of which we now know are supported by sigma hole interactions), the term "sigma hole" was still over three decades away from being linked to the phenomenon with which we associate it today [3].

Indeed, one of the categories of structures discussed by Hassel in a published version of his Nobel Lecture [7] is a three-dimensional assembly of molecular strings made up of weakly bound alternating units of 1,4 dioxan and X_2 molecules, and a three-monomer segment of one such string, where X = Cl, is shown in Figure 11.7.

In the system shown in Figure 11.7, the O—Cl separation (2.76 Å) is well below the sum of their van der Waals radii (1.52 Å (O) + 1.75 Å (Cl) = 3.27 Å), and – as expected from our contemporary understanding of halogen bonding – the O—Cl−Cl bond is close to 180° as the Cl−Cl bond orients itself to allow the sigma hole on each Cl center to align and interact as optimally as possible with one of the lone pairs on the adjacent O center. Unsurprisingly, perhaps, no fluorine analog – i.e. no extended

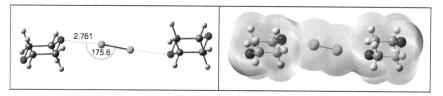

Figure 11.7 Structure of an optimized segment of a chain of alternating 1,4 dioxan and Cl_2 units, with the corresponding electrostatic potential (ESP) map beside it. The bond distance and angle are in angstrom and degree units, respectively.

solid made of an assembly of 1,4-dioxan and F_2 'polymer' chains – was reported. The O—F halogen bond is expected to be exceptionally weak in such a solid were it to be prepared. After all, as we show in Figure 11.5, the computed $V_{s,max}$ at F in the F_2 molecule is only ~54% of that obtained for Cl in Cl_2. And – based simply on the continuous increase in $V_{s,max}(X)$ as X gets larger – the O—X interactions are expected to strengthen going from X = Cl to X = I. In commenting on structural trends in the 1,4-dioxan + X_2 extended systems, Hassel pointed out that, "A comparison between the oxygen-halogen [O—X] separations and of the interhalogen [X—X] bond lengths in the three 1,4-dioxan adducts leads to the conclusion that the former increases rather slowly from chlorine to iodine, indicating a certain degree of compensation of the effect of larger halogen radius by the increase in [O—X] charge-transfer bond strength." And the X_2 bond distances do increase somewhat relative to the free X_2 units due evidently to charge transfer from O to the empty antibonding orbital of the X_2 molecule [6]. So the existence of these 1,4-dioxan + X_2 crystal structures may be accounted for in large part by a partnership of charge transfer and electrostatic contributions in stabilizing the observed O—X halogen bonds.

Figure 11.8 shows three model cases of a string of molecules stabilized by halogen bonding: chains of alternating NC—CN and XC≡CX units. They are simplified variations on chains identified in a crystal structure (co-crystals of diiodobutadiyne and linear alkyl nitrile chains) reported by Sun et al. [39]. The slow change in the N—X contacts as X gets larger echoes the observation of Hassel that we quoted above, and the structures show some more subtle features of the bonding as well: e.g. the C≡C triple bond in the XC≡CX molecular unit get slightly longer going from X = Cl to Br and I, implying some growing donation of electron density from p orbitals on X into C≡C anti-bonding orbitals as X gets larger. The reason for the very slight curvature in the chain for X = I is not clear.

In the past several decades, since the 1969 Nobel Prize celebration of charge transfer interactions, many other intriguing (confirmed and posited) instances of halogen bonding interactions have been reported. These sub-covalent R—X—Y types of interactions in which the halogen is the electrophilic site have been identified in contexts spanning materials science, crystal engineering, and biochemistry [24, 31, 40–42]. Halogen bonding is implicated, for instance, in the binding of thyroxine (the thyroid hormone also known as tetraiodo-thyronine or "T_4") to certain proteins [31]. And a role has been posited for halogen bonding as well in the initiation of chalcogen (sulfur or selenium)-mediated abstraction of an iodine atom from T_4 to generate the more active triiodo – (T_3) form [43, 44].

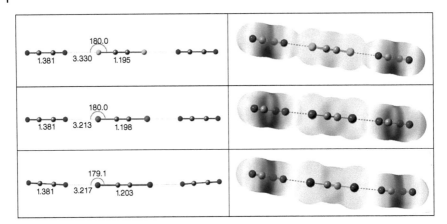

Figure 11.8 Structure of an optimized segment of a chain of alternating NC−CN and X_2 units (for X = Cl, Br, and I), with the corresponding electrostatic potential (ESP) maps beside them. The bond distances and angles are in angstrom and degree units, respectively.

11.2.2.3 The Sigma Hole and the Whole Story

Although various analyses of the energy contributions to the net interaction have attracted controversy, the existence of the halogen bonding *per se* is not in question, and the use of the term halogen bonding is as valid as the use of the more familiar hydrogen bonding. Yet, as others have pointed out repeatedly, both of those categories of interactions may also be described as sigma hole interactions or sigma hole bonding [45, 46]. Hydrogen bonding will probably continue to be treated as a separate category for both historical reasons and because an R−H−Y hydrogen bond is typically stronger than the analogous R−X−Y halogen bond. Yet, the use of the term sigma hole bonding offers the advantage of stemming any further proliferation (and we say more about this later on) of element- or group-based names for such interactions. "Lithium bonding" has already been used, for instance [47, 48], to describe the R−Li−Y type interaction, which is the same basic phenomenon as hydrogen and halogen bonding – with R polarizing Li and fostering thus a sigma hole type interaction with the base, Y. Since the sigma hole is a fundamental feature of the electron density distribution at atomic centers involved in the R−M−Y type interactions described so far in this work, the description of such M−Y contacts as sigma hole interaction is empirically grounded and has predictive power. What that name does not span explicitly, however, is the whole story of the bonding. Charge transfer, dispersion, and other contributions to bonding, beyond any electrostatic interactions between the sigma hole and the donor electrons, can play important roles in stabilizing the weak R−M−Y bonds that tend to be labeled nowadays simply as sigma hole interactions. A term that relieves the sigma hole, therefore, of the need to account fully for the overall bonding interaction but acknowledges its significance in stabilizing such complexes is '*sigma hole-supported*' interactions, and we draw on that language in this chapter.

11.2.3 Chalcogens

Like the group 17 compounds, it is traditionally expected that group 16 atomic centers in molecules will be nucleophilic rather than electrophilic. Yet bonding patterns to the contrary have been observed in which group 16 atomic centers exhibit significant electrophilic tendencies. Indeed, it has now been shown unequivocally that if they are sufficiently polarized, positive sigma holes can be induced on group 16 atomic sites in molecules – i.e. on the chalcogen center, opposite the bond to a polarizing substituent – and such sigma holes can lead to bonding interactions (so-called chalcogen bonds) [48, 49]. As is the case for polarized halogen atoms in molecules, sigma hole supported interactions (R—M—Y) involving chalcogen atoms tend to strengthen as the group 16 atom gets larger and as any charge transfer (from the base, Y, to the chalcogen center) becomes more substantial.

The presence of sigma holes on divalent chalcogen compounds is clear from the computed ESPs shown in Figure 11.9. In that Figure, we show the ESP maps for the chalcogen difluoride, MF_2, molecules, and we also include the perhaps more experimentally relevant perfluorodimethyl ether system and its heavier analogs plus examples of bound F_2M—NH_3 structures. These model cases are included here for illustrative purposes, but several other examples of chalcogen bonding have been identified or posited in the literature [48, 49].[5]

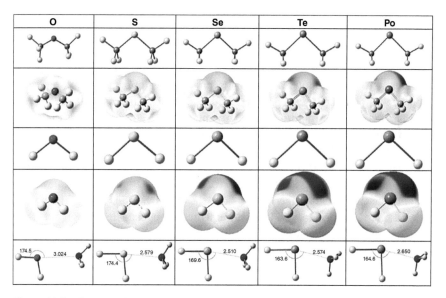

Figure 11.9 Structure of optimized MF_2 and $M(CF_3)_2$ molecules (for M = O, S, Se, Te, and Po) and corresponding electrostatic potential (ESP) maps. Structural parameters for the optimized F_2M—NH_3 complexes are included as well (with bond distances in angstrom units and bond angles in degrees).

---

5 See also the suggestion in Ref. [43] that chalcogen bonding may play a role in mediating thyroid hormone function in the body.

The ESP maps shown in Figure 11.9 can all be compared with each other since, as we mentioned earlier, all of the ESP maps shown in this chapter have been plotted on the same iso-density surface and employ the same ESP color scale. The sigma holes of interest for us in this case (partially visible in the MF_2 and $M(CF_3)_2$ ESP maps in Figure 11.9 as the bluest regions on each M surface opposite the R—M bonds) become more prominent (more intensely blue) as the M atom gets larger. This is a visual indication of the growing electrophilicity of the sigma hole, which eventually expands into a band across the top of the ESP maps as it intensifies going from left to right (from O to Po) in Figure 11.9.

Of note, this increasing electrophilicity of the sigma hole is reflected in the decrease in the M—N distance in the optimized F_2M—NH_3 complexes, even as the atoms get larger going from M = O to M = Se. After M = Se, that contraction is successfully countered by the ever-increasing atomic radius of M going down the group, but the influence of the increasingly positive sigma hole shows up still in the very slow increase in the M—N contacts going from Se to Po in Figure 11.9. There are two lone pairs on each of the group 16 M centers, but they appear to be substantially contracted in the MF_2 molecules, as is probably expected: in line with Bent's rule [50], the lone pair hybrid orbitals on M in MF_2 sacrifice p character to the M orbitals involved in the M—F bonds since F is very electronegative, so the former are expected to be somewhat richer in s character and more contracted. The lone pairs are less prominent, therefore, on the 0.001 au surface than they are in H_2O, for example (Figure 11.2), but they are there, subdued.

11.2.4 Pnictogens

The presence of sigma holes on group 15 (pnictogen) atomic centers in molecules opposite very polar bonds, such as N—F and P—F bonds, has been confirmed as well [51]. The pyramidal structure of the trivalent group 15 species makes it relatively easy to identify sigma holes at the top of the ESP maps in Figure 11.10. In those maps, the sigma holes on M are opposite the M—F bonds, around an extension of that bond through the M center.

As M gets larger, going left to right across Figure 11.10, the sigma holes merge to dominate the whole surface of M above the MF_3 pyramid, even if the positive extrema on the surface are still located immediately opposite the M—F bond.

To confirm the location and assess the influence of positive extrema on the molecular surfaces, we consider here a series of simple sigma hole supported pairs involving each MF_3 molecule and ammonia, the latter being our base of choice for most of the model complexes that we will consider throughout this chapter.

An undergraduate chemistry student might anticipate perhaps that NH_3 and NF_3 will form a hydrogen bond involving one of the H centers of the ammonia molecule and a lone pair on the N or one of the F centers of NF_3. Yet we find that a local minimum energy sigma hole supported complex shown on the bottom left in Figure 11.10 is possible as well – a minimum energy (pnictogen bonding) structure in which the lone pair of NH_3 engages in a bonding interaction with the N center in NF_3.

Remarkably, this interaction (if we can make an inference based solely on the geometrical evidence without an energy decomposition analysis) transitions from

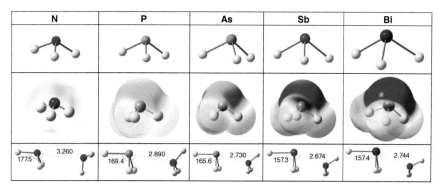

Figure 11.10 Structure of optimized MF$_3$ molecules (for M = N, P, As, Sb, and Bi) and corresponding electrostatic potential (ESP) maps. Structural parameters for the optimized F$_3$M—NH$_3$ complexes are included as well (with bond distances in angstrom units and bond angles in degrees).

a weak electrostatic interaction to increasing charge transfer as the F$_3$M—NH$_3$ distance contracts and gets progressively shorter all the way from M = N to M = Sb even as the atomic radius is increasing! Only as we go from Sb to Bi does the continued increase in the atomic radius of M manage apparently to outstrip the influence of the M—N attraction enough to enforce an increase in the minimum M—N distance, and then only barely (see Figure 11.10).

The progression in the F—M—N bond angles shown in Figure 11.10 tells an intriguing story. It is pointed out regularly in discussions of halogen bonding, for example, that the interactions are usually very directional and the R—M—Y bond angles are close to 180°. That is what we have reported here, too, but for sigma hole supported interactions to central atoms – such as groups 15 and 16 atoms – with two or more pre-existing covalently bonded substituents (such as the F atoms in their MF$_3$ and MF$_2$ molecules), substantial deviations from a linear R—M—Y arrangement may be observed. That is partly because the base, Y, can interact strongly sometimes with other parts of the R–M molecule as well. In the case of the pnictogen fluorides, the structural evidence suggests that the direct F–M—NH$_3$ interaction is disrupted to open up the possibility of side-on F—H hydrogen bonding between two hydrogen atoms on ammonia and the two F sites of the MF$_3$ molecule that are not associated with the compromised (F—M—N) pnictogen bond. And other secondary interactions of that sort may also lead to noticeable deviations from what would otherwise be linear sigma-hole-type interactions to main group central atoms. As we will see in the next section, the match in symmetry between the ammonia molecule and the triangular face of the tetrahedral MF$_4$ molecules of group 14 atoms obviates the need for any significant deviations from linearity for secondary interactions of the types just mentioned to take place.

11.2.5 Tetrels

The S$_N$2 reaction in carbon chemistry and specifically the formation of the five-coordinate S$_N$2 transition state structure (Figure 11.11) have been discussed

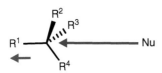

Figure 11.11 Model of the nucleophilic attack in S_N2 reactions at tetravalent centers.

in the research literature and organic chemistry textbooks for decades [52–54].[6] ESP maps of the surfaces of C centers with electron-withdrawing substituents reveals the presence of a sigma hole precisely where the incoming nucleophilic substituent (Nu) approaches the tetravalent C center, opposite the "R^1–C" bond in S_N2 reactions where R^1 is the leaving group (Figure 11.12). Of particular relevance to this discussion, it has been shown that the ESP at the C center (coincident as it is with the R^1–C σ^* orbital) plays an apparently decisive role in bimolecular nucleophilic substitution reactions [55–57].

Indeed, computational evidence suggests that covalent bond formation (and leaving group elimination) through S_N2 type reactions (at tetracoordinate C as well as Si and Ge centers) may be preceded in some cases by a relatively weak five-coordinate minimum energy arrangement – a tetrel (sigma hole) bond [58]. The nature of the bonding and thus the separation between the central atom and the base (nucleophile, Y) in the minimum energy five-coordinate, R_4M—Y, species can vary substantially, however, depending on the identity of the base as well as the four substituents on M [58–60]. The identity of the R groups is decisive for both the energy of the lowest unoccupied molecular orbital (LUMO) on MR_4 – for charge

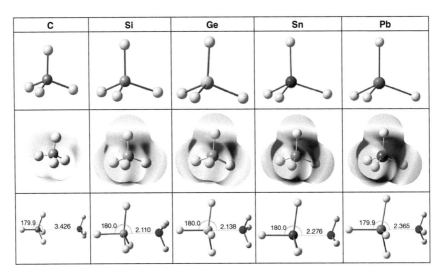

Figure 11.12 Structure of optimized MF_4 molecules (for M = C, Si, Ge, Sn, and Pb) and corresponding electrostatic potential (ESP) maps. Structural parameters for the optimized F_4M–NH_3 complexes are included as well (with bond distances in angstrom units and bond angles in degrees).

---

6 These references trace the origins of the ideas back to Walden and subsequent contributions from Hughes, Ingold, Bateman, and others.

transfer from the base – as well as the magnitude of the potentials induced at the sigma holes on the M center. And (along with the identity of Y itself) the identity of the substituents (for M = Si or Ge, for instance) can be decisive for whether a double minimum (i.e. both a noncovalent R$_4$M—Y and a dative R$_4$M·Y complex with a much shorter M—Y bond) exist on the potential energy surface.

As an aside, a substantial sensitivity to the identity of R and Y has even been noted for rates of S$_N$2 reactions. Brauman, Olmstead, and Lieder identified, for instance, apparent reversals in the relative leaving group ability of Cl$^-$ versus Br$^-$ depending on the identity of the nucleophile and concluded that "it is impossible to establish any general scale on the order of either nucleophilicity or leaving group ability without direct reference to the specific reaction [54]."

For M = C, computationally identified pentavalent instances of tetrel type bonding tend to feature long and weak C—Base contacts, but MR$_4$·(NH$_3$)$_n$ type species (for M = Si and Ge, R = F and n = 1 at least) have been prepared (possibly unintentionally, early on) [61] and examined in the chemical literature for years [62–68]. Those investigations and others using alkylamines as bases precede the sigma hole concept, and the M—N separations in MR$_4$·Y compounds where M is below C in group 14 are usually much shorter and stronger than the weak pairs formed when M = C – hence the use of the dot (MF$_4$·Y) notation, which is normally reserved for dative bonding. Nonetheless, the location of the coordinating NH$_3$ in MF$_4$·NH$_3$ adducts, for instance, coincides precisely with the location of the sigma hole opposite one of the F—M bonds. The molecule has a locally trigonal bipyramidal structure, with the Si ← NH$_3$ bond immediately opposite one of the Si—F bonds while the other three Si—F bonds move apart into a trigonal planar arrangement in the middle of the molecule.

As in other groups of the periodic table, the sigma hole on M tends to get stronger as M gets larger, more polarizable, and less electronegative (Figure 11.12). By Pauling's and Mulliken's (valence state) electronegativity schemes, the electronegativity does not decrease uniformly going down groups 13 or 14. But that irregularity does not seem to lead to any detour from the expected trend.

The sigma hole strengthens continuously going from carbon to lead in group 14, and this is reflected in the rapid contraction of the M—N distance going from C to Si as we mentioned above and the remarkably slow increase going from Si to Pb despite the substantial increase in atomic radius of the M atom. As our research group and others have shown, the interaction moves closer to simple F$_4$M ← NH$_3$ dative bonding at and below period 3 (M = Si to Pb), but there is strong evidence that the bonding is supported significantly by the presence of the sigma hole on M [59].

Interestingly, experimental data on the higher SiF$_4$·(NH$_3$)$_2$ adduct show that the two ammonia molecules are opposite each other (*trans*) in a locally octahedral arrangement around the Si center [69]. This arrangement does not imply, however, that the influence of the sigma hole is negligible. The presence of two sigma-hole maxima on M above and below the plane of the square planar SiF$_4$ fragment of the octahedral hypervalent SiF$_4$·(NH$_3$)$_2$ complex favors this positioning of the NH$_3$ units above and below (i.e. perpendicular to) the SiF$_4$ plane. And a surprisingly similar pattern is observed, as we will show presently, in the hypercoordinate group 13 species.

11.2.6 Triels

In the case of the optimized trigonal planar group 13 MF$_3$ molecules, the maximum of the positive potential on the isodensity surface is not opposite the R—M bonds but perpendicular to them, above and below the plane of the molecule (see Figure 11.13).

That situation complicates things a bit since – probably due to the dominance of halogen bonding in the broader discussion of sigma hole supported bonding – sigma holes are usually expected to appear opposite to the polar M—R bond along the bond axis, not perpendicular to it. But, as we saw for hydrogen bonding, for example, a large area of positive potentials can arise on an atomic surface due to a substantial shift of electron density from an atomic center (M in the case of the group 13 MR$_3$ molecules) toward the M—R bonding region. Instead of having only one polar bond, as is the case for H in H—F, however, the polarized M center in the group 13 MF$_3$ molecule has three coplanar polar M—F bonds. Each of those bonds strip away electron density from M in three directions, and – with no lone pair on M to mask the impact of that depletion – it turns out that the most exposed parts of that polarized M center in the planar MF$_3$ molecule are the regions above and below M in the center of the molecule. So those two regions of the iso-density surface are almost expected to have relatively high positive surface potentials. In that regard, therefore, it is not surprising that the potential maxima arise where they do on trivalent group 13 atomic

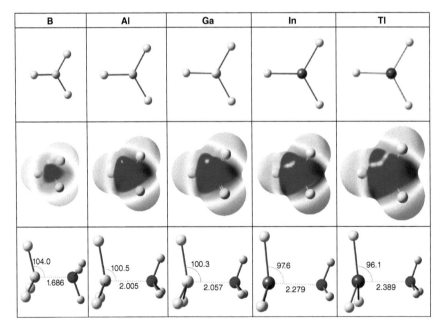

Figure 11.13 Structure of optimized MF$_3$ molecules (for M = B, Al, Ga, In, and Tl) and corresponding electrostatic potential (ESP) maps. Structural parameters for the optimized F$_3$M—NH$_3$ complexes are included as well (with bond distances in angstrom units and bond angles in degrees).

centers in the MF$_3$ molecules. And that is so even if the electrostatic potentials on M immediately opposite the M—F bonds, where sigma holes are traditionally expected to arise, are also positive.

Rather than being an exception relative to what we see in the group 17–14 systems, therefore, the group 13 case may be seen as the first instance (going right to left in the main group) where that exposed extrema of positive potential does not coincide with the extension of the M—R bond axis. Yet, these positive extrema are also the results of the electron-withdrawing effects of R in the M—R sigma bonds of the MR$_3$ molecules. With the electron density around M localized primarily in the three coplanar M—R bonds, the center of the molecule is left to betray the fullest impact of a substantial polarization of the hypovalent M center (Figure 11.13).

The two positive extrema on the surface of the planar trivalent group 13 MR$_3$ molecule happen to coincide with the empty p orbital of the group 13 M atom in the molecule. That orbital is perpendicular to the plane of the group 13 MR$_3$ molecule and allows such molecules to be viable Lewis acids for the formation of dative adducts for which BH$_3$·NH$_3$ is the classic example. And, although adducts such as BH$_3$·NR$_3$ or BF$_3$·NR$_3$ [69–71] are expected to be stabilized primarily by charge transfer, the formation of weaker triel bonds in R$_3$M—Y addition pairs is favored as well by the presence of the positive potentials on M in the MR$_3$ molecules [72]. Double-well potentials have been identified for certain complexes (such as BCl$_3$—NCH) where two minimum energy structures are located for the same acid–base pair: one structure with a short and strong bond and another with a weaker noncovalent (classical sigma-hole-type) bond (with computed B—N distances of 1.628 and 2.817 Å, respectively) [73].

Grabowski has proposed the term π-hole (pi-hole) for these types of interactions where the region of positive potential on the surface coincides with a p orbital or the π-system of a molecule rather than following the classical pattern for sigma holes along the extension of the polar M—R bond axis [74]. To be clear, however, the areas of positive potential on the surfaces of these group 13 MR$_3$ molecules, arise, like all of the sigma holes considered in previous sections, from the polarization of M by sigma bonds. That the exposed part of the polarized M surface coincides with an empty p orbital does not confer on it any particular π characteristic. Interactions between these triel molecules and a base directed toward the p-orbital coincident with these regions of positive potential can be quite strong [72] and the minimum energy R$_3$M—Y separations are short compared to analogous sigma hole interactions (with the same base) to groups 14 – 17 centers, where the sigma hole does not coincide with a low-energy empty valence orbital. Compare, for example, F$_3$B·NH$_3$ and F$_4$C—NH$_3$ in Figures 11.12 and 11.13! We were unable to find a double-well potential for either of these two pairs, so the structures shown in those figures seem to be their optimal triel and tetrel bonding arrangements.

For the triel systems, as for other pairs that we have discussed so far, it is important to analyze energetic and geometrical properties to acknowledge the role of secondary interactions beyond the primary M—Base interactions in the overall bonding. We pointed out that those influences are consequential for the structures of chalcogen- and pnictogen-bonded complexes of NH$_3$. In the case of tetrel- and triel-bonding, however, the triangular faces of the tetrahedral MF_4 structure and the trigonal planar MF_3 structure, respectively, each share a threefold axis with NH_3. So, it is possible for strong secondary repulsive or attractive interactions (e.g. between the F sites on MF_4 or MF_3 and H sites on NH_3) in those symmetrical F_4M—NH_3 and F_3M—NH_3 complexes to occur without disrupting the direct alignment of the M σ hole and the N lone pair (Figures 11.12 and 11.13). In the MF_3 triel complexes themselves, other secondary effects, such as intra-molecular changes in any π-delocalization of electron density from the F lone pairs into the empty orbital of M in MF_3 itself as the F_3M—NH_3 complex is formed [75], will also not show up in the symmetry of the F_3M—NH_3 complex. But those influences may impact the energetics of the complexes in drastic ways [76]. So, the fact that $V_{s,max}$ at B in the isolated BF_3 molecule is greater than $V_{s,max}$ at B in the isolated BH_3 molecule, for instance, does not guarantee (considering the various possible secondary interactions) that the binding energies of their complexes with a given base (e.g. F_3B—NH_3 and H_3B—NH_3) will vary accordingly [73]. Strong positive potentials can promote dative bonding and dominate noncovalent interactions, but, even in the latter case, sigma holes are rarely the solitary drivers of the structure or energetics of the bonding. Secondary effects can have substantial consequences.

Another case where the exposed region of positive potential on the isodensity surface is not coincident with the M—R bond axis is that of the linear group 2 dihalides, as we will see presently. In that case, the two M—R bonds are opposite each other. So, no sigma hole is available on M opposite the M—R bonds. Instead, a belt of electron depletion arises around the M center due to the direct polarization of electron density to the left and to the right away from the M center and toward the two M—R bonding regions.

11.3 Related Interactions Elsewhere in the Main Group

11.3.1 Group 2

The nature of acid–base interactions with group 2 atomic centers has received some attention in the literature as well. Since several of the group 2 dihalides are linear (though some of them are known to be bent) [77, 78], any region of induced positive potential below one M—R bond is hidden by the other M—R bond. If the polarization of the M center is extensive, however, the evidence of that polarization on the isodensity surface presents itself as a belt of positive potential around the M center, as shown in Figure 11.14.

The M centers in divalent group 2 molecules have two empty p orbitals in the plane perpendicular to the principal axis and together have the same symmetry (a torus around the M center) as the sigma belt visible in Figure 11.14. So, here, as in the case of triel bonding, the interactions of bases with the central atom can be quite strong and tend toward dative bonding with substantial charge transfer contributions. The presence of this belt of positive charge is an intriguing feature of group 2 molecules with polar bonds to electron-withdrawing R groups and offers insights for tuning

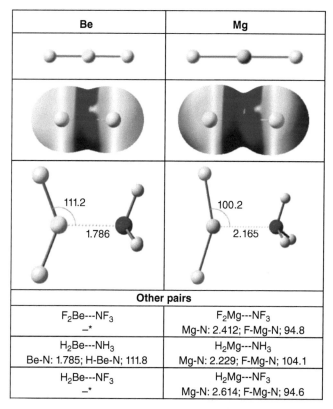

Figure 11.14 Structure of optimized MF$_2$ molecules (for M = Be and Mg) and corresponding electrostatic potential (ESP) maps. Structural parameters for the optimized F$_2$M—NH$_3$ complexes are included. For comparison, the same parameters for other R$_2$M—NR$_3$ complexes (for R = H or F) are provided as well (with bond distances in angstrom units, and bond angles in degrees). *No optimized Be—N structure was obtained for M = Be with the NF$_3$ base.

the structure and reactivity of those compounds. We include in Figure 11.14 key structural parameters for the optimized F$_2$M—NH$_3$ complexes for M = Be and Mg. For comparison, we also include the analogous parameters for other pairs formed by MR$_2$ and NR$_3$ molecules for R = H and F. The minimum energy structures feature rather short M ← N contacts for most of those combinations, except for the cases that combine the least polarizable group 2 atom (M = Be) and the weak base (NF$_3$) – see Figure 11.14.

The history of hypovalent group 2 compounds in organic and other areas of chemistry is rich, with Grignard reagents (R–Mg–X where R is an organic substituent) being among the best known. For those reagents, the formation of dative bonds, supported evidently by the presence of the positive potential on M, plays a significant role in the reaction mechanism [79, 80].

Given that coordination complexes to hypovalent group 2 compounds are known to feature short dative M-Base bonds close to or well below 2.0 Å for M = Be and

Mg, most with clear evidence of charge transfer (such as M—R bond elongation and a deviation of MR$_2$ from linearity [see Figure 11.14]), is it meaningful to classify any group 2 M-Base interaction as "alkali earth bonds" in the sense of halogen or even tetrel bonds? A case has been made in "Beryllium Bonds, Do They Exist?" [81] and elsewhere that there are some features analogous to traditional sigma hole bonds (such as the involvement of the antibonding R–M σ* orbitals of the bent MR$_2$ fragment in the R$_2$M-Base complex) that are worth acknowledging [81–83].

11.3.2 Group 1

We have already met in this chapter an instance of sigma hole-type interactions with group 1 atomic centers – the so-called lithium bond [47]. Similar polarization is observed for the heavier group 1 fluorides, and similar interactions to those of halogen bonding may be expected since, like hydrogen and the halides, the group 1 metals are typically terminal or bridging atoms in molecules. As for the alkali earth metals, less work has been done specifically on their involvement of group 1 atoms in sigma hole supported bonding, but both lithium (dating back to 1970 at least) [47, 84–87] and sodium bonding [83, 88] have been considered and named as such in the literature.

Dimers of the group 1 halides are known to form anti-parallel rhomboidal global minimum energy structures that are generally interpreted as an optimization of dipole–dipole interactions as well as the linear X–M—X-M alternative as a local minimum [84, 87, 89]. That local minimum is a favorable dipole–dipole arrangement as well, but it is also an example of what might be called an "alkali bond," and the X–M—NH$_3$ examples shown in Figure 11.15 qualify in that regard as well.

Along with the more extensively studied sigma hole interactions involving group 14–17 sites as acceptors, work continues in the exploration, improved classification, and exploitation of the noncovalent interactions that may be possible with groups 1, 2, and 13 atoms [83, 90].

11.3.3 Group 18

The elements of group 18 carry the largely well-deserved reputation of being unreactive, but compounds of the heavier atoms in the group have been known for decades [91–93]. Weak sigma-hole-type interactions with group 18 atomic centers dubbed aerogen bonds have been identified computationally and in published crystal structures [94, 95]. The sigma holes induced on the Kr and Xe centers in one of those molecules – MF$_2$O – are shown in Figure 11.16. In those systems, the expansive sigma hole is clearly visible opposite the O=M bonds.

N is a slightly larger atom than F, but the exceptionally long M—N bond distances (2.669 and 2.857 Å) compared, for example, to average M—F bonds in the molecules (1.882 and 1.986 Å, respectively) are consistent with expectations for noncovalent sigma hole supported M—N bonding. Such noncovalent interactions can be relatively strong, however (comparable to halogen and hydrogen bonds), and with the

11.3 Related Interactions Elsewhere in the Main Group | 307

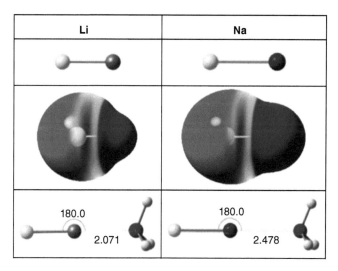

Figure 11.15 Structure of optimized MF molecules (for M = Li and Na) and corresponding electrostatic potential (ESP) maps. Structural parameters for the optimized FM—NH$_3$ complexes are included as well (with bond distances in angstrom units, and bond angles in degrees).

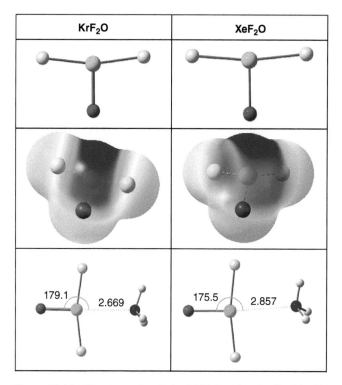

Figure 11.16 Structure of optimized MF$_2$O molecules (for M = Kr and Xe) and corresponding electrostatic potential (ESP) maps. Structural parameters for the optimized F$_2$OM—NH$_3$ complexes are included as well (with bond distances in angstrom units, and bond angles in degrees).

sigma hole extending like a bridge over the M center, it is apparently possible for a base to interact with the M center opposite the O=M bonds and perpendicular to it, comparable (in the latter case) to the so-called π hole bond [96]. Although many group 18 compounds are reactive and explosive, the chemistry of these compounds is rather well developed and continues to attract experimental and computational interest [97], and interest in group 18 noncovalent interactions is expected to grow.

11.4 Contested Interpretations

In the long history of hydrogen bonding, the question of the origins of that interaction, with electrostatics and charge transfer as primary claimants, has attracted a substantial degree of attention [12]. And the same continues to be true for halogen bonding and other sigma-hole-type interactions highlighted in this discussion. The latter exchange has been particularly lively, with theoretically grounded analyses claiming the primacy of electrostatic [98–101] or the significance of charge transfer [102–104] contributions (supported in either case by dispersion and other stabilizing contributions and offset partly by Pauli repulsion). One metaphorical red card has even been issued in the process [101], but the evidence is increasingly clear that the predominant contribution depends to a great extent on the specific identity of both the base and the acid involved in the interaction [105]. Changes in the substituents on either the acid or the base can substantially impact the geometry and energetics of sigma hole supported complexes. As a simple case in point, the I—N separation in the $F_3Pb-I-N(CH_3)_3$ halogen bond was found computationally to be 2.520 Å, but it is noticeably longer (3.328 Å) in $F_3Pb-I-NF_3$ [59, 106], and similar variations in the nature of noncovalent interactions when the atomic centers involved are unchanged but their substituents or the chemical environments are different have been noted repeatedly.

11.5 Conclusions

There is sometimes a lot in a name. What we call things can narrow our view to see in them only what the names convey, even when there is much more there. Yet the names that we give to chemical phenomena can be useful in affirming too the dominant contributions of certain influences on chemical bonding and even in drawing deserved attention to important but previously unrecognized or unacknowledged features that offer deeper insight and opportunities for predictions. The sigma hole concept [3, 8, 9] has borne much fruit over the past few decades and builds on many more decades of work on the consequences of inductive effects and electronegativity differences for chemical structure and reactivity. A survey of the sigma hole phenomenon across the main group has been provided here to support new entrants to the field in developing their own understanding of the scope of interactions supported by sigma holes and the range of incarnations of induced positive potentials on the surfaces of main group compounds. Notwithstanding the associated proliferation of column-based names (which now extends, unfortunately, even to the d-block) [10] to create sub-categories for sigma hole supported interactions and the introduction of additional terms such as π-holes as distinguished from σ-holes, the sigma hole concept has fed new energy into the study of noncovalent interactions. It has motivated renewed interest in hydrogen bonding, cooperativity of weak interactions, the intentional incorporation of weak interactions such as halogen bonds in crystal engineering and materials science, and critical (re)assessments of structural data for biomolecules that have led to the identification of roles for sigma-hole-type interactions in biochemistry.

We employ the term "sigma hole supported" interactions here to pull all of these interactions under one umbrella. Depending on the identity of the Lewis acid and Lewis base fragments involved in the interaction (including the identity of the substituents on the actual atomic centers involved in the interaction), the evidence suggests that the roles of electrostatics, charge transfer, and other contributions to the bonding can vary greatly. There have been some controversies around the question of which interactions dominate in particular contexts, but the growing understanding of these interactions will continue, we expect, to energize efforts to apply or capitalize on them in new, interesting, and practical ways. Exciting developments are ahead.

Acknowledgment

Work reported in this chapter was supported by the National Science Foundation NSF-RUI Award (CHE-2055119) and NSF-MRI Grants CHE-0958696 (University of Richmond) and CHE-1662030 (the MERCURY consortium). K.J.D. acknowledges the support of the Henry Dreyfus Teacher-Scholar Awards Program (TH-16-015). The support of the University of Richmond is also gratefully acknowledged.

References

1. Ogburn, W.F. and Thomas, D. (1922). Are inventions inevitable? A note on social evolution. *Poli. Sci. Quart.* 37: 83–98.
2. Tambolo, L. and Cevolani, G. (2021). Multiple discoveries, inevitability, and scientific realism. *Stud. Hist. Philos. Sci. A* 90: 30–38.
3. Clark, T., Hennemann, M., Murray, J.S., and Politzer, P. (2007). Halogen bonding: the σ-hole. *J. Mol. Model.* 13 (2): 291–296.
4. Brinck, T., Murray, J.S., and Politzer, P. (1992). Surface electrostatic potentials of halogenated methanes as indicators of directional intermolecular interactions. *Int. J. Quant. Chem.* 44: 57–64.
5. Bent, H.A. (1968). Structural chemistry of donor-acceptor interactions. *Chem. Rev.* 68: 587–648.

6 Hassel, O. (1970). Nobel Lecture, June 9, 1970 Structural aspects of interatomic charge-transfer bonding. *NobelPrize.org*. https://www.nobelprize.org/prizes/chemistry/1969/hassel/lecture (accessed 04 December 2023).

7 Hassel, O. (1970). Structural aspects of interatomic charge-transfer bonding. *Science* 170: 497–502.

8 Murray, J.S., Lane, P., and Politzer, P. (2009). Expansion of the sigma-hole concept. *J. Mol. Model.* 15: 723–729.

9 Clark, T. (2013). σ-Holes. *Wiley Interdiscip. Rev. Comput. Mol. Sci.* 3: 13–20.

10 Alkorta, I., Elguero, J., and Frontera, A. (2020). Not only hydrogen bonds: other noncovalent interactions. *Crystals* 10: 180(1-29).

11 Clark, T. (2023). How deeply should we analyze non-covalent interactions? *J. Mol. Model.* 29 (3): 66.

12 van der Lubbe, S.C.C. and Fonseca Guerra, C. (2019). The nature of hydrogen bonds: a delineation of the role of different energy components on hydrogen bond strengths and lengths. *Chem. Asian J.* 14: 2760–2769.

13 Shaik, S., Danovich, D., and Zare, R.N. (2023). Valence bond theory allows a generalized description of hydrogen bonding. *J. Am. Chem. Soc.* 145: 20132–20140.

14 Frisch, M.J., Trucks, G.W., Schlegel, H.B. et al. (2016). *Gaussian 16, Rev. B.01*. Wallingford, CT: Gaussian, Inc.

15 Chai, J.-D. and Head-Gordon, M. (2008). Long-range corrected hybrid density functionals with damped atom-atom dispersion corrections. *Phys. Chem. Chem. Phys.* 10: 6615–6620.

16 Kendall, R.A., Dunning, T.H. Jr., and Harrison, R.J. (1992). Electron affinities of the first-row atoms revisited. Systematic basis sets and wave functions. *J. Chem. Phys.* 96: 6796–6806.

17 Energy-consistent Pseudopotentials of the Stuttgart/Cologne Group. http://www.tc.uni-koeln.de/PP/clickpse.en.html (accessed 08 December 2023).

18 Chemcraft. Graphical software for visualization of quantum chemistry computations. www.chemcraftprog.com (accessed 08 December 2023).

19 Dennington, R., Keith, T.A., and Millam, J.M. (2016). *GaussView 6.0*. Shawnee Mission, KS: Semichem Inc.

20 Lu, T. and Chen, F. (2012). Multiwfn: a multifunctional wavefunction analyzer. *J. Comput. Chem.* 33: 580–592.

21 Lu, T. and Chen, F. (2012). Quantitative analysis of molecular surface based on improved Marching Tetrahedra algorithm. *J. Mol. Graphics Modell.* 38: 314–323.

22 Petruševski, V.M. and Cvetković, J. (2018). On the 'true position' of hydrogen in the periodic table. *Found. Chem.* 20 (3): 251–260.

23 Murray, J.S. and Politzer, P. (2011). The electrostatic potential: an overview. *Wiley Interdiscip. Rev. Comput. Mol. Sci.* 1: 153–163.

24 Politzer, P. and Murray, J.S. (2013). Halogen bonding: an interim discussion. *ChemPhysChem* 14 (2): 278–294.

25 Donald, K.J., Wittmaack, B.K., and Crigger, C. (2010). Tuning σ-holes: charge redistribution in the heavy (group 14) analogues of simple and mixed

halomethanes can impose strong propensities for halogen bonding. *J. Phys. Chem. A* 114: 7213.

26 Martire, D.E., Sheridan, J.P., King, J.W., and O'Donnell, S.E. (1976). Thermodynamics of molecular association. 9. An NMR study of hydrogen bonding of chloroform and bromoform to di-n-octyl ether, di-n-octyl thioether, and di-n-octylmethylamine. *J. Am. Chem. Soc.* 98 (11): 3101–3106.

27 Dumas, J.M., Geron, C., Peurichard, H., and Gomel, M. (1976). MX_4-organic base interactions (M = C, Si; X = Cl, Br). Study of the influence of the central element and the halogen. *Bull. Soc. Chim. Fr.* (5–6): 720–728.

28 Dumas, J.-M., Kern, M., and Janier-Dubry, J.L. (1976). Cryoscopic and calorimetric study of MX_4-polar organic base interactions (M = C, SI, X = Cl, Br)—influence of element and of halogen. *Bull. Soc. Chim. Fr.* 11 (1): 1785–1790.

29 Dumas, J.-M., Peurichard, H., and Gomel, M. (1978). CX_4...base interactions as models of weak charge-transfer interactions: comparison with strong charge-transfer and hydrogen-bond interactions. *J. Chem. Res.* 2: 54–55.

30 Parker, A.J., Stewart, J., Donald, K.J., and Parish, C.A. (2012). Halogen bonding in DNA base pairs. *J. Am. Chem. Soc.* 134: 5165–5172.

31 Wilcken, R., Zimmermann, M.O., Lange, A. et al. (2013). Principles and applications of halogen bonding in medicinal chemistry and chemical biology. *J. Med. Chem.* 56: 1363–1388.

32 Politzer, P., Murray, J.S., and Concha, M.C. (2007). Halogen bonding and the design of new materials: organic bromides, chlorides and perhaps even fluorides as donors. *J. Mol. Model.* 13: 643–650.

33 Wang, Y.-H., Lu, Y.-X., Zou, J.-W., and Yu, Q.-S. (2008). Use of ab initio calculations to provide insights into the strength and nature of interfluorine interactions. *Int. J. Quantum Chem.* 108: 1083–1089.

34 Mulliken, R.S. (1950). Structures of complexes formed by halogen molecules with aromatic and with oxygenated solvents. *J. Am. Chem. Soc.* 72: 600–608.

35 Hildebrand, J.H. and Glascock, B.L. (1909). The color of iodine solutions. *J. Am. Chem. Soc.* 31: 26–31.

36 Guthrie, F. (1863). On the iodide of iodammonium. *J. Chem. Soc.* 16: 239–244.

37 Chattaway, F.D. (1896). The constitution of the so-called "nitrogen iodide". *J. chem. Soc. Trans.* 69: 1572–1583.

38 Remsen, I. and Norris, J.F. (1896). The action of halogens on the methylamines. *Am. Chem. J.* 18: 90–95.

39 Sun, A., Lauher, J.W., and Goroff, N.S. (2006). Preparation of poly(diiododiacetylene), an ordered conjugated polymer of carbon and iodine. *Science* 312: 1030–1034.

40 Auffinger, P., Hays, F.A., Westhof, E., and Ho, P.S. (2004). Halogen bonds in biological molecules. *Proc. Nat. Acad. Sci.* 101: 16789–16794.

41 Metrangolo, P., Meyer, F., Pilati, T. et al. (2008). Halogen bonding in supramolecular chemistry. *Angew. Chem. Int. Ed.* 47 (33): 6114–6127.

42 Scholfield, M.R., Vander Zanden, C.M., Carter, M., and Ho, P.S. (2013). Halogen bonding (X-bonding): a biological perspective. *Protein Sci.* 22: 139–152.

43 Manna, D. and Mugesh, G. (2012). Regioselective deiodination of thyroxine by iodothyronine deiodinase mimics: an unusual mechanistic pathway involving cooperative chalcogen and halogen bonding. *J. Am. Chem. Soc.* 134: 4269–4279.

44 Bayse, C.A. and Rafferty, E.R. (2010). Is halogen bonding the basis for iodothyronine deiodinase activity? *Inorg. Chem.* 49: 5365–5367.

45 Wolters, L.P. and Bickelhaupt, F.M. (2012). Halogen bonding versus hydrogen bonding: a molecular orbital perspective. *ChemistryOpen* 1: 96–105.

46 Politzer, P., Murray, J.S., and Clark, T. (2015). σ-Hole bonding: a physical interpretation. *Top. Curr. Chem.* 358: 19–42.

47 Shigorin, D.N. (1959). Infra-red absorption spectra study of H-bonding and of metal-element bonding. *Spectrochim. Acta* 14: 198–212.

48 Wang, W., Ji, B., and Zhang, Y. (2009). Chalcogen bond: a sister noncovalent bond to halogen bond. *J. Phys. Chem. A* 113: 8132–8135.

49 Scilabra, P., Terraneo, G., and Resnati, G. (2019). The chalcogen bond in crystalline solids: a world parallel to halogen bond. *Acc. Chem. Res.* 52 (5): 1313–1324.

50 Bent, H.A. (1961). An appraisal of valence-bond structures and hybridization in compounds of the first-row elements. *Chem. Rev.* 61: 275–311.

51 Zahn, S., Frank, R., Hey-Hawkins, E., and Kirchner, B. (2011). Pnicogen bonds: a new molecular linker? *Chem. Euro. J.* 17 (22): 6034–6038.

52 Deasy, C.L. (1945). The Walden inversion in nucleophilic aliphatic substitution reactions. *J. Chem. Ed.* 22: 82–83.

53 Ingold, C.K. (1969). *Structure and Mechanism in Organic Chemistry*, 2e. Ithaca, NY: Cornell University Press. (see Chapter 7).

54 Brauman, J.I., Olmstead, W.N., and Lieder, C.A. (1974). Gas-phase nucleophilic displacement reactions. *J. Am. Chem. Soc.* 96: 4030–4031.

55 Galabov, B., Nikolova, V., Wilke, J.J. et al. (2008). Origin of the S_N2 benzylic effect. *J. Am. Chem. Soc.* 130: 9887–9896.

56 Uggerud, E. (2009). Steric and electronic effects in S_N2 reactions. *Pure Appl. Chem.* 81: 709–717.

57 Hamlin, T.A., Swart, M., and Bickelhaupt, F.M. (2018). Nucleophilic substitution (S_N2): dependence on nucleophile, leaving group, central atom, substituents, and solvent. *ChemPhysChem* 19: 1315–1330.

58 Grabowski, S.J. (2014). Tetrel bond–σ-hole bond as a preliminary stage of the S_N2 reaction. *Phys. Chem. Chem. Phys.* 16: 1824–1834.

59 Donald, K.J. and Tawfik, M. (2013). The weak helps the strong: sigma-holes and the stability of MF_4·base complexes. *J. Phys. Chem. A* 117: 14176–14183.

60 Donald, K.J., Befekadu, E., and Prasad, S. (2017). Coordination and insertion: competitive channels for Borylene reactions. *J. Phys. Chem. A* 121: 8982–8994.

61 Davy, J. (1812). An account of some experiments on different combinations of fluoric acid. *Philos. Trans. R. Soc. Lond.* 102: 352–369.

62 Beattie, I.R. and Ozin, G.A. (1970). Vibrational spectra, vibrational analysis, and shapes of some 1:1 and 1:2 addition compounds of group IV tetrahalides with trimethylamine and trimethylphosphine. *J. Chem. Soc. A* 370–377.

63 Ault, B.S. (1981). Matrix-isolation studies of Lewis acid/base interactions: infrared spectra of the 1:1 adduct $SiF_4 \cdot NH_3$. *Inorg. Chem.* 20: 2817–2822.

64 Lorenz, T.J. and Ault, B.S. (1982). Matrix-isolation studies of Lewis acid-base interactions. 2. 1/1 adduct of tetrafluorosilane with methyl-substituted amines. *Inorg. Chem.* 21: 1758–1761.

65 McNair, A.M. and Ault, B.S. (1982). Matrix-isolation studies of Lewis acid-base interactions. 3. Infrared spectra of 1/1 tetrafluorogermane-amine complexes. *Inorg. Chem.* 21: 1762–1765.

66 Marsden, C.J. (1983). Structure and energetics of tetrafluorosilane-ammonia ($SiF_4 \cdot NH_3$) ab initio molecular orbital calculations. *Inorg. Chem.* 22 (22): 3177–3178.

67 Walther, A.M. and Ault, B.S. (1984). Infrared matrix isolation study of intermediate molecular complexes: complexes of tetrafluorogermane with oxygen-containing bases. *Inorg. Chem.* 23: 3892–3897.

68 Ruoff, R.S., Emilsson, T., Jaman, A.I. et al. (1992). Rotational spectra, dipole moment, and structure of the $SiF_4–NH_3$ dimer. *J. Chem. Phys.* 96: 3441–3446.

69 Kraus, F. and Baer, S.A. (2010). Higher ammoniates of BF_3 and SiF_4: syntheses, crystal structures, and theoretical calculations. *ZAAC* 636 (2): 414–422.

70 Kraus, C.A. and Brown, E.H. (1929). Studies relating to boron. I. Reaction of boron trifluoride with ammonia and alkylamines. *J. Am. Chem. Soc.* 51 (9): 2690–2696.

71 Brown, H.C. and Johnson, S. (1954). Molecular addition compounds. I. The interaction of ammonia with ammonia-boron trifluoride at low temperatures. *J. Am. Chem. Soc.* 76: 1978–1979.

72 Grabowski, S.J. (2014). Boron and other triel Lewis acid centers: from hypovalency to hypervalency. *ChemPhysChem* 15 (14): 2985–2993.

73 Grabowski, S.J. (2020). The nature of triel bonds, a case of B and Al centres bonded with electron rich sites. *Molecules* 25 (1–15): 2703.

74 Grabowski, S.J. (2015). π-Hole bonds: boron and aluminum Lewis acid centers. *ChemPhysChem* 16 (7): 1470–1479.

75 Frenking, G., Fau, S., Marchand, C.M., and Grützmacher, H. (1997). The π-donor ability of the halogens in cations and neutral molecules. A theoretical study of AX_3^+, AH_2X^+, YX_3, and YH_2X (A = C, Si, Ge, Sn, Pb; Y = B, Al, Ga, In, Tl; X = F, Cl, Br, I). *J. Am. Chem. Soc.* 119 (28): 6648–6655.

76 Donald, K.J., Prasad, S., and Wilson, K. (2021). Group 14 central atoms and halogen bonding in different dielectric environments: how germanium outperforms silicon. *ChemPlusChem* 86: 1387–1396.

77 Buchler, A. and Klemperer, W. (1958). Infrared spectra of the alkaline-earth halides. I. Beryllium fluoride, beryllium chloride, and magnesium chloride. *J. Chem. Phys.* 29: 121–123.

78 Wharton, L., Berg, R.A., and Klemperer, W. (1963). Geometry of the alkaline earth dihalides. *J. Chem. Phys.* 39: 2023–2031.

79 Miller, J., Gregoriou, G., and Mosher, H.S. (1961). Relative rates of Grignard addition and reduction reactions 1-3. *J. Am. Chem. Soc.* 83 (19): 3966–3971.

80 Blomberg, C. (1996). Chapter 11: Mechanisms of reactions of Grignard reagents. In: *Handbook of Grignard Reagents* (ed. G.S. Silverman and P.E. Rakita), 219–249. CRC Press.

81 Yáñez, M., Sanz, P., Mó, O. et al. (2009). Beryllium bonds, do they exist? *J. Chem. Theory Comput.* 5 (10): 2763–2771.

82 Montero-Campillo, M.M., Sanz, P., Mó, O. et al. (2018). Alkaline-earth (Be, Mg and Ca) bonds at the origin of huge acidity enhancements. *Phys. Chem. Chem. Phys.* 20 (4): 2413–2420.

83 Das, A. and Arunan, E. (2022). Non-covalent bonds in group 1 and group 2 elements: the 'alkalene bond'. *Phys. Chem. Chem. Phys.* 24 (47): 28913–28922.

84 Kollman, P.A., Liebman, J.F., and Allen, L.C. (1970). Lithium bond. *J. Am. Chem. Soc.* 92: 1142–1150.

85 Latajka, Z. and Scheiner, S. (1984). Ab initio comparison of H bonds and Li bonds. Complexes of LiF, LiCl, HF, and HCl with NH_3. *J. Chem. Phys.* 81: 4014–4017.

86 Szczęśniak, M.M., Latajka, Z., Piecuch, P. et al. (1985). Theoretical studies of lithium bonding in lithium chloride/aliphatic amine complexes. *Chem. Phys.* 94: 55–63.

87 Sannigrahi, A.B. (1986). The lithium bond. *J. Chem. Educ.* 63 (10): 843.

88 Esrafili, M.D. and Mohammadirad, N. (2014). Halogen bond interactions enhanced by sodium bonds — theoretical evidence for cooperative and substitution effects in NCX···NCNa···NCY complexes (X = F, Cl, Br, I; Y = H, F, OH). *Can. J. Chem.* 92: 653–658.

89 Welch, D.O., Lazareth, O.W., Dienes, G.J., and Hatcher, R.D. (2008). Alkali halide molecules: configurations and molecular characteristics of dimers and trimers. *J. Chem. Phys.* 64 (2): 835–839.

90 Niu, Z., McDowell, S.A.C., and Li, Q. (2023). Triel bonds with Au atoms as electron donors. *ChemPhysChem* 24 (6): e202200748.

91 Bartlett, N. (1962). Xenon hexafluoroplatinate(v) $Xe+[PtF6]^-$. *Proc. Chem. Soc.* 6: 218.

92 Claassen, H.H., Selig, H., and Malm, J.G. (1962). Xenon tetrafluoride. *J. Am. Chem. Soc.* 84 (18): 3593–3593.

93 Slivnik, J. (1962). Synthesis of XeF_6. *Croatica Chem. Acta* 34: 253.

94 Brock, D.S., Bilir, V., Mercier, H.P.A., and Schrobilgen, G.J. (2007). $XeOF_2$, $F_2OXeN\equiv CCH_3$, and $XeOF_2 \cdot nHF$: rare examples of Xe(IV) oxide fluorides. *J. Am. Chem. Soc.* 129 (12): 3598–3611.

95 Bauzá, A. and Frontera, A. (2015). Aerogen bonding interaction: a new supramolecular force? *Angew. Chem. Int. Ed.* 54 (25): 7340–7343.

96 Zierkiewicz, W., Michalczyk, M., and Scheiner, S. (2018). Aerogen bonds formed between $AeOF_2$ (Ae = Kr, Xe) and diazines: comparisons between σ-hole and π-hole complexes. *Phys. Chem. Chem. Phys.* 20 (7): 4676–4687.

97 Grochala, W. (2007). Atypical compounds of gases, which have been called 'noble'. *Chem. Soc. Rev.* 36 (10): 1632–1655.

98 Stone, A.J. (2013). Are halogen bonded structures electrostatically driven? *J. Am. Chem. Soc.* 135 (18): 7005–7009.

99 Politzer, P., Murray, J.S., and Clark, T. (2010). Halogen bonding: an electrostatically-driven highly directional noncovalent interaction. *Phys. Chem. Chem. Phys.* 12: 7748.

100 Řezáč, J. and de la Lande, A. (2017). On the role of charge transfer in halogen bonding. *Phys. Chem. Chem. Phys.* 19 (1): 791–803.

101 Brinck, T. and Borrfors, A.N. (2019). Electrostatics and polarization determine the strength of the halogen bond: a red card for charge transfer. *J. Mol. Model.* 25 (5): 125.

102 Wang, C., Danovich, D., Mo, Y., and Shaik, S. (2014). On the nature of the halogen bond. *J. Chem. Theory Comput.* 10 (9): 3726–3737.

103 Mo, Y., Danovich, D., and Shaik, S. (2022). The roles of charge transfer and polarization in non-covalent interactions: a perspective from ab initio valence bond methods. *J. Mol. Model.* 28 (9): 274.

104 Inscoe, B., Rathnayake, H., and Mo, Y. (2021). Role of charge transfer in halogen bonding. *J. Phys. Chem. A* 125 (14): 2944–2953.

105 Thirman, J., Engelage, E., Huber, S.M., and Head-Gordon, M. (2018). Characterizing the interplay of Pauli repulsion, electrostatics, dispersion and charge transfer in halogen bonding with energy decomposition analysis. *Phys. Chem. Chem. Phys.* 20 (2): 905–915.

106 Tawfik, M. and Donald, K.J. (2014). Halogen bonding: unifying perspectives on organic and inorganic cases. *J. Phys. Chem. A* 118: 10090–10100.

12

On the Generalization of Marcus Theory for Two-State Photophysical Processes

Chao-Ping Hsu[1,2] and Chou-Hsun Yang[1]

[1]*Academia Sinica, Institute of Chemistry, 128 Sec. 2 Academia Road, Taipei 115, Taiwan*
[2]*National Center of Theoretical Sciences, Division of Physics, 1 Sec. 4 Roosevelt Road, Taipei, 106, Taiwan*

12.1 Introduction

Many photophysical processes include changes in electronic states. When two molecules (or fragments) are brought together, a change in the charge state, excitation energy, or spin may occur, leading to electron transfer (ET), singlet-energy transfer (SET), or processes involving spin exchange accompanied by energy transfer, such as triplet-energy transfer (TET), singlet fission (SF), and triplet–triplet annihilation (TTA). In predicting rates of such processes, the golden rule rate has been useful, especially under weak coupling conditions. Rate theories beyond the golden rule have been developed as well [1–3]. For SET, Coulomb coupling can be significant, and as a result, the coherent dynamics are typically beyond golden rule theory. Predicting SET dynamics with advanced theories or computational schemes has been an active research field [4–6]. Golden-rule-based rate expressions have long been discussed for processes with spin exchange [7, 8]. For TET, SF, and TTA, the nature of weak coupling allows a generalization from Golden-rule rates.

The early development of ET rate theory has laid a solid foundation for Golden-rule-based theories in condensed phases. The Marcus theory employed classical statistical treatments, and quantum treatments needed for high-frequency modes were developed by Levich and Jortner [9–11]. Since the two-state models are similar, ET rate expression has been widely generalized to other two-state processes like TET and SF [12–20], and both classical and quantum expressions have been employed.

Meanwhile, ET theory was developed with a change of the charge state in the system, which induces a dielectric response in the polar solvent, feeding back to the system's energetics [21]. Therefore, the influence of polarization of the environment and other nuclear degrees of freedom, so-called "system-bath coupling," or "electron–phonon coupling" in condensed matter theories [22, 23], tends to be quite large. On the other hand, TET, SF, and TTA are energy and spin-exchange processes. It is expected that the environment moves little, in a much shorter range, as the process proceeds. Internal degrees of freedom in molecules or fragments involved can

Exploring Chemical Concepts Through Theory and Computation, First Edition. Edited by Shubin Liu.
© 2024 WILEY-VCH GmbH. Published 2024 by WILEY-VCH GmbH.

also move, leading to high-frequency contributions to system-bath coupling. The difference in treatment of system-bath coupling can lead to orders-of-magnitude differences in predicted rates. Therefore, when directly generalizing ET theory to describe TET, SF, and TTA, it is necessary to carefully consider the physical origin of system-bath coupling and to choose a suitable rate expression.

In this work, we review quantum and classical treatments of ET rates, together with the Fröster theory [24] for SET, also a useful form of golden-rule-based rate theories. With the mathematical equivalence established among rate expressions, we discuss its application to general two-state processes, especially for TET, SF, and TTA.

12.2 The Golden Rule Rate Expression

In two-state transition problems, theoretical models allow descriptions of dynamics as well as the rate of transition [1, 23]. The simplest rate expression is Fermi's golden rule [25, 26], which is rooted in principle of perturbation theory. The Fermi's golden rule can be derived from time-dependent perturbation theory. It can be expressed as [25]:

$$k = \frac{2\pi}{\hbar}|V_{if}|^2 \delta(E_i - E_f) \qquad (12.1)$$

Here, k represents the transition rate, $V_{if} (\equiv \langle \psi_i | V | \psi_f \rangle)$ is the matrix element of the perturbation operator V between the initial state ψ_i and the final state ψ_f, and $\delta(E_i - E_f)$ indicates an energy-conserving condition. In practice, when dealing with systems having a collection of states, such as those in a condensed phase, there are continuous distributions of states; thus, the δ function becomes a measure of the density of states, the Franck–Condon weighted density-of-state (FCWD) [27].

$$k = \frac{2\pi}{\hbar}|V_{if}|^2 f(E_f) \qquad (12.2)$$

where f denotes the FCWD, reflecting the distribution of initial states when the reaction occurs, and its Franck–Condon factor weighted overlap with the available final state density. It can be written as [11, 28]

$$f(E_f) = \sum_{\mu,\nu} p(E_{i\mu}) \langle \Theta_{i\mu} | \Theta_{f\nu} \rangle \delta(E_{i\mu} - E_{f\nu}) \qquad (12.3)$$

where $p(E_{i\mu})$ is the distribution of initial states,

$$p(E_{i\mu}) = \frac{\exp(-E_{i\mu})/k_B T}{\sum_{\mu'} \exp(-E_{i\mu'})/k_B T} \qquad (12.4)$$

In Eq. (12.3), we have assumed that both the initial and final states have nuclear vibrations. That is, these states are composed of an electronic wavefunction $\Psi_{i(f)}$ and a vibration state $\mu(\nu)$, whose wavefunction is denoted with $\Theta_{i\mu(f\nu)}$. While the dependence of the nuclear position of V_{if} can be further characterized as discussed in the literature [29–31], in the present work, we took Condon's approximation and treated V_{if} as a constant of nuclear coordinates [32]. We also assumed that

$$E_{i\mu} = E_i^0 + \sum_k \left(\mu_k + \frac{1}{2}\right)\hbar\omega_k \qquad (12.5)$$

$$E_{f\nu} = E_f^0 + \sum_k \left(\nu_k + \frac{1}{2}\right)\hbar\omega_k \qquad (12.6)$$

In principle, μ should be treated as a vector of vibrational quantum numbers of all modes $\mu = (\mu_1, \mu_2, \ldots)^T$, and so is ν. One can also define the energy difference between the bottom of the potential energy of initial and final states, E_f^0 and E_i^0 as ΔE_0,[1]

$$\Delta E_0 \equiv E_f^0 - E_i^0 \qquad (12.7)$$

In this fashion, $f(E_f)$ defined in Eq. (12.3) can be re-written as $f(\Delta E_0)$, which represent the overlap density of states as a function of the difference of initial and final states.

The electronic states i and f impose a different force field on the nuclei, leading to different force field parameters, such as equilibrium configurations and force constants. In further evaluating Eq. (12.3), it is common to assume that the two relevant states are influenced by a set of harmonic oscillators having the same frequencies but with different displacements Δ_k. In other words, the potentials of the two states are,

$$V_i(q) = \sum_k \frac{1}{2}\hbar\omega_k q_k^2 \qquad (12.8)$$

$$V_f(q) = \sum_k \frac{1}{2}\hbar\omega_k(q_k - \Delta_k)^2 + \Delta E_0 = V_i(q) - \sum_k \hbar\omega_k \Delta_k q_k + \lambda + \Delta E_0 \qquad (12.9)$$

where q_k is the coordinate of harmonic mode k, in dimensionless units. These displacements add a linear term to the harmonic potential, and they give rise to the *reorganization energy*, commonly denoted as λ:

$$\lambda_k = \frac{1}{2}\hbar\omega_k \Delta_k^2 \qquad (12.10)$$

$$\lambda = \sum_k \lambda_k \qquad (12.11)$$

λ is an important characteristic for the potential energy surface. It is the energy required from the equilibrium nuclear position for one state to the equilibrium position for another state, along one of the two potential energy surfaces. Since the frequencies of the two states are assumed to be the same, λ obtained from either the initial or the final states would be the same. Another related quantity is the Huang–Rhys factor S [33],

$$S_k = \lambda_k/\hbar\omega_k$$
$$= \frac{1}{2}\Delta_k^2 \qquad (12.12)$$

---

1 Rigorously speaking, the zero-point energy (ZPE) should be included in ΔE_0. However since we are going to assume that the vibrational frequencies $\{\omega_k\}$ are the same for the two states, the energy difference with ZPE is identical to the one without ZPE.

Therefore, FCWD is determined by the ZPE difference ΔE_0 (or ΔG_0), frequencies of vibration that influence the system $\{\omega_k\}$, and the extent of their influence, which is the amount of coordinate shift, characterized either by their contribution to the reorganization energy λ_k or by the dimensionless Huang–Rhys factor S_k. It is also often described with the spectral density function $J(\omega)$ defined as:

$$J(\omega) = \frac{\pi}{2} \sum_k \hbar \omega_k^2 \Delta_k^2 \delta(\omega - \omega_k) \tag{12.13}$$

In further evaluating the FCWD and the subsequent transition-rate expression, it is important to note whether frequencies of important modes (those with large displacements) have frequencies much lower than the thermal energy $k_B T$, with T being the temperature. If so, a classical statistical mechanical treatment would be sufficient, or otherwise, a quantum mechanical treatment would be necessary, as further outlined below.

12.2.1 The Marcus Theory: The Classical Treatment

In the Marcus theory, nuclear degrees of freedom are treated with classical statistics. It can be shown that

$$f(\Delta E_0) = \frac{1}{\sqrt{4\pi \lambda k_B T}} \exp[-(\Delta E_0 + \lambda)^2 / 4\lambda k_B T] \tag{12.14}$$

leading to the Marcus theory,

$$k = \frac{2\pi}{\hbar} \frac{|V_{if}|^2}{\sqrt{4\pi \lambda k_B T}} \exp[-(\Delta E_0 + \lambda)^2 / 4\lambda k_B T] \tag{12.15}$$

The derivation for Eq. (12.15) can be from the Landau–Zener theory [34], by generalization of the transition state theory [35], or by a second-order expansion on time-evolution of density matrix with a spin-boson model [35, 36], as discussed extensively in the literature, where we refer the interested readers.

12.2.2 The Marcus–Levich–Jortner Expression: A Quantum Expression for High-Frequency Modes

When there high-frequency vibrational modes contribute to the reorganization energy, a quantum mechanical treatment is necessary. The boundary of "low" and "high" frequency is generally the thermal energy, $k_B T$, which is approximately 200 cm^{-1}, or 25 meV at room temperature. In a condensed phase, among all possible nuclear modes that can affect the two states, there is always a set of low-frequency modes involving collective motion of the environment. Effects of intramolecular, high-frequency modes may also exist. So a practical, general treatment should include both. Detailed derivations of such a separation can be found in Refs. [11, 28, 37] reported a detailed derivation of such a separation. With Eqs. (12.5) and (12.6), we first separate the vibrational energy into a set of low-frequency, solvation-like modes s and higher frequency, intramolecular modes from the total energy difference ΔE

$$\Delta E = E_{i\mu} - E_{f\nu}$$
$$= \Delta E_0 + E_{fs} - E_{is} + E_{fv} - E_{iv}$$
$$= \Delta E_0 + \Delta E_s + \Delta E_v \tag{12.16}$$

where

$$E_{is} = \sum_{k \in s} \left(\mu_k + \frac{1}{2}\right)\hbar\omega_k \tag{12.17}$$

$$E_{iv} = \sum_{k \notin s} \left(\mu_k + \frac{1}{2}\right)\hbar\omega_k \tag{12.18}$$

and a similar definition applies to the final state. The delta function can also be separated

$$\delta(\Delta E) = \int_{-\infty}^{\infty} \delta(\Delta E_0 + \Delta E_s - x)\delta(\Delta E_v + x)dx \tag{12.19}$$

where the energy partition of x is allowed between s and v sets of modes. The FCWD becomes,

$$f(\Delta E_0) = \int_{-\infty}^{\infty} dx F_s(\Delta E_0 - x)F_v(\Delta E_v + x) \tag{12.20}$$

For $F_s(\Delta E_0 - x)$, the effects of low-frequency, solvation-like vibration modes are included, which is given by the expression in the classical Marcus theory [21, 28]:

$$F_s(\Delta E_0 - x) = \frac{1}{\sqrt{4\pi \lambda_s k_B T}} \exp[-(\Delta E_0 - x + \lambda_s)^2 / 4\lambda_s k_B T] \tag{12.21}$$

where λ_s denotes the contribution of solvent modes to the reorganization energy, and $F_v(x)$ is for high-frequency modes,

$$F_v(x) = \frac{1}{Z_v} \sum_{\mu_1=0 v_1=0}^{\infty}\sum_{}^{\infty} \cdots \sum_{\mu_N=0 v_N=0}^{\infty}\sum_{} \prod_{k \notin s} |\langle \Theta_{\mu_k,i}|\Theta_{v_k,f}\rangle|^2$$
$$\times \exp[-E_{iv}/k_B T]\delta(E_{fv} - E_{iv} + x) \tag{12.22}$$

where $\Theta_{\mu_k,i}$ is the vibrational wave function of mode k, quantum number μ_k in state i. In practice, since $\hbar\omega_k \gg k_B T$, $\exp(-\hbar\omega_k/k_B T) \ll 1$, it is practical to assume that in initial states, vibration modes are in their ground states, $\mu_k = 0$. With harmonic oscillator, $E_{vib,f} = \sum_k (v_k + 1/2)\hbar\omega_k$, and $|\langle 0|v\rangle|^2 = S^v/v! \exp(-S)$, and we have

$$f(\Delta E) = \frac{1}{\sqrt{4\pi \lambda_s k_B T}} \exp\left(-\sum_{k \notin s} S_k\right)$$
$$\times \sum_{v_1=0}^{\infty}\sum_{v_2=0}^{\infty} \cdots \sum_{v_N=0}^{\infty} \prod_{k \notin s} \frac{S_k^{v_k}}{v_k!} \exp\left[-\left(\Delta E_0 + \lambda_s + \sum_k \hbar\omega_k v_k\right)^2 / 4\lambda_s k_B T\right]$$
$$\tag{12.23}$$

In practice, the expression in Eq. (12.23) is often further simplified for one high-frequency vibrational mode k:

$$f(\Delta E) = \frac{1}{\sqrt{4\pi \lambda_s k_B T}} \exp(-S_k)$$

$$\times \sum_{v_k=0}^{\infty} \frac{S^{v_k}}{v_k!} \exp[-(\Delta E_0 + \lambda_s + v_k \hbar \omega_k)^2 / 4\lambda_s k_B T] \quad (12.24)$$

Reduction to one mode has been discussed extensively in the literature (e.g. [38]). This is a practical approach: the more high-frequency modes included, the more undetermined parameters the model requires, increasing the chance of obtaining the right results but not the right physical insights. Moreover, as discussed below, in many cases, it is possible to determine parameters for FCWD from fittings to experimental spectra. Since most room-temperature condensed-phase spectra are with significant line broadening, fitting such broadened spectra with many parameters for multiple modes is rather impractical, and instead, a single-mode progression model can recover the FCWD effectively [38–41], on which we focus in the following discussion.

12.2.3 The Föster Theory: Separating Donor and Acceptor Parts in FCWD

Another way to dissect the golden rule rate is to separate contributions from donor and acceptor molecules or fragments. It is convenient to do so especially in the case of SET, where the FCWD can be derived approximately from experimentally measurable spectra of the molecules involved. Such an approach is used directly in the Föster theory [24] and it can be generalized to account for the rate for TET and SF [14, 16, 17, 27, 42].

By assuming that the system can be separated into a donor and an acceptor without much interaction between them, the energy is broken down to the energy of donor and acceptor:

$$E_i = E^D_{i\mu^D} + E^A_{i\mu^A} = \sum_k \left(\mu^D_k + \frac{1}{2}\right)\hbar\omega^D_k + \sum_{k'} \left(\mu^A_{k'} + \frac{1}{2}\right)\hbar\omega^A_{k'} \quad (12.25)$$

$$E_f = E^D_{fv^D} + E^A_{fv^A} = \sum_k \left(v^D_k + \frac{1}{2}\right)\hbar\omega^D_k + \Delta E^D_0 + \sum_{k'} \left(v^A_{k'} + \frac{1}{2}\right)\hbar\omega^A_{k'} + \Delta E^A_0$$

$$(12.26)$$

Under the Condon approximation, and assuming independent nuclear motion in the donor and acceptor, we can rewrite Eqs. (12.2) and (12.3) as [27]

$$k_{if} = \frac{2\pi}{\hbar}|H_{if}|^2 \sum_{\mu^D, v^D \in D\mu^A, v^A \in A} p\left(E^D_{i\mu^D}\right) p\left(E^A_{i\mu^A}\right) \left|\langle \Theta_{i\mu^D}|\Theta_{fv^D}\rangle\langle \Theta_{i\mu^A}|\Theta_{fv^A}\rangle\right|^2$$

$$\times \delta\left(\Delta E^D_{fv^D, i\mu^D} + \Delta E^A_{fv^A, i\mu^A}\right) \quad (12.27)$$

where $\Delta E^{D(A)}_{fv,i\mu} = E^{D(A)}_{fv} - E^{D(A)}_{i\mu}$ is the difference of vibronic energy for the donor (or acceptor) from the initial to the final state. The state energies, $\varepsilon_{i\mu}$ and ε_{fv}, are assumed to be the sum of donor and acceptor energies in their corresponding vibronic states:

$$\varepsilon_{i\mu} = E_{i\mu}^D + E_{i\mu'}^A \tag{12.28}$$

and similar definition for $E_{fv(v')}^{D(A)}$ can be generalized. $\Theta_{i\mu}^{D(A)}$ are vibration wavefunctions for one of the fragments in the corresponding electronic and vibration state. The transition rate k_{if} in Eq. (12.27) can be written as

$$k_{if} = \frac{2\pi}{\hbar}|H_{if}|^2 \int_{-\infty}^{\infty} dE f_D(E) f_A(E) \tag{12.29}$$

where $f_D(E), f_A(E)$ are

$$f_D(E) = \sum_{\mu}\sum_{v} p\left(E_{i\mu}^D\right) \left|\langle\Theta_{i\mu^D}|\Theta_{fv^D}\rangle\right|^2 \delta\left(E + \Delta E_{fv,i\mu}^D\right) \tag{12.30}$$

$$f_A(E) = \sum_{\mu'}\sum_{v'} p\left(E_{i\mu'}^A\right) \left|\langle\Theta_{i\mu'^A}|\Theta_{fv'^A}\rangle\right|^2 \delta\left(E - \Delta E_{fv',i\mu'}^A\right) \tag{12.31}$$

which are densities of states weighted by vibrational Franck–Condon factors, or FCWD functions for the donor and the acceptor, respectively.

We can further separate modes affecting donor or acceptor into high- and low-frequency sets, again using thermal energy $k_B T$ as the boundary. Following the same procedure as in Eqs. (12.19)–(12.24), it is possible to show that

$$f_D(E) = \frac{1}{\sqrt{4\pi\lambda_{Ds}k_B T}} \exp(-S_D)$$
$$\times \sum_{v_D=0}^{\infty} \frac{S_D^{v_D}}{v_D!} \exp\left[-(E + \Delta E_0^D + \lambda_{Ds} + v_D\hbar\omega_D)^2/4\lambda_{Ds}k_B T\right], \tag{12.32}$$

$$f_A(E) = \frac{1}{\sqrt{4\pi\lambda_{As}k_B T}} \exp(-S_A)$$
$$\times \sum_{v_A=0}^{\infty} \frac{S_A^{v_A}}{v_A!} \exp\left[-(E - \Delta E_0^A - \lambda_{As} - v_A\hbar\omega_A)^2/4\lambda_{As}k_B T\right] \tag{12.33}$$

Therefore, the rate is now an overlap between two density-of-state functions, which can be modeled using experimental spectra, if available, constituting the foundation of Föster theory for SET. In the case in which one high-frequency mode dominates, $f_{D/A}(E)$ is a sum of Gaussian peaks broadened by $\sqrt{2\lambda_{D_s}k_B T}$ or $\sqrt{2\lambda_{A_s}k_B T}$, with relative heights $S_{D/A}^v/v!$, forming a vibronic progression.

With donor–acceptor separation, it is desirable to further combine Eqs. (12.32) and (12.33) with Eq. (12.20) to see if it can lead to Eq. (12.23), which is a rather good exercise. Following Eq. (12.29) with Eqs. (12.32) and (12.33), integration of E can be performed with a Gaussian integration, as,

$$I \equiv \int_{-\infty}^{\infty} dE f_D(E) f_A(E)$$

$$= \frac{1}{4\pi k_B T \sqrt{\lambda_{Ds}\lambda_{As}}} \sum_{v_D=0}^{\infty}\sum_{v_A=0}^{\infty} \frac{S_D^{v_D} S_A^{v_A}}{v_D! v_A!} G(v_D, v_A) \tag{12.34}$$

where $G(v_D, v_A)$ is the integration of donor and acceptor vibronic bands:

$$G(v_D, v_A) \equiv \int_{-\infty}^{\infty} dE \exp\left[-\frac{(E+\varepsilon_D)^2}{4\lambda_{Ds}k_B T} - \frac{(E-\varepsilon_A)^2}{4\lambda_{As}k_B T}\right] \quad (12.35)$$

For the sake of simplicity, we define,

$$\varepsilon_D \equiv \Delta E_0^D + \lambda_{Ds} + v_D \hbar \omega_D \quad (12.36)$$

$$\varepsilon_A \equiv \Delta E_0^A + \lambda_{As} + v_A \hbar \omega_A \quad (12.37)$$

It can be shown that,

$$G(v_D, v_A) = \int_{-\infty}^{\infty} dE \exp\left\{-\frac{1}{4k_B T}\left[\left(\frac{1}{\lambda_{Ds}} + \frac{1}{\lambda_{As}}\right)E^2 + 2E\left(\frac{\varepsilon_D}{\lambda_{Ds}} - \frac{\varepsilon_A}{\lambda_{As}}\right)\right.\right.$$
$$\left.\left.+ \frac{\varepsilon_D^2}{\lambda_{Ds}} + \frac{\varepsilon_A^2}{\lambda_{As}}\right]\right\}$$

$$= \int_{-\infty}^{\infty} dE \exp\left\{-\frac{1}{4k_B T}\left(\frac{\lambda_{Ds}+\lambda_{As}}{\lambda_{Ds}\lambda_{As}}\right)\left[\left(E + \frac{\lambda_{As}\varepsilon_D - \lambda_{Ds}\varepsilon_A}{\lambda_{Ds}+\lambda_{As}}\right)^2\right.\right.$$
$$\left.\left.- \left(\frac{\lambda_{As}\varepsilon_D - \lambda_{Ds}\varepsilon_A}{\lambda_{Ds}+\lambda_{As}}\right)^2 + \frac{\lambda_{Ds}\varepsilon_A^2 + \lambda_{As}\varepsilon_D^2}{\lambda_{Ds}+\lambda_{As}}\right]\right\} \quad (12.38)$$

The sum of the last two terms in the exponential function is

$$-\left(\frac{\lambda_{As}\varepsilon_D - \lambda_{Ds}\varepsilon_A}{\lambda_{Ds}+\lambda_{As}}\right)^2 + \frac{\lambda_{Ds}\varepsilon_A^2 + \lambda_{As}\varepsilon_D^2}{\lambda_{Ds}+\lambda_{As}} = \frac{\lambda_{Ds}\lambda_{As}(\varepsilon_D + \varepsilon_A)^2}{(\lambda_{Ds}+\lambda_{As})^2} \quad (12.39)$$

Using Gaussian integration,

$$\int_{-\infty}^{\infty} dx e^{-ax^2} = \sqrt{\frac{\pi}{a}}$$

the integration of E can be performed with the first term in the exponential function in Eq. (12.39). Then we have,

$$G(v_D, v_A) = \sqrt{\frac{4\pi k_B T \lambda_{Ds}\lambda_{As}}{\lambda_{Ds}+\lambda_{As}}} \exp\left[-\frac{(\varepsilon_D+\varepsilon_A)^2}{4k_B T(\lambda_{Ds}+\lambda_{As})}\right] \quad (12.40)$$

The integral of f_D and f_A in Eq. (12.35) becomes,

$$I = \frac{1}{\sqrt{4\pi k_B T(\lambda_{Ds}+\lambda_{As})}} \sum_{v_D=0}^{\infty}\sum_{v_A=0}^{\infty} \frac{S_D^{v_D} S_A^{v_A}}{v_D! v_A!} \exp\left[-\frac{(\varepsilon_D+\varepsilon_A)^2}{4k_B T(\lambda_{Ds}+\lambda_{As})}\right] \quad (12.41)$$

Setting

$$v = v_D + v_A \quad (12.42)$$

$$\lambda_s = \lambda_{Ds} + \lambda_{As} \quad (12.43)$$

and assuming that frequencies of major modes $\omega_D = \omega_A \equiv \omega$, we have

$$\varepsilon_D + \varepsilon_A = \Delta E_0 + \lambda_s + v\hbar\omega \quad (12.44)$$

and,

$$I = \frac{1}{\sqrt{4\pi\lambda_s k_B T}} \sum_{v=0}^{\infty} \left(\sum_{v_D=0}^{v} \frac{S_D^{v_D} S_A^{v-v_D}}{v_D!(v-v_D)!} \right) \exp\left[-\frac{(\Delta E_0 + \lambda_s + v\hbar\omega)^2}{4k_B T \lambda_s} \right] \quad (12.45)$$

With binomial expansion, and setting $S \equiv S_D + S_A$, we have,

$$S^v = (S_D + S_A)^v = \sum_{v_D=0}^{v} v! \frac{S_D^{v_D} S_A^{v-v_D}}{v_D!(v-v_D)!} \quad (12.46)$$

The FCWD can be written as,

$$I = \frac{1}{\sqrt{4\pi\lambda_s k_B T}} \sum_{v=0}^{\infty} \frac{S^v}{v!} \exp\left[-\frac{(\Delta E_0 + \lambda_s + v\hbar\omega)^2}{4k_B T \lambda_s} \right] \quad (12.47)$$

which is identical to Eq. (12.24). The corresponding Golden-rule rate, expressed with a high-frequency mode, is then,

$$k_{if} = \frac{2\pi}{\hbar} \frac{|H_{if}|^2}{\sqrt{4\pi\lambda_s k_B T}} \sum_{v=0}^{\infty} \frac{S^v}{v!} \exp\left[-\frac{(\Delta E_0 + \lambda_s + v\hbar\omega)^2}{4k_B T \lambda_s} \right] \quad (12.48)$$

Therefore, the overlap spectral approach originating from the Föster theory is mathematically equivalent to the Marcus–Levich–Jortner quantum ET rate. This gives us the freedom to utilize these rate expressions for the two-state processes discussed in the present work. In other words, it is legitimate to use ET rate expressions for SET, TET, SF, or TTA processes, provided that the proper parameter values and the suitable variant of the rate expression are employed, which we further elaborate in Section 12.3.

12.3 Application

The physical natures of factors determining FCWD, such as S and λ_s, are dependent on properties of the actual system in applying the theory. Here we discuss three cases: ET, SET, in which a direct application of Föster theory is useful, and a general case in which other energy and spin-exchange process, such as TET, SF, and TTA.

12.3.1 Electron Transfer

In ET, a change of charge distribution caused strong system-bath coupling, leading to typical dielectric solvation in λ_{out}, as illustrated in Figure 12.1. For example, the standard Marcus theory has the donor and acceptor in spherical cavities [43, 44]:

$$\lambda_{out} = (\Delta e)^2 \left(\frac{1}{2r_D} + \frac{1}{2r_A} - \frac{1}{R_{DA}} \right) \left(\frac{1}{\epsilon_{op}} - \frac{1}{\epsilon_s} \right) \quad (12.49)$$

where Δe is the amount of charge transferred in the reaction; $r_{D(A)}$ is the radius of the donor (acceptor), when a spherical cavity is used to model the molecule or fragment; R_{DA} is the center-to-center distance between the donor and acceptor; ϵ_{op} is the optical dielectric constant; and ϵ_s is the static dielectric constant for the solvent. The

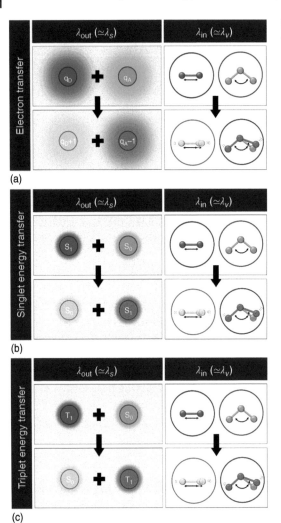

Figure 12.1 The origin and size of various sources of reorganization energies, λ_out and λ_in, in (a) ET, (b) SET, and (c) TET.

expression in Eq. (12.49) was derived from a model in which the dielectric solvation energy change was estimated from two independent spheres (infinitely separated) containing the donor and acceptor, and changes in the interaction energy between the two (the term involving $1/R_\text{DA}$). Typical values of λ_out are 0.5–1 eV for polar solvents, about 2–5 times larger than typical λ_in, often regarded as 0.1–0.2 eV [39, 41, 45], as calculated using quantum chemistry.

We note that λ_out is mainly in the low-frequency region [46, 47] as the vibrations of polar solvents contribute only marginally, while important λ_in is mostly in the frequencies higher than thermal energy. In order to proceed with the FCWD expression given in the Section 12.2.3, we further assume that $\lambda_s \simeq \lambda_\text{out}$, and $\lambda_v \simeq \lambda_\text{in}$.

FCWDs with classical and quantum treatments are included in the top row of Figure 12.2. In this case, it is seen that ET with a large λ_s, the classical FCWD (Eq. (12.14)) is quite close to that of quantum expression (Eq. (12.24)).

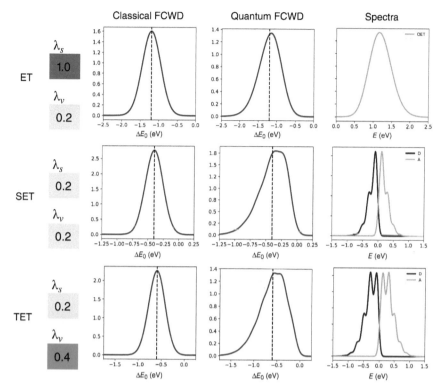

Figure 12.2 FCWD for 3 combinations of reorganization energy, as shown, mimicking ET, SET, and TET situations. For each case, the classical FCWD (left, Eq. (12.14)) and quantum FCWD (middle, Eq. (12.24)) as a function of ΔE_0 are shown, in units of eV^{-1}, simulated with the progression frequency, $\hbar\omega$, as 0.2 eV at 300 K. For ET, the simulated optical electron transfer (OET) absorption spectrum is included to the right (with Eq. (12.33)). For SET and TET, a symmetric separation into donor emission (Eq. (12.32)) and acceptor absorption (Eq. (12.33)) spectra (arbitrary units) are shown in the right panels, with ΔE set to zero, without loss of generality. Values of λ_v and λ_s employed are as listed to the left of each row.

It is rare to treat the FCWD of an ET rate as an overlap integral of spectra. In principle, a similar spectral concept is applicable. It has been reported for ET occurring with a optical transition, a situation termed "optical electron transfer" (OET) [48, 49]. Due to the large λ_s, the absorption spectra for such an ET transition can be characterized as a red-shifted broad band with a low extinction coefficient, usually without any visible vibronic progression, as depicted in the top-right panel in Figure 12.2.

12.3.2 SET: Using Spectra for FCWD

For SET, the Förster theory has been the most useful way to predict the rate, which takes the donor emission and acceptor absorption spectra for FCWD. In this case, experimental optical spectra are used, providing reliable information about the FCWD of system.

SET involves singlet excited states of a donor and an acceptor.

$$D^*(S_1) + A(S_0) \longrightarrow D(S_0) + A^*(S_1) \tag{12.50}$$

The break-down of FCWD to those of donor and acceptor is a convenient approach, as it connects the Förster theory with the Marcus–Levich–Jortner theory. For SET, Förster theory takes the (experimental) optical spectra. In this case, the donor emission,

$$D^*(S_1) \longrightarrow D(S_0) \tag{12.51}$$

and acceptor absorption

$$A(S_0) \longrightarrow A^*(S_1) \tag{12.52}$$

are the two separate processes, allowing us to calculate FCWD as the overlap of the two spectra. The spectral function $f_D(E)$, can be derived from the donor emission $A_{if}^D(E)$, normalized after scaling down by the E^3 dependence in the Einstein's spontaneous emission coefficient [17, 50],

$$f_D(E) = \frac{A_{if}^D(E)/E^3}{\int_{-\infty}^{\infty} dE\, A_{if}^D(E)/E^3} \tag{12.53}$$

and $f_A(E)$ can be similarly derived from acceptor absorption spectra, $\sigma_{if}^A(E)$, with first-order dependence on E in the extinction coefficient,

$$f_A(E) = \frac{\sigma_{if}^A(E)/E}{\int_{-\infty}^{\infty} dE\, \sigma_{if}^A(E)/E} \tag{12.54}$$

In Eqs. (12.53) and (12.54), we have used the relationship [51]:

$$\sum_\mu \sum_\nu p\left(\omega_{i\mu}^{D(A)}\right) \left|\langle \Theta_{i\mu D(A)} | \Theta_{f\nu D(A)} \rangle\right|^2 = 1 \tag{12.55}$$

It is also possible to consider the rate of SET from a generalized ET rate. Without changing the charge states, SET can be generally regarded as a charge-neutral event. Therefore, the interaction of the system and its surroundings is quite different from that of ET: the polarization effect of the environment is usually not present, and a smaller outer reorganization energy (λ_{out} and thus, λ_s) is expected. On the contrary, vibrational states in the donor and acceptor fragments are likely to be changed with the electronic excitation, leading to a contribution from the inner-shell reorganization energy λ_{in} (and λ_v as well) (Figure 12.1b). Moreover, since the excitation could affect the bonding strength as $\pi - \pi^*$ excitation is often involved in many organic, aromatic systems, alternate stretching of aromatic bonds can be shifted significantly, leading to high-frequency contributions to the inner-shell reorganization energy.

Therefore, in the generalized form of ET rate theories for SET, the Marcus–Levich–Jortner expression with a quantum treatment for high-frequency modes is far better than classical Marcus Theory. As seen in the middle row of Figure 12.2, the quantum FCWD (center panel) is significantly broadened compared to classical one (left panel), indicating a wider energetic window available for reaction, and estimation with either quantum or classical expressions could lead to drastically different conclusions.

12.3.3 TET and Other Energy Transfer Process with Spin Exchange

TET, SF, and TTA involve an exchange of spin and energy. TET can generally be processed like,

$$D^*(T_1) + A(S_0) \longrightarrow D(S_0) + A^*(T_1) \tag{12.56}$$

and for SF it is,

$$D^*(S_1) + A(S_0) \longrightarrow D^*(T_1) + A^*(T_1) \tag{12.57}$$

We note that triplet–triplet annihilation (TTA) is the reverse process of SF,

$$D^*(T_1) + A^*(T_1) \longrightarrow D(S_0) + A^*(S_1) \tag{12.58}$$

which is also frequently studied with methodologies similar to SF.

Other than minor charge transfer (CT) components mixed in the diabatic states involved, as reported in the literature [14, 16, 20, 52, 53], the process is mostly neutral in its charge distribution. Without the electric interaction, the interaction of the system (donor and acceptor) with the surrounding environment is not too strong. Again, TET, SF, and TTA are expected to be cases with low λ_{out}, and high-frequency components λ_{in} arising from the $\pi - \pi^*$ transitions can be important (Figure 12.1c). Therefore, when considering the rate by generalizing Marcus ET rate theory, the Marcus–Levich–Jortner quantum expression is a much better choice than the classical Marcus theory. Moreover, the spectral overlap approach for SET can be also generalized to TET, SF, and TTA, but triplet states are involved in the process.

In adapting the Förster theory for the rates of TET, SF, and TTA, the problem is that experimental absorption and emission spectra involving triplet states are not as available. In SF and TTA, spectra between S_1 and T_1 of the donor would be necessary, but these are not available from typical experimental works. Specifically, for TET, the "emission" and "absorption" spectra needed are

$$D^*(T_1) \longrightarrow D(S_0) \tag{12.59}$$

$$A(S_0) \longrightarrow A^*(T_1) \tag{12.60}$$

and for SF, they are

$$D^*(S_1) \longrightarrow D^*(T_1) \tag{12.61}$$

$$A(S_0) \longrightarrow A^*(T_1) \tag{12.62}$$

or TTA

$$D^*(T_1) \longrightarrow D(S_0) \tag{12.63}$$

$$A^*(T_1) \longrightarrow A^*(S_1) \tag{12.64}$$

Such spectra can be simulated with sensible approximations. With quantum chemistry computation, optimization for the structure of corresponding states in each fragment can be performed first, followed by vibrational analyses. In this way, corresponding $S_{D(A)}$ and $\omega_{D(A)}$ values can be obtained for D and A. To determine E_0, experimental results, when available, are a far better source than typical density functional theory (DFT) calculations. In addition to typical problems for DFT to estimate state energies, it is also hard to capture effects of a packing condensed phase environment that is typical of optical-electronic processes. If regular spectra are available (for transition between singlet states), the width of progression bands can be adapted to determine λ_s, an assumption that the outer-sphere, low-frequency reorganization energy is similar in $S_1 \longleftrightarrow S_0$ as in $T_1 \longleftrightarrow S_{0(1)}$.

In this way, to the best of our knowledge, the FCWD can be derived, and the final rates of TET, SF, or TTA can be estimated. A set of simulated quantum and classical FCWDs, with spectral functions are included in Figure 12.2. Here we assumed that the λ_v is larger than that of SET in transitions involving triplet states, according to our previous estimation [16]. Again, quantum and classical FCWDs can be quite different, indicating the need to employ the quantum vibronic effect when estimating the rates.

12.4 Conclusion

In modeling rates of photophysical processes involving transitions between electronic states, the simplest approach is to use the golden rule. It is important to consider the nature of the transition, and to properly determine the suitable rate expression, especially for the reorganization energy and its corresponding frequency of motion. For energy transfer or spin- and energy-exchange events such as SF and TTA, in most cases, the states involved are largely charge-neutral, and a quantum rate expression with a high-frequency component would be much more suitable than classical Marcus theory.

Acknowledgments

We gratefully acknowledge support from Academia Sinica and the National Science and Technology Council of Taiwan through project 112-2123-M-001-002 and 111-2123-M-001-003.

References

1 Song, X. and Stuchebrukhov, A.A. (1993). Outer-sphere electron transfer in polar solvents: quantum scaling of strongly interacting systems. *J. Chem. Phys.* 99(2): 969–978. https://doi.org/10.1063/1.465310.

2 Bixon, M. and Jortner, J. (1968). Intramolecular radiationless transitions. *J. Chem. Phys.* 48(2): 715–726. https://doi.org/10.1063/1.1668703.

3 Zusman, L.D. (1980). Outer-sphere electron transfer in polar solvents. *Chem. Phys.* 49(2): 295–304. https://doi.org/10.1016/0301-0104(80)85267-0.

4 Dani, R., Kundu, S., and Makri, N. (2023). Coherence maps and flow of excitation energy in the bacterial light harvesting complex 2. *J. Phys. Chem. Lett.* 14(16): 3835–3843. https://doi.org/10.1021/acs.jpclett.3c00670.

5 Hwang-Fu, Y.-H., Chen, W., and Cheng, Y.-C. (2015). A coherent modified Redfield theory for excitation energy transfer in molecular aggregates. *Chem. Phys.* 447: 46–53. https://doi.org/10.1016/j.chemphys.2014.11.026.

6 Redfield, A.G. (1965). The theory of relaxation processes. In: *Advances in Magnetic and Optical Resonance*, Advances in Magnetic Resonance, vol. 1 (ed. J.S. Waugh), 1–32. Academic Press https://doi.org/10.1016/B978-1-4832-3114-3.50007-6.

7 Closs, G.L., Johnson, M.D., Miller, J.R., and Piotrowiak, P. (1989). A connection between intramolecular long-range electron, hole, and triplet energy transfers. *J. Am. Chem. Soc.* 111(10): 3751–3753. https://doi.org/10.1021/ja00192a044.

8 Haberkorn, R., Michel-Beyerle, M.E., and Marcus, R.A. (1979). On spin-exchange and electron-transfer rates in bacterial photosynthesis. *Proc. Natl. Acad. Sci. U.S.A.* 76(9): 4185–4188. https://doi.org/10.1073/pnas.76.9.4185.

9 Jortner, J. (1976). Temperature dependent activation energy for electron transfer between biological molecules. *J. Chem. Phys.* 64(12): 4860–4867. https://doi.org/10.1063/1.432142.

10 Levich, V.G., Dogonadze, R.R., German, E.D. et al. (1970). Theory of homogeneous reactions involving proton transfer. *Electrochim. Acta* 15(2): 353–367. https://doi.org/10.1016/0013-4686(70)80027-5.

11 Ulstrup, J. and Jortner, J. (1975). The effect of intramolecular quantum modes on free energy relationships for electron transfer reactions. *J. Chem. Phys.* 63(10): 4358–4368. https://doi.org/10.1063/1.431152.

12 Azarias, C., Russo, R., Cupellini, L. et al. (2017). Modeling excitation energy transfer in multi-BODIPY architectures. *Phys. Chem. Chem. Phys.* 19(9): 6443–6453. https://doi.org/10.1039/C7CP00427C.

13 Ito, A. and Meyer, T.J. (2012). The Golden Rule. Application for fun and profit in electron transfer, energy transfer, and excited-state decay. *Phys. Chem. Chem. Phys.* 14(40): 13731. https://doi.org/10.1039/c2cp41658a.

14 Lin, H.-H., Kue, K.Y., Claudio, G.C., and Hsu, C.-P. (2019). First principle prediction of intramolecular singlet fission and triplet triplet annihilation rates. *J. Chem. Theory Comput.* 15(4): 2246–2253. https://doi.org/10.1021/acs.jctc.8b01185.

15 Ullrich, T., Munz, D., and Guldi, D.M. (2021). Unconventional singlet fission materials. *Chem. Soc. Rev.* 50(5): 3485–3518. https://doi.org/10.1039/D0CS01433H. Authors' argument is because quantum tunneling which is different yours.

16 Yang, C.-H. and Hsu, C.-P. (2015). First-principle characterization for singlet fission couplings. *J. Phys. Chem. Lett.* 6(10): 1925–1929. https://doi.org/10.1021/acs.jpclett.5b00437.

17 You, Z.-Q. and Hsu, C.-P. (2011). Ab inito study on triplet excitation energy transfer in photosynthetic light-harvesting complexes. *J. Phys. Chem. A* 115(16): 4092–4100. https://doi.org/10.1021/jp200200x.

18 Ha, D.-G., Wan, R., Kim, C.A. et al. (2022). Exchange controlled triplet fusion in metal–organic frameworks. *Nat. Mater.* 21(11): 1275–1281. https://doi.org/10.1038/s41563-022-01368-1.

19 Renaud, N., Sherratt, P.A., and Ratner, M.A. (2013). Mapping the relation between stacking geometries and singlet fission yield in a class of organic crystals. *J. Phys. Chem. Lett.* 4(7): 1065–1069. https://doi.org/10.1021/jz400176m.

20 Yost, S.R., Lee, J., Wilson, M.W.B. et al. (2014). A transferable model for singlet-fission kinetics. *Nat. Chem.* 6(6): 492–497. https://doi.org/10.1038/nchem.1945.

21 Marcus, R.A. (1956). On the theory of oxidation-reduction reactions involving electron transfer. I. *J. Chem. Phys.* 24(5): 966–978. https://doi.org/10.1063/1.1742723.

22 Holstein, T. (1959). Studies of polaron motion: Part I. The molecular-crystal model. *Ann. Phys.* 8: 325.

23 Leggett, A.J., Chakravarty, S., Dorsey, A.T. et al. (1987). Dynamics of the dissipative two-state system. *Rev. Mod. Phys.* 59(1): 1–85. https://doi.org/10.1103/RevModPhys.59.1.

24 Förster, Th. (1948). Zwischenmolekulare Energiewanderung und Fluoreszenz. *Ann. Phys.* 437(1-2): 55–75. https://doi.org/10.1002/andp.19484370105.

25 Fermi, E. (1974). *Nuclear Physics: A Course given by Enrico Fermi at the University of Chicago*. Chicago: University of Chicago Press.

26 Dirac, P.A.M. and Bohr, N.H.D. (1997). The quantum theory of the emission and absorption of radiation. *Proc. R. Soc. London, Ser. A, Contain. Pap. Math. Phys. Charact.* 114(767): 243–265. https://doi.org/10.1098/rspa.1927.0039.

27 Lin, S.H. (1973). On the theory of non-radiative transfer of electronic excitation. *Proc. R. Soc. London, Ser. A, Math. Phys. Sci.* 335(1600): 51–66.

28 Levich, V.G. (1966). Present state of the theory of oxidation-reduction in solution (bulk and electrode reactions). In: *Advances in Electrochemistry and Electrochemical Engineering*, vol. 4 (ed. P.N. Delahay), 249–371. New York: Interscience Publishers.

29 Wang, Y.-S., Wang, C.-I., Yang, C.-H., and Hsu, C.-P. (2023). Machine-learned dynamic disorder of electron transfer coupling. *J. Chem. Phys.* 159(3): 034103. https://doi.org/10.1063/5.0155377.

30 Medvedev, E.S. and Stuchebrukhov, A.A. (1997). Inelastic tunneling in long-distance biological electron transfer reactions. *J. Chem. Phys.* 107(10): 3821–3831. https://doi.org/10.1063/1.474741.

31 Troisi, A., Ratner, M.A., and Zimmt, M.B. (2004). Dynamic nature of the intramolecular electronic coupling mediated by a solvent molecule: a computational study. *J. Am. Chem. Soc.* 126(7): 2215–2224. https://doi.org/10.1021/ja038905a.

32 Condon, E.U. (1928). Nuclear motions associated with electron transitions in diatomic molecules. *Phys. Rev.* 32(6): 858–872. https://doi.org/10.1103/PhysRev.32.858.

33 Huang, K. and Rhys, A. (1950). Theory of light absorption and non-radiative transitions in f-centres. *Proc. R. Soc. London, Ser. A, Math. Phys. Sci.* 204(1078): 406–423.

34 Zener, C. (1997). Non-adiabatic crossing of energy levels. *Proc. R. Soc. London, Ser. A, Contain. Pap. Math. Phys. Charact.* 137(833): 696–702. https://doi.org/10.1098/rspa.1932.0165.

35 Tokmakoff, A. (2018). 15.5: Marcus Theory for Electron Transfer. https://chem.libretexts.org/Bookshelves/Physical_and_Theoretical_Chemistry_Textbook_Maps/Time_Dependent_Quantum_Mechanics_and_Spectroscopy_(Tokmakoff)/15\LY1\textbackslash%3A_Energy_and_Charge_Transfer/15.05\LY1\textbackslash%3A_Marcus_Theory_for_Electron_Transfer (accessed 5 January 2024).

36 Lawrence, J.E., Fletcher, T., Lindoy, L.P., and Manolopoulos, D.E. (2019). On the calculation of quantum mechanical electron transfer rates. *J. Chem. Phys.* 151(11): 114119. https://doi.org/10.1063/1.5116800.

37 Heller, E.R. and Richardson, J.O. (2020). Semiclassical instanton formulation of Marcus–Levich–Jortner theory. *J. Chem. Phys.* 152(24): 244117. https://doi.org/10.1063/5.0013521.

38 Stehr, V., Fink, R.F., Tafipolski, M. et al. (2016). Comparison of different rate constant expressions for the prediction of charge and energy transport in oligoacenes. *WIREs Comput. Mol. Sci.* 6(6): 694–720. https://doi.org/10.1002/wcms.1273.

39 Bussolotti, F., Han, S.W., Honda, Y., and Friedlein, R. (2009). Phase-dependent electronic properties of monolayer and multilayer anthracene films on graphite [0001] surfaces. *Phys. Rev. B* 79(24): 245410. https://doi.org/10.1103/PhysRevB.79.245410.

40 Kera, S., Yamane, H., and Ueno, N. (2009). First-principles measurements of charge mobility in organic semiconductors: valence hole–vibration coupling in organic ultrathin films. *Prog. Surf. Sci.* 84(5): 135–154. https://doi.org/10.1016/j.progsurf.2009.03.002.

41 Yamane, H., Nagamatsu, S., Fukagawa, H. et al. (2005). Hole-vibration coupling of the highest occupied state in pentacene thin films. *Phys. Rev. B* 72(15): 153412. https://doi.org/10.1103/PhysRevB.72.153412.

42 Dexter, D.L. (2004). A theory of sensitized luminescence in solids. *J. Chem. Phys.* 21(5): 836–850. https://doi.org/10.1063/1.1699044.

43 Marcus, R.A. (1982). The second R. A. Robinson memorial lecture. Electron, proton and related transfers. *Faraday Discuss. Chem. Soc.* 74: 7–15. https://doi.org/10.1039/DC9827400007.

44 Marcus, R.A. (1965). On the theory of electron-transfer reactions. VI. Unified treatment for homogeneous and electrode reactions. *J. Chem. Phys.* 43(2): 679–701. https://doi.org/10.1063/1.1696792.

45 Coropceanu, V., Malagoli, M., Da Silva Filho, D.A. et al. (2002). Hole- and electron-vibrational couplings in oligoacene crystals: intramolecular contributions. *Phys. Rev. Lett.* 89(27): 275503. https://doi.org/10.1103/PhysRevLett.89.275503.

46 Hsu, C.-P., Georgievskii, Y., and Marcus, R.A. (1998). Time-dependent fluorescence spectra of large molecules in polar solvents. *J. Phys. Chem. A* 102(16): 2658–2666. https://doi.org/10.1021/jp980255n.

47 Hsu, C.-P., Song, X., and Marcus, R.A. (1997). Time-dependent stokes shift and its calculation from solvent dielectric dispersion data. *J. Phys. Chem. B* 101(14): 2546–2551. https://doi.org/10.1021/jp9630885.

48 Blasse, G. (1991). Optical electron transfer between metal ions and its consequences. In: *Complex Chemistry, Structure and Bonding*, vol. 76, 153–187. Berlin, Heidelberg: Springer-Verlag. ISBN: 978-3-540-46631-4. https://doi.org/10.1007/3-540-53499-7_3.

49 Hupp, J.T. and Weydert, J. (1987). Optical electron transfer in mixed solvents. Major energetic effects from unsymmetrical secondary coordination. *Inorg. Chem.* 26(16): 2657–2660. https://doi.org/10.1021/ic00263a021.

50 Hilborn, R.C. (1982). Einstein coefficients, cross sections, f values, dipole moments, and all that. *Am. J. Phys.* 50(11): 982–986. https://doi.org/10.1119/1.12937.

51 Lin, S.H. (1968). Spectral band shape of absorption and emission of molecules in dense media. *Theor. Chim. Acta* 10(4): 301–310. https://doi.org/10.1007/BF00526493.

52 Feng, X., Luzanov, A.V., and Krylov, A.I. (2013). Fission of entangled spins: an electronic structure perspective. *J. Phys. Chem. Lett.* 4(22): 3845–3852. https://doi.org/10.1021/jz402122m.

53 Smith, M.B. and Michl, J. (2013). Recent advances in singlet fission. *Annu. Rev. Phys. Chem.* 64(1): 361–386. https://doi.org/10.1146/annurev-physchem-040412-110130.

13

Computational Modeling of CO_2 Reduction and Conversion via Heterogeneous and Homogeneous Catalysis

Yue Zhang[1], Lin Zhang[1], Denghui Ma[1], Xinrui Cao[2], and Zexing Cao[1]

[1] Xiamen University, State Key Laboratory of Physical Chemistry of Solid Surfaces and Fujian Provincial Key Laboratory of Theoretical and Computational Chemistry, College of Chemistry and Chemical Engineering, No. 422, Siming South Road, Xiamen 360015, China
[2] Xiamen University, Department of Physics and Collaborative Innovation Center for Optoelectronic Semiconductors and Efficient Devices, Fujian Provincial Key Laboratory of Theoretical and Computational Chemistry, No. 422, Siming South Road, Xiamen 361005, China

13.1 Introduction

Carbon dioxide (CO_2) is one of the main greenhouse gases, and the continuous increase of its atmospheric concentration has a severe impact on environmental change and global warming. Nevertheless, as an important C_1 resource, CO_2 can be transformed into usable chemical feedstocks and liquid fuels through photocatalysis, electrocatalysis, and thermocatalysis [1–5]. Photocatalytic CO_2 reduction provides an ideal way to convert renewable solar energy into chemical energy, but this strategy still faces many challenges in the stability and activity of photocatalysts, as well as in the use of noble metals and organic electron sacrificial agents. CO_2 electroreduction can directly transform electric energy to chemical energy, although it may suffer from low selectivity, high overpotential, and low faradaic efficiency. Thermocatalytic hydrogenation of CO_2, using hydrogen sources generated by renewable energy, is also quite a promising mode for current CO_2 utilization.

Generally, formic acid (HCOOH) and methanol (CH_3OH) from CO_2 reduction are more suitable for transport and storage as C_1 liquid fuels, and as important platform molecules, they can not only be used for the synthesis of other chemicals but also serve as promising candidates for H_2 storage and raw materials for fuel cells. However, due to the extremely high thermodynamic stability and kinetic inertia, the activation and catalytic transformation of CO_2 are usually quite energy-demanding and challenging. In the past decades, much effort has been made to tackle the inertness of CO_2 in its hydrogenation reduction, and many novel low-dimensional, nanomaterial-supported, atomically dispersed metal catalysts were developed [6–10]. In particular, two-dimensional (2D) nanomaterials, such as graphene, N-doped carbon sheets, boron nitride sheets, and graphitic carbon nitride (g-C_3N_4), are often used as substrates for single-atom catalysts (SACs)

Exploring Chemical Concepts Through Theory and Computation, First Edition. Edited by Shubin Liu.
© 2024 WILEY-VCH GmbH. Published 2024 by WILEY-VCH GmbH.

toward CO_2 reduction reactions, owing to their specific geometric structures and unique electronic properties. In general, precious metal catalysts (e.g. Rh, Ru, Pd, Pt, Au, and so on) for CO_2 conversion are relatively efficient but costly. Although SACs maximize the use of noble metal atoms [11–15], their effective replacement by nonprecious metals or even nonmetals is highly required for the development of high-performance catalysts toward CO_2 conversion, which is reminiscent of 2D materials confining single atoms for the catalytic reduction of CO_2.

Except for the hydrogenation reduction, the catalytic conversion of CO_2 into useful organic compounds has received considerable attention in the chemical fixation of CO_2 [16–18]. The coupling reaction of CO_2 and epoxides was demonstrated to be the most sustainable route for the generation of cyclic carbonates [19–23] since this cycloaddition is 100% atom economy reaction and the cyclic carbonate products promise wide application perspective due to their high boiling point, low toxicity, good hydrophilicity, and biodegradability. The synthesis of cyclic carbonates through cycloaddition generally involves three steps – ring opening, CO_2 insertion, and ring closure – which require the use of effective catalysts and cocatalysts. In past decades, many kinds of catalysts have been developed for the selective coupling of CO_2 and epoxide in homogeneous and heterogeneous catalysis [16–23].

So far, lots of review articles have summarized recent advances in CO_2 capture, utilization, and storage [24–30]. Here, several types of low-dimensional, nanomaterial-supported, atomically dispersed metal and nonmetal catalysts for electrocatalytic and thermocatalytic reduction of CO_2, as well as metal macrocycles for the coupling of CO_2 with epoxide, were constructed and characterized computationally. Detailed reaction mechanisms and related correlationships between structure and activity were explored, based on extensive quantum-mechanical calculations and molecular dynamics (MD simulations on homogeneous and heterogeneous catalytic transformation of CO_2.

13.2 Computational Methods

With the rapid development of computational methodologies and computer technologies, electronic structure calculation and MD simulations are widely used to explore the mechanistic aspects of homogeneous and heterogeneous catalysis and facilitate the rational design of multifunctional catalysts [27, 31–33]. In the past decade, the massive accumulation of data and huge advances in data-driven science also opened a new avenue for chemists to interpret and predict complex reactions in catalysis using machine learning (ML) as a novel tool [34–39].

In computational heterogeneous catalysis here, all calculations were performed by the Vienna ab initio simulation package (VASP) [40] with the spin polarization density functional theory (DFT). The electron exchange correlation functional was described by the GGA-PBE functional [41], and the Van der Waals interactions were evaluated by using the DFT-D3 method of Grimme [42]. The cutoff energy of plane-wave basis sets was set to be 450–500 eV, and the K-point grid was set to $5 \times 5 \times 1$ generated with the Monkhorst–Pack method. The convergence tolerances of total energy and atomic force for geometry optimization are 10^{-5} eV and 0.02 eV/Å, respectively. The climbing image nudged elastic band (CINEB) [43] and dimer method [44] were used to locate the transition state (TS), and the force threshold was set at 0.05 eV/Å. A vacuum distance of 20 Å was used to avoid periodic interlayer interaction. Bader charge analysis [45] was employed to describe the charge transfer performance. The solvent effect was evaluated by using the implicit water solvent model in VASPsol [46]. All TS structures were verified by vibrational frequency analyses. The structural stability was evaluated by ab initio molecular dynamics (AIMD) simulations and phonon spectrum calculations, where AIMD simulations were carried out in the canonical ensemble with a constant particle number, constant volume and a temperature (NVT) by using the Nose–Hoover thermostat method [47] with a time step of 1 fs under different temperatures.

The binding energy (ΔE_b) of the metal atom anchored on the substrate material is calculated as:

$$\Delta E_b = E_{M@Substrate} - (E_{Substrate} + E_M)$$

where $E_{M@Substrate}$, $E_{Substrate}$, and E_M are the energies of M@Substrate, the substrate, and the single metal atom, respectively. The dissolution potential (U_{dis}) [48] of M in M@Substrate at pH = 0 is calculated by:

$$U_{dis} = U_M^0 + [E_{M,bulk} - (E_{M@Substrate} - E_{Substrate})]/ne$$

where U_M^0 is the standard dissolution potential of M, n is the number of electrons involved in the dissolution, and e is the unit charge.

The Gibbs free energy change (ΔG) for elementary steps of CO_2 electroreduction reaction (CO_2ER) can be estimated by using the computational hydrogen electrode (CHE) model [49], which is expressed as:

$$\Delta G = \Delta E + \Delta ZPE + \int C_p dT - T\Delta S + \Delta G_{pH} + \Delta E_{sol}$$

where ΔE, ΔZPE, $\int C_p dT$, and $T\Delta S$ are the electron energy difference, zero-point energy change, enthalpy change, and entropy change, respectively. ΔG_{pH} is expressed as $\Delta G_{pH} = k_B T \ln 10 \times pH$, and here, the pH value was assumed to be 0. ΔE_{sol} is the energy correction for the solvation effect. The limiting potential (U_L) for CO_2ER is obtained from the potential-limiting step with the maximum free energy change (ΔG_{MAX}) defined as $U_L = -\Delta G_{MAX}/e^-$. The overpotential ($\eta$) of CO_2ER is the difference between the equilibrium potential (U_0) and U_L.

For homogeneous catalytic conversion of CO_2, all calculations were performed by the Gaussian 09 program [50], and different functionals have been tested and considered. In full geometry optimization and single-point energy calculations, different-sized basis sets were chosen, in consideration of the balance between the computational cost and accuracy. The stable structures of reactants, intermediates, and products are identified by all positive frequencies, while the TSs are characterized by only one imaginary frequency. The minimum energy path (MEP) is constructed by the intrinsic reaction coordinate (IRC) method to verify if the TS correlates with two desired minima on the potential energy surface. In consideration of the overestimation of the entropic contribution to the Gibbs free energy, from the gas-phase calculations for the reaction step with different numbers of reactant and product molecules in the condensed phase, a correction of -2.6 (or 2.6) kcal/mol ($T = 298.15$ K) was applied to calibrate the relative free energies for the 2 : 1 (or 1 : 2) transformation according to the free volume theory [51–53].

13.3 Activation and Reduction of CO_2

Nonpolar linear CO_2 is extremely thermodynamically stable, and its activation requires electron capture or injection since it has relatively low-lying π^* antibonding orbitals to accommodate extra electrons. Generally, the catalytic reduction of CO_2 can be achieved with the use of homogeneous and heterogeneous catalysts. In heterogeneous catalysis, atomically dispersed metal catalysts have received considerable attention owing to their excellent catalytic activity and the full utilization of metal atoms [6–15]. Of these, low-dimensional nanomaterials, including one-dimensional nanotubes and 2D nanosheets, have emerged as promising candidates for the catalyst substrate in electrocatalytic and thermocatalytic reduction of CO_2 to liquid fuels and platform chemicals, owing to their large specific surface area and unique physical and chemical properties [54, 55]. By tuning the surface structure and charge distribution of 2D materials, the interaction of reactants and intermediates with the active site can be modified, thereby affecting the catalytic activity. Here, we summarize our recent theoretical studies of atomically dispersed metal and silicon atoms on carbon-based BN, and MXene materials for the electrochemical and thermocatalytic reduction of CO_2, as well as plausible mechanisms and various strategies to further improve stability, selectivity, and activity of low-dimensional nanomaterial catalysts.

13.3.1 Computational Catalyst Design

Carbon-based materials, such as graphene-based, carbon nanotubes, carbon-nitrogen-based (e.g. CN, C_2N, and $g-C_3N_4$), boron nitride (BN)-based materials, and MXene materials, are currently the most prevalent type of low-dimensional nanomaterials. Here, heteroatom doping, configuration modification, and an external electric field are used to regulate the properties of these carbon-based low-dimensional materials, and their application to CO_2 activation and reduction has been investigated through extensive first-principles calculations. Several types of low-dimensional nanomaterial catalysts considered here are depicted in Figure 13.1, and intrinsic mechanisms for the enhancement of CO_2 activation and conversion efficiency, and the effects of doping atom and doping concentration, substrate curvature, steric hindrance, and external electric field, have been explored.

13.3.1.1 Doping of Metal and Nonmetal Atoms

By incorporating heteroatoms, such as transition metal and main group elements, into the substrate material, the surface chemical activity of the low dimensional material can be triggered due to electronegativity difference and charge transfer

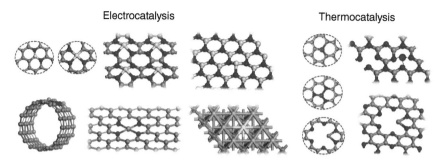

Figure 13.1 Atomically dispersed catalysts designed for electrocatalytic and thermocatalytic reduction of CO_2.

between the substrate and heteroatoms. The introduction of heteroatoms may change the electronic and magnetic properties of the material, which becomes a feasible method to develop new materials with better or specific physical and chemical properties. Herein, we constructed carbon-based and MXenes materials by embedding the single transition-metal atom into low-dimensional Mo_2C and N-doped carbon materials [56, 57], and the single Si atom into BN and graphene-based sheets [58–60]. First-principles calculations and AIMD simulations show that these 2D materials with atomically dispersed moieties are predicted to have high thermodynamic and dynamic stability, and heteroatom doping can modify the d-band or p-band center and the electronic feature of the active site, resulting in better activation of CO_2 and modulation of the catalytic performance.

13.3.1.2 Structural Modification

The curvature of the support material, local coordination environment of the active site, and steric hindrance of low-dimensional nanomaterial substrates can regulate the electronic and structural properties of atomically dispersed catalysts and their activities toward CO_2 reduction. We evaluated the curvature effect of carbon-based support materials on CO_2ER catalyzed by the atomically dispersed Cu and Fe on N-doped nanotubes and N-doped graphene [57], and the local bonding effect of single Si sites for CO_2 activation, capture, and electroreduction on monolayer materials [60], and notable structural modification effects were found.

Based on the valence-shell electron structural character of main group elements Al and N, we designed several heterogeneous frustrated Lewis acid/base pair (FLP) catalysts for CO_2 activation and hydrogenation. Moreover, using the unique pore structure of C_3N_4 monolayers, we constructed the nonprecious metal single-atom-embedded graphitic s-triazine-based C_3N_4 catalysts for hydrogenation reduction of CO_2. These 2D materials as single-atom catalysts (SACs) are predicted to have high thermal and dynamic stability, as well as high activity toward CO_2 hydrogenation to value-added chemicals [61, 62].

13.3.1.3 Application of an External Electric Field

External electric fields (EEF) can regulate the adsorption of CO_2 on the surface of catalysts, and the tunable, easy-to-operate, and low-energy characteristics of EEFs may be well coordinated with 2D nanomaterials for further tuning of the catalytic performance. For example, application of an external electric field and introduction of excess electrons to the 2D SiN_4C_4 sheet may remarkably promote activation and capture of CO_2, while other gas molecules of CH_4, N_2, and H_2 are almost not influenced, which may facilitate the separation and capture of CO_2 from these gas mixtures [59]. In the catalytic coupling of CH_4 with CO_2 into acetic acid by a Rh(I)-substituted human carbonic anhydrase, the application of the oriented EEF can also remarkably reduce the free energy span [63].

13.3.2 Electrocatalytic Reduction of CO_2

The CO_2ER is a promising method for converting CO_2 into liquid fuels and chemicals under mild ambient conditions, such as C_{1-2} products of CO, HCOOH, CH_3OH, CH_4, C_2H_4, and CH_3CH_2OH. CO_2ER effectively converts the electricity generated from renewable energy sources into carbon energy, mitigating greenhouse gas emissions. The half electrochemical reactions and corresponding thermodynamic equilibrium potentials for CO_2ER are compiled into Figure 13.2a, and plausible pathways for CO_2ER to several C_1/C_2 products on the electrode surface are presented in Figure 13.2b.

In principle, the proton-assisted multielectron transfer process is thermodynamically more favorable than the direct reduction of CO_2 to CO_2^- by one electron. As shown in Figure 13.2b, the simplest process for CO_2ER to C_1 products CO and HCOOH consists of two proton-coupled electron transfer steps. Other C_1/C_2 products, involving more protons and electrons, can be produced through more complicated reaction steps. However, hydrogen evolution reduction (HER) is competitive with CO_2ER (Figure 13.2a), which may limit the overall catalytic efficiency. Accordingly, an ideal catalyst should facilitate proton–electron transfer with relatively low energy barriers. Additionally, it is also essential for the effective conversion of CO_2 to enhance the ability to capture electrons of its antibonding orbitals.

Graphene is a chemically inert 2D material with low catalytic activity for CO_2ER, and it can be activated by doping with heteroatoms such as transition metals, noble metals, and nonmetallic elements. For example, recent calculations reveal that the main group single-atom Si-embedded N-doped graphene nanosheet and the SiN_4C_4 monolayer with the planar tetracoordinate silicon (ptSi) can effectively activate CO_2 under ambient temperatures, exhibiting good activity toward CO_2ER [59, 60]. Furthermore, CO_2 activation at the atomically dispersed Si site strongly depends on its local bonding environment [58, 60]. It is interesting to find that the more N atoms coordinate with Si in both SiN_xC_{3-x} and SiN_xC_{4-x} sheets, the more significant the CO_2 activation is, as shown in Figure 13.3a. Notably, the three-coordinated Si of SiN_xC_{3-x} acts as an electron donor in CO_2 activation, and its activity is dominated by both its electron population and the Si-p_z band center. On the contrary, the four-coordinated Si of SiN_xC_{4-x} behaves as an electron shuttle for the electron transfer from the SiN_xC_{4-x} framework to CO_2, where the deeper Si-p_z band center is beneficial to the activation of CO_2, differing from the conventional p-band center model for the p-block element.

13.3 Activation and Reduction of CO_2

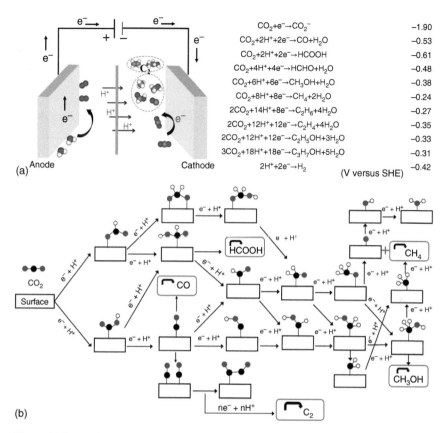

Figure 13.2 Half electrochemical reactions for CO_2ER (a) and plausible pathways for CO_2ER to C_1/C_2 products on the electrode surface (b).

Predicted relative free energy diagrams for CO_2ER to CO, HCOOH, CH_3OH, and CH_4 on SiN_3C_0 are given in Figure 13.3b. The calculated limit potentials for the generation of CO, HCOOH, and CH_3OH/CH_4 are −0.95, −0.54, and −0.59 V, respectively, suggesting that SiN_3C_0 as the electrocatalyst for CO_2ER is more efficient than Cu–NC and Cu(111) surfaces. Furthermore, activation and chemisorption of CO_2 on the SiN_4C_4 sheet are almost barrier-free and favorable energetically. The electroreduction of CO_2 to HCOOH, CH_3OH, and CH_4 on the SiN_4C_4 sheet has low limiting potentials of about −0.46 V.

CO_2ER on single-atom metal sites embedded in N-doped-graphene (M@N-Gr, M = Cu, and Fe) and carbon nanotubes (M@N-CNT) has been explored by extensive first-principles calculations in combination with the computational hydrogen electrode model [57]. The results showed that the catalytic activity of the single-Fe-atom catalysts for generating CO, HCOOH, and CH_3OH was higher than that of the single-Cu-atom catalysts. Additionally, the limiting potentials for C_1 products on the high-curvature Cu@N-CNT are significantly lower, compared to Cu@N-Gr. However, the curvature effect was less pronounced for single-Fe-atom catalysts.

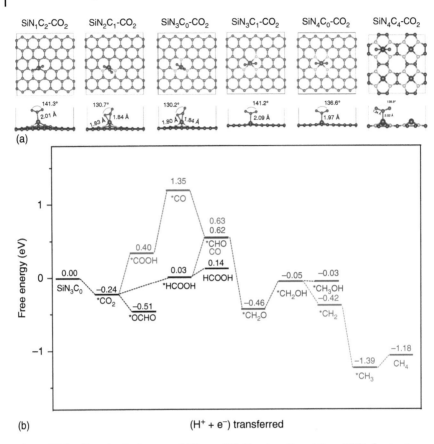

Figure 13.3 Chemisorbed states of CO_2 on SiN_xC_{n-x} ($n = 3$ and 4) and SiN_4C_4 sheets (a) and free energy diagrams for CO_2ER to CO, HCOOH, CH_3OH, and CH_4 on SiN_3C_0 (b).

2D transition-metal carbide materials (MXenes) have great potential in the fields of energy utilization and catalysis. Note that Mo_2C electrocatalysts can reduce CO_2 to CH_4 with low overpotentials, but also accelerate the HER. By embedding the single transition-metal atom into Mo_2C MXenes, we constructed a series of transition metal-embedded Mo_2C (M@Mo_2C) catalysts, and plausible mechanisms and pathways for CO_2ER to C_1/C_2 products on M@Mo_2C were investigated by first-principles calculations [56]. The computational results indicated that M@Mo_2C (M = Cr, Mn, Fe, and Co) catalysts showed higher activity and selectivity toward CO_2ER than HER, compared to the pristine Mo_2C. Notably, Fe@Mo_2C and Co@Mo_2C are quite promising for CO_2ER to CH_4 or CH_3OH with low limiting potential. Fe@Mo_2C, on the other hand, has a smaller $E_B(OH)$, breaking the inherent linear scaling of binding energies of key intermediates, and is predicted to be more active for CO_2ER to C_2 products C_2H_4 and CH_3CH_2OH.

13.3.3 Hydrogenation Reduction of CO_2

Heterogeneous catalytic hydrogenation of CO_2 is also quite appealing in reducing atmospheric CO_2 emissions and accelerating the optimal utilization of carbon

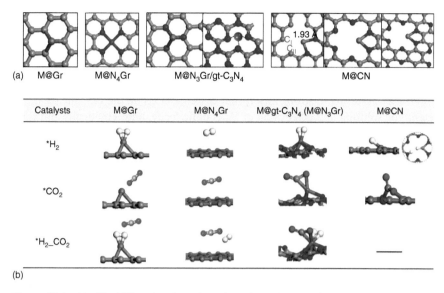

Figure 13.4 Modified 2D carbon-based catalysts (a) and adsorption configurations of H_2, CO_2, and their coadsorption at the active site (b).

resources. Among C_1 molecules, CH_3OH and $HCOOH$ are key chemical feedstocks and excellent liquid fuels that are easy to store and transport, and their preparation for CO_2 reduction has been receiving considerable attention [64]. However, the efficient hydrogenation of CO_2 to $HCOOH$, CH_3OH, and CH_4 still faces tremendous challenges due to the chemical inertness of CO_2 and the notable energy demand for the reduction of the highest oxidation state of carbon [65], and thus the development of novel catalysts with high activity to tackle the inertness of CO_2 is highly required. Ideally, the catalyst should be capable of adsorbing and activating both CO_2 and H_2 simultaneously and facilitating the formation of hydrogenated products. Here, we reviewed the most recent computational modeling of CO_2 hydrogenation on atomically dispersed metal catalysts supported by 2D carbon-based materials, including graphene, N-doped graphene, and C_3N_4 and C_2N sheets, as shown in Figure 13.4, and the reaction mechanisms and activities for the thermocatalytic reduction of CO_2 on atomically dispersed metal catalysts of M@Gr, M@N_4Gr, M@N_3Gr, M@gt-C_3N_4, and M@CN were discussed, based on extensive first-principles calculations.

It is well recognized that the surface defective sites and coordination environments of 2D carbon-based materials play vital roles in fabricating the atomically dispersed catalysts for the activation and hydrogenation of CO_2. The computational results demonstrated a stronger adsorption of H_2 than CO_2 on SACs supported on the doped and defective graphene (M@Gr, M = Cu, Ru, and Pd), along with an efficient activation of H_2 [66–68]. Similarly, in the coadsorption of H_2 and CO_2, CO_2 is physically adsorbed around the chemisorbed H_2.

Both M@gt-C_3N_4 and Co@N_3-Gr can effectively activate both CO_2 and H_2 and moreover, the coadsorption of H_2 and CO_2 on these SACs is energetically more favorable than their individual adsorption configurations [62, 69, 70]. On the contrary,

Co@N$_4$-Gr exhibits weak surface reactivity, and H$_2$ and CO$_2$ are only physically adsorbed on the surface [69]. According to the local structural environment of the anchored main-group single atom M (M = B, Al, and Ga), three types of M@NC catalysts with the surface frustrated Lewis pair (FLPs) were constructed, as shown in Figure 13.4. Interestingly, these M@NC catalysts can activate both CO$_2$ and H$_2$ remarkably, resulting in strong adsorption of CO$_2$ and H$_2$ on the FLP site [61]. The heterolytic splitting and adsorption of H$_2$ is quite facile, both kinetically and thermodynamically, superior to CO$_2$ adsorption on the surface.

Catalytic reaction mechanisms usually depend on the components of atomically dispersed catalysts, such as loaded metal atoms, support materials, and their combinatorial construction forms. Two types of reaction mechanisms are generally concerned, based on the initial state of reactants on the catalyst, i.e. Eley–Rideal (ER) mechanism and Langmuir–Hinshelwood (LH) mechanism, as shown in Figure 13.5. CO$_2$ hydrogenation on M@gt-C$_3$N$_4$ (M = Mn, Fe, Co, Ni, Cu, and Mo) follows the LH mechanism [62], where the primary reaction pathways include the HCOO* pathway to produce HCOOH, the reverse water-gas-shift (RWGS) pathway to generate CO, and the mixed pathway (RWGS + CO-hydro) to yield CH$_3$OH. Based on the coadsorption configuration H$_2$*_CO$_2$*, the reaction of H$_2$* with CO$_2$* may form HCOO* and COOH* intermediates, and HCOO* can be further hydrogenated to HCOOH*. Alternatively, the H adatoms from the adsorbed activated H$_2$ consecutively attack the O atom of adsorbed CO$_2$ to generate CO and H$_2$O (RWGS pathway). Note that Co@gt-C$_3$N$_4$ exhibits the highest activity. Since the CO* desorption is difficult, the nascent CO* from the RWGS process tends to further hydrogenate into CH$_3$OH from the coadsorption state of CO and H$_2$ at the Co site, i.e. CO_H$_2$* intermediate. Subsequently, the consecutive coupling of CO* with the hydrogen atom forms CH$_3$OH*. The overall pathway for CO$_2$ hydrogenation to HCOOH and CH$_3$OH on Co@gt-C$_3$N$_4$ is shown in Figure 13.5.

The ER mechanism for CO$_2$ hydrogenation usually includes (i) H$_2$ chemisorption or heterolytic splitting of H$_2$ into a metal-bound hydride and a proton accommodated by the metal-embedded C/N defect site of carbon materials; (ii) insertion of CO$_2$ into M-H to form HCOO*_H*/COOH*_H* intermediates; and (iii) hydrogenation of HCOO*/COOH* to HCOOH* or HCOOH* and H*_H* intermediates through H adatom or the second H$_2$ molecule. We note that CO$_2$ hydrogenation on several heterogeneous frustrated Lewis acid/base pair (FLP) catalysts (Al@N-Gr-1, Al@N-Gr-2, and Al@C$_2$N) resembles the ER mechanism [61], and the hydrogenation reaction proceeds through a three-step mechanism, i.e. heterolytic dissociation of H$_2$ to Al-H and C/N-H, the Al-bound hydride transfer to CO$_2$ to yield HCOO*, and HCOO* hydrogenation arising from the second H$_2$ molecule to HCOOH*. Among these FLP catalysts, Al@N-Gr-2 exhibits the best catalytic performance and the rate-determining step is the hydrogenation of HCOO* to HCOOH* arising from the second H$_2$ molecule, with an activation energy of 0.59 eV.

On the other hand, CO$_2$ hydrogenation on Ga@N-Gr-2 follows a two-step mechanism consisting of H$_2$ dissociation and subsequent hydrogen transfer, with an energy barrier of 0.83 eV for the rate-determining step. Similarly, the hydrogenation of CO$_2$ on M@Gr generally involves the dissociation of adsorbed H$_2$, the interaction

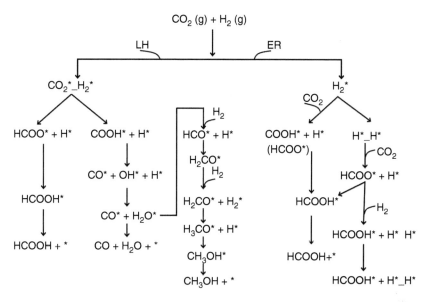

Figure 13.5 The proposed reaction networks for CO_2 hydrogenation through LH and ER mechanisms.

of H adatom with CO_2 to form HCOO*, and the formation of HCOOH* [68, 69]. Notably, the diverse characteristics of metal atoms can lead to the participation of the modified carbon support in the catalytic process through various pathways of H_2 adsorption and activation.

13.4 Catalytic Coupling of CO_2 with CH_4

Emission reduction and catalytic conversion of CH_4 and CO_2 have been increasingly attracting great concern in society, and their direct coupling to platform chemicals will provide one of the important measures to mitigate the environmental impact caused by CH_4 and CO_2 as main greenhouse gases. Given the chemical inertness of both CH_4 and CO_2, it remains quite difficult to achieve efficient catalytic conversion under mild conditions, and tremendous efforts have been made to tackle such challenges, both academically and industrially [71–74]. Traditional routes have mainly focused on CO_2 reforming of CH_4 into synthesis gas (i.e. $CO + H_2$) since the production of syngas is of high economic benefits. However, the preparation and utilization of syngas are still confronted with some problems, such as the involvement of multi-step post-treatment processes in desorption, storage, and transport of products, harsh requirements, and narrow selection spaces for catalysts, as well as security. Alternatively, the catalytic conversion of CH_4 and CO_2 to acetic acid and acetone through the C–C coupling is highly demanded in agriculture and pharmaceutical industries [75].

Biotransformation in the specific reaction environment may open up a promising pathway for chemically unfavorable reaction processes. For example, recent calculations indicate that the catalytic coupling of CH_4 with CO_2 into acetic acid catalyzed by a rhodium(I)-substituted human carbonic anhydrase [Rh(hCAII)] can be achieved under the oriented external electric field (OEEF) [63]. Figure 13.6 shows the relative free energy profiles for the entire catalytic conversion of CH_4 and CO_2 into CH_3COOH, without the application of OEEFs, and under an external electric field of $F_x = +0.0075$ au along the direction of the C–C formation of the rate-determining step in this transformation. Generally, the catalytic reaction is composed of three steps: (i) CH_4 activation and coordination of CO_2; (ii) CO_2 insertion and C–C coupling; (iii) CH_3COOH formation and desorption. As Figure 13.6 shows, the CO_2 insertion coupled with the C–C bond formation is the rate-determining step for the catalytic transformation of CH_4 and CO_2 into acetic acid. The application of an oriented external electric field ($F_x = +0.0075$ au) can significantly reduce the free energy barrier from 37.4 to 19.3 kcal/mol, suggesting that the introduction of an oriented external electric field could regulate the local electric field inside the active site of the enzyme and the spatial configuration of protein skeletons and related residues to a certain extent, which makes the improvement of the catalytic ability become possible.

Besides, the heterogeneous catalytic co-conversion of CH_4 and CO_2 into acetic acid and acetone also receives considerable attention [76–79], both experimentally and theoretically. First-principles calculations and microkinetics analysis show that $(ZnO)_3/In_2O_3(110)$ exhibits better activity towards the direct coupling of CO_2 and CH_4, compared to the interface of Ga_2O_3 and $(ZrO_2)_3$ supported on $In_2O_3(110)$, where the interface between the oxides plays a key role in the activation of CH_4 and CO_2 simultaneously [76]. Actually, the primitive ZnO is inactive for the coupling of CH_4 and CO_2, but the transition-metal doping can significantly improve the activity of ZnO. For example, the doping ZnO with Fe (Fe/ZnO) can remarkably promote the production of acetic acid from the co-conversion of CH_4 and CO_2, and such wonderful strong synergy of the Fe/ZnO catalyst has been identified by combining first-principles calculations and in situ DRIFTS experiments [71].

Most recently, catalytic conversion of CO_2 and CH_4 to acetic acid and acetone on the doped $In_2O_3(110)$ surfaces were explored by extensive first-principles calculations [80], and the Ga or Al substitution for the single In atom at the active oxygen vacancy site of $In_2O_3(110)$ can stabilize the reaction species and reduce the free energy barrier of the rate-limiting C–H activation for the coupling of CO_2 and CH_4 to acetic acid. The calculations reveal that the metal doping lowers the energy level of partially empty s and p orbitals of In_1 at the oxygen vacancy site and manipulates its electronic properties, resulting in activity improvement (see in Figure 13.7). The stable intermediate with the newly-formed CH_3COO* has the available In site for subsequent CH_4 activation, which may initiate the direct C–C coupling of CH_3COO* and CH_3* to yield C3 species on the doped $In_2O_3(110)$. These findings suggest that the metal doping of the active oxygen vacancy opens an avenue for carbon-chain growth through heterogeneously catalytic coupling of CO_2 and CH_4 [80].

As an ideal platform for the rational design of atomically accurate heterogeneous catalysts, metal–organic frameworks (MOFs) have received great attention in catalysis, owing to the high-density active sites available, good stability, high

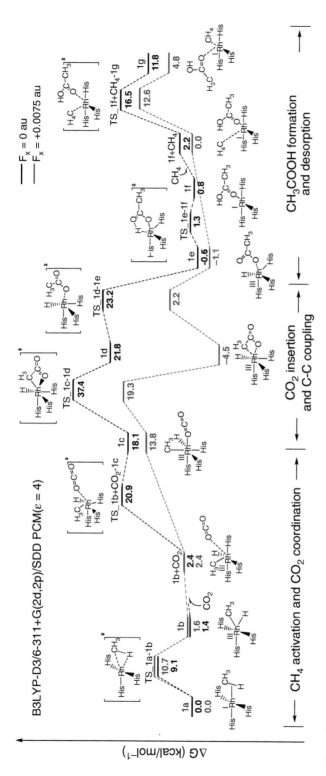

Figure 13.6 Predicted free energy profiles for the C–C coupling of CH_4 and CO_2 to CH_3COOH under no electric field (black) and an electric field with $F_x = +0.0075$ au (red) along the direction of the C–C bond by B3LYP-D3 calculations.

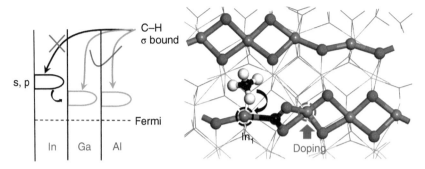

Figure 13.7 Diagram of energy level of the In_1 atom in coadsorption configurations of CH_4 and CO_2 on the undoped and doped $In_2O_3(110)$ surfaces.

accessibility, and chemical compatibility [81]. Very recently, computational design and mechanistic characterization of metal-alkoxide-functionalized MOF UiO-67 with well-defined dual sites for co-conversion of CO_2 and CH_4 into acetic acid were reported [79], and the metal dual-site UiO-67s can activate CO_2 and CH_4 cooperatively to promote production of acetic acid, compared to the single-site counterparts.

13.5 Homogeneous Catalytic Conversion of CO_2

Homogeneous catalysis for CO_2 conversion into high value-added chemicals, such as urea, HCOOH, methyl alcohol, and cyclic carbonates, also provides an effective means for optimal utilization of CO_2 under mild conditions. However, owing to the extremely high thermodynamic stability of nonpolar CO_2, its chemical fixation posed a challenge for energy requirement and proper use of catalyst. More specifically, the organometallic reaction mediated by transition metal complexes has been widely used for homogeneous catalytic transformation of CO_2, and coordination modes of CO_2 to a metal center and catalytic mechanisms have been well discussed and documented [82]. In past decades, various homogeneous catalysts [2, 5, 83–85], such as metal complexes, frustrated Lewis-pairs (FLPs), and N-heterocyclic carbenes (NHCs), were developed, and these catalytic systems play essential roles in facilitating the transformation of CO_2.

13.5.1 Catalytic CO_2 Fixation into Cyclic Carbonates

Cyclic carbonates are essential chemicals and raw materials, widely used in organic synthesis, fuel additives, and aprotic solvents. CO_2 has been considered a useful and sustainable feedstock for the production of cyclic carbonates because of its high atom utilization efficiency and low by-product production rate. In particular, the coupling reaction between CO_2 and epoxides, especially for propylene oxide (PO), has been identified as the most sustainable route.

Extensive DFT calculations have been used to elucidate the coupling mechanism of CO_2 with epoxide catalyzed by metal-porphyrin and -corrole complexes [86],

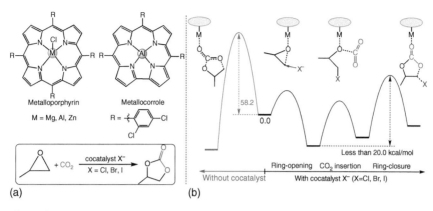

Figure 13.8 The cycloaddition reaction of CO_2 and PO catalyzed by metal-porphyrin and -corrole complexes (a), and the corresponding relative Gibbs free energy profiles (in kcal/mol) with and without the presence of cocatalyst ligand X^- (b).

and we examined how the *meso*-substitution of the macro ring and the presence of a cocatalyst can impact the reaction (see Figure 13.8). The computational results show that the cycloaddition reaction catalyzed by substituted and unsubstituted porphyrin- and corrole-based complexes follows a similar multistep mechanism, involving ring-opening, CO_2-insertion, and ring-closure. Generally, ring-opening and -closure steps experience relatively high free energy barriers, which can be used to evaluate the catalytic activity of metalloporphyrin complexes. In addition, the Lewis acidic metal center of the catalyst and the nucleophilic X^- (X = Cl, Br, I) ligand of the cocatalyst exert synergistically an influence on the initial ring-opening of epoxide. Overall, the presence of X^- may lower the free energy span of the whole reaction to below 20 kcal/mol. Although the ring-opening step is less influenced by the *meso*-substitution of the porphyrin and corrole rings with aryl and chlorinated aryl groups or resulting in a slight increase in the free energy barriers, the remarkable enhancement of binding interactions among catalyst, cocatalyst, and reactant can be achieved, which may greatly drive the initial reaction.

The catalytic coupling of CO_2 with PO by the bifunctional catalysts IL-M(TPP) and IL-M(Cor), fabricated by combining metal macrocycles (metalloporphyrin and metallocorrole) with imidazolium ionic liquid (see Figure 13.9), was explored, both experimentally and theoretically [87–89]. DFT calculations identified that the metal site of macrocycles in these catalysts serves as the catalytic center, except for the COOH-IL-Zn(TPP), and the Br^- anion acts as the nucleophile to facilitate the ring opening of PO. The synergistic effect between the electrophilic metal center and the flexible nucleophilic Br^- of the bifunctional catalysts is responsible for the high catalytic activity. In these bifunctional catalysts, the cooperation between the metal macrocycle and its bonded ionic liquid plays an important role in reducing the free energy barrier of the rate-determining step with the ring opening of PO. Accordingly, the metal macrocycles functionalized with imidazolium bromide are quite promising catalysts for CO_2 conversion, and their catalytic activity could be regulated by the metal center substitution, ligand structure modification, and countercation addition.

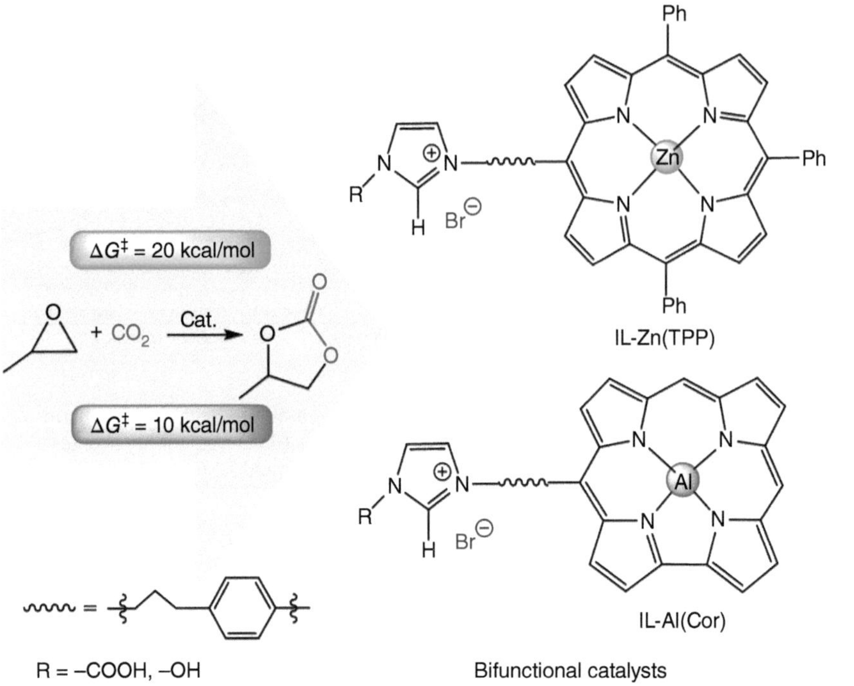

Figure 13.9 The cycloaddition reaction of CO_2 and PO catalyzed by bifunctional catalysts IL-Zn(TPP) and IL-Al(Cor).

13.5.2 CO_2 Hydrogenation Catalyzed by Metal PNP-Pincer Complexes

On the other hand, the homogeneous catalytic reduction of CO_2 also offers a feasible way of synthesizing valuable chemicals, including HCOOH, CH_3OH, and derivatives thereof, as well as other hydrogen storage products. As shown in Figure 13.10a, DFT calculations have been employed to investigate plausible mechanisms of three cascade cycles for the hydrogenation of CO_2 to CH_3OH catalyzed by a tri-coordinated Mn-PNP complex **^{1}Mn** [Mn(Ph$_2$PCH$_2$SiMe$_2$)$_2$NH(CO)$_2$Br] [90]. The H_2-splitting process was predicted to be the rate-determining step of each catalytic cycle, and the free energetic spans of the whole reaction in gas phase, water, toluene, and tetrahydrofuran are 27.1, 21.3, 20.8, and 20.4 kcal/mol, respectively. Such relatively low energy spans suggest that this manganese complex could be a promising catalyst for CO_2 hydrogenation into CH_3OH under mild conditions using H_2 as a hydrogen source.

Note that a major obstacle in base-assisted CO_2 hydrogenation reactions requests for an "external" sacrificial base to facilitate H_2 splitting and product release. This can be efficiently averted by introducing a secondary amine as an "internal base" into the pincer ligand, as shown in Figure 13.10b [91]. The mechanistic analysis and reaction energetics calculations on both tetra-coordinated PNPN-pincer complexes **^{2}Mn** and **^{3}Fe** reveal that the secondary amine moiety in-built in the ligand backbone plays a crucial role in the heterolytic H_2-splitting. Interestingly, the ruthenium

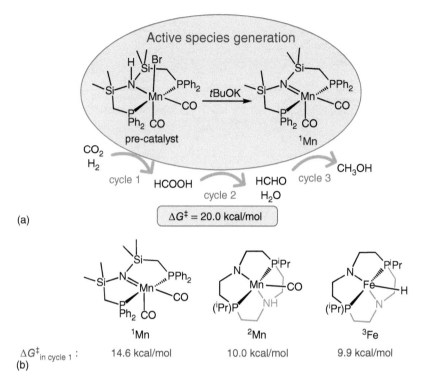

Figure 13.10 Proposed active catalytic species ^{1}Mn for the CO_2 hydrogenation to methanol with three catalytic cycles and the predicted free energetic span (in kcal/mol) of the whole reaction is given (a), and the predicted free energy barriers for the hydrogenation of CO_2 to HCOOH (cycle 1) catalyzed by different catalysts (b).

pincer complexes not only catalyze the hydrogenation of CO_2 to CH_3OH under mild conditions, but also trigger the dehydrogenation of CH_3OH, formaldehyde, and isopropanol to provide the hydrogen resource [92, 93]. These findings provide a basis for exploring the catalytic potential of a wide range of metal-pincer complexes and the rational design of novel catalysts for CO_2 reduction.

More recently, plausible pathways for CO_2 hydroboration into methyl boronate catalyzed by the Mn-PNP complex were investigated by DFT calculations, and both carbonyl association and carbonyl dissociation mechanisms were proposed, as shown in Figure 13.11 [94]. The predicted free energy spans of the carbonyl association and carbonyl dissociation mechanisms are 27.0 and 27.3 kcal/mol, respectively. The carbonyl release experiences a substantially high energy barrier of 35.2 kcal/mol for the rate-determining step, which may suppress the occurrence of the carbonyl dissociation mechanism. In addition, the effect of different trans ligands on the catalytic activity of manganese complex was explored, and the Mn^{I-} anionic complex with an H ligand exhibits the highest catalytic activity with a free energetic span of 21.0 kcal/mol. These computational results not only demonstrate the feasibility of homogeneous reduction of CO_2 to CH_3OH and its derivatives by using Mn-based pincer catalysts but also provide a basis for the design of metal catalysts for CO_2 conversion.

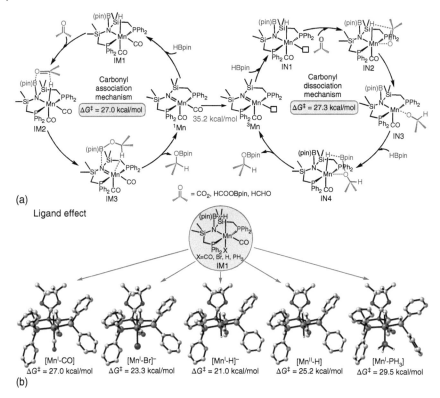

Figure 13.11 Proposed reaction mechanisms of the CO_2 hydroboration to methyl boronate catalyzed by the Mn-PNP complex (a), and geometric structures of complex **IM1** with different X ligands (X = CO, Br, H, PH_3) and the related free energy spans (in kcal/mol) for the carbonyl association mechanism are given (b). Less relevant H atoms are omitted for clarity.

13.6 Conclusion and Outlook

We summarized our recent work on computational modeling of CO_2 conversion and reduction through heterogeneous and homogeneous catalysis. Based on the N-doped graphene (N-Gr-1 and N-Gr-2), porous carbon nitride (C_2N) and graphitic s-triazine-based C_3N_4 (gt-C_3N_4) 2D materials, and the frustrated Lewis acid/base pair (FLP) concept, we constructed various single-atom catalysts (SACs) for CO_2 hydrogenation, where the solid FLP catalysts Al@N-Gr-2 and Ga@N-Gr-2, as well as the Co single-atom-embedded gt-C_3N_4 (Co@gt-C_3N_4) are predicted to be quite promising SACs for CO_2 hydrogenation to high value-added chemicals. The computationally designed SiN_4C_4 monolayer with the planar tetracoordinate silicon (ptSi) exhibits good activity toward CO_2 activation and electroreduction to C_1 products, and the application of an external electric field can facilitate CO_2 activation and capture under mild conditions. The local bonding environment of single Si sites may modify the electron mechanism for CO_2 activation, resulting in different linear correlations between the activity and the p-band center. These computational studies provide an alternative solution for CO_2 reduction and conversion by designing novel and effective single-atom catalysts based on nonprecious metals and main group elements. Most recently, the emergence and development of heterogeneous dual single-atom catalysts opens a new avenue for the construction of high-performance catalysts for CO_2 transformation under ambient conditions.

In the homogeneous reactive capture of CO_2, except for metal macrocycles functionalized with the ionic liquid or coupled with cocatalyst, the frustrated Lewis-pair (FLP) and N-heterocyclic carbene (NHC) catalysts also received considerable attention. Especially, phosphine-borane FLPs can effectively catalyze the copolymerization of CO_2 and PO, and their catalytic performance may be well manipulated by the electronic and steric properties of Lewis bases and acids. The most recently synthesized tetrakis(NHC)-diboron(0) compound has two highly reactive electrons in the π^* antibonding orbital and exhibits remarkable double single-electron-transfer (SET) reactivity, which can activate CO_2 through the SET mechanism to yield CO_2 radical anion $CO_2^{\bullet-}$. The continuous accumulation of atomic-level understanding of the catalytic mechanism for CO_2 conversion may facilitate the rapid development of CO_2 chemistry and carbon-neutral processes.

Acknowledgments

This work was supported by the National Science Foundation of China (21933009 and 22073076).

References

1 Wu, Q.J., Liang, J., Huang, Y.B. et al. (2022). Thermo-, electro-, and photocatalytic CO_2 conversion to value-added products over porous metal/covalent organic frameworks. *Acc. Chem. Res.* 55: 2978–2997.

2 Fors, S.A. and Malapit, C.A. (2023). Homogeneous catalysis for the conversion of CO_2, CO, CH_3OH, and CH_4 to C_{2+} chemicals via C–C bond formation. *ACS Catal.* 13: 4231–4249.

3 Qiu, L.Q., Yao, X.Y., Zhang, Y.K. et al. (2023). Advancements and challenges in reductive conversion of carbon dioxide via thermo-/photocatalysis. *J. Org. Chem.* 88 (8): 4942–4964.

4 Zhang, S., Fan, Q., Xia, R. et al. (2020). CO_2 reduction: from homogeneous to heterogeneous electrocatalysis. *Acc. Chem. Res.* 53: 255–264.

5 Siegel, R.E., Pattanayak, S., and Berben, L.A. (2023). Reactive capture of CO_2: opportunities and challenges. *ACS Catal.* 13: 766–784.

6 Zhai, S.L., Sun, J.K., Sun, L. et al. (2023). Heteronuclear dual single-atom catalysts for ambient conversion of CO_2 from air to formate. *ACS Catal.* 13: 3915–3924.

7 Zhu, H.W., Liu, S., Yu, J.H. et al. (2023). Computational screening of effective g-C_3N_4 based single atom electrocatalysts for the selective conversion of CO_2. *Nanoscale* 15: 8416–8423.

8 Yan, C., Liu, Y.L., Zeng, Q.Y. et al. (2023). 2D Nanomaterial supported single-metal atoms for heterogeneous photo/electrocatalysis. *Adv. Funct. Mater.* 33: 2210837.

9 Franco, F., Rettenmaier, C., Jeon, H.S. et al. (2020). Transition metal-based catalysts for the electrochemical CO_2 reduction: from atoms and molecules to nanostructured materials. *Chem. Soc. Rev.* 49: 6884–6946.

10 Lin, X.L., Ng, S.F., and Ong, W.J. (2022). Coordinating single-atom catalysts on two-dimensional nanomaterials: a paradigm towards bolstered photocatalytic energy conversion. *Coord. Chem. Rev.* 471: 214743.

11 Yang, X.F., Wang, A.Q., Qiao, B.T. et al. (2013). Single-atom catalysts: a new frontier in heterogeneous catalysis. *Acc. Chem. Res.* 46: 1740–1748.

12 Ji, S.F., Chen, Y.J., Wang, X.L. et al. (2020). Chemical synthesis of single atomic site catalysts. *Chem. Rev.* 120: 11900–11955.

13 Liang, R., Du, X.R., Huang, Y.K. et al. (2020). Single-atom catalysts based on the metal–oxide interaction. *Chem. Rev.* 120: 11986–12043.

14 Sarma, B.B., Maurer, F., Doronkin, D.E. et al. (2023). Design of single-atom catalysts and tracking their fate using operando and advanced X-ray spectroscopic tools. *Chem. Rev.* 123: 379–444.

15 Zhou, H.Y., Zhang, X., Liang, J.X. et al. (2020). Theoretical understandings of graphene-based metal single-atom catalysts: stability and catalytic performance. *Chem. Rev.* 120: 12315–12341.

16 Lidston, C.A.L., Severson, S.M., Abel, B.A. et al. (2022). Multifunctional catalysts for ring-opening copolymerizations. *ACS Catal.* 12: 11037–11070.

17 Centeno-Pedrazo, A., Perez-Arce, J., Freixa, Z. et al. (2023). Catalytic systems for the effective fixation of CO_2 into epoxidized vegetable oils and derivates to obtain biobased cyclic carbonates as precursors for greener polymers. *Indus. Eng. Chem. Res.* 62: 3428–3443.

18 Yang, G.W., Zhang, Y.Y., Xie, R. et al. (2020). Scalable bifunctional organoboron catalysts for copolymerization of CO_2 and epoxides with unprecedented efficiency. *J. Am. Chem. Soc.* 142: 12245–12255.

19 Shaikh, R.R., Pornpraprom, S., and D'Elia, V. (2018). Catalytic strategies for the cycloaddition of pure, diluted, and waste CO_2 to epoxides under ambient conditions. *ACS Catal.* 8: 419–450.

20 Ullah, H., Ullah, Z., Khattak, Z.A.K. et al. (2023). Formation of value-added cyclic carbonates by coupling of epoxides and CO_2 by ruthenium pincer hydrazone complexes under atmospheric pressure. *Energy Fuels* 37: 2178–2187.

21 Liu, N., Xie, Y.F., Wang, C. et al. (2018). Cooperative multifunctional organocatalysts for ambient conversion of carbon dioxide into cyclic carbonates. *ACS Catal.* 8: 9945–9957.

22 Gao, Z.Y., Liang, L., Zhang, X. et al. (2021). Facile one-pot synthesis of Zn/Mg-MOF-74 with unsaturated coordination metal centers for efficient CO_2 adsorption and conversion to cyclic carbonates. *ACS Appl. Mater. Interfaces* 13: 61334–61345.

23 Martin, C., Fiorani, G., and Kleij, A.W. (2015). Recent advances in the catalytic preparation of cyclic organic carbonates. *ACS Catal.* 5: 1353–1370.

24 Dunstan, M.T., Donat, F., Bork, A.H. et al. (2021). CO_2 capture at medium to high temperature using solid oxide-based sorbents: fundamental aspects, mechanistic insights, and recent advances. *Chem. Rev.* 121: 12681–12745.

25 Yang, X., Rees, R.J., Conway, W. et al. (2017). Computational modeling and simulation of CO_2 capture by aqueous amines. *Chem. Rev.* 117: 9524–9593.

26 Gao, W.L., Liang, S.Y., Wang, R.J. et al. (2020). Industrial carbon dioxide capture and utilization: state of the art and future challenges. *Chem. Soc. Rev.* 49: 8584–8686.

27 Li, L., Li, X.D., Sun, Y.F. et al. (2022). Rational design of electrocatalytic carbon dioxide reduction for a zero-carbon network. *Chem. Soc. Rev.* 51: 1234–1252.

28 Wei, J., Yao, R.W., Han, Y. et al. (2021). Towards the development of the emerging process of CO_2 heterogenous hydrogenation into high-value unsaturated heavy hydrocarbons. *Chem. Soc. Rev.* 50: 10764–10805.

29 Xia, Q., Zhang, K.E., Zheng, T.T. et al. (2023). Integration of CO_2 capture and electrochemical conversion. *ACS Energy Lett.* 8: 2840–2857.

30 Bierbaumer, S., Nattermann, M., Schulz, L. et al. (2023). Enzymatic conversion of CO_2: from natural to artificial utilization. *Chem. Rev.* 123: 5702–5754.

31 Jafarzadeh, M. and Daasbjerg, K. (2023). Rational design of heterogeneous dual-atom catalysts for CO_2 electroreduction reactions. *ACS Appl. Energy Mater.* 6: 6851–6882.

32 Yoon, Y.J., You, H.M., Kim, H.J. et al. (2022). Computational catalyst design for dry reforming of methane: a review. *Energy Fuels* 36: 9844–9865.

33 Yu, S.Y., Zhang, C., and Yang, H. (2023). Two-dimensional metal nanostructures: from theoretical understanding to experiment. *Chem. Rev.* 123: 3443–3492.

34 Chen, H.H., Zheng, Y.Z., Li, J.L. et al. (2023). AI for nanomaterials development in clean energy and carbon capture, utilization and storage (CCUS). *ACS Nano* 17: 9763–9792.

35 Mai, H.X., Le, T.C., Chen, D.H. et al. (2022). Machine learning for electrocatalyst and photocatalyst design and discovery. *Chem. Rev.* 122: 13478–13515.

36 Ma, S.C. and Liu, Z.P. (2020). Machine learning for atomic simulation and activity prediction in heterogeneous catalysis: current status and future. *ACS Catal.* 10: 13213–13226.

37 Mazheika, A., Wang, Y.G., Valero, R. et al. (2022). Artificial-intelligence-driven discovery of catalyst genes with application to CO_2 activation on semiconductor oxides. *Nat. Commun.* 13: 419.

38 Takahashi, K., Ohyama, J., Nishimura, S. et al. (2023). Catalysts informatics: paradigm shift towards data-driven catalyst design. *Chem. Commun.* 59: 2222–2238.

39 Zhu, A., Jiang, K.Z., Chen, B. et al. (2023). Data-driven design of electrocatalysts: principle, progress, and perspective. *J. Mater. Chem. A* 11: 3849–3870.

40 Kresse, G. and Furthmuller, J. (1996). Efficient iterative schemes for ab initio total-energy calculations using a plane-wave basis set. *Phys. Rev. B* 54: 11169–11186.

41 Perdew, J.P., Burke, K., and Ernzerhof, M. (1996). Generalized gradient approximation made simple. *Phys. Rev. Lett.* 77: 3865–3868.

42 Grimme, S., Antony, J., Ehrlich, S. et al. (2010). A consistent and accurate ab initio parametrization of density functional dispersion correction (DFT-D) for the 94 elements H-Pu. *J. Chem. Phys.* 132: 154104–154122.

43 Henkelman, G., Uberuaga, B.P., and Hennig, J. (2000). A climbing image nudged elastic band method for finding saddle points and minimum energy paths. *J. Chem. Phys.* 113: 9901–9904.

44 Henkelman, G. and Jonsson, J. (1999). A dimer method for finding saddle points on high dimensional potential surfaces using only first derivatives. *J. Chem. Phys.* 111: 7010–7022.

45 Bader, R.F.W. (1985). Atoms in molecules. *Acc. Chem. Res.* 18: 9–15.

46 Mathew, K., Kolluru, V.S.C., and Mula, S. (2019). Implicit self-consistent electrolyte model in plane wave density-functional theory. *J. Chem. Phys.* 151: 234101–234108.

47 Nosé, S. (1984). A unified formulation of the constant temperature molecular-dynamics methods. *J. Chem. Phys.* 81: 511–519.

48 Greeley, J. and Nørskov, J.K. (2007). Electrochemical dissolution of surface alloys in acids: thermodynamic trends from first principles calculations. *Electrochim. Acta* 52: 5829–5836.

49 Nørskov, J.K., Rossmeisl, J., and Logadottir, A. (2004). Origin of the overpotential for oxygen reduction at a fuel-cell cathode. *J. Phys. Chem. B* 108: 17886–17892.

50 Frisch, M.J., Trucks, G.W., Schlegel, H.B., et al. (2010). Gaussian 09, Revision B.01. Gaussian, Inc.: Wallingford, CT.

51 Berson, S.W. (1982). *The Foundations of Chemical Kinetics*. Malabar, FL: Krieger.

52 Okuno, Y. (1997). Theoretical investigation of the mechanism of the Baeyer-Villiger reaction in nonpolar solvents. *Chem. Eur. J.* 3: 212–218.

53 Schoenebeck, F. and Houk, K.N. (2010). Ligand-controlled regioselectivity in palladium-catalyzed cross coupling reactions. *J. Am. Chem. Soc.* 132: 2496–2497.

54 Voiry, D., Shin, H., Loh, K. et al. (2018). Low-dimensional catalysts for hydrogen evolution and CO_2 reduction. *Nat. Rev. Chem.* 2: 0105.

55 Mihet, M., Dan, M., and Lazar, M.D. (2022). CO_2 hydrogenation catalyzed by graphene-based materials. *Molecule* 27: 3367–3373.

56 Zhang, Y. and Cao, Z.X. (2021). Tuning the activity of molybdenum carbide MXenes for CO_2 electroreduction by embedding the single transition-metal atom. *J. Phys. Chem. C* 125: 13331–13342.

57 Zhang, Y., Fang, L., and Cao, Z.X. (2020). Atomically dispersed Cu and Fe on N-doped carbon materials for CO_2 electroreduction: insight into the curvature effect on activity and selectivity. *RSC Adv.* 10: 43075–43084.

58 Fang, L. and Cao, Z.X. (2021). CO_2 activation and capture on a Si-doped h-BN sheet: insight into the local bonding effect of single Si sites. *J. Phys. Chem. C* 125: 5048–5055.

59 Fang, L., Zhang, C.Y., Cao, X.R., and Cao, Z.X. (2020). Tackling the inertness of CO_2: facile activation and electroreduction on the metal-free SiN_4C_4 monolayer sheet. *J. Phys. Chem. C* 124: 18660–18669.

60 Fang, L. and Cao, Z.X. (2021). CO_2 activation at atomically dispersed Si sites of N-doped graphenes: insight into distinct electron mechanisms from first-principles calculations. *AIP Adv.* 11: 115302.

61 Zhang, Y., Mo, Y.R., and Cao, Z.X. (2021). Rational design of main group metal-embedded nitrogen-doped carbon materials as frustrated Lewis pair catalysts for CO_2 hydrogenation to formic acid. *ACS Appl. Mater. Interfaces* 14: 1002–1014.

62 Zhang, Y., Cao, X.R., and Cao, Z.X. (2022). Unraveling the catalytic performance of the nonprecious metal single-atom-embedded graphitic s-triazine-based C_3N_4 for CO_2 hydrogenation. *ACS Appl. Mater. Interfaces* 14: 35844–35853.

63 Ma, D.H., Xie, H.J., and Cao, Z.X. (2020). Catalytic coupling of CH_4 with CO_2 and CO by a modified human carbonic anhydrase combined with oriented external electric fields: mechanistic insights from DFT calculations. *Organometallics* 39: 4657–4666.

64 Jiang, X., Nie, X.W., Guo, X.W. et al. (2020). Recent advances in carbon dioxide hydrogenation to methanol via heterogeneous catalysis. *Chem. Rev.* 120: 7984–8034.

65 Alvarez, A., Bansode, A., Urakawa, A. et al. (2017). Challenges in the greener production of formates/formic acid, methanol, and DME by beterogeneously catalyzed CO_2 hydrogenation processes. *Chem. Rev.* 117: 9804–9838.

66 Sirijaraensrea, J. and Limtrakul, J. (2016). Hydrogenation of CO_2 to formic acid over a Cu-embedded graphene: a DFT study. *Appl. Surf. Sci.* 364: 241–248.

67 Sredojevic, D., Sljivancanin, Z., Brothers, E. et al. (2018). Formic acid synthesis by CO_2 hydrogenation over single-atom catalysts based on Ru and Cu embedded in graphene. *ChemistrySelect* 3: 2631–2637.

68 Ali, S., Iqbal, R., Khan, A. et al. (2021). Stability and catalytic performance of single-atom catalysts supported on doped and defective graphene for CO_2 hydrogenation to formic acid: a first-principles study. *ACS Appl. Nano Mater.* 4: 6893–6902.

69 Esrafili, M.D. and Nejadebrahimi, B. (2019). Theoretical insights into hydrogenation of CO_2 to formic acid over a single Co atom incorporated nitrogen-doped graphene: a DFT study. *Appl. Surf. Sci.* 475: 363–371.

70 Poldorn, P., Wongnongwa, Y., Mudchimo, T. et al. (2021). Theoretical insights into catalytic CO_2 hydrogenation over single-atom (Fe or Ni) incorporated nitrogen-doped graphene. *J. CO2 Util.* 48: 101532.

71 Nie, X., Ren, X., Tu, C. et al. (2020). Computational and experimental identification of strong synergy of the Fe/ZnO catalyst in promoting acetic acid synthesis from CH_4 and CO_2. *Chem. Commun.* 56: 3983–3986.

72 Puliyalil, H., Jurkovic, D.L., Dasireddy, V.D.B.C. et al. (2018). A review of plasma-assisted catalytic conversion of gaseous carbon dioxide and methane into value-added platform chemicals and fuels. *RSC Adv.* 8: 27481–27508.

73 Ban, T., Yu, X.Y., Kang, H.Z. et al. (2022). Design of single-atom and frustrated-Lewis-pair dual active sites for direct conversion of CH_4 and CO_2 to acetic acid. *J. Catal.* 408: 206–215.

74 Zhao, Y., Cui, C., Han, J. et al. (2016). Direct C-C coupling of CO_2 and the methyl group from CH_4 activation through facile insertion of CO_2 into $Zn-CH_3$ sigma-bond. *J. Am. Chem. Soc.* 138: 10191–10198.

75 Rahman, M.S. and Xu, Y. (2023). Acetate formation on metals via CH_4 carboxylation by CO_2: a DFT study. *Catal. Today* 416: 113891.

76 Zhao, Y., Wang, H., Han, J. et al. (2019). Simultaneous activation of CH_4 and CO_2 for concerted C-C coupling at oxide–oxide interfaces. *ACS Catal.* 9: 3187–3197.

77 Li, Y.F., Zheng, K., Shen, Y. et al. (2023). Acetic acid production from CH_4 and CO_2 via synergistic catalysis between Pd particles and oxygen vacancies generated in ZrO_2. *J. Phys. Chem. C* 127: 5841–5854.

78 Tu, C.Y., Nie, X.W., and Chen, J.G.G. (2021). Insight into acetic acid synthesis from the reaction of CH_4 and CO_2. *ACS Catal.* 11: 3384–3401.

79 Yang, K. and Jiang, J. (2022). Rational design of metal-alkoxide-functionalized metal-organic frameworks for synergistic dual activation of CH_4 and CO_2 toward acetic acid synthesis. *ACS Appl. Mater. Interfaces* 14: 52979–52992.

80 Ma, D.H. and Cao, Z.X. (2022). Electron regulation of single indium atoms at the active oxygen vacancy of $In_2O_3(110)$ for production of acetic acid and acetone through direct coupling of CH_4 with CO_2. *Chem. Asian J.* 17: e202101383.

81 Bavykina, A., Kolobov, N., Khan, I.S. et al. (2020). Metal-organic frameworks in heterogeneous catalysis: recent progress, new trends, and future perspectives. *Chem. Rev.* 120: 8468–8535.

82 Fan, T., Chen, X.H., and Lin, Z.Y. (2012). Theoretical studies of reactions of carbon dioxide mediated and catalysed by transition metal complexes. *Chem. Commun.* 48: 10808–10828.

83 Piccirilli, L., Rabell, B., Padilla, R. et al. (2023). Versatile CO_2 hydrogenation-dehydrogenation catalysis with a Ru–PNP/ionic liquid system. *J. Am. Chem. Soc.* 145: 5655–5663.

84 Fan, J., Koh, A.-P., and Zhou, J.S. (2023). Tetrakis(N-heterocyclic carbene)-diboron(0): double single electron-transfer reactivity. *J. Am. Chem. Soc.* 145: 11669–11677.

85 Wang, Y., Liu, Z.H., Guo, W.Q. et al. (2023). Phosphine-borane frustrated Lewis pairs for metal-free CO_2/epoxide copolymerization. *Macromolecules* 56: 4901–4909.

86 Li, P. and Cao, Z.X. (2018). Catalytic preparation of cyclic carbonates from CO_2 and epoxides by metal–porphyrin and – corrole complexes: insight into effects of cocatalyst and *meso*-substitution. *Organometallics* 37: 406–414.

87 Li, P. and Cao, Z.X. (2019). Catalytic coupling of CO_2 with epoxide by metal macrocycles functionalized with imidazolium bromide: insights into the mechanism and activity regulation from density functional calculations. *Dalton Trans.* 48: 1344–1350.

88 Chen, Y.J., Luo, R.C., Yang, Z. et al. (2018). Imidazolium-based ionic liquid decorated zinc porphyrin catalyst for converting CO_2 into five-membered heterocyclic molecules. *Sustain. Energy Fuels* 2: 125–132.

89 Maeda, C., Shimonishi, R., Miyazaki, J.Y. et al. (2016). Highly active and robust metalloporphyrin catalysts for the synthesis of cyclic carbonates from a broad range of epoxides and carbon dioxide. *Chem. Eur. J.* 22: 6556–6563.

90 Zhang, L., Pu, M., and Lei, M. (2021). Hydrogenation of CO_2 to methanol catalyzed by a manganese pincer complex: insights into the mechanism and solvent effect. *Dalton Trans.* 50: 7348–7355.

91 Moni, S. and Mondal, B. (2023). Correlation between key steps and hydricity in CO_2 hydrogenation catalysed by non-noble metal PNP-pincer complexes. *Catalysts* 13: 592.

92 Zhou, Y., Zhao, Y.Q., Shi, X.F. et al. (2022). A theoretical study on the hydrogenation of CO_2 to methanol catalyzed by ruthenium pincer complexes. *Dalton Trans.* 51: 10020–10028.

93 Lei, M., Pan, Y.H., and Ma, X.L. (2015). The nature of hydrogen production from aqucous-phase methanol dehydrogenation with ruthenium pincer complexes under mild conditions. *Eur. J. Inorg. Chem.* 794–803.

94 Zhang, L., Zhao, Y.Q., Liu, C. et al. (2022). Hydroboration of CO_2 to methyl boronate catalyzed by a manganese pincer complex: insights into the reaction mechanism and ligand effect. *Inorg. Chem.* 61: 5616–5625.

14

Excited States in Conceptual DFT

Frédéric Guégan[1], Guillaume Hoffmann[2], Henry Chermette[2], and Christophe Morell[2]

[1]Université de Poitiers – CNRS, IC2MP UMR 7285, 4, rue Michel Brunet TSA, 51106–86073 Cedex 9 Poitiers, France
[2]Université de Lyon, Institut des Sciences Analytiques, UMR 5280, CNRS, Université Lyon 1 - 5, rue de la Doua, F-69100 Villeurbanne, France

14.1 Introduction

The initial grounds of CDFT (conceptual Density Functional Theory) are Hohenberg and Kohn theorems [1]. Because of this, CDFT is usually seen as a ground-state theory. In this chapter, we propose to first show that the very conceptual framework used in CDFT (the so-called "perturbative perspective") is already inviting one to consider CDFT as a theory involving not only the ground state but excited states as well. From this point onwards, two main directions can be drawn: using excited states to highlight the chemical properties in the ground state and using excited states to probe excited states.

Most research efforts have been dedicated to the first idea, as one may expect from Rayleigh Schrödinger perturbation theory [2]; it will be the first section of this chapter. The second idea has been less explored, although it will certainly gain more interest in the following years, in line with the concomitant development of both computational resources, appropriate numerical models for the description of excited states, and experimental photochemical setups. This will be the topic of the second section of this chapter.

14.2 Exploring Ground State Properties Thanks to Excited States

14.2.1 Context and Justification

At the core of conceptual DFT lies the idea of perturbative responses: chemical properties are probed through the study of the responses of the studied system under careful designed perturbations, usually a variation in the electron count (probing

Exploring Chemical Concepts Through Theory and Computation, First Edition. Edited by Shubin Liu.
© 2024 WILEY-VCH GmbH. Published 2024 by WILEY-VCH GmbH.

redox processes) or a change in the external potential (approach of a reagent, change of geometry) [3–5].

Perturbation theory is thus underlying in every conceptual DFT approach. Let us focus in the following on time-independent perturbation theory (one often reasons in terms of stationary states). We will consider a chemical system represented by the Hamiltonian $\mathcal{H}$, associated to energies and wavefunctions E_n and ψ_n, respectively. For the sake of simplicity, we will here consider nondegenerate states. By using the Rayleigh–Schrödinger formalism [2], it is possible to express the change in energy and wavefunction to the application of a small perturbation $\delta\mathcal{H}$. For instance, at the first order the perturbed wavefunction of the ground state ($n = 0$) writes

$$\psi_0 = c_0 \left(\psi_0^{(0)} + \sum_{i \neq 0} \frac{\langle \psi_i^{(0)} | \delta\mathcal{H} | \psi_0^{(0)} \rangle}{E_0 - E_i} \psi_i^{(0)} \right) \quad (14.1)$$

where c_0 is a normalization constant, and (0) an index referring to the unperturbed states.

It is quite plain from this expression that the responses to perturbation in the ground state of any given system can be expanded over the set of excited states, which in turn suggests *valuable insight on the chemical properties in the ground state can be inferred from the analysis of the excited states*. We propose in the following to show how this can be actually achieved.

14.2.2 Chemical Hardness Revisited

A. Nagy was among the first to propose the use of excited state-related quantities to evaluate ground state properties. More specifically, she proceeded to show how chemical hardness could be revisited and identified to the first excitation energy [6].

The approach relies on the so-called ensemble state theory [7–16]: instead of considering an isolated eigenstate of the Hamiltonian, one may alternatively consider an ensemble of M states with different weights w_i (such that $w_0 \geq w_1 \geq \cdots \geq 0$), and try to minimize the associated ensemble energy E

$$E = \sum_{i=0}^{M} w_i E_i \quad (14.2)$$

By doing so, one may eventually define an ensemble electron density, which is the simple weighted superimposition of the electron densities of each state,

$$\rho(\mathbf{r}) = \sum_{i=0}^{M} w_i \rho_i(\mathbf{r}) \quad (14.3)$$

and generalized Kohn–Sham equations [17] can be used to relate this density to a set of Kohn–Sham orbitals, displaying noninteger occupations.

By doing so in the case of an ensemble constituted by the ground and first excited state, one may then show that the excitation energy can be written according to the generalized Kohn–Sham orbitals energies ε_i:

$$E_1 - E_0 = \varepsilon_{N+1} - \varepsilon_N + \frac{\partial E_{xc}[\rho]}{\partial w} \quad (14.4)$$

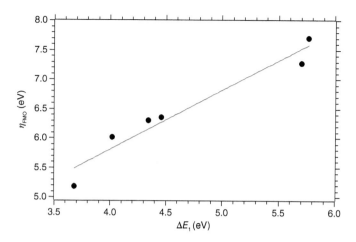

Figure 14.1 Correlation between the first excitation energy ΔE_1 and the frozen orbital approximation of hardness η_{FMO} for a series of carbonylated compounds, computed at the (TD)B3LYP/def2-TZVP level of theory. The red solid line corresponds to the best linear fit, $\eta_{FMO} = \Delta E_1 + 1.8$ ($R^2 = 0.94$).

with w the weight of the first excited state, ε_N the energy of the orbital which would be the last occupied in the ground state, ε_{N+1} the energy of the first vacant one, and E_{xc} the exchange-correlation energy for the ensemble state.

When $w \to 0$, generalized Kohn–Sham orbitals tend to those of the ground state, and the first term in Eq. (14.4) then matches the frozen orbital approximation of hardness, $\eta \approx \varepsilon_{N+1} - \varepsilon_N$. It was then proposed that the first excitation energy could serve as an estimation of chemical hardness. Interestingly, it was observed that excitation energies and finite difference approximation of hardness (that is, $\eta \approx I - A$) are correlated, although excitation energies are almost always found lower than $I - A$ (indicating the exchange-correlation contribution in Eq. (14.4) is seemingly always negative). The same inequation holds also for the frozen orbital approximation of hardness η_{FMO} (difference between the energies of the Lowest Unoccupied Molecular Orbital, LUMO, and the Highest Occupied Molecular Orbital, HOMO)

$$\eta_{FMO} = \varepsilon_{LUMO} - \varepsilon_{HOMO} \qquad (14.5)$$

We illustrate this in Figure 14.1, in the case of a series of carbonylated compounds,[1] computed at equilibrium geometry at the (TD)B3LYP/def2-TZVP level of theory using Gaussian 16.A01. A rather satisfactory linear relationship is observed, with a slope equal to 1.0 and an offset of 1.8 eV ($R^2 = 0.94$).

In the end, one may then expect to evaluate chemical hardness (at least relative chemical hardness) by the mere computation of the first excitation energy. This can be quite advantageous in the case of quite electron-rich molecules, for which the calculation of electron affinities is far from being trivial (self-ionizing anions).

---

[1] Namely, H_2CO, CH_3CHO, CH_3COCH_3, $HCOOCH_3$, and $HCON(CH_3)_2$.

14.2.3 State-Specific Dual Descriptors

The same issue of self-ionizing anions led to the development of another excited-state variation of a well-established CDFT quantity: the dual descriptor [18], defined as

$$\Delta f(\mathbf{r}) = \left(\frac{\partial^2 \rho(\mathbf{r})}{\partial N^2}\right)_{v(\mathbf{r})} \tag{14.6}$$

This descriptor, fitted to describe local philicity, can indeed be evaluated according to

$$\Delta f(\mathbf{r}) = \rho_{N+1}(\mathbf{r}) + \rho_{N-1}(\mathbf{r}) - 2\rho_N(\mathbf{r}) \approx \rho_{\text{LUMO}}(\mathbf{r}) - \rho_{\text{HOMO}}(\mathbf{r}) \tag{14.7}$$

under, respectively, the finite difference limit (indices N, $N+1$, and $N-1$ referring, respectively, to the initial system and vertical anion and cation) and the frozen orbital scheme (indices referring to the Kohn–Sham orbital densities). As can be inferred, self-ionizing anions preclude the use of the first expression (and sometimes even the second, as the LUMO is ill-defined). The frozen orbital approximation on the other hand may be too crude to offer a valuable insight on reactivity/selectivity, as it neglects orbital relaxation – which may be quite large in electron-rich systems.

In 2013 [19], it was proposed to describe the evolution of the electron density along a given chemical reaction (when changing a reaction coordinate from one point R to another one P), which could be expressed as a weighted sum of electron density reorganization under excitation (at point R):

$$\Delta\rho_{R\to P,0}(\mathbf{r}) = \sum_{i\geq 0} \alpha_i \left[\rho_{R,i}(\mathbf{r}) - \rho_{R,0}(\mathbf{r})\right] = \sum_{i\geq 0} \alpha_i \Delta f_{R,i}(\mathbf{r}) \tag{14.8}$$

Although the explicit formulas for the weights are not known, it may be expected that high-energy excited states will not contribute significantly to the reactivity in the ground state, hence most of the chemical properties should be retrieved within the very first excited states. This in fact meets a precedent proposition by Pearson [20]. Indeed, by the study of the UV-visible spectra of organic compounds, he noticed a trend between the first excitation energy and reactivity: the lower that energy, the more reactive, suggesting most information about the ground state reactivity is contained in the first excited state.

Let us now focus on the contribution from this first excited state. Assuming this excited state is reasonably reproduced by a HOMO $\to$ LUMO excitation, the electron density reorganization in Eq. (14.8) will boil down (dropping the "R" index) to

$$\Delta f_1(\mathbf{r}) = \rho_1(\mathbf{r}) - \rho_0(\mathbf{r}) \approx \rho_{\text{LUMO}}(\mathbf{r}) - \rho_{\text{HOMO}}(\mathbf{r}) \tag{14.9}$$

As one may note, under these conditions the right-hand side of Eq. (14.9) is directly the frozen orbital approximation of the dual descriptor. From this, it was proposed to qualify the electron density reorganization Δf_i the ith state-specific dual descriptor (SSDD).

In many instances, the first state-specific dual descriptor affords a satisfactory depiction of reactivity/selectivity. This is noticeably the case when the frozen orbital scheme itself is efficient, but not only. For instance, it was shown in the case of isoquinoline that usual reactivity/selectivity descriptors, including the frozen orbital

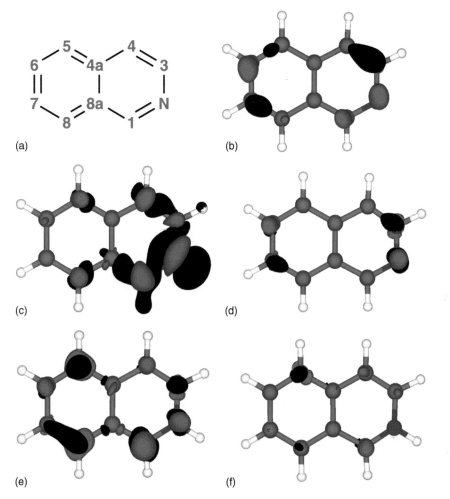

Figure 14.2 Isosurfaces of the frozen orbital approximation of the DD (b), of the first (c) and second (d) SSDD for the isoquinoline molecule, and of the first (e) and second (f) SSDD for the isoquinolium ion, computed at the (TD)B3LYP/def2-TZVP level of theory. Isovalue: 0.005 a.u. Positive values are depicted in red and negative values in black. The atom numbering is also reminded (a). Isosurfaces were produced using VESTA 3.5.8. Source: Adapted from [22].

approximation of the dual descriptor, fail at matching the observed trends in aromatic electrophilic substitution (that is, a pronounced reactivity at position 5, followed by position 8). Interestingly, as shown on Figure 14.2, in this example the first state-specific dual descriptor hints at a pronounced nucleophilicity on the nitrogen atom, as one might in fact expect, owing to its lone pair. Experimentally, aromatic electrophilic substitution reactions are known to proceed under quite strong acidic (either Brønsted or Lewis) conditions [21], and the first SSDD actually indicates that the first reaction to occur should be at the nitrogen, either a protonation or complexation of a Lewis acid. Computing the first SSDD for the isoquinolium ion, one

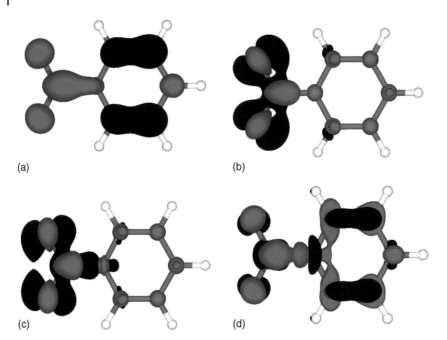

Figure 14.3 Isosurfaces of the frozen orbital approximation of the DD (a) of the first (b), second (c), and third (d) SSDD for the nitrobenzene molecule, computed at the (TD)B3LYP/def2-TZVP level of theory. Isovalue: 0.005 a.u. Positive values are depicted in red and negative values in black.

then directly retrieves the experimental results, as nucleophilicity is now principally observed at the C5 and C8 positions.

In some other cases, more excitations need to be considered to retrieve a satisfactory depiction of reactivity and selectivity. This is for instance the case for the simple nitrobenzene molecule. Dewar [23], and later on Langenaeker et al. [24], pointed out that frontier molecular orbital theory and Fukui functions usually misrepresent the order of reactivity of this molecule in electrophilic substitutions. Holleman rules [21] indeed indicate that the meta position is expected to be the most reactive one, while almost no reactivity should be observed in ortho and para positions. As can be seen in Figure 14.3, the frozen-orbital approximation of the DD indeed hints at nucleophilicity both at the ortho and meta position, with a slight (but not marked) preference over the latter. Conversely, the first and second SSDD do not display any nucleophilicity on the endocyclic carbon atoms, but rather electrophilicity in ortho and para positions. This already helps to understand the lack of reactivity on these positions but does not allow to understand the possibility of reaction in meta. This only becomes apparent in the third SSDD, which is quite reminiscent of the frozen orbital approximation of the DD. Overall, the meta position is expected to be the least electrophilic one and thus to be associated to reactivity. One may here even wonder whether the fact that one has to go up to the third excited state to observe endocyclic nucleophilicity could not be seen as a marker of a lesser reactivity, which is experimentally known to occur with nitro groups (deactivating with respect to the SEAr reaction).

14.2.4 Polarization Interaction

In the previous illustrations, the perturbative framework remained somehow implicit. In 2019 [25], the present authors proposed to explicitly use Rayleigh–Schrödinger perturbation theory to probe the responses of chemical systems to perturbations in their external potential. The starting point of the development is thus known: we assume the set of energies E_n and eigenvectors ψ_n of the Hamiltonian are known, and we wish to study the responses of these when an external perturbation $\delta \mathcal{H} = \delta v(\mathbf{r})$ is invoked. We will assume the ground state is nondegenerate (which is desirable if we are to use DFT calculations).

As we saw earlier, the perturbed ground state wavefunction at first order writes as

$$\psi_0 = c_0 \left(\psi_0^{(0)} + \sum_{i \neq 0} \frac{\langle \psi_i^{(0)} | \delta \mathcal{H} | \psi_0^{(0)} \rangle}{E_0 - E_i} \psi_i^{(0)} \right) \tag{14.1}$$

where the exponents refer to the unperturbed solutions of the Schrödinger equation. From this, we may express the reorganization of the ground state electron density induced by perturbation at first order: it will simply reduce to the difference between the expectation value of the density operator in ψ_0,[2]

$$\rho_{\text{pert}}(\mathbf{r}) = \langle \psi_0 | \hat{\rho}(\mathbf{r}) | \psi_0 \rangle$$

$$= c_0^2 \left(\langle \psi_0^{(0)} | \hat{\rho} | \psi_0^{(0)} \rangle + 2 \sum_{i \neq 0} \frac{\langle \psi_i^{(0)} | \delta v(\mathbf{r}) | \psi_0^{(0)} \rangle}{E_0 - E_i} \langle \psi_0^{(0)} | \hat{\rho} | \psi_i^{(0)} \rangle \right) \tag{14.10}$$

and that in $\psi_0^{(0)}$,

$$\rho_0(\mathbf{r}) = \langle \psi_0^{(0)} | \hat{\rho} | \psi_0^{(0)} \rangle \tag{14.11}$$

If we now assume the perturbation remains small, in a first approach we may expect that c_0^2 will be very close to unity, hence one may simplify the electron density reorganization at first order according to

$$\delta \rho(\mathbf{r}) \approx 2 \sum_{i \neq 0} \frac{\langle \psi_i^{(0)} | \delta v(\mathbf{r}) | \psi_0^{(0)} \rangle}{E_0 - E_i} \rho_0^i(\mathbf{r}) \tag{14.12}$$

$$= \sum_{i \neq 0} \left[\frac{2}{E_0 - E_i} \int \rho_0^i(\mathbf{r}) \delta v(\mathbf{r}) d\mathbf{r} \right] \rho_0^i(\mathbf{r}) \tag{14.13}$$

where ρ_0^i are the so-called transition densities (which one may compute from Time-dependent DFT (TDDFT) calculations). Because they integrate to zero over the whole space, so does the density reorganization – as expected since we implicitly supposed the system under study is closed.

---

2 In the following expression, we used the hermiticity to factorize the integrals involving both the ground and excited states, and consistency in perturbation orders to remove integrals involving two different excited states.

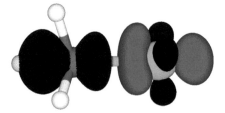

Figure 14.4 Isosurface of the first-order electron density polarization for a molecule of CH_3Cl perturbed by a negative −0.1 e point charge placed behind the C−Cl bond. Isovalue: 0.0005 a.u. The positive regions are depicted in red and negative in black. The perturbing point charge is depicted by a small blue sphere. Calculation was performed using a Python/Orbkit adaptation of our original Fortran program, truncating the sum on 50 excited states.

If $\delta v(\mathbf{r})$ is chosen to reproduce the effect of the approach of a reagent, $\delta \rho(\mathbf{r})$ will thus translate the electron density polarization induced by the perturbation: how electron density responds, at first order, to the electric field induced by the incoming species. We illustrate in Figure 14.4 a typical example: the polarization of the density of the chloromethane molecule induced by the approach of a negative point charge (here −0.1 a.u.) at the back of the C−Cl bond. As one may note, the principal effect of this perturbation is to promote the transfer of electron density from the carbon atom to the chlorine one, along the C−Cl axis. Perturbation is here weakening the C−Cl bond, and we observe the expected electron density movement for a nucleophilic substitution reaction.

We present in Figure 14.5 the results we obtained for another prototypical example: the acroleine molecule. In this unsaturated carbonyl system, addition of a nucleophile can occur either on the carbon atom involved in the carbonyl function or on the β position (extremal carbon). Experimental evidence indicate that hard nucleophiles (in the sense of the Hard and Soft Acids and Bases (HSAB) theory)

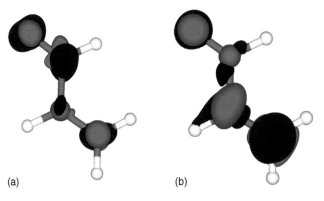

(a)   (b)

Figure 14.5 Isosurface of the first-order electron density polarization for an acrolein molecule perturbed by a negative −0.1 e point charge placed at the position of the nucleus of the carbonyl (a) and β (b) carbon atoms. Isovalue of 0.0005 a.u. The positive regions are depicted in red and negative in black. Calculation was performed using a Python/Orbkit adaptation of our original Fortran program, truncating the sum on 50 excited states.

react on the carbonyl position, while soft nucleophiles react on the β position. We thus probed the response of electron density to a perturbation by a negative point charge located on these two carbon atom nuclei. As expected, in the direct vicinity of the perturbation, a depletion of electron density is observed (because of electrostatic repulsion). It may be noted that electron density in both cases relocates over the molecule, but interesting larger volumes are observed when the perturbation is applied on the β carbon atom. This suggests a stronger polarization response in that case (which one may in fact have expected from simple mesomery): this position will likely be more polarizable, hence softer in the sense of Pearson. We thus appear to meet experiment here.

It is possible to obtain a more reliable quantification by evaluating how many electrons are in fact impacted by the phenomenon,

$$\delta N = \int_{\forall \mathbf{r} | \delta\rho(\mathbf{r})>0} \delta\rho(\mathbf{r})d\mathbf{r} = -\int_{\forall \mathbf{r} | \delta\rho(\mathbf{r})<0} \delta\rho(\mathbf{r})d\mathbf{r} = \frac{1}{2}\int |\delta\rho(\mathbf{r})|d\mathbf{r} \qquad (14.14)$$

Quantification can also be attained by the study of energy responses to such a perturbation:

$$\delta E^{(1)} = \left\langle \psi_0^{(0)} | \delta v(\mathbf{r}) | \psi_0^{(0)} \right\rangle = \int \rho_0(\mathbf{r})\delta v(\mathbf{r})d\mathbf{r} \qquad (14.15)$$

$$\delta E^{(2)} = \sum_{i \neq 0} \frac{\left|\left\langle \psi_i^{(0)} | \delta v(\mathbf{r}) | \psi_0^{(0)} \right\rangle\right|^2}{E_i - E_0} = \sum_{i \neq 0} \frac{1}{E_i - E_0}\left[\int \rho_0^i(\mathbf{r})\delta v(\mathbf{r})d\mathbf{r}\right]^2 \qquad (14.16)$$

$$= \frac{1}{2}\int \delta\rho(\mathbf{r})\delta v(\mathbf{r})d\mathbf{r} \qquad (14.17)$$

As one may note, the first-order response does not in fact involve any excited state. It translates the electron density contribution to the electrostatic interaction energy with the incoming reagent (contribution to the molecular electrostatic potential at the location of the perturber if it is a point charge). The second-order contribution on the other hand relies on an expansion over all excited states. Being always negative, it translates the stabilization undergone by the system by polarizing its electron density in the field of the incoming reagent.

Coming back to example of acrolein, perturbed by negative point charges (−0.1 a.u.) located, respectively, on the carbonyl or β carbon atom nuclei, we obtain after calculation that the polarization energy at second order is -95 and -184×10^{-6} a.u. Similarly, the shifted fractions of electron are, respectively, of 101×10^{-4} and 150×10^{-4} a.u. We thus retrieve our initial conclusion: the β position is much more polarizable (thus softer) than the carbonyl one, in line with experimental evidences.

Another approach can be used to reach quantification. Looking back at Eq. (14.1), the perturbation coefficients can be seen as probabilities of finding the system in the excited state i as a consequence of the application of the perturbation in the ground state. If we now consider a large number of replicas of the system under study, these coefficients describe a statistical distribution of replicas in every excited states. The width of distribution can then be quantified using the Gibbs–Shannon entropy

$$S = -k_B \sum_{i=0} c_i^2 \ln c_i^2 \qquad (14.18)$$

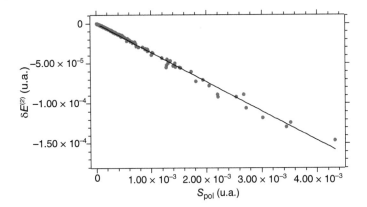

Figure 14.6 Evolution of the second order energy variation $\delta E^{(2)}$ induced by a point charge perturbation (−0.1 a.u.) against entropy S_{pol}, for successive perturbations at 400 randomly selected positions, in the case of a chloromethane molecule. The straight line depicts the best linear fit $\delta E^{(2)} = -0.0364(2) S_{pol}$ ($R^2 = 0.993$).

with

$$c_0 = 1 - \sum_{i \neq 0} \frac{\left|\langle \psi_i^{(0)} | \delta H | \psi_0^{(0)} \rangle\right|^2}{(E_0 - E_i)^2}, \quad c_i = \frac{\left|\langle \psi_i^{(0)} | \delta H | \psi_0^{(0)} \rangle\right|^2}{(E_0 - E_i)^2} \quad (14.19)$$

Obviously, before perturbation the entropy will be zero ($c_0 = 1$ and the other coefficients are zero), so S can also be seen as an entropy variation induced by perturbation. From the previous expression, we may see that large responses to perturbation should be associated to large values of S: the entropy variation should thus be a measure of softness, as energy is.

If we come back to example of acrolein, we satisfactorily find that the entropy variation induced by a perturbation on the nucleus of carbon atom β results in an entropy variation of 62×10^{-4} a.u., while a perturbation on the carbonyl position results in a variation of 31×10^{-4} a.u.

In fact, in all studied cases so far, a strong linearity was reported between the second-order energy variation and entropy for point charge perturbations, regardless of their position in space [26]. This is for instance visible in Figure 14.6, where the evolution of the second-order energy variation $\delta E^{(2)}$ against entropy S_{pol} is depicted in the case of a chloromethane molecule perturbed by a −0.1 a.u. point charge, successively applied to 400 random positions.

From this, it appears that a polarization temperature can be defined by analogy with statistical mechanics,

$$T_{pol} = \frac{\delta E^{(2)}}{S} \quad (14.20)$$

Perturbation can thus be viewed as a thermodynamic heat exchange between a heat source (the perturber) and the system under study.

14.3 Exploring the Reactivity of Excited States with Excited States

In the first section of this chapter (14.2), we focused on exploring the reactivity and selectivity in the ground state through the excited states. One may alternatively try and study the reactivity and selectivity in excited states themselves. This is the topic of this section.

14.3.1 Local Chemical Potential

A conceptually simple approach to probe chemical reactivity and selectivity in the excited states was proposed in 2009 by one of the present authors [27]. At the root of this development is the long-known relation between the chemical potential, the external potential, and the universal functional electron density derivative in the ground state:

$$\mu = \left(\frac{\delta E}{\delta \rho(\mathbf{r})}\right)_N = v(\mathbf{r}) + \left(\frac{\delta F_{HK}}{\delta \rho(\mathbf{r})}\right) \tag{14.21}$$

This simple equation is remarkable as it indicates the chemical potential, a global quantity, is the sum of two local functions. In fact, the constancy of μ is a consequence of the minimization of the energy functional. As such, application of any trial density (other than the actual ground state density) to the energy functional should result in a nonconstant chemical potential, and the associated shape in real space of this function $\mu(\mathbf{r})$ should provide insight on the way electron density should distort in order to tend to the ground state density.

Feeding a vertically excited state electron density ρ_k to the energy functional should thus provide clue as to how electron density will relax from the excited state to the ground state at constant geometry. In fact, the local chemical potential can be written in this case as

$$\mu_k(\mathbf{r}) = v(\mathbf{r}) + \left(\frac{\delta F_{HK}}{\delta \rho(\mathbf{r})}\right)[\rho_k(\mathbf{r})] \tag{14.22}$$

One may here further expand through a Taylor development using

$$\rho_k(\mathbf{r}) = \rho_0(\mathbf{r}) + \Delta f_k(\mathbf{r}) \tag{14.23}$$

where Δf_k represents the k-th state-specific dual descriptor (electron density difference between the k-th excited and ground states). Doing so, we have up to second order

$$\left(\frac{\delta F_{HK}}{\delta \rho(\mathbf{r})}\right)[\rho_k(\mathbf{r})] = \left(\frac{\delta F_{HK}}{\delta \rho(\mathbf{r})}\right)[\rho_0(\mathbf{r}) + \Delta f_k(\mathbf{r})]$$

$$= \left(\frac{\delta F_{HK}}{\delta \rho(\mathbf{r})}\right)[\rho_0(\mathbf{r})] + \int \left(\frac{\delta^2 F_{HK}}{\delta \rho(\mathbf{r})\delta \rho(\mathbf{r}')}\right)[\rho_0(\mathbf{r})]\Delta f_k(\mathbf{r}')d\mathbf{r}' \tag{14.24}$$

The first term can be written as $\mu - v(\mathbf{r})$, while the second-order derivative of F_{KH} is the hardness kernel. If we assume the Coulombic contribution in this kernel is predominant over the other ones (i.e. kinetic, exchange, and correlation), one may approximate it as $\dfrac{1}{|\mathbf{r} - \mathbf{r}'|}$, hence we can eventually write

$$\mu_k(\mathbf{r}) \approx \mu + \int \frac{\Delta f_k(\mathbf{r}')}{|\mathbf{r} - \mathbf{r}'|} d\mathbf{r}'$$

Overall, the local chemical potential can be written as the sum of the ground state μ and a local contribution. We may already note that this is expected to be correct only when the second-order truncation of the Taylor development is reasonable, hence mostly for low-lying excited states (the excited state HK energy resembling the ground state one).

Let us focus on the local component: its sign will give indication on the direction of the electron flow to apply in order to tend to the stationary conditions (stability). Where Δf_k is positive, the chemical potential associated to the excited state density will be larger than that of the ground state. Hence when returning to the ground state, such a site is expected to lose electron density. Conversely, a site associated to $\Delta f_k < 0$ will present a lower chemical potential in the excited state, hence electron density should flow to this position when returning to the ground state.

As such, we may note that in low-lying excited states we can predict the philicity based only on the state-specific dual descriptors (or electron density differences with the ground state), and that the association between the philicity and the sign of the SSDD is reversed compared to the ground state case [27, 28].

We present in Figure 14.7 an example of application of this idea in the case of the Diels–Alder [2+2] photo-cycloaddition of ethylene on itself.

Experimentally, such a reaction will occur when one reagent molecule is excited by light and reacts with another reagent in the ground state. Following Kasha's rule [29], the reactive excited state is expected to be the first excited state, so the relevant SSDD to study here should be the first one. As one can note, the 0.001 a.u. isosurface of the first SSDD is indeed complying with our expectation: the red volumes pointing outside of the molecule will be associated to electrophilicity for the ground state

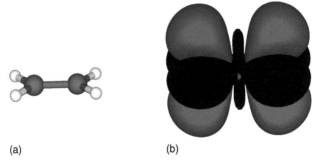

(a)   (b)

Figure 14.7 Isosurface of the first SSDD for the ethylene molecule (b), computed at the (TD)B3LYP/def2-TZVP level of theory. Isovalue: 0.001 a.u. Positive values are depicted in red and negative values in black. The orientation of the molecule is given in (a).

molecule and to nucleophilicity for the excited one, and conversely the black domain developing in the middle of the C–C bond will be associated to nucleophilicity in the ground state and electrophilicity in the excited state. As such, the suprafacial [2+2] addition is expected to optimize the overlap of domains with opposite philicities, hence the reaction should proceed.

14.3.2 Polarization in the Excited States

As we saw, the previous development allows to probe the reactivity and selectivity in low-lying excited states. And what about higher energy excited states?

For these, we may here too consider the Rayleigh–Schrödinger perturbation framework. Indeed, the perturbation equations are in principle applicable to *any eigenstate* of the system. Let us consider the first order perturbation expansion of the wavefunction of one excited state:

$$\psi_k = \psi_k^{(0)} + \sum_{j \neq k} \frac{\langle \psi_k | \mathcal{H} | \psi_j \rangle}{E_k - E_j} \psi_j^{(0)} \qquad (14.25)$$

Following the same approach we unfolded in Section 14.2.4, we may express the electron density in this perturbed excited state

$$\rho_k(\mathbf{r}) = \langle \psi_k | \hat{\rho} | \psi_k \rangle$$

$$\approx \left(\langle \psi_k^{(0)} | \hat{\rho} | \psi_k^{(0)} \rangle + 2 \sum_{j \neq k} \frac{\langle \psi_j^{(0)} | \delta v(\mathbf{r}) | \psi_k^{(0)} \rangle}{E_k - E_j} \langle \psi_j^{(0)} | \hat{\rho} | \psi_k^{(0)} \rangle \right)$$

$$= \rho_k^{(0)}(\mathbf{r}) + 2 \sum_{j \neq k} \frac{1}{E_k - E_j} \left[\int \rho_k^j(\mathbf{r}) \delta v(\mathbf{r} d\mathbf{r} \right] \rho_k^j(\mathbf{r}) \qquad (14.26)$$

where we once again considered a weak perturbation in the external potential and a nondegenerate state.

We can here too express the electron density variation in the excited state induced by perturbation,

$$\delta \rho_k(\mathbf{r}) = 2 \sum_{j \neq k} \frac{1}{E_k - E_j} \left[\int \rho_k^j(\mathbf{r}) \delta v(\mathbf{r} d\mathbf{r} \right] \rho_k^j(\mathbf{r}) \qquad (14.27)$$

which can then be used to evaluate the evolution of this state energy at second order according to

$$\delta E_k^{(2)} = \int \delta \rho_k(\mathbf{r}) \delta v(\mathbf{r}) d\mathbf{r} \qquad (14.28)$$

and also quantify the total electron displacement induced by perturbation,

$$\delta N_k = \frac{1}{2} \int |\delta \rho_k(\mathbf{r})| d\mathbf{r} \qquad (14.29)$$

The prefactors in Eq. (14.27) can also be used to evaluate the entropy variation associated to the change in distribution of the eigenstate induced by the perturbation,

$$\delta S_k = -\sum_{j \neq k} c_j^2 \ln c_j^2 - \left(1 - \sum_{k \neq j} c_j^2\right) \ln \left(1 - \sum_{k \neq j} c_j^2\right) \qquad (14.30)$$

Overall, the same polarization descriptors that we developed in the ground state can be formulated in the excited states.

Now a question remains: how can one compute these descriptors? As in the ground state case, the central quantities to evaluate are the transition densities between states j and k. Recalling these are the integrals of a monoelectronic operator over the two eigenfunctions ψ_j and ψ_k, and since these can be written as a linear combination of monoexcited Slater determinants relying on the same set of orbitals (through TDDFT calculations for instance), it should be possible to compute these transition densities by using the Slater–Condon rules [30, 31]. Three cases appear:

- the considered Slater determinants in ψ_j and ψ_k are the same, thus their contribution to the transition density is the electronic density associated to the Slater determinant, multiplied by its coefficient in both wavefunctions;
- they differ by one orbital only, thus one can be seen as a monoexcitation of the other one. In that case the contribution is the product of the two differing orbitals, once again multiplied by the coefficient in both wavefunctions;
- they differ by more than one orbital; in that case they can be seen as being more than singly excited configurations of one another, and their contribution is zero.

Naturally, this method is only expected to be efficient when the excited state themselves are reasonably described. In the case of TDDFT calculations, one may already infer that calculations will become tainted with significant error as the excited state energies become large. Nevertheless, as indicated before most photochemical reactions are expected to occur in the very first excited states, since internal conversion is expected to occur rapidly, thus the proposed methodology is expected to meet some success in the future.

We propose here to consider the case of the butadiene molecule, and to compare the polarization of electron density induced by a point charge perturbation in the ground and first excited states. We will here focus on the response to a perturbation on one terminal carbon atom. As one may note, in both cases we retrieve the expected depletion at the perturbed C atom site (here on the right-hand side). Differences are mostly apparent on the relocation sites. In the ground state, we observe a relocation of electron density on carbon atoms number 2 and 4, which can be rationalized by invoking mesomery (curvy arrows). In the first excited state, we note that electron density relocation is mostly concentrated on carbon atom number 2, but also spreads on C3, while C4 presents a depletion volume. A drastic change is thus observed, which relates to the difference of electronic structure of butadiene in the ground and first excited state.

A careful analysis of the first excited state wavefunction shows it principally consists in the promotion of electrons from the HOMO to the LUMO. These orbitals are in fact essentially resembling those one would expect to observe in the framework of the old – but still useful – Hückel theory: the HOMO is bonding between C1 and C2 and C3 and C4, and antibonding between C2 and C3. As such we can see this MO as describing the presence of both double bonds in the ground state. The LUMO on the other hand is antibonding between C1 and C2, bonding between C2 and C3, and antibonding between C3 and C4. Promotion of electrons from the HOMO to the

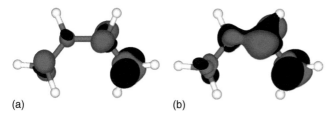

Figure 14.8 Isosurface of the first-order electron density polarization for a butadiene molecule perturbed by a negative −0.1 e point charge placed at the position of the nucleus of a terminal carbon atom in the ground (a) and first excited (b) states. Isovalue of 0.00001 a.u. The positive regions are depicted in red and negative in black. Calculation was performed using a Python/Orbkit adaptation of our original Fortran program, truncating the sum on 50 excited states.

LUMO can then be seen as relocating them from the external double bonds to the central bond. We may thus conceive the electronic configuration of butadiene in the first excited state as being a diradical, with two unpaired electrons on the terminal C atoms and a double bond linking the central C atoms. With such a Lewis picture in mind, we can then make more sense of our observation on the electron density polarization: the similarity of behavior of both terminals and central carbon atoms can be expected.

We may nevertheless note two additional features that deserve a final comment. First, the volumes in Figure 14.8 are larger in the excited state case, as we could have expected since excited states are more polarizable than ground state. Second, we note that in the excited state case the contributions on the left-hand side of the molecule are seemingly weaker. Indeed, if the isovalue is increased it is possible to mask all contributions from the associated carbon atoms: the main features of the polarization density are centered on the perturbed C atom and its direct neighbor. This tends to suggest that conjugation between the two sides of the molecule is diminished in the excited state. It would be interesting to see if this can be retrieved in other electron density bases analyses, noticeably the electron localization function (ELF), which has recently been studied by the group of C. Cardenas and coworkers [32].

Another interesting matter to consider would be the study of the existence of a polarization temperature in the excited states. First evidences seem to suggest it may still be derived; if so, how do polarization temperatures compare between ground and excited states?

14.4 Conclusion

Conceptual Density Functional Theory can nowadays be seen as a perturbation theory. It starts from the chemical reactants and tries to predict the features of the transition state by evaluating the response of the starting reagents to perturbations. Therefore, the fact that all eigenstates help to get a better understanding of chemical reactivity and selectivity should not come as a surprise. In this chapter, we have tried to review ancient and recent research studies in conceptual DFT in which excited states are the primary ingredient, either for characterizing the reactivity of the ground state or the explaining the behavior of excited states themselves. Up to now, most efforts have been dedicated to the use of excited states to characterize reactivity and selectivity in the ground state. On the other hand, Conceptual DFT in excited states has been paid very little attention so far. The only real attempt has consisted to propose the use of a local chemical potential, but recent developments opened new possibilities to extend the field to photochemistry, hopefully affording to cross one of the identified CDFT frontiers [33].

References

1 Hohenberg, P. and Kohn, W. (1964). Inhomogeneous electron gas. *Phys. Rev.* 136: B864.
2 Schrödinger, E. (1926). Quantisierung als eigenwertproblem. *Ann. Phys.* 385: 437–490.
3 Chermette, H. (1999). Chemical reactivity indexes in density functional theory. *J. Comput. Chem.* 20 (1): 129–154.
4 Geerlings, P., De Proft, F., and Langenaeker, W. (2003). Conceptual density functional theory. *Chem. Rev.* 103 (5): 1793–1873.
5 Liu, S. (ed.) (2022). *Conceptual Density Functional Theory*. Wiley. https://doi.org/10.1002/9783527829941.
6 Nagy, Á. (2005). Hardness and excitation energy. *J. Chem. Sci.* 117 (5): 437–440.
7 Perdew, J.P., Parr, R.G., Levy, M., and Balduz, J.L. (1982). Density-functional theory for fractional particle number: derivative discontinuities of the energy. *Phys. Rev. Lett.* 49: 1691–1694. https://doi.org/10.1103/PhysRevLett.49.1691.
8 Cárdenas, C., Ayers, P.W., and Cedillo, A. (2011). Reactivity indicators for degenerate states in the density-functional theoretic chemical reactivity theory. *J. Chem. Phys.* 134 (17): 174103. https://doi.org/10.1063/1.3585610.
9 Franco-Pérez, M., Ayers, P.W., and Gázquez, J.L. (2016). Average electronic energy is the central quantity in conceptual chemical reactivity theory. *Theor. Chem. Acc.* 135 (8): https://doi.org/10.1007/s00214-016-1961-2.
10 Miranda-Quintana, R.A. and Ayers, P.W. (2016). Fractional electron number, temperature, and perturbations in chemical reactions. *Phys. Chem. Chem. Phys.* 18 (22): 15070–15080. https://doi.org/10.1039/c6cp00939e.
11 Franco-Pérez, M., Heidar-Zadeh, F., Ayers, P.W. et al. (2017). Going beyond the three-state ensemble model: the electronic chemical potential and Fukui function for the general case. *Phys. Chem. Chem. Phys.* 19 (18): 11588–11602. https://doi.org/10.1039/c7cp00224f.
12 Franco-Pérez, M. (2019). An electronic temperature definition for the reactive electronic species: conciliating practical approaches in conceptual chemical reactivity theory with a rigorous ensemble formulation. *J. Chem. Phys.* 151 (7): https://doi.org/10.1063/1.5096561.
13 Gázquez, J.L., Franco-Pérez, M., Ayers, P.W., and Vela, A. (2018). Temperature-dependent approach to chemical reactivity concepts in density

functional theory. *Int. J. Quantum Chem.* 119 (2): https://doi.org/10.1002/qua.25797.

14 Miranda-Quintana, R.A., Franco-Pérez, M., Gázquez, J.L. et al. (2018). Chemical hardness: temperature dependent definitions and reactivity principles. *J. Chem. Phys.* 149 (12): https://doi.org/10.1063/1.5040889.

15 Franco-Pérez, M., Gázquez, J.L., Ayers, P.W., and Vela, A. (2020). Temperature-dependent approach to electronic charge transfer. *J. Phys. Chem. A* 124 (26): 5465–5473. https://doi.org/10.1021/acs.jpca.0c02496.

16 Franco-Pérez, M., Polanco-Ramírez, C.A., Gázquez, J.L. et al. (2020). Study of organic reactions using chemical reactivity descriptors derived through a temperature-dependent approach. *Theor. Chem. Acc.* 139 (3): https://doi.org/10.1007/s00214-020-2557-4.

17 Kohn, W. and Sham, L.J. (1965). Self-consistent equations including exchange and correlation effects. *Phys. Rev.* 140 (4A): A1133–A1138. https://doi.org/10.1103/physrev.140.a1133.

18 Morell, C., Grand, A., and Toro-Labbé, A. (2005). New dual descriptor for chemical reactivity. *J. Phys. Chem. A* 109 (1): 205–212.

19 Tognetti, V., Morell, C., Ayers, P.W. et al. (2013). A proposal for an extended dual descriptor: a possible solution when frontier molecular orbital theory fails. *Phys. Chem. Chem. Phys.* 15: 14465–14475.

20 Pearson, R.G. (1988). Electronic spectra and chemical reactivity. *J. Am. Chem. Soc.* 110 (7): 2092–2097.

21 Carey, F.A. and Sundberg, R.J. (2001). Aromatic substitution reactions. In: *Part B: Reactions and Synthesis*, 693–745. Berlin, Heidelberg: Springer-Verlag https://doi.org/10.1007/978-3-662-39510-3\LY1\textbackslash.11.

22 Momma, K. and Izumi, F. (2011). VESTA for three-dimensional visualization of crystal, volumetric and morphology data. *J. Appl. Crystallogr.* 44 (6): 1272–1276. https://doi.org/10.1107/s0021889811038970.

23 Dewar, M.J.S. (1989). A critique of frontier orbital theory. *J. Mol. Struct. THEOCHEM* 200: 301–323. https://doi.org/10.1016/0166-1280(89)85062-6.

24 Langenaeker, W., Demel, K., and Geerlings, P. (1991). Quantum-chemical study of the Fukui function as a reactivity index. *J. Mol. Struct. THEOCHEM* 234: 329–342. https://doi.org/10.1016/0166-1280(91)89021-r.

25 Guégan, F., Pigeon, T., De Proft, F. et al. (2020). Understanding chemical selectivity through well selected excited states. *J. Phys. Chem. A* 124 (4): 633–641.

26 Guégan, F., Tognetti, V., Martínez-Araya, J.I. et al. (2020). A statistical thermodynamics view of electron density polarisation: application to chemical selectivity. *Phys. Chem. Chem. Phys.* 22: 23553–23562.

27 Morell, C., Labet, V., Grand, A. et al. (2009). Characterization of the chemical behavior of the low excited states through a local chemical potential. *J. Chem. Theory Comput.* 5 (9): 2274–2283.

28 Morell, C., Labet, V., Ayers, P.W. et al. (2011). Use of the dual potential to rationalize the occurrence of some DNA Lesions (pyrimidic dimers). *J. Phys. Chem. A* 115 (27): 8032–8040.

29 Kasha, M. (1950). Characterization of electronic transitions in complex molecules. *Discuss. Faraday Soc.* 9: 14–19.
30 Slater, J.C. (1929). The theory of complex spectra. *Phys. Rev.* 34: 1293–1322. https://doi.org/10.1103/PhysRev.34.1293.
31 Condon, E.U. (1930). The theory of complex spectra. *Phys. Rev.* 36: 1121–1133. https://doi.org/10.1103/PhysRev.36.1121.
32 Echeverri, A., Gallegos, M., Gómez, T. et al. (2023). Calculation of the ELF in the excited state with single-determinant methods. *J. Chem. Phys.* 158 (17): https://doi.org/10.1063/5.0142918.
33 Geerlings, P., Chamorro, E., Chattaraj, P.K. et al. (2020). Conceptual density functional theory: status, prospects, issues. *Theor. Chem. Acc.* 139 (2): 36.

15

Modeling the Photophysical Processes of Organic Molecular Aggregates with Inclusion of Intermolecular Interactions and Vibronic Couplings

WanZhen Liang, Yu-Chen Wang, Shishi Feng, and Yi Zhao

Xiamen University, College of Chemistry and Chemical Engineering, State Key Laboratory of Physical Chemistry of Solid Surfaces, iChEM, Fujian Provincial Key Laboratory of Theoretical and Computational Chemistry, 42 Siming South Road, Xiamen 361005, People's Republic of China

15.1 Introduction

Molecular photophysics concerns molecular electronic excited-state dynamics including both radiative (fluorescence and phosphorescence) and non-radiative (internal conversion and intersystem crossing) pathways. As a molecule is at its electronic excited state by photoexcitation, the excitation energy is immediately dissipated into vibrations and rotations via internal conversion until the bottom of the lowest excited state is reached, and then fluorescence emits from this state. The intersystem crossing by changing spin multiplicity may further access lower triplet states, leading to phosphorescence. Although these photophysical processes are well understood by Kasha's rule [1], they may become dramatically different as the molecules aggregate together [2, 3]. Typical examples include aggregation-caused quenching [4] where the molecule is emissive in the solution but nonluminescent in the aggregate state, and its opposite situation coined as aggregation-induced emission [5]. Understanding these curious photo-physical features of molecular aggregates is extremely important in designing light-emitting diodes, thin-film transistors, and photovoltaics.

The aggregation behavior of different organic molecules is quite different because of the flexibility of the organic molecular framework and the weak van der Waals interaction between molecules, which makes the same molecule often exhibit different photophysical behaviors in different aggregate states. In many molecular films, due to the disordered molecular arrangement and weak intermolecular excitonic couplings, the photophysical processes are dominated by the monomer, and the aggregate effect is considered as the environment one. With respect to this simplified model, the environment-induced structural change and intramolecular electron–phonon coupling in the monomer successfully explain the dramatic switch between the radiative and non-radiative pathways in aggregates [6].

In well-arranged molecular aggregates, such as molecular crystals, however, the intermolecular excitonic couplings are quite strong. In this case, concerned by

Exploring Chemical Concepts Through Theory and Computation, First Edition. Edited by Shubin Liu.
© 2024 WILEY-VCH GmbH. Published 2024 by WILEY-VCH GmbH.

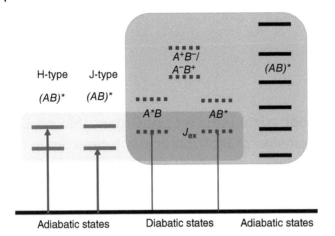

Figure 15.1 Schematic representation of the adiabatic $(AB)^*$ (black solid lines) and quasi-diabatic A^*B, AB^*, A^+B^- and A^-B^+ (dashed lines) excited states for a dimer AB, where the states in yellow and blue boxes represent the original Kasha's exciton model and extended exciton model including CT states, respectively. The vertical arrow lines represent the states with the dipole-allowed features.

the chapter, photoexcitation may generate a delocalized excited state and result in significant collective molecular effects. Either coherent or incoherent excitation energy transfer (EET) and charge transfer (CT) are dependent on the competing influences of the exciton bandwidth and thermal fluctuations/disorder. The Kasha exciton model has been extensively used to describe the pure electronic states of molecular aggregates due to its simplicity and its computational feasibility [2, 3]. In such a case, crystal states, also known as the delocalized Frenkel excitons (FEs) states, are constructed from superpositions of localized exciton (LE) states, in which electron and hole are confined to the same chromophore. FE model is a feasible one to predict the spectra of multichromophores when the chromophores have nonoverlapping charge distributions. Take a dimer AB as an example, as shown in Figure 15.1 within the purple box. The intermolecular excitonic coupling J_{ex} between two LE states A^*B and AB^* (the transition dipole–dipole coupling at the condition that two monomers are enough far from each other) leads to two splitting adiabatic excited states $(AB)^*$, one higher and the other lower than the energy of original monomeric excited state. According to Kasha exciton model, the "face-to-face" transition dipole arrangement performs an allowed transition from the ground state to the higher excited state, resulting in a blue-shifted absorption peak. This type of aggregate is called an H-aggregate, in which the radiative transition to the ground state is suppressed because the electrons in the higher excited state can quickly relax to the lower excited state. When the monomers are described as packing in a "head-to-tail" way, the transition from the ground state to the lower excited state is allowed, resulting in a red-shifted absorption peak and an enhanced emission. This class of aggregates is called the J-aggregates (shown in the left side in Figure 15.1).

For closely-stacking organic molecules, the noncovalent intermolecular interactions may give rise to the significant overlap between the frontier molecular orbitals on adjacent chromophores, leading to through-space charge delocalization along the stacking direction. The intermolecular charge-transfer excitations (CTEs), in which electron and hole are located on adjacent chromophores, may play an important role on both the spectra and exciton dynamics [7–10]. More complex exciton models are therefore required to account for the intermolecular CT states, such as A^+B^- and A^-B^+ in a molecular dimer AB, as shown within the blue box in Figure 15.1. In the latter case, there exist complex photophysical processes in adiabatic states of $(AB)^*$ because of the mixing between various kinds of LE and CT states [8, 11].

In realistic molecular aggregates, optical excitation is essentially comprised both electronic and vibrational degrees of freedom (DOFs) because organic molecular assemblies are commonly flexible in the sense that photophysics is accompanied by significant nuclear rearrangements within the constituent chromophores. It is thus essential to incorporate both the vibronic and exciton couplings to have a quantitative explanation of the experimental results. In many previous descriptions of photophysical processes, either intermolecular exciton couplings or electron–phonon couplings were taken as a perturbation to derive the analytical expressions for spectral simulations and EET/CT rate's calculations [8, 11, 12], and the properties of quasi-diabatic electronic states were commonly fitted from the experimental measurements. The analytical expressions are extremely useful for understanding the structure–property relationships in photophysical processes. However, in realistic organic molecular aggregates, both the exciton couplings and electron–phonon couplings may not be treated as a perturbation, and they should be considered on an equal footing. Many photophysical processes, such as EET and CT, intramolecular and intermolecular internal conversion, intersystem crossing, and singlet fission, are entangled together to generate quantum coherence and interference among those pathways, which may lead to new spectral mechanisms and characteristics.

In this chapter, we first describe the possible electronic structure theories for getting the energies of quasi-diabatic states and their couplings and display the fragment particle–hole density (FPHD) method recently proposed by us as an example [13]. We then summarize the methods for calculating electron–phonon couplings, and exhibit a non-Markovian stochastic Schrödinger equation (NMSSE) [14–17] for quantum dynamic simulations. The possible applications and challenges are further discussed.

15.2 Theoretical Approaches

15.2.1 Model Hamiltonian

Due to the complexity and considerable sizes of interacting units in the molecular aggregates, the extended exciton model combined with the quantum master equation or stochastic Schrödinger equation approaches are usually adopted to model the molecular aggregates and photosynthetic complexes. The model Hamiltonian for a general situation involving various kinds of photophysical processes (EET, CT, singlet fission, etc.) can be written as

$$\hat{H} = \hat{H}_e + \hat{H}_{ph} + \hat{H}_{e-ph} \tag{15.1}$$

Here, H_e, H_{ph}, and H_{e-ph} represent the electronic Hamiltonian, the phonon Hamiltonian, and the electron–phonon interaction, respectively. The electronic system of interest has been assumed to couple to atomic nuclei and environmental DOFs, or a "bath" held at thermal equilibrium. All nuclear DOFs are treated on the same footing, namely, as a heat bath. The bath is described as the set of (intra- and intermolecular) vibrations that couple to the electronic excitations.

The electronic Hamiltonian is generally expressed as

$$\hat{H}_e = \sum_{nm} E_{nm} |n\rangle\langle m| \tag{15.2}$$

where $|n\rangle$ represents the nth electronic state with the state energy E_{nn}, and $E_{nm}(n \neq m)$ is the electronic coupling between $|n\rangle$ and $|m\rangle$. In photophysical processes of aggregates, the breakage of chemical bonds is not generally involved in and the nuclear positions usually oscillate slightly around the equilibrium positions. Therefore, a collection of harmonic oscillators is suitable to describe the nuclear motions, and the bilinear electron–phonon couplings are a good approximation to characterize the position distortions induced by electronic excitation and the nuclear-dependent electronic couplings. Concretely speaking, the vibrational Hamiltonian is expressed as

$$\hat{H}_{ph} = \sum_{\mu} \omega_\mu \left(\hat{b}^\dagger_\mu \hat{b}_\mu + \frac{1}{2} \right) \tag{15.3}$$

where $\hat{b}^\dagger_\mu$ and $\hat{b}_\mu$ represent the creation and annihilation operators of the μth vibrational mode, respectively, with the frequency ω_μ. The electron–phonon interaction is given by

$$\hat{H}_{e-ph} = \sum_{\mu} \hat{X}_\mu \otimes (\hat{b}^\dagger_\mu + \hat{b}_\mu) \tag{15.4}$$

with

$$\hat{X}_\mu = \sum_{nm} g_{nm\mu} |n\rangle\langle m| \tag{15.5}$$

Here, $\hat{X}_\mu$ is an electronic operator determining the concrete coupling form between electronic states and the μth phonon mode. $g_{nm\mu}$ is the electron–phonon interaction coefficients, characterizing the strength (and phase) of the μth mode coupled to specific electronic states. $g_{nn\mu}$ and $g_{nm\mu}(n \neq m)$ correspond to the local and nonlocal electron–phonon interactions, respectively.

Having established the basis of a general model Hamiltonian, it is helpful to discuss its specific form particularly tailored for the EET and CT processes. The relevant electronic states involve the LE states, CT states, and charge-separated (CS) states. For an N-monomer aggregate, the electronic Hamiltonian has the following form

$$\hat{H}_e = \sum_{nmi} E_{n,m;L_i} |n, m; L_i\rangle\langle n, m; L_i|$$
$$+ \sum_{nmi} \sum_{n'm'i'}{}' V_{nm,n'm';L_i L_{i'}} |n, m; L_i\rangle\langle n', m'; L_{i'}| \tag{15.6}$$

Here, $|n, m; L_i\rangle$ represents the electronic excited state. For $n \neq m$, it represents the excited state that the nth monomer is in the cationic state and the mth monomer is in the anionic state simultaneously, for $n = m$, it corresponds to a LE state with the nth monomer being excited, and all other monomers in the aggregate are all in the ground state. L_i orders the electronic states for a given $|n, m\rangle$ state in terms of the energy. $E_{n,m;L_i}$ is the vertical excitation energy of the state $|n, m; L_i\rangle$ and $V_{nm,n'm';L_i L_{i'}}$ corresponds to the electronic coupling between states $|n, m; L_i\rangle$ and $|n', m'; L_{i'}\rangle$. In the second row of Eq. (15.6), the superscript prime beside the summation notation indicates that the terms where $i = i'$, $n = n'$ and $m = m'$ are excluded. Under this notation convention, $|n, n; L_i\rangle$ and $|n, n \pm 1; L_i\rangle$ represent the LE state and the bound CT state, respectively, whereas $|n, m; L_i\rangle$ with $|n - m| \geq 2$ denotes the CS state.

The vibrational Hamiltonian is described as

$$\hat{H}_{ph} = \sum_n \sum_j \omega_j \left(\hat{b}^\dagger_{nj} \hat{b}_{nj} + \frac{1}{2} \right) \tag{15.7}$$

where $\hat{b}^\dagger_{nj}$ and $\hat{b}_{nj}$ represent the creation and annihilation operator of the jth vibrational mode in the nth monomer with the frequency of ω_j, respectively. Finally, the electron–phonon interaction term is given by

$$\hat{H}_{e\text{-}ph} = \sum_n \sum_j \hat{x}_{nj} \otimes (\hat{b}^\dagger_{nj} + \hat{b}_{nj}) \tag{15.8}$$

with

$$\hat{x}_{nj} = \sum_i g^{ex}_{j,L_i} |n, n; L_i\rangle\langle n, n; L_i| + \sum_i \sum_{m \neq n} g^+_{j,L_i} |n, m; L_i\rangle\langle n, m; L_i|$$
$$+ \sum_i \sum_{m \neq n} g^-_{j,L_i} |m, n; L_i\rangle\langle m, n; L_i| \tag{15.9}$$

where $\hat{x}_{nj}$ corresponds to the electron–phonon interaction operator coupled with the different electronic states. g^{ex}_{j,L_i}, g^+_{j,L_i}, and g^-_{j,L_i} are the electron–phonon interaction coefficients of the jth vibrational mode associated with the ith excited, cationic, and anionic state of the monomer, respectively.

15.2.2 Parameterizing Electronic Excited-State Hamiltonian

To obtain H_e, we have to know the energies of quasi-diabatic electronic excited states and the couplings among them, i.e. the following electronic-state Hamiltonian in the diabatic representation for a typical aggregate should be constructed

$$H_e = \begin{bmatrix} E_{11} & J_{12} & \cdots & J_{1n} \\ J_{21} & E_{22} & \cdots & J_{2n} \\ \cdots & \cdots & \cdots & \cdots \\ J_{n1} & J_{n2} & \cdots & E_{nn} \end{bmatrix} \equiv \sum_{n,m} E_{nm} |n\rangle\langle m| \tag{15.10}$$

where n and m index quasi-diabatic states (LE, CT, etc.), the diagonal element E_{nn} is the nth quasi-diabatic state energy, and the off-diagonal element $J_{nm}(\equiv E_{nm})$ is the electronic coupling between the nth and mth states. In general, conventional electronic structure theories – which begin with the adiabatic representation under the Born–Oppenheimer approximation – calculate the adiabatic excited-state quantities. Consequently, these theories cannot be straightforwardly utilized to construct the Hamiltonian of Eq. (15.10). To acquire quasi-diabatic states, several electronic structure approaches have been proposed, either by imposing constraints on electron densities or by using chemically intuitive methods to construct localized configurations. These methods include the constrained density functional theory [18–20], the valence bond theory [21–23], and the multistate density functional theory [24–26]. Such approaches are thriving [27–29], and an appropriate descriptor of electronic localized properties is crucial for the construction of correct quasi-diabatic states.

Despite the aforementioned methods, a more popular spirit to constructing quasi-diabatic states involves utilizing adiabatic excited-state calculations, subsequently employing an adiabatic-to-diabatic (ATD) unitary transformation. There are two kinds of categories following such methodology. One category adopts the concept of requiring the quasi-diabatic wavefunction to be as smooth as possible in terms of nuclear coordinates, i.e. the requirement of the configurational uniformity [30]. Methods falling into this category include the conventional block diagonalization [31, 32] and the fourfold way [33–35]. The projection approach [36–38] may also be classified in this group. The other category requires the smooth condition for various physical properties [39]. In this kind of methods, the most well-known one is the generalized Mulliken–Hush method [40] for electron transfer, which adopts the dipole moment as an indicator to maximize the distance between two charge centers. Extensions to multiple charge centers [41–43] and to include quadruple moments [44, 45] have also been proposed. Sharing the same spirit, methods like the fragment charge difference [46] (FCD) and its variants [47–51], along with other recently proposed schemes [52, 53], construct quasi-diabatic states using fragment-based quantities such as the charge difference or excitation difference. While these methods have achieved great success in many complex systems, there are still certain factors that may limit their practical applications. For instance, they may be more applicable to CT or EET but not both, or may be best suited for specific types of adiabatic wave functions, or may involve complex implementations, or may be limited to dimer systems.

Apart from the above methods, there is a newly proposed ATD diabatization scheme called FPHD [13]. The primary idea of FPHD is to find an optimal unitary transformation of the adiabatic excited states to maximize the localization of particle and hole densities in terms of predefined molecular fragments. Compared to other ATD methods, the FPHD scheme has several appearing merits. It applies to both CT and EET processes in multichromophoric systems with multiple charge and excitation centers, it can be combined with various electronic structure theories, and it is able to simultaneously treat up to hundreds of electronic states within a single diabatization procedure in a nearly black-box manner. In the following, we will first outline the theoretical foundation of FPHD and then take FPHD as a typical protocol to demonstrate how to parameterize the electronic Hamiltonian equation (15.10) from first-principles calculations.

Under the ATD formalism, the electronic Hamiltonian in the diabatic representation is obtained via a simple unitary transformation as follows

$$\hat{H}_e = \hat{U}^T \hat{H}_e^a \hat{U} \tag{15.11}$$

provided that the ATD transformation matrix $\hat{U}$ is known, where H_e^a represents a diagonal matrix with the adiabatic excited-state energies as elements. To construct the optimal transform matrix $\hat{U}$, the core issue is to find an appropriate descriptor for quasi-diabatic states. In the FPHD method, the desired quasi-diabatic states should have the property of maximally localized particle and hole densities on the fragments (monomers) which are divided in advance from the aggregate. Figure 15.2 is a schematic diagram that illustrates the localization procedure, taking a four-monomer aggregate as an example. The properties of the adiabatic excited states of the aggregate can be easily obtained with conventional electronic structure calculations. The first row of Figure 15.2 displays the calculated hole density and particle (electron) density for several adiabatic excited states S_n. We denote the corresponding adiabatic wave functions as $|\Psi^{(n)}\rangle$. The one-particle transition density between $|\Psi^{(n)}\rangle$ and a reference state (the ground state) $|\Psi_{ref}\rangle$ can be defined by

$$\gamma^{(n)}(r, r') = \sum_{pq} D_{pq}^{(n)} \varphi_p(r) \varphi_q(r') \tag{15.12}$$

Here, p and q denote the pth and qth spin-polarized molecular orbitals, respectively, with $\varphi_p(r)$ and $\varphi_q(r)$ being the spatial part of the spin–orbital wave functions. $D_{pq}^{(n)} = \langle \Psi_{ref}|\hat{a}_p^\dagger \hat{a}_q|\Psi^{(n)}\rangle$ is the element of the transition density matrix, with $\hat{a}_p^\dagger$ and $\hat{a}_q$ being the creation and annihilation operators of the pth and qth molecular orbitals, respectively. $\gamma^{(n)}(r, r')$ represents the amplitude that an electron is moved from r' to r during the $|\Psi_{ref}\rangle \rightarrow |\Psi^{(n)}\rangle$ transition. As such, the hole density and

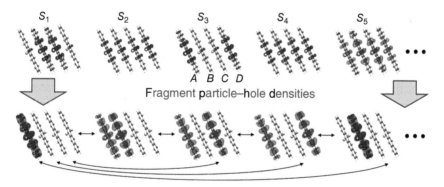

Figure 15.2 Schematic representation of the adiabatic excited states S_i (top) of aggregate ABCD, and quasi-diabatic states A*BCD, A⁻B⁺CD and A⁻BC⁺D, A⁻BCD⁺ and AB*CD (bottom from left to right sides). The blue and red clouds represent hole density and particle (electron) density. Source: Wang et al. [13]/Adapted with permission of American Chemical Society.

particle density in this transition can be expressed as $\rho_n^h(r) = \int |\gamma^{(n)}(r,r')|^2 dr'$ and $\rho_n^e(r) = \int |\gamma^{(n)}(r',r)|^2 dr'$, respectively. We can integrate $\rho_n^h(r)$ and $\rho_n^e(r)$ over specific spatial regions to obtain the number of particles and holes located on each fragment. For the numerical benefit, we introduce the fragment hole number matrix $\hat{N}^{h,A}$ and the fragment particle number matrix $\hat{N}^{e,A}$ in terms of the fragment A, with their elements being defined as

$$N_{nm}^{h,A} = \sum_{pq} \left[\hat{D}^{(n)}(\hat{D}^{(m)})^T\right]_{pq} S_{pq}^{(A)} \tag{15.13}$$

and

$$N_{nm}^{e,A} = \sum_{pq} \left[(\hat{D}^{(n)})^T \hat{D}^{(m)}\right]_{pq} S_{pq}^{(A)} \tag{15.14}$$

where $S_{pq}^{(A)} = \int_{r \in A} \varphi_p(r) \varphi_q(r) dr$ is the partial overlap integral between $\varphi_p(r)$ and $\varphi_q(r)$ over the spatial region of the fragment A. It can be shown [13] that the diagonal elements of $\hat{N}^{h,A}$ and $\hat{N}^{p,A}$ represent the concrete number of holes and particles locating in fragment A, respectively. Numerically, various partition schemes such as Mulliken [54], Hirshfeld [55], or the Löwdin population analysis [56] can be adopted to calculate $S_{pq}^{(A)}$. The representation transformations of the fragment number matrices satisfy the following relations:

$$\begin{cases} \bar{N}^{h,A} = \hat{U}^T \hat{N}^{h,A} \hat{U} \\ \bar{N}^{e,A} = \hat{U}^T \hat{N}^{e,A} \hat{U} \end{cases} \tag{15.15}$$

where $\bar{N}^{h,A}$ and $\bar{N}^{p,A}$ are the fragment hole and particle matrices in the diabatic representation. Equations (15.13) and (15.14) in conjunction with Eq. (15.15) offer an elegant and highly efficient way to calculate the fragment-based matrices for any electronic state. To obtain the ATD unitary transformation matrix $\hat{U}$, we simultaneously diagonalize all of the fragment number matrices by the Jacobi sweep algorithm [57], and identify the nature of each quasi-diabatic state according to the eigenvalues of the matrices. For instance, as depicted in the second row of Figure 15.2, the quasi-diabatic states are A^*BCD, A^-B^+CD, A^-BC^+D, A^-BCD^+, and AB^*CD. In these states, the LE states correspond to the situation where the non-zero eigenvalues of particle and hole number matrices are in the same fragment. Conversely, the CT states correspond to those situations where they reside in different fragments. Then, the quasi-diabatic Hamiltonian can be obtained via Eq. (15.11). Moreover, other parameters in the diabatic representation, such as transition dipole moments, can also be obtained via a similar unitary transformation.

The FPHD approach, being efficient in calculations, simple for implementation, and flexible to combine with various kinds of electronic structure theories, serves as an important theoretical tool in the investigation of photoinduced CT and EET processes. It is noted that a closely related approach called the fragment excitation difference [47] (FED) uses the sum of the attachment and detachment densities [58] to distinguish LE states with different excitation centers, and works quite well with the configuration interaction singles (CIS) and time-dependent density functional theory within the Tamm–Dancoff approximation (TDA-time-dependent density functional theory [TDDFT]). However, for a higher-level electronic structure theory, the calculation of attachment and detachment densities involves cumbersome eigendecomposition of the difference density matrix [50] and thus severely hinders the application of the FED method. On the contrary, the particle and hole densities adopted in the FPHD scheme are calculated based on the transition density matrix, and their calculations follow the same procedure irrespective of the electronic structure theories being used.

15.2.3 Calculating Electron–Phonon Couplings

The interaction between electronic and vibrational DOFs plays a crucial role in modeling exciton dynamics and spectra, which enters the exciton model in a form of coordinate-dependent state energies and electronic couplings, and is represented by the spectral density. However, it remains a challenge to model the spectral density because the number of DOFs in a complex aggregate grows exponentially with the number of physical particles. Furthermore, the electron–phonon (e–p) interactions in realistic organic materials are quite complex and intricately dependent on intramolecular motions, intermolecular couplings, stacking motifs, and crystal environment, among other factors. To simplify the problem, the parameterization of $\hat{H}_{ph}$ and $\hat{H}_{e-ph}$ is often based on the monomer model and the Condon approximation, which not only significantly reduces the complexity but also provides a good approximation.

We demonstrate this protocol with the model Hamiltonian presented in Eqs. (15.7)–(15.9), which is suitable for the description of CT and EET processes in molecular aggregates. Since Eq. (15.7) has assumed that all the monomers in the aggregate share the same set of vibrational frequencies, it is straightforward to obtain the frequencies and the corresponding normal modes by diagonalizing the Hessian matrix of the monomer at its ground state equilibrium geometry. The environmental effects can be accounted for via either a solvent model or a quantum mechanics/molecular mechanics model.

For the parameterization of the electron–phonon interaction coefficients, one must explicitly calculate properties of the associated excited, cationic, or anionic states of the monomer, depending on the status of the monomer in the quasi-diabatic states. Under the Condon approximation, the nonlocal electron–phonon interaction coefficients $g_{nm\mu}$ in Eq. (15.5) are zero and do not enter Eq. (15.9). On the other hand, the local ones (i.e. g^{ex}_{j,L_i}, g^{+}_{j,L_i} and g^{-}_{j,L_i} in Eq. (15.9) are directly related to the reorganization energies and can be obtained from the corresponding normal-mode coordinate shifts between the electronic ground state and the relevant electronic states, as shown in Figure 15.3. Taking g^{ex}_j as a concrete example (the parameterization of g^{+}_{j,L_i} and g^{-}_{j,L_i} follows similar procedures), where we have omitted the subscript L_i and explicitly considered one of the excited states (usually the lowest-lying one) of the monomer. Then we have

$$g^{ex}_j = \sqrt{\frac{\omega_j^3}{2}} \Delta Q_j \qquad (15.16)$$

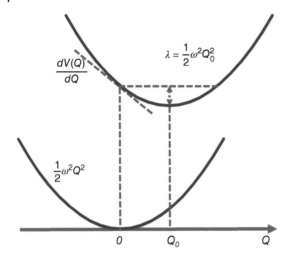

Figure 15.3 Schematic representation of the one-dimensional Harmonic potential energy surface and reorganization energy of nuclear motion.

where ΔQ_j is the jth normal mode shift between the electronic ground and excited states. The shift is also connected to the mode-resolved reorganization energy λ_j and Huang–Rhys factor S_j via $\lambda_j = \frac{1}{2}\omega_j^2 \Delta Q_j^2$ and $S_j = \frac{1}{2}\omega_j \Delta Q_j^2$.

To get $\Delta \vec{Q}$, one may start from the following expression

$$\vec{q}_e = \vec{q}_g + \Delta \vec{q} \tag{15.17}$$

where $\vec{q}_e$ and $\vec{q}_g$ represent the mass-weighted Cartesian coordinates of the excited and ground states, respectively, and $\Delta \vec{q}$ is the corresponding shift. The normal mode coordinates $\vec{Q}$ can be transformed from Cartesian coordinates $\vec{q}$ as [59]

$$\vec{Q} = L\vec{q} \tag{15.18}$$

where L is the transform matrix, which can be obtained from the diagonalization of the Hessian matrix K by

$$L^T K L = \omega^2 \tag{15.19}$$

with ω being the frequencies. By using Eq. (15.18), Eq. (15.17) can be cast into

$$\vec{Q}_e = L_e L_g^T \vec{Q}_g + L_e \Delta \vec{q} \tag{15.20}$$

where L_e and L_g correspond to the transform matrix for the excited and ground states, respectively, and $L_e L_g^T$ is called the Duschinsky rotation matrix, which is approximated to unit in the calculations. Therefore, the normal mode shifts can be obtained [59] via $\Delta \vec{Q} = L_e \Delta \vec{q}$.

Alternatively, under the assumption that the frequency and coordinate direction of each normal mode are the same in two electronic states, one can adopt the vertical gradient technique [60–62], in which g_j^{ex} factor is calculated by

$$g_j^{ex} = \sqrt{\frac{1}{2\omega_j} \frac{\partial E(\vec{Q})}{\partial Q_j}} \tag{15.21}$$

where $E(\vec{Q})$ is the potential energy of the excited state, Q_j is the coordinate of the jth normal mode, and the partial derivative is taken at the ground state equilibrium geometry. The schematic picture of the potential gradient is shown in Figure 15.3. The vertical gradient technique is usually much more efficient than the rotation technique because the former one avoids direct geometry optimization and second-order energy derivative calculations of excited states.

In the fitting of experimental measurements, an effective normal mode to mimic multi-modes is commonly introduced. From the rate calculations of CT and EET processes based on the Fermi's golden rule, we can get the effective-mode parameters from multi-mode properties [63]. The effective frequency is given by

$$\omega_{\text{eff}}^2 = \sum_j (\lambda_j \omega_j^2)/\lambda \tag{15.22}$$

where $\lambda = \sum_j \lambda_j$ is the total reorganization energy. The effective Huang–Rhys factor S_{eff} is

$$S_{\text{eff}} = \lambda/\omega_{\text{eff}} \tag{15.23}$$

15.2.4 Propagating the Photophysical Dynamics

In pure electronic states, the photophysical processes are easily obtained by solving time-dependent Schrödinger equation with Hamiltonian of Eq. (15.10). The dynamics with 10 thousands of electronic states can be easily solved by straightforwardly diagonalizing Hamiltonian. Concretely, at a given time t, the wavefunction can be calculated by

$$\psi(t) = \hat{T} e^{-i\hat{E}t} \hat{T}^\dagger \psi(0) \tag{15.24}$$

where $\hat{E}$ is a diagonal matrix with the elements being the eigenvalues of $\hat{H}_e$, and $\hat{T}$ is transform matrix for the diagonalization (that is, $\hat{E} = \hat{T}^\dagger \hat{H}_e \hat{T}$). However, straightforwardly solving the Schrödinger equation becomes nearly formidable when the vibrational motions of aggregates are incorporated. This is primarily because it necessitates dealing with nuclear DOFs in dynamics simulations, which, given the current computational capacity, are still limited to fewer than tens of DOFs. By reducing the number of basis vectors within controllable numerical errors, celebrated approaches have been developed, such as the numerical renormalization group [64, 65], density matrix renormalization group [66], time evolving density matrix using orthogonal polynomials algorithms [67], and multi-layer multi-configuration time-dependent Hartree (ML-MCTDH) method [68].

In the photophysical processes of aggregates, it is often the case that the primary interest lies in the dynamics of the electronic states. Consequently, the vibrational DOF are typically traced over. Such reduced dynamics can be formally calculated using the path integral technique [69]. However, this operation also brings about new challenges, as the influence functional is nonlocal in both spatial and temporal domains. To overcome this problem, Makri have originally proposed an efficient tensor multiplication scheme that preserves the long-time stability on the foundation of an improved quasi-adiabatic path integral (QUAPI) discretization [70]. The QUAPI's applications to complex systems are hindered due to the fact that the computational efforts grow dramatically with the increase of system size and memory time, and further extensions have been made [71]. An alternative powerful approach that was originally established within the path-integral formalism is the non-perturbative and non-Markovian hierarchical equations of motion (HEOM) [72]. Pioneered by Tanimura and Kubo in their seminal work [73], great efforts from different contributors have been devoted to derive a numerically exact version of HEOM [74–76]. For a comprehensive introduction of HEOM, the readers can refer to [72]. Given all these past developments, HEOM has become a standard approach in the field of open quantum systems nowadays.

Another disparate strategy is to perform complete stochastic unraveling to the influence functional by the introduction of complex Gaussian stochastic fields, which results in the stochastic Liouville–von Neumann equation [77–83]. This approach is not limited to the form of the spectral density function and can be directly decomposed into a forward and a backward stochastic Schrödinger equations in the case of a pure initial state of the system. However, it may suffer from a severe stochastic convergence problem in long-time simulations or when the system-bath interaction is not weak. Recently, several density-matrix approaches [84–88] that combine the hierarchical expansion and stochastic unraveling techniques have also been proposed, which inherit both merits of HEOM and the stochastic Liouville–von Newmann equation, in the sense that the resulting hierarchical structure is simpler and the stochastic convergence property is better.

One less satisfactory feature of density-matrix approaches is that the number of elements that need to be calculated scales quadratically with the system size. As a counterpart of density-matrix approaches, wavefunction-based methods have also been playing a prominent role in the dynamics of open quantum systems due to the beneficial linear scaling of Hilbert space with respect to the system size. Inspired by the spirits of mixed deterministic-stochastic scheme, Suess et al. [89] applied the hierarchical expansion technique to the functional derivative term of non-Markovian quantum state diffusion and proposed a numerically exact approach named hierarchy of pure states (HOPS), which is a closed set of stochastic Schrödinger equations with hierarchical structure. Whereafter, at about the same time, Song et al. [90] and Ke and Zhao [91] independently derived two similar approaches. In the new approaches, temperature effects are mostly or entirely accounted for by the stochastic fields such that the resulting equations of motion are more stable at finite temperatures as compared with HOPS. Soon afterward, with a similar spirit, Hartmann et al. [92, 93] improved the original HOPS by including temperature effects via a stochastic contribution to the system Hamiltonian.

As the magnitudes of electron–phonon couplings are smaller than or similar to those of electronic couplings, one may use partial perturbations for electron–phonon couplings to establish cheaper NMSSE [15, 94], applicable on an equal footing for any kind of spectral density function and for the quantum dynamics of large-scale systems. NMSSE thus becomes very suitable for investigating photophysical dynamics of aggregates. Here, we outline the working equation of NMSSE proposed in our group [14–17].

15.2 Theoretical Approaches

We start from the reduced density matrix of electronic states at time t

$$\hat{\rho}_e(t) = \text{Tr}_{\text{ph}} \left\{ e^{-i\hat{H}t} |\psi_0\rangle\langle\psi_0| \otimes \rho_{\text{ph}}(T) e^{i\hat{H}t} \right\} \tag{15.25}$$

where $|\psi_0\rangle$ is the initial state of the electronic DOF, $\rho_{\text{ph}}(T) = e^{-\beta\hat{H}_{\text{ph}}}/Z_{\text{ph}}$, $\beta = 1/k_BT$ with T being the temperature, $Z_{\text{ph}} = \text{Tr}\{e^{-\beta\hat{H}_{\text{ph}}}\}$, and Tr_{ph} represents to take the trace over the phonon DOFs. In NMSSEs, the effect of vibrational motions on the exciton dynamics is incorporated with phonon-induced stochastic fields as well as non-Markovian terms. After disentangling the forward and backward propagation, we obtain the general expression of NMSSE

$$i\frac{\partial}{\partial t}|\psi_\xi(t)\rangle = \hat{H}_\xi(t)|\psi_\xi(t)\rangle \tag{15.26}$$

with

$$\hat{H}_\xi(t) = \hat{H}_e + \hat{V}_\xi(t) + \hat{V}_{\text{mem}}(t) \tag{15.27}$$

where $\hat{V}_\xi$ and $\hat{V}_{\text{mem}}$ are stochastic and memory terms, respectively. For the general model Hamiltonian given in Eqs. (15.1)–(15.5), we have

$$\hat{V}_\xi(t) = \sum_\mu \xi_\mu(t) \hat{X}_\mu \tag{15.28}$$

and

$$\hat{V}_{\text{mem}}(t) = -i \int_0^t d\tau \sum_\mu \hat{X}_\mu \hat{X}_\mu(-\tau) \alpha_\mu(\tau) \tag{15.29}$$

Here, $\xi_\mu(t)$ is the mode-specific stochastic field which satisfies the following statistical properties

$$\begin{cases} \langle \xi_\mu(t)\xi^*_{\mu'}(t')\rangle = \delta_{\mu\mu'}(\bar{n}_\mu e^{-i\omega_\mu(t-t')} + (\bar{n}_\mu + 1)e^{i\omega_\mu(t-t')}) \\ \langle \xi_\mu(t)\xi_{\mu'}(t')\rangle = \delta_{\mu\mu'}\sqrt{2\bar{n}_\mu(\bar{n}_\mu+1)}\cos\omega_\mu(t-t') \end{cases} \tag{15.30}$$

where $\bar{n}_\mu \equiv 1/(e^{\beta\omega_\mu}-1)$, $\langle \xi_\mu(t)\xi^*_{\mu'}(t')\rangle$ accounts for the correlation between the forward and backward paths in the reduced density matrix, whereas $\langle \xi_\mu(t)\xi_{\mu'}(t')\rangle$ is obligated to the correlation within the same path. These stochastic fields can be efficiently generated by

$$\xi_\mu(t) = \sqrt{\frac{\bar{n}_\mu+1}{2}}(\gamma^1_\mu + i\gamma^2_\mu)e^{i\omega_\mu t} + \sqrt{\frac{\bar{n}_\mu}{2}}(\gamma^1_\mu - i\gamma^2_\mu)e^{-i\omega_\mu t} \tag{15.31}$$

where γ^1_μ and γ^2_μ are independent Gaussian white noises that obey $\langle \gamma^i_\mu \rangle = 0$ and $\langle \gamma^i_\mu \gamma^{i'}_{\mu'}\rangle = \delta_{\mu\mu'}\delta_{ii'}$. In the memory term, $\hat{X}_\mu(-\tau) = \hat{U}_0(\tau)\hat{X}_\mu \hat{U}^\dagger_0(\tau)$ with $\hat{U}_0(\tau) = e^{-i\hat{H}_{el}\tau}$, and

$$\alpha_\mu(\tau) = \tanh\frac{\beta\omega_\mu}{4}\cos\omega_\mu\tau - i\sin\omega_\mu\tau \tag{15.32}$$

Substituting Eq. (15.5) into Eq. (15.29), we obtain a more explicit expression for the memory term

$$\hat{V}_{\text{mem}}(t) = -i\int_0^t d\tau \sum_\mu \sum_{nm}\sum_{n'm'} g_{nm\mu}g_{n'm'\mu}|n\rangle\langle m|\hat{U}_0(\tau)|n'\rangle\langle m'|\hat{U}^\dagger_0(\tau)\alpha_\mu(\tau) \tag{15.33}$$

As a special of case Eqs. (15.1)–(15.5), the model Hamiltonian given in Eqs. (15.6)–(15.9) corresponds to the following form of NMSSE

$$i\frac{\partial}{\partial t}|\psi_\xi(t)\rangle = \left[\hat{H}_e + \sum_{nj}\xi_{nj}(t)\hat{x}_{nj} - i\sum_{nj}\hat{x}_{nj}\int_0^t \hat{x}_{nj}(-\tau)\alpha_j^{res}(\tau)d\tau\right]|\psi_\xi(t)\rangle \quad (15.34)$$

Here, $\hat{x}_{nj}(-\tau) = \hat{U}_0(\tau)\hat{x}_{nj}\hat{U}_0(\tau)^\dagger$. The residual correlation function is

$$\alpha_j^{res}(t) = \tanh\frac{\beta\omega_j}{4}\cos\omega_j t - i\sin\omega_j t \quad (15.35)$$

$\xi_{nj}(t)$ is the time-dependent stochastic field satisfying statistical properties similar to Eq. (15.30), and can be generated via

$$\xi_{nj}(t) = \sqrt{\frac{\bar{n}_j + 1}{2}}(\gamma_{nj}^1 + i\gamma_{nj}^2)e^{i\omega_j t} + \sqrt{\frac{\bar{n}_j}{2}}(\gamma_{nj}^1 - i\gamma_{nj}^2)e^{-i\omega_j t} \quad (15.36)$$

where γ_{nj}^1 and γ_{nj}^2 are independent Gaussian white noises that obey $\langle\gamma_{nj}^i\rangle = 0$ and $\langle\gamma_{nj}^i\gamma_{n'j'}^{i'}\rangle = \delta_{nn'}\delta_{jj'}\delta_{ii'}$.

The time-dependent wave function $|\psi_\xi(t)\rangle$ is then obtained by solving Eqs. (15.26)–(15.29) (or Eq. (15.34), according to the model Hamiltonian adopted) with standard algorithms like the fourth-order Runge–Kutta method, with the initial condition $|\psi_\xi(0)\rangle = |\psi_0\rangle$. The reduced density matrix of the electronic DOF is obtained after averaging over the stochastic wavefunction ensemble

$$\hat{\rho}_e(t) = M[|\psi_\xi(t)\rangle\langle\psi_\xi(t)|]_\xi \quad (15.37)$$

where $M[\cdot]_\xi$ represents the average over the stochastic fields. It is noted that the norm of $\hat{\rho}_e(t)$ is usually not conserved due to the complex stochastic field and non-Hermitian integral term in Eqs. (15.26) and (15.34), and an artificial normalization to this reduced density matrix is usually required after the average step.

15.2.5 Numerical Examples

Once the parameterized model Hamiltonian Eqs. (15.1)–(15.5) (or Eqs. (15.6)–(15.9)) is available, we can calculate the time-dependent wavefunction and time-dependent observables according to Eqs. (15.26)–(15.29) (or Eq. (15.34)). In the following, we show numerical examples for the calculations of spectra and carrier dynamics in organic zinc phthalocyanine (ZnPc) aggregates [95–97].

15.2.5.1 Absorption Spectra of Aggregates

The linear absorption spectra correspond to Fourier transform of time-dependent correlation function of transition dipole moment. For the model Hamiltonian given in Eqs. (15.6)–(15.9), the transition dipole moment operator can be expressed as

$$\hat{\mu} = \sum_{nmi}\vec{\mu}_{n,m;L_i}(|n,m;L_i\rangle\langle g| + |g\rangle\langle n,m;L_i|) \quad (15.38)$$

where $|g\rangle$ is the electronic ground state of the aggregate, and $\vec{\mu}_{n,m;L_i}$ is the transition dipole moment between $|g\rangle$ and $|n,m;L_i\rangle$. The absorption spectra are given by

$$D(\omega) = \frac{1}{2\pi} \Re \int_0^\infty C_{\mu\mu}(t) e^{i\omega t} dt \tag{15.39}$$

where $\Re$ represents to take the real part. Within the NMSSE framework, the α component ($\alpha = x, y, z$) of the dipole autocorrelation function is calculated by

$$C_{\mu\mu}^\alpha(t) = M\big[\langle g|\hat{\mu}_\alpha|\psi_\xi(t)\rangle\big]_\xi \tag{15.40}$$

with the initial condition

$$|\psi_\xi(0)\rangle = \hat{\mu}_\alpha|g\rangle \tag{15.41}$$

where $\hat{\mu}_\alpha$ is the projection of $\hat{\mu}$ along the α direction. The total autocorrelation function is obtained by $C_{\mu\mu}(t) = \sum_{\alpha=x,y,z} C_{\mu\mu}^\alpha(t)$.

In concrete examples, we focus on the absorption spectra of ZnPc crystals of α and β phases [98], and their crystal structures are shown in Figure 15.4. Experimentally, both phases exhibit two absorption peaks in the Q-band region, but it is remarkable that the relative intensities of the two peaks are reversed [99–105], as shown in Figure 15.5. Furthermore, the evidence from the time-resolved photoluminescence spectra have indicated that the ultrafast exciton dynamics occurring right after the photoexcitation in the two phases also exhibit significant differences [106]. To reveal the corresponding relationship between the structure and photophysical processes, we first use the FPHD method to construct quasi-diabatic states from adiabatic electronic excited states of α and β phases.

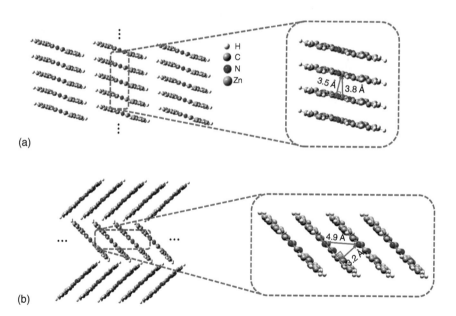

Figure 15.4 Crystal stacking structures of (a) α-phase (brickstone) and (b) β-phase (herringbone) ZnPc aggregates. The two magnification boxes present the structures of the ZnPc tetramers extracted from the corresponding aggregates. Source: Feng et al. [97]/Royal Society of Chemistry.

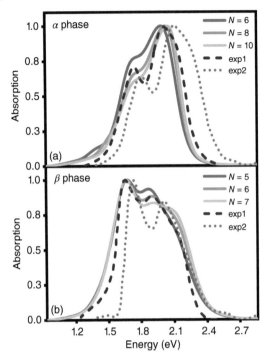

Figure 15.5 Normalized vibrationally resolved absorption spectra of (a) α-phase and (b) β-phase ZnPc aggregates with different aggregation lengths. The gray-doted and black-dashed lines are experimental results extracted from Refs. [105] and [104], respectively. The calculated spectra are red-shifted by 0.461 and 0.303 eV for the α and β phases, respectively, for better comparison to experimental measurements. The spectra are broadened using the Gaussian line shape function with a full width of 0.05 eV. Source: Feng et al. [97]/Royal Society of Chemistry.

The fruitful features in the absorption spectra in two phases can be explained from the interplay between the bright LE states and the CT states. Along with the formation of the aggregate, due to the interference between LE states, the absorption spectra will be dominated by a single peak corresponding to the (delocalized) bright FE state. As the dark CT state takes part in, the two states may mix together to constitute two new eigenstates, which causes the splitting of the original single absorption peak [8]. The energies and oscillator strengths of the new states, which correspond to the positions and intensities of the absorption peaks, respectively, strongly depend on the couplings and the energy alignment between the FE and CT states. It is known that for an H-aggregate with positive excitonic couplings, the energy of the FE state gradually increases as the aggregation length increases, and eventually arrives at a stationary value of twice the excitonic coupling. Moreover, since the CT state is usually transition forbidden, the peak intensities are determined by the FE proportions of the corresponding eigenstates, that is, the higher FE proportion, the stronger peak intensity.

To simulate the photoinduced exciton dynamics, one should have knowledge about the initial adiabatic electronic wavepacket created by the incident light. However, direct TDDFT calculations for the aggregates including thousands of atoms still meet great challenges currently. This difficulty can be overcome by utilizing the localization nature of quasi-diabatic states. Concretely speaking, we first calculate the adiabatic excited states of the aggregate with a few of monomers. The quasi-diabatic states and the diabatic Hamiltonian for this model aggregate can be obtained by a subsequent FPHD calculation. Since the quasi-diabatic states are

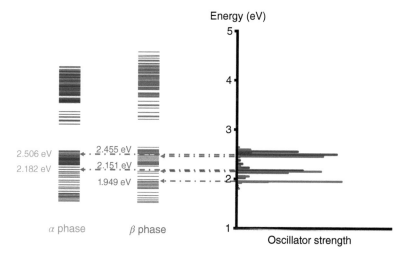

Figure 15.6 Energy level diagrams and oscillator strengths of the adiabatic electronic states of the α-phase and β-phase aggregates formed from 11 monomers. Source: Feng et al. [97]/Royal Society of Chemistry.

well-localized, their electronic couplings rapidly vanish as the distance of exciton (or charge) transfer increases. As such, we can directly extend the aggregate to a desired length by simply using the periodicity of the diabatic Hamiltonian and neglecting the electronic couplings for transfer distances larger than a threshold value. After diagonalizing the constructed Hamiltonian, we acquire the adiabatic excited states for more complex aggregates.

Figure 15.6 shows the adiabatic energy level diagrams and oscillator strengths of two phases with incorporation of 11 monomers, which are obtained based on the adiabatic results of 4 monomers by TDDFT calculations [97]. Among all of the adiabatic states, we choose those states with the largest oscillator strengths as the initial dynamic states to imitate the photoexcitation process. For the α phase, the two selected excitation energies (denoted as E_{opt} in the following) are 2.182 and 2.506 eV, corresponding to the low-energy and high-energy absorption peaks in the Q-band region, respectively. As for the β phase, the three adiabatic states with the energies of 1.949, 2.151, and 2.455 eV are selected. The first one represents the lowest-lying L_2-FE peak, whereas the other two correspond to the two L_1-CT-mediated peaks.

It is noted that there exist large energy gaps in the energy level diagrams of both phases. The adiabatic states above the gap are mainly constituted by the CS states, whereas those below the gap are mostly formed by the LE and CT states. Thereby, free carriers can be hardly generated by the photoexcitation from the Q-band region, as has been confirmed in our previous work [96].

The population evolution of different states in the α and β phases with 11 monomers are presented in Figures 15.7 and 15.8, respectively. In the α phase, as has already been discussed previously [96], the optical excitation from the Q-band region can give rise to strong mixing between the LE and CT states. At the initial

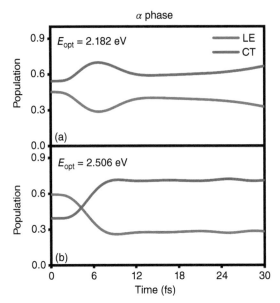

Figure 15.7 Population evolution of the LE and CT states after optical excitation in the α-phase aggregate with excitation energies of (a) 2.182 eV and (b) 2.506 eV. Source: Feng et al. [96]/Adapted with permission from American Chemical Society.

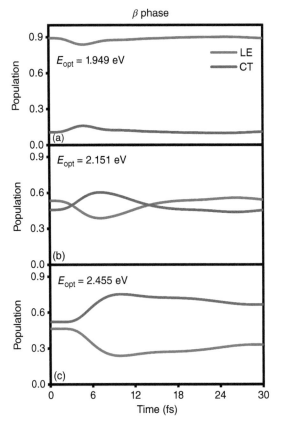

Figure 15.8 Population evolution of the LE and CT states after optical excitation in the β-phase aggregate with excitation energies of (a) 1.949 eV, (b) 2.151 eV, and (c) 2.455 eV. Source: Feng et al. [97]/Royal Society of Chemistry.

time, the proportion of the LE state is smaller than that of the CT state for the lower excitation energy (2.182 eV), whereas the opposite situation occurs for the higher one (2.506 eV). Nonetheless, after a rapid population exchange process between the LE and CT states within the first 10 fs, the aggregates seem to fall into some static states where the population distributions do not obviously evolve. For the β phase, similar interplay between the LE and CT states arises for the photoexcitation of the two L_1-CT-mediated peaks, as can be seen in Figure 15.8b,c. However, for the optical transition of the L_2-FE peak shown in Figure 15.8a, the LE and CT states only weakly mix with each other, and the population evolution curves are rather flat. The above dynamical results are in line with previous analyses of the absorption spectra.

It should be addressed that those ultrafast dynamics within 30 fs already determine the properties of absorption spectra, but they cannot be straightforwardly measured experimentally because the individual initial state is a single adiabatic excited state that can be only excited by continuous-wave (CW) laser without the time resolution. However, the realistic time-resolved dynamics can be simulated with an initial state including a collective of adiabatic states generated by ultrafast laser pulse, and the NMSSE can predict the corresponding dephasing and relaxation processes for experimental measurements.

15.3 Concluding Remarks

In summary, theoretical unveiling of photophysical processes in molecular aggregates demands both excited-state electronic structure calculations and quantum dynamic simulations. The former offers the properties of quasi-diabatic states, such as the excited-state energies, electronic couplings and electron–phonon interaction coefficients, and the latter exhibits the quantum dynamics of carriers. Although we have displayed the FPHD and quantum chemical calculations based on the vertical gradient approximation for constructing Hamiltonian and a NMSSE for carrier dynamics, there are many challenges to be addressed. For instance, the work should include establishing the quantitative relationship between the internal harmonic vibration coupling, anharmonic, Herzberg–Teller effects and the radiation and nonradiation processes, and proposing the efficient way to design high-performance photovoltaic aggregates. Quantum coherence and interference between channels should be clarified occurring in the radiation and nonradiation processes, which may lead to new luminous mechanisms and luminous properties. The different photophysical processes should be clarified from coherent light (laser) and incoherent light (sun) excitation. The effects of energy transfer, charge transfer, doping, and structural disorder on photophysics should be investigated. In the calculations of time-dependent correlation functions under the thermal equilibrium distribution of excited states (for instance, fluorescence and phosphorescence), quantum dynamics approaches need further extension to efficiently propagate imaginary and real-time dynamics for complex aggregates simultaneously.

Building an excitonic Hamiltonian is difficult because it includes a lot of microscopic physical parameters which are closely related with the intramolecular geometric and electronic structures, the intermolecular mutual orientation and distance, and the electron–phonon interactions. Traditionally, those parameters are determined by fitting the experimental spectroscopies so that the construction of effective model Hamiltonian is fully established on the basis of experimental measurements. This semiempirical fitting method may produce several sets of fitting parameters based on the same experimental results, leading to uncertainties to our understanding of the microscopic mechanism. In addition, for new materials, there are no experimental data, this semiempirical fitting method will not have predictive role on the material's properties in practice. To achieve in-depth understanding of the experimental phenomena and microscopic processes, it is thus essential to build the effective Hamiltonian with respect to the high-level excite-state electronic structure methods, which determine the quality of the excited-state simulations of molecular aggregates. There are many state-of-the-art excite-state electronic structure methods, ranged from *ab initio* wavefunction-based methods to the TDDFT. For a medium and even large-sized molecular system, the most widely used method is TDDFT although it can produce spurious results for some electronic states due to its single-reference character and the approximations introduced in the numerical implementation. For some of the processes relevant to aggregates, such as singlet fission and triplet–triplet annihilation, both single and multiple excitations are included. In this case, more sophisticated diabatization schemes based on high-level electronic structure calculations are usually required to obtain the accurate energies of the quasi-diabatic states and the electronic couplings [107].

Many important photo-initialized processes take place in the realistic and complicated condensed-phase environment such as the different solvents, the protein microscopic environment or the metal surface. The aggregate-environmental interaction results in the exchanges of energies and charges between the system and the environment, subsequently affecting the system's properties and dynamics. For example, for an aggregate in condensed-phase environment, the site energies, electronic couplings, and exciton coherence lengths may be profoundly different in the condensed phase than those in the gas phase, and these details ultimately dictate the photophysical and photochemical processes. It is thus essential to develop computational methods which can take into account the environmental effect, including possible electronic couplings to other chromophores.

Acknowledgments

We first acknowledge the essential contributions of several former and present group members to this chapter: in particular, Prof. Xinxin Zhong, Dr. Yaling Ke, Dr. Man Lian, and Mr. Xunkun Huang. This work is supported by the National Science Foundation of China (Grant Nos. 21833006, 22033006, 22173074).

References

1. Kasha, M. (1950). Characterization of electronic transitions in complex molecules. *Discuss. Faraday Soc.* 9: 14–19.
2. Kasha, M. (1963). Energy transfer mechanisms and the molecular exciton model for molecular aggregates. *Radiat. Res.* 20: 55–70.
3. Davydov, A.S. (1964). The theory of molecular excitons. *Sov. Phys. Usp.* 7: 145–178.
4. Birks, J.B. (1970). *Photophysics of Aromatic Molecules*. London: Wiley-Interscience.
5. Mei, J., Leung, N.L.C., Kwok, R.T.K. et al. (2015). Aggregation-induced emission: together we shine, united we soar! *Chem. Rev.* 115: 11718–11940.
6. Peng, Q. and Shuai, Z. (2021). Molecular mechanism of aggregation-induced emission. *Aggregate* 2: e91.
7. Gao, F., Zhao, Y., and Liang, W. (2011). Vibronic spectra of perylene bisimide oligomers: effects of intermolecular charge-transfer excitation and conformational flexibility. *J. Phys. Chem. B* 115: 2699–2708.
8. Hestand, N.J. and Spano, F.C. (2018). Expanded theory of H- and J-molecular aggregates: the effects of vibronic coupling and intermolecular charge transfer. *Chem. Rev.* 118: 7069–7163.
9. Hernández, F.J. and Crespo-Otero, R. (2023). Modeling excited states of molecular organic aggregates for optoelectronics. *Ann. Rev. Phys. Chem.* 74: 547–571.
10. Liang, W. and Wu, W. (2013). Theory and algorithms for the excited states of large molecules and molecular aggregates. *Sci. China Chem.* 56: 1267–1270.
11. Bardeen, C.J. (2014). The structure and dynamics of molecular excitons. *Annu. Rev. Phys. Chem.* 65: 127–148.
12. Deshmukh, A.P., Geue, N., Bradbury, N.C. et al. (2022). Bridging the gap between H- and J-aggregates: classification and supramolecular tunability for excitonic band structures in two-dimensional molecular aggregates. *Chem. Phys. Rev.* 3: 021401.
13. Wang, Y.-C., Feng, S., Liang, W., and Zhao, Y. (2021). Electronic couplings for photoinduced charge transfer and excitation energy transfer based on fragment particle–hole densities. *J. Phys. Chem. Lett.* 12: 1032–1039.
14. Zhong, X. and Zhao, Y. (2013). Non-Markovian stochastic Schrödinger equation at finite temperatures for charge carrier dynamics in organic crystals. *J. Chem. Phys.* 138: 014111.
15. Ke, Y. and Zhao, Y. (2017). Perturbation expansions of stochastic wavefunctions for open quantum systems. *J. Chem. Phys.* 147: 184103.
16. Wang, Y.-C. and Zhao, Y. (2020). The hierarchical stochastic Schrödinger equations: theory and applications. *Chin. J. Chem. Phys.* 33: 653–667.
17. Lian, M., Wang, Y.-C., Peng, S., and Zhao, Y. (2022). Photo-induced ultrafast electron dynamics in anatase and rutile TiO_2: effects of electron-phonon interaction. *Chin. J. Chem. Phys.* 35: 270–280.
18. Wu, Q. and Van Voorhis, T. (2005). Direct optimization method to study constrained systems within density-functional theory. *Phys. Rev. A* 72: 024502.

19 Van Voorhis, T., Kowalczyk, T., Kaduk, B. et al. (2010). The diabatic picture of electron transfer, reaction barriers, and molecular dynamics. *Annu. Rev. Phys. Chem.* 61: 149–170.

20 Kaduk, B., Kowalczyk, T., and Van Voorhis, T. (2012). Constrained density functional theory. *Chem. Rev.* 112: 321–370.

21 Song, L. and Gao, J. (2008). On the construction of diabatic and adiabatic potential energy surfaces based on ab initio valence bond theory. *J. Phys. Chem. A* 112: 12925–12935.

22 Lin, X., Liu, X., Ying, F. et al. (2018). Explicit construction of diabatic state and its application to the direct evaluation of electronic coupling. *J. Chem. Phys.* 149: 044112.

23 Zhang, Y., Su, P., and Lasorne, B. (2020). A novel valence-bond-based automatic diabatization method by compression. *J. Phys. Chem. Lett.* 11: 5295–5301.

24 Cembran, A., Song, L., Mo, Y., and Gao, J. (2009). Block-localized density functional theory (BLDFT), diabatic coupling, and their use in valence bond theory for representing reactive potential energy surfaces. *J. Chem. Theory Comput.* 5: 2702–2716.

25 Ren, H., Provorse, M.R., Bao, P. et al. (2016). Multistate density functional theory for effective diabatic electronic coupling. *J. Phys. Chem. Lett.* 7: 2286–2293.

26 Grofe, A., Qu, Z., Truhlar, D.G. et al. (2017). Diabatic-At-construction method for diabatic and adiabatic ground and excited states based on multistate density functional theory. *J. Chem. Theory Comput.* 13: 1176–1187.

27 Lu, Y. and Gao, J. (2022). Multistate density functional theory of excited states. *J. Phys. Chem. Lett.* 13: 7762–7769.

28 Ahart, C.S., Rosso, K.M., and Blumberger, J. (2022). Implementation and validation of constrained density functional theory forces in the CP2K package. *J. Chem. Theory Comput.* 18: 4438–4446.

29 Wang, K., Xie, Z., Luo, Z., and Ma, H. (2022). Low-scaling excited state calculation using the block interaction product state. *J. Phys. Chem. Lett.* 13: 462–470.

30 Ruedenberg, K. and Atchity, G.J. (1993). A quantum chemical determination of diabatic states. *J. Chem. Phys.* 99: 3799–3803.

31 Pacher, T., Cederbaum, L.S., and Köppel, H. (1988). Approximately diabatic states from block diagonalization of the electronic Hamiltonian. *J. Chem. Phys.* 89: 7367–7381.

32 Domcke, W. and Woywod, C. (1993). Direct construction of diabatic states in the CASSCF approach. Application to the conical intersection of the $^1A_2$ and $^1B_1$ excited states of ozone. *Chem. Phys. Lett.* 216: 362–368.

33 Nakamura, H. and Truhlar, D.G. (2001). The direct calculation of diabatic states based on configurational uniformity. *J. Chem. Phys.* 115: 10353–10372.

34 Nakamura, H. and Truhlar, D.G. (2002). Direct diabatization of electronic states by the fourfold way. II. Dynamical correlation and rearrangement processes. *J. Chem. Phys.* 117: 5576–5593.

35 Nakamura, H. and Truhlar, D.G. (2003). Extension of the fourfold way for calculation of global diabatic potential energy surfaces of complex, multiarrangement, non-Born–Oppenheimer systems: application to HNCO (S,S$_1$). *J. Chem. Phys.* 118: 6816–6829.

36 Tamura, H., Burghardt, I., and Tsukada, M. (2011). Exciton dissociation at thiophene/fullerene interfaces: the electronic structures and quantum dynamics. *J. Phys. Chem. C* 115: 10205–10210.

37 Tamura, H. (2016). Diabatization for time-dependent density functional theory: exciton transfers and related conical intersections. *J. Phys. Chem. A* 120: 9341–9347.

38 Xie, Y., Jiang, S., Zheng, J., and Lan, Z. (2017). Construction of vibronic diabatic Hamiltonian for excited-state electron and energy transfer processes. *J. Phys. Chem. A* 121: 9567–9578.

39 Kryachko, E.S. and Yarkony, D.R. (1999). Diabatic bases and molecular properties. *Int. J. Quantum Chem.* 76: 9.

40 Cave, R.J. and Newton, M.D. (1996). Generalization of the Mulliken–Hush treatment for the calculation of electron transfer matrix elements. *Chem. Phys. Lett.* 249: 15–19.

41 Subotnik, J.E., Yeganeh, S., Cave, R.J., and Ratner, M.A. (2008). Constructing diabatic states from adiabatic states: extending generalized Mulliken–Hush to multiple charge centers with boys localization. *J. Chem. Phys.* 129: 244101.

42 Subotnik, J.E., Cave, R.J., Steele, R.P., and Shenvi, N. (2009). The initial and final states of electron and energy transfer processes: diabatization as motivated by system-solvent interactions. *J. Chem. Phys.* 130: 234102.

43 Subotnik, J.E., Vura-Weis, J., Sodt, A.J., and Ratner, M.A. (2010). Predicting accurate electronic excitation transfer rates via Marcus theory with Boys or Edmiston–Ruedenberg localized diabatization. *J. Phys. Chem. A* 114: 8665–8675.

44 Hoyer, C.E., Xu, X., Ma, D. et al. (2014). Diabatization based on the dipole and quadrupole: the DQ method. *J. Chem. Phys.* 141: 114104.

45 Hoyer, C.E., Parker, K., Gagliardi, L., and Truhlar, D.G. (2016). The DQ and DQϕ electronic structure diabatization methods: validation for general applications. *J. Chem. Phys.* 144: 194101.

46 Voityuk, A.A. and Rüsch, N. (2002). Fragment charge difference method for estimating donor-acceptor electronic coupling: application to DNA π-stacks. *J. Chem. Phys.* 117: 5607–5616.

47 Hsu, C.-P., You, Z.-Q., and Chen, H.-C. (2008). Characterization of the short-range couplings in excitation energy transfer. *J. Phys. Chem. C* 112: 1204–1212.

48 You, Z.-Q. and Hsu, C.-P. (2010). The fragment spin difference scheme for triplet-triplet energy transfer coupling. *J. Chem. Phys.* 133: 074105.

49 Yang, C.-H. and Hsu, C.-P. (2013). A multi-state fragment charge difference approach for diabatic states in electron transfer: extension and automation. *J. Chem. Phys.* 139: 154104.

50 Kue, K.Y., Claudio, G.C., and Hsu, C.-P. (2018). Hamiltonian-independent generalization of the fragment excitation difference scheme. *J. Chem. Theory Comput.* 14: 1304–1310.

51 Voityuk, A.A. (2014). Fragment transition density method to calculate electronic coupling for excitation energy transfer. *J. Chem. Phys.* 140: 244117.

52 Liu, W., Lunkenheimer, B., Settels, V. et al. (2015). A general ansatz for constructing quasi-diabatic states in electronically excited aggregated systems. *J. Chem. Phys.* 143: 084106.

53 Voityuk, A.A. (2017). Electronic couplings for photoinduced electron transfer and excitation energy transfer computed using excited states of noninteracting molecules. *J. Phys. Chem. A* 121: 5414–5419.

54 Mulliken, R.S. (1955). Electronic population analysis on LCAO-MO molecular wave functions. I. *J. Chem. Phys.* 23: 1833–1840.

55 Hirshfeld, F.L. (1977). Bonded-atom fragments for describing molecular charge densities. *Theor. Chim. Acta* 44: 129–138.

56 Löwdin, P. (1950). On the non-orthogonality problem connected with the use of atomic wave functions in the theory of molecules and crystals. *J. Chem. Phys.* 18: 365–375.

57 Cardoso, J.-F. and Souloumiac, A. (1996). Jacobi angles for simultaneous diagonalization. *SIAM J. Matrix Anal. Appl.* 17: 161–164.

58 Head-Gordon, M., Grana, A.M., Maurice, D., and White, C.A. (1995). Analysis of electronic transitions as the difference of electron attachment and detachment densities. *J. Phys. Chem.* 99: 14261–14270.

59 Liang, W., Zhao, Y., Sun, J. et al. (2006). Electronic excitation of polyfluorenes: a theoretical study. *J. Phys. Chem. B* 110: 9908–9915.

60 Santoro, F., Cappelli, C., and Barone, V. (2011). Effective time-independent calculations of vibrational resonance Raman spectra of isolated and solvated molecules including Duschinsky and Herzberg–Teller effects. *J. Chem. Theory Comput.* 7: 1824–1839.

61 Avila Ferrer, F., Barone, V., Cappelli, C., and Santoro, F. (2013). Duschinsky, Herzberg–Teller, and multiple electronic resonance interferential effects in resonance Raman spectra and excitation profiles. The case of pyrene. *J. Chem. Theory Comput.* 9: 3597–3611.

62 Ma, H., Zhao, Y., and Liang, W. (2014). Assessment of mode-mixing and Herzberg–Teller effects on two-photon absorption and resonance hyper-Raman spectra from a time-dependent approach. *J. Chem. Phys.* 140: 094107.

63 Zhao, Y., Liang, W., and Nakamura, H. (2006). Semiclassical treatment of thermally activated electron transfer in the intermediate to strong electronic coupling regime under the fast dielectric relaxation. *J. Phys. Chem. A* 110: 8204–8212.

64 Bulla, R., Tong, N.-H., and Vojta, M. (2003). Numerical renormalization group for bosonic systems and application to the sub-ohmic spin-boson model. *Phys. Rev. Lett.* 91: 170601.

65 Bulla, R., Lee, H.-J., Tong, N.-H., and Vojta, M. (2005). Numerical renormalization group for quantum impurities in a bosonic bath. *Phys. Rev. B* 71: 045122.

66 White, S.R. and Feiguin, A.E. (2004). Real-time evolution using the density matrix renormalization group. *Phys. Rev. Lett.* 93: 076401.
67 Chin, A.W., Rivas, Á., Huelga, S.F., and Plenio, M.B. (2010). Exact mapping between system-reservoir quantum models and semi-infinite discrete chains using orthogonal polynomials. *J. Math. Phys.* 51: 092109.
68 Wang, H. and Thoss, M. (2003). Multilayer formulation of the multiconfiguration time-dependent Hartree theory. *J. Chem. Phys.* 119: 1289–1299.
69 Feynman, R.P. and Vernon, F.L. (1963). The theory of a general quantum system interacting with a linear dissipative system. *Ann. Phys.* 24: 118–173.
70 Makri, N. (1995). Numerical path integral techniques for long time dynamics of quantum dissipative systems. *J. Math. Phys.* 36: 2430–2457.
71 Makri, N. (2020). Small matrix path integral for system-bath dynamics. *J. Chem. Theory Comput.* 16: 4038–4049.
72 Tanimura, Y. (2020). Numerically "exact" approach to open quantum dynamics: the hierarchical equations of motion (HEOM). *J. Chem. Phys.* 153: 020901.
73 Tanimura, Y. and Kubo, R. (1989). Time evolution of a quantum system in contact with a nearly Gaussian-Markoffian noise bath. *J. Phys. Soc. Jpn.* 58: 101–114.
74 Yan, Y.-A., Yang, F., Liu, Y., and Shao, J. (2004). Hierarchical approach based on stochastic decoupling to dissipative systems. *Chem. Phys. Lett.* 395: 216–221.
75 Xu, R.-X., Cui, P., Li, X.-Q. et al. (2005). Exact quantum master equation via the calculus on path integrals. *J. Chem. Phys.* 122: 041103.
76 Ishizaki, A. and Tanimura, Y. (2005). Quantum dynamics of system strongly coupled to low-temperature colored noise bath: reduced hierarchy equations approach. *J. Phys. Soc. Jpn.* 74: 3131–3134.
77 Cao, J., Ungar, L.W., and Voth, G.A. (1996). A novel method for simulating quantum dissipative systems. *J. Chem. Phys.* 104: 4189–4197.
78 Stockburger, J.T. and Grabert, H. (2002). Exact c-number representation of non-Markovian quantum dissipation. *Phys. Rev. Lett.* 88: 170407.
79 Stockburger, J.T. (2004). Simulating spin-boson dynamics with stochastic Liouville–von Neumann equations. *Chem. Phys.* 296: 159–169.
80 Shao, J. (2004). Decoupling quantum dissipation interaction via stochastic fields. *J. Chem. Phys.* 120: 5053–5056.
81 Chen, X., Cao, J., and Silbey, R.J. (2013). A novel construction of complex-valued Gaussian processes with arbitrary spectral densities and its application to excitation energy transfer. *J. Chem. Phys.* 138: 224104.
82 Yan, Y.-A. and Shao, J. (2016). Stochastic description of quantum Brownian dynamics. *Front. Phys.* 11: 1–24.
83 Yan, Y.-A. and Shao, J. (2018). Equivalence of stochastic formulations and master equations for open systems. *Phys. Rev. A* 97: 042126.
84 Tanimura, Y. (2006). Stochastic Liouville, Langevin, Fokker–Planck, and master equation approaches to quantum dissipative systems. *J. Phys. Soc. Jpn.* 75: 082001.
85 Zhou, Y., Yan, Y., and Shao, J. (2005). Stochastic simulation of quantum dissipative dynamics. *Europhys. Lett.* 72: 334.

86 Zhou, Y. and Shao, J. (2008). Solving the spin-boson model of strong dissipation with flexible random-deterministic scheme. *J. Chem. Phys.* 128: 034106.

87 Moix, J.M. and Cao, J. (2013). A hybrid stochastic hierarchy equations of motion approach to treat the low temperature dynamics of non-Markovian open quantum systems. *J. Chem. Phys.* 139: 134106.

88 Zhu, L., Liu, H., and Shi, Q. (2013). A new method to account for the difference between classical and quantum baths in quantum dissipative dynamics. *New J. Phys.* 15: 095020.

89 Suess, D., Eisfeld, A., and Strunz, W. (2014). Hierarchy of stochastic pure states for open quantum system dynamics. *Phys. Rev. Lett.* 113: 150403.

90 Song, K., Song, L., and Shi, Q. (2016). An alternative realization of the exact non-Markovian stochastic Schrödinger equation. *J. Chem. Phys.* 144: 224105.

91 Ke, Y. and Zhao, Y. (2016). Hierarchy of forward-backward stochastic Schrödinger equation. *J. Chem. Phys.* 145: 024101.

92 Hartmann, R. and Strunz, W.T. (2017). Exact open quantum system dynamics using the hierarchy of pure states (HOPS). *J. Chem. Theory Comput.* 13: 5834–5845.

93 Hartmann, R., Werther, M., Grossmann, F., and Strunz, W.T. (2019). Exact open quantum system dynamics: optimal frequency vs time representation of bath correlations. *J. Chem. Phys.* 150: 234105.

94 Gaspard, P. and Nagaoka, M. (1999). Slippage of initial conditions for the Redfield master equation. *J. Chem. Phys.* 111: 5668–5675.

95 Feng, S., Wang, Y.-C., Ke, Y. et al. (2020). Effect of charge-transfer states on the vibrationally resolved absorption spectra and exciton dynamics in ZnPc aggregates: simulations from a non-Makovian stochastic Schrödinger equation. *J. Chem. Phys.* 153: 034116.

96 Feng, S., Wang, Y.-C., Liang, W., and Zhao, Y. (2021). Vibrationally resolved absorption spectra and exciton dynamics in zinc phthalocyanine aggregates: effects of aggregation lengths and remote exciton transfer. *J. Phys. Chem. A* 125: 2932–2943.

97 Feng, S., Wang, Y.-C., Liang, W., and Zhao, Y. (2022). Vibrationally resolved absorption spectra and ultrafast exciton dynamics in α-phase and β-phase zinc phthalocyanine aggregates. *Phys. Chem. Chem. Phys.* 24: 2974–2987.

98 Cranston, R.R. and Lessard, B.H. (2021). Metal phthalocyanines: thin-film formation, microstructure, and physical properties. *RSC Adv.* 11: 21716–21737.

99 Qiu, Y., Chen, P., and Liu, M. (2008). Interfacial assembly of an achiral zinc phthalocyanine at the air/water interface: a surface pressure dependent aggregation and supramolecular chirality. *Langmuir* 24: 7200–7207.

100 Zanfolim, A.A., Volpati, D., Olivati, C.A. et al. (2010). Structural and electric-optical properties of zinc phthalocyanine evaporated thin films: temperature and thickness effects. *J. Phys. Chem. C* 114: 12290–12299.

101 Roy, D., Das, N.M., Shakti, N., and Gupta, P.S. (2014). Comparative study of optical, structural and electrical properties of zinc phthalocyanine Langmuir–Blodgett thin film on annealing. *RSC Adv.* 4: 42514–42522.

102 Roy, D., Das, N.M., Gupta, M. et al. (2014). Optical and surface morphology study of zinc phthalocyanine Langmuir Blodgett thin film. *AIP Conf. Proc.* 1591: 968.

103 Ahn, H. and Chu, T.-C. (2016). Annealing-induced phase transition in zinc phthalocyanine ultrathin films. *Opt. Mater. Express* 6: 3586–3593.

104 Roy, D., Das, N.M., Gupta, M., and Gupta, P.S. (2016). Study of polymorphism of ZnPc LB thin film on annealing. *AIP Conf. Proc.* 1731: 030007.

105 Shahiduzzaman, M., Horikawa, T., Hirayama, T. et al. (2020). Switchable crystal phase and orientation of evaporated zinc phthalocyanine films for efficient organic photovoltaics. *J. Phys. Chem. C* 124: 21338–21345.

106 Kato, M., Nakaya, M., Matoba, Y. et al. (2020). Morphological and optical properties of α- and β-phase zinc phthalocyanine thin films for application to organic photovoltaic cells. *J. Chem. Phys.* 153: 144704.

107 Wang, Y.-C., Feng, S., Kong, Y. et al. (2023). Electronic couplings for singlet fission processes based on the fragment particle-hole densities. *J. Chem. Theory Comput.* 19: 3900–3914. https://doi.org/10.1021/acs.jctc.3c00243.

16

Duality of Conjugated π Electrons

Yirong Mo

University of North Carolina at Greensboro, Joint School of Nanoscience & Nanoengineering, Department of Nanoscience, 2907 E Gate City Blv, Greensboro, NC 27401, USA

Key Points/Objectives Box

- Stabilizing π conjugation
- Pauli repulsion among π electron pairs
- Duality of π conjugated electrons
- Intramolecular multi-bond strain
- Block-localized wavefunction (BLW) method
- Strictly electron-localized (resonance) state
- Hypothetical linear molecule HBBBBH
- CN bond distances in nitrobenzene ($C_6H_5NO_2$) and aniline ($C_6H_5NH_2$)
- Long NN bond distance in dinitrogen tetroxide (N_2O_4)

16.1 Introduction

16.1.1 Conjugated Systems and the Concept of Conjugation

A conjugated system often refers to a planar molecule of connected p orbitals with delocalized π electrons. Typical conjugated systems, such as acyclic butadiene C_4H_6 and cyclic benzene C_6H_6, are molecules with multiple bonds that are separated by single bonds. Compared with nonconjugated or saturated molecules, conjugated systems are characterized by shortened single-bond distances and stretched multiple-bond distances. While carbon with sp² hybridization is the major atom participating in conjugations, other main group atoms such as nitrogen, oxygen, halogen, or even transition metals can also get involved in conjugations. The most prominent example of conjugation is benzene, which exhibits equal carbon–carbon bond distances, and its extraordinary stability leads to the concept of aromaticity and Huckel's rule [1]. The infinite extension of benzene results in graphene, which is an

---

Exploring Chemical Concepts Through Theory and Computation, First Edition. Edited by Shubin Liu.
© 2024 WILEY-VCH GmbH. Published 2024 by WILEY-VCH GmbH.

isolated single layer of carbon hexagons consisting of sp² hybridized carbon–carbon bonding with π-electron clouds. Currently, graphene and related two-dimensional (2D) materials, including hexagonal boron nitride and molybdenum disulfide, have drawn intensive attention in the development of novel catalysts and devices at the atomic limit and in many other technological fields, largely due to the conjugated π electrons [2]. Other widely used conjugated systems include conducting polymers and organic pigments in organoluminescent devices and dye-sensitized solar cells.

In Chemistry, a molecule is often illustrated with a Lewis structure, which is a much simplified representation of all valence shell electrons in a molecule [3]. A pair of valence electrons is either shared between two atoms and thus forms a chemical bond, or solely owned by one atom as an electron lone pair [4, 5]. Due to the localization nature of all valence electrons, a Lewis structure K can well describe a saturated molecule like alkanes, and its corresponding wavefunction can be expressed with a Heitler–London–Slater–Pauling (HLSP) function within the valence bond (VB) theory [6–10] as

$$\Phi_K = \hat{A}(\phi_{1,2}\phi_{3,4}\cdots\phi_{2n-1,2n}\varphi_{2n+1}\alpha(2n+1)\cdots\varphi_N\alpha(N)) \tag{16.1}$$

where $\hat{A}$ is the antisymmetrizer and $\phi_{i,j}$ is a bond function corresponding to the bond between orbitals φ_i and φ_j (or a lone pair if $\varphi_i = \varphi_j$)

$$\phi_{i,j} = \hat{A}\{\varphi_i\varphi_j[\alpha(i)\beta(j) - \beta(i)\alpha(j)]\} \tag{16.2}$$

Equation (16.1) corresponds to a molecule of $N = 2n + 2S$ electrons (n is the number of chemical bonds or electron pairs and S is the spin quantum number), and thus there are $2S$ singly occupied orbitals from φ_{2n+1} to φ_N. Note that, for the sake of simplicity, we omit the normalization constants in the above equations. As each bond function (Eq. 16.2) can be expanded into 2 Slater determinants, a HLSP comprises 2^n Slater determinants.

For a conjugated system, however, one Lewis structure cannot even correctly describe the electronic structure of the molecule. A typical example is benzene, which can be expressed with two equivalent Kekulé structures as

Pauling envisioned that more than one Lewis structure or resonance structure is needed to elucidate molecular structures and properties and subsequently proposed the resonance theory [6]. According to the resonance theory, the actual normal state of a molecule is represented not by a single Lewis (resonance) structure but by a combination of several distinct resonance structures. Translated to the mathematical formula, the resonance theory (or more generally, the VB theory) expresses each resonance structure with an HLSP function and the overall many-electron wavefunction as a linear combination of HLSP functions

$$\Psi = \sum_K C_K \Phi_K \tag{16.3}$$

where the coefficients $\{C_K\}$ are self-consistently determined by minimizing the energy of Ψ. As a matter of fact, Eq. (16.3) is similar to the configuration interaction (CI) method within the molecular orbital (MO) theory, and the calculated energy of the overall wavefunction $E(\Psi)$, which is a resonance hybrid, must be lower than the energy of any individual resonance structure $E(\Phi_K)$. Accordingly, conjugation (resonance) energy can be derived "by subtracting the actual energy of the molecule in question from that of the most stable contributing structure," [11] or

$$\Delta E_{conj} = E(\Phi_K) - E(\Psi) \tag{16.4}$$

where we assume that the most stable resonance structure is K.

Thus, from the theoretical point of view, conjugation must be a stabilizing force by definition, and a conjugated system must be stabilized by conjugation (resonance). This is the prevailing understanding and interpretation of conjugation.

16.1.2 Alternative Proposal from Rogers

However, resonance structures are hypothetical and not experimental observables. Besides, although a resonance state can be well described within the VB theory (Eq. 16.1), the popular MO theory and the density function theory (DFT) have difficulties in defining electron-localized states for references (i.e. the most stable resonance structure) in the way the VB theory does [12]. This is due to the different strategies in MO/DFT and VB theories. MO/DFT theory adopts MOs that are delocalized over the whole system and constrained to be orthogonal, whereas VB theory uses nonorthogonal atomic orbitals to construct HLSPs for resonance structures. Both the lack of direct experimental evidence and the incapability of constructing resonance states with the MO/DFT methods result in persistent controversies in estimating the conjugation energy, notably in the determination of aromaticity in conjugated rings [13]. A typical approach in MO/DFT methods is the use of reference molecules without or with limited conjugation, and these reference molecules are real. With reference molecules, various isodesmic and homodesmotic model reactions are consequently designed to estimate resonance energies [14]. But it has been known that other effects, such as strain, hyperconjugation, Coulomb repulsion imbalance, and uncompensated van der Waals attractions, are involved in reference systems [15, 16]. In other words, choosing different reference systems often leads to very different results.

Experimentally, Kistiakowsky et al. first measured and noticed the differences in the hydrogenation heats of carbon–carbon double bonds in substituted and/or conjugated systems [17, 18]. For instance, the hydrogenation heat of butadiene is 57.1 kcal/mol, less than two times the hydrogenation heat of 1-butene (30.3 kcal/mol), and the difference (3.5 kcal/mol) is the extra stabilization due to the resonance between two double bonds in the former, referred to as the Kistiakowsky resonance energy. The process can be expressed as the sequential hydrogenations (in kcal/mol) [18]

$$CH_2 = CH - CH = CH_2 \xrightarrow[+H_2]{-26.8} CH_2 = CH - C_2H_5 \xrightarrow[+H_2]{-30.3} C_4H_{10} \tag{16.5}$$

The above process can be simplified with the following isodesmic reaction:

$$CH_2=CH-CH=CH_2 + C_4H_{10} \rightarrow 2CH_2=CH-C_2H_5 + 3.5 \text{ kcal/mol} \quad (16.6)$$

Obviously, in the above reaction, we take 1-butene as the reference molecule. But there is a $\sigma \rightarrow \pi^*$ interaction from the ethyl group to the double bond, or hyperconjugation, in 1-butene. Hyperconjugation was unknown at the time when Kistiakowsky proposed his idea, as it was proposed a few years later by Mulliken et al. in 1941 [19]. A better selection of the reference molecule is thus ethylene, which is free of any hyperconjugative interaction

$$CH_2=CH-CH=CH_2 + 2C_2H_6 \rightarrow 2CH_2=CH_2 + C_4H_{10} + 8.5 \text{ kcal/mol} \quad (16.7)$$

Equations (16.6) and (16.7) show that different reference molecules can lead to very different energy values for conjugation stabilization. Following early Kistiakowsky's definition [17], Rogers et al. computationally studied the stepwise hydrogenation of 1,3-butadiyne as

$$CH\equiv C-C\equiv CH \xrightarrow[+H_2]{-69.6} CH\equiv C-C_2H_5 \xrightarrow[+H_2]{-69.6} C_4H_{10} \quad (16.8)$$

and concluded that there is no conjugation stabilization in this molecule [20, 21]. In terms of isodesmic reactions, it is

$$CH\equiv C-C\equiv CH + C_4H_{10} \rightarrow 2CH\equiv C-C_2H_5 + 0.0 \text{ kcal/mol} \quad (16.9)$$

where 1-butyne is the reference molecule.

Unsurprisingly, Rogers' claim that the conjugation stabilization in 1,3-butadiyne is zero received instant objections [22–24]. Jarowski et al. pointed out that there is significant hyperconjugation from the ethyl group to the triple bond (i.e. $\sigma \rightarrow \pi^*$) in 1-butyne [22]. This stabilizing force leads to an underestimation of conjugation with Kistiakowsky's definition. Similar to Eq. (16.7), if we use ethyne as the reference molecule for the following isodemic reaction:

$$CH\equiv C-C\equiv CH + 2C_2H_6 \rightarrow 2CH\equiv CH + C_4H_{10} + 10.0 \text{ kcal/mol} \quad (16.10)$$

we would get a value of 10.0 kcal/mol for the conjugation stability in 1,3-butadiyne. This is close to rather than twice the value for butadiene with reference to ethylene (8.5 kcal/mol in Eq. 16.7). Based on the resonance theory, we would expect that the thermodynamic conjugation stabilization in 1,3-butadiyne is two times the quantity in 1,3-butadiene, as demonstrated first by Kollmar, who derived the resonance stabilization energies in 1,3-butadiene and 1,3-butadiyne as 9.7 and 19.1 kcal/mol, respectively, by replacing their π MO's with the π MO's of ethylene and ethyne, respectively [25]. Based on the high-level computations of several isomerization reactions, Jarowski et al. predicted the conjugation energies of 9.3 kcal/mol for diynes and 8.2 kcal/mol for dienes [22]. Alternatively, based on the energy decomposition analysis between two fragments (•C≡CH and •CH=CH$_2$), Cappel et al.

estimated the conjugative stabilization in 1,3-butadiyne (45.0 kcal/mol) to be about twice the value in 1,3-butadiene (19.5 kcal/mol) [23].

Interestingly, Roger et al. continued their work and showed the cases of 2,3-butanedione and cyanogen where the conjugation is even destabilizing [26], e.g.

$$CH_3CH_2CH_2CH_3 \xrightarrow[+1/2O_2-H_2]{-27.0} CH_3COCH_2CH_3 \xrightarrow[+1/2O_2-H_2]{-21.1} CH_3COCOCH_3 \tag{16.11}$$

$$N \equiv C - C \equiv N \xrightarrow[+H_2]{-44.0} N \equiv C - CH_2NH_2 \xrightarrow[+H_2]{-32.6} H_2NCH_2CH_2NH_2 \tag{16.12}$$

Rogers proposed that the lack of overall thermodynamic stabilization in polyynes is due to the repulsions among the six electrons of each triple bond. But we note that this kind of interaction similarly exists in ethyne, and the extra stabilization in 1,3-butadiyne with reference to ethyne is still only 10.0 kcal/mol (Eq. 16.10).

16.1.3 Origin of the Disparity

The key to the above controversies is whether conjugation must be stabilizing or sometimes it could be destabilizing.

We believe that the controversies over Rogers' claim come from the different interpretations of conjugation from the experimental and theoretical perspectives. Consequently, the question to be addressed is what the exact cause of the difference between theoretical and experimental conjugation energies is. While the theoretical conjugation energy seems well defined by Eq. (16.4), it is not realistic in general due to the overwhelming popularity of the MO/DFT methods, as these methods cannot define a hypothetical electron-localized resonance state, where conjugation is strictly forbidden.

For the example of butadiene, the conjugation from the VB perspective can be described with two covalent resonance structures (**I** and **II**) as

$$CH_2=CH-CH=CH_2 \xleftrightarrow{\text{Resonance}} CH_2-CH=CH-CH_2$$
$$\text{I} \qquad\qquad\qquad \text{II}$$

Obviously, structure **I** is the much more stable resonance structure, and conjugation results from the minor contribution of resonance structure **II** with a very long bond in the dashed line. The participation of **II** leads to the stability of butadiene with reference to the covalent state **I**. In other words, conjugation is a stabilizing force. From the MO perspective, however, conjugation occurs from intramolecular electron transfers from one moiety to the other in the nonconjugated structure **I**. If we assume the carbon–carbon π bonds in **I** as 1π and the corresponding anti-bonds as $1\pi^*$, the conjugation can be illustrated in Figure 16.1a, where A and B refer to the left and right carbon–carbon π bonds in **I**. But there is considerable Pauli repulsion

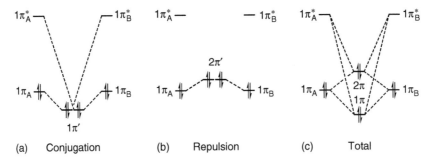

Figure 16.1 Interactions between two π bonds: (a) attractive conjugative interactions, (b) Pauli repulsion, and (c) the summation of both attractive and repulsive interactions. Source: Reproduced from Ref. [27] with permission from the Royal Society of Chemistry.

between the two adjacent π bonds, which coexists with the conjugative interactions. Interestingly, the repulsion among σ bonds has been well recognized even in the case of ethane, where the rotation barrier from the staggered to the eclipsed conformer is ascribed to the steric repulsion between two methyl groups (thus among carbon-hydrogen bonds). But for conjugated systems, the Pauli repulsion among adjacent π bonds has been little mentioned until our work. This repulsion can be schematized in Figure 16.1b. Both stabilizing conjugation and destabilizing Pauli repulsion coexist and reflect the duality of conjugated π electrons. As these two opposing forces are usually coupled and cannot be evaluated individually, both the experimental and MO/DFT computational data involve both forces (Figure 16.1c). In contrast, the VB treatment (Eq. 16.4) does not involve Pauli repulsion.

In conclusion, the controversies over Rogers' claims and the disagreements among very different aromatic stabilization energies largely result from the neglection of the π–π Pauli repulsion, in our opinion.

16.2 The New Concept of Intramolecular Multibond Strain

To recognize the π–π Pauli repulsion in conjugated systems, we proposed the concept of intramolecular multi-bond strain [27, 28]. This destabilizing strain exists in all conjugated systems and influences the molecular structure and properties implicitly. Shaik, Hiberty, and their coworkers once studied the π-π repulsion in the discussion of π-distortivity in benzene and proposed that if there were no π electronic delocalization (conjugation), instead of a D_{6h} geometry, the molecule would prefer a D_{3h} geometry with alternating long and short carbon–carbon bonds [29–31]. In other words, the π electrons "possess a global distortive tendency." We assume that this distortive tendency primarily comes from the π–π repulsion or the intramolecular multi-bond strain.

Molecular strain is certainly nothing new, as Baeyer proposed the strain theory as early as 1885 [32, 33]. Molecular strain is particularly known in small organic rings and plays a significant role in the preference of molecular conformations and reactivity. Like the evaluation of aromatic stabilization, there have been many proposals for the evaluation of strain energy. Strain energy is often derived as the difference between the experimental heat of formation for a strained molecule and the expected heat of formation for a hypothetical strain-free molecule with the same number of atoms, which is derived from group contributions of additivity methods or ab initio computations [34–38]. Molecular strain can also be analyzed in terms of bond distances, bond angles, torsional angles, and noncovalent interactions (NCIs). In general, molecular strain can be classified as either ring strain, such as in cyclopropane, or steric strain, such as in eclipsed ethane [39–42]. Now, there is a new type of molecular strain, namely the intramolecular multi-bond strain. With the new concept, the subsequent two key questions are: (i) what the magnitude of the π–π repulsion is, and (ii) whether there is any direct or indirect experimental evidence to support this new concept.

16.3 Theoretical Method

To evaluate the π–π repulsion within a conjugated system, we need to derive the electron-localized state, or the most stable resonance structure first. Obviously, this can be achieved with ab initio VB methods (i.e. Eq. 16.1). Over the years, we developed a simplified variant of VB theory, namely the block-localized wavefunction (BLW) method [43–46]. The BLW method combines the advantages of both MO/DFT and VB theories. It not only has MO/DFT computational efficiency but is also able to provide physical insights in terms of VB theory. The BLW method can generate optimal geometries and spectral properties for strictly localized (diabatic) molecular structures, and results can be verified indirectly by numerous experimental data. It can probe the bonding nature, analyze the substitution effect, and clarify the exact roles of charge transfer, polarization, and π resonance in intra- or intermolecular interactions, ultimately elucidate the correlations between molecular structures and properties and mechanisms of molecular interactions and recognitions [42, 47–58].

The BLW method starts from Eq. (16.2) where each orbital function is composed of two Slater determinants in VB theory. In MO theory, however, it is assumed that a bonding electron pair occupies the same orbital, and thus the bond function can be reduced to only one Slater determinant as

$$\phi_{2i-1,2i} = \hat{A}\{\varphi_i'\varphi_i'[\alpha(2i-1)\beta(2i)]\} \tag{16.13}$$

where $\{\varphi_i'\}$ are delocalized and orthogonal, in contrast to $\{\varphi_i\}$ in VB that is localized and nonorthogonal. The hybrid usage of Eqs. (16.2) and (16.13) in Φ_K (Eq. 16.1) leads to the GVB method proposed by Goddard [59, 60], which retains the VB form for one or a few focused bonds (perfect-pairs, Eq. (16.2)) but accommodates the remaining electrons with orthogonal and doubly occupied MOs (Eq. 16.13). The further introduction of the strong orthogonality condition between bond orbitals and MOs remarkably reduces the computational demand for GVB calculations.

Alternatively, the doubly occupied bond orbital in Eq. (16.12) can be expanded only with the atomic orbitals (or basis functions) of bonding atoms [25, 61–70]. In such a way, bond orbitals are local and nonorthogonal, reflecting the combination of the VB and MO methods. Since 1996, we have been working on the proposal of a BLW for a strictly electron-localized state. In the BLW approach, we assume that the total electrons and primitive basis functions are partitioned into k subgroups (blocks). This assumption is consistent with conventional VB ideas. In detail, we assume that the ith subspace consists of $\{\chi_{i\mu}, \mu = 1, 2,...m_i\}$ basis functions and accommodates n_i electrons (for the sake of simplicity, here we assume that n_i is an even number). Clearly, for a Lewis structure, every two electrons form a subspace. However, we generalize the definition of resonance structures and allow a subspace to have any number of electrons, depending on the research question to be addressed. The block-localized MOs for the ith subspace $\{\phi_{ij}, j = 1, 2,...m_i\}$ are expanded in terms of basis functions $\{\chi_{i\mu}\}$ only, i.e.

$$\phi_{ij} = \sum_{\mu=1}^{m_i} C_{ij\mu} \chi_{i\mu} \tag{16.14}$$

Subsequently, the BLW is defined using one Slater determinant and in the closed-shell case ($S = 0$) it is expressed as

$$\Psi_R^{BLW} = M_R(N!)^{-1/2} \det | \phi_{11}^2 \phi_{12}^2 \cdots \phi_{1\frac{n_1}{2}}^2 \phi_{21}^2 \cdots \phi_{i1}^2 \cdots \phi_{i\frac{n_i}{2}}^2 \cdots \phi_{k\frac{n_k}{2}}^2 | \tag{16.15}$$

where M_R is the normalization constant. Following the MO methods, we constrain the orbitals in the same subspace to be orthogonal, but orbitals belonging to different subspaces are free to overlap and thus nonorthogonal in general. As such, the BLW method combines the advantages of both MO and VB theories. The corresponding energy of the above BLW can be computed as

$$E^{BLW} = \langle \Psi_R^{BLW} | H | \Psi_R^{BLW} \rangle = \sum_{\mu=1}^{m} \sum_{\nu=1}^{m} d_{\mu\nu} h_{\mu\nu} + \sum_{\mu=1}^{m} \sum_{\nu=1}^{m} d_{\mu\nu} F_{\mu\nu} \tag{16.16}$$

where m is the total number of basis functions, $h_{\mu\nu}$ and $F_{\mu\nu}$ are elements of the usual one-electron and the Fock matrices, and $d_{\mu\nu}$ is an element of the density matrix, $D = C(C^+SC)^{-1}C^+$ (S is the overlap matrix of the basis functions).

The key step in the BLW computations is the optimization of orbitals, which can be accomplished using successive Jacobi rotation [43] or Gianinettia et al.'s algorithm [69, 70]. The latter generates coupled Roothaan-like equations, and each equation corresponds to a block. The first-order derivative of the energy with respect to nuclear coordinates $\{q_i\}$ directly takes the form in conventional HF theory, and for closed-shell HF, it is [71]

$$\frac{\partial E^{BLW}}{\partial q_i} = 2 \sum_{\mu\nu}^{m} d_{\mu\nu} \frac{\partial h_{\mu\nu}}{\partial q_i} + \sum_{\mu\nu\rho\sigma}^{m} [2 d_{\mu\nu} d_{\rho\sigma} - d_{\mu\rho} d_{\nu\sigma}] \frac{\partial (\mu\nu | \rho\sigma)}{\partial q_i} - 2 \sum_{\mu\nu}^{m} W_{\mu\nu} \frac{\partial S_{\mu\nu}}{\partial q_i} \tag{16.17}$$

where $W_{\mu\nu}$ is a Lagrange variable. After the analytical first-order derivatives, the second-order derivatives can be computed numerically. Thus, both geometry optimizations and spectral computations for an electron-localized state BLW are feasible. Due to the low computational cost and considerable incorporation of electron correlation, DFT methods provide a sound basis for the development of computational strategies for studying potential energy surfaces, dynamics, various response functions, spectroscopy, excited states, and more. Currently, various DFT methods are the most popular computational tools. In DFT, the self-consistent Kohn–Sham (KS) procedure is strictly analogous to the Hartree–Fock–Roothaan SCF procedure, except that the HF exchange potential is replaced by a DFT exchange–correlation (XC) potential. Thus, the orbital equations of DFT have the same forms as those in HF theory except with a different Fock matrix F' as

$$F' = h + J + F^{XC} \tag{16.18}$$

where h is the one-electron Hamiltonian matrix and J is the Coulomb matrix. The elements of XC matrix F^{XC} can be evaluated by a one-electron integral involving local electron spin densities (within the local spin density or LSD methods) or by an integral involving electron density and its gradients (within the generalized gradient approximation or GGA methods). Thus, it is straightforward to implement the BLW idea into the KS-DFT algorithm if we keep all the equations unchanged except that the Fock matrix therein is replaced by a DFT one (F^{XC_α} in Eq. 16.18). As the density matrix, $D = C(C^+SC)^{-1}C^+$, satisfies the symmetry ($D^T = D$), rank ($\mathrm{Tr}(DS) = N$), and idempotency ($DSD = D$) conditions, the electron density is given as follows:

$$\rho(\mathbf{r}) = \sum_{\mu\nu}^{m} |\chi_\mu(\mathbf{r}) > d_{\mu\nu} < \chi_\nu(\mathbf{r})| = \chi(\mathbf{r}) D \chi^T(\mathbf{r}) \tag{16.19}$$

Using the one-particle density matrix and electron density computed from the nonorthogonal KS orbitals, we can express the ground state energy identically as that in the case of orthogonal KS orbitals [46]:

$$E[\rho(\mathbf{r})] = \mathrm{Tr}(\mathbf{DH}) + \frac{1}{2}\mathrm{Tr}(\mathbf{DJD}) + E_{xc}[\rho(\mathbf{r})] + E_{nuc} \tag{16.20}$$

where E_{nuc} is the Coulomb energy of the nuclei and $E_{xc}[\rho(\mathbf{r})]$ is the XC energy functional.

The applications of the BLW method can be best exemplified with the Kekulé structure of benzene (Figure 16.2). The BLW method can generate not only its energy but also its optimal geometry. Within the BLW method, the Kekulé structure can be defined with four blocks, including three carbon–carbon π bonds and the rest σ framework (Figure 16.2a). At the B3LYP/6-311+G(d,p) level, we obtain the optimal single and double bond lengths in the optimal Kekulé structure 1.325 and 1.523 Å, respectively. These values are comparable to the bond lengths in ethylene (1.339 Å) and ethane (1.536 Å). While the compression energy (29.0 kcal/mol) is very close to the empirical estimate (30 kcal/mol) [72], the resonance energy (63.2 kcal/mol) is much higher than the experimental value (36 kcal/mol) [17]. But it has been known that the hyperconjugation in the reference molecule cyclohexene was not considered by Kistiakowsky at that time, and the best "corrected" experimental value would be 65 kcal/mol [73]. This value is very close to our BLW estimate of 63.2 kcal/mol.

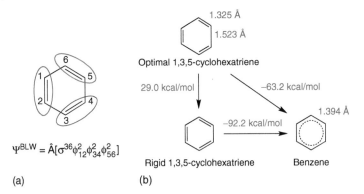

Figure 16.2 (a) BLW definition for the Kekulé structure and (b) computational results for benzene at the B3LYP/6-311+G(d,p) level.

16.4 Computational Analysis of the Concept of Intramolecular Multibond Strain

With the BLW method, we can now derive the wavefunction for the strictly localized covalent resonance structure **I** of butadiene or the Kekulé structure of benzene. But it remains a challenge to evaluate the π–π repulsion therein. Recalling that the conformational change such as the rotation of a methyl group in ethane is usually used to compute the steric repulsion changes, we wonder whether we can evaluate the π–π repulsion changes by altering their relative orientations. In such a way, at least we can sense the magnitude of the π–π repulsion. Here we explore the lower limits of π–π repulsion with a model molecule B_4H_2. For comparison, we also study the σ–σ and σ–π repulsions.

16.4.1 π–π Repulsion in Linear Model Molecule HBBBBH (B_4H_2)

Boron atom is electron-deficient with three valence electrons. If we allow a boron atom to form two σ bonds with others in a linear structure, there would be two p (p_x and p_y if we assume the molecule is in the z-axis) orbitals left with only one π electron. The π electron has the flexibility to occupy any of the two p orbitals. Consequently, we can design a linear molecule B_4H_2 (HB=B–B=BH). Though this is a hypothetical molecule yet, it offers an interesting opportunity to directly probe the π–π repulsion with two localized π pairs either parallelly or perpendicularly arranged, as shown in Figure 16.3. The ground state of linear B_4H_2 involves two 4-center-2-electron (Π_4^2) bonds, which leads to the shortening of the central BB bond to 1.486 Å, even shorter than the terminal BB bonds (1.528 Å). If we strictly localize each π electron pair on two terminal BB bonds with the BLW method, we can derive the localized state with two perpendicular π bonds (up on the right side of Figure 16.3). In this electron-localized state, the central single BB bond distance is stretched to 1.623 Å, but the terminal BB double bonds change little (1.526 Å).

From the localized state with two perpendicular π bonds, we rotate one π bond by 90° and make the two π bonds parallel (comparable to the eclipsed structure of

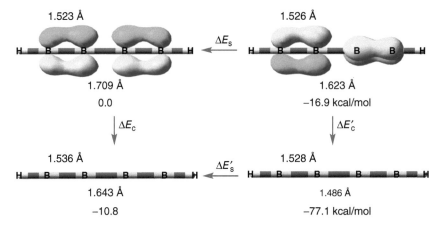

Figure 16.3 Energetic and structural changes along the rotation and conjugation of interacting moieties from parallel to perpendicular orientation for B_4H_2 at the B3LYP/6-311+G(d,p) level.

ethane, up on the left side of Figure 16.3). In the process of rotation, remarkably, we observe a barrier of 16.9 kcal/mol. This is a significant value as it highlights the lower limit for the π–π repulsion. The intramolecular multi-bond strain also noticeably stretches the central single BB bond by 0.086 Å to 1.709 Å, while the terminal BB double bonds shorten slightly by 0.003 Å. This surprisingly high repulsion signifies the strain in all conjugated systems. The conjugation between the π bonds, much like in butadiene, stabilizes the system by 10.8 kcal/mol and shortens the central bond by 0.066 Å to 1.643 Å. Notably, since the π–π resonance stabilization in the Π_4^4 state of B_4H_2 is less than the π–π repulsion, this molecule would exhibit "destabilizing conjugation" like in Eq. (16.11). But we clarify that conjugation must be stabilizing by definition, and the intramolecular multi-bond strain needs to be excluded in the estimate of conjugation stability.

16.4.2 σ–σ Repulsion in Model Molecule B_2H_4

With the π–π repulsion energy larger than 16.9 kcal/mol, we are curious how strong the σ–σ repulsion could be. Often, we use the rotation barrier of ethane as an example to illustrate the steric effect among σ bonds. But the rotation from the energy-minimum staggered conformer to the energy-maximum eclipsed conformer costs only 2.95–2.80 kcal/mol [74], and surely there is still strong repulsion remaining in the eclipsed structure with CH bonds separated by 60°. We assume that B_2H_4 [75, 76] may be a better case to measure the lower limit of σ–σ repulsion.

In the planar B_2H_4, there is no π electron in p orbitals. By keeping these p orbitals strictly vacant, we rotate one BH_2 group to the perpendicular conformation (Figure 16.4). The rotation is accompanied by the shortening of the single BB bond from 1.745 to 1.688 Å, or by 0.057 Å, and the lowering of the energy by 7.7 kcal/mol. Here, the energy change in this rotation process should be solely ascribed to the relief of the σ–σ repulsion between BH bonds (ΔE_s in Figure 16.4). This energetics

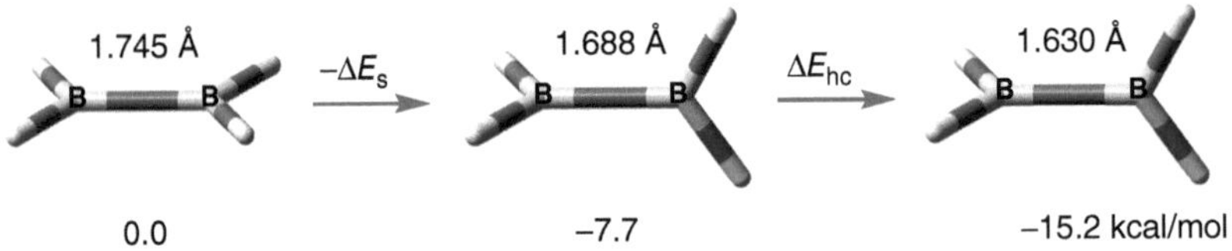

Figure 16.4 Energetic and structural changes along the rotation and hyperconjugation of interacting moieties from planar to perpendicular conformers of B_2H_4 at the B3LYP/6-311+G(d,p) level. Source: Reproduced from Ref. [27] with permission from the Royal Society of Chemistry.

is considerably higher than the rotation barrier in ethane (2.7 kcal/mol at the same B3LYP/6-311+G(d,p) theoretical level), as there is still significant repulsion even in its staggered conformation. We note that ΔE_s is contributed by both Pauli repulsion and electrostatic repulsion, as the BH bond is slightly polar. The introduction of the σ–π* hyperconjugative interaction would further shorten the BB bond to 1.630 Å and stabilize the perpendicular conformation by 7.5 kcal/mol. In total, the rotation barrier in B_2H_4 is 15.2 kcal/mol.

The much lower σ–σ repulsion in B_2H_4 (7.7 kcal/mol) than the π–π repulsion in B_4H_2 (16.9 kcal/mol) comes from the fact that the BH σ bonds in B_2H_4 are farther away than the π bonds in B_4H_2, leading to the lower overlap between BH σ bonds in B_2H_4. The subsequent question is, how about the σ–π repulsion, which coexists with the hyperconjugation?

16.4.3 σ–π Repulsion in Model Molecule B_3H_3

We continue to use boron compounds for the study of the σ–π interactions and adopt the model system B_3H_3 (Figure 16.5), which is a planar molecule with two π electrons forming a double BB bond. The π bond can be in the same plane with the

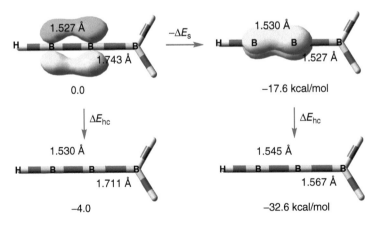

Figure 16.5 Energetic and structural changes along the rotation and conjugation of interacting moieties from parallel to perpendicular orientation for B_3H_3 at the B3LYP/6-311+G(d,p) level.

terminal BH_2 group (up left in Figure 16.5), or out of the planar and perpendicular plane with the terminal BH_2 group (upright in Figure 16.5). Obviously, the latter is relieved of the σ–π repulsion, and computations show that it is more stable by 17.6 kcal/mol. This magnitude is very close to the π–π repulsion. The strong σ–π repulsion is characterized by the very long BB single bond (1.743 Å), and the rotation of the π bond significantly reduces this single bond by 0.216 Å to 1.527 Å. Of course, we can allow the hyperconjugative interactions in the in-plane conformation, which stabilizes the system further by only 4.0 kcal/mol, or both the conjugation (3-center-2-electron bond) and the hyperconjugation (σ → p*) in the perpendicular conformation, which considerably stabilizes the system by 15.0 kcal/mol. The ground state of B_3H_3 involves two π electrons among three boron atoms, and the Π_3^2 bond is perpendicular to the molecular frame.

16.4.4 Conjugation and Repulsion in Butadiene, Butadiyne, Cyanogen, and α-dicarbonyl

The above studies on the π–π, σ–σ, and σ–π interactions with model systems confirm the significant Pauli repulsion among bonds in a molecule, and their magnitudes may be beyond our initial expectations. However, evaluating the Pauli repulsion within molecules remains a challenge. For conjugated systems, although we are currently unable to compute the exact magnitude of π–π repulsion, we can use the BLW method to examine the impact of conjugation on geometries and energetics. The results can be used to infer the π–π repulsion therein. The conjugated systems studied here include butadiene (**1**), butadiyne (**2**), cyanogen (**3**), and α-dicarbonyl (**4**), with bond distances labeled with R_1 and R_2.

$$R_1 \quad R_2$$
$$CH_2=CH-CH=CH_2$$
1

$$R_1 \quad R_2$$
$$CH\equiv C-C\equiv CH$$
2

$$R_1 \quad R_2$$
$$N\equiv C-C\equiv N$$
3

$$R_1 \quad R_2$$
$$O=CH-CH=O$$
4

We first compare the resonance in the two most typical conjugated systems, butadiene **1** and butadiyne **2**. Geometry optimizations with the regular DFT and the BLW methods (Table 16.1) showed that the localization of π electrons on their respective multiple bonds considerably stretches the central carbon–carbon bonds by 0.071 and 0.101 Å, respectively, for **1** and **2**. The central carbon–carbon bond distances in the BLW optimal geometries reflect the Csp–Csp (1.465 Å) and Csp^2–Csp^2 (1.528 Å) single σ bonds. The deactivation of conjugation modestly shortens the double and triple bond lengths, which are essentially the same as the bond distances in ethylene (1.329 Å) and acetylene (1.199 Å) at the same B3LYP/6-311+G(d,p) level. Within the resonance theory, there are two types of resonance (conjugation) energies, namely vertical resonance energy (VRE) and

Table 16.1 Major optimal bond distances (Å) in delocalized (DFT) and localized (BLW) states and the computed resonance energies (VRE and ARE, in kcal/mol) compared with experimental resonance energies (ERE, in kcal/mol) and the force constant (kcal/Å^2).

Molecule	Method	R_1	R_2	VRE	ARE	ERE	k
1	DFT	1.338	1.457	14.5	12.6	8.5	382
	BLW	1.326	1.528	10.9			
2	DFT	1.207	1.364	32.9	27.0	10.0	589
	BLW	1.194	1.465	22.2			
3	DFT	1.155	1.376	27.2	22.7	−4.3[a]	576
	BLW	1.145	1.467	18.9			
4	DFT	1.203	1.529	6.3	5.6	4.3[a]	380
	BLW	1.196	1.583	4.9			

a) Reaction enthalpy at 298 K.

adiabatic resonance energy (ARE). The former is the energy difference between DFT and BLW computations at the same geometry, while the latter is the energy difference between the optimal delocalized state (i.e. DFT optimization) and the optimal localized state (i.e. BLW optimization). Figure 16.6 shows the change of VRE along the central carbon–carbon bond distance in the four conjugated systems studied here. There is an excellent exponential correlation between VRE and the central bond distance in each molecule.

At the DFT-optimized geometries, the VRE in butadiyne (32.9 kcal/mol) is a little more than two times the value in butadiene (14.5 kcal/mol). This is in agreement

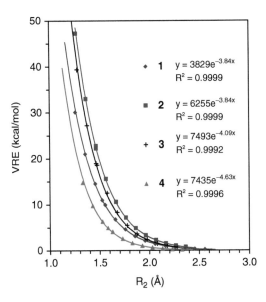

Figure 16.6 The exponential correlation of vertical resonance energy (VRE) with the central CC bond distance (R_2) in butadiene (**1**), butadiyne (**2**), cyanogen (**3**), and α-dicarbonyl (**4**) at the B3LYP/6-311+G(d,p) theoretical level. Source: Reproduced from Ref. [27] with permission from the Royal Society of Chemistry.

with the studies by Kollmar [25], and Cappel et al. [23], and consistent with the facts that the central bond in butadiyne is shorter than in butadiene, and there are two Π_4^4 bonds in the former but only one in the latter. The ARE is supposed to be comparable to experimental resonance energies (EREs) with reference to individual multiple bonds such as ethylene and acetylene. However, isodesmic reactions show that the EREs for butadiene (8.5 kcal/mol in Eq. 16.7) and butadiyne (10.0 kcal/mol in Eq. 16.10) are not only similar but also considerably lower than theoretical AREs (12.6 and 27.0 kcal/mol for **1** and **2**). If we take the difference between ARE and ERE as the steric contribution, the steric repulsion in butadiene and butadiyne are 4.1 and 18.8 kcal/mol respectively. The latter is much more than two times the former, due to the much-shortened central carbon–carbon bond distance, and the repulsive force increases exponentially along the distance [77]. It should be noted that our AREs are very much close to the evaluations of conjugation stabilization (14.8 and 27.1 kcal/mol) in butadiene and butadiyne by Wodrich et al., who reinterpreted the differences in the hydrogenations of the first and second multi-bonds in Eqs. (16.5) and (16.8) after introducing the "protobranching" concept [73].

Pauli repulsion is believed to result from the overlap of electron densities [77]. One way to estimate such intramolecular repulsion ΔE_s is the use of the compression energy [11], which is the difference between VRE at the optimal DFT structure and ARE and reflects the energy cost for the structural change (dominated by the central carbon–carbon single bond variation ΔR_2) when conjugation is deactivated. Consequently, we propose a force constant k to evaluate and compare the magnitude of the intramolecular steric repulsion

$$k = \frac{\Delta E_s}{[\Delta R_2]^2} = \frac{\text{VRE-ARE}}{[R_2(\text{BLW}) - R_2(\text{DFT})]^2} \tag{16.21}$$

Table 16.1 lists the k values for the four conjugated molecules. As expected, butadiyne has a much higher k value than butadiene. Of course, k measures the change of repulsion with distance, i.e. the repulsive force, rather than the absolute energetic value of the steric repulsion. While the π–π repulsion in considerably offsets the theoretical resonance energy (ARE) and leads to the ERE, the repulsion in butadiyne is much stronger than in butadiene. In the end, both molecules exhibit comparable ERE's, and interestingly, as found by Rogers et al. [20, 21], there is zero thermodynamic conjugation stabilization in butadiyne.

We continue to analyze cyanogen (**3**) and α-dicarbonyl (**4**), where Rogers et al. showed even the thermodynamic destabilization [26]. While the conjugation stabilization in **3** is lower than in **2** by 4–6 kcal/mol, the conjugation in **4** is similarly weaker than in **1**. The reduced conjugation results in the central carbon–carbon bond in cyanogen being 0.012 Å longer than that in butadiyne. But when π electrons are localized, the optimal central bond lengths are quite similar (1.465 versus 1.467 Å). This implies a stronger repulsion between triple bonds in cyanogen than in butadiyne. In fact, if we use the isodesmic reaction (Eq. 16.22) to evaluate the ERE, we find that the ERE is even negative at the G3 theoretical level. This is consistent with Eq. (16.11), where the hydrogenation of the first cyano group is more exothermic than the hydrogenation of the second cyano group.

$$N \equiv C - C \equiv N + 2CH_3NH_2 \rightarrow 2HCN + H_2NCH_2 - CH_2NH_2 - 4.3 \text{ kcal/mol} \quad (16.22)$$

$$HCO - CHO + 2CH_3OH \rightarrow 2H_2CO + CH_2OH - CH_2OH + 4.3 \text{ kcal/mol} \quad (16.23)$$

Different from **3**, α-dicarbonyl exhibits limited conjugation stabilization, as shown in Eq. (16.13). Notably, we observed much stretched central carbon–carbon bond distance (1.529 Å). The location of π electrons further stretches the bond to 1.583 Å. We believe that the dipole–dipole repulsion between the carbonyl groups plays a role here.

16.5 Experimental Evidence

Above, we proposed that there is a duality of π electrons in conjugated systems. Apart from the much-familiar π conjugation, which is a stabilizing force, there is destabilizing π–π repulsion, which is largely neglected and is the primary culprit for never-ending controversies over conjugation stability, notably aromaticity in conjugated rings. Here we discuss a few experimental clues for the π–π repulsion [28].

16.5.1 Longer Carbon–Nitrogen Bond in Nitrobenzene than in Aniline

Substituents can influence the properties and reactivity of benzene. For instance, an amino substituent remarkably increases the rate of electrophilic substitution, while a nitro substituent decreases the benzene ring's reactivity significantly. This substituent effect has been intuitively summarized by Hammett's substituent constant σ which measures the electronic effect of replacing H with a given substituent [78, 79]. The resonance between the benzene ring and its substituent can be shown as

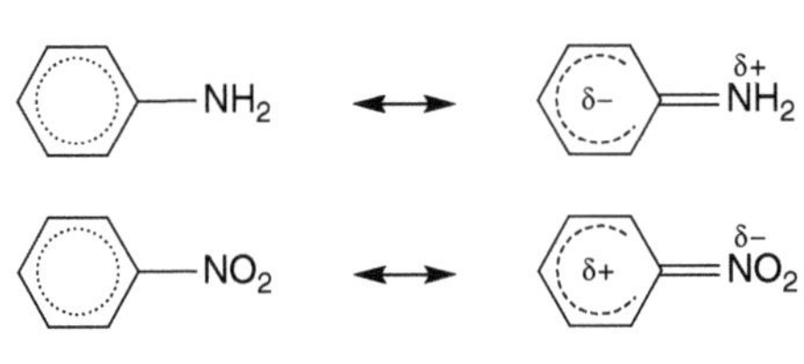

In the above scheme, the left resonance structures (note that they are not classical Lewis structures anymore) correspond to diabatic electron-localized states before any electron is delocalized between the benzene ring and a substituent, while the right ones are a diabatic states with electrons already delocalized (transferred). While both amino and nitro exhibit similar inductive effects, they have contrasting behaviors for their π conjugation capability. The right resonance structures indicate that $-NH_2$ is an electron-donating substituent with a negative σ (−0.66 at the *para* position of aniline), and $-NO_2$ is an electron-withdrawing substituent with a positive σ (+0.778 at the *para* position of nitrobenzene). Based on the resonance picture, it would be expected that the C−N bonds in both systems have certain double bond property. However, gas-phase electron diffraction experiments show a significant disparity for the C−N bond lengths between aniline (1.407 Å) [80] and nitrobenzene (1.486 Å) [81]. For comparison, the C−N bonds in methylamine (1.472 Å [82]) and nitromethane (1.488 Å [83]) are similar and close to the bond in nitrobenzene. As bond lengths generally correlate with bond energies [84, 85], the much longer bond length in nitrobenzene implies that the C−N bond is even weaker than the single C−N bond in methylamine. In aniline, the amino group is slightly pyramidal, and the nitrogen takes a hybridization mode between sp^3 and sp^2, due to the conjugation between the benzene ring and the nitrogen lone pair. The latter also results in a low inversion barrier (1–2 kcal/mol) for the amino group [86, 87], compared with 4–5 kcal/mol in aliphatic amines and 5.0 kcal/mol in ammonia [88]. In contrast, nitrobenzene is planar, and the nitrogen adopts a sp^2 hybridization mode. In general, it is well recognized that the increasing *s* component in a hybrid orbital leads to a shorter bond, as observed in C−H bond lengths in ethane (1.096 Å), ethene (1.085 Å), and ethyne (1.061 Å) [89]. Thus, the much shorter C−N bond in aniline than in nitrobenzene is puzzling and requires an explanation.

To probe the origin of the disparity in aniline and nitrobenzene and differentiate the σ inductive effect and the π resonance, it is essential to get the optimal bond lengths when the conjugation over C−N bonds in these systems is hypothetically and completely quenched. Table 16.2 summarizes the geometry optimizations at the B3LYP/6-311+G(d,p) level. For comparison, we also listed the data for dithionitrobenzene. Regular DFT computations result in the C−N bond lengths in aniline (1.399 Å) and nitrobenzene (1.481 Å), which are very close to experimental data (1.407 Å [80] and 1.486 Å [81]). Due to the conjugation over the C−N bond and the subsequent reduction of the nitrogen lone pair and NH bonds, the amino group in aniline is highly flattened, with the CNH bond angle of 115.6° and the CCNH dihedral angle of 24.5°. Natural population analysis (NPA) [90] also shows that the π-electron population in the ring increases to 6.143 e in aniline but decreases to 5.933 e in nitrobenzene. This may imply that the magnitude of conjugation in nitrobenzene is much less than in aniline. This is verified by the BLW computations as shown in Table 16.2.

Table 16.2 Optimal C−N bond distances (Å) and conjugation energies (kcal/mol) at the B3LYP/6-311+G(d,p) level.

Molecule	Method	R(C−N)	VRE	ARE
C_6H_5-NH_2	DFT	1.399	19.3	15.2
	BLW	1.491	/	
C_6H_5-NO_2	DFT	1.481	10.8	8.8
	BLW	1.565	/	
C_6H_5-NS_2	DFT	1.472	13.4	10.3
	BLW	1.579	/	

By strictly localizing the six π electrons in the benzene ring and quenching the conjugation over the C–N bond, BLW geometry optimizations led to stretched C–N bonds. Conjugation shortens C–N bonds by 0.092 and 0.084 Å in aniline and nitrobenzene, respectively. Without conjugation, as expected, the pyramidality of the amino group in aniline would be very much like in ammonia, as the BLW optimization results in the CNH bond angle of 108.7° and the dihedral angle of 114.4°, compared with 107.9° and 116.4° for ammonia at the same level. It should be noted that the C−N single bond in aniline (1.491 Å) is longer than the bond length in methylamine (1.465 Å). Two factors may be involved here, including the repulsion between the nitrogen lone pair and the π cloud of benzene in aniline and the hyperconjugative interaction between the methyl and amino groups in methylamine.

What is the Primary cause for the difference in C−N bond distances in aniline and nitrobenzene, which is little changed with or without the conjugation over the C−N bonds? We propose that it is the π–π repulsion responsible for the stretched C−N bond in nitrobenzene compared with aniline. This π–π repulsion is essentially a kind of steric effect and is composed of both Pauli exchange repulsion and electrostatic interaction. To verify our proposal, we first examine the change of the C−N bond distance with the rotation of the nitro group with all π electrons localized on either the benzene ring or the nitro group. Indeed, we can see the shortening of the C−N bond in the perpendicular nitrobenzene from 1.565 to 1.557 Å (Figure 16.7), due to the relief of the π–π Pauli repulsion. But this change is very modest as there is π–σ Pauli repulsion in the perpendicular structure, and the energy even goes up by 5.1 kcal/mol with the removal of the stabilizing C–H⋯O interactions. Considering the delocalization of the lone pair electrons on N to more electronegative oxygen atoms, we speculate that the resonance of the π electrons within the nitro group generates a significant dipole, and the π–π electrostatic repulsion may play a big role in stretching the C–N bond. To prove this hypothesis, we performed BLW optimizations by deactivating the π resonance within the nitro group and found that there

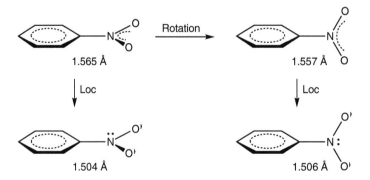

Figure 16.7 Variation of the central C−N bond in nitrobenzene with the rotation of the nitro group and the localization of π electrons on the nitro group based on BLW optimizations. Source: Reproduced from Ref. [28] with permission from the Royal Society of Chemistry.

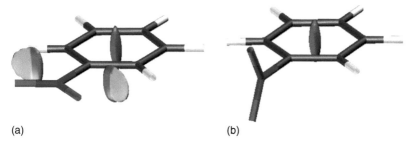

(a)   (b)

Figure 16.8 Gradient isosurfaces with a value of 0.6 au for (a) planar and (b) perpendicular nitrobenzene. The surfaces are colored on a green-red scale (no blue here) according to values of sign(λ_2)ρ, ranging from −0.02 to 0.02 au. Green indicates van der Waals attractive interaction, and red indicates steric repulsion. For details, see Refs. [91, 92]. Source: Reproduced from Ref. [28] with permission from the Royal Society of Chemistry.

is a significant shortening of the C—N bond in both planar and perpendicular conformations, as shown in Figure 16.7. In fact, the values (1.504 and 1.506 Å) are now very much close to the bond distance in aniline (1.491 Å, see Table 16.2).

The π–π repulsion within nitrobenzene can be intuitively demonstrated with the NCIs plot based on the electron density and its reduced gradient [91, 92]. With the Multiwfn program [93], we generate the NCI plot (Figure 16.8), which shows the repulsion between the nitro group and the benzene ring (in red), as well as the attraction between the H and O atoms (in green), in the planar structure. In the perpendicular structure, however, both forces disappear.

16.5.2 Abnormally Long Nitrogen–Nitrogen Bond in Dinitrogen Tetroxide (N$_2$O$_4$)

N$_2$O$_4$ is a very unique molecule with an extremely long (experimental value 1.756 Å) and weak nitrogen–nitrogen bond [94]. B3LYP/6-311+G(d,p) optimization leads to a bond distance of 1.809 Å, indicating the impact of electron correlation which is not our focus. We argue that this very long nitrogen–nitrogen bond is due to the enhanced π–π repulsion. Interestingly, there is negligible conjugation over the N—N bond as the deactivation of the conjugation only slightly stretches the bond by 0.024–1.833 Å. Dinitrogen tetroxide prefers a planar geometry due to the through-space coupling between oxygen atoms rather than the conjugation.

By rotating one nitro group around the N—N bond, the central bond is shortened slightly by 0.016 Å due to the relief of the π–π repulsion, though the cost for the decoupling of O···O interactions is quite high (9.7 kcal/mol, see Figure 16.9). The deactivation of the π conjugation over the N—N bond stretches the bond to 1.833 Å. Remarkably, if the π conjugation within nitro groups is shut down, the π dipole–dipole electrostatic repulsion would be eliminated, and the N—N bond would be significantly shortened to 1.539 Å. The rotation of one nitro group would further relieve the π–π Pauli exchange repulsion and shorten the N—N bond to 1.471 Å, and even more evidently, the perpendicular structure would be more stable than the planar structure by 5.9 kcal/mol!

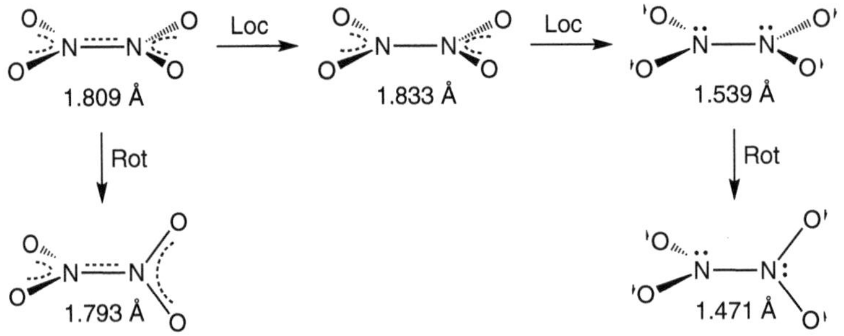

Figure 16.9 Evolution of the central nitrogen–nitrogen bond distance with the gradual localization of π electrons and the rotation of one nitro group based on BLW optimizations. Source: Reproduced from Ref. [28] with permission from the Royal Society of Chemistry.

Considering the high electronegativity of oxygen, one may wonder whether the σ induction plays a role in N_2O_4. To verify the role of induction for the C−N bond in nitrobenzene, we performed similar computations of dithionitrobenzene as shown in Table 16.2. While the conjugation over the C−N bond in dithionitrobenzene is slightly enhanced compared with nitrobenzene, the replacement of oxygen by sulfur only slightly shortens the C−N bond by 0.009 Å. This is evidenced by the population analysis (5.923 e in the benzene ring of dithionitrobenzene) and the resonance energies (Table 16.2).

16.6 Summary

In this chapter, we discussed the duality of π electrons in conjugated systems. While researchers often focus on the electron delocalization (conjugation) side of π electrons, there is considerable π–π repulsion that is often neglected, and we believe that this neglection is the cause for many controversies related to conjugation, including the aromaticity in conjugated rings. To emphasize this π–π repulsion, we proposed the concept of intramolecular multi-bond strain. Using the model molecule HBBBBH, we demonstrated that the parallel arrangement of the two strictly localized π pairs is less stable by 16.9 kcal/mol than the perpendicular arrangement. This significant value informs us of the very strong π–π repulsion in conjugated systems, which coexists with the stabilizing conjugation.

References

1 von Ragué Schleyer, P. (2001). Introduction: Aromaticity. *Chem. Rev.* 101: 1115–1118. https://doi.org/10.1021/cr0103221.
2 Novoselov, K.S., Geim, A.K., Morozov, S.V. et al. (2004). Electric field effect in atomically thin carbon films. *Science* 306: 666–669. https://doi.org/10.1126/science.1102896.
3 Lewis, G.N. (1916). The atom and the molecule. *J. Am. Chem. Soc.* 38: 762–785.

4 Shaik, S. (2007). The Lewis legacy: the chemical bond—a territory and heartland of chemistry. *J. Comput. Chem.* 28: 51–61. https://doi.org/10.1002/jcc.20517.

5 Zhao, L., Hermann, M., Schwarz, W.H.E., and Frenking, G. (2019). The Lewis electron-pair bonding model: modern energy decomposition analysis. *Nat. Rev. Chem.* 3: 48–63. https://doi.org/10.1038/s41570-018-0060-4.

6 Pauling, L.C. (1960). *The Nature of the Chemical Bond*, 3e. Cornell University Press.

7 Cooper, D. (2002). *Valence Bond Theory*, vol. 10. Amsterdam: Elsevier.

8 Gallup, G.A. (2002). *Valence Bond Methods: Theory and Applications*. Cambridge University Press.

9 Shaik, S.S. and Hiberty, P.C. (2007). *A Chemist's Guide to Valence Bond Theory*. Wiley.

10 Wu, W., Su, P., Shaik, S., and Hiberty, P.C. (2011). Classical valence bond approach by modern methods. *Chem. Rev.* 111: 7557–7593.

11 Wheland, G.W. (1944). *The Theory of Resonance: The Theory of Resonance and its Application to Organic Chemistry*. Wiley.

12 Dewar, M.J.S. and Schmeising, H.N. (1959). A re-evaluation of conjugation and hyperconjugation: the effects of changes in hybridisation on carbon bonds. *Tetrahedron* 5: 166–178.

13 Cyrański, M.K., Krygowski, T.M., Katritzky, A.R., and von Ragué Schleyer, P. (2002). To what extent can aromaticity be defined uniquely? *J. Org. Chem.* 67: 1333–1338. https://doi.org/10.1021/jo016255s.

14 Hehre, W.J., Radom, L., von Ragué Schleyer, P., and Pople, J.A. (1986). *Ab Initio Molecular Orbital Theory*. Wiley.

15 Cyrański, M.K., von Ragué Schleyer, P., Krygowski, T.M. et al. (2003). Facts and artifacts about aromatic stability estimation. *Tetrahedron* 59: 1657–1665. https://doi.org/10.1016/S0040-4020(03)00137-6.

16 Slayden, S.W. and Liebman, J.F. (2001). The energetics of aromatic hydrocarbons: an experimental thermochemical perspective. *Chem. Rev.* 101: 1541–1566. https://doi.org/10.1021/cr990324.

17 Kistiakowsky, G.B., Ruhoff, J.R., Smith, H.A., and Vaughan, W.E. (1936). Heats of organic reactions. IV. Hydrogenation of some dienes and of benzene. *J. Am. Chem. Soc.* 58: 146–153. https://doi.org/10.1021/ja01292a043.

18 Conant, J.B. and Kistiakowsky, G.B. (1937). Energy changes involved in the addition reactions of unsaturated hydrocarbons. *Chem. Rev.* 20: 181–194. https://doi.org/10.1021/cr60066a002.

19 Mulliken, R.S., Rieke, C.A., and Brown, W.G. (1941). Hyperconjugation*. *J. Am. Chem. Soc.* 63: 41–56. https://doi.org/10.1021/ja01846a008.

20 Rogers, D.W., Matsunaga, N., Zavitsas, A.A. et al. (2003). The conjugation stabilization of 1,3-butadiyne is zero. *Org. Lett.* 5: 2373–2375. https://doi.org/10.1021/ol030019h.

21 Rogers, D.W., Matsunaga, N., McLafferty, F.J. et al. (2004). On the lack of conjugation stabilization in polyynes (polyacetylenes). *J. Org. Chem.* 69: 7143–7147. https://doi.org/10.1021/jo049390o.

22 Jarowski, P.D., Wodrich, M.D., Wannere, C.S. et al. (2004). How large is the conjugative stabilization of diynes? *J. Am. Chem. Soc.* 126: 15036–15037. https://doi.org/10.1021/ja046432h.

23 Cappel, D., Tüllmann, S., Krapp, A., and Frenking, G. (2005). Direct estimate of the conjugative and hyperconjugative stabilization in diynes, dienes, and related compounds. *Angew. Chem. Int. Ed.* 44: 3617–3620. https://doi.org/10.1002/anie.200500452.

24 Feixas, F., Matito, E., Poater, J., and Solà, M. (2011). Understanding conjugation and hyperconjugation from electronic delocalization measures. *J. Phys. Chem. A* 115: 13104–13113. https://doi.org/10.1021/jp205152n.

25 Kollmar, H. (1979). Direct calculation of resoanance energies of conjugated hydrocarbons with ab initio MO methods. *J. Am. Chem. Soc.* 101: 4832–4840.

26 Zavitsas, A.A., Rogers, D.W., and Matsunaga, N. (2011). Destabilization of conjugated systems of α-dicarbonyls and of cyanogen. *Aust. J. Chem.* 64: 390–393. https://doi.org/10.1071/CH10394.

27 Mo, Y., Zhang, H., Su, P. et al. (2016). Intramolecular multi-bond strain: the unrecognized side of the dichotomy of conjugated systems. *Chem. Sci.* 7: 5872–5878. https://doi.org/10.1039/C6SC00454G.

28 Zhang, H., Jiang, X., Wu, W., and Mo, Y. (2016). Electron conjugation versus π–π repulsion in substituted benzenes: why the carbon–nitrogen bond in nitrobenzene is longer than in aniline. *Phys. Chem. Chem. Phys.* 18: 11821–11828. https://doi.org/10.1039/C6CP00471G.

29 Shaik, S.S. and Hiberty, P.C. (1985). When does electronic delocalization become a driving force of molecular shape and stability? 1. The aromatic sextet. *J. Am. Chem. Soc.* 107: 3089–3095. https://doi.org/10.1021/ja00297a013.

30 Hiberty, P.C., Danovich, D., Shurki, A., and Shaik, S. (1995). Why does benzene possess a d6h symmetry? A quasiclassical state approach for probing .pi.-bonding and delocalization energies. *J. Am. Chem. Soc.* 117: 7760–7768. https://doi.org/10.1021/ja00134a022.

31 Shaik, S., Shurki, A., Danovich, D., and Hiberty, P.C. (2001). A different story of π-delocalization the distortivity of π-electrons and its chemical manifestations. *Chem. Rev.* 101: 1501–1540. https://doi.org/10.1021/cr990363l.

32 Baeyer, A. (1885). Ueber Polyacetylenverbindungen. *Ber. Dtsch. Chem. Ges.* 18: 2269–2281. https://doi.org/10.1002/cber.18850180296.

33 Wiberg, K.B. (1986). The concept of strain in organic chemistry. *Angew. Chem. Int. Ed. English* 25: 312–322. https://doi.org/10.1002/anie.198603121.

34 Dudev, T. and Lim, C. (1998). Ring strain energies from ab initio calculations. *J. Am. Chem. Soc.* 120: 4450–4458. https://doi.org/10.1021/ja973895x.

35 Barić, D. and Maksić, Z.B. (2005). On the origin of Baeyer strain in molecules – an ab initio and DFT analysis. *Theor. Chem. Acc.* 114: 222–228. https://doi.org/10.1007/s00214-005-0664-x.

36 Sellers, B.D., James, N.C., and Gobbi, A. (2017). A comparison of quantum and molecular mechanical methods to estimate strain energy in druglike fragments. *J. Chem. Information Model.* 57: 1265–1275. https://doi.org/10.1021/acs.jcim.6b00614.

37 Franklin, J.L. (1949). Prediction of heat and free energies of organic compounds. *Indus. Eng. Chem.* 41: 1070–1076. https://doi.org/10.1021/ie50473a041.

38 Engler, E.M., Andose, J.D., and Schleyer, P.V.R. (1973). Critical evaluation of molecular mechanics. *J. Am. Chem. Soc.* 95: 8005–8025. https://doi.org/10.1021/ja00805a012.

39 Mo, Y. and Gao, J. (2007). Theoretical analysis of the rotational barrier of ethane. *Acc. Chem. Res.* 40: 113–119. https://doi.org/10.1021/ar068073w.

40 Mo, Y., Wu, W., Song, L. et al. (2004). The magnitude of hyperconjugation in ethane: a perspective from ab initio valence bond theory. *Angew. Chem. Int. Ed.* 43: 1986–1990. https://doi.org/10.1002/anie.200352931.

41 Pitzer, R.M. (1983). The barrier to internal rotation in ethane. *Acc. Chem. Res.* 16: 207–210.

42 Mo, Y. (2011). Rotational barriers in alkanes. *Wiley Interdiscip. Rev. Comput. Mol. Sci.* 1: 164–171.

43 Mo, Y. and Peyerimhoff, S.D. (1998). Theoretical analysis of electronic delocalization. *J. Chem. Phys.* 109: 1687–1697.

44 Mo, Y., Gao, J., and Peyerimhoff, S.D. (2000). Energy decomposition analysis of intermolecular interactions using a block-localized wave function approach. *J. Chem. Phys.* 112: 5530–5538. http://dx.doi.org/10.1063/1.481185.

45 Mo, Y., Bao, P., and Gao, J. (2011). Intermolecular interaction energy decomposition based on block-localized wavefunction and block-localized density functional theory. *Phys. Chem. Chem. Phys.* 13: 6760–6775.

46 Mo, Y., Song, L., and Lin, Y. (2007). Block-localized wavefunction (BLW) method at the density functional theory (DFT) level. *J. Phys. Chem. A* 111: 8291–8301.

47 Mo, Y. (2010). Computational evidence that hyperconjugative interactions are not responsible for the anomeric effect. *Nat. Chem.* 2: 666–671.

48 Wang, C., Chen, Z., Wu, W., and Mo, Y. (2013). How the generalized anomeric effect influences the conformational preference? *Chem. Eur. J.* 19: 1436–1444.

49 Wang, C., Danovich, D., Shaik, S., and Mo, Y. (2017). A unified theory for the blue- and red-shifting phenomena in hydrogen and halogen bonds. *J. Chem. Theory Comput.* 13: 1626–1637.

50 Wang, C. and Mo, Y. (2019). Classical electrostatic interaction is the origin for blue-shifting halogen bonds. *Inorg. Chem.* 58: 8577–8586.

51 Mo, Y., Wang, C., Guan, L. et al. (2014). On the nature of blue-shifting hydrogen bonds. *Chem. Eur. J.* 20: 8444–8452.

52 Lin, X., Jiang, X., Wu, W., and Mo, Y. (2018). A direct proof of the resonance-impaired hydrogen bond (RIHB) concept. *Chem. Eur. J.* 24: 1053–1056.

53 Lin, X. and Mo, Y. (2022). On the bonding nature in the crystalline tri-thorium cluster: core-shell syngenetic σ-aromaticity. *Angew. Chem. Int. Ed.* https://doi.org/10.1002/anie.202209658.

54 Lin, X., Wu, W., and Mo, Y. (2019). How resonance modulates multiply hydrogen bonding in self-assembled systems. *J. Org. Chem.* 84: 14805–14815.

55 Lin, X., Wu, W., and Mo, Y. (2020). A theoretical perspective of agostic interactions in early transition metal compounds. *Coord. Chem. Rev.* 419: 213401.

56 Zhang, H., Cao, Z., Wu, W., and Mo, Y. (2018). The transition-metal-like behavior of $B_2(NHC)_2$ in the activation of CO: HOMO-LUMO swap without photoinduction. *Angew. Chem. Int. Ed.* 57: 13076–13081.

57 Zhang, H., Yuan, R., Wu, W., and Mo, Y. (2020). Two push-pull channels enhance the dinitrogen activation by borylene compounds. *Chem. Eur. J.* 26: 2619–2625.

58 Inscoe, B., Rathnayaki, H., and Mo, Y. (2021). The role of charge transfer in halogen bonding. *J. Phys. Chem. A* 125: 2944–2953.

59 Goddard, W.A. III, (1967). Improved quantum theory of many-electron systems. I. Construction of eigenfunctions of S_2 which satisfy Paul's principle. *Phys. Rev.* 157: 73–80.

60 Bobrowicz, F.W. and Goddard, W.A. III, (1977). *Methods of Electronic Structure Theory* (ed. H.F. Schaefer III,), 79–127. New York: Springer.

61 Mulliken, R.S. and Parr, R.G. (1951). Linear-combination-of-atomic-orbital-molecular-orbital computation of resonance energies of benzene and butadiene, with general analysis of theoretical versus thermochemical resonance energies. *J. Chem. Phys.* 19: 1271–1278.

62 Sovers, O.J., Kern, C.W., Pitzer, R.M., and Karplus, M. (1968). Bond-function analysis of rotational barriers: ethane. *J. Chem. Phys.* 49: 2592–2599.

63 Stoll, H. and Preuss, H. (1977). On the direct calculation of localized HF orbitals in molecule clusters, layers and solids. *Theor. Chim. Acta* 48: 11–21.

64 Stoll, H., Wagenblast, G., and Preuss, H. (1980). On the use of local basis sets for localized molecular orbitals. *Theor. Chim. Acta* 57: 169–178.

65 Daudey, J.P., Trinquier, G., Barthelat, J.C., and Malrieu, J.P. (1980). Decisive role of p-conjugation in the central bond length shortening of butadiene. *Tetrahedron* 36: 3399–3401.

66 Mehler, E.L. (1977). Self-consistent, nonorthogonal group function approximation for polyatomic systems. I. Closed shells. *J. Chem. Phys.* 67: 2728–2739.

67 Mehler, E.L. (1981). Self-consistent, nonorthogonal group function approximation for polyatomic systems. II. Analysis of noncovalent interactions. *J. Chem. Phys.* 74: 6298–6306.

68 Fülscher, M.P. and Mehler, E.L. (1981). Self-consistent, nonorthogonal group function approximation. III. Approaches for modeling intermolecular interactions. *J. Comp. Chem.* 12: 811–828.

69 Gianinetti, E., Raimondi, M., and Tornaghi, E. (1996). Modification of the Roothaan equations to exclude BSSE from molecular interaction calculations. *Int. J. Quantum Chem.* 60: 157–166.

70 Gianinetti, E., Vandoni, I., Famulari, A., and Raimondi, M. (1998). Extension of the SCF-MI method to the case of K fragments one of which is an open-shell system. *Adv. Quantum Chem.* 31: 251–266.

71 Famulari, A., Gianinetti, E., Raimondi, M., and Sironi, M. (1998). Implementation of gradient-optimization algorithms and force constant computations in BSSE-free direct and conventional SCF approaches. *Int. J. Quantum Chem.* 69: 151–158.

72 Coulson, C.A. and Altmann, S.L. (1952). Compressional energy and resonance energy. *Trans. Faraday Soc.* 48: 293–302.

73 Wodrich, M.D., Wannere, C.S., Mo, Y. et al. (2007). The concept of protobranching and its many paradigm shifting implications for energy evaluations. *Chem. Euro. J.* 13: 7731–7744. https://doi.org/10.1002/chem.200700602.

74 Ercolani, G. (2005). Determination of the rotational barrier in ethane by vibrational spectroscopy and statistical thermodynamics. *J. Chem. Educ.* 82: 1703. https://doi.org/10.1021/ed082p1703.

75 Demachy, I. and Volatron, F. (1994). Hyperconjugation versus steric effects: Ab Initio study of the B2D4 systems (D=H, CH_3, NH_2, OH, F, Cl). *J. Phys. Chem.* 98: 10728–10734.

76 Mo, Y. and Lin, Z. (1996). Theoretical study of conjugation, hyperconjugation, and steric effect in B2D4 (D=H, F, OH, NH_2, and CH_3). *J. Chem. Phys.* 105: 1046–1051.

77 Rackers, J.A. and Ponder, J.W. (2019). Classical Pauli repulsion: an anisotropic, atomic multipole model. *J. Chem. Phys.* 150: 084104. https://doi.org/10.1063/1.5081060.

78 Hammett, L.P. (1935). Some relations between reaction rates and equilibrium constants. *Chem. Rev.* 17: 125–136. https://doi.org/10.1021/cr60056a010.

79 Hansch, C., Leo, A., and Taft, R.W. (1991). A survey of Hammett substituent constants and resonance and field parameters. *Chem. Rev.* 91: 165–195. https://doi.org/10.1021/cr00002a004.

80 Schultz, G., Portalone, G., Ramondo, F. et al. (1996). Molecular structure of aniline in the gaseous phase: a concerted study by electron diffraction and ab initio molecular orbital calculations. *Struct. Chem.* 7: 59–71. https://doi.org/10.1007/BF02275450.

81 Domenicano, A., Schultz, G., Hargittai, I. et al. (1990). Molecular structure of nitrobenzene in the planar and orthogonal conformations. *Struct. Chem.* 1: 107–122. https://doi.org/10.1007/BF00675790.

82 Iijima, T., Jimbo, H., and Taguchi, M. (1986). The molecular structure of methylamine in the vapour phase. *J. Mol. Struct.* 144: 381–383. https://doi.org/10.1016/0022-2860(86)85017-7.

83 Cox, A.P. (1983). Recent microwave studies of internal rotation and molecular structure. *J. Mol. Struct.* 97: 61–76. https://doi.org/10.1016/0022-2860(83)90178-3.

84 Zavitsas, A.A. (2003). The relation between bond lengths and dissociation energies of carbon–carbon bonds. *J. Phys. Chem. A* 107: 897–898. https://doi.org/10.1021/jp0269367.

85 Blanksby, S.J. and Ellison, G.B. (2003). Bond dissociation energies of organic molecules. *Acc. Chem. Res.* 36: 255–263. https://doi.org/10.1021/ar020230d.

86 Bock, C.W., George, P., and Trachtman, M. (1986). A molecular orbital study of nitrogen inversion in aniline with extensive geometry optimization. *Theor. Chim. Acta* 69: 235–245. https://doi.org/10.1007/BF00526422.

87 Wang, Y., Saebø, S., and Pittman, C.U. (1993). The structure of aniline by ab initio studies. *J. Mol. Struct. THEOCHEM* 281: 91–98. https://doi.org/10.1016/0166-1280(93)87064-K.

88 Léonard, C., Carter, S., and Handy, N.C. (2003). The barrier to inversion of ammonia. *Chem. Phys. Lett.* 370: 360–365. https://doi.org/10.1016/S0009-2614(03)00107-6.

89 Delley, B. (1991). Analytic energy derivatives in the numerical local-density-functional approach. *J. Chem. Phys.* 94: 7245–7250. https://doi.org/10.1063/1.460208.

90 Reed, A.E., Weinstock, R.B., and Weinhold, F. (1985). Natural population analysis. *J. Chem. Phys.* 83: 735–746. https://doi.org/10.1063/1.449486.

91 Contreras-García, J., Johnson, E.R., Keinan, S. et al. (2011). NCIPLOT: a program for plotting non-covalent interaction regions. *J. Chem. Theory Comput.* 7: 625–632. https://doi.org/10.1021/ct100641a.

92 Johnson, E.R., Keinan, S., Mori-Sánchez, P. et al. (2010). Revealing noncovalent interactions. *J. Am. Chem. Soc.* 132: 6498–6506. https://doi.org/10.1021/ja100936w.

93 Lu, T. and Chen, F. (2012). Multiwfn: a multifunctional wavefunction analyzer. *J. Comput. Chem.* 33: 580–592. https://doi.org/10.1002/jcc.22885.

94 Wesolowski, S.S., Fermann, J.T., Crawford, T.D., and Schaefer, H.F. III, (1997). The weakly bound dinitrogen tetroxide molecule: high level single reference wavefunctions are good enough. *J. Chem. Phys.* 106: 7178–7184. https://doi.org/10.1063/1.473679.

17

Energy Decomposition Analysis and Its Applications

Peifeng Su

Xiamen University, The State Key Laboratory of Physical Chemistry of Solid Surfaces, Fujian Provincial Key Laboratory of Theoretical and Computational Chemistry, and College of Chemistry and Chemical Engineering, Department of Chemistry, No. 422, Siming South Road, Xiamen, Fujian 361005, China

17.1 Introduction

Intermolecular interactions are responsible for various important processes in physics, chemistry, and life sciences. In the textbook of general chemistry, intermolecular interactions are recognized as the following types: electrostatic, steric, induction, and dispersion. For example, electrostatic interaction can be charge–charge interaction, dipole–dipole interaction, etc., while dispersion is defined as simultaneous dipole–dipole interactions. Considering the quantum mechanical properties of multi-electron systems, these concepts can be used to understand but not to quantify the intermolecular interactions between chemically meaningful fragments ranging from covalent bonds to weak noncovalent interactions.

As shown in Figure 17.1, energy decomposition analysis (EDA) is a tool that translates the classical concepts related to intermolecular interactions into the language of quantum mechanics [1–9]. By using anti-symmetrical wavefunction, the EDA method divides the total interaction energy into several physically meaningful components, and then quantitatively predicts the properties of interactions that cannot be determined experimentally. From the table in Figure 17.1, electrostatic, polarization, and dispersion are included in the EDA results. The three EDA terms are defined in forms that have an analogous physical meaning to the classical concepts. The additional term is ΔE^{exrep}, or called Pauli repulsion, denoting the quantum mechanical properties of the electrons. This table shows that EDA can explore the nature of intermolecular interactions and chemical bonds. In detail, in the C—C covalent bond in CH_3CH_3, the polarization energy term is the largest, showing the role of orbital relaxation in the formation of the single covalent bond. For the ionic bond (Na^+Cl^-), however, the electrostatic term dominates the cation–anion bonding. For the typical hydrogen bond in the water dimer, the electrostatic interaction is the most important, and the polarization term plays a secondary role. As can be seen from the last row, the π...π interactions in the

Exploring Chemical Concepts Through Theory and Computation, First Edition. Edited by Shubin Liu.
© 2024 WILEY-VCH GmbH. Published 2024 by WILEY-VCH GmbH.

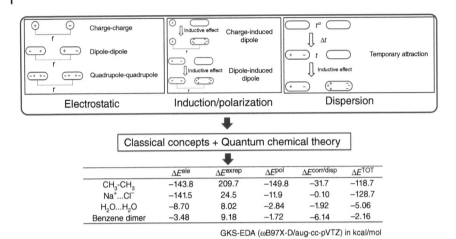

Figure 17.1 Energy decomposition analysis methods and their applications to selected examples.

benzene dimer are dominated by the correlation/dispersion term, showing the vdW character.

Based on quantum mechanical calculations, EDA methods have been widely employed in intermolecular interactions and chemical bonds [1–9]. EDA methods are proposed based on single determinant molecular orbital (MO) methods, Kohn–Sham density functional theory (KS-DFT), or multi-reference wavefunction methods.

17.1.1 Single-Determinant MO-Based EDA

In the early stage, EDA was proposed based on the framework of single-determinant MO methods. Popular MO-based EDA methods include the Kitaura–Morokuma (KM)–EDA [10], the restricted variational space (RVS)–EDA [11, 12], constrained space orbital variations (CSOVs)–EDA [13, 14], block-localized wavefunction (BLW)–EDA [15, 16], absolutely localized molecular orbital (ALMO)–EDA [17, 18], localized molecular orbital (LMO)–EDA method [19], the symmetry adapted perturbation theory (SAPT), etc.

Among them, KM-EDA is the first EDA method, which employs restricted Hartree–Fock (RHF) wavefunction to explore the close-shell noncovalent interactions [10]. It decomposes total interaction energy into electrostatic, polarization, charge transfer (CT), exchange, and MIX. The first four components are computed by the diagonalization of additional Fock matrix in which the appropriate matrix elements are set equal to zero, while the MIX term is computed as a difference between the total interaction energy and the sum of the first four contributions.

LMO-EDA provides analyses for open or closed shell-interacting systems in the gas phase with restricted, restricted-open, or unrestricted types of HF orbitals [19]. It divides the total interaction energy into electrostatic, exchange-repulsion, polarization, and dispersion terms.

SAPT method is the widely used perturbation EDA method, starting from HF orbitals of monomers [2, 5, 6]. This method employs perturbation theory with double perturbation operators, i.e. intra-monomer correlation operators and intermolecular interaction operators, to explore the nature of noncovalent interactions. The EDA terms in SAPT include electrostatic, exchange, induction, and dispersion. The advantage of SAPT is its ability to rigorously quantify dispersion energy.

Single-determinant MO-based EDA methods usually employ post-HF methods, such as MP2, and CCSD/CCSD(T), to consider the dynamic correlation. However, it should be noted that CISD is not suitable for EDA because it is not a size-consistent method. To ensure computational accuracy, post-HF calculations require large basis sets, for example, aug-cc-pVTZ, leading to high computational demands.

17.1.2 DFT-Based EDA

Currently, DFT-based EDA methods are widely used because of the fine balance between computational efficiency and accuracy. In particular, ETS–EDA is a widely used variational EDA approach proposed by Ziegler and Rauk [20]. It decomposes total interaction energy into electrostatics, Pauli, and orbital terms. In conjunction with natural orbital chemical valence (NOCV) [21], ETS–NOCV has been utilized to investigate the energy and charge of various molecular interactions, particularly chemical bonds [8, 22–25].

GKS–EDA is a variational EDA method based on KS-DFT calculations [26]. Since 2014, this method has been extensively applied in various strong chemical bonds and noncovalent interactions with restricted restricted open-shell, and unrestricted KS orbitals. GKS–EDA can also be efficiently extended to analyze intermolecular interactions in radical–radical interactions and intermolecular interactions with excited states, which are introduced in this chapter.

DFT–SAPT (or SAPT(DFT)) is the variant of SAPT based on KS–DFT [27, 28]. In DFT–SAPT, the MOs in the HF method are replaced by KS orbitals to obtain individual interaction terms. The intramolecular electron correlation effect is accounted for by the KS–DFT functional, while the intermolecular electron correlation effect is treated with perturbation techniques. By employing the density fitting technique, DFT–SAPT is affordable for systems containing more than 100 atoms.

In 2007, Liu proposed the DFT steric energy to divide the total DFT energy into the independent contributions of steric, electrostatic, and quantum effects [29]. Thereafter, by using the steric concept, Liu and the coworkers developed the EDA method, called DFTs–EDA, to explore the physical origin of various chemical systems [30]. With the comparisons between the results of CSOV–EDA and DFTs–EDA, they revealed the linear relationship between the polarization and CT from CSOV–EDA and the steric and Pauli energies from DFTs–EDA.

Recently, the analysis results of three EDA methods, GKS–EDA, ETS–EDA, and DFT–SAPT, were extensively assessed for various intermolecular interactions [31]. The EDA terms in the three methods were grouped into four categories: electrostatics, exchange-repulsion/Pauli/exchange, polarization/orbital/induction, and correlation/dispersion terms. A total of 1092 noncovalent interaction complexes in a series of standard sets were used to test. It was concluded that despite the different basis sets and different running platforms (or programs), the results of the three EDA methods are comparable. In general, except for the dispersion term, all the EDA terms in the three methods are in excellent agreement. The correlation/dispersion term in GKS–EDA is comparable with the dispersion term in DFT–SAPT.

Besides KS-DFT-based EDA methods, several EDA schemes have been introduced within the framework of semi-empirical DFT and tight-binding-based density functional theory (DFTB) [32, 33]. Lately, a general tight-binding-based EDA scheme for intermolecular interactions has been proposed [34, 35]. In DFTB–EDA, the total interaction energy is divided into frozen, polarization, and dispersion terms. DFTB–EDA is capable of performing interaction analysis with all self-consistent charge-type DFTB methods, including SCC–DFTB2/3 and GFN1/2–xTB, despite their different formulas and parametrization schemes.

17.1.3 Multireference Wavefunction-Based EDA

In principle, the single-determinant MO or KS-DFT EDA methods mentioned above are hard to handle intermolecular interactions with the character of static correlation. Meanwhile, multi-reference (or configuration) wavefunction-based EDA methods can be expected to overcome this difficulty. In the 1980s, Bernardi and Robb proposed a multi-configuration self-consistent-field (MCSCF)-based EDA scheme [36], which divides total interaction energy into the sum of electrostatic and exchange repulsion, valence CT, the sum of orbital polarization and CT, and configuration interaction terms. It was applied to the rotational barrier in ethane, the dimerization of methylene, the reaction of methylene and silylene, etc. Sequentially, several analogous EDA methods have been proposed [37–41].

Lately, SAPT(MC) method has been extended to consider molecular complexes with localized excitons. SAPT(MC) describes electrostatic, exchange, induction, and dispersion interaction energy components in a rigorous manner. Currently, SAPT(MC) is limited to second-order terms in the intermolecular interaction operator [42].

Compared to multi-reference MO methods, one of the advantages of classical valence bond (VB) theory is its intuitive wavefunction, which is expressed as a linear combination of chemically meaningful VB structures with localized VB orbitals [43–45]. Recently, several VB theory-based analysis schemes have been proposed for CH…HC interactions, halogen bonds, blue-shift hydrogen bonds, and so on [40, 46, 47]. In these studies, the origin of noncovalent intermolecular interactions can be revealed with the compact VB wavefunction.

Lately, a VB-based EDA method, called VB–EDA, has been presented for a unified description of strong covalent bonds, charge-shift bonds, and cation–π interactions. By VB–EDA, total interaction energy can be decomposed into frozen, quasi-resonance, reference state switch, and polarization terms. Different from MO and DFT-based EDA methods, with reference state switch and quasi-resonance terms, VB-EDA is able to explore the role of reference state and covalent-ionic mixing effects, which is helpful for providing insights into the electron-pair bonds [48].

Suffering from the expensive computational cost, multi-reference wavefunction-based EDA methods are difficult for large or medium-sized systems. For intermolecular interactions in which static correlation is important, an alternative solution is to develop the EDA method with the idea of broken symmetry (BS) and unrestricted density functional theory, which will be introduced in the next section of this chapter.

17.1.4 Definitions of EDA Terms

Generally speaking, there are four kinds of popular EDA terms in popular EDA methods.

1. Electrostatic term, which accounts for Coulomb interactions with monomer's wavefunctions.
2. Pauli/exchange-repulsion, which arises from the antisymmetrized and renormalized form of monomer's wavefunctions.
3. Orbital/Polarization/Induction/Charge transfer, which denotes the orbital interactions between monomers.
4. Dispersion/Correlation, which corresponds to the electronic correlations.

Most EDA methods contain the first three types of terms/components, while SAPT, LMO–EDA, and GKS–EDA explicitly include the correlation/dispersion term. The frozen term in several EDA methods, such as BLW–EDA and ALMO–EDA, is a combination of electrostatic and Pauli repulsion contributions.

In this chapter, we introduce the fundamentals of the GKS–EDA family and its recent applications by our research group. It is worthwhile to point out that we do not intend to exhaustively summarize all the GKS–EDA applications. The recent review of GKS–EDA [9], which focuses on the methods and applications before 2020, is recommended to readers who may have interests in metal–ligand interactions, noncovalent interactions in various environments, intramolecular interactions, etc.

17.2 Methodology

17.2.1 GKS–EDA

In the generalized Kohn–Sham theory [49], the KS-DFT Fock operator is defined as the summation of noninteracting electron kinetics T, Coulomb repulsion v_J, exact exchange potential, and the GKS correlation potential with the Hartree–Fock–Kohn–Sham formulism:

$$\left(h + v_J + v_X^{HF} + v_C^{GKS}\right)\psi_i = \varepsilon_i \psi_i \tag{17.1}$$

Here $h = T + v_{ne}$, where v_{ne} is the electron-nuclear attraction potential, and v_X^{HF} is the exact HF exchange potential calculated with KS orbitals. Ψ_i and ε_i denote the occupied KS orbitals and corresponding energies. The correlation potential and energy can be expressed as:

$$v_C^{GKS} = \frac{\delta E_C^{GKS}}{\delta \rho} \tag{17.2}$$

$$E_C^{GKS}(\rho) = E_{XC}(\rho) - E_X^{HF} \tag{17.3}$$

In the GKS–EDA method, the total interaction energy is decomposed into energy components as the following [26]:

$$\Delta E^{TOT} = \Delta E^{ele} + \Delta E^{exrep} + \Delta E^{pol} + \Delta E^{corr} + \Delta E^{disp} \tag{17.4}$$

In Equation (17.4), the electrostatics term ΔE^{ele} refers to Coulomb interactions between electrons and nucleus from different monomers, which can be defined as:

$$\Delta E^{ele} = \langle \Psi_0 | f^0 | \Psi_0 \rangle - \sum_M \langle \Psi_M | f^0 | \Psi_M \rangle + \Delta V_{NN} \tag{17.5}$$

$$f^0 = h + \frac{1}{2}(v_J + v_X^{HF}) \tag{17.6}$$

where Ψ_0 is the direct product of monomers' wavefunctions, while Ψ_M is the wavefunction of monomer M (the number of monomers, which can be two or more). ΔV_{NN} is the nuclear repulsion energy among monomers.

Especially for the two-body interaction between monomers A and B (M = A or B), electrostatic interaction can be written as:

$$\Delta E^{ele} = \sum_{a \in A, b \in B} \frac{Z_a Z_b}{R_{ab}} - \sum_{a \in A} \int \frac{\rho_B(\mathbf{r}) Z_a}{|\mathbf{R}_a - \mathbf{r}|} d\mathbf{r} - \sum_{b \in B} \int \frac{\rho_A(\mathbf{r}) Z_b}{|\mathbf{R}_b - \mathbf{r}|} d\mathbf{r}$$
$$+ \int \frac{\rho_A(\mathbf{r}_1) \rho_B(\mathbf{r}_2)}{r_{12}} d\mathbf{r}_1 d\mathbf{r}_2 \tag{17.7}$$

Here ρ_A and ρ_B are the electron densities of monomers A and B, respectively, $\mathbf{R}_a$ and $\mathbf{R}_b$ represent the vectors of nuclei a with charge Z_a and nuclei b with charge Z_b located at monomers A and B, respectively, while $\mathbf{r}_{12}$ is the distance between electron 1 at monomer A and electron 2 at monomer B. It is noted that the definition is the same as the corresponding terms in SAPT/DFT–SAPT and ETS–EDA.

The exchange-repulsion term ΔE^{exrep} arises from enforcing the direct product of monomers' wavefunctions, Ψ_0, to obey the antisymmetry and orthonormalization requirement:

$$\Delta E^{exrep} = \langle \Psi_{ASN} | f^0 | \Psi_{ASN} \rangle - \langle \Psi_0 | f^0 | \Psi_0 \rangle \tag{17.8}$$

Here Ψ_{ASN} is the antisymmetrization and normalization of Ψ_0. This term is an energy penalty, which imposes the Pauli exclusion principle, preventing electrons of the same spin from ever being in the same place. It is noted that ΔE^{exrep} is completely determined by the monomers' wavefunctions, while ΔE^{Pauli} in ETS–EDA also includes the contribution from the variation of E_{XC} energy with the antisymmetrization and normalization.

The polarization term ΔE^{pol} is defined as the energy variation with a self-consistent field (SCF) procedure,

$$\Delta E^{pol} = \langle \Psi_S | f^0 | \Psi_S \rangle - \langle \Psi_{ASN} | f^0 | \Psi_{ASN} \rangle \quad (17.9)$$

Here Ψ_S is the supermolecular wavefunction. ΔE^{pol} term in GKS–EDA contains the CT interactions between monomers. Among the definitions for EDA interaction terms, CT is a highly controversial topic. The CT term is very sensitive to the scale of the basis set. With a complete basis set, CT and polarization are indistinguishable [50]. In reality, some EDA methods explicitly include CT, while others, including GKS–EDA, ETS–EDA, and DFT–SAPT, do not.

The correlation term ΔE^{corr} is defined as the difference of GKS correlation energy E_C^{GKS} from monomers to a supermolecule. When a hybrid DFT functional is applied, this term is defined as:

$$\Delta E^{corr} = E_C^{GKS}(\rho_S) - \sum_M E_C^{GKS}(\rho_M)$$

$$= (1-a)\left\{ \left[E_X(\rho_S^\alpha, \rho_S^\beta) - E_X^{HF}(S) \right] - \sum_M \left[E_X(\rho_M^\alpha, \rho_M^\beta) - E_X^{HF}(M) \right] \right\}$$

$$+ \left\{ E_C(\rho_S^\alpha, \rho_S^\beta) - \sum_M E_C(\rho_M^\alpha, \rho_M^\beta) \right\} \quad (17.10)$$

Here a is the hybrid coefficient, showing the portion of HF exact change in the hybrid DFT functional, while $E_X^{HF}(S)$ and $E_X^{HF}(M)$ are the exact exchange energies of supermolecule and monomer M, respectively, which are computed from KS determinants.

Therefore, ΔE^{corr} term is the linear combination of two contributions: the first one is the difference between the DFT E_X energy and the exact exchange energy, while the other one is the contribution from the DFT E_C energy. The former can be regarded as the static correlation in several kinds of literature. For example, Fogueri et al. [51] and Martin et al. [52]. employed the exchange energy difference as the diagnostic for static correlation. Meanwhile, the latter one is considered as the dynamic correlation provided by DFT functional. For example, Lie and Clementi [53] and Wu and co-workers [54] used E_C energy to improve the results of VBSCF and HF, respectively. Given the fact that the functionals of E_X and E_C cannot be separately treated, these two contributions should be grouped together to denote the correlation effect from DFT calculations.

If Grimme's DFT-D dispersion correction [55, 56] is employed, an additional term ΔE^{disp} will be introduced as:

$$\Delta E^{disp} = E_S^{disp} - \sum_M E_M^{disp} \quad (17.11)$$

The introduction of dispersion correction aims to improve the description of DFT functionals for vdW interactions. Because recently developed DFT functionals, especially for hybrid meta-GGA functionals such as M06-2X or range-separated functional ωB97X-D, can describe vdW interactions, correlation terms in these functionals tend to mimic the contribution of dispersion. In practice, ΔE^{corr} and ΔE^{disp} are often combined as the $\Delta E^{corr/disp}$ term, called the correlation/dispersion term, to consider the long-range and short-range dispersion (or correlation) contributions.

By default, GKS–EDA calculations are performed based on the monomers' electron densities that depend on the monomers' point groups. For atomic or radical fragments, GKS–EDA considers the orientations of the unpaired electrons in different monomers by using guess orbitals.

GKS–EDA can be naturally extended to intermolecular interactions in a solvated environment. Combined with the implicit solvation model, the total free energy can be decomposed into [26]:

$$\Delta G^{TOT} = \Delta G^{ele} + \Delta G^{exrep} + \Delta G^{pol} + \Delta G^{desol} + \Delta G^{corr} + \Delta G^{disp} \quad (17.12)$$

Here, the desolvation term ΔG^{desol} accounts for the free energy penalty by the environment due to monomers' interaction. For the calculation of the desolvation term, the monomers' cavities are constructed by the fixed points with variable areas (FIXPVA) scheme proposed by Su and Li [57], which ensures that the total interaction energy and individual EDA terms change smoothly along with the whole potential energy surface. Moreover, for the definition of cavity in GKS–EDA, if the distances of monomers are smaller than a certain criterion of the cavity, they are enclosed in one cavity; if not, they are in their isolated cavities. Thus, GKS–EDA considers the cavity construction error in the implicit solvation model for intermolecular interactions in a solvated environment. In 2018, GKS–EDA was used to explore the physical origin of a series of neutral halogen bonds X'–X…Y (X'–X = BrF, ClF, I_2, Br_2, and Cl_2; Y = pyridine, NH_3, H_2S, HCN, H_2O, and dimethyl ether) in various environments. It was shown that the cavity construction error makes differences in the computations of the binding strengths [58].

17.2.2 GKS–EDA(BS)

GKS–EDA(BS) was developed to treat the interactions in open-shell singlet molecular systems based on the BS unrestricted density functional theory (BS-UDFT) [59]. Because unrestricted Kohn–Sham orbitals are not the eigenfunctions of the spin operator S^2, the use of spin decontamination schemes is required. Various schemes have been presented to address the problem of spin contamination. In the spin projection approximation proposed by Yamaguchi and coworkers [60–62], the ground state (GS) energy can be expressed as:

$$E_{GS} = (1 + c)E_{BS} - cE_{HS}, \quad (17.13)$$

where E_{BS} and E_{HS} are the total energies in the BS singlet state and high spin (HS) state, respectively. c is defined as:

$$c = \frac{\langle \hat{S}^2 \rangle_{BS}}{\langle \hat{S}^2 \rangle_{HS} - \langle \hat{S}^2 \rangle_{BS}}. \quad (17.14)$$

Accordingly, the total interaction energy at the ground state of the interacting system is expressed as:

$$\Delta E_{GS}^{TOT} = (1 + c)\, \Delta E_{BS}^{TOT} - c\Delta E_{HS}^{TOT}, \tag{17.15}$$

where ΔE_{BS}^{TOT} and ΔE_{HS}^{TOT} are the total interaction energies in BS state and HS state, respectively:

$$\Delta E_{BS}^{TOT} = E_{BS} - E_A - E_B; \tag{17.16}$$

$$\Delta E_{HS}^{TOT} = E_{HS} - E_A - E_B. \tag{17.17}$$

The total interaction energies in the BS state and HS state are both spin-contaminated. In GKS–EDA(BS), the spin contamination in the HS state is neglected.

The interaction energy at a high spin state, ΔE_{HS}^{TOT}, can be obtained by a GSK-EDA calculation with parallel spins of monomers, which shows that ΔE_{HS}^{TOT} can be decomposed as:

$$\Delta E_{HS}^{TOT} = \Delta E_{HS}^{ele} + \Delta E_{HS}^{exrep} + \Delta E_{HS}^{pol} + \Delta E_{HS}^{corr} + \Delta E_{HS}^{disp}, \tag{17.18}$$

To compute ΔE_{BS}^{TOT}, the BS form ϕ_0^{BS} is constructed from the direct product of the monomers' wavefunctions. Unpaired electrons with different spins are located at different monomers. For example, for the interacting system in open-shell singlet state with two unpaired electrons, α electron is located at monomer A, while β electron is in monomer B.

With the BS wavefunction, ΔE_{BS}^{TOT} is decomposed as:

$$\Delta E_{BS}^{TOT} = \Delta E_{BS}^{ele} + \Delta E_{BS}^{exrep} + \Delta E_{BS}^{pol} + \Delta E_{BS}^{corr} + \Delta E_{BS}^{disp}. \tag{17.19}$$

Therefore, the total interaction energy in open-shell singlet state is decomposed as [59]:

$$\Delta E_{GS}^{TOT} = \Delta E_{GS}^{ele} + \Delta E_{GS}^{exrep} + \Delta E_{GS}^{pol} + \Delta E_{GS}^{corr} + \Delta E_{GS}^{disp}. \tag{17.20}$$

where:

$$\Delta E_{GS}^{X} = (1 + c)\, \Delta E_{BS}^{X} - c\Delta E_{HS}^{X} \tag{17.21}$$

X = electrostatic, exchange–repulsion, polarization, correlation, or dispersion. If $c = 0$, these definitions will reduce to the regular GKS–EDA formulas.

17.2.3 GKS–EDA(TD)

Excited-state intermolecular interactions are important in photophysical processes and photochemical reactions. Lately, GKS–EDA(TD) was proposed to analyze intermolecular interactions in excited states, in which one monomer is in excited states based on TD–DFT [63]. Suppose that a supermolecule contains several monomers, where one monomer (denoted as monomer A) is in a singly excited state and the other monomers B in their ground state, and the TD–DFT excitation energies of monomer A and supermolecule are ω_A and ω_S, respectively. The KS-wavefunction of monomer(s) B is denoted as Φ_B, while the corresponding wavefunctions of monomer A and supermolecule can be expressed as:

$$\Psi_A^* = \sum_{i,a \in A} c_i^a \Phi_i^a, \tag{17.22}$$

$$\Psi_S^* = \sum_{i,a \in S} d_i^a \Theta_i^a. \tag{17.23}$$

Here, the asterisk in the superscript of wavefunction Ψ represents excited states; index i represents occupied orbitals while a represents unoccupied orbitals; c_i^a and d_i^a are the coefficients of excited configurations for monomer A (Φ_i^a) and supermolecule S (Θ_i^a), respectively. A pseudo-direct-product wavefunction is defined as Ψ_{DP}^*:

$$\Psi_{DP}^* = \sum_{i,a \in A} c_i^a \cdot \Phi_i^a \cdot \Phi_B. \tag{17.24}$$

In GKS–EDA(TD), the total interaction energy in an excited state, ΔE^{TOT*}, is decomposed into these energy components: electrostatic, exchange-repulsion, polarization, correlation, and dispersion:

$$\Delta E^{TOT*} = \Delta E^{ele*} + \Delta E^{exrep*} + \Delta E^{pol*} + \Delta E^{corr*} + \Delta E^{disp}. \tag{17.25}$$

The asterisk denotes the interaction term for the excited state. The definition of dispersion in GKS–EDA(TD) is the same as in GKS–EDA. This means that the GKS–EDA(TD) method does not account for the excited-state dispersion because dispersion correction is not considered for TD-DFT calculations at present. Currently, in GKS–EDA(TD), dispersion is only treated at the ground state density level [63].

The procedure of GKS–EDA(TD) is summarized as:

(1) TD-DFT calculations of the supermolecule and monomer A are performed. According to the excitation energies and the orbitals involved in the transitions, the excited states of the supermolecule and monomer A are identified. The excitation type of monomer A should be the same as the noncovalently bound supermolecule.
(2) With the assigned excited state, GKS–EDA(TD) calculations are carried out with a modified procedure of GKS–EDA.

17.3 Applications of GKS–EDA

GKS–EDA has been widely used in various bonding and nonbonding interactions with all kinds of popular functionals. Lately, a fast and multipurpose EDA program, called XEDA, was developed for GKS–EDA calculations [64]. Here we show the latest applications of GKS–EDA for chemical bonds, noncovalent interactions with various spin states, and excited states. All the calculations were performed with the XEDA program.

17.3.1 Strong Chemical Bonds

The GKS–EDA results of the covalent, dative, and transition metal–ligand bonds are shown in Table 17.1. In general, it can be found that the GKS–EDA results with the hybrid functional PBE0 are analogous to those with pure GGA functional PBE. In strong chemical bonds, different from the results for noncovalent interactions, the role of the polarization term is highlighted.

17.3.1.1 Covalent Bonds in Diatomic Molecules

For diatomic molecules such as N_2, CO, BF, and P_2, the total interaction energies of these chemical bonds are dominated by the polarization terms, while the electrostatic terms are secondary. The percentage contribution of polarization in the total attractive interaction ($\Delta E^{pol}/(\Delta E^{ele} + \Delta E^{pol} + \Delta E^{corr}) \times 100\%$) ranges from 42.2% to 64.7% by PBE0 and from 40.9% to 64.1% by PBE.

By PBE0, the value of the polarization term for N_2 is −602.1 kcal/mol, smaller than the value of −686.3 kcal/mol in CO. Except ΔE^{pol}, all the EDA terms for CO are smaller than the corresponding terms for N_2, and it can be found that ΔE^{pol} is the main contribution to the large binding strength of CO compared to N_2. The large polarization term arises from the existence of lone pair delocalization from O to C.

Compared with N_2 and P_2, the electrostatics term in the P≡P bonding is close to the polarization term. The contribution of the polarization term in the P≡P bonding is smaller than N_2, showing that the orbital interaction in P_2 is weaker than N_2.

Table 17.1 Results of the chemical bonds (covalent, dative, and transition metal–ligand) by GKS–EDA (PBE0/cc-pVTZ, PBE0/cc-pVTZ, in kcal/mol).

	GKS–EDA (PBE0/cc-pVTZ)					GKS–EDA (PBE/cc-pVTZ)				
	ΔE^{ele}	ΔE^{exrep}	ΔE^{pol}	ΔE^{corr}	ΔE^{tot}	ΔE^{ele}	ΔE^{exrep}	ΔE^{pol}	ΔE^{corr}	ΔE^{tot}
N_2	−320.99	812.46	−602.06	−113.67	−224.26	−318.29	817.39	−607.40	−134.55	−242.85
CO	−288.76	805.42	−686.33	−84.90	−254.56	−285.62	807.9	−689.48	−101.23	−268.43
BF	−220.96	524.83	−437.39	−45.27	−178.78	−218.01	521.78	−434.69	−55.09	−186.01
P_2	−182.14	338.15	−189.41	−77.62	−111.01	−181.71	339.85	−188.65	−90.94	−121.45
CH≡CH	−141.96	229.62	−256.55	−94.64	−263.52	−142.19	233.85	−259.32	−107.19	−274.85
CH_2=CH_2	−180.99	269.15	−210.24	−60.29	−182.37	−179.41	272.66	−214.20	−67.75	−188.70
CH_3–CH_3	−134.28	204.13	−153.22	−30.40	−113.76	−133.54	207.24	−156.55	−31.97	−114.81
CH_3–NH_2	−163.23	338.06	−238.28	−34.75	−98.20	−162.96	340.98	−240.78	−37.78	−100.53
BH_3–NH_3	−79.51	118.09	−73.57	−13.34	−48.34	−79.69	121.66	−76.19	−13.70	−47.93
BH_3–PH_3	−59.33	123.01	−85.10	−19.69	−41.11	−58.38	124.31	−85.71	−22.31	−42.08
$(CO)_5$W–CO	−83.21	121.99	−58.51	−20.18	−39.91	−85.61	125.59	−60.46	−20.08	−40.57
$(CO)_5$W–BF	−126.48	162.45	−75.27	−18.98	−58.27	−124.63	161.92	−78.88	−15.82	−57.41
$(CO)_5$W–C_2H_2	−74.78	114.49	−59.17	−14.62	−34.07	−73.84	115.83	−62.97	−10.86	−31.84
$(CO)_5$W–C_2H_4	−71.44	103.03	−48.78	−15.04	−32.22	−70.25	103.83	−52.29	−11.34	−30.06

17.3.1.2 Covalent Bonds (X—Y Bonds) in XH_n-YH_n Molecules (X = C or B; Y = C, N or P, n = 2 or 3)

For the X—Y chemical bonds in XH_n-YH_n molecules, analogous to those in diatomic molecules, the total interaction energies are dominated by the polarization/orbital terms. From C_2H_6, C_2H_4 to C_2H_2, the increase in total interaction energy can be mainly attributed to the enlargement of the polarization term. Compared with CH_3-CH_3, the low binding energy of the C—N bond in CH_3-NH_2 is due to the large exchange-repulsion/Pauli term.

For the B—N and B—P bonds in BH_3-NH_3 and BH_3-PH_3, GKS–EDA indicates that the electrostatics term is the largest contribution in BH_3-NH_3 while the polarization term is the most important in BH_3-PH_3. The BH_3-PH_3 bond is weaker than BH_3-NH_3 because of the smaller electrostatics term.

17.3.1.3 Transition Metal–Ligand Bonds in $(CO)_5W\cdots L$ (L = CO, BF, C_2H_2, and C_2H_4) Complexes

For the metal–ligand bonds in the four $(CO)_5W\cdots L$ complexes, in general, the metal–ligand bonds are dominated by electrostatics terms. Among these metal–ligand bonds, the $(CO)_5W\cdots BF$ interaction energy is the strongest, which can be attributed to the large electrostatics and polarization terms. Different from BF and CO, C_2H_2, and C_2H_4 form the π-type metal–ligand bonds with the $(CO)_5W$ group. According to all the EDA results, the π ligand bonds are weaker than the σ ones (CO and BF) because of the smaller electrostatics terms.

In general, for these strong chemical bonds, compared with the results for the $(CO)_5W\cdots L$ complexes by ETS–EDA [22–24], GKS–EDA, and ETS–EDA can provide consistent results. Different from ETS–EDA, GKS–EDA is capable of explicitly demonstrating the role of correlation terms.

17.3.2 Radical-Pairing Interactions

Mechanically interlocked molecules (MIMs) are composed of multiple components connected by mechanical bonds, which can be described as an entanglement in space that prevents two parts of a molecule from separating. Cyclobis(paraquat-p-phenylene) ($CBPQT^{4+}$) has been widely used in MIMs because of its unique redox properties [65, 66]. It is known that different charge states of CBPQT have different multiplicities. Noncovalent interactions between $CBPQT^{n+}$ (n = 1–3) and radical recognition unit (RU) belong to radical–radical interactions. Recently, the interactions between $CBPQT^{n+}$ (n = 0–4) and a series of recognition units (RUs) were investigated by GKS–EDA(BS) [67]. Here the radical recognition unit $BIPY^{\bullet+}$ was selected to show the radical-pairing interactions in MIMs.

The $BIPY^{\bullet+}\cdots CBPQT^{n+}$ interaction can be radical (host)–radical (RU) interactions or radical (host)–close shell (RU) interactions with the variation of n. For example, if n = 4, there are no π electrons in frontier orbitals of $CBPQT^{4+}$, while for n = 3, there are two unpaired electrons located at $BIPY^{\bullet+}$ and $CBPQT^{3+}$ separately. As such, $BIPY^{\bullet+}\cdots CBPQT^{4+}$ is known as radical–close shell interactions, while $BIPY^{\bullet+}\cdots CBPQT^{3+}$ belongs to radical–radical interactions. It can be suggested

Table 17.2 The GKS–EDA results of BIPY•+···CBPQT^{n+} complexes obtained by ωB97X-D/6-31+G* in MeCN solvent (kcal/mol) [67].

Complexes	ΔG^{ele}	ΔG^{exrep}	ΔG^{pol}	ΔG^{desol}	$\Delta G^{disp/corr}$	ΔG^{TOT}
BIPY•+···CBPQT^{4+}	187.20	31.77	−8.09	−186.23	−43.82	−19.17
BIPY•+···CBPQT•$^{3+}$	135.20	37.84	−11.37	−138.20	−51.08	−27.61
BIPY•+···CBPQT$^{2(•+)}$(T)	83.16	42.48	−11.37	−93.51	−54.53	−33.77
BIPY•+···CBPQT$^{2(•+)}$(S)	83.86	41.82	−10.13	−93.22	−51.00	−28.67
BIPY•+···CBPQT•+	31.16	47.04	−14.18	−42.96	−56.91	−35.84
BIPY•+···CBPQT^{0}	−19.77	44.44	−13.18	3.43	−51.35	−36.42

that the orbital relaxation, which contains the effects of CT and induction, is unimportant in BIPY•+···CBPQT^{n+}.

Table 17.2 collects the GKS–EDA results for the BIPY•+···CBPQT^{n+} interactions. For BIPY•+···CBPQT^{n+}, the total interaction energy ranges from −19.17 to −36.42 kcal/mol. The interaction energy of BIPY•+···CBPQT$^{2(•+)}$, −33.8 kcal/mol, is close to the result of −29.6 kcal/mol in literature [68]. In general, for the BIPY•+···CBPQT^{n+} interaction, the electrostatic and desolvation terms are sensitive to the variation of n value. The role of correlation/dispersion is also important. In agreement with our expectations, the contribution of the polarization term ranges from −8.09 to −14.18 kcal/mol, which is the smallest among all the EDA terms.

The origin of BIPY•+···CBPQT^{n+} is different from BIPY•$^{2+}$···CBPQT^{n+}, which can be donor–acceptor interactions or close shell (host)–radical (RU) interactions depending on the n value. For example, there is donor–acceptor interaction in BIPY•$^{2+}$···CBPQT^{4+}. The great geometrical relaxation in this complex can be attributed to the polarization term due to electron transfer between the host and guest [67].

As such, the interactions between the radical recognition unit and CBPQT^{n+} always have enough binding strength because of the correlation and dispersion terms. Small polarization terms indicate the small electron transfer between the host and guest, which is beneficial for the movement in MIM. This understanding is useful for the design of new recognition sites and multi-stable molecular switches.

17.3.3 Noncovalent Interactions in Excited States

17.3.3.1 Hydrogen Bonds Between Aromatic Heterocycles and H_2O

As shown in Figure 17.2, pyridazine, pyrimidine, and pyrazine form the N···H−O hydrogen bonds with H_2O. In the ground state, the distance between nitrogen and hydrogen atoms forming hydrogen bonds increases sequentially (pyridazine-H_2O: 1.975 Å; pyrimidine-H_2O: 2.001 Å; pyrazine-H_2O: 2.011 Å). The GKS–EDA results show that the intermolecular interactions of the three complexes in the ground state are dominated by electrostatics and polarization, which conform to the characteristics of hydrogen bonds. Large bonding distance corresponds to the small

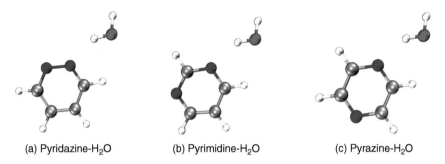

(a) Pyridazine-H_2O (b) Pyrimidine-H_2O (c) Pyrazine-H_2O

Figure 17.2 Pyridazine-H_2O (a), pyrimidine-H_2O (b), pyrazine-H_2O (c).

Table 17.3 GKS–EDA/GKS–EDA(TD) results for pyridazine-H_2O, pyrimidine-H_2O, and pyrazine-H_2O with the $n \rightarrow \pi^*$ transition at CAM-B3LYP/6-311++G(d,p) level (in kcal/mol).

		ΔE^{ele}	ΔE^{exrep}	ΔE^{pol}	ΔE^{corr}	ΔE^{TOT}
Pyridazine-H_2O	S_0	−11.21	11.61	−3.82	−3.23	−7.55
	S_1	−2.94	9.26	−5.52	−4.04	−3.24
Pyrimidine-H_2O	S_0	−10.81	10.75	−3.64	−2.91	−6.61
	S_1	−3.50	8.96	−4.67	−3.67	−2.88
Pyrazine-H_2O	S_0	−10.13	10.24	−3.26	−3.01	−6.16
	S_1	−6.12	9.16	−4.48	−2.91	−4.36

total interaction energies and the small electrostatic, exchange–repulsion, and polarization terms.

For the excited state, the TD-DFT calculations at the CAM-B3LYP/6-311++G** level show that the S1 states of these complexes belong to the $n \rightarrow \pi^*$ transitions, which are mainly related to the transit orbitals in heterocyclic molecules. GKS–EDA(TD) results in Table 17.3 show that the binding strengths of the hydrogen bonds in the $n \rightarrow \pi^*$ excited states are greatly weakened, which can be contributed by the decrease of the electrostatic term. It is because with the excitation, the lone pair of electrons on the nitrogen atoms get involved in the formation of hydrogen bond transitions to π^* orbitals, resulting in a significant weakening of electrostatic interaction, as can be seen from Figure 17.3. It is noted that different from the electrostatic and exchange–repulsion terms, the polarization and correlation terms in the excited state are slightly enhanced compared to those in the ground state.

17.3.3.2 Base Pair Interactions with Excited States

Figure 17.4 shows the Waston–Crick type AT and GC base pairs (denoted as AT(WC) and GC(WC) in the following). In their ground states (S_0), the GKS–EDA results show that the intermolecular interactions in AT(WC) and GC(WC) are dominated by electrostatic and polarization terms.

The first (S_1) and the second excited state (S_2) of AT(WC) originate from the $n \rightarrow \pi^*$ excitation and $\pi \rightarrow \pi^*$ excitation, respectively. The transition orbitals are localized in

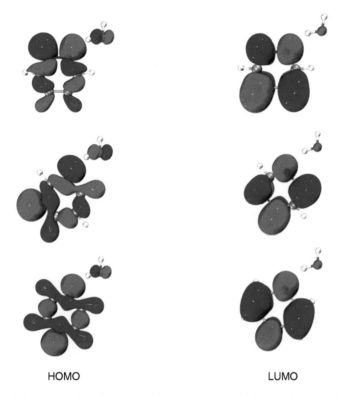

            HOMO   LUMO

Figure 17.3 HOMOs and LUMOs of pyridazine-H_2O, pyrimidine-H_2O, and pyrazine-H_2O.

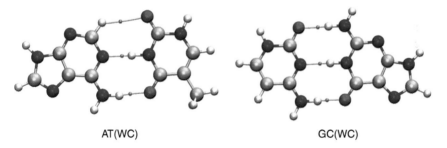

            AT(WC)   GC(WC)

Figure 17.4 Waston–Crick AT and GC base pairs.

T. As shown in Figure 17.5, the GKS–EDA(TD) results show that with the $n \rightarrow \pi^*$ transition ($S_0 \rightarrow S_1$), the hydrogen bond in AT(WC) is weakened, which can be interpreted as the dramatic decrease of the polarization term. This is due to the fact that the orbitals involved in the $n \rightarrow \pi^*$ excitation are related to the hydrogen bonding interaction between A and T. Therefore, in the S_1 state, the origin of the interaction in AT(WC) is governed by dispersion and correlation term, different from that in the ground state.

For the $\pi \rightarrow \pi^*$ excitation ($S_0 \rightarrow S_2$) of AT(WC), the GKS–EDA results reveal that the nature of intermolecular interactions remains almost unchanged because the

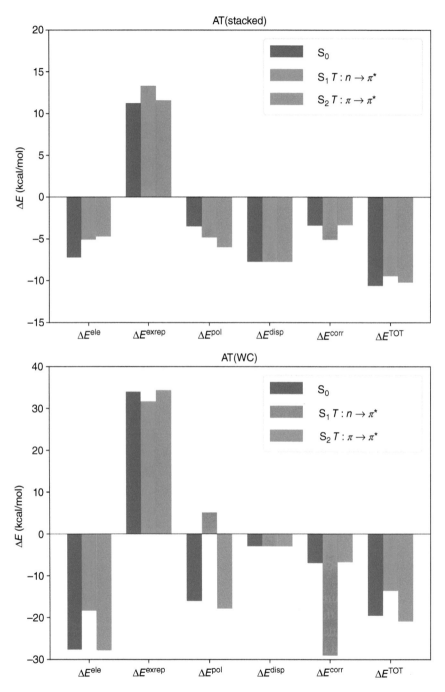

Figure 17.5 GKS–EDA(TD) results for intermolecular interactions in AT and CG with the lowest excited states.

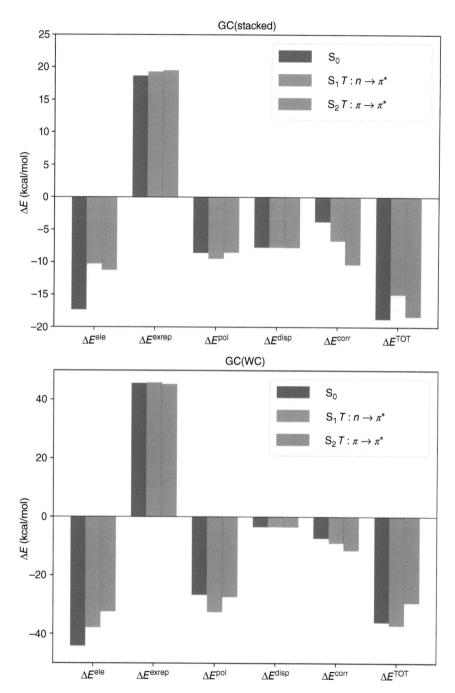

Figure 17.5 (Continued)

transit orbitals are not involved in the hydrogen bonding interaction. For the same reason, for the first and second excited states of GC(WC), the GKS–EDA(TD) results are similar to those in the ground state.

Therefore, the relation between the geometrical configuration and the electron transition is revealed. It has been found that the origin of the intermolecular interaction in the excited state can be quite different from the ground state. In detail, for the Waston–Crick type base pair, the variation of the overall interaction from the ground state to the excited state is sensitive to the $n \to \pi^*$ transition.

17.4 Conclusion

In this chapter, after a brief survey of EDA methods, the fundamentals and applications of the GKS–EDA family are introduced. It is shown that GKS–EDA method can be successfully used for strong chemical bonds and noncovalent interactions with different spin states and electronic states:

1. Combined with fragmentation scheme to study intramolecular interactions.
2. Combined with implicit solvation methods to investigate intermolecular interactions in a solvated environment.
3. Combined with the BS DFT to study intermolecular interactions with open shell singlet states.
4. Combined with TD-DFT to study intermolecular interactions with singly excited states.

The first two strategies have been mentioned in the published review [9], while the last two are discussed in this chapter.

Acknowledgments

This project is supported by the National Natural Science Foundation of China (No. 22173076), Science and Technology Projects of Innovation Laboratory for Sciences and Technologies of Energy Materials of Fujian Province (IKKEM) (No: RD2022070103), the Research Funds of National Program for Top Students Training in Basic Disciplines (No. 20222107), and Fujian Province College Education Research Funds (No. FBJG20220159). The author thanks the student Longxiang Yan for his help in data curation and visualization.

References

1 Stone, A.J. (1996). *The Theory of Intermolecular Forces*. Oxford: Oxford University Press.
2 Jeziorski, B., Moszynski, R., and Szalewicz, K. (1994). Perturbation theory approach to intermolecular potential energy surfaces of van der Waals complexes. *Chem. Rev.* 94 (7): 1887–1930.

3 Bickelhaupt, F.M. and Baerends, E.J. (2000). Kohn-Sham density functional theory: predicting and understanding chemistry. *Rev. Comput. Chem.* 15: 1–86.

4 Hohenstein, E.G. and Sherrill, C.D. (2012). Wavefunction methods for noncovalent interactions. *Wiley Interdiscip. Rev. Comput. Mol. Sci.* 2 (2): 304–326.

5 Szalewicz, K. (2012). Symmetry-adapted perturbation theory of intermolecular forces. *Wiley Interdiscip. Rev. Comput. Mol. Sci.* 2 (2): 254–272.

6 Jansen, G. (2014). Symmetry-adapted perturbation theory based on density functional theory for noncovalent interactions. *Wiley Interdiscip. Rev. Comput. Mol. Sci.* 4 (2): 127–144.

7 Phipps, M.J., Fox, T., Tautermann, C.S., and Skylaris, C.-K. (2015). Energy decomposition analysis approaches and their evaluation on prototypical protein–drug interaction patterns. *Chem. Soc. Rev.* 44 (10): 3177–3211.

8 Zhao, L., von Hopffgarten, M., Andrada, D.M., and Frenking, G. (2018). Energy decomposition analysis. *Wiley Interdiscip. Rev. Comput. Mol. Sci.* 8 (3): e1345.

9 Su, P., Tang, Z., and Wu, W. (2020). Generalized Kohn-Sham energy decomposition analysis and its applications. *Wiley Interdiscip. Rev. Comput. Mol. Sci.* 10 (5): e1460.

10 Kitaura, K. and Morokuma, K. (1976). A new energy decomposition scheme for molecular interactions within the Hartree-Fock approximation. *Int. J. Quantum Chem.* 10: 325–340.

11 Stevens, W.J. and Fink, W.H. (1987). Frozen fragment reduced variational space analysis of hydrogen bonding interactions. Application to the water dimer. *Chem. Phys. Lett.* 139 (1): 15–22.

12 Chen, W. and Gordon, M.S. (1996). Energy decomposition analyses for many-body interaction and applications to water complexes. *J. Phys. Chem.* 100 (34): 14316–14328.

13 Bagus, P.S., Hermann, K., and Bauschlicher, C.W. Jr. (1984). A new analysis of charge transfer and polarization for ligand–metal bonding: model studies of Al_4CO and Al_4NH_3. *J. Chem. Phys.* 80 (9): 4378–4386.

14 Bagus, P.S. and Illas, F. (1992). Decomposition of the chemisorption bond by constrained variations: order of the variations and construction of the variational spaces. *J. Chem. Phys.* 96 (12): 8962–8970.

15 Mo, Y., Gao, J., and Peyerimhoff, S.D. (2000). Energy decomposition analysis of intermolecular interactions using a block-localized wave function approach. *J. Chem. Phys.* 112 (13): 5530–5538.

16 Mo, Y., Bao, P., and Gao, J. (2011). Energy decomposition analysis based on a block-localized wavefunction and multistate density functional theory. *Phys. Chem. Chem. Phys.* 13 (15): 6760–6775.

17 Khaliullin, R.Z., Cobar, E.A., Lochan, R.C. et al. (2007). Unravelling the origin of intermolecular interactions using absolutely localized molecular orbitals. *J. Phys. Chem. A* 111 (36): 8753–8765.

18 Mao, Y., Horn, P.R., and Head-Gordon, M. (2017). Energy decomposition analysis in an adiabatic picture. *Phys. Chem. Chem. Phys.* 19 (8): 5944–5958.

19 Su, P. and Li, H. (2009). Energy decomposition analysis of covalent bonds and intermolecular interactions. *J. Chem. Phys.* 131 (1): 014102.

20 Ziegler, T. and Rauk, A. (1977). On the calculation of bonding energies by the Hartree Fock Slater method. *Theor. Chim. Acta* 46: 1–10.

21 Mitoraj, M.P., Michalak, A., and Ziegler, T. (2009). A combined charge and energy decomposition scheme for bond analysis. *J. Chem. Theory Comput.* 5 (4): 962–975.

22 Frenking, G., Wichmann, K., Fröhlich, N. et al. (2003). Towards a rigorously defined quantum chemical analysis of the chemical bond in donor–acceptor complexes. *Coord. Chem. Rev.* 238–239: 55–82.

23 Bessac, F. and Frenking, G. (2006). Chemical bonding in phosphane and amine complexes of main group elements and transition metals. *Inorg. Chem.* 45 (17): 6956–6964.

24 Jerabek, P. and Frenking, G. (2014). Comparative bonding analysis of N2 and P2 versus tetrahedral N4 and P4. *Theor. Chem. Acc.* 133 (3): 1447.

25 Zhao, L., Hermann, M., Schwarz, W.H.E., and Frenking, G. (2019). The Lewis electron-pair bonding model: modern energy decomposition analysis. *Nat. Chem. Rev.* 3 (1): 48–63.

26 Su, P., Jiang, Z., Chen, Z., and Wu, W. (2014). Energy decomposition scheme based on the generalized Kohn–Sham scheme. *J. Phys. Chem. A* 118 (13): 2531–2542.

27 Heßelmann, A., Jansen, G., and Schütz, M. (2004). Density-functional theory-symmetry-adapted intermolecular perturbation theory with density fitting: a new efficient method to study intermolecular interaction energies. *J. Chem. Phys.* 122 (1): 014103.

28 Misquitta, A.J. and Szalewicz, K. (2005). Symmetry-adapted perturbation-theory calculations of intermolecular forces employing density-functional description of monomers. *J. Chem. Phys.* 122 (21): 214109.

29 Liu, S. (2007). Steric effect: a quantitative description from density functional theory. *J. Chem. Phys.* (24): 126.

30 Fang, D., Piquemal, J.-P., Liu, S., and Cisneros, G.A. (2014). DFT-steric-based energy decomposition analysis of intermolecular interactions. *Theor. Chem. Acc.* 133: 1–14.

31 Xu, Y., Zhang, S., Wu, W., and Su, P. (2023). Assessments of DFT-based energy decomposition analysis methods for intermolecular interactions. *J. Chem. Phys.* 158 (12): 124116.

32 Miriyala, V.M. and Řezáč, J. (2017). Description of non-covalent interactions in SCC-DFTB methods. *J. Comput. Chem.* 38 (10): 688–697.

33 Fedorov, D.G. and Kitaura, K. (2018). Pair interaction energy decomposition analysis for density functional theory and density-functional tight-binding with an evaluation of energy fluctuations in molecular dynamics. *J. Phys. Chem. A* 122 (6): 1781–1795.

34 Xu, Y., Zhang, S., Lindahl, E. et al. (2022). A general tight-binding based energy decomposition analysis scheme for intermolecular interactions in large molecules. *J. Chem. Phys.* 157 (3): 034104.

35 Xu, Y., Friedman, R., Wu, W., and Su, P. (2021). Understanding intermolecular interactions of large systems in ground state and excited state by using density functional based tight binding methods. *J. Chem. Phys.* 154 (19): 194106.

36 Bernardi, F. and Robb, M.A. (1983). A multi-reference approach to energy decomposition for molecular interactions. *Mol. Phys.* 48 (6): 1345–1355.

37 Bernardi, F., Bottoni, A., and Robb, M.A. (1984). A multireference energy decomposition scheme with respect to fragment valence states. *Theoret. Chem. Acta* 64 (4): 259–263.

38 Bauschlicher, C.W. Jr., Bagus, P.S., Nelin, C.J., and Roos, B.O. (1986). The nature of the bonding In XCO for X=Fe, Ni, and Cu. *J. Chem. Phys.* 85 (1): 354–364.

39 Cardozo, T.M. and Nascimento, M.A.C. (2009). Energy partitioning for generalized product functions: the interference contribution to the energy of generalized valence bond and spin coupled wave functions. *J. Chem. Phys.* 130 (10): 104102.

40 Danovich, D., Shaik, S., Neese, F. et al. (2013). Understanding the nature of the CH···HC interactions in alkanes. *J. Chem. Theory Comput.* 9 (4): 1977–1991.

41 Zhang, Y., Chen, S., Ying, F. et al. (2018). Valence bond based energy decomposition analysis scheme and its application to cation–π interactions. *J. Phys. Chem. A* 122 (27): 5886–5894.

42 Jangrouei, M.R., Krzeminska, A., Hapka, M. et al. (2022). Dispersion interactions in exciton-localized states. Theory and applications to ?-?* and n-?* excited states. *J. Chem. Theory Comput.* 18 (6): 3497–3511.

43 Wu, W., Su, P., Shaik, S., and Hiberty, P.C. (2011). Classical valence bond approach by modern methods. *Chem. Rev.* 111 (11): 7557–7593.

44 Hiberty, P. and Shaik, S. (2008). *A Chemist's Guide to Valence Bond Theory*. Hoboken, NJ: Wiley.

45 Su, P. and Wu, W. (2013). Ab initio nonorthogonal valence bond methods. *Wiley Interdiscip. Rev. Comput. Mol. Sci.* 3 (3): 56–68.

46 Wang, C., Danovich, D., Shaik, S., and Mo, Y. (2017). A unified theory for the blue- and red-shifting phenomena in hydrogen and halogen bonds. *J. Chem. Theory Comput.* 13 (4): 1626–1637.

47 Chang, X., Zhang, Y., Weng, X. et al. (2016). Red-shifting versus blue-shifting hydrogen bonds: a perspective from ab initio valence bond theory. *J. Phys. Chem. A* 120 (17): 2749–2756.

48 Zhang, Y., Wu, X., Su, P., and Wu, W. (2022). Exploring the nature of electron-pair bonds: an energy decomposition analysis perspective. *J. Phys. Condens. Matter* 34 (29): 294004.

49 Seidl, A., Görling, A., Vogl, P. et al. (1996). Generalized Kohn-Sham schemes and the band-gap problem. *Phys. Rev. B* 53 (7): 3764–3774.

50 Stone, A.J. and Misquitta, A.J. (2009). Charge-transfer in symmetry-adapted perturbation theory. *Chem. Phys. Lett.* 473 (1): 201–205.

51 Fogueri, U.R., Kozuch, S., Karton, A., and Martin, J.M.L. (2013). A simple DFT-based diagnostic for nondynamical correlation. *Theor. Chem. Acc.* 132 (1): 1291.

52 Martin, J.M.L., Santra, G., and Semidalas, E. (2022). An exchange-based diagnostic for static correlation. *AIP Conf. Proc.* 2611 (1): 020014.

53 Lie, G.C. and Clementi, E. (1974). Study of the electronic structure of molecules. XXI. Correlation energy corrections as a functional of the Hartree-Fock density and its application to the hydrides of the second row atoms. *J. Chem. Phys.* 60 (4): 1275–1287.

54 Ying, F., Su, P., Chen, Z. et al. (2012). DFVB: a density-functional-based valence bond method. *J. Chem. Theory Comput.* 8 (5): 1608–1615.

55 Grimme, S., Antony, J., Ehrlich, S., and Krieg, H. (2010). A consistent and accurate ab initio parametrization of density functional dispersion correction (DFT-D) for the 94 elements H-Pu. *J. Chem. Phys.* 132 (15): 154104.

56 Grimme, S., Ehrlich, S., and Goerigk, L. (2011). Effect of the damping function in dispersion corrected density functional theory. *J. Comput. Chem.* 32 (7): 1456–1465.

57 Su, P. and Li, H. (2009). Continuous and smooth potential energy surface for conductorlike screening solvation model using fixed points with variable areas. *J. Chem. Phys.* 130 (7): 074109–074113.

58 Shen, D., Su, P., and Wu, W. (2018). What kind of neutral halogen bonds can be modulated by solvent effects? *Phys. Chem. Chem. Phys.* 20 (41): 26126–26139.

59 Tang, Z., Jiang, Z., Chen, H. et al. (2019). Energy decomposition analysis based on broken symmetry unrestricted density functional theory. *J. Chem. Phys.* 151 (24): 244106.

60 Yamaguchi, K., Takahara, Y., and Fueno, T. (1986). Ab-initio molecular orbital studies of structure and reactivity of transition metal-OXO compounds. In: *Applied Quantum Chemistry* (ed. V.H. Smith Jr., H.F. Schaefer III, and K. Morokuma), 155–184. Dordrecht, Holland: D. Reidel Publishing Company.

61 Yamaguchi, K., Okumura, M., Takada, K., and Yamanaka, S. (1993). Instability in chemical bonds. II. Theoretical studies of exchange-coupled open-shell systems. *Int. J. Quantum Chem.* 48 (S27): 501–515.

62 Yamanaka, S., Kawakami, T., Nagao, H., and Yamaguchi, K. (1994). Effective exchange integrals for open-shell species by density functional methods. *Chem. Phys. Lett.* 231 (1): 25–33.

63 Tang, Z., Shao, B., Wu, W., and Su, P. (2023). Energy decomposition analysis methods for intermolecular interactions with excited states. *Phys. Chem. Chem. Phys.* 25 (27): 18139–18148.

64 Tang, Z., Song, Y., Zhang, S. et al. (2021). XEDA, a fast and multipurpose energy decomposition analysis program. *J. Comput. Chem.* 42 (32): 2341–2351.

65 Stoddart, J.F. (2017). Mechanically interlocked molecules (MIMs)—molecular shuttles, switches, and machines (Nobel lecture). *Angew. Chem. Int. Ed. Engl.* 56 (37): 11094–11125.

66 Cai, K., Zhang, L., Astumian, R.D., and Stoddart, J.F. (2021). Radical-pairing-induced molecular assembly and motion. *Nat. Rev. Chem.* 5 (7): 447–465.

67 Wang, W., Wu, W., and Su, P. (2023). Radical pairing interactions and donor–acceptor interactions in cyclobis(paraquat-p-phenylene) inclusion complexes. *Molecules* 28 (5): 2057.

68 Trabolsi, A., Khashab, N., Fahrenbach, A.C. et al. (2010). Radically enhanced molecular recognition. *Nat. Chem.* 2 (1): 42–49.

18

Chemical Concepts in Solids

Peter C. Müller[1], David Schnieders[1], and Richard Dronskowski[1,2,3]

[1] RWTH Aachen University, Institute of Inorganic Chemistry, Landoltweg 1a, 52056 Aachen, Germany
[2] RWTH Aachen University, Jülich-Aachen Research Alliance (JARA-CSD), 52056 Aachen, Germany
[3] Shenzhen Polytechnic University, Hoffmann Institute of Advanced Materials, 7098 Liuxian Blvd, Nanshan District, 518055 Shenzhen, China

18.1 The Three Schisms of Solid-State Chemistry

Even in the twenty-first century, a newcomer to the field of solid-state chemistry or the quantum chemistry of the solid state will sooner or later witness several astonishing schisms in how theory and experiment are complementing each other – or sometimes not so – in providing essential insight or help us understand what is going and what might be the next scientific target.

Most solid-state chemists have been heavily influenced by the ionic notion covered in traditional textbooks or lectures simply because archetype solid-state materials such as NaCl and MgO are prone to be oversimplified as purely ionic species. Hence, the notion of ionicity massively prevails in "explaining" the stability of periodic solids, simply because combinations of Na^+ and Cl^- or Mg^{2+} and O^{2-} allude to the validity of the octet rule, and then one simply sums up their electrostatic Madelung energies by the Coulomb law as if we were still living in pre-1926 times when wave mechanics had not been introduced [1]. It is, of course, attractive to do so since such rather naive models work nicely here and there. Also, when virtually *everything* is explained electrostatically, there is a lot of error cancellation, and the method may even turn out quantitatively "correct." Let's not forget, however, that Hellmann already showed that systems only composed of potential energy are inherently unstable [2].

An ionicitistic view of the solid-state world has other drawbacks, for example, the bizarre idea that most solid-state materials do not contain covalent bonds such as in molecules, so interacting with quantum chemists specialized in exactly that is difficult, if not impossible. This communication problem has been severe, even counterproductive [3], and we trust that it has now been overcome in the early twenty-first century. Likewise, metallicity, the somewhat mysterious special case of covalency with a plethora of atoms and an undersupply of electrons [4], has been usually left to the metallurgists or solid-state physicists, and they, while they were and still are

Exploring Chemical Concepts Through Theory and Computation, First Edition. Edited by Shubin Liu.
© 2024 WILEY-VCH GmbH. Published 2024 by WILEY-VCH GmbH.

interested in many exciting solid-state topics, will probably ignore chemical bonding, at least it is less important to them. So, there is a schism as regards how we look at solids or group them in the first place.

Compared with any sizeable molecule, a crystal is infinitely larger. Hence, an intelligent quantum chemist trained in molecules could, of course, handle a covalently bonded periodic solid such as carbon in the diamond structure where every C atom bonds covalently to four other C atoms, just like in many hydrocarbons. Then, however, the theorist would need to include Bloch's theorem [5] because the periodic species is (almost) infinite in size, not finite like in typical small molecules. This immediately translates into the necessity of a spatially infinite basis set, totally delocalized, no longer localized like for a molecule, although moving between a fully delocalized and a fully localized representation is possible, in principle and also in practice, but only recently (see Section 18.4). So, second, there is yet another schism, this time as regards the sheer size of the matter.

And there is yet another incompatibility with respect to molecular quantum chemistry and the solid state, in particular, because the majority of the elements comes in the form of metals, and this does not go well with the quantum-chemical workhorse, Hartree–Fock theory. Despite the usual Hartree–Fock adolation in quantum-chemistry textbooks, already in the very beginning of quantum mechanics it was found that an exact solution of the Hartree–Fock equations for simple metals (the so-called Sommerfeld model) arrives at a ground state with an electronic density-of-states of zero at the Fermi level, the worst thinkable result [6]. Clearly, electronic exchange (important for molecules) is less critical than electronic correlation (important for solids) in the solid state, so there is a technical barrier to be broken, eventually solved by the ingenious invention of density-functional theory, which came out of the solid [7, 8]. Even here, the solid-state chemist chooses LDA or GGA-type functionals whenever possible, with a Hubbard U correction if needed. The hybrid functionals that are so popular in the domain of molecular quantum chemistry are less frequently used in the periodic solid. So, there is a technical schism in terms of electron–electron interactions; at least, there was one in the past.

Because of the oversimplifying notion to cover the solid state electrostatically, that is, the tendency to isolate the field from the molecular field, theoretical difficulties in using a fitting basis set which was entirely different from the molecular approach, and also the existence of the metallic state which poses problems to traditional, wave-function-based approaches in *ab initio* theory, it is probably fair to say that the coverage of chemical bonding in solids is less developed than in molecules, all because of unfitting technical or computational tools, lack of communication between chemistry and physics, and possibly insufficient interest [6].

That being the case, at present, a significant number of quantum-chemical analyses of solids are carried out using density-based schemes despite the fact that the most crucial component for such analyses – the orbital *phases* – are not available. Likewise, a density-of-states (DOS) diagram contains zero bonding information, as oddly as that may sound. DOS leads nowhere as regards chemical bonding, unfortunately. But times are changing, chemical-bonding analyses of solids do not have to be inferior to their molecular cousins, and this book, as well as this chapter, is vivid proof. And we are going to cover that in what follows.

18.2 Bloch's Theorem

In order to describe the infinitely large solid-state structures alluded to in Section 18.1, all the solid-state theorists must use Bloch's theorem [5], the most important law of electronic-structure theory for periodic solids, which only rests on translational invariance, nothing more. It simply defines the translationally invariant wave function $\Psi_\mathbf{k}$ of any periodic (crystalline) system as

$$\Psi_\mathbf{k}(\mathbf{r} + \mathbf{T}) = \Psi_\mathbf{k}(\mathbf{r})e^{i\mathbf{k}\mathbf{T}} \tag{18.1}$$

and tells us that the solution to the Schrödinger (or Kohn–Sham) equation(s) within one particular unit cell may be dubbed $\Psi_\mathbf{k}(\mathbf{r})$ and, therefore, depends on a new quantum number $\mathbf{k}$, which stems from reciprocal space. And if we want to know how a such-defined wave function looks like after some translational vector $\mathbf{T}$ has been applied to it, one only needs to multiply the original wave function $\Psi_\mathbf{k}(\mathbf{r})$ with a periodic modulation which reads $e^{i\mathbf{k}\mathbf{T}}$, the phase factor which is unknown in molecular quantum chemistry (but somehow corresponds to the mixing coefficients in LCAO theory, as a theoretical chemist would immediately guess). Because of the phase factor, Felix Bloch did not only make band-structure calculations possible, but he also found why the valence electrons do not "bounce" into the atoms – instead, they "ride" on the $\mathbf{k}$-dependent symmetry-adapted wave function similar to the "riding" electrons in the π molecular orbitals of benzene. Nonetheless, this phase information is lost by square-integration toward the densities, and not without consequences: phase and nodal information of the wave function, crucial to chemical bonding analysis, is lost, and we will discuss practical implications in Section 18.5.

To give an instructive picture of the Bloch theorem, let us construct the simplest one-dimensional case, that is, an infinite crystal composed of only H atoms. Here, the atomic wave function inside the one-atom unit cell is just the 1s atomic orbital of H, and all vectors k have become scalars. At the zone center Γ of the Brillouin zone (which is the unit cell of reciprocal space), the k quantum number equals zero such that the exponential factor is unity for whatever translation T. Hence, the wave function is still 1s at T, at $3T$, at $-5T$, and so forth. If the translation T equals the lattice parameter, or, likewise, the H–H distance dubbed a, the crystal wave function, regardless of whatever normalization, simply reads

$$...1s(-3a) + 1s(-2a) + 1s(-a) + 1s(0) + 1s(a) + 1s(2a) + 1s(3a)...$$

but for the zone edge $\mathbf{X}$, the quantum number k equals π/a, so the exponential modulation e^{ikT} alternates between $e^{i\frac{\pi}{a}a} = e^{\pi i} = -1$ or $e^{i\frac{\pi}{a}2a} = e^{2\pi i} = 1$ for odd or even translations, so the crystal wave function then reads

$$\cdots - 1s(-3a) + 1s(-2a) - 1s(-a) + 1s(0) - 1s(a) + 1s(2a) - 1s(3a) \cdots$$

Clearly, at the zone center Γ, we have found the bonding combination of all 1s orbitals, the extended equivalent of the bonding σ_g molecular orbital of H_2, and at the zone edge **X**, there is the antibonding combination of all 1s orbitals, the periodic equivalent of the antibonding σ_u^* molecular orbital. In between those two high symmetry points, there exist infinitely many more points in reciprocal space, each defining different linear combinations of the infinitely numbered interacting 1s orbitals. If we were to weaken the interactions between the 1s orbitals by increasing H–H distances, the energetic difference between bonding and antibonding levels would decrease as well, so the band width would get smaller, or less disperse, in terms of solid-state language. If we follow this *gedankenexperiment* to the extreme, we would end up with an infinite number of noninteracting hydrogen atoms such that the band structure would collapse to a single discrete energy level representing the 1s orbital of the hydrogen atom. That being said, the periodic modulation of the crystal wave function resembles the LCAO mixing coefficient from molecular orbital theory, as alluded to before. It is almost trivial to plot the band structure of the one-dimensional hydrogen chain next to the density-of-states (see Figure 18.1b). At Γ, the states are close to each other, so the band has a flat incline and the DOS has a maximum value. With increasing energy, the levels widen so the band gets steeper and the DOS decreases. At the **X** point, the levels come closer again, the band flattens out and the DOS increases. It is rather instructive to compare the band structure with the electronic structures of rings of H atoms differing in the number of atoms, see Figure 18.1a. As the molecule grows from H_2 to H_3 to H_4 to H_6 to H_{20}, the MO levels gradually approach the band of the infinite H chain.

The above example, a pictorial introduction into band-structure theory, goes back to Hoffmann [3], and its ingenuity is given by the ability to link molecular and

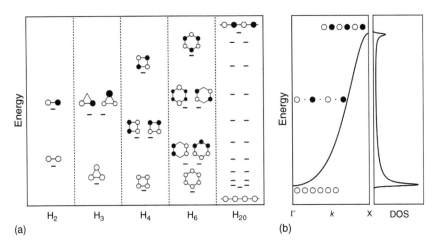

Figure 18.1 (a) Transition from the linear dihydrogen molecule over cyclic H_3, H_4, H_6, and H_{20} to an infinitely long hydrogen chain. For every hydrogen atom added, there is an additional molecular orbital, ultimately leading to an energy band that no longer has discrete levels. (b) Band structure and density of states (DOS) of a linear chain of H-atoms. In all cases, the distance between two hydrogens is set to 1 Å.

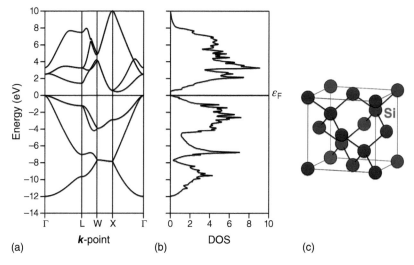

Figure 18.2 (a) Band structure, (b) density of states (DOS) and (c) unit cell of silicon in the diamond structure. Source: Adapted from [9] De Gruyter.

solid-state language, almost without any activation barrier. Nonetheless, we need to mess up things because, for real, three-dimensional materials, the situation gets a little more complicated. For example, the three-dimensional silicon crystal exhibits a far more complex band structure, as shown in Figure 18.2. Because this band structure does not allow for an easy interpretation as the one given in Figure 18.1, further interpretational tools need to be applied, and we will describe them in great detail in Section 18.4. Given a little training in solid-state theory, however, the reader will realize that we took a different path in the likewise three-dimensional Brillouin zone (the unit cell of reciprocal space, as said before), and we walked from the zone center Γ toward **L**, then toward **W**, then toward **X**, and back to Γ. Clearly, Si is an indirect semiconductor because the lowest part of the conduction band occurs at a different **k** point than the highest part of the valence band. The band gap (see DOS in Figure 18.2b) is about 0.5 eV, slightly too small compared to experiment (about 1.1 eV), a "failure" of the exchange–correlation functional, which was not developed for that purpose.

Not only the spatial dimensions, but also the heavier atoms introduce huge complexity into the electronic structure. Let us now imagine a one-dimensional Na crystal and its valence orbital 3s filled with one electron. How does the extended wave function look like? This is shown in Figure 18.3, in which the crystal orbital or one-electron band $\Psi(\mathbf{k}, \mathbf{r})$ has been schematically sketched as shown for the zone edge **X**.

Quite obviously, the band has a dual character because it looks – if we forget about the wiggles close to the atoms – like a simple sine or cosine function oscillating from atom to atom since we are at the zone edge. That being said, the band is very much plane-wave-like, as also expected from Bloch's theorem, whose periodic modulation by translation is nothing but a plane wave. Close to the atom, the band gets more

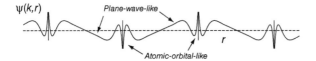

Figure 18.3 3s-like wave function of a one-dimensional crystal made up of Na atoms at the zone edge **X**. Source: Dronskowski [6]/Adapted from John Wiley and Sons.

nodal as it must become more atomic-orbital-like; note that the 3s is an orthogonal function to the lower-lying 2p, 2s, and 1s functions, hence orthonormality enforces nodality. Alternatively expressed, the wiggles reflect the high kinetic energy of the 3s electron. Understanding the duality of the periodic wave function or band (nodal close to the nucleus and sine-like in-between), that is to say, its general behavior, is a prerequisite to realize how periodic electronic-structure calculations, often going under the name band-structure calculations, must proceed, at least in theory.

18.3 Basis Sets

A moment reflection of Figure 18.3 yields that there are at least three fundamentally different ways to approximate the optimum periodic wave function, or, alternatively expressed, the basis-set choice if we allude to the language of molecular quantum chemistry. What can be learned for the one-dimensional case also translates into two and three dimensions without any additional difficulty.

For the molecular quantum chemist, it seems quite natural and straightforward to totally forget about the plane-wave character of the periodic wave function (one-electron band) and to approximate it like he or she is doing it for molecules, namely by a linear superposition of atomic-like wave functions or orbitals, but still considering Bloch's theorem. So, instead of combining atomic orbitals ϕ_μ to yield molecular orbitals (LCAO-MO), one may express the crystal orbitals $\psi_{j\mathbf{k}}$ according to a similar principle, that is [6]

$$\psi_{j\mathbf{k}}(\mathbf{r}) = \sum_{\mu \mathbf{T}} c_{\mu j \mathbf{k}}^{\text{LCAO}} e^{i\mathbf{k}\cdot\mathbf{T}} \phi_\mu(\mathbf{r} - \mathbf{T}) \tag{18.2}$$

Here, the ϕ_μ run over all atoms in the unit cell and yield $\psi_{j\mathbf{k}}$ given the right LCAO coefficients and a periodic modulation $e^{i\mathbf{k}\cdot\mathbf{T}}$. Exactly the same approach has been carried out in solid-state physics for a long time already, and here it goes under the name "tight binding" for the simple reason that the atom's electrons are thought of as being "tightly bonded" to the atom. In other words, the atomic wave functions are considered good (enough) approximations for the crystal orbital, just like in molecular quantum chemistry. The tight-binding approach comes in different varieties, mostly using semi-empirical parameterizations of the Hamiltonian, and it can be extremely effective, not only because of the minimum size of the basis set but also because it is almost trivial to project any periodic property to the atom. When overlap is included, tight-binding theory is equivalent to extended Hückel theory in chemistry (for example, the YAeHMOP code [10] using Slater-type orbitals). Still, *ab initio* variants of tight binding are also available, for example, using Hartree–Fock theory (the CRYSTAL code using Gaussian-type orbitals [11]), at least as long as materials are not metallic. If atomic orbitals are good approximations for molecules, why shouldn't they be for periodic solids as well? In fact, the natural habitat of the periodic wave function, the reciprocal space, allows for analytical evaluation of the integrals over Slater-type orbitals needed within the quantum chemical Hamilton matrices, while molecular codes need to introduce additional numerical approximations such as B-splines [12, 13].

The exactly opposite strategy, which is the *de facto* standard of the community, measured by plenty of citations of the most relevant codes, focuses on the plane-wave parts only and gives up all the nodal atomic nature, as strangely as that may sound. Before discussing how the method operates, one needs to highlight the enormous advantage of plane waves: they are the natural wave functions (according to the Bloch theorem, plane waves are built into it) for periodic solids, spatially infinite in principle and totally delocalized. If that can be carried out (and it can), two significant advantages are immediately apparent. In a plane-wave basis, all Hellmann–Feynman forces are *exact* because the method does not – in principle – suffer from whatever kind of basis-set superposition error well known from molecular quantum chemistry. Second, it is rather trivial to improve the basis-set quality, namely from an increasing energy cutoff (squared frequencies) for the plane waves so that a larger basis has more waves with "shorter" wavelengths. The shorter they are, the better the basis set. Mathematically, one simply writes down a sum of plane waves up to a certain reciprocal lattice vector $\mathbf{G}$, as given in

$$\Psi_\mathbf{k}(\mathbf{r}) = \frac{e^{i\mathbf{k}\mathbf{r}}}{\sqrt{\Omega}} \sum_{\mathbf{G}=0}^{\infty} c_\mathbf{k}(\mathbf{G}) e^{i\mathbf{G}\cdot\mathbf{r}} \qquad (18.3)$$

and then the energy cutoff can be formulated as $\frac{\hbar^2}{2m}|\mathbf{k}+\mathbf{G}|^2 \leq E_{cut}$, very simple indeed [6]. There is a caveat, though, and it relates to the nodality mentioned in the very beginning: in case the atomic nodal nature should also expressed by plane waves, the energy cutoff would rather have to become far too high, resulting in an enormous number of plane waves, in particular for heavy elements with highly nodal valence functions. Instead, one simply throws away the core orbitals, replaces the true nuclear potential by a much softer "pseudopotential" [14] sometimes also dubbed "effective core potential," and then gets rid of all nodality because the resulting pseudo valence function no longer has these, then easily expressed by a relatively small number of plane waves. The projector-augmented wave (PAW) strategy by Blöchl [15] extends the pseudopotential approach and allows for the reconstruction of an all-electron wave function, because the pseudopotential "rides" on an all-electron representation of the atom, even when iterating toward self-consistency, so a lot of arbitrariness is removed forever from pseudopotential theory.

The mighty plane-wave-pseudopotential combination, in particular when carried out using flexible all-electron-derived PAW pseudopotentials (see Section 18.4), makes us understand why the vast majority of periodic electronic-structure calculations is presently performed in such a way, and powerful computer programs such as VASP [16–20], Quantum ESPRESSO [21–23], ABINIT [24–27], and CASTEP [28] confirm this finding. The method is mathematically simple, has enormous technical advantages, and it is also quite fast. It pays off to use the natural Bloch functions when dealing with crystalline materials! There is also a big disadvantage, of course, namely the loss of all chemical information: the atoms are gone, and so are the atomic orbitals. If one is not interested in chemistry, life stays simple, but in the other case, however, chemical-bonding information must be extracted via an in-between step, as shown in the sequel.

Before we do so, let us recall the third strategy to generate a high-quality wave function, which – in terms of accuracy – is clearly the best but also most complicated. Those *cellular* methods solve a radial Schrödinger equation numerically close to the atoms (and yield "partial waves" = numerical atomic orbitals) and glue the resulting atomic wave functions to a plane wave in the outer between-atoms region. Slater's ingenious augmented plane-wave (APW) method must be mentioned [29], later "linearized" to Andersen's linear muffin-tin orbital (LMTO) and linear augmented-plane wave (LAPW) approaches [30]. While LAPW is still the accuracy benchmark of solid-state theory, the simpler and much faster LMTO has given birth to chemical-bonding analysis within an *ab initio* framework. Why is that? LMTO established friendship between theoretical physics and chemistry of the solid state, so to speak, because it is based on *orbitals*. Almost needless to say, the dominating pseudopotential theory of today (PAW) solely rests on the principles of all-electron LMTO and LAPW theories, it actually stems from the same school.

18.4 Interpretational Tools

Knowing the chemical-bonding situation is crucial for understanding and even predicting the stability and reactivity of a given compound. This relationship has been known to and successfully applied by molecular chemists for decades, leading to the design of new target molecules and synthetic pathways. While the same principles also apply to the solid state, their application is not as trivial, although solid-state chemists have successfully interpreted a plethora of materials using the ionic notion. The main difference between a molecule and a crystal is that the latter features translational invariance, leading to plane waves being the natural choice of wave function. While they have the advantage of low computational cost, they are delocalized over the entire crystal and, therefore, carry no local information that is crucial for bonding analysis. Thus, a quantum-chemical understanding of a crystal is impossible at this point, and one can only look at its physical properties: the total energy can be compared between related structures, allowing one to judge relative stabilities. The density of states, as well as the spatially resolved electron density, are especially liked by physicists as they allude to "observables," but they do not provide any bonding information *per se*. The reason is that the precious phase information of the wave function is totally missing, and one needs to differentiate between bonding and antibonding states, by using the phases.

At this point, the mindful reader may ask whether we have to make a trade-off here. Can we either exploit the low computational cost of plane waves or benefit from the quantum-chemical information of a local-orbital basis, which, on the other hand, would be exceedingly expensive to compute? The answer is most certainly No!

In principle, both approaches arrive at the same overall wave function, so it should be possible to convert one of them into the other, and indeed, this so-called projection scheme was introduced quite a while ago, based on the idea to unitarily transform a given plane-wave wave function into a specified set of local (atomic) orbitals [31, 32]. The LCAO coefficient matrix elements are obtained from the transfer matrix elements $T_{\mu j}[\mathbf{k}]$ between the local orbital ϕ and the plane-wave (PW) band $\tilde{\psi}$:

$$T_{\mu j \mathbf{k}} = \left\langle \phi_{\mu \mathbf{k}} \middle| \tilde{\psi}_{j \mathbf{k}} \right\rangle \tag{18.4}$$

Using this formalism, it is possible to extract the LCAO coefficients by solving linear algebraic expressions [13], and thereby regain the locality from a delocalized wave function. The schematic procedure can be seen in Figure 18.4. An auxiliary local atomic function "probes" the locality of the plane wave (Figure 18.4a). In case the overlap integral between the two is close to unity, the plane wave is local (and s-like as in Figure 18.4b). If the integral is (close to) zero, the plane wave does not show the correct (p-like) angular-momentum character, see Figure 18.4c.

When going from a pure PW to a PAW wave function, additional transfer matrices have to be evaluated as the wave function inside the PAW spheres differs from the plane-wave setup, although the general idea remains the same. Once evaluated, the transfer matrix can be used to determine the local Hamiltonian. Interested readers may refer to the literature [31–34].

The aforementioned projection scheme has been conveniently included into the Local-Orbital Basis Suite Towards Electronic-Structure Reconstruction (LOBSTER) computer program. Its general workflow is shown in Figure 18.5. The procedure starts with a PAW calculation using either VASP, Quantum ESPRESSO or ABINIT followed by a projection onto a local STO basis because then, local quantities allow for a chemical bonding analysis using three descriptors that are introduced in the following section.

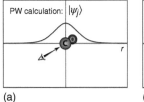
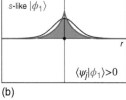
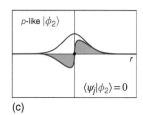

Figure 18.4 Schematic procedure of the projection (a) from plane waves (PW) to LCAO. The overlap between the band function $|\psi_j\rangle$ and atomic orbitals $|\phi_n\rangle$ is evaluated. In this case, the PW band coincides with a carbon 2s orbital in the CO molecule (b) but not with a carbon 2p orbital (c). Source: Deringer et al. [13]/Adapted from American Chemical Society.

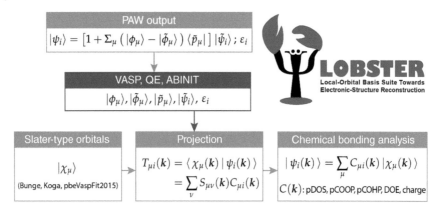

Figure 18.5 Schematic workflow of the LOBSTER program. Starting with a (delocalized) PAW wave function, Slater-type orbitals are used as a basis set for the projection onto atomic orbitals. With this local basis, a chemical bonding analysis becomes possible by means of, for example, projected COHP (pCOHP) and wave-function-based charges. Source: Nelson et al. [35]. John Wiley & Sons.

To repeat: after having transformed the entire delocalized electronic structure into another representation described by a local-orbital basis, it is possible to perform any analysis that is available for finite molecules. As a first tool, let us consider Mulliken's population analysis [36–39], which determines an orbital's gross population GP from its net population NP and overlap population OP:

$$\text{GP}_\mu = \text{NP}_\mu + \frac{1}{2}\sum_\nu \text{OP}_{\mu\nu} \tag{18.5}$$

$$\text{NP}_\mu = \sum_{j\mathbf{k}} f_j c^2_{\mu j \mathbf{k}} w_\mathbf{k} \tag{18.6}$$

$$\text{OP}_{\mu\nu} = \sum_{j\mathbf{k}} f_j c^*_{\mu j \mathbf{k}} c_{\nu j \mathbf{k}} S_{\mu\nu\mathbf{k}} w_\mathbf{k} \tag{18.7}$$

μ and ν represent the atomic orbitals, j is the (crystal) orbital, and f_j are the occupation numbers. Or, alternatively,

$$\text{GP}_\mu = \sum_\nu \int P_{\mu\nu\mathbf{k}} S_{\mu\nu\mathbf{k}} d\mathbf{k} \tag{18.8}$$

with the density-matrix elements

$$P_{\mu\nu\mathbf{k}} = \sum_j f_{j\mathbf{k}} c^*_{\mu j \mathbf{k}} c_{\nu j \mathbf{k}}. \tag{18.9}$$

While the Mulliken population analysis shows significant interpretational advantages as regards to density-based partitioning schemes both in terms of computational speed and quality of the results, it may – in certain cases – lead to nonphysical orbital occupations such as values smaller than 0 and larger than 2, conflicting with the Pauli principle. Also, the symmetric splitting of the overlap population is at least questionable, less so for elements with similar electronegativity but increasingly for polarized and ionic compounds. Both issues can be addressed by applying Löwdin's symmetric orthonormalization (LSO) [40] to the basis. By doing so, all occupations are strictly limited to the range of $0 \leq f_j \leq 2$, and the overlap matrix becomes the identity matrix, so there is no overlap matrix to begin with. The gross population can then simply be determined from the on-site density matrix elements.

$$GP_\mu^{LSO} = \int P_{\mu\mu k}^{LSO} d\mathbf{k} \tag{18.10}$$

To make the density matrix approach energy-resolved, one simply needs to use the respective coefficient-matrix elements and replace the orbital (or, in this case, band) occupation by a delta distribution as in Eq. (18.11), so we are left with a density-of-states matrix approach. The result is the orbital-projected or local DOS, to be regarded as an energy-resolved gross population. It can be formulated using the original as well as the orthonormalized basis.

$$LDOS_\mu(E) = \sum_{(v,j)} \int \text{Re}(c_{\mu j k}^* c_{v j k}) S_{\mu v k} \delta(E - \varepsilon_{j\mathbf{k}}) d\mathbf{k} \tag{18.11}$$

$$LDOS_\mu^{LSO}(E) = \sum_{(v,j)} \int \text{Re}(c_{\mu j k}^{*\,LSO} c_{v j k}^{LSO}) \delta(E - \varepsilon_{j\mathbf{k}}) d\mathbf{k} \tag{18.12}$$

These orbital gross populations can be easily transferred into quantum-chemical charges by subtracting the atomic gross population from the element's valence electron count. Doing so gives rise to the first kind of interatomic interaction, usually referred to as the ionic bond.

$$q = N - \sum_\mu GP_\mu \tag{18.13}$$

Ionicity has been subject of research ever since the advent of solid-state structural analysis, naturally so, since among the simplest structures we find the alkali halides. They tend to crystallize in the highly symmetric rock-salt structure that can easily be "explained" by a simple electrostatic approach. Assuming ideal integer charges of ±1 seems to be a reasonable choice as the difference in electronegativity is the largest in this kind of compounds, in particular when quantum-chemical alternatives had not been developed yet. The energy between two point charges q_i and q_j is proportional to their reciprocal distance d_{ij}, and a negative sign means an attraction while a positive sign indicates repulsion.

$$E_{ij} \propto \frac{q_i q_j}{d_{ij}} \tag{18.14}$$

For a finite system, say, a molecule, the electrostatic interaction strength can simply be determined following this approach but for a solid the translational symmetry generates a periodic crystal field that goes by the name Madelung [41]. As the electrostatic potential is proportional to $\frac{1}{d}$, it asymptotically approaches zero but never completely vanishes. This means that an ion in a given unit cell "feels" the presence of every other ion in the entire crystal, and in order to quantify the electrostatic lattice energy, it would be necessary to determine the sum over all possible ion–ion combinations, infinite in theory, so convergence is never reached. To enforce convergence, multiple mathematical tricks have been suggested and in 1921 already, Ewald [42] proposed a method to split the total electrostatic potential into two separate potentials that converge independently from each other.

$$v_{\text{total}}(j) = v_1(j) + v_2(j) \tag{18.15}$$

The first potential v_1 determines the long-range interactions in reciprocal space, while the second potential v_2 considers the close-range interactions in real space. Since v_1 and v_2 converge independently from each other, so does their sum, and the total electrostatic (Madelung) energy of a given crystal is easily determined:

$$\varepsilon_M = \frac{1}{2} \sum_j q_j v_{\text{total}}(j) \tag{18.16}$$

While the ionic bond plays an important role in the solid state, molecular chemistry is dominated by covalent bonding, also found in "infinite" covalently bonded crystals such as elemental diamond. It goes without saying that molecular chemists have been using the overlap and Hamilton matrix elements since Mulliken's population analysis was published in 1955, but it took almost three decades until the latter was generalized to the solid state. In 1983, Hughbanks and Hoffmann [43] introduced the crystal orbital overlap population (COOP) in the framework of extended Hückel theory, and it can be understood as an overlap-weighted DOS.

$$\text{COOP}(E) = S_{\mu\nu} \sum_j \int \text{Re}(c^*_{\mu j \mathbf{k}} c_{\nu j \mathbf{k}}) \delta(E - \varepsilon_{j\mathbf{k}}) d\mathbf{k} \tag{18.17}$$

If integrated to the Fermi level, a scalar quantity for the bond strength is achieved. The integrated COOP dubbed ICOOP equals the number of electrons occupying a certain bond, just like Mulliken's overlap population. Only 10 years later, the crystal orbital Hamilton population was proposed by one [44] of us in the framework of DFT. It is an energy-partitioning scheme that attributes the band energy to individual orbital–orbital interactions.

$$\text{COHP}(E) = H_{\mu\nu} \sum_j \int \text{Re}(c^*_{\mu j \mathbf{k}} c_{\nu j \mathbf{k}}) \delta(E - \varepsilon_{j\mathbf{k}}) d\mathbf{k} \tag{18.18}$$

Because DFT is variational (unlike extended Hückel), COHP focuses on the Hamiltonian, not on the overlap. There has been some debate concerning the comparability of COHP for strongly differing crystal structures. The reason lies in the definition of the crystal's electrostatic potential as derived from its volume as well as its surface – both have no meaning in a perfectly translationally invariant system. Thus, the electrostatic potential is usually defined arbitrarily in common DFT codes. While the total energy of a neutrally charged system is invariant toward the electrostatic potential as the same energy shift is applied to positive as well as negative charges and both cancel each other, this is no longer true for a charged system. Since the band energy refers to the electrons, an inherently charged system, COHP is not trivial to be put into context for different crystal structures. When employing an orthonormal basis, this issue does not arise in the first place [45, 46], however, because the Hamilton matrix element is given as

$$H_{\mu\nu} = \langle \phi_\mu | \hat{H} | \phi_\nu \rangle \tag{18.19}$$

Within first-order perturbation theory, and for an orthonormal basis, we arrive at the following expression:

$$H_{\mu\nu}^{V_0} = \left\langle \phi_\mu \left| \hat{H} + V_0 \right| \phi_\nu \right\rangle = \left\langle \phi_\mu \left| \hat{H} \right| \phi_\nu \right\rangle + \left\langle \phi_\mu \left| V_0 \right| \phi_\nu \right\rangle = H_{\mu\nu} + S_{\mu\nu} V_0 = H_{\mu\nu} \quad (18.20)$$

So, the off-site (bonding) matrix elements are not affected by the perturbation. For onsite interactions, this is not true. Here, the overlap matrix element is not zero, therefore $H_{\mu\mu}$ is affected by the shift of V_0.

$$H_{\mu\mu}^{V_0} = \left\langle \phi_\mu | \hat{H} + V_0 | \phi_\mu \right\rangle = \left\langle \phi_\mu | \hat{H} | \phi_\mu \right\rangle + \left\langle \phi_\mu | V_0 | \phi_\mu \right\rangle = H_{\mu\mu} + V_0 \quad (18.21)$$

Both COOP and COHP have been generalized to the PAW formalism, but to discern them from their original definitions, we refer to them as *projected* quantities, namely pCOOP and pCOHP. In most cases, they look like mirror images differing by the plus/minus sign. Figure 18.6 depicts the DOS and projected COHP analysis for elemental Si: in the COHP, 3p–3p, 3s–3p, and 3s–3s bonding interactions are shown, and almost all of them point to the right below the Fermi level, so those are *attractive* Si–Si interactions. In contrast, repulsive ones show up in the unoccupied region, in the conduction band. Because of improper 3s–3p spatial fit (unlike in C), the 3p and 3s contributions are even energetically apart, so there isn't much "sp³" hybridization. To make COHP look similar to COOP, it carries a minus sign, so the bonding levels go to the right, a silent agreement in solid-state bonding theory.

In addition to COOP and COHP, a third bonding descriptor was introduced in 2021 and dubbed crystal orbital bond index (COBI) [47]. Its general idea resembles the bond index originally proposed by Wiberg within the complete neglect of differential overlap (CNDO) approximation [48], later generalized by Mayer for

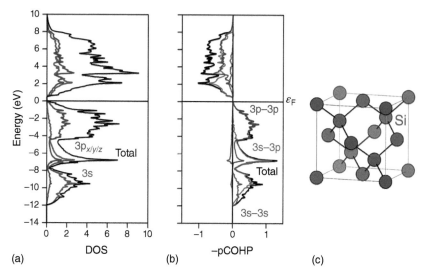

Figure 18.6 (a) Orbital-projected DOS, (b) orbitalwise projected COHP for the closest Si–Si contact, and (c) unit cell of silicon in the diamond structure.

nonorthonormal bases [49]. For a bond between an atom A and an atom B the bond indices read

$$\mathrm{BI}_{\mathrm{Wiberg}} = \sum_{\substack{\mu \in A \\ \nu \in B}} |P_{\mu\nu}|^2; \quad \mathrm{BI}_{\mathrm{Mayer}} = \sum_{\substack{\mu \in A \\ \nu \in B}} P_{\mu\nu} P_{\nu\mu} \quad (18.22)$$

Combining the general idea of COOP/COHP with the Wiberg/Mayer bond indices leads to the definition of COBI. Therefore, the overlap/Hamilton matrix element has to be replaced by the respective density-matrix element, leading to

$$\mathrm{COBI}(E) = P_{\mu\nu} \sum_{j} \int \mathrm{Re}(c^*_{\mu j \mathbf{k}} c_{\nu j \mathbf{k}}) \delta(E - \varepsilon_{j\mathbf{k}}) \mathrm{d}\mathbf{k} \quad (18.23)$$

Using COBI and its energy integral, the ICOBI, gives access to covalent bond orders in solid-state materials – a useful quantity that remained elusive up to this point. While COBI is analogous to COOP and COHP in terms of qualitative interpretation, COBI excels by the simplicity of its quantitative analysis. As the energy integral simply equals the bond order, it does not need a reference point, unlike ICOOP and ICOHP. This can nicely be demonstrated by considering three archetypical solid-state phases shown in Figure 18.7. On the left-hand side, there is carbon in the diamond type. Here, each carbon atom forms four σ-bonds that have a bond order of one. Indeed, the ICOBI of 0.95 nicely mirrors this qualitative and simplistic prediction. Naturally, the charges of each C atom are zero for an elemental compound. Silicon dioxide in the α-quartz structure (middle) is an example of covalent and ionic bonding being present at the same time. The ICOBI of 0.76 is still large enough for the Si–O bond to be considered covalent, and the Löwdin charges of −0.90 e for O and 1.81 e for Si indicate significant charge transfer. This localization is also apparent from the rather sharp peaks of the COBI curve. In Figure 18.7 c (right), the famous representative of ionicity is shown: rock salt. Here, the Löwdin charges of ±0.66 e indicate a

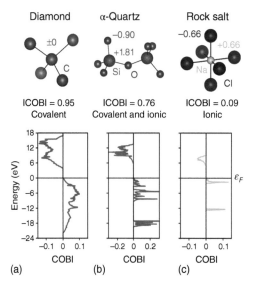

Figure 18.7 Direct coordination environment and chemical bonding descriptors for three archetypical phases, that is, (a) diamond, (b) SiO_2, and (c) NaCl. Source: Müller et al. [47].

strong impact of electrostatic interactions, while the ICOBI (0.09) shows that covalency is negligible. The localization is also visible from the distinct peaks of the COBI plot.

In addition to conventional two-center (2c) bonds, COBI can also highlight multi-center bonding. To do so, the product of density-matrix elements is expanded involving three atomic orbitals μ, ν, and χ according to

$$\text{BI}_{\mu\nu\chi} = (P_{\mu\nu}P_{\nu\chi}P_{\chi\mu}) \tag{18.24}$$

Generalizing the multicenter bond index to the solid state is analogous to the two-center case but the density-matrix elements are added to the product in front of the summation.

$$\text{COBI}^{(3)}_{\mu\nu\chi} = P_{\mu\nu}P_{\nu\chi}\sum_j \int \text{Re}(c^*_{\chi jk}c_{\mu jk})\delta(E - \varepsilon_{jk})\text{d}\mathbf{k} \tag{18.25}$$

While the integral value of the multicenter COBI is invariant to the order of orbitals μ, ν, and χ, this assertion does not apply to its energy-resolved form. As the energy dependence is introduced by the delta distribution in Eq. (18.25), the plot thus achieved only mirrors the energy dependence of the μ–χ interaction while the other two are lost. In order to solve this issue, one needs to sum over all cyclic permutations of the orbitals μ, ν, and χ according to Eq. (18.26).

$$\text{COBI}^{(3)} = \text{COBI}^{(3)}_{\mu\nu\chi} + \text{COBI}^{(3)}_{\nu\chi\mu} + \text{COBI}^{(3)}_{\chi\mu\nu} \tag{18.26}$$

The original multicenter bond index was developed by two groups independently from each other [50, 51] and successfully applied to three-center bonds as they appear in boranes such as B_2H_6 as well as hypervalent compounds like XeF_2. As a result, positive three-center bond indices indicate an electron-*deficient* compound while negative values are indicative of an electron-*rich* bonding mechanism. For illustration, a quantitative COBI analysis is depicted in Figure 18.8 using XeF_2

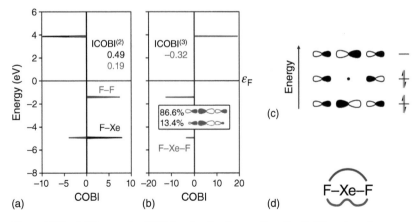

Figure 18.8 (a) Two-center COBI for the Xe–F as well as the F–F bonds and (b) three-center COBI for the F–Xe–F bonds in XeF_2. (c) Simple Pimentel-style MO-diagram for XeF_2 and (d) structure of the XeF_2 molecule. The color code of bonds follows the curves in subfigure (a) and (b). Source: Müller et al. [47].

with 10 involved electrons, violating the octet rule. The two-center bonds between xenon and fluorine have a bond order of almost 0.5, which perfectly matches the expectation from the simple Pimentel-style MO diagram in Figure 18.8c; here, the lowest-lying MO is the only one contributing to the Xe–F bond with an in-phase constructive interference between all three *p*-orbitals. Alternatively expressed, the electron pair is somewhat delocalized over both Xe–F contacts, resulting in a "half" bond order, often stressed in the literature but without getting enough attention. Interestingly, there is also an F–F interaction with an ICOBI of 0.19, only surprising at first sight. This is because the *p*-orbitals of F in the second MO in Figure 18.8c (with a node at the central Xe atom) are engaged in a bonding interaction, albeit at a range exceeding common bonding distances. Analyzing the three-center COBI reveals the origin of this long-distance bond, with two significant peaks in the COBI$^{(3)}$ plot as well as a relatively large integral value of −0.32. As alluded to before, the negative sign is indicative of an electron-rich bonding mechanism as it appears in the three-center four-electron (3c4e) bond in XeF$_2$.

18.5 Applications

By tradition, the three fundamental bonding types, covalency, metallicity, and ionicity, are conveniently summarized in the van Arkel–Ketelaar triangle [53, 54] depicted in Figure 18.9. It sets the average electronegativity (EN) into context with the difference in electronegativities, hence the bonding types separate into different regions. For a large EN difference, one or more electrons move from an energetically high-lying "cationic" atomic orbital into a low-lying "anionic" orbital, so there is charge separation, and quantum mechanics becomes almost classical, such that electrostatic interactions prevail: ionic bonding results. In the lower part, for (almost) no EN difference, we find true interference of wave functions, however, so there are covalent molecules in the lower right corner and metals in the lower

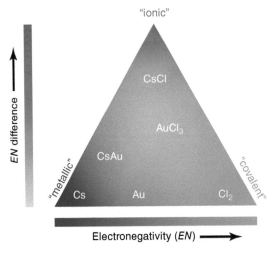

Figure 18.9 van Arkel–Ketelaar triangle of bonding. Source: Reproduced from [52] by kind permission of Elsevier.

left corner, another variant of covalency but with too few electrons delocalized over too many atoms. In the following section, we will walk along the coordinates of the triangle, showing materials with various types of bonding situations.

Let us start with the form of bonding found in the right corner of the triangle, often seen in molecules, for example, the covalent electron pair bond as it appears in elemental carbon. Here, each carbon atom is coordinated tetrahedrally and forms a single bond to each of its neighbors (cf. Figure 18.7a). Graphite, on the other hand, consists of hexagonal graphene sheets in which each atom forms three σ-bonds. Additionally, there is a completely delocalized π-system that is responsible for the electrical conductivity and metallic character in graphite. Figure 18.10a and b show the DOS and two-center COBI curves for graphite. As expected, the region of σ-bonding for a C–C bond is at lower energy than the π-contribution. Interestingly, the σ-bond order has an ICOBI of 0.94, which is almost the same as for diamond. The π-bond order is 0.27, a value that is slightly lower than the simple qualitative value of 0.33 to be expected if the π-electron is distributed over the three nearest-neighbor bonds that each atom contributes to. If we go from the 2D-periodic sheets of graphite to a 1D-periodic system, we arrive at a carbon nanotube, a wrapped graphene layer, shown in Figure 18.10c and d. It, therefore, does not surprise us that the integral values of COBI are almost unaffected by the decrease in dimensionality: the σ bond order increases slightly in the third decimal digit, leading to a differently rounded value but the π bond order stays exactly the same. The only qualitative difference between both systems lies in the shape of the DOS and COBI curves, a function of "curvature." By decreasing the dimensionality, the states become more localized and, therefore, the curves are "sharper" as the

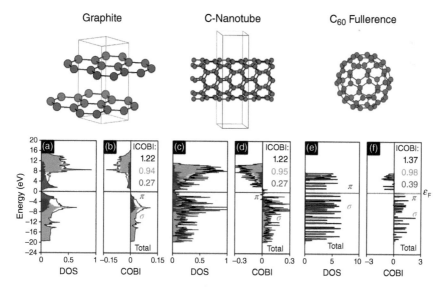

Figure 18.10 Density-of-states (a, c, e) and COBI curves (b, d, f) for graphite (a, b), a carbon nanotube (c, d) and the C_{60} Buckminster fullerene. DOS and COBI for σ bonds are shown in gray, π bonds are shown in red. Note the increasing discretization with decreasing dimensionality of the periodicity.

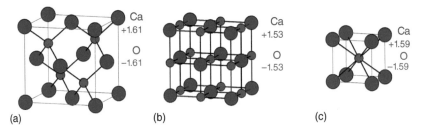

Figure 18.11 Hypothetical polymorphs of CaO, starting with the [ZnS] type in (a). While the [NaCl] structure (b) is stable under standard conditions, [CsCl] can be achieved at high pressure (c).

electronic bands approach the discrete energy levels known from molecules. This discretization can be extended even further if we consider an isolated molecule, namely the C_{60} Buckminster fullerene shown in Figure 18.10e and f. Here, the picture changes significantly and the molecule only possesses discrete energy levels. Structurally, the formation of a sphere requires a "bending" of the rather stiff C_6 rings, which is achieved by the formation of five-membered rings in addition to the six-membered rings that are present in graphite and the nanotube. As a consequence, there are two different types of bonds: a slightly longer (1.45 Å) along the fivefold rings and a shorter one (1.40 Å) that is along the sixfold rings. As the 1.40 Å bond is shorter than the bond in graphite, we observe an increase in the ICOBI value, which is mainly due to an increased π-contribution as seen in Figure 18.10f.

Having analyzed elemental polymorphs (also dubbed allotropes) that move between metallic and covalent bonding in the van Arkel–Ketelaar triangle as a function of structure, we now investigate the interplay of ionic and covalent bonding. An archetypical ionic compound is CaO, crystallizing in the [NaCl] structure type under standard conditions. Aside from this experimentally verified structure, two other polymorphs are conceivable: the [CsCl] (which is known at high pressure) and [ZnS] structure types, shown in Figure 18.11.

All three structure types have been calculated using DFT methods, and DFT agrees with experiment regarding the most stable polymorph: the [NaCl] structure. But how do these structures differ in terms of covalent and ionic bonding? A chemical bonding analysis using LOBSTER can tell us, and the results are summarized in Table 18.1. Obviously, the ground state is lowest in total energy, apart by about 96 kJ/mol from the high-pressure [CsCl] and by about 37 kJ/mol from the low-pressure [ZnS] polymorphs. Although the ionic archetype [NaCl] is usually alluded to ionicity, the Löwdin charges are *lowest* for [NaCl], so CaO is most covalent in the rock-salt structure. This is also reflected from the Madelung energies, this time calculated using wave-function-based charges (but not fictitious ones such as +2 and −2), also revealing the strongest covalency (weakest ionicity) for [NaCl]-like CaO. The same finding is given more directly from the bond orders (ICOBI) over the next-nearest coordinating atoms. For [NaCl], CaO is most covalent.

Now, we want to look at the interplay of changing ionic and covalent bonding during compositional change, namely by replacing the chalcogenide in CaO with

18.5 Applications

Table 18.1 Total and Madelung energies and ICOBI per polyhedron for CaO in [ZnS], [NaCl], and [CsCl].

Structure type	[ZnS]	[NaCl]	[CsCl]
ΔE (kJ/mol)	+37.1	0	+95.7
$E_{Madelung}$ (kJ/mol)	−2630	−2351	−2404
Σ ICOBI	0.56	0.60	0.56

The total energies are relative to [NaCl], which is stable under standard conditions. The Madelung energies are given with negative numbers, on purpose, as opposed to the usual solid-state chemical definition.

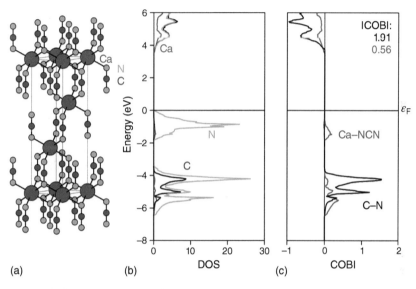

Figure 18.12 Structure (a), DOS (b), and COBI (c) for CaNCN.

the more covalent pseudo chalcogenide NCN^{2-}, a complex anion. Figure 18.12 displays crystal structure, projected DOS and COBI for the archetypical carbodiimide CaNCN. The anionic character of NCN^{2-} is visible from the occupied levels below ε_F dominated by N and C, and N is charged −0.92 e, so significantly less charged than O in CaO. While it is rather trivial to detect the covalent N=C bond inside NCN^{2-} by means of ICOBI = 1.91 (so, a double bond), the covalency of Ca–N is quantified by ICOBI = 0.56 for the entire octahedron, and it is in the same order of magnitude as found for CaO, even a little lower, surprising at first sight. Likewise, the charge on Ca is slightly increased from +1.53 e (CaO) to +1.65 e (CaNCN), so both compounds do resemble each other cationic charge-wise, unsurprisingly so, as NCN^{2-} is often referred to as *pseudo* chalcogenide. The change in the interplay of covalent and ionic bonding is clearly revealed, however, by the change of the Madelung energy, which amounts to −2152 kJ/mol^{-1} for CaNCN, lowered by almost 9% compared to CaO. Even though there is a slightly larger anion–cation

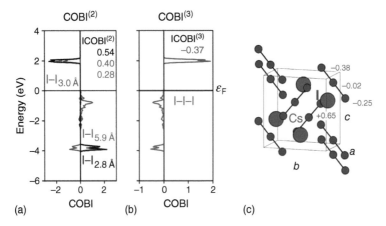

Figure 18.13 (a) Two-center and (b) three-center crystal orbital bond index for the I_3^- complex anion in CsI_3. (c) Crystal structure of CsI_3. Adapted from [47] by kind permission of the authors.

charge separation, judged by the Löwdin charge on Ca, the Coulombic interactions are weaker due to the covalency found within the complex anion, and the smaller Madelung energy.

Another complex anion is the triiodide I_3^- anion. It appears in the solid-state compound CsI_3 shown in Figure 18.13. Unsurprisingly, the bonds between the central iodine atom and the terminal atoms are rather short (2.8 Å and 3.0 Å) and their ICOBI is approx. 0.5, which is exactly in line with the 3c4e bond that can also be formulated as two "half" bonds. Not too surprisingly, I_3^- is isoelectronic with XeF_2, see discussion before. In addition to that, there exists a long-ranged contact with a distance of 5.9 Å, between the terminal iodine atoms, and it has an ICOBI of 0.28 (blue curve in Figure 18.13a). What may contradict chemical intuition, but only at first sight, can be attributed to the 3c4e bond in the molecule as quantified by the $COBI^{(3)}$ in Figure 18.13b. The integral value of -0.37 is indicative of an electron-rich bond, just like in XeF_2 ($ICOBI^{(3)} = -0.32$), and it is amongst the largest negative values that can be achieved. Interestingly, the $ICOBI^{(3)}$ can also be used to examine higher-order bonds.

The next group we consider is located somewhere in the middle of the bottom part of the bond triangle. Phase-change materials (PCM) have very low ionic contributions (for example, the EN difference in GeTe, an archetype phase, is almost zero on the Pauling scale), so the bonding is both a little delocalized and at the same time directed and largely covalent. The exact nature of the chemical bonding situation has been vividly discussed in the past. Based on a lattice-dynamics-based approach, it was found that there are surprisingly strong force constants in the solid-state mixture $GeTe-Sb_2Te_3$ (so-called "GST") system at very large distances [56], seemingly contradicting chemical intuition if "regular" covalency would be at play. These strong interactions follow linear atomic chains, just as we have seen in the triiodide anion, in Figure 18.14, albeit the number of atoms involved is by far larger in the PCM. As these compounds violate the octet rule by having

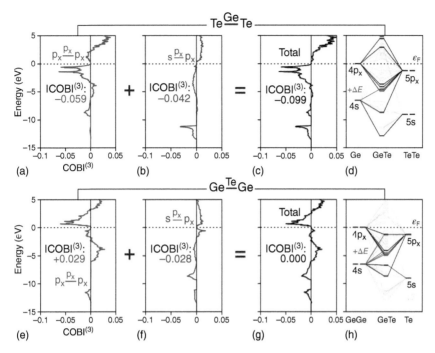

Figure 18.14 (a) and (b) Orbitalwise COBI$^{(3)}$ curves, (c) total COBI$^{(3)}$ plot and (d) simplified MO diagram for the Te$\stackrel{Ge}{=}$Te fragment in β-GeTe. (e)–(h) The same, but for the Ge$\stackrel{Te}{=}$Ge fragment. Source: Reproduced from [55] by kind permission of John Wiley and Sons.

too many electrons, similar to triiodide and XeF$_2$, it seems logical to postulate an analogous bonding mechanism for the GST family. Here, the idea is that in GeTe (with 10 valence electrons, just like XeF$_2$), the central Ge^{2+} with a 4s^2 "lone pair" is the cause of breaking the octet rule, and it mediates three-center bonding of the type Te$\stackrel{Ge}{=}$Te. Indeed, we find a significant three-center COBI as well as ICOBI for the Te$\stackrel{Ge}{=}$Te fragment in β-GeTe, a polymorph that crystallizes in the rock-salt structure. Figure 18.14a–c reveal the presence of an electron-rich multicenter bond in this fragment, as seen by the negative ICOBI$^{(3)}$. The simplified MO interaction diagram projected from plane waves in Figure 18.14d shows the same result. Interestingly, the other Ge$\stackrel{Te}{=}$Ge combination has a *negligible* ICOBI$^{(3)}$ value despite showing a strong stiffness in lattice-dynamic simulations. As can be visualized in Figure 18.14e and f, the contributions of $p_x\stackrel{p_x}{=}p_x$ and $s\stackrel{p_x}{=}p_x$ have roughly the same absolute value but with a different sign, so they cancel each other, leading to the curve in Figure 18.14g that has a non-zero COBI$^{(3)}$ plot but a zero integral value. The same result is displayed in the simplified MO interaction diagram in Figure 18.14h. Here, there are a lot of destabilizing interactions alongside the stabilizing ones. In analogy to lone pairs that are formed by bonding and anti-bonding AO interactions, we can call the Ge$\stackrel{Te}{=}$Ge fragment a three-center analog to the classical lone pair [55]. That is to say, the chemical bonding in GST materials is covalent in nature but, due to the too

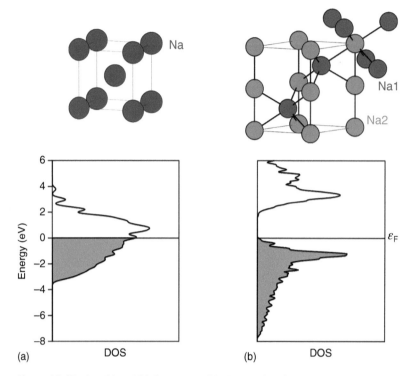

Figure 18.15 bcc (a) and high pressure (b) phases of sodium metal and their DOS curves.

many electrons and violation of the octet rule, involves multicenter interactions known from molecules (say, XeF_2) but is not a fundamentally different bonding type [57].

Let's get fully metallic: the bcc type of metallic Na is not spectacular since this is the true ground state. As shown in Figure 18.15, the DOS plot of bcc Na resembles the Sommerfeld model of almost "free" electrons, also found for other simple metals of the alkali or alkaline-earth groups, both boring and fundamental, too. At very high pressure, however, sodium metal shows a strange behavior: while usually pressure is expected to make everything metallic, in this case, sodium undergoes a phase transition from the metallic bcc structure over a close-packed fcc structure to a *transparent* high-pressure phase of [NiAs] type that is, in fact, an insulator [58]. The chemical interpretation of this rather puzzling phenomenon is quite simple: sodium first undergoes a phase transition from the bcc to the fcc structure to optimize the space-filling using equally sized atoms, a necessity of high pressure. When even more pressure is applied, there is no way to further optimize space-filling using equally sized atoms, but there is a way by *differently* sized atoms, e.g., sodium anions and cations. In this scenario, the anions would adopt a close-packed structure (in this case: hcp), while the cations would occupy the empty sites within the anion lattice. The result is the experimentally observed [NiAs] structure, and indeed, LOBSTER evaluates the charges to be ± 0.11 e, supporting the interpretation of the unusual phase transition. As a consequence, a band gap opens in the DOS figure for transparent high-pressure sodium. Although issues with the STO basis under these extreme pressures suggest treating the numerics of the Löwdin population analysis with care, the qualitative picture holds true: a charge separation into anions and cations occurs under extreme pressures, so a metal becomes a salt.

18.6 Summary

Density-functional theory together with the pseudopotential approximation has allowed for calculating the electronic structures of almost every thinkable solid-state material to an astonishingly high degree of accuracy, amazingly enough. Even though translational invariance suggests totally delocalized plane-wave-like basis sets to be the optimum choice for periodic systems, thereby nicely conforming with the fundamental Bloch theorem, local chemical-bonding information can nowadays be also extracted from such delocalized electronic-structure calculations, very similar to the "molecular" case, fortunately enough. To do so, one needs to unitarily transform the maximally delocalized wave functions into a localized representation, a standard automatic procedure using the LOBSTER computer code. Given such a localized, inherently quantum-chemical picture in terms of atomic orbitals, wave-function-based descriptors such as Crystal Orbital Hamilton Population (COHP) and Crystal Orbital Bond Index (COBI) may serve the busy quantum chemist interested in crystalline material. The latter technique, targeting the bond order, also allows to detect multicenter bonding in electron-rich solids. Likewise, atomic charges are automatically generated (that is, "fall off" from the theory) and help to understand the interplay between ionicity and covalency in whatever crystalline material (archetype binaries, elemental allotropes, nanomaterials, high-pressure phases, etc.) After far too many years of operating theoretically isolated from each other, first-principles molecular and solid-state quantum chemistries are finally reaching out to each other, a fortunate development.

References

1 Greenwood, N.N. (1968). *Ionic Crystals, Lattice Defects and Nonstoichiometry*. London: Butterworth.
2 Hellmann, H. (1937). *Einführung in die Quantenchemie*. Leipzig, Wien: Franz Deuticke.
3 Hoffmann, R. (1988). *Solids and Surfaces: A Chemist's View of Bonding in Extended Structures*. Weinheim, New York: VCH Publishers.
4 Allen, L.C. and Burdett, J.K. (1995). The metallic bond—dead or alive? A comment and a reply. *Angew. Chem. Int. Ed. Engl.* 34(18): 2003.
5 Bloch, F. (1929). Über die Quantenmechanik der Elektronen in Kristallgittern. *Z. Phys.* 52(7-8): 555–600.
6 Dronskowski, R. (2005). *Computational Chemistry of Solid State Materials*. Weinheim, Germany: Wiley-VCH.

7 Hohenberg, P. and Kohn, W. (1964). Inhomogeneous electron gas. *Phys. Rev.* 136(3B): B864.
8 Kohn, W. and Sham, L.J. (1965). Self-consistent equations including exchange and correlation effects. *Phys. Rev.* 140(4A): A1133.
9 Dronskowski, R. (2023). *Chemical Bonding from Plane Waves via Atomic Orbitals*. Berlin, Boston, MA: De Gruyter.
10 Landrum, G. and Glassey, W. (2004). *Yet Another Extended Hückel Molecular Orbital Package (YAeHMOP)*. Ithaca, NY: Cornell University.
11 Erba, A., Desmarais, J.K., Casassa, S. et al. (2023). CRYSTAL23: A program for computational solid state physics and chemistry. *J. Chem. Theory Comput.* 19(20): 6891–6932.
12 Niehaus, T.A., López, R., and Rico, J.F. (2008). Efficient evaluation of the Fourier transform over products of Slater-type orbitals on different centers. *J. Phys. Chem. A* 41(48): 485205.
13 Deringer, V.L., Tchougréeff, A.L., and Dronskowski, R. (2011). Crystal orbital Hamilton population (COHP) analysis as projected from plane-wave basis sets. *J. Phys. Chem. A* 115(21): 5461–5466.
14 Hellmann, H. (1935). A new approximation method in the problem of many electrons. *J. Chem. Phys.* 3(1): 61–61.
15 Blöchl, P.E. (1994). Projector augmented-wave method. *Phys. Rev. B* 50(24): 17953.
16 Kresse, G. and Hafner, J. (1993). Ab initio molecular dynamics for liquid metals. *Phys. Rev. B* 47(1): 558.
17 Kresse, G. and Hafner, J. (1994). Ab initio molecular-dynamics simulation of the liquid-metal–amorphous-semiconductor transition in germanium. *Phys. Rev. B* 49(20): 14251.
18 Kresse, G. and Furthmüller, J. (1996). Efficiency of ab-initio total energy calculations for metals and semiconductors using a plane-wave basis set. *Comput. Mater. Sci.* 6(1): 15–50.
19 Kresse, G. and Furthmüller, J. (1996). Efficient iterative schemes for ab initio total-energy calculations using a plane-wave basis set. *Phys. Rev. B* 54(16): 11169.
20 Kresse, G. and Joubert, D. (1999). From ultrasoft pseudopotentials to the projector augmented-wave method. *Phys. Rev. B* 59(3): 1758.
21 Giannozzi, P., Baroni, S., Bonini, N. et al. (2009). Quantum ESPRESSO: a modular and open-source software project for quantum simulations of materials. *J. Phys. Condens. Matter* 21(39): 395502.
22 Giannozzi, P., Andreussi, O., Brumme, T. et al. (2017). Advanced capabilities for materials modelling with quantum ESPRESSO. *J. Phys. Condens. Matter* 29(46): 465901.
23 Giannozzi, P., Baseggio, O., Bonfà, P. et al. (2020). Quantum ESPRESSO toward the exascale. *J. Chem. Phys.* 152(15): 154105.
24 Gonze, X., Beuken, J.M., Caracas, R. et al. (2002). First-principles computation of material properties: the ABINIT software project. *Comput. Mater. Sci.* 25(3): 478–492.

25 Torrent, M., Jollet, F., Bottin, F. et al. (2008). Implementation of the projector augmented-wave method in the ABINIT code: application to the study of iron under pressure. *Comput. Mater. Sci.* 42(2): 337–351.

26 Romero, A.H., Allan, D.C., Amadon, B. et al. (2020). ABINIT: Overview and focus on selected capabilities. *J. Chem. Phys.* 152(12): 124102.

27 Gonze, X., Amadon, B., Antonius, G. et al. (2020). The ABINIT project: impact, environment and recent developments. *Comput. Phys. Commun.* 248: 107042.

28 Clark, S.J., Segall, M.D., Pickard, C.J. et al. (2005). First principles methods using CASTEP. *Z. Kristallogr. –Cryst. Mater.* 220(5-6): 567–570.

29 Slater, J.C. (1937). Wave functions in a periodic potential. *Phys. Rev.* 51(10): 846–851.

30 Andersen, O.K. (1975). Linear methods in band theory. *Phys. Rev. B* 12(8): 3060.

31 Chadi, D.J. (1977). Localized-orbital description of wave functions and energy bands in semiconductors. *Phys. Rev. B* 16(8): 3572.

32 Sánchez-Portal, D., Artacho, E., and Soler, J.M. (1995). Projection of plane-wave calculations into atomic orbitals. *Solid State Commun.* 95(10): 685–690.

33 Maintz, S., Deringer, V.L., Tchougréeff, A.L., and Dronskowski, R. (2013). Analytic projection from plane-wave and PAW wavefunctions and application to chemical-bonding analysis in solids. *J. Comput. Chem.* 34(29): 2557–2567.

34 Maintz, S., Deringer, V.L., Tchougréeff, A.L., and Dronskowski, R. (2016). LOBSTER: A tool to extract chemical bonding from plane-wave based DFT. *J. Comput. Chem.* 37(11): 1030–1035.

35 Nelson, R., Ertural, C., George, J. et al. (2020). LOBSTER: Local orbital projections, atomic charges, and chemical-bonding analysis from projector-augmented-wave-based density-functional theory. *J. Comput. Chem.* 41(21): 1931–1940.

36 Mulliken, R.S. (1955). Electronic population analysis on LCAO–MO molecular wave functions. I. *J. Chem. Phys.* 23(10): 1833–1840.

37 Mulliken, R.S. (1955). Electronic population analysis on LCAO–MO molecular wave functions. II. Overlap populations, bond orders, and covalent bond energies. *J. Chem. Phys.* 23(10): 1841–1846.

38 Mulliken, R.S. (1955). Electronic population analysis on LCAO–MO molecular wave functions. III. Effects of hybridization on overlap and gross AO populations. *J. Chem. Phys.* 23(12): 2338–2342.

39 Mulliken, R.S. (1955). Electronic population analysis on LCAO–MO molecular wave functions. IV. Bonding and antibonding in LCAO and valence-bond theories. *J. Chem. Phys.* 23(12): 2343–2346.

40 Löwdin, P.O. (1970). On the nonorthogonality problem. *Adv. Quantum Chem.* 5: 185–199.

41 Madelung, E. (1918). Das elektrische Feld in Systemen von regelmäßig angeordneten Punktladungen. *Phys. Z.* 19: 524–533.

42 Ewald, P.P. (1921). Die Berechnung optischer und elektrostatischer Gitterpotentiale. *Ann. Phys.* 369(3): 253–287.

43 Hughbanks, T. and Hoffmann, R. (1983). Chains of trans-edge-sharing molybdenum octahedra: metal–metal bonding in extended systems. *J. Am. Chem. Soc.* 105(11): 3528–3537.

44 Dronskowski, R. and Blöchl, P.E. (1993). Crystal orbital Hamilton populations (COHP). Energy-resolved visualization of chemical bonding in solids based on density-functional calculations. *J. Phys. Chem.* 97(33): 8617–8624.

45 Müller, P.C. (2019). *Quantenchemische Deskriptoren zur Kartierung von Funktions-materialien. Master Thesis*. RWTH Aachen University.

46 Nelson, R., Ertural, C., Müller, P.C., and Dronskowski, R. (2022). Chemical bonding with plane waves. In: *Comprehensive Inorganic Chemistry III*, vol. 3 (ed. J. Reedijk and K. Poeppelmeier), 141–201. Elsevier.

47 Müller, P.C., Ertural, C., Hempelmann, J., and Dronskowski, R. (2021). Crystal orbital bond index: covalent bond orders in solids. *J. Phys. Chem. C* 125(14): 7959–7970.

48 Wiberg, K.B. (1968). Application of the Pople-Santry-Segal CNDO method to the cyclopropylcarbinyl and cyclobutyl cation and to bicyclobutane. *Tetrahedron* 24(3): 1083–1096.

49 Mayer, I. (1983). Charge, bond order and valence in the Ab initio SCF theory. *Chem. Phys. Lett.* 97(3): 270–274.

50 Sannigrahi, A.B. and Kar, T. (1990). Three-center bond index. *Chem. Phys. Lett.* 173(5-6): 569–572.

51 Giambiagi, M., de Giambiagi, M.S., and Mundim, K.C. (1990). Definition of a multicenter bond index. *Struct. Chem.* 1: 423–427.

52 Deringer, V.L. and Dronskowski, R. (2013). Computational methods for solids. In: *Comprehensive Inorganic Chemistry II*, vol. 9 (ed. J. Reedijk and K. Poeppelmeier), 59–87. Oxford: Elsevier.

53 van Arkel, A.E. and Swallow, J.C. (1956). *Molecules and Crystals in Inorganic Chemistry*, vol. 2. London: Butterworths.

54 Ketelaar, J.A.A. (1958). *Chemical Constitution. An Introduction to the Theory of the Chemical Bond*, vol. 2. Amsterdam: Elsevier.

55 Hempelmann, J., Müller, P.C., Ertural, C., and Dronskowski, R. (2022). The orbital origins of chemical bonding in Ge–Sb–Te phase-change materials. *Angew. Chem. Int. Ed.* 61(17): e202115778.

56 Hempelmann, J., Müller, P.C., Konze, P.M. et al. (2021). Long-range forces in rock-salt-type tellurides and how they mirror the underlying chemical bonding. *Adv. Mater.* 33(37): 2100163.

57 Jones, R.O., Elliott, S.R., and Dronskowski, R. (2023). The myth of "metavalency" in phase-change materials. *Adv. Mater.* 35(30): 2300836.

58 Ma, Y., Eremets, M., Oganov, A.R. et al. (2009). Transparent dense sodium. *Nature* 458(7235): 182–185.

19

Toward Interpretable Machine Learning Models for Predicting Spectroscopy, Catalysis, and Reactions

Jun Jiang[1] and Shubin Liu[2,3]

[1] *University of Science and Technology of China, Key Laboratory of Precision and Intelligent Chemistry, School of Chemistry and Materials Science, Hefei 230026 Anhui, China*
[2] *University of North Carolina, Research Computing Center, Chapel Hill NC 27599-3420, USA*
[3] *University of North Carolina, Department of Chemistry, Chapel Hill NC 27599-3290, USA*

19.1 Introduction

Machine learning (ML) is a subset of artificial intelligence (AI) to develop algorithms and statistical models to make fast, reliable, and robust predictions. It enables computers to learn and make predictions without being explicitly programmed. Instead, ML allows systems to automatically learn from the features of datasets and improve their performance over time. This is achieved through various computational techniques such as supervised learning, unsupervised learning, reinforcement learning, and deep learning. ML has been enjoying explosive popularity and vast applications in the recent literature of many disciplines including, but not limited to, theoretical and computational chemistry [1–7]. It is the intention of this chapter to briefly introduce this new subject to our readers, highlight a few of our recent ML applications to predict spectroscopy, catalysis, and reactions [8–17], and relate them to chemical understandings and conceptual frameworks from the perspective of interpretable ML models [18–22].

19.2 ML in a Nutshell

In ML, the goal is to build empirical/statistical models that can learn from the features of quality datasets and then make robust predictions for the new data [23]. In this regard, to build a viable ML model, three ingredients, i.e. datasets, features, and algorithms, are mandatory. ML features refer to the attributes of datasets that can be employed to train ML algorithms. ML algorithms learn patterns and establish relationships between the features and the target variable and use this knowledge to make generalizations and predictions for the new data.

ML datasets are collections of data used to train ML models. They are often big in size with millions and even billions of datapoints and in a variety of formats,

Exploring Chemical Concepts Through Theory and Computation, First Edition. Edited by Shubin Liu.
© 2024 WILEY-VCH GmbH. Published 2024 by WILEY-VCH GmbH.

Table 19.1 Samples of ML databases in theoretical and computational chemistry.

DATASET	Brief description	URL link
AFLOWLIB	Property database of over 625K materials	http://www.aflowlib.org
ANI-1	DFT dataset with 20M conformations for 57.5K small organic molecules	https://github.com/isayev/ANI1_dataset
ANI-1x/ANI-1ccx	5M DFT, 500K data points at CCSD(T)/CBS level of theory	https://github.com/aiqm/ANI1x_datasets
CMR	Computational materials repository	https://cmr.fysik.dtu.dk
FreeSolv	Hydration-free energies for neutral molecules in water	http://www.escholarship.org/uc/item/6sd403pz
Materials project	structural, electronic, and energetic data for over 500k compounds	https://www.materialsproject.org
MD17	MD trajectories for 1M conformational geometries	http://www.sgdml.org
MoleculeNet	Properties of over 700k compounds	https://moleculenet.org/
Open Catalyst Project	1.2M structures 250 M DFT calculations	https://opencatalystproject.org
OQMD	DFT predicted results for over 200K crystal structures	http://oqmd.org
PubChemQC PM6	221M structures at PM6 level	https://nakatamaho.riken.jp/pubchemqc.riken.jp/
PubChemQC	3M structure at DFT + 2M TD-DFT	https://nakatamaho.riken.jp/pubchemqc.riken.jp/b3lyp_pm6_datasets.html
QM9	Organic molecules up to 23 atoms with 7165 molecules	http://quantum-machine.org/datasets

including numerical, text, images, audio, and video. ML datasets are essential for training ML models. Without them, ML models cannot be trained, and thus they are useless in making predictions. ML datasets should be both representative and high quality. Otherwise, ML models based on them are inaccurate and not robust because they will not be able to generalize well to predict new data. There are many free ML datasets available online, such as MNIST, CIFAR-10, ImageNet, and TensorFlow Datasets. When an ML dataset is chosen, the following three factors should be taken into careful consideration, size, diversity, and quality. In Table 19.1, we list a handful of widely used datasets in theoretical and computational chemistry.

ML features can be any measurable quantity or observable property accessible from the dataset, such as numerical values (e.g. age, temperature, or income), categorical labels (e.g. gender, color, or type of vehicle), molecular properties (e.g. charge, dipole, or frontier orbital), text data (e.g. word and sentence), time series (e.g. stock

Table 19.2 Examples of ML descriptors used in theoretical and computational chemistry.

Descriptors	N-body nature
Atom-centered symmetry functions (ASCF)	1,2,3-body terms, cutoff
Atomic cluster expansion	1,2-body terms
Bag of bonds (BoB)	1,2-body terms
Cormorant	1,2-body terms, hierarchical
Coulomb matrix (CM)	1,2-body terms
Deep potential-smooth edition (DeepPot-SE)	1,2,3-body terms
Ewald sum matrix	1,2-body terms
Faber–Christensen–Huang–Lilienfeld (FCHL)	1,2,3-body terms
Invariant many-body interaction descriptor (MBI)	1,2,3-body terms
Many-body tensor representation (MBTR)	1,2,3-body terms
MPNN, SchNet	1,2-body terms, hierarchical
Overlap matrix	1,2-body terms
REMatch	1,2-body terms
Smooth overlap of atomic positions (SOAP)	Density-based, SO(3) rotational group
Spectrum of London and Axilrod-Teller-Muto potential (SLATM)	1,2,3,4-body terms
Symmetric Gradient Domain Machine Learning (sGDML)	1,2-body terms

price or sensor reading over time), or even image pixels. In ML, it is routine practice to generate thousands or even millions of features. There are several common techniques to select the most relevant features for training an ML algorithm, such as univariate selection (e.g. chi-square test, ANOVA, or correlation coefficient), recursive feature elimination (RFE), L1 regularization (Lasso), decision tree-based method (e.g. random forests [RFs] and gradient boosting), and principal component analysis (PCA). Table 19.2 illustrates examples of some commonly used ML features and their many-body nature. In Section 19.3, we will introduce some descriptors with the chemistry origin as ML features from our recent applications to predict spectroscopy, catalytical properties, and chemical reactions [8–17].

The third key ingredient of ML is the algorithms applied to train the dataset. They include:

- Linear Regression: This algorithm is used for solving regression problems, where the target variable is continuous.
- Logistic Regression: It is used for classification problems, where the target variable is categorical.
- Decision Trees: These are tree-like models where each internal node represents a feature, each branch represents a decision rule, and each leaf node represents the outcome.

- Random Forest: It is an ensemble learning method that combines multiple decision trees to make predictions.
- Support Vector Machines (SVM): This algorithm is used for both regression and classification tasks. It works by finding the best hyperplane that separates the data into different classes.
- Naive Bayes: It is a probabilistic classifier based on Bayes' Theorem. It assumes that features are conditionally independent given the class variable.
- K-Nearest Neighbors (KNN): This algorithm classifies new instances based on similarity measures, where k is the number of nearest neighbors to consider.
- Neural Networks (NN): These models are inspired by the human brain and consist of interconnected nodes or "neurons" that process and transmit information.

Most ML algorithms in the recent literature of theoretical and computational chemistry are based on NN. The key difference between regression analysis and ML is that regression aims to reconstruct the function that goes through a set of *known* data points with the lowest error, whereas ML aims to identify functions to *predict* interpolations between data points and thus minimize the prediction error for new data points that might appear later. In the formulation of typical regression analysis, to fit a dataset $\mathbf{x}$ to find the model function $\widehat{f}$ so that $\mathbf{y} = \widehat{f}(\mathbf{x})$ is satisfied as much as possible, we minimize the square loss (cost function) $\mathcal{L}(\widehat{f}(\mathbf{x}), \mathbf{y})$ with $\mathcal{L}(\widehat{f}(\mathbf{x}), \mathbf{y}) = (\widehat{f}(\mathbf{x}) - \mathbf{y})^2$, via the least-square optimization process. This process can be written as [23]:

$$\widehat{f} = \arg\min_{f \in F} \left[\sum_{i}^{M} \mathcal{L}(\widehat{f}(\mathbf{x}_i), \mathbf{y}_i) \right].$$

In ML algorithms, however, typical ML models to be trained are in the following formulation by introducing an additional *regularization* term $\Gamma(\Theta)$ in the cost function to avoid overfitting of the data, especially when the trained and test data are much varying [23], a popular choice of which is LASSO (least absolute shrinkage and selection operator).

$$\widehat{f} = \arg\min_{f \in F} \left[\sum_{i}^{M} \mathcal{L}(\widehat{f}(\mathbf{x}_i), \mathbf{y}_i) + ||\Gamma(\Theta)||^2 \right]$$

where Γ is a matrix that defines "simplicity" with regard to the model parameters Θ. This additional regularization term is necessary in ML because it influences the selection of candidate models by introducing additional properties to promote *generalization*. Usually, $\Gamma = \lambda \mathbf{I}$ (where $\mathbf{I}$ is the identity matrix and $\lambda > 0$) is chosen to simply favor a small L^2-norm on the parameters, such that the solution does not rely on individual input features (i.e. descriptors) too strongly. This approach is called Tikhonov *regularization*.

Standard ML models obtained in this manner are intrinsically nonlinear and, in most cases, highly uninterpretable. This is because, besides the nonlinearity, ML models often contain excessive number of parameters. It can be hard to gain physiochemical insights into the modeled relationship using a small set of simple rules. Techniques have emerged to make ML models interpretable, such as interpretable ML and explainable AI (XAI) [24, 25]. Interpretable ML is a branch of ML that focuses on making ML models understandable to humans. This contrasts with typical black-box ML models, which are opaque and cannot be easily explained. There are many reasons why interpretability is important in ML. It can help us to understand how the model works and why it makes robust predictions. It can also be helpful to debug models, identify potential biases, and ensure the fairness and robustness of model predictions. There are different methods available in the literature to make ML models interpretable in terms of, for instance, feature importance, domain specificity, and rule-based modeling. The interpretability of ML models can also be determined by ML features. Interpretable ML models as a subset of more general ML models are what we will focus on and discuss in the chapter.

ML has become an increasingly important tool in theoretical and computational chemistry in recent years [1–7]. Main applications have been found in property predictions of stability, reactivity, and solubility, reaction predictions to design new synthetic pathways and understand reaction mechanisms, structure predictions of molecular geometry, electronic structure, and vibrational frequencies, and discoveries of new materials and new drugs. Common features utilized in ML models, as illustrated in Table 19.2, are topological (e.g. number of atoms, number of bonds, and connectivity of the atoms), geometric (e.g. bond lengths, bond angles, and molecular volume), electronic (e.g. ionization potential, electron affinity, and HOMO-LUMO gap), and quantum mechanical (e.g. molecular energy, dipole moment, and polarizability). In this chapter, based on our work in the past few years, we present a list of chemistry-based descriptors as ML features and then highlight a few ML applications in predicting spectroscopy, catalysts, and chemical reactions [8–17]. This chapter is by no means a comprehensive review of the recent literature on this topic. Several excellent reviews can be found elsewhere [1, 6, 7]. Rather, it is our humble intention to bring this subject to the attention of this group of the physical chemistry community, where chemical understanding from the perspective of theory and computation is of particular interest.

19.3 Chemistry-Based Descriptors as ML Features

Using quantities from ab initio electronic structure calculations of molecules and solids alike as the descriptors to evaluate physiochemical properties has a long history in the literature. The linear free energy relationship (LFRE) in physical organic chemistry is one example. QSAR/QSPR (quantitative structure–activity relationship/quantitative structure–property relationships) in medicinal chemistry and pharmacy is another one. In recent studies, for example, we employed molecular electrostatic potential and valence atomic orbital energies to accurately evaluate molecular acidity [26] and utilized Hirshfeld charge and information gain to quantify electrophilicity and nucleophilicity [27], the capability of molecules to accept and donate electrons. These empirical models are often linear and involve only one or a few descriptors, so they are by no means ML models per se. However, these descriptors have well-defined physiochemical origin, so they can be viewed as the precursors of interpretable ML models.

As illustrated in Table 19.2, there can be a variety of possible chemistry-based descriptors as ML features. Molecular descriptors can be classified into two

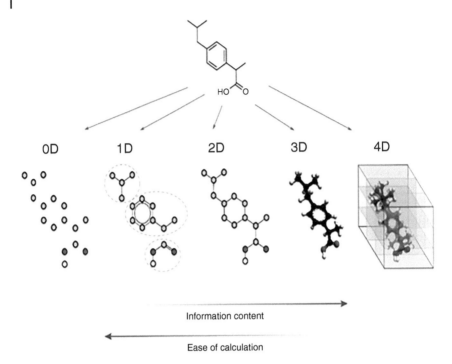

Figure 19.1 Five (0D–4D) dimensions of chemistry-based theoretical descriptors. Source: Grisoni et al. [28]/Springer Nature.

categories, experimental and theoretical. Experimental descriptors are any physicochemical properties from experimental measurements. One example of such experimental descriptors is the partition coefficient, which is defined as the ratio of the concentrations of a compound in a mixture of two nonmiscible solvents at equilibrium. Theoretical descriptors are derived from the theoretical or computational representation of molecules. This category of descriptors can be divided into five classes according to their dimensionality, from 0D (zero-dimension) to 4D (four-dimension), of the information they provide. An example of these five-dimensional descriptors for a simple organic molecule is shown in Figure 19.1.

Earlier, a different partition set of the theoretical descriptors was proposed by us, as shown in Figure 19.2, from our efforts of designing catalysts and predicting catalytical reactivity and selectivity using ML models. It consists of 14 descriptors, which can be categorized into 5 families, including structural, electrostatic, covalency, electron occupancy, and exchange interaction. In what follows, according to our work as well as recent studies on catalysis, in particular, from other groups that are interested in ML models, we present chemistry-based theoretical descriptors in a little different manner.

19.3.1 Intrinsic Atomic Property Descriptors

Wang and coworkers [30] recently combined density functional theory (DFT) calculations and ML techniques to develop a simple and universal class of

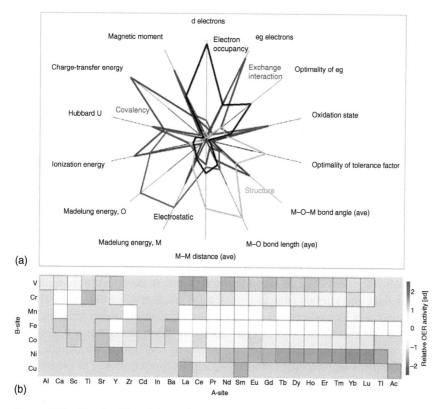

Figure 19.2 Five families of theoretical descriptors from our recent work. Source: Reproduced with permission of Ref. [29]. Copyright 2023, Wiley.

descriptors based on inherent atomic properties (e.g. electronegativity, electron type, and number). With this class of descriptors as ML features, they designed 2D materials supporting dual atom electrocatalysts (DACs@2D) that possess superior reactivity and selectivity. For a catalytic metal atom M interacting with the set of coordination atoms X, the proposed descriptor φ was expressed as $\varphi = (\chi_M + \Sigma \chi_x) + N_{d/p}$, in which $\chi_M + \Sigma \chi_x$ and $N_{d/p}$ denote the coordination environment defined by atomic electronegativity values χ and the number of d or p electrons on the metal atom, respectively. This kind of descriptor has been applied to evaluate the complicated interfacial effects on the electrochemical reduction reactions involving CO_2, O_2, and N_2 reduction reactions, and design new catalysts with superior activity and selectivity, viz. CuCr/g-C_3N_4 for CH_4 and CuSn/N-BN for HCOOH (Figure 19.3).

19.3.2 Electronic and Structural Property Descriptors

Li et al. [31] combined DFT-calculated d-band features of active sites and an ANN model to learn and predict adsorption energies of *CO and *OH on alloy surfaces for the rapid screening of bimetallic catalysts. The d-band features used include filling

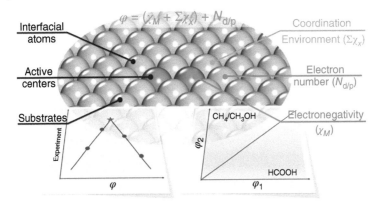

Figure 19.3 Intrinsic atomic property descriptors for heterogeneous catalysis. Source: Reproduced with permission from Ref. [30]. Copyright 2022, American Chemical Society.

(zeroth moment up to the Fermi level), center (first moment relative to the Fermi level), width (square root of the second central moment), skewness (third standardized moment), and kurtosis (fourth standardized moment). In addition, local electronegativity was also selected to assess the contribution of sp-electron density to adsorption energies. To enhance interpretability for ML models, *feature importance* analysis was carried out using normalized sensitivity coefficients, which reflect the degree of dependency of a target property on a given input feature. The existence of linear dependence between the six primary features and the *CO/*OH adsorption energies (Figure 19.4, inset) confirms that d-states and sp-states act as distinct governing factors of *CO and *OH adsorption on bimetallic catalyst surfaces, in good agreement with our chemical intuition.

Various geometric descriptors have been developed in heterogeneous catalyses, such as bond length, rotational angle, smooth overlap of atomic positions (SOAP), many-body tensor representation (MBTR), atom-centered symmetry functions (ACSF), Coulomb matrix (CM), coordination number, and so on. MBTR and CM are global descriptors based on tensor representation and coulomb repulsion, respectively, whereas SOAP and ACSF are local descriptors. SOAP, MBTR, ACSF, and CM were applied to predict the hydrogen adsorption energy on the surface of nanoclusters MoS_2 and AuCu, and SOAP was found to perform significantly better than others.

19.3.3 Multilevel Attention Mechanisms-Identified Descriptors

Ma and coworkers [32] have developed a multilevel attention graph convolution neural network (MA-GCNN) that has been applied to predict the energetics of hydroxyapatite nanoparticles and quantum chemical properties of organic molecules. For a given molecule, the weights assigned to different effects of adjacent atoms on the central atom are called "attention." In contrast to other attention algorithms that use "attention" at a single step, MA-GCNN applies multilevel attention at every message-passing step, gradually capturing the influence of

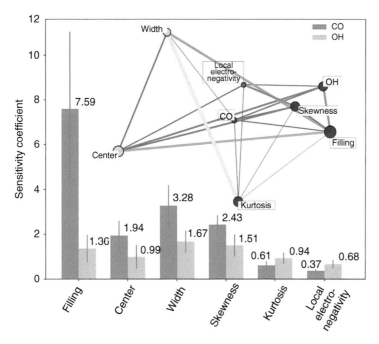

Figure 19.4 Feature importance scores for *CO and *OH adsorption energies based on sensitivity analysis of ANN models. The inset shows the linear dependence among input features and adsorption energies. Node size is proportional to the degree of linear dependence of one variable on the others. Source: Li et al. [31]/The Royal Society of Chemistry.

different atomic nodes at each time step. By using MA-GCNN, Gu et al. identified the important roles played by hydrogen bonding (HB) interactions and metal coordination (metal acidity) in predicting reaction energies of the NRR in metal-zeolites, which motivated them to select HB features and local acidity as the descriptors of constructing interpretable ML models. It was used to predict the energy changes (ΔE: the relative energy to free N_2, $\Delta\Delta E$: energy difference between two successive steps) of the NRR process. The results of the attention mechanism for the three main hydrogenation and dehydrogenation steps gave large weights (red color) to the metal centers (M, e.g. Ti), intermediates (NN*, NNH*, NH_2*, NH_3*, NNH_3*, N*), and HB interactions between the H atoms of intermediates and O atoms of channels, indicating the important role played by the HB interactions in the nitrogen reduction reaction (NRR) process (Figure 19.5).

19.3.4 SISSO Method-constructed Descriptors

Data-driven descriptors constructed by the compressed sensing method SISSO (sure-independence screening and sparsifying operator) are also effective in predicting physiochemical properties and catalytic reactivity. Based on a combinatorial pool of features and mathematical operators, SISSO can handle high-dimensional and nonlinear relationships and identify the best descriptor

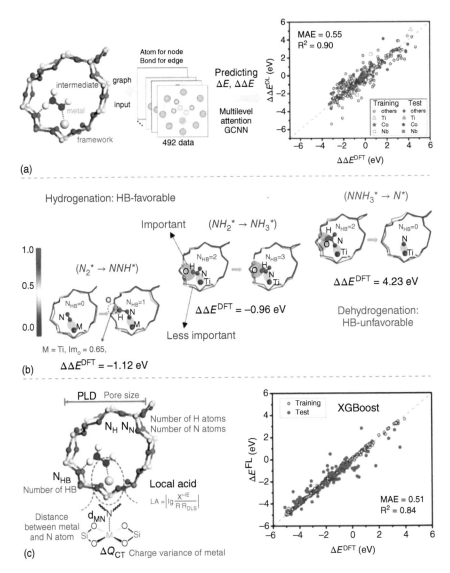

Figure 19.5 (a) A flowchart of ΔE (relative energy to free N_2) and $\Delta\Delta E$ (energy difference between two successive steps) prediction by a multilevel attention graph convolutional neural network (GCNN) applied to NRR metal-zeolite catalysts. (b) The multilevel attention mechanism reveals that hydrogen bonding (HB) interactions favor the hydrogenation step but disfavor the dehydrogenation step. DFT calculated $\Delta\Delta E$ values for these steps are given. Red and blue colors denote relative weights of 1 and 0, respectively. (c) Descriptors used for predicting ΔE and the results predicted by the XGBoost model. MAE, mean absolute error; PLD, pore largest diameter; N_{HB}, number of hydrogen bonds. Source: Mou et al. (2022)/The Royal Society of Chemistry.

out of an immensity of even billions of candidates. Combining DFT calculations and SISSO, Han et al. [33] reported a fast yet reliable high-throughput method for screening more than five thousand single-atom-alloy catalysts (SAACs) of hydrogenation reactions. This method reduced the computational cost by at least one thousand times compared with a pure DFT approach. Their screening criteria included the hydrogen binding energy, the H_2 dissociation energy barrier, and the guest-atom segregation energy evaluated in the presence of adsorbed hydrogen. Accurate predictions were obtained by SISSO-derived descriptors that were constructed based on only 19 primary features of the host surfaces and guest single atoms. SISSO-constructed descriptors are usually complex analytic formulas, reflecting the complexity of the relationships between the primary features and the target properties. While potentially interpretable, SISSO by itself does not provide a straightforward way of evaluating the relative importance of different features for identifying desirable changes in target properties. However, the data-mining method such as subgroup discovery (SGD) has proved useful for facilitating a physiochemical understanding of SISSO descriptors. Given a dataset and a target property, the SGD algorithm identifies local patterns that maximize or minimize a quality function and describes them as an intersection of simple inequalities involving a defined set of features and a set of adjacent data cluster borders (a_1, a_2, ...), for example, "(feature1 < a_1) AND (feature2 > a_2) AND ...." From this, SGD identifies both the most important subgroups and the relevant primary features for a given target property. Through a qualitative analysis of complex SISSO descriptors by SGD, Han et al. revealed the actuating mechanisms for desirable changes in the target properties, for example, increasing the catalyst's stability and reducing the reaction barrier, in terms of basic features of the material.

19.3.5 Spectral Descriptors

The electric dipole moment parameters provide both quantitative and spatial information about electron distributions and are computationally accessible. In a recent study [34], with a NN ML technique, the adsorption energy (E_{ads}) and charge transfer Δe of CO and NO adsorption on an Au(111) surface were first investigated based on calculations of 10 000 adsorption configurations, generated by changing the adsorption angles φ (from 0° to 90°) and θ (from 0° to 180°) in small steps and then performing static calculations directly. Two descriptors were extracted from the surface-dipole (d_{sur}) and molecule-dipole (d_{mol}) moments, that is, the dipole–dipole interaction potential energy (V_{dd}) and the angle α between d_{sur} and d_{mol}, quantities which have been shown to have a significant impact on charge/energy transfer. In addition, two descriptors widely used in catalysis research, work function (WF) and d-band center (ε_d), were also included. The four selected descriptors more accurately predicted E_{ads} and Δe of CO and NO adsorption on an Au(111) surface compared to DFT calculations. Feature importance analysis revealed the effectiveness of the dipole-related descriptor α for predicting E_{ads} and Δe as it had the highest importance among the four descriptors.

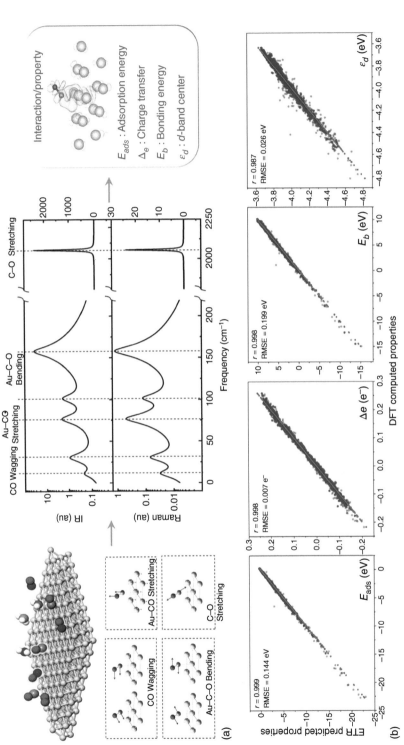

Figure 19.6 Using spectroscopy-based features to predict surface–adsorbate interactions. (a) Schematic illustration of using spectral features as descriptors to predict molecular adsorption: left panel: cartoon representation of a typical surface–adsorbate system (top) and typical vibrational modes (bottom), central panel: exemplary IR and Raman spectra of such a system and assignment of its six key vibrational modes, right panel: target properties to be predicted. (b) Comparison of DFT computed (red dots) and ETR predicted (blue dots) properties (E_{ads}, Δe, E_b, and ε_d) for CO@Au(111). Input parameters are vibrational frequencies and IR and Raman intensities of the six most relevant vibrations. All negative frequencies, if any, are set to zero. Source: Wang et al. [34]/American Chemical Society.

Molecular dipoles are very difficult to measure experimentally but have a strong association with vibrational spectra, such as infrared (IR) and Raman spectra, which can be measured experimentally, simulated theoretically, and even predicted by ML tools. We have demonstrated that surface–adsorbate interaction properties including adsorption energy and charge transfer can be quantitatively determined directly from IR and Raman spectroscopic signals of adsorbates. As shown in Figure 19.6, six vibrational modes most relevant to CO adsorption (two bending modes ω_1 and ω_2, two wagging modes ω_3 and ω_4, one weak metal–C stretching mode ω_5, and one strong C–O stretching mode ω_6) and their corresponding IR/Raman intensities (I and R) were selected as the input features to describe target properties including E_{ads}, Δe, C–O bond energy (E_b), and the d-band center of the metal surface (ε_d). The application of these 18 vibrational spectral features combined with ML extra-trees regression (ETR) was used to accurately predict E_{ads}, Δe, E_b, and ε_d for the CO@Au(111)/Ag(111) system. Moreover, by employing SISSO, the machine-learned spectrum–property relationships can be described by mathematical formulas, with adsorbate spectral features being the variables. Interestingly, these vibrational feature-based formulas have generalizability to a series of new surface–adsorbate systems, including new metals, binary alloys, and high-entropy alloys, with excellent predictive ability, typically with r values greater than 0.8 and many exceeding 0.9. It was also found that these spectroscopy-based formulas allow the separation of contributions from substrate and adsorbate: the variables I, R, and ω are adsorbate spectral signals, while the parameters a, b, c, and d are constants related to intrinsic characteristics of the substrate. The development of spectral features as catalytic descriptors to establish quantitative spectrum–property relationships opens a new avenue for investigations of catalytic activity, circumventing the difficulties in learning detailed geometric structures of complex catalysts. These spectroscopic descriptors were recently applied to recognize protein secondary structures as well. Other applications are possible.

19.4 Selected ML Applications

ML has greatly advanced the application of quantum modeling and chemical physics for determining electronic structures, energetics, reaction activities, drug discovery, and materials design. Extensive efforts in applying ML techniques have been reported for correlating spectral features with descriptors from molecular structures and other properties. For example, photoionization, X-ray absorption (XAS), UV–visible, IR, and nuclear magnetic resonance (NMR) spectra can now be predicted by ML models based on either detailed geometrical structure (i.e. 3D coordinates) or abstract structural descriptors only containing atomic connectivity. Here, we highlight a few of our recent applications to predict spectroscopy, catalytical properties, and chemical reactions from the perspective of interpretable ML models using descriptors with the well-defined physiochemical origin.

19.4.1 ML Prediction of IR/Raman Spectroscopy

Vibrational spectroscopy is one of the most applied techniques for determining molecular structures. Conventional approaches to obtain spectroscopy often involve experimental measurement or first-principles simulations to establish the one-to-one spectrum–structure relationship. ML protocols and models to predict IR/Raman spectra were recently made possible using spectral fingerprints with local molecular structures, which are not only quick and accurate in predicting IR absorption and Raman spectra but, more importantly, also enable structure recognition of chemical groups from vibrational spectral features. In addition, ML models demonstrate reliable transferability, implying its likelihood of being extended to other spectral or nonspectral characteristics. In one of our recent studies [35], it was done with QM9 dataset with 21 950 molecules (Figure 19.7). Molecules containing –OH and –C=O groups were employed to construct NN models. Radial symmetry functions (RSFs) and angular symmetry functions (ASFs) were employed

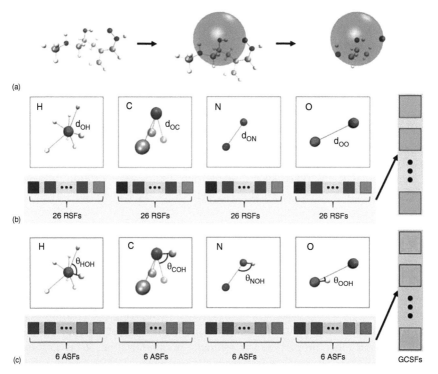

Figure 19.7 Schematic of the symmetry function descriptor for encoding the local chemical environment into feature vectors. (a) An appropriate cutoff radius for the hydroxyl group is selected, only atoms within the cutoff sphere are considered. (b) Radial symmetry functions (RSF) centered on the hydroxyl oxygen atom. 26 RSFs are used to encode the contribution of atoms of each element H, C, N, and O. There are 104 (26 × 4) elements in total arising from the radial part. (c) Similar to (b), but for angular symmetry functions (ASFs), with six ASFs for each element. Source: Reproduced with permission from Ref. [35]. Copyright 2021.

as descriptors to account for two-body and three-body interactions between atoms. ML predictions compared with DFT results are the B3LYP/6-31G(2df,p) of level (Figure 19.8).

Recently, we also applied the same idea to predict IR spectra for proteins [36]. The ML protocol is shown in Figure 19.9. We adopt a divide-and-conquer strategy to treat the amide I vibrations of the whole protein. The Frenkel exciton model is employed to construct a vibrational model Hamiltonian, in which the diagonal elements are the frequency (ω_i) of the ith amide I oscillator, and the off-diagonal elements represent the coupling between two oscillators i and j. For a pair of non-neighboring oscillators, since the distances between oscillators are greater than their sizes, the coupling is calculated with the dipole approximation. The

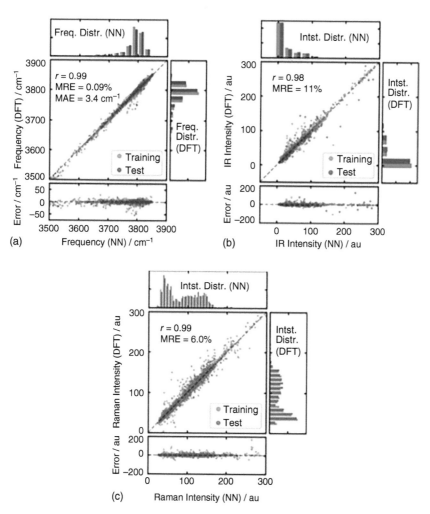

Figure 19.8 Correlation plots of (a) NN and DFT frequencies, (b) IR, and (c) Raman intensities for −OH stretching vibrations. Top and right panels depict distributions, while bottom panel shows errors. Source: Ren et al. [35]/Elsevier.

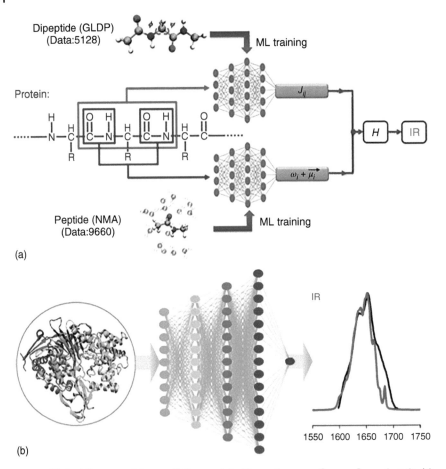

Figure 19.9 ML protocol for predicting protein IR spectroscopy. Source: Reproduced with permission from Ref. [36]. Copyright 2021, American Chemical Society.

N-methylacetamide (NMA) molecule was taken as the model system for NN training. For the vibrational couplings between two neighboring peptide bonds, we approximated by the glycine dipeptide. We employed CM as the descriptor for these predictions.

19.4.2 ML Prediction of Surface-Enhanced Raman Spectroscopy

Surface-enhanced Raman spectroscopy (SERS) is a powerful analytical tool for probing interfacial structures *in situ* at the molecular level. First-principles prediction of SERS is a long-standing challenge because of the diversified interfacial structures. We developed a cost-effective ML RF model that can predict SERS signals of a *trans*-1,2-bis (4-pyridyl) ethylene (BPE) molecule adsorbed on a gold substrate [10]. We employed the RF technique and geometric descriptors extracted from quantum chemistry simulations of thousands of ab initio molecular dynamics conformations (Figure 19.10). The ML protocol can predict SERS frequencies and

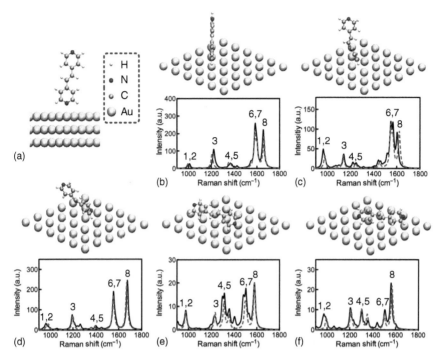

Figure 19.10 (a) Atomic model of BPE on Au(111) surface. The BPE/Au configuration taken at 0 (b), 1 ps (c), 2 ps (d), and 3 ps (e) of AIMD evolution and the stable structure (f) from DFT optimization, together with DFT-calculated (solid black line) and the ML-predicted SERS spectra (dashed red line). Source: Reproduced with permission from Ref. [10]. Copyright 2019, American Chemical Society.

intensities that are comparable with DFT and experimental results (Figure 19.11). It can also predict SERS responses of the molecule on different surfaces, or under external fields of electric fields and solvent environment, demonstrating its decent transferability and applicability.

19.4.3 ML Prediction of Ultraviolet Absorption Spectroscopy for Proteins

Ultraviolet (UV) absorption spectra are commonly used for characterizing the global structure of proteins. However, the theoretical interpretation of UV spectra is hindered by the large number of required expensive ab initio calculations of excited states spanning a huge conformation space. We established an ML protocol for far-UV (FUV) spectra of proteins [37], which can predict FUV spectra of proteins with comparable accuracy to DFT calculations but with three to four orders of magnitude in computational cost reduction (Figure 19.12). It has exhibited excellent predictive power and transferability that can be used to probe structural mutations and protein folding/unfolding pathways. We employed internal coordinates, embedded density, and converted Cartesian coordinates as the molecular descriptors for the input layer of a deep-learning NN containing three

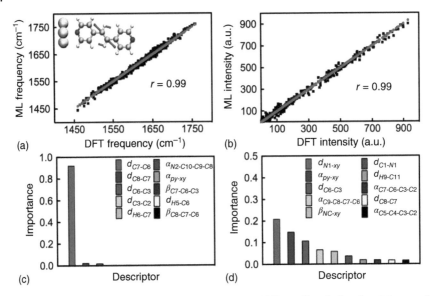

Figure 19.11 (a,b) Comparison of DFT-computed and ML-predicted vibrational frequencies and Raman intensities for mode 8 (shown in the inset of (a)). The Pearson correlation coefficient (r) of ML reflects the agreement. (c,d) Descriptor importance analysis of frequencies and Raman intensities. Source: Reproduced with permission from Ref. [10]. Copyright 2019, American Chemical Society.

hidden layers (with 32, 64, and 128 neurons, respectively) and L2 regularization (Figure 19.13).

19.4.4 ML Prediction of Protein Circular Dichroism Spectra

Proteins with different secondary structures have distinctive signatures in ECD (electronic circular dichroism) spectra, making them useful in studying protein dynamics such as folding/unfolding and binding events, yet the challenge of predicting electric and magnetic transition dipole moments poses a significant barrier to overcome. To tackle the issue, we designed ML models [38] with ordinary pure geometry-based descriptors replaced by embedded density descriptors (Figure 19.14). These models can simulate protein CD spectra with nearly four orders of magnitude faster than conventional first-principles simulations, and the predicted CD spectra are in reasonable agreement with experimental results (Figure 19.15). We also predicted a series of CD spectra for a Trp-cage protein along its folding pathway, suggesting that our ML model can be applied for real-time CD spectroscopy study of protein dynamics.

19.4.5 ML Prediction of Bond Dissociation Energy

Bond dissociation energy (BDE) as a basic thermodynamic property of a molecule is a quantitative measure of the strength of its chemical bonds. It can be either determined experimentally by pyrolysis, kinetics, and electrochemistry, or obtained

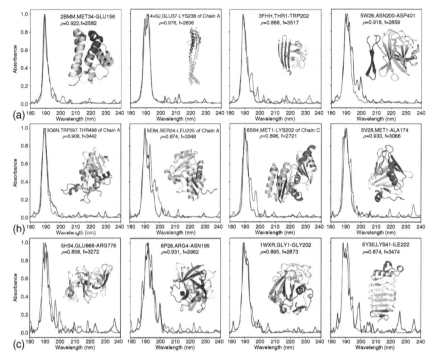

Figure 19.12 FUV spectra of 12 proteins (a: α-helix, β-sheet, b,c: α-helix + β-sheet) were calculated with the DFT/TDDFT (black curves) and ML (red curves) methods. Source: Reproduced with permission from Ref. [37]. Copyright 2021, American Chemical Society.

theoretically with computational approaches. These methods are usually costly and inefficient, posing challenges for large-scale high-throughput screening of materials. We built an ML model with six geometric descriptors (Figure 19.16) and a fully connected neural network (FCNN) and RF (Figure 19.17) to reliably and efficiently predict BDE values for organic carbonyls.

19.4.6 ML Predictions of Catalytical Properties

Using ML techniques and spectroscopic descriptors, we recently established quantitative spectrum–property relationships. Key interaction properties of substrate–adsorbate systems, including adsorption energy and charge transfer, are quantitatively determined from IR/Raman spectroscopic signals of the adsorbates. These ML-based spectrum–property relationships are physically interpretable and therefore transferrable to a series of metal/alloy surfaces. We studied four systems, CO or NO adsorption on pristine Au or Ag substrates. The IR/Raman spectra of small molecules CO or NO derived from first-principles calculations were selected as descriptors, which have six vibrational modes (Figure 19.18): one strong stretching (v_{C-O} or v_{N-O}), one weak stretching (v_{M-C} or v_{M-N}), two bending (v_{bend}), and two wagging modes ($v_{wagging}$). Explicit vibrational modes enabled us to associate spectroscopic descriptors with the specific vibrational

500 | *19 Toward Interpretable Machine Learning Models*

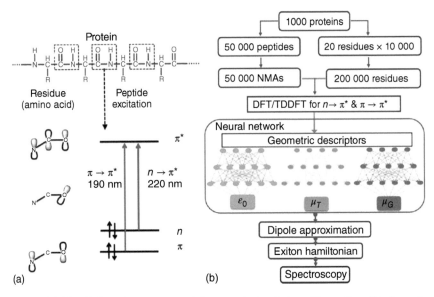

Figure 19.13 (a) Protein structure and the two electronic transitions of peptide ($n \to \pi^*$ and $\pi \to \pi^*$) which contribute to FUV adsorption. (b) Schematic of the machine-learning protocol for FUV protein spectroscopy. Source: Reproduced with permission from Ref. [37]. Copyright 2021, American Chemical Society.

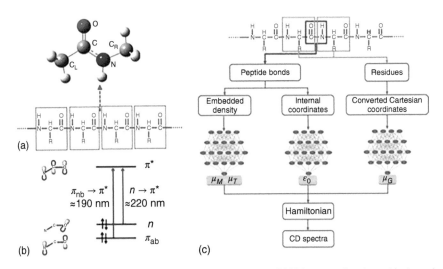

Figure 19.14 (a) NMA structure and protein structure. (b) Valence molecular orbitals and two electronic transitions of the peptide bond which are $n \to \pi^*$ or $\pi \to \pi^*$ transitions. (c) Machine learning protocol for predicting protein CD spectra. Source: Reproduced with permission from Ref. [38]. Copyright 2021, American Chemical Society.

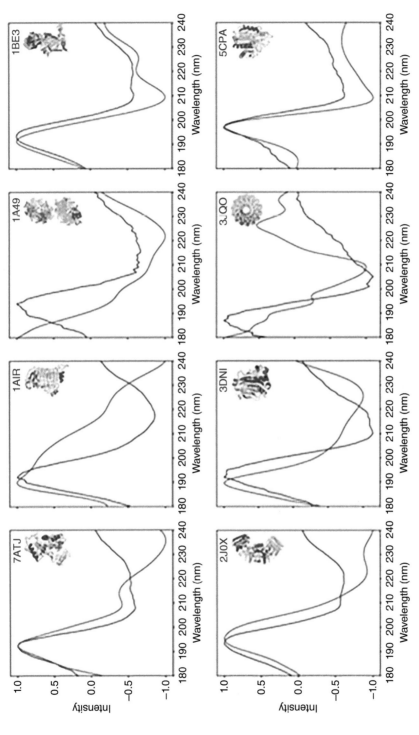

Figure 19.15 Experimental (black curves) and ML predicted (red curves) CD spectra of different types of proteins. Intensity is scaled to have the same maximum intensity for each panel. Source: Reproduced with permission from Ref. [38]. Copyright 2021, American Chemical Society.

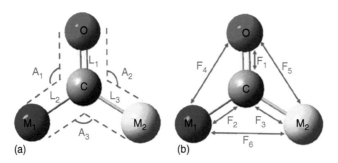

Figure 19.16 Feature descriptors. (a) L_1, L_2, and L_3 are bond lengths, and A_1, A_2, and A_3 are bond angles. (b) F_1–F_6 are the Coulomb forces. (O) Oxygen, (C) carbon, (M_1, M_2) ortho groups. Here, M_1 and M_2 are determined by the atomic order of SMILES format. Source: Reproduced with permission from Ref. [39]. Copyright 2020, American Chemical Society.

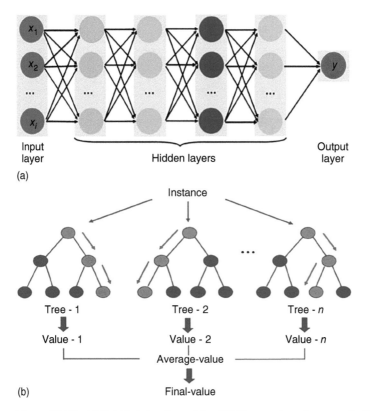

Figure 19.17 (a) Detailed structure of the NN. The number of neurons in the input layer is 6. There are four hidden layers, and the number of neurons in each layer is 100, 150, 150, and 100, respectively. The number of neurons in the output layer is 1. (b) Random forest model structure with 240 decision trees. Source: Reproduced with permission from Ref. [39]. Copyright 2020, American Chemical Society.

behavior of small molecules, making spectroscopic descriptors physically interpretable. For better interpretability, glass-box methods were employed with SISSO (Figure 19.18). We observed that the analytical expressions generated by SISSO primarily relied on high-frequency vibrational modes and often did not require information from low-frequency vibrational modes. The analytical expressions derived from SISSO were successfully applied to make predictions for 34 new catalytic systems (Figure 19.19). Their outstanding transferability suggests that these spectral descriptors have captured the underlying physiochemical insights. The analytical expressions derived from spectroscopic descriptors enable the separation of substrate and adsorbate contributions and thus make the ML models interpretable.

19.4.7 ML Predictions for Imperfect and Small Chemistry Data

The rapid development of ML techniques has brought a new paradigm of data-intelligence-driven research in chemistry. However, the mainstream ML methods often provide only "black box" predictive models that are good within homogenous datasets. They have to rely on the supply of large training datasets for transfer learning to heterogeneous datasets. Unfortunately, high-quality data are often difficult to obtain in chemical research. Limited by experimental and computational capabilities and costs, chemical data are often sparse or even lacking to meet the data needs of intelligent models. To address this challenge, establishing interpretable ML models provides a promising solution [40]. We took the task of predicting the performance metrics (adsorption energy and charge transfer) of copper-based metal–organic frameworks (MOFs) catalysts as an example and developed interpretable ML models based on the SISSO algorithm using IR/Raman spectral features as descriptors (Figure 19.20a).

As shown in Figure 19.20b, when the training dataset has more than 500 samples, the prediction performance of the NN model remains at an acceptably high level. However, when the number of samples continues to decrease, the prediction performance of NN model decreases dramatically, which is a common problem for black-box ML models. In contrast, the predictive capability of the SISSO mathematical formula model remains almost unchanged at high levels as the decrease of training dataset. The advantage of interpretable mathematical formulas on small datasets will be reflected in the future prediction of expensive and sparse experimental samples. Similarly, when adding incorrect data to the training set, the prediction performance of NN model immediately deteriorates, while the SISSO formula is more robust and error-tolerant, which can be attributed to the fact that fewer parameters in the mathematical formula allow the model to avoid falling into the trap of incorrect data points. Overall, a spectroscopy–property relationship appeared as a "glass box" ML model with explicit mathematical expressions. Such "glass box" model breaks the dependence of "black box" models on large datasets, and shows excellent generalization capability, even with small amounts of training data and even with partial errors in the data. The approach toward interpretable

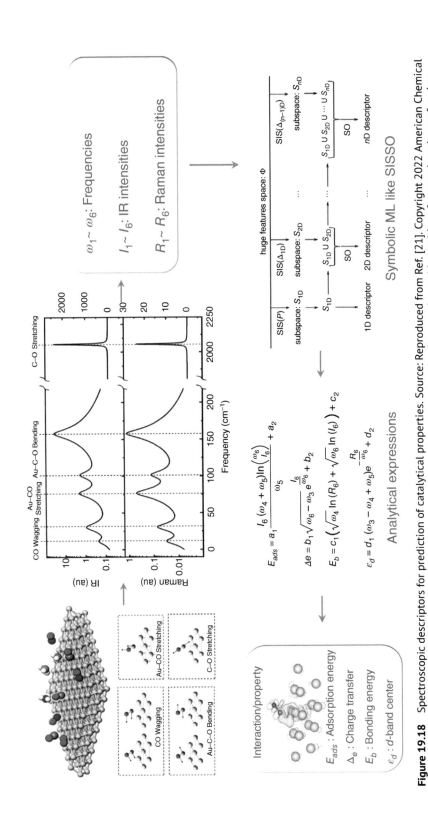

Figure 19.18 Spectroscopic descriptors for prediction of catalytical properties. Source: Reproduced from Ref. [21]. Copyright 2022 American Chemical Society. The workflow includes collecting experimental/computational spectra, extracting frequencies and intensities of spectral peaks, performing symbolic ML like SISSO, obtaining analytical expressions, and predicting catalytic interactions or properties. Source: Wang et al. [17]/American Physical Society.

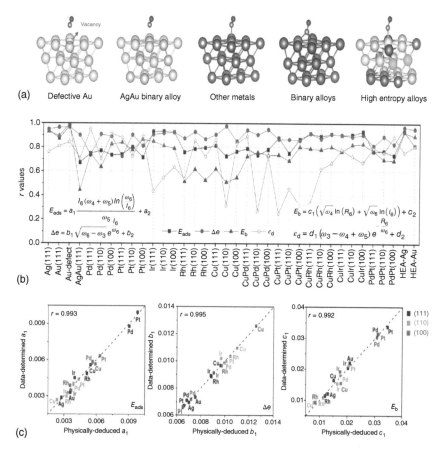

Figure 19.19 Advantages of interpretable ML model with spectroscopic descriptors. Source: Reproduced from Ref. [21]. Copyright 2022 American Chemical Society. (a) Examples of 34 surface–adsorbate systems. (b) Predication power of SISSO formulas trained from the source domain transfer (CO@Ag(111) and CO@Au(111)) to the target domains (34 surface–adsorbate systems). The r represents the Pearson correlation coefficient. (c) The parameters can also be derived from several intrinsic properties of metal substrates.

intelligent models will greatly enhance the applicability of ML methods in the field of chemistry. In future, we expect to use theoretical big data to generate interpretable pre-trained models and then rely on small experimental data for transfer learning to build a theory-experiment collaboration model for complex systems to facilitate *in situ* and real-time analysis of real catalytic systems.

19.4.8 ML Predictions in Reactions and Retrosynthesis

To apply ML techniques to predict chemical reactions, especially for the purpose of retrosynthesis, we recently introduced chemical information, including NMR chemical shifts, bond energies, catalysts, and solvents into the descriptor of molecules and reactions and into molecular graphs to represent molecules

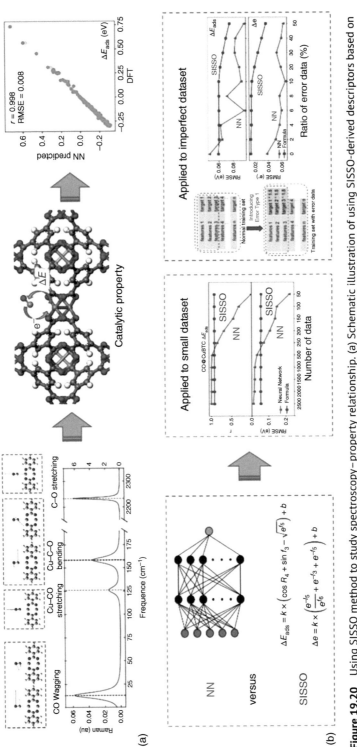

Figure 19.20 Using SISSO method to study spectroscopy–property relationship. (a) Schematic illustration of using SISSO-derived descriptors based on vibrational spectral features to predict catalytic properties such as adsorption energy and transfer charge. Left panel: vibrational spectra of a typical adsorbate (CO) – substrate (CuBTC cluster) system and assignment of its six key vibrational modes, central panel: target properties to be predicted, right panel: comparison of DFT computed and ML predicted catalytic property values. (b) Comparison of "black box" model (NN) and "glass box" model (SISSO). Left panel: schematic illustration of NN model and SISSO formulas, central panel: the comparation results with small dataset, right panel: the comparation results with imperfect dataset. Source: Reproduced with permission from Ref. [40]. Copyright 2023, AAAS.

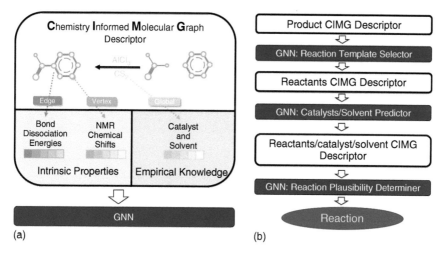

Figure 19.21 (a) Data structure of CIMG as reaction descriptors used in GNN models. (b) Overall workflow for single-step retrosynthesis: for any given product, finding an appropriate reaction leading to it. Source: Janet and Kulik [41]/Proceedings of the National Academy of Sciences PNAS.

and reactions, and constructed a retrosynthesis planning model. We designed a chemistry-informed molecular graph (CIMG) to describe chemical reactions (Figure 19.21). A collection of key information that is most relevant to chemical reactions is integrated in CIMG:NMR chemical shifts as vertex features, bond dissociation energies as edge features, and solvent/catalyst information as global features. For any given compound as a target, a product CIMG is generated and exploited by a graph neural network (GNN) model to choose reaction template(s) leading to this product. A reactant CIMG is then inferred and used in two GNN models to select the appropriate catalyst and solvent, respectively. Finally, a fourth GNN model compares the two CIMG descriptors to check the plausibility of the proposed reaction. A reaction vector is obtained for every molecule in training these models. The chemical wisdom of reaction propensity contained in the pretrained reaction vectors is exploited to automatically categorize molecules/reactions and to accelerate Monte Carlo tree search (MCTS) for multistep retrosynthesis planning.

Our model, trained with a dataset of 1.4 million reaction data, achieved a top 50 accuracy of 0.94 for reaction template selection, a top 10 accuracy of 0.93 for catalyst prediction, and a top 10 accuracy of 0.89 for solvent prediction. The introduction of chemical information greatly enhances the accuracy, reliability, and efficiency of both single- and multi-step reaction planning. Examples are shown in Figure 19.22.

19.5 Concluding Remarks

To wrap up, in this contribution, we outlined a few chemistry-based descriptors utilized as ML features in the recent literature and highlighted their applications in

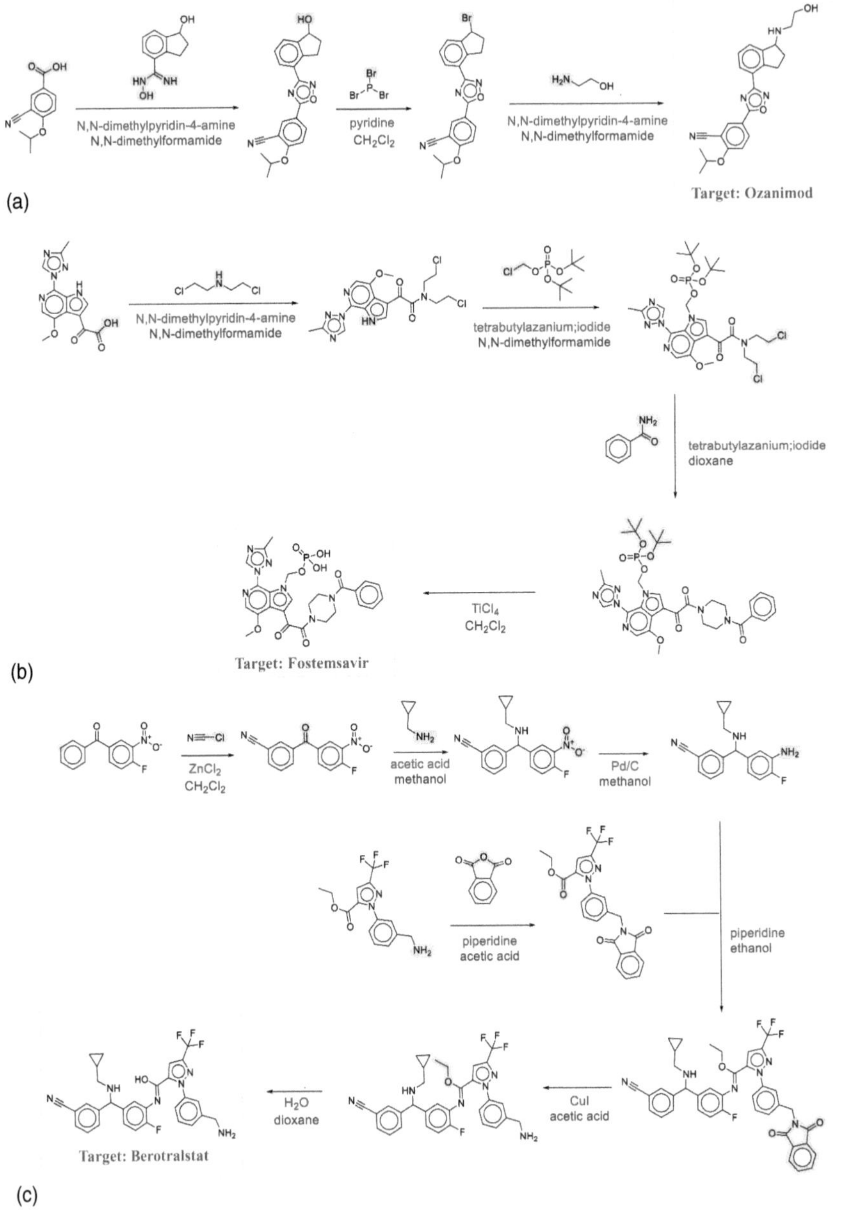

Figure 19.22 Examples of complete retrosynthesis planning. (a) The predicted synthesis path of Ozanimod. (b) The predicted synthesis path of Fostemsavir. (c) The predicted synthesis path of Berotralstat. Source: Janet and Kulik [41]/Proceedings of the National Academy of Sciences PNAS.

predicting spectroscopy, catalysis, and chemical reactions from our recent studies. It is without question that ML is picking up steam and becoming one of the most rigorous and widely applied techniques in many disciplines including theoretical and computational chemistry. We are certain that the examples illustrated above confirmed the applicability, variety, and diversity of such applications.

Even though ML models are often excessively parametric, intrinsically nonlinear, and thus highly uninterpretable per se, it is still possible to garner chemical understandings and physical insights from ML models. This is done through a few ML techniques such as feature importance and rule-based modeling. Our past experiences outlined above indicate that chemistry-based descriptors from either experimental measurements or computational simulations when utilized as ML features could play a paramount role in this regard. We presented a few examples of such descriptors above to illustrate their usefulness and effectiveness in spectroscopic and catalytical applications. This kind of chemistry-based descriptors keeps coming up in the literature and it is certainly not exhausted yet.

Looking ahead, we anticipate that two issues in ML need to be continuously and carefully addressed. The first one is the availability and accessibility of quality databases. We do have decent chemistry-oriented databases online or from the literature, but their size and quality are not adequate. The second yet more important issue is the descriptors themselves. The presently available descriptors in the literature are, in most cases, ad hoc in nature. Is it possible to have a theory-guided approach to generate necessary yet sufficient chemistry or physics-based descriptors as ML features? How to craft a conceptual framework to thoroughly develop these descriptors for applications of different purposes is still a grand challenge, in our opinion. From this very point of view, we think the subject of ML fits in very well with the main theme of this book.

Acknowledgments

J.J. acknowledges financial support from the Innovation Program for Quantum Science and Technology (2021ZD0303303), the National Key Research and Development Program of China (2018YFA0208603), the CAS Project for Young Scientists in Basic Research (YSBR-005), and the National Natural Science Foundation of China (22025304, 22033007, 22203082, and 12227901).

References

1 Keith, J.A., Vassilev-Galindo, V., Cheng, B. et al. (2021). Combining machine learning and computational chemistry for predictive insights into chemical systems. *Chem. Rev.* 121: 9816–9872.
2 Duan, C., Nandy, A., Adamji, H. et al. (2022). Machine learning models predict calculation outcomes with the transferability necessary for computational catalysis. *J. Chem. Theory Comp.* 18: 4282–4292.

3 Dral, P.O. (2020). Quantum chemistry in the age of machine learning. *J. Phys. Chem. Lett.* 11: 2336–2347.

4 Goh, G.B., Hodas, N.O., and Vishnu, A. (2017). Deep learning for computational chemistry. *J. Comp. Chem.* 38: 1291–1307.

5 Goldman, B.B. and Walters, W.P. (2006). Machine learning in computational chemistry. *Ann. Rep. Comp. Chem.* 2: 127–140.

6 Mai, H., Le, T.C., Chen, D. et al. (2022). Machine learning for electrocatalyst and photocatalyst design and discovery. *Chem. Rev.* 122 (16): 13478–13515.

7 Andersen, M. and Reuter, K. (2021). Adsorption enthalpies for catalysis modeling through machine-learned descriptors. *Acc. Chem. Res.* 54 (12): 2741–2749.

8 Wang, X., Zhang, G., Yang, L. et al. (2018). Material descriptors for photocatalyst/catalyst design. *Wiley Interdiscip. Rev. Comput. Mol. Sci.* 8 (5): e1369.

9 Li, Q.-K., Li, X.-F., Zhang, G., and Jiang, J. (2018). Cooperative spin transition of monodispersed FeN_3 sites within graphene induced by CO adsorption. *J. Am. Chem. Soc.* 140 (45): 15149–15152.

10 Hu, W., Ye, S., Zhang, Y. et al. (2019). Machine learning protocol for surface-enhanced Raman spectroscopy. *J. Phys. Chem. Lett.* 10: 6026–6031.

11 Tang, S., Dang, Q., Liu, T. et al. (2020). Realizing a not-strong-not-weak polarization electric field in single-atom catalysts sandwiched by boron nitride and graphene sheets for efficient nitrogen fixation. *J. Am. Chem. Soc.* 142 (45): 19308–19315.

12 Huang, Y., Yang, T., Yu, H. et al. (2020). Theoretical calculation of hydrogen generation and delivery via photocatalytic water splitting in boron-carbon-nitride nanotube/metal cluster hybrid. *ACS Appl. Mater. Interfaces* 12 (43): 48684–48690.

13 Wang, X., Ye, S., Hu, W. et al. (2020). Electric dipole descriptor for machine learning prediction of catalyst surface-molecular adsorbate interactions. *J. Am. Chem. Soc.* 142 (17): 7737–7743.

14 Dang, Q., Tang, S., Liu, T. et al. (2021). Regulating electronic spin moments of single-atom catalyst sites via single-atom promoter tuning on s-vacancy MoS_2 for efficient nitrogen fixation. *J. Phys. Chem. Lett.* 12 (34): 8355–8362.

15 Jia, C., Wang, Q., Yang, J. et al. (2022). Toward rational design of dual-metal-site catalysts: catalytic descriptor exploration. *ACS Catal.* 12 (6): 3420–3429.

16 Zhong, W., Qiu, Y., Shen, H. et al. (2021). Electronic spin moment as a catalytic descriptor for Fe single-atom catalysts supported on C_2N. *J. Am. Chem. Soc.* 143 (11): 4405–4413.

17 Wang, S. and Jiang, J. (2023). Interpretable catalysis models using machine learning with spectroscopic descriptors. *ACS Catal.* 13: 74328–77436.

18 Segler, M.H., Preuss, M., and Waller, M.P. (2018). Planning chemical syntheses with deep neural networks and symbolic AI. *Nature* 555 (7698): 604–610.

19 Esterhuizen, J.A., Goldsmith, B.R., and Linic, S. (2020). Interpretable machine learning for knowledge generation in heterogeneous catalysis. *Nat. Catal.* 5 (3): 175–184.

20 Senior, A.W., Evans, R., Jumper, J. et al. (2020). Improved protein structure prediction using potentials from deep learning. *Nature* 577 (7792): 706–710.

21 Wang, X., Jiang, S., Hu, W. et al. (2022). Quantitatively determining surface-adsorbate properties from vibrational spectroscopy with interpretable machine learning. *J. Am. Chem. Soc.* 144 (35): 16069–16076.

22 Li, H., Jiao, Y., Davey, K., and Qiao, S.Z. (2023). Data-driven machine learning for understanding surface structures of heterogeneous catalysts. *Angew. Chem. Int. Ed.* 62: e202216383.

23 Janet, J.P. and Kulik, H.J. (2020). *Machine learning in Chemistry*. American Chemical Society https://doi.org/10.1021/acs.infocus.7e4001.

24 Samek, W., Montavon, G., Vedaldi, A. et al. (ed.) (2019). *Explainable AI: Interpreting, Explaining and Visualizing Deep Learning*, 1e. Springer.

25 Molnar, C. (2022). *Interpretable Machine Learning: A Guide for Making Black Box Models Explainable*. Independently published (February 28, 2022). ISBN 978-0-244-76852-2.

26 Liu, S.B., Schauer, C.K., and Pedersen, L.G. (2009). Molecular acidity: a quantitative conceptual density functional theory description. *J. Chem. Phys.* 131: 164107.

27 Liu, S.B., Rong, C.Y., and Lu, T. (2014). Information conservation principle determines electrophilicity, nucleophilicity, and regioselectivity. *J. Phys. Chem. A* 118: 3698–3704.

28 Grisoni, F., Ballabio, D., Todeschini, R., and Consonni, V. (2018). Molecular descriptors for structure–activity applications: a hands-on approach. In: *Computational Toxicology. Methods in Molecular Biology*, vol. 1800 (ed. O. Nicolotti). New York, NY: Humana Press.

29 Mou, L.-H., Han, T.T., Smith, P.E.S. et al. (2023). Machine learning descriptors for data-driven catalysis study. *Adv. Sci.* 10: 2301020.

30 Ren, C., Lu, S., Wu, Y. et al. (2022). A universal descriptor for complicated interfacial effects on electrochemical reduction reactions. *J. Am. Chem. Soc.* 144: 12874.

31 Li, Z., Wang, S., Chin, W.S. et al. (2017). High-throughput screening of bimetallic catalysts enabled by machine learning. *J. Mater. Chem. A* 5: 24131–24138.

32 Gu, Y., Zhu, Q., Liu, Z. et al. (2022). Nitrogen reduction reaction energy and pathways in metal-zeolites: deep learning and explainable machine learning with local acidity and hydrogen bonding features. *J. Mater. Chem. A* 10: 14976–14988.

33 Han, Z.K., Sarker, D., Ouyang, R. et al. (2021). Single-atom alloy catalysts designed by first-principles calculations and artificial intelligence. *Nat. Commun.* 12: 1833.

34 Wang, X., Jiang, S., Hu, W. et al. (2022). Quantitatively determining surface-adsorbate properties from vibrational spectroscopy with interpretable machine learning. *J. Am. Chem. Soc.* 144: 16069–16070.

35 Ren, H., Li, H., Zhang, Q. et al. (2021). A machine learning vibrational spectroscopy protocol for spectrum prediction and spectrum-based structure recognition. *Fundamental Res.* 1: 488–494.

36 Ye, S., Zhong, K., Zhang, J. et al. (2020). A machine learning protocol for predicting protein infrared spectra. *J. Am. Chem. Soc.* 142: 19071–19077.

37 Zhang, J., Ye, Z., Zhong, K. et al. (2021). A machine-learning protocol for ultraviolet protein-backbone absorption spectroscopy under environmental fluctuations. *J. Phys. Chem. B* 125: 6171–6178.

38 Zhao, L., Zhang, J., Zhang, Y. et al. (2021). Accurate machine learning prediction of protein circular dichroism spectra with embedded density descriptors. *J. Am. Chem. Soc. Au* 1: 2377–2384.

39 Yu, H., Wang, Y., Wang, X. et al. (2020). Using machine learning to predict the dissociation energy of organic carbonyls. *J. Phys. Chem. A* 124: 3844–3850.

40 Chong, Y., Huo, Y., Jiang, S. et al. (2023). Machine learning of spectra-property relationship for imperfect and small chemistry data. *Proc. Natl. Acad. Sci.* 120: e2220789120.

41 Zhang, B., Zhang, X., Du, W. et al. (2022). Chemistry-informed molecular graph as reaction descriptor for machine-learned retrosynthesis planning. *Proc. Natl. Acad. Sci.* 119: e2212711119.

20

Learning Design Rules for Catalysts Through Computational Chemistry and Machine Learning

Aditya Nandy[1,2] and Heather J. Kulik[1,2]

[1] Massachusetts Institute of Technology, Department of Chemical Engineering, 77 Massachusetts Ave, Rm 66-464, Cambridge, MA 02139, USA
[2] Massachusetts Institute of Technology, Department of Chemistry, Cambridge, MA 02139, USA

20.1 Computational Catalysis

20.1.1 Catalyst Design with Density Functional Theory

Computational modeling is widely used in both predicting fleeting reactive intermediates in catalysis and deducing mechanisms as well as in the design of new catalysts. Density functional theory (DFT) is a first-principles theory that enables the prediction of chemical trends from computational modeling. Since the 1990s, improvements in exchange-correlation functionals that are used in approximate DFT, increases in computing power, and increasingly user-friendly software have led to the widescale adoption of DFT. The balanced tradeoff between accuracy and efficiency in DFT has led to its indispensable role in understanding the reactivity of homogeneous (e.g. transition metal complexes [1, 2]) and heterogeneous (e.g., alloys [3] and surfaces [4]) catalysts. In particular, DFT enables detailed mechanistic studies and analysis of trends over large chemical spaces beyond what is tractable with experiments [5]. Quantum chemical analyses of individual reaction steps provide insights into important catalyst design factors. The roles of strain [6, 7], ligand chemistry [8], and electron configuration [9], which are challenging to deconvolute experimentally, have all been quantified via DFT studies. DFT has been used in tandem with experimental data to determine the reactive intermediates that cause specific reactivity patterns [10–12].

Terminal metal-oxo moieties (i.e. an oxygen atom bound on a metal) are frequently invoked in C–H activation catalysis, since they hydroxylate inert C–H bonds in alkane substrates such as methane or ethane [13, 14]. For this challenging reaction, DFT has been used to decouple the influence of different variables on individual reaction steps [15, 16]. Hydrogen atom transfer (HAT), where a proton and electron are transferred from the substrate to a target metal-oxo moiety, has been a significant focus for many studies on C–H activation [17, 18] (Figure 20.1). Understanding the separate roles of redox potentials (which quantify the energetics

Exploring Chemical Concepts Through Theory and Computation, First Edition. Edited by Shubin Liu.
© 2024 WILEY-VCH GmbH. Published 2024 by WILEY-VCH GmbH.

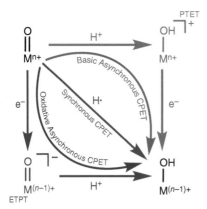

Figure 20.1 A thermodynamic square diagram for proton-coupled electron transfer (PCET). The stepwise reactions (ETPT or PTET) are shown on the edges of the square and the concerted proton-electron transfer (CPET) is shown on the diagonal, with distinct asynchronous behavior (i.e. PT occurring faster than ET or vice versa) shown as curved arrows. Source: Reproduced with permission from Ref. [19]. Copyright 2019, American Chemical Society, Washington, DC.

of electron transfer) and pK_as (which quantify proton transfer) in governing bond dissociation free energies has paved the way for understanding design principles that have led to metal-oxo compounds with improved reactivity [15, 19]. Analysis of the elementary processes in HAT enables the identification of the nature of the electron transfer step, either taking place in a concerted fashion with proton transfer or stepwise [19] (Figure 20.1). From a fundamental understanding of these reactivity factors, novel transition metal complex ligand designs have focused on making metal-oxo complexes that display more favorable C–H bond activation kinetics [20, 21].

DFT enables the discovery of reactivity and property trends across groups and periods of the periodic table. For instance, DFT studies on metal-oxo and metal-nitrido moieties have found that these compounds are characterized by distinct thermodynamics and kinetics. In comparison to metal-nitrido compounds, metal-oxo compounds more readily activate inert C–H bonds during HAT, even with the same ligands and metal coordination geometry [22]. The relationship between DFT-computed kinetic barrier heights and HAT enthalpies indicates a strong Brønsted–Evans–Polanyi (BEP) relation that is independent of the C–H activating moiety [22] (Figure 20.2). We recently showed that this same BEP relation holds for metal-oxo moieties independent of metal-local distortion [7] or oxidation and spin state [9], with a unity slope, indicating that there is a one-to-one relationship between a change in reaction thermodynamics and activation energies. DFT study of metal–ligand bonds enables a detailed understanding of hypothesized trends, such as the metal-oxo wall theory that states that higher d-filling in a tetragonal geometry prevents formation of metal-oxo double bonds and therefore metal-oxo moieties [7, 23, 24]. Recent DFT studies have extended this analysis to metal-fluoro moieties that do not form metal–ligand multiple bonds and are thus unable to activate strong substrate C–H bonds, concluding that a fluoro wall is unlikely to exist in contrast to the oxo wall [23]. Beyond individual reaction steps, DFT sheds light on the transferability of reactivity trends across periods in the periodic table, revealing where design principles derived from one row of the periodic table are appropriately applied to a new set of metals [25]. For instance, the spin-splitting energies of 3d and 4d transition metal complexes with the same

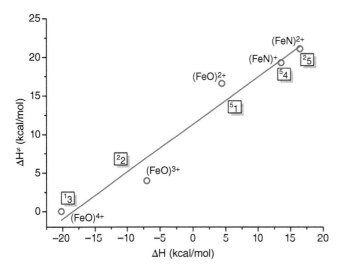

Figure 20.2 A Brønsted–Evans–Polanyi relationship relating activation energy barriers for C–H activation ($\Delta H^{\ddagger}$) to reaction enthalpies (ΔH) for various iron-oxo and iron-nitrido moieties. Spin and charge states for five distinct systems are compared and indicated inset: the baseline number corresponds to an intermediate number from the original Ref. [22] and the superscript number corresponds to the spin multiplicity of the intermediate. All complexes have four equatorial NH_3 ligands and a distal hydroxo. Source: Reproduced with permission from Ref. [22]. Copyright 2014, Royal Society of Chemistry.

ligands are intimately related, enabling the prediction of 4d spin-splitting energies from those of the corresponding 3d transition metal complexes [25, 26]. Developing more transferable trends across the periodic table leads to screening with reduced computational cost rather than requiring the explicit calculation of properties for systems with increasing numbers of electrons.

Reactivity principles such as hard–soft acid–base theory (HSAB [27]), the maximum hardness principle (MHP [28]), minimum electrophilicity principle (MEP [29]), and "|Δμ| big is good" (DMB [30]) all enable better descriptors and interpretability for quantitative structure–property relationships (QSPRs). Combined with DFT, HSAB has been used to understand ligand binding at transition metal centers [31]. Similarly, the MHP and MEP are both derived from DFT electron densities and have been used to rationalize reactivity of Lewis acid catalysts for Diels–Alder reactions [32, 33]. In these studies, the MHP predicts the major regioisomer of the reaction but fails to predict the reaction direction; in contrast, MEP is able to predict both. Since the chemical potential of a system is defined as the derivative of the energy with respect to total number of electrons, the DMB principle rationalizes reactivity by maximizing the change in electronegativity. These reactivity principles provide a quantitative link between a computed quantity and a reaction outcome, facilitating interpretability.

New concepts have arisen from DFT studies of reactivity trends. For instance, when protons and electrons are transferred in proton-coupled electron transfer events, the relative energetics of proton and electron transfer have led to the development of the asynchronicity factor, which quantifies whether PCET is proton transfer driven or electron transfer driven [34]. A new thermodynamic factor, coined as frustration [35], was identified based on DFT calculations of Gibbs free energy profiles on experimentally known catalysts for C–H activation. When this frustration factor was incorporated into a Marcus-theory model, it could be used to quantitatively predict experimental hydrogen atom abstraction energies [35]. Analyzing the kinetic energy distribution in a C–H activation transition state with DFT enables prediction of post-C–H activation reaction selectivity where C–H activation is not the turnover-determining transition state and thus is not rate limiting [36]. Similarly, when there are confinement factors that influence catalysis, such as the reaction occurring in a small zeolite pore, DFT enables quantification of the spatial proximity of the reactants to each other and to the wall of the zeolite pore [37, 38]. Insights from fundamental chemical bonding concepts such as molecular orbital theory lead to a powerful conceptual understanding from DFT, which provides new criteria for molecular design [39].

20.1.2 Mechanistic Modeling: The Degree of Rate Control and Energetic Span Model

DFT data in tandem with experimental kinetic studies enable determination of reaction mechanism and quantification of the corresponding energy landscape. At the same time, DFT allows the discovery of alternate reactivity pathways that explain off-cycle intermediate formation and side-product formation. Experimental kinetic measurements on catalysts are typically obtained as observed overall reaction rates from which rate laws and rate constants are derived. This rate information is not immediately comparable to energetics obtained from DFT. Enthalpic contributions computed with DFT as the sum of electronic energies and zero-point vibrational energies at 0 K must be supplemented with relevant entropic (i.e. vibrational, rotational, and translational) contributions at higher temperatures. To derive reaction rates from DFT-computed energetics, the energetics (i.e. activation energies) must be converted to elementary rate constants that can be used as inputs into a microkinetic model. These microkinetic models, which are typically a series of coupled differential equations, enable direct comparisons to experimental observables such as catalytic turnover frequencies or expected overall reaction rates. Simplifications to this procedure enable a more straightforward comparison of DFT-computed energetics to reaction rates observed in experiment.

The concept of the degree of rate control (DRC) has been used to identify key steps in reaction energy landscapes [40–43]. The DRC concept is particularly powerful due to its ability to be realized both computationally and experimentally. In particular, when the rate-determining step is identified by experiment to be a single elementary step (i.e. intermediate and corresponding transition state), that step has a DRC of 1, with other steps contributing minimally [42]. Here, isotope labeling studies that quantify a kinetic isotope effect identify individual steps that are rate limiting. When the DRC of a given step over a catalytic cycle is established, factors that influence the energetics of that step or reactive intermediate, e.g. noncovalent interactions, sterically bulky functional groups, donating or withdrawing groups, or different ligand scaffold, may be used for catalyst design. The DRC provides a quantitative, simplified metric that unites the kinetics of elementary steps with observed rate laws.

The energetic span model [43, 44] is closely related to the DRC approach and offers further simplification of microkinetic modeling. The energetic span model more directly employs DFT-computed energetics to draw conclusions on turnover frequencies, facilitating high-throughput virtual screening of catalysts where the mechanism is known (Figure 20.3). In particular, the most widely applied simplification of the energetic span model identifies that turnover frequencies are governed by two specific states that dictate the energy span of the catalyst [43, 44]: (i) the turnover-determining transition state (TDTS), and (ii) the turnover-determining intermediate (TDI) (Figure 20.3). The energetic span model driven by DFT-computed energetics facilitates ligand design beyond Edisonian approaches typically realized by experiment or small-scale computation. The assumption implicit in the energetic span model is that the TDTS and TDI do not change as the environment around the active site changes. The energetic span

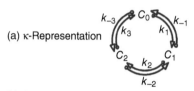

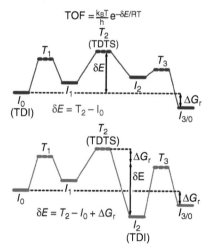

Figure 20.3 An example of a catalytic cycle is shown in the k- (a) and E-representations (b). The turnover frequency (TOF) is calculated as a function of the energetic span, δE, denoted inset for two scenarios: (i) when the turnover-determining intermediate (TDI) precedes the turnover-determining transition state (TDTS) and (ii) when the TDTS precedes the TDI. Source: Reproduced with permission from Ref. [44]. Copyright 2011, American Chemical Society, Washington, DC.

tailoring strategy has been successful for screening catalysts for CO_2 hydration [45] and Rh sandwich catalysts [46] that are useful for [2+2+2] cyclotrimerization to convert alkynes into benzene. In these cases, the mechanism was already known and was assumed to be independent of ligand choice. When multiple transition states or intermediates govern catalysis (i.e. the DRC of individual steps is not unity including when there is branching from multiple pathways that a catalyst can pass through), this assumption may be an oversimplification. When a single TDTS and single TDI govern catalysis, changes to catalysts that systematically reduce this energy span are desirable for improved catalyst design because a reduced energy span leads to increased turnover frequencies and thus a more active catalyst [45]. More recently, the energetic span model has been applied to CO oxidation on Pt atoms dispersed on TiO_2 to distinguish between multiple proposed mechanisms [47]. Additionally, it has also been applied to electrocatalytic reactions [48]. The quantitative nature of this model makes it a useful tool for catalyst design.

20.1.3 The Utility of Scaling Relationships for Catalyst Design

Exhaustive study with DFT becomes cost-prohibitive when a large number of catalysts need to be screened, because high-throughput virtual screening scales in proportion to both the number of distinct systems and the number of reactive intermediates. Here, using scaling relationships that relate energetics of intermediates to each other (i.e. linear free energy relationships, or LFERs) and thermodynamics to kinetics (i.e. BEP relationships) assists with determination of the rate-determining steps, and thus the construction of both thermodynamic and kinetic volcano plots [49, 50] (Figure 20.4). In particular, LFERs relate energetics of reactive intermediates to each other and BEPs relate the energetics of reactive intermediates to computed activation energies. The BEP allows researchers to avoid explicit calculation of time-consuming transition states. Together, LFERs and BEP relations estimate the relationships between energetics of catalyzed steps, including where the largest kinetic barriers and deepest thermodynamic traps are. In turn, these scaling relations produce estimates of catalyst turnover frequencies based on an intuitive and easy-to-compute "descriptor" variable, typically the binding energy of a single atom [51]. These relationships are expected because the bond strengths between a catalyst and a set of related species, such as that of a single bound oxygen atom and a hydroxyl moiety, should be proportional to one another. The Sabatier principle, which was introduced years before the advent of DFT, proposes that the best catalyst will be the one that binds an adsorbate atom neither too tightly nor too weakly. Thus, the comparison between the binding strength descriptor and catalyst turnover frequency leads to a plot that contains a maximum, which is the Sabatier maximum. These so-called "volcano plots" enable rapid high-throughput virtual screening when LFERs and BEP relations are known because the descriptor variable can be used to identify catalysts that are at the top of the volcano. In the limit of a fixed mechanism, volcano plots establish theoretical optimal performance for a designed catalyst. In realistic catalyst systems, the turnover-determining states may indeed change and must be accounted for.

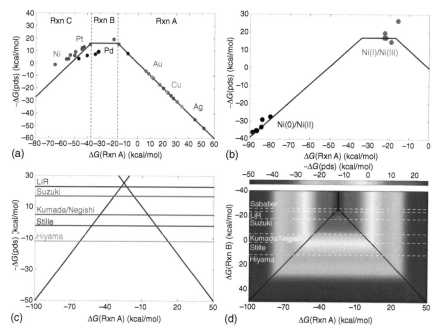

Figure 20.4 (a) Thermodynamic volcano plots for the Suzuki–Miyaura C–C cross-coupling reaction where different reaction mechanisms (Rxn A, B, or C) are rate determining in different parts of the plot. (b) The influence of oxidation state is highlighted on the volcano plot. (c) Changes in volcano plots due to different reaction energies for transmetalation for different catalytic systems. (d) Energetic relationship between three reaction steps for different catalytic systems. Source: Reproduced with permission from Ref. [49]. Copyright 2021, American Chemical Society, Washington, DC.

Since the early 1900s, phenomenological LFERs from experimental data have been used with substituted variants of substrates to understand reactivity trends [52]. These trends, initially developed for organic chemistry to understand how functional groups on a substrate influence reactivity, eventually saw applications in homogeneous [49, 50] and heterogeneous [4, 53, 54] catalysis. DFT studies have quantified these reactivity trends without the need for phenomenological experimental studies [51]. Independent of whether LFERs are developed with experimental or computational data, neither the "descriptor" nor the linear relationship is known in advance. Therefore, the LFER must be developed for a set of systems prior to application to new systems (Figure 20.5). When a LFER developed on one set of compounds generalizes to new systems, it forgoes the need for exhaustive DFT calculations of reactive intermediates by predicting the relative energetics of new catalysts based on a single descriptor. As an example LFER, the metal-oxo formation energy (i.e. the energy required to bind an oxygen atom) is typically used as a descriptor to predict HAT thermodynamics [55, 56]. At times, introducing structure specificity to LFERs is essential for predicting scaling behavior. For instance, the influence of a generalized coordination number has been incorporated into scaling behavior for nanoparticle catalysts [57]. However, since

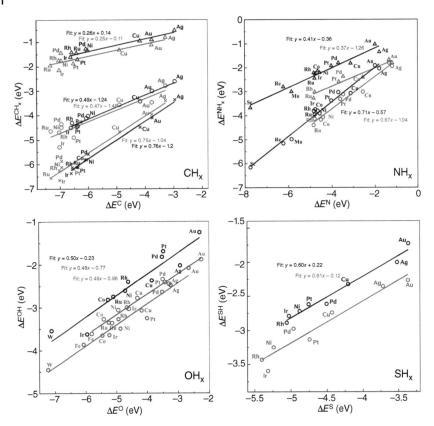

Figure 20.5 Adsorption energies for adsorbate atoms (i.e. C, N, O, or S, respectively) versus CH_x (crosses: $x = 1$, circles: $x = 2$, triangles: $x = 3$; top left), NH_x (circles: $x = 1$, triangles: $x = 2$; top right), OH_x (circles: $x = 1$; bottom left), and SH_x (circles: $x = 1$; bottom right) intermediates on the metal surfaces labeled on each point. Black and red refer to close-packed and stepped transition metal surfaces, respectively. For OH_x, the fcc(100) structure is shown in blue. Best-fit lines are shown. In all cases, adsorption energies are defined as the lowest energy structure minus the sum of the adsorbate and clean surface. Source: Reproduced with permission from Ref. [54]. Copyright 2007, American Physical Society.

LFERs are constructed upon the concept of bond-order conservation [54, 58] (i.e. the bond order of the adsorbate atom will remain constant independent of the reactive intermediate), deviations from bond-order conservation lead to broken LFERs [7, 9, 59]. Thus, introducing specific interactions, i.e. noncovalent interactions, distinct metal–ligand bonding, or external electric and magnetic fields, that disrupt bond order conservation also cause catalyst energetics to deviate from LFERs.

Explicitly computing transition states with DFT is expensive, due to the cost of computing a Hessian matrix and identifying a transition state structure with a single imaginary frequency. If BEP relations hold across a set of systems, the need to compute a transition state structure is avoided, as it is substituted with the cheaper thermodynamic descriptor variable that is easier to calculate [60]. Indeed, BEP relations have also been utilized experimentally to correlate easy-to-obtain

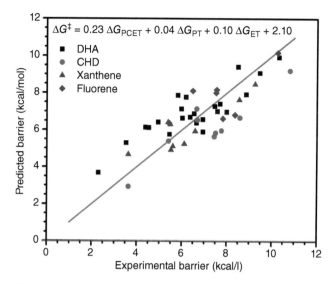

Figure 20.6 Multiple linear regression analysis for experimental barrier heights for PCET-mediated C–H activation relative to barrier height predictions from ΔG_{PCET}, ΔG_{PT}, and ΔG_{ET}. The formula for predictions is denoted inset. The gray line indicates parity. Source: Reproduced with permission from Ref. [64]. Copyright 2021, Royal Society of Chemistry.

properties to catalytic activity [17]. For specific reactions, such as CO_2 reduction, the properties of a metal center have been used to predict the effect of solvation on reactivity [61] and selectivity [62]. In other reactions, such as C–H activation, BEP relations have been seen to hold independent of catalyst identity [55], geometry [7], oxidation, and spin state [9]. When mechanisms differ, however, a single BEP relation does not describe all data [63]. For C–H activation, recent studies have found that HAT is universally thermodynamically controlled, indicating that BEP relations will work well for catalyst screening [64] (Figure 20.6). Specifically, the slope of the BEP relation also indicates the "tunability" of the kinetics. If the slope of the BEP relation is less than unity, then altering the thermodynamics does not have as large of an influence on the kinetics. In select cases, such as HAT, where the slope of the BEP is unity, changing the thermodynamics leads to an equivalent change in kinetics [5, 7].

In the limit of strong scaling (i.e. a linear relationship is predictive and holds tightly with little spread), thermodynamic and kinetic volcano plots are useful tools for understanding and accelerating tuning of catalyst reactivity. Thermodynamic volcano plots are useful for reactions that are thermodynamically limited, where knowledge of reaction energies is sufficient to guide which will be the most reactive catalyst, an approximation that is often applied in electrochemistry. For instance, thermodynamic volcanoes indicate the change in the thermodynamic reaction steps that govern catalysis as the descriptor variable varies. By combining LFERs, it is evident that different reaction steps will become thermodynamically limiting depending on substrate binding strength. In contrast to thermodynamic volcano plots, kinetic volcano plots identify the kinetically limiting step for a given reaction, which will often vary within a single mechanism depending on the value of the descriptor variable. Kinetic volcano plots have been used to understand regioselectivity in catalysis, because the regioselectivity is governed by relative activation energies (i.e. kinetics) [49]. While volcano plots are most useful for understanding trends in reactivity versus binding strength, they are also useful for accelerating calculations. For example, the validity of BEP relations is implicitly assumed during a thermodynamics-led screen, which reduces the cost of screening by eliminating catalysts that do not enable favorable thermodynamics. Nevertheless, BEP relations should still be determined beforehand for these catalysts. LFERs and BEPs paired with thermodynamic or kinetic volcano plots, respectively, provide a simplification of catalyst activity to a single descriptor that enhances understanding and also simultaneously accelerates computational screening.

20.1.4 Quantifying Active Site Environments with DFT

There are numerous cases where LFERs and BEP relations are unable to accurately predict reactivity. These include the presence of structural distortions or noncovalent interactions as well as when collective attributes of the intermediate species in the reactive site exhibit behavior distinct from their constituent atoms. In these examples, altering the relative energetics of one intermediate does not affect others, leading to broken scaling behavior. One mode of LFER disruption is mechanical strain, which has been exerted on homogeneous [7] and heterogeneous [6] catalysts. The disruption of LFERs based on strain facilitates improved catalyst design but limits the applicability of strong scaling relations. In homogeneous catalysis, varying the ligand chemistry with different degrees of ligand donation to the metal (i.e. the metal–ligand bond covalent character and strength) leads to scaling behavior along a LFER, whereas out-of-plane metal distortion breaks scaling entirely (Figure 20.7). Similarly, geometric distortion facilitates distinct scaling by reactivity pathway (i.e. σ versus π C–H activation) on transition metal complexes [65]. In single-atom catalysts (SACs), increased distortion in select reactive intermediates promotes strong deviations from scaling behavior [66]. In heterogeneous catalysis, uniaxial strain causes broken scaling behavior [6]. In both of these cases, strain destabilizes specific d orbitals without influencing all of them equivalently. Because transition metals typically use their partially filled d orbitals to store electrons in bonds during catalysis – thus enabling reactivity – the preferential destabilization of specific d orbitals dictates the energetics of the highest occupied molecular orbital (HOMO) level. Correspondingly, when reactivity depends on more than one of these d orbitals (i.e. the gap between two d orbitals instead of the HOMO level alone), distortion influences reactivity significantly. Thus, LFERs and BEP relations that do not incorporate structural parameters will fail to account for distortion effects that influence catalysis.

Although strain is one way to break the thermodynamic scaling that limits catalyst performance, there are many other ways to induce interactions that influence individual reactive intermediates and thus disrupt scaling. For instance, open-shell active sites that form multiple metal–ligand bonds (i.e. metal–ligand bonds with bond order greater than unity) with specific substrates (e.g. oxo [8, 9] and nitrido [22]

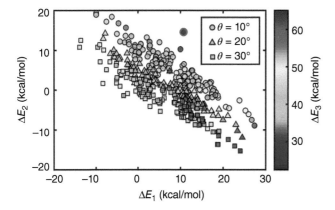

Figure 20.7 Oxo formation energy (ΔE_1) versus hydrogen atom transfer energy (ΔE_2) for various minimal model catalysts, grouped by distortion angle (θ). Symbols are colored according to the methanol release energy (ΔE_3) as indicated by the inset color bar. One outlier is included and denoted with a red border. Source: Reproduced with permission from Ref. [7]. Copyright 2018, American Chemical Society, Washington, DC.

moieties) lead to broken scaling behavior. In particular, each electron configuration demonstrates distinct scaling behavior, indicating that using a single LFER over all different metals, oxidation states, spin states, and ligands leads to bias against specific subsets of data [9, 55] (Figure 20.8). Additionally, introducing noncovalent interactions that stabilize one moiety (e.g. a metal-oxo moiety [7, 20, 21]) that is not present in all reactive intermediates gives rise to broken scaling behavior. Specifically, introducing hydrogen bond donors that stabilize a metal-oxo moiety more than a metal-hydroxo moiety improves relative reaction energetics of distinct reaction steps in cases where a scaling trade-off normally holds (i.e. between the energy to form the metal-oxo intermediate and the HAT step). Therefore, careful analysis of geometric and electronic structures of individual reaction intermediates produces opportunities to interrupt scaling.

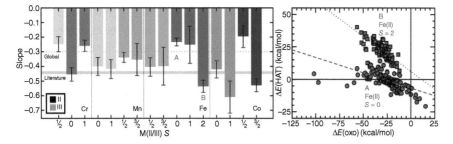

Figure 20.8 (left) Slopes and standard errors for linear free energy relationships between ΔE(oxo) and ΔE(HAT) across over 1000 catalysts in 16 distinct metal-oxidation-spin state combinations. The range of literature [55, 56] slopes are shown as a tan line. The fit through all data is shown as a gray dotted line. (right) Representative datasets are used to determine linear free energy relationship slopes. Fits through subsets of data are shown as dashed and dotted lines, respectively. Source: Reproduced with permission from Ref. [9]. Copyright 2020, American Chemical Society, Washington, DC.

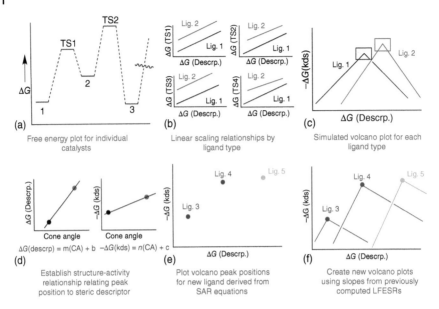

Figure 20.9 Procedure to construct structure–activity relationships for catalysts when linear free energy relationships and Brønsted–Evans–Polanyi relationships hold. (a) First, a subset of catalysts are simulated. (b) Linear free energy relationships and Brønsted–Evans–Polanyi relationships are established. (c) Volcano plots are constructed from thermodynamic and kinetic scaling relationships. (d) A simple descriptor is identified to reduce screening costs. (e) Hypothetical volcano positions based on new ligands are established. (f) New volcano plots are made for the computed data. Source: Reproduced with permission from Ref. [67]. Copyright 2016, Royal Society of Chemistry.

Even when moderate scaling holds (i.e. a linear relationship captures most of the trends in the data), structural variations, such as the previously described out-of-plane distortion example, frequently interrupt scaling. In transition metal complexes, variations in ligand properties can alter the metal-local environment. Ligands range from small to large, with the size of the ligand influencing the accessibility of the active site. This active site accessibility, termed "steric bulk," is not incorporated in LFERs. Therefore, efforts to use LFERs in homogeneous catalysis have utilized ligand-specific LFERs, thus reducing LFER errors at the cost of limiting generalizability to new ligands [67] (Figure 20.9). Since the structure and flexibility of ligands both influence catalysis, there have been significant efforts to quantify these effects. Sterimol [68] descriptors are a set of geometric features that quantify steric bulk of ligands and have seen use predominantly in small-molecule screening for therapeutic drug discovery. In combination with multiple linear regression (MLR) models, Sterimol descriptors facilitate the development of structure-dependent LFERs for transition metal complexes that incorporate ligand influence on active site environments [63]. Despite the incorporation of structural factors [57] in scaling, dynamic factors such as nanoparticle cluster size changing with reactive intermediate have been shown to disrupt linear scaling [69] since these factors selectively influence individual reaction intermediates.

When scaling behavior is not strong (i.e. a linear relation does not describe all of the available data), the assumption of scaling leads to best-in-class catalysts being overlooked [9, 65]. To overcome this challenge, recent efforts have incorporated data-driven methods to build more complex descriptor-based relationships that extend beyond one or two descriptors [70, 71]. In particular, these efforts have focused on selecting descriptors based on statistical techniques such as sure independence screening and sparsifying operator (SISSO), a dimensionality reduction technique that can identify a composite set of features that minimize the error of a scaling relation. SISSO has been used to identify the essential descriptors to predict scaling relationships for reaction intermediates in OER electrocatalysis on transition metal oxides [70, 71]. For reactions that occur on open-shell transition metals, such as partial methane oxidation under the radical rebound mechanism, LFERs are weak [5, 8]. The relationship between initial oxo formation and subsequent steps of the reaction energy landscape is not strong, and the HAT (i.e. C–H bond activation) and methanol release (i.e. to return the catalyst to its resting state) steps are very decoupled from each other [9]. Thus, to avoid the computation of descriptors entirely, we directly predicted reaction energies from catalyst connectivity [5, 8]. This strategy enabled us to identify catalysts with properties that are missed when LFERs are used (Figure 20.10). By eliminating dependence on LFERs, we discovered catalysts that are oxidatively stable but favorably form metal-oxo intermediates, and we also found examples of catalysts that activate C–H bonds but release methanol. In both of these scenarios, assuming a strong scaling relationship is invalid and limits the discovery of catalysts that display optimal properties.

20.1.5 Method Sensitivity in DFT: Influence on Catalysis

A key ingredient in computational catalysis is the selection of the appropriate density functional approximation (DFA) [2]. DFAs are often distinguished by which rung of a hierarchical "Jacob's ladder" they belong to, with increasing complexity moving from local to nonlocal descriptions of exchange and correlation [72]. Nevertheless, in open-shell transition metal chemistry, it is not universally known whether increasing complexity increases accuracy as well [1, 73, 74]. This is important because DFAs from various rungs of Jacob's ladder can lead to distinct predictions of kinetics [75], thermodynamics [76], or mechanism [77]. Understanding DFA sensitivity for catalysis helps to uncover how functional choice influences conclusions on turnover and thus catalyst design. Practical implementations of approximate DFT typically utilize either localized atomic basis sets or plane wave basis sets. The type of basis set influences DFA selection from different rungs of "Jacob's ladder." In homogeneous catalysis, where localized-basis set DFT is pragmatic, global hybrid DFAs that incorporate fractions of Hartree–Fock (HF) exchange are practical. In contrast, calculations with global hybrid DFAs in plane-wave DFT are significantly more expensive than the same calculations with generalized gradient approximation (GGA) DFAs and are often cost-prohibitive [76]. Unfortunately, trade-offs inherent in DFA performance often mean that improving the density may worsen the energetics or that improved energetics are obtained for one step while they are worsened for another [75, 76, 78].

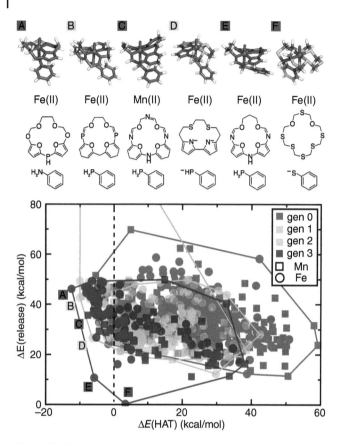

Figure 20.10 Pareto-optimal transition metal catalysts from a high-throughput virtual screen that simultaneously optimizes $\Delta E(\text{HAT})$ and $\Delta E(\text{release})$ on the methane-to-methanol energy landscape. Compounds are colored by generation and shaped by metal identity. Pareto-optimal compounds and their corresponding metals and ligands are shown with letters. Source: Reproduced with permission from Ref. [5]. Copyright 2022, American Chemical Society, Washington, DC.

The use of a Hubbard U correction, in an approach referred to as DFT + U, in combination with semi-local DFT, has been utilized to enforce piecewise linearity of the energy with respect to fractional electron removal or addition without increasing computational cost [79, 80]. Piecewise linearity requires derivative discontinuity at integer electron occupations and represents the correct physics, which is missing from many DFAs [79] (Figure 20.11). Thus, supplementing GGA DFAs with a tunable Hubbard U parameter systematically improves energetics at the cost of semilocal DFT. Despite improving energetics, the U parameter in DFT + U alone is unable to tune multiple properties of a catalyst to be more accurate at the same time. In particular, DFT + U with atom-centered projectors fails when there is strong metal–ligand rehybridization [76, 82, 83]. Within the DFT + U framework, improving the energetics may lead to a simultaneous worsening of the geometry or other properties. Adaptations to DFT + U in the judiciously modified

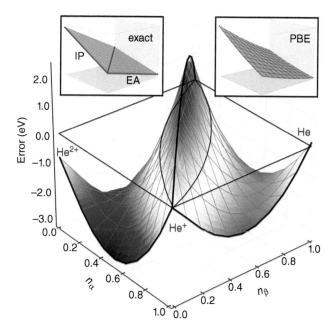

Figure 20.11 The flat-plane error profile (center) for the PBE functional (top right) relative to experiment (top left). Energies are aligned at He$^+$. Positive errors are shown in blue and negative errors in red. Axes refer to fractional spin-up electron (n_α) and fractional spin-down electron (n_β). Source: Reproduced with permission from Ref. [81]. Copyright 2017, American Institute of Physics.

DFT framework improve the description of the electronic structure for open-shell transition metals by recovering piecewise linearity with nonempirical coefficients [84]. Thus, systematic improvements in catalyst predictions within DFT + U require careful consideration of the metal–ligand bonding and the states to which the U corrections are applied.

Although DFT + U has seen wide use in plane wave basis set DFT, ensemble approaches have also been used to reduce errors. For instance, the Bayesian error estimation functional with van der Waals (BEEF-vdW) correlation [85], trained on an ensemble of DFAs, provides error estimates for computed quantities. For the task of understanding ammonia synthesis, the BEEF-vdW functional and its corresponding error estimates have highlighted that synthesis rates relative to a standard catalyst are more accurate than absolute rate predictions [86]. Indeed, these predictions improve catalyst design even when DFT does not accurately predict absolute rates because they identify the catalysts that will outperform the current state-of-the-art [86]. In recent work, we have developed a data-driven approach where we use an ensemble of 23 DFAs from across "Jacob's ladder" to understand the agreement between DFAs [87]. By limiting ourselves to areas of the chemical space where DFAs agree, we identify experimentally realizable spin crossover complexes that have not yet been tested (Figure 20.12). This strategy has been extended beyond spin crossover complexes and has successfully identified

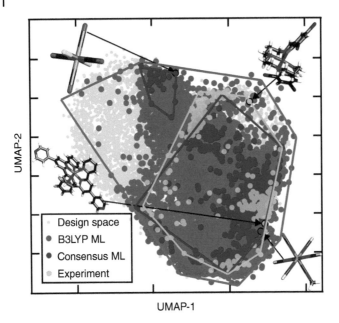

Figure 20.12 Visualization of a design space of nearly 200 000 transition metal complexes as well as predicted spin crossover complexes by B3LYP (red), consensus machine learning with 23 density functional approximations (blue), and experiment (green). Convex hulls are shown for each set of data. Representative experimental complexes are shown (refcodes YANLAC and JUMPEO) in addition to representative design space complexes ($Cr(III)(CO)_4(F^-)_2$ and $Co(II)(NCS^-)_4(NCO^-)(C_2H_5N)$). Atoms are colored as follows: orange for Fe, purple for Mn, pink for Cr, blue for N, red for O, gray for C, cyan for F, dark red for Br, and white for H. Source: Reproduced with permission from Ref. [87]. Copyright 2021, Royal Society of Chemistry.

light-harvesting complexes [88]. Therefore, we would expect similar approaches to work well in catalysis where functional sensitivity influences conclusions [5, 75, 77].

Within global hybrid DFT, the fraction of HF exchange in a DFA is typically the most important tunable parameter for open-shell transition metal catalysis [26, 75, 77]. The value of HF exchange, denoted as a_{HF}, is typically tuned within a previously developed DFA by DFT practitioners to reproduce experimental quantities or recover piecewise linearity for specific test molecules. DFAs have HF exchange values ranging from 0% to 50%. Understanding the variation in properties with respect to HF exchange (i.e. the HF exchange sensitivity) indicates when material properties are strongly sensitive to DFA [26, 77]. We have found that DFA choice alters conclusions on relative turnover frequencies for C–H activation catalysts, since varying HF exchange changes the energy span, instead of shifting the TDTS and TDI equivalently [75]. Additionally, we have found that HF exchange sensitivity changes the overall landscape for methane oxidation, with increased HF exchange uniformly making oxo formation less favorable and HAT more favorable [77]. Correspondingly, we see little variation in relative energetics between catalysts with different metal centers, oxidation states, and spin states. Despite relative energetics being preserved with HF exchange variation, mechanisms differ significantly.

Analysis over more than 1000 C–H activation catalysts reveals that the reaction mechanism changes with greater HF exchange fractions [77]. In particular, reactions move from spin-forbidden (i.e. where the ground spin state changes over the energy landscape) to increasingly spin-allowed (i.e. where the ground state spin is the same across all intermediates). Taken together, HF exchange sensitivity leads to distinct scaling relations and thus distinct volcano plots. Therefore, before using an LFER or volcano plot to carry out a large-scale screen, a researcher should understand the extent to which DFA sensitivity will impact the type of lead compounds suggested and what mechanisms will be favored. Beyond a sensitivity approach for DFA uncertainty, we have developed a recommender system that recommends a DFA for a given molecule and property in a system-dependent manner [89]. Our data-driven approach avoids the systematic biases that are present with ad hoc method selection.

20.2 Machine Learning (ML) in Catalysis

20.2.1 Utilizing ML in Catalysis for Improved Design

Structure–property mappings are relationships that provide guidance on how chemical composition of molecules or materials relates to observed properties. These can be heuristic or mathematical in nature. In the field of machine learning (ML), "supervised learning" represents a broad class of techniques for building structure–property relationships if we have available data regarding molecules and their properties. In supervised learning, we build a mathematical model that can map between these structures and properties by fitting a mathematical model capable of obtaining "y" (the property) from "x" (the features). All such data-driven ML models require a representation to learn from [1]. This representation, most commonly referred to as a "featurization," indicates how molecules are represented as numerical vectors [90–93]. Thus, a powerful featurization or one that encodes known chemical trends is of paramount importance in developing QSPRs [25].

Featurizations may be discrete or continuous and represent a key way by which chemical intuition is imparted to ML models for chemistry [92, 93]. The appropriate level of complexity of a featurization depends on the scale of the data at hand, the difficulty of the learning task, and whether or not interpretability of the features is important to the end use [1]. Additionally, the cost of the featurization is an important factor in the scalability of a ML model: it is essential that featurizations are low cost (i.e. do not require an experiment or a lot of calculation time from DFT to compute) so that screening efforts are readily extended to large chemical spaces [5, 88, 94]. For example, composition or graph-based descriptors can be used to train ML models to predict ionization potentials or redox potentials [93], HOMO energies [95], or catalyst reaction energetics [8] at low cost. These low-cost featurizations then allow for rapid evaluation of properties across larger spaces. Even when featurizations are more computationally demanding, they may still be motivated when they reduce the relative computational cost. For instance, ML models trained on a feature set consisting of multi-reference (MR) diagnostics predict when a DFT calculation is acceptable for property evaluation. However, these diagnostics require wavefunction theory and are expensive to obtain over large data sets. Thus, reducing the computational cost by using DFT-computed diagnostics and explicit geometric information enables scalability to large chemical spaces [96, 97]. Developing representations that generalize to chemically dissimilar spaces has advantages in understanding trends across the periodic table such as when design rules from 3d transition metal complexes are applicable to 4d counterparts [25, 98, 99].

Geometry-free or graph-based featurizations are a widely used class of featurization due to their low cost and their connection to skeleton structures frequently used by chemists in mechanistic study [1, 93, 100]. Since molecular connectivity graphs (i.e. what atoms are connected to what other atoms) are inexpensive to obtain and do not require explicit information about geometries, they do not require any structural data and readily generalize to large chemical spaces when properties are less sensitive to geometric variation. All connectivity graphs contain nodes (atoms) and edges (bonds). Thus, molecular graphs always contain information regarding atom identity on the nodes. When bond order is known, edges can be weighted with bond order information [101]. Connectivity graphs encode information about coordination number (i.e. how many ligand-binding sites a metal contains) but not coordination geometry (i.e. if a metal complex is in a square planar versus tetrahedral geometry). Correspondingly, connectivity-only graph representations fail to faithfully distinguish between conformers, connectivity isomers, and stereoisomers but are able to distinguish between molecules that have distinct connectivity. Moreau–Broto autocorrelations are a representative example of a graph-based representation that has been used for therapeutic drug discovery [102, 103]. In this featurization, sums of products of atomwise properties (i.e. products of nuclear charges or electronegativities) are collected at various bond distances to fingerprint each molecule. In an extension of this representation known as revised autocorrelations (RACs), the incorporation of sums of differences captures chemical trends in transition metal complexes [93] (Figure 20.13). Furthermore, supplementing this representation with heuristics such as group number enables the representation to work to describe trends across the periodic table [25].

At times, building a representation that matches our chemical intuition is not possible. Here, ML models may take graphs as inputs and learn a representation during ML model training. When graphs are used as inputs to ML models, they must be "zero-padded" to account for differences in molecular size [104]. Chemical information is learned during training with "graph convolutions" or "message passing" in graph neural networks [105] (GNNs). Graph convolutions are filters that slide over the molecular graph and relate neighbors that are a certain number of bonds away from each other. In contrast, message passing is when nearest-neighbor nodes influence molecular predictions. GNNs that learn representations from graph inputs must be permutation invariant because the order of the nodes in the graph should not influence property prediction [106] (i.e. the same graph in a different order should produce the same output). The concept that a change in input order will lead to an equivalent change in output order is known as "equivariance" and is necessary for optimal GNN training [106]. Together, GNNs have been successful at

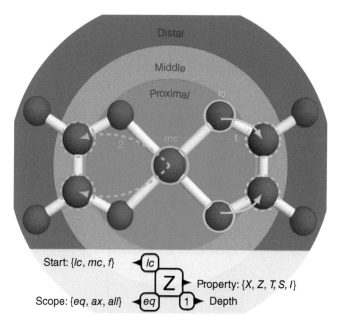

Figure 20.13 Representation of revised autocorrelation (RAC) depths in the equatorial plane of an Fe(oxalate)$_2$ transition metal complex. Atoms are colored according to their chemical identity: carbon in gray, oxygen in red, and iron in blue. The bond-path distance of features containing these atoms is categorized by shaded concentric circles as metal-local (proximal) colored in red, metal-distal (distal) colored in blue, and middle colored in green. Different scopes such as the metal-centered (mc) or ligand-centered (lc) scopes are shown, in addition to various depths (i.e. lc-depth-1 and mc-depth-2). Source: Reproduced with permission from Ref. [93]. Copyright 2017, American Chemical Society, Washington, DC.

both predicting node-level quantities such as partial charges [107] and graph-level quantities such as energies from quantum mechanics or solubility [108].

In certain material classes, such as porous materials, geometric information may be essential. For instance, metal–organic frameworks (MOFs) have large surface areas and pore volumes that render them useful as functional materials [109]. Here, geometric fingerprints are essential for quantifying pore shape and size [110, 111]. These representations are more computationally expensive than graph-based representations because they require 3D coordinate information. Broad geometric descriptions of pore size may fail to accurately describe pore shape. Persistent homology is a technique that has been used to more accurately represent pore shape and volume by generating "barcodes" for each material based on a point cloud on the surface of the MOF [112, 113] (Figure 20.14). Since the barcodes derived from persistent homology represent the shape well but do not represent a consistent length feature vector, methods to convolve these features with a filter (i.e. use a graph convolution) enable ML models to be trained from this data [114]. Beyond pore size and shape metrics, many representations encode molecular geometries. For instance, atom-centered symmetry functions [115], RACs supplemented with geometric information [96, 98] (i.e. Coulomb-decay

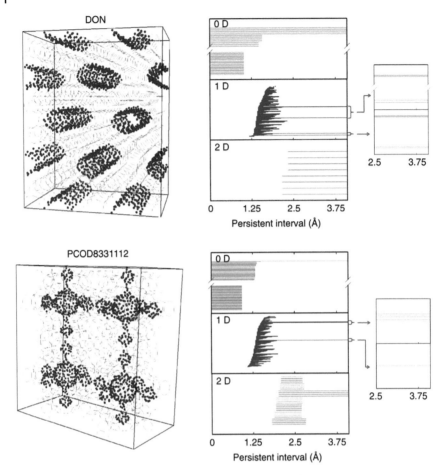

Figure 20.14 Examples of persistent homology fingerprint "barcodes" on two porous materials. For each structure (left), a fingerprint is developed based on the combination of 0D, 1D, and 2D barcodes (middle), with the details of the 1D barcode shown (right). The fingerprint is developed by taking random dots on the pore surface and growing them to construct the barcodes at each dimensionality. Source: Lee et al. [113]/Springer Nature/CC by 4.0.

RACs, or CD-RACs), or Coulomb matrices (i.e. a matrix of distance-dependent interaction strengths based on Coulomb's law) [116] are all featurizations that encode geometries and distinguish between conformers. Thus, when geometric information is cheaply attainable, these representations may produce improved performance relative to a graph representation.

Aside from learning molecular properties from connectivities and structures, ML models have been trained to accelerate the sampling and determination of equilibrium geometries. For instance, ML-derived force fields and neural network potentials enable geometry optimization at DFT accuracy and low cost, thus reducing the difference between first-principles and classical methods [117–119]. ML-derived force fields have been used to address challenging geometry

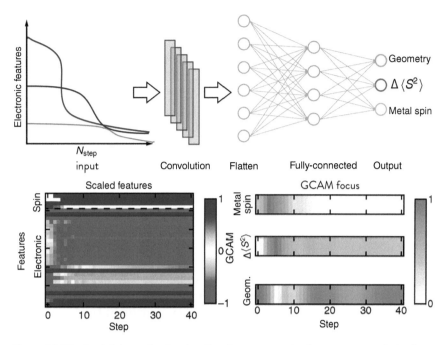

Figure 20.15 (top) Schematic of a classifier that uses on-the-fly geometry optimization steps (i.e. electronic structure features) as inputs to a convolutional neural network to classify whether a calculation is converging toward an intended geometry (i.e. octahedral) and toward a desired electronic state (i.e. one that is not spin-contaminated). (Bottom left) Example of electronic structure features by step of geometry optimization. (Bottom right) Distinct convolutional neural network focus for different classification tasks. Source: Reproduced with permission from Ref. [121]. Copyright 2021, American Chemical Society, Washington, DC.

optimizations in systems that are large and contain many non-covalently interacting parts [118]. In transition metal chemistry, geometry optimization is plagued with convergence to the incorrect electronic state or unintended geometry [120]. We have developed convolutional neural networks (CNNs) that monitor the progress of geometry optimizations and terminate them when they are unproductive. These "dynamic classifiers," which identify whether calculations are likely to be productive or not based on on-the-fly electronic structure information, save computational resources by terminating unproductive simulations [120] (Figure 20.15). The featurization for these models is time-series data that fingerprints the electronic structure near the metal. Although this featurization is much more expensive than a graph-based representation because it requires that several steps of a geometry optimization be carried out, it is more transferable. We found that our models are readily transferable to different reactive intermediates with distinct metals and ligands, indicating that the ML models learned generalizable trends that cause calculation failures [122]. Analysis of model focus using gradient-weighted class activation map (Grad-CAM), an approach that uncovers model focus on the trajectory data, unveils generalizable trends independent of reactive intermediate [121, 123].

20.2.2 The Role of ML in the Limits of Strong and Weak Scaling Relationships

ML surrogate models have been an indispensable tool for catalyst screening due to their low-cost evaluation of properties of millions of candidate compounds [5]. When scaling relations are strong, a descriptor variable produces an accurate estimate of the catalyst energy landscape [49]. However, even calculating a descriptor variable can become cost-prohibitive when expanding to large multimillion compound spaces [124]. Because descriptor variables for volcano plots frequently require one or more DFT calculations, efforts in homogeneous catalysis have focused on using supervised learning approaches to obtain a descriptor variable from a geometry-dependent featurization (e.g. the "bag of bonds" [125] representation, which fingerprints variations in bond lengths between two atoms) on a single reactive intermediate [126]. To reduce the cost of the geometry-dependent featurization, geometric properties (i.e. the 3D molecular structure) are obtained with DFT using a small 3-21G basis set [126]. A similar strategy where a geometry-dependent fingerprint was developed on a single reactive intermediate was utilized for the discovery of new heterogeneous electrocatalysts for the CO_2 reduction reaction and hydrogen evolution reaction where scaling relations were strong [127].

When scaling relations are weak, ML-computed descriptor variables have limited utility since they do not generate predictive energy landscapes. Instead, distinct scaling relations must be developed for each metal to screen diverse chemical spaces [9, 128]. To avoid this computationally expensive step, direct prediction of reaction energies and corresponding energy landscapes is more effective [5, 8]. Here, feature selection methods coupled with graph representations (e.g. RACs) identify the molecular characteristics that are most essential for predicting a given property [93]. For partial methane oxidation to methanol, forming a stable metal-oxo intermediate is frequently a challenge [8, 51]. Thus, scaling relations have been developed to relate resting state HOMO levels to metal-oxo formation energies for Fe catalysts [51]. By coupling the direct prediction of reaction energies to feature selection, we revealed that the scaling relation developed on high-spin Fe does not extend to other metals, oxidation states, spin states, or ligands. ML model analysis helps to pinpoint why the traditional descriptor-based approach would fail. Feature analysis reveals that this failure of scaling is because the metal-oxo formation energy is governed by more metal-local features (i.e. the identity of the metal, its electron configuration, and the metal-coordinating atoms) whereas the HOMO level is dictated by metal-distal features (i.e. functional groups far away from the metal) (Figure 20.16). Thus, this descriptor-based relation only holds if the metal-local environment is fixed and only functional groups are varied [8]. Furthermore, since scaling relations systematically fail in open-shell transition metal chemistry for partial methane oxidation, ML provides a way to discover ligands that display Pareto-optimal reaction energetics where a catalyst is able to activate methane but also release methanol [5]. Here, judicious design strategies are essential for new insights and guidance for synthetically accessible ligands. We recombined subunits (e.g. pyrroles or furans) of existing ligands such as porphyrins

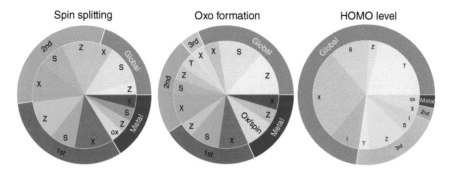

Figure 20.16 Pie charts of the selected features for predicting spin-splitting energetics (left), oxo formation energetics (middle), and HOMO level (right). Features are grouped by the most distant atoms, with the metal represented in blue, the first coordination shell (i.e. one bond away from the metal) in red, the second coordination shell in green, the third coordination shell in orange, and global features in gray. Within each color, the features (Z, T, S, I, or χ) are denoted and oxidation/spin states are assigned to the metal. The selected features for oxo formation are more similar to those for spin splitting than HOMO level. Source: Reproduced with permission from Refs. [1, 8]. Copyright 2019 and 2021, American Chemical Society, Washington, DC.

to construct a space of realistic macrocycles. By training separate ML models for HAT and methanol release energetics, we found needle-in-the-haystack ligands that generate methane oxidation catalysts. Over a set of over 16 million catalysts, we found that using anionic axial ligands facilitates C–H activation while still permitting methanol release to return the catalyst to its resting state.

In some cases, DFT-computed transition state barrier height differences predict selectivity and thus may be used as a figure of merit for screening [129]. After benchmark experimental studies found quantitative agreement between DFT-computed barrier height differences and 1-octene reaction selectivity over 1-hexene for Cr oligomerization catalysts, ML was combined with DFT data to understand key transition metal complex features that resulted in greater barrier height differences and thus selectivity [130]. These efforts indicated that Cr–N distances and other key geometric variations of the metal with respect to the ligands governed barrier height differences. Correspondingly, three generations of ligand exploration led to new Cr oligomerization catalysts that have superior 1-octene selectivity (Figure 20.17). When a single feature cannot be readily identified, unsupervised approaches such as principal component analysis or singular value decomposition are also effective in identifying descriptors that are combinations of many factors [131]. When transition states are necessary for catalyst evaluation but are challenging to locate, nonlinear ML models such as random forests [132] have been used to identify key bond-breaking and bond-forming distances [133]. These distances are useful for generating guessed transition state structures without the need for expensive transition state search methods such as nudged elastic band.

MLR models that rely on precise geometric descriptors provide a path forward for interpretable data-driven insights when specific target quantities are challenging to obtain and no general relations are known [134]. For instance, enantioselectivity is

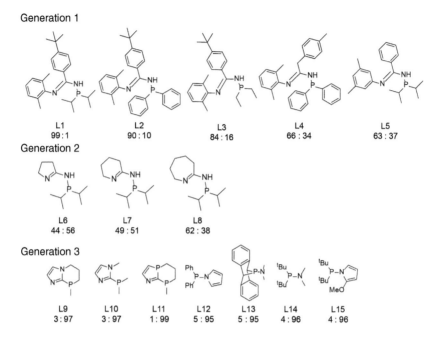
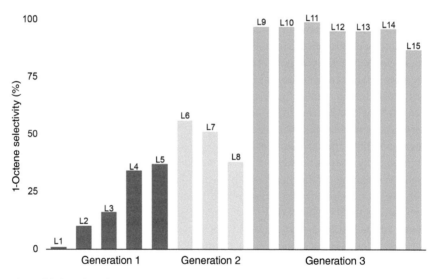

Figure 20.17 Chemical structures, predicted 1-octene selectivity, and evolution by generation over a space of ligands used to optimize Cr oligomerization catalysts. Each generation is colored distinctly. Source: Reproduced with permission from Ref. [130]. Copyright 2020, Royal Society of Chemistry.

difficult to compute by DFT because small differences in barrier heights distinguish catalysts that are enantioselective [135]. DFT remains useful because it can be used to obtain descriptors that enable screening once an MLR model has been fit to experimental data. Quantum mechanical (i.e. natural bond orbital charges [136]) and geometric (i.e. Sterimol [137] or buried volume [138]) descriptors that quantify a transition metal complex have been used to predict these experimental properties accurately. MLR models have the added benefit of interpretability because the magnitude and sign of MLR coefficients indicate the direction in which predictions are influenced.

20.2.3 Uncertainty Quantification in ML and Improved Model Performance

One of the key challenges of all ML models is whether the model remains predictive (i.e. generalizes) on new compounds beyond those in the training set. A related challenge is whether we can easily tell when the model is failing to generalize on new compounds. To evaluate where models are most "confident," we require a quantitative uncertainty metric that indicates where models are applicable and where they are not [100, 139]. In linear regression, a simple "feature space distance" (i.e. the Euclidean distance between a new point and the training data in the feature space) indicates when a prediction would require extrapolation beyond the training data. These distance metrics may fail to faithfully measure similarity in chemical discovery where either the relative importance of features is not equal or when used for nonlinear ML models that alter the mapping of property space with respect to the input features. For nonlinear ML models, efforts to quantify uncertainty include the use of ensemble models trained on the same data (i.e. testing multiple models and their deviations in predictions) or Monte Carlo dropout [140] on neural network weights (i.e. removing path dependence in a neural network).

Gaussian process (GP) models [141] naturally provide estimates of uncertainty, making them a useful tool for chemical discovery [142, 143]. Similarly, distance to training data provides a natural measure of uncertainty in kernel-based models such as kernel ridge regression (KRR) where significant deviation from the training data will result in a prediction of the training data average [93]. Although uncertainty quantification (UQ) is built into GP models, it is less developed for neural networks, which are widely employed in molecular discovery tasks. Neural networks are an example of a highly nonlinear ML model because they take in a representation and use nonlinear activation functions to distort the inputs until a structured "latent space" is formed, where data is subject to either regression or classification. Thus, the power of a neural network comes from its ability to learn the weights and biases that transform the input data. Thus, distance to the training data in the latent space is a natural UQ metric [100] we have found to outperform distance-based metrics that measured distance to the training data in the feature space [92] (Figure 20.18). We found that the distance in latent space metric quantitatively tracks with model prediction errors and enables us to systematically reduce prediction errors by considering uncertainty. Additionally, the latent space UQ metric behaves as a de facto active

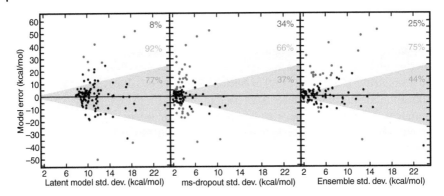

Figure 20.18 Performance of three different uncertainty metrics: latent model standard deviation in energy units (left), Monte Carlo dropout standard deviation in energy units (middle), and ensemble standard deviation in energy units (right) on a 116-molecule test set not seen by the artificial neural network model. Green and yellow zones represent one or two standard deviations, respectively, with the percentage of points lying between them indicated as a percentage for each metric. Source: Reproduced with permission from Ref. [100]. Copyright 2019, Royal Society of Chemistry.

learning criterion, indicating areas of chemical space where the existing ML models do not generalize well [5, 88, 94]. Utilizing this UQ metric to select new data to ingest into model training has led to improved model performance over time, enabling us to train ML models that generalize more readily over larger chemical spaces.

A careful understanding of feature distributions and overlap across different sets of training data indicate opportunities for more generalizable models. For instance, returning to the discussion of the dynamic classifiers we constructed to monitor on-the-fly geometry optimizations and terminate unproductive calculations reveals lessons for computational catalysis [120]. For these dynamic classifiers, we found that classifiers trained on a single reactive intermediate readily generalize to other reactive intermediates [121, 122]. This behavior is attributed to comparable electronic structure features, which are inputs to the model, across reactive intermediates. Similarly, introducing group number as a RAC feature enables models to generalize predictions across rows of the periodic table [25]. When evaluating whether DFT is applicable for molecular property prediction, it is typical to calculate a MR diagnostic [98]. Certain MR diagnostics [97] are more transferable between inorganic chemistry (i.e. transition metal complexes) and organic chemistry (i.e. small molecules). For instance, the natural orbital occupations of the HOMO computed with Møller–Plesset second-order perturbation theory [144] (n_{HOMO}[MP2]) generalizes better than other MR diagnostics such as T_1, derived from coupled-cluster theory [145], due to the overlap in values between inorganic and organic compounds [98] (Figure 20.19). Similarly, using electronic structure information from Hartree–Fock (e.g. molecular orbital features) enables transferable prediction of post-Hartree–Fock correlation energies across organic molecules of distinct chemical compositions [146–148]. Thus, careful representation selection for ML models that leads to improved data overlap also leads to more transferable predictions.

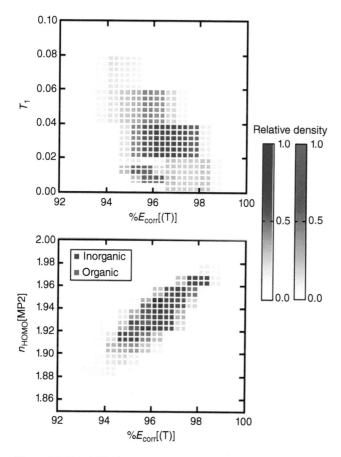

Figure 20.19 A 2D histogram of multireference diagnostics: (top) T_1 and $\%E_{corr}[(T)]$ and (bottom) $n_{HOMO}[MP2]$ and $\%E_{corr}[(T)]$. Over 10 000 transition metal complexes (labeled inorganic, blue) are compared to over 12 000 distorted organic molecules. Source: Reproduced with permission from Ref. [98]. Copyright 2022, Royal Society of Chemistry.

20.2.4 The Role of Optimization Algorithms in Combination with ML Models

ML models trained on cheap-to-compute descriptors reduce the cost of molecular property prediction by orders of magnitude, from days to milliseconds. This, in turn, enables ML to be combined with optimization tools such as genetic algorithms (GAs) [90, 95, 149] or Bayesian methods with active learning [5, 88, 94] for rapid chemical space exploration. GAs are widely used evolutionary algorithms that provide a means for global exploration in chemical spaces that may be intractable for exhaustive screening and can also be used with full property evaluation from DFT or experiment as long as a target property (i.e. the fitness function) that is optimized can be evaluated. In a GA, each molecule or material is assigned a series of "genes" (i.e. the characteristics that define that molecule). In transition metal chemical space, genes can encode information about the metal, oxidation state, spin state, and ligand

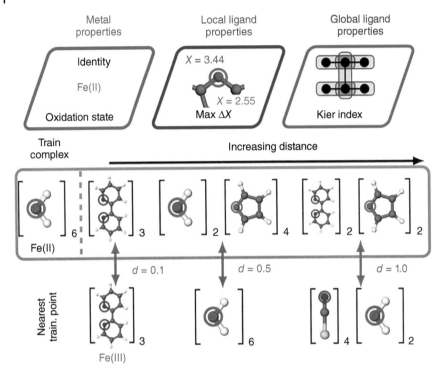

Figure 20.20 (top) Schematic representing the features used in a machine learning model to predict transition metal complex properties during ML-led exploration. (middle) Representative ligands that may be paired with Fe(II) to make a transition metal complex. (bottom) Nearest-neighbor compounds to representative transition metal complexes and their corresponding distances are shown. Source: Reproduced with permission from Ref. [90]. Copyright 2018, American Chemical Society, Washington, DC.

chemistry. We evaluate each set of genes with an ML model that produces a property prediction and is subsequently scored using a fitness function, such as the difference between an energy evaluated from ML or DFT and the target energy [90]. High-fitness compounds then undergo "mutation" (i.e. exploration) and "crossover" (i.e. exploitation between two high-fitness "parent" genes) events that provide a new set of "offspring" genes to test. When the GA fitness is evaluated with an ML model, it can be important to modify the fitness function to avoid making predictions on points with high "distance" from the training data (i.e. that are distinct from model training data, Figure 20.20). Over multiple generations of simulated evolution, we showed it was possible to discover compounds with targeted properties [90]. In addition to transition metal complexes, GAs coupled with ML models have shown utility for identifying over 100 state-of-the-art MOFs for methane storage that show working capacities beyond the current world-record performance [149].

Exhaustive screening with Bayesian methods provides an alternative to the stochastic nature of exploration and exploitation in GAs. Over large multimillion molecule spaces, this exhaustive screening is made possible by graph-based descriptors (e.g. RACs [93]) that are readily obtained with near-zero cost [5, 88, 94]. For

Figure 20.21 Schematic of efficient global optimization (EGO) combined with active learning to improve surrogate models. A large design space is first reduced to a tractable number by performing *k*-medoids clustering and setting up DFT calculations. Upon completion of said calculations, the surrogate model is retrained, EGO is carried out, and a new round of clustering occurs. This process is repeated until there are no more lead compounds or model improvements. Source: Reproduced with permission from Ref. [94]. Copyright 2020, American Chemical Society, Washington, DC.

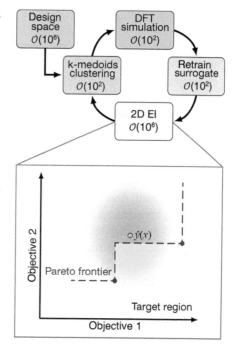

instance, efficient global optimization (EGO) with an expected improvement (EI) criterion helps balance exploration and exploitation [94, 150–153]. The EI criterion factors in both the uncertainty in the prediction as well as the improvement a new molecule is expected to make. This EI criterion has seen extensions into two dimensions [154] where it has shown utility for multi-objective Pareto optimization (Figure 20.21). A Pareto front is defined as the region of chemical space where there is an optimal trade-off between two competing objectives (i.e. one property cannot be improved without worsening the other). Active learning, as is used in 2D-EI, is the process of improving ML models with new information over time. For active learning to be effective, we must utilize UQ to identify the data points that will add the most information to the model [153, 155]. Therefore, incorporating these new points into the training data leads to improved model generalization. Since uncertainty is incorporated into the EI criterion, retraining models with points selected from EI improves model performance over multiple generations [5, 88, 94]. When extending beyond two dimensions, EGO methods can use expected hyper-volume improvement criteria that simultaneously optimize competing objectives [153].

With the 2D-EI criterion, we have discovered transition metal complexes that show the best tradeoff between solubility and redox potential [94], methane oxidation catalysts that activate a methane C–H bond but also release methanol [5], and light-harvesting complexes that have low MR character but desired optical properties [88]. These discoveries typically correspond to a two-to-three order of magnitude speed-up over random search, allowing researchers to identify which chemical design principles are most relevant for the optimal materials on a much faster timescale. For instance, we evaluated our improved ML models over a large design space for methane oxidation catalysis after performing active learning. From this analysis, we found that anionic axial ligands in a tetragonal ligand field environment promote methane C–H bond activation while also facilitating methanol release. This behavior arises since neutral axial ligands generate more electropositive metals that bind methanol more tightly [5]. Therefore, active learning and model predictions provide a way to reveal "hidden trends" within data that provide design insights.

20.2.5 Leveraging Experimental Data for Molecular Design

Efforts to use ML-driven discovery methods with experimental data are typically limited by small data set sizes [156]. This data is rarely systematically curated and frequently contains anthropogenic bias that results in local exploration of materials [157] (i.e. small changes on state-of-the-art compounds). Although many databases exist for curated experimental structures of transition metal complexes [158, 159] and MOFs [160, 161], they rarely contain experimental properties pertaining to those structures. Correspondingly, first-principles calculations are typically used to understand variations in properties such as band gaps in MOFs [162] or frontier orbital energetics in transition metal complexes [163] as computed on experimentally characterized structures. In select cases, curated experimental data uncovers unique trends in the influence of synthesis conditions [164–167] or additives on materials property outcomes [168] or materials stability [169].

Due to the availability of structural data and the influence of metal electronic structure (i.e. electron configuration) on metal–ligand bonds, we studied the spin-state assignment of Fe(II) transition metal complexes in the Cambridge Structural Database [101]. For iron complexes, oxidation and spin state are determined by Mössbauer spectroscopy, which requires isolation of the compound and a corresponding spectroscopic study. ML methods trained on graph representations can be used to predict geometries, an easier learning task than predicting the experimental spin state. Because geometries show minimal variation by DFA choice, bond length predictions are also much less sensitive to DFA than spin-splitting energy predictions [93, 170]. Therefore, predicting the DFT bond lengths for a complex with a given set of ligands in a specific oxidation and spin state enables facile assignment of the ground state spin, avoiding the need for a Mössbauer experiment (Figure 20.22). This zero-cost strategy produces a way to confidently assign ground state spin when a compound is isolable, but it is not possible to perform a spectroscopic study. Beyond mononuclear iron complexes, structural data sheds light on metal–metal bonding. In bimetallic transition metal complexes, metal–metal interactions produce emergent phenomena where the complex does not behave like the sum of its parts [171]. Despite numerous efforts, DFT calculations have failed to explain these emergent phenomena that produce multiple redox events for a bimetallic complex [172, 173] whereas the mononuclear counterpart only undergoes a single redox event (Figure 20.23). Our recent work leveraged over 300 experimental structures to instead predict the formal shortness ratio of the metal–metal bonds to within 5% from connectivity [99]. Feature analysis on this

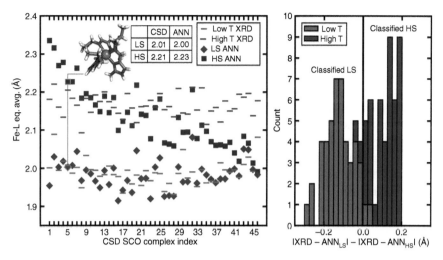

Figure 20.22 (left) Average equatorial bond lengths for Fe compounds, with artificial neural network predictions shown as diamonds and squares and experimental data shown as lines. A representative structure (refcode: BAKGUR) is shown, with Fe in orange, N in blue, C in gray, and H in white. (right) A histogram of neural network prediction deviations from experiment. Compounds to the left of the zero line are identified as low-spin in contrast to the high-spin on the right of the zero line. Source: Reproduced with permission from Ref. [101]. Copyright 2020, American Chemical Society, Washington, DC.

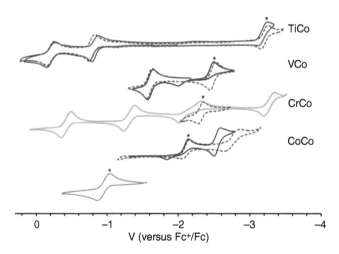

Figure 20.23 Cyclic voltammograms of various bimetallic systems under the presence of nitrogen (colored line) or argon (dotted line). The Co/Al combination contains only a single redox event in contrast to other metal–metal combinations with 3d metals. Source: Reproduced with permission from Ref. [171]. Copyright 2015, American Chemical Society, Washington, DC.

data indicates that formal shortness ratio prediction is transferable across periods. Similarly, we were able to predict the oxidation potential to 0.38 V on unseen complexes over a nearly 3 V range from connectivity [99]. As experimental data collection efforts grow, we expect our approach to generalize over larger chemical spaces with different molecular properties.

In mixed-metal–oxide materials where emergent phenomena are more challenging to decipher, single-source high-throughput experimentation has generated new insights on the design of a catalyst with tunable transparency, catalytic activity, and stability in strong electrolytes [174]. In complex mixed-metal–oxide chemical spaces, compositional variation induces optical property variation [175]. Over a set of nearly 179 000 distinct materials, convolutional and deep neural networks are able to predict the UV–vis absorption spectrum from a microscope image without the need for a first-principles time-dependent DFT (TDDFT) calculation that would typically be necessary to predict a spectrum. Similarly, deep learning models also solve the converse problem by generating the image of a hypothetical material given its UV–vis absorption spectrum [175]. Aside from mixed-metal–oxides, single-source experimental data have been used to develop ML models that predict MOF water stability [176]. Due to limited data quantities, the generalizability of these models remains unknown.

When single-source data are challenging to obtain, data curation from the extant experimental literature provides an opportunity to harness community knowledge. Indeed, utilizing natural language processing (NLP) tools reveals that synthesis conditions influence materials property outcomes for zeolites [164, 165, 167] and carbon nanotubes [166]. Similarly, these strategies have uncovered the role of additives such as organic structure-directing agents on zeolite formation [168]. In other porous materials such as MOFs, analysis has shown that surface areas grow exponentially with time, indicating that synthesis methods have improved systematically for commonly studied MOFs [177]. To be usable for practical applications, MOFs must be able to withstand heat (i.e. thermal stability) and have solvent removed from their pores after synthesis (i.e. activation stability). By mining experimental thermal and activation stability from the extant literature, we trained ML models to predict whether or not MOFs will be thermally stable and stable upon activation [169, 178]. This strategy enabled zero-cost predictions on how to improve MOF stability (Figure 20.24). Using models trained from this data, we constructed a new hypothetical MOF database comprising recombinations of synthesized MOFs that are likely to be stable [179]. As a final example, ML has enabled verification of oxidation state labels assigned to transition metal complexes, where heuristics may be incorrect or inconsistent [180]. Collective knowledge unlocks information from the literature that no single scientist or group would be able to gather alone.

Community feedback is of paramount importance to improve ML predictions, especially when subjectivity influences the data [156]. Here, community-based active learning represents a way in which domain experts are able to "vote" on whether predictions are accurate [178]. Strategies to utilize domain expertise have led to the discovery of new synthesizability metrics [181] and molecules with

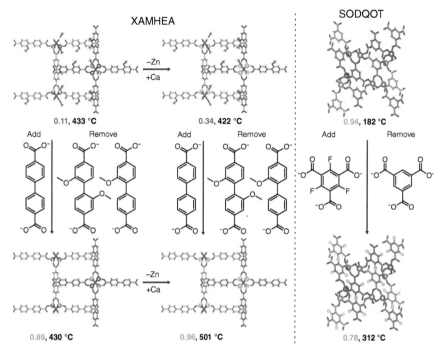

Figure 20.24 Representative zero-cost example of model-informed MOF engineering to improve stability. Model predictions for activation stability (between 0 and 1) and thermal stability (floating point value) are shown, with activation stability values below 0.5 (i.e. not stable) shown in red, and those above or equal to 0.5 (i.e. stable) shown in green. Changes to the linkers and metals lead to tangible differences in stability. Source: Nandy et al. [169]/American Chemical Society.

state-of-the-art properties [182]. Additionally, community-based studies on MOF color metrics indicate that qualitative color assignments are imprecise [183]. Along with our extracted stability data for MOFs, we recently released a web interface for feedback on MOF stability predictions by our data-driven models [178]. On this platform, users are able to deposit new data, comment on model predictions on new systems, and improve the fidelity of the training data. Over time, the models will continue to improve and improve user confidence in the predictions generated by ML regarding candidate materials for catalysis.

20.3 Summary

Over the past two decades, DFT has played an essential role in understanding mechanism and reactivity for various homogeneous and heterogeneous catalytic systems. Concepts such as LFERs and BEP relationships, coupled with DFT-computed descriptor variables, have facilitated large-scale screening over catalyst composition space to identify high-activity catalysts. When reaction intermediates are challenging to isolate, DFT has played an essential role in identifying the roles of variables such as distortion or spin and oxidation state. With an increase in data quantity and fidelity, DFT method selection has become more systematic, ensuring increasingly faithful comparison to experimental observables.

ML has paved the way for obtaining DFT-quality predictions on materials at a fraction of the cost, reducing the time to identifying lead candidates. Interpretable representations of catalysts and materials have facilitated data-driven insights. For catalysis, data-driven methods provide a path forward when LFERs and BEP relations break down. Combining UQ with ML models and merging them with optimization algorithms has unlocked large-scale chemical space exploration over millions and billions of molecules and materials. These large-scale explorations have led to extraction of hidden trends that are difficult to analyze over small data sets. As more experimental data becomes available, ML on experimental data or a multi-fidelity approach combining computational and experimental data becomes increasingly promising. The catalysts needed to tackle global challenges may have not yet been identified because they violate the assumptions baked into current screening protocols. ML acceleration will enable sophisticated searches over wide swaths of chemical space and will allow us to relax our assumptions in order to discover the catalysts of the future.

References

1 Nandy, A., Duan, C., Taylor, M.G. et al. (2021). Computational discovery of transition-metal complexes: from high-throughput screening to machine learning. *Chem. Rev.* 121(16): 9927–10000.

2 Vogiatzis, K.D., Polynski, M.V., Kirkland, J.K. et al. (2018). Computational approach to molecular catalysis by 3d transition metals: challenges and opportunities. *Chem. Rev.* 119(4): 2453–2523.

3 Rostamikia, G. and Janik, M.J. (2010). Direct borohydride oxidation: mechanism determination and design of alloy catalysts guided by density functional theory. *Energy Environ. Sci.* 3(9): 1262.

4 Nørskov, J.K., Abild-Pedersen, F., Studt, F., and Bligaard, T. (2011). Density functional theory in surface chemistry and catalysis. *Proc. Natl. Acad. Sci.* 108(3): 937–943.

5 Nandy, A., Duan, C., Goffinet, C., and Kulik, H.J. (2022). New strategies for direct methane-to-methanol conversion from active learning exploration of 16 million catalysts. *JACS Au* 2(5): 1200–1213.

6 Khorshidi, A., Violet, J., Hashemi, J., and Peterson, A.A. (2018). How strain can break the scaling relations of catalysis. *Nat. Catal.* 1(4): 263–268.

7 Gani, T.Z.H. and Kulik, H.J. (2018). Understanding and breaking scaling relations in single-site catalysis: methane to methanol conversion by $Fe^{IV}=O$. *ACS Catal.* 8(2): 975–986.

8 Nandy, A., Zhu, J., Janet, J.P. et al. (2019). Machine learning accelerates the discovery of design rules and exceptions in stable metal–oxo intermediate formation. *ACS Catal.* 9(9): 8243–8255.

9 Nandy, A. and Kulik, H.J. (2020). Why conventional design rules for C–H activation fail for open-shell transition-metal catalysts. *ACS Catal.* 10(24): 15033–15047.

10 Kjaergaard, C.H., Qayyum, M.F., Wong, S.D. et al. (2014). Spectroscopic and computational insight into the activation of O_2 by the mononuclear Cu center in polysaccharide monooxygenases. *Proc. Natl. Acad. Sci.* 111(24): 8797–8802.

11 Dey, A., Peng, Y., Broderick, W.E. et al. (2011). S K-edge XAS and DFT calculations on SAM dependent pyruvate formate-lyase activating enzyme: nature of interaction between the Fe_4S_4 cluster and SAM and its role in reactivity. *J. Am. Chem. Soc.* 133(46): 18656–18662.

12 Heyer, A.J., Plessers, D., Braun, A. et al. (2022). Methane activation by a mononuclear copper active site in the zeolite mordenite: effect of metal nuclearity on reactivity. *J. Am. Chem. Soc.* 144(42): 19305–19316.

13 Ravi, M., Ranocchiari, M., and van Bokhoven, J.A. (2017). The direct catalytic oxidation of methane to methanol-a critical assessment. *Angew. Chem. Int. Ed.* 56(52): 16464–16483.

14 Nandy, A., Adamji, H., Kastner, D.W. et al. (2022). Using computational chemistry to reveal nature's blueprints for single-site catalysis of C–H activation. *ACS Catal.* 12(15): 9281–9306.

15 Usharani, D., Lacy, D.C., Borovik, A.S., and Shaik, S. (2013). Dichotomous hydrogen atom transfer vs proton-coupled electron transfer during activation of X–H bonds (X = C, N, O) by nonheme iron–oxo complexes of variable basicity. *J. Am. Chem. Soc.* 135(45): 17090–17104.

16 Klein, J.E.M.N. and Knizia, G. (2018). cPCET versus HAT: a direct theoretical method for distinguishing X–H bond-activation mechanisms. *Angew. Chem. Int. Ed.* 57(37): 11913–11917.

17 Mayer, J.M. (2004). Proton-coupled electron transfer: a reaction Chemist's view. *Annu. Rev. Phys. Chem.* 55(1): 363–390.

18 Hammes-Schiffer, S. and Stuchebrukhov, A.A. (2010). Theory of coupled electron and proton transfer reactions. *Chem. Rev.* 110(12): 6939–6960.

19 Goetz, M.K. and Anderson, J.S. (2019). Experimental evidence for pK_a-driven asynchronicity in C–H activation by a terminal Co(III)–oxo complex. *J. Am. Chem. Soc.* 141(9): 4051–4062.

20 Oswald, V.F., Lee, J.L., Biswas, S. et al. (2020). Effects of noncovalent interactions on high-spin Fe(IV)–oxido complexes. *J. Am. Chem. Soc.* 142(27): 11804–11817.

21 Borovik, A.S. (2004). Bioinspired hydrogen bond motifs in ligand design: the role of noncovalent interactions in metal ion mediated activation of dioxygen. *Acc. Chem. Res.* 38(1): 54–61.

22 Geng, C., Ye, S., and Neese, F. (2014). Does a higher metal oxidation state necessarily imply higher reactivity toward H-atom transfer? A computational study of C–H bond oxidation by high-valent iron-oxo and -nitrido complexes. *Dalton Trans.* 43(16): 6079–6086.

23 Rolfes, J.D., van Gastel, M., and Neese, F. (2020). Where is the fluoro wall?: a quantum chemical investigation. *Inorg. Chem.* 59(2): 1556–1565.

24 Winkler, J.R. and Gray, H.B. (2011). Electronic structures of oxo-metal ions. In: *Molecular Electronic Structures of Transition Metal Complexes I Structure and Bonding*, vol. 142 (ed. D. Mingos, P. Day, and J. Dahl), 17–28. Heidelberg, Berlin: Springer.

25 Harper, D.R., Nandy, A., Arunachalam, N. et al. (2022). Representations and strategies for transferable machine learning improve model performance in chemical discovery. *J. Chem. Phys.* 156(7): 074101.

26 Nandy, A., Chu, D.B.K., Harper, D.R. et al. (2020). Large-scale comparison of 3d and 4d transition metal complexes illuminates the reduced effect of exchange on second-row spin-state energetics. *Phys. Chem. Chem. Phys.* 22(34): 19326–19341.

27 Miranda-Quintana, R.A. (2017). Note: the minimum electrophilicity and the hard/soft acid/base principles. *J. Chem. Phys.* 146(4): 046101.

28 Chattaraj, P.K., Liu, G.H., and Parr, R.G. (1995). The maximum hardness principle in the gyftopoulos-hatsopoulos three-level model for an atomic or molecular species and its positive and negative ions. *Chem. Phys. Lett.* 237(1–2): 171–176.

29 Chamorro, E., Chattaraj, P.K., and Fuentealba, P. (2003). Variation of the electrophilicity index along the reaction path. *J. Phys. Chem. A* 107(36): 7068–7072.

30 Miranda-Quintana, R.A., Heidar-Zadeh, F., and Ayers, P.W. (2018). Elementary derivation of the "$|\Delta\mu|$ big is good" rule. *J. Phys. Chem. Lett.* 9(15): 4344–4348.

31 Novikov, A.S. (2015). 1,3-dipolar cycloaddition of nitrones to transition metal-bound isocyanides: DFT and HSAB principle theoretical model together with analysis of vibrational spectra. *J. Organomet. Chem.* 797: 8–12.

32 Geerlings, P., Chamorro, E., Chattaraj, P.K. et al. (2020). Conceptual density functional theory: status, prospects, issues. *Theor. Chem. Acc.* 139(2): 36.

33 Xia, Y., Yin, D., Rong, C. et al. (2008). Impact of Lewis acids on Diels–Alder reaction reactivity: a conceptual density functional theory study. *J. Phys. Chem. A* 112(40): 9970–9977.

34 Bím, D., Maldonado-Domínguez, M., Rulíšek, L., and Srnec, M. (2018). Beyond the classical thermodynamic contributions to hydrogen atom abstraction reactivity. *Proc. Natl. Acad. Sci.* 115(44): 10287–10294.

35 Maldonado-Domínguez, M. and Srnec, M. (2022). H-atom abstraction reactivity through the lens of asynchronicity and frustration with their counteracting effects on barriers. *Inorg. Chem.* 61(47): 18811–18822.

36 Maldonado-Domínguez, M. and Srnec, M. (2020). Understanding and predicting post H-atom abstraction selectivity through reactive mode composition factor analysis. *J. Am. Chem. Soc.* 142(8): 3947–3958.

37 Szécsényi, Á., Khramenkova, E., Chernyshov, I.Y. et al. (2019). Breaking linear scaling relationships with secondary interactions in confined space: a case study of methane oxidation by Fe/ZSM-5 zeolite. *ACS Catal.* 9(10): 9276–9284.

38 Goltl, F., Michel, C., Andrikopoulos, P.C. et al. (2016). Computationally exploring confinement effects in the methane-to-methanol conversion over iron-oxo centers in zeolites. *ACS Catal.* 6(12): 8404–8409.

39 Betley, T.A., Wu, Q., Van Voorhis, T., and Nocera, D.G. (2008). Electronic design criteria for O–O bond formation via metal–oxo complexes. *Inorg. Chem.* 47(6): 1849–1861.

40 Campbell, C.T. (1994). Future directions and industrial perspectives micro- and macro-kinetics: their relationship in heterogeneous catalysis. *Top. Catal.* 1(3–4): 353–366.

41 Campbell, C.T. (2001). Finding the rate-determining step in a mechanism. *J. Catal.* 204(2): 520–524.

42 Campbell, C.T. (2017). The degree of rate control: a powerful tool for catalysis research. *ACS Catal.* 7(4): 2770–2779.

43 Kozuch, S. and Shaik, S. (2010). How to conceptualize catalytic cycles? The energetic span model. *Acc. Chem. Res.* 44(2): 101–110.

44 Kozuch, S. and Martin, J.M.L. (2011). What makes for a bad catalytic cycle? A theoretical study on the Suzuki–Miyaura reaction within the energetic span model. *ACS Catal.* 1(4): 246–253.

45 Kim, J.Y. and Kulik, H.J. (2018). When is ligand pK_a a good descriptor for catalyst energetics? In search of optimal CO_2 hydration catalysts. *J. Phys. Chem. A* 122(18): 4579–4590.

46 Orian, L., Wolters, L.P., and Bickelhaupt, F.M. (2013). In silico design of heteroaromatic half-sandwich Rh^I catalysts for acetylene [2+2+2] cyclotrimerization: evidence of a reverse Indenyl effect. *Chem. Eur. J.* 19(40): 13337–13347.

47 Bac, S. and Mallikarjun, S.S. (2022). CO oxidation with atomically dispersed catalysts: insights from the energetic span model. *ACS Catal.* 12(3): 2064–2076.

48 Chen, J., Chen, Y., Li, P. et al. (2018). Energetic span as a rate-determining term for electrocatalytic volcanos. *ACS Catal.* 8(11): 10590–10598.

49 Wodrich, M.D., Sawatlon, B., Busch, M., and Corminboeuf, C. (2021). The genesis of molecular volcano plots. *Acc. Chem. Res.* 54(5): 1107–1117.

50 Wodrich, M.D., Sawatlon, B., Solel, E. et al. (2019). Activity-based screening of homogeneous catalysts through the rapid assessment of theoretically derived turnover frequencies. *ACS Catal.* 9(6): 5716–5725.

51 Liao, P., Getman, R.B., and Snurr, R.Q. (2017). Optimizing open iron sites in metal–organic frameworks for ethane oxidation: a first-principles study. *ACS Appl. Mater. Interfaces* 9(39): 33484–33492.

52 Hammett, L.P. (1938). Linear free energy relationships in rate and equilibrium phenomena. *Trans. Faraday Soc.* 34: 156.

53 Nørskov, J.K., Bligaard, T., Rossmeisl, J., and Christensen, C.H. (2009). Towards the computational design of solid catalysts. *Nat. Chem.* 1(1): 37–46.

54 Abild-Pedersen, F., Greeley, J., Studt, F. et al. (2007). Scaling properties of adsorption energies for hydrogen containing molecules on transition-metal surfaces. *Phys. Rev. Lett.* 99(1): 016105.

55 Latimer, A.A., Kulkarni, A.R., Aljama, H. et al. (2017). Understanding trends in C-H bond activation in heterogeneous catalysis. *Nat. Mater.* 16(2): 225–229.

56 Rosen, A.S., Notestein, J.M., and Snurr, R.Q. (2019). Structure–activity relationships that identify metal–organic framework catalysts for methane activation. *ACS Catal.* 9(4): 3576–3587.

57 Calle-Vallejo, F., Loffreda, D., Koper, M.T.M., and Sautet, P. (2015). Introducing structural sensitivity into adsorption–energy scaling relations by means of coordination numbers. *Nat. Chem.* 7(5): 403–410.

58 Shustorovich, E. (1988). Analysis of CO hydrogenation pathways using the bond-order-conservation method. *J. Catal.* 113(2): 341–352.

59 Pérez-Ramírez, J. and López, N. (2019). Strategies to break linear scaling relationships. *Nat. Catal.* 2(11): 971–976.

60 Nørskov, J.K., Bligaard, T., Logadottir, A. et al. (2002). Universality in heterogeneous catalysis. *J. Catal.* 209(2): 275–278.

61 Tsay, C., Livesay, B.N., Ruelas, S., and Yang, J.Y. (2015). Solvation effects on transition metal hydricity. *J. Am. Chem. Soc.* 137(44): 14114–14121.

62 Ceballos, B.M. and Yang, J.Y. (2018). Directing the reactivity of metal hydrides for selective CO_2 reduction. *Proc. Natl. Acad. Sci.* 115(50): 12686–12691.

63 Lan, Z. and Mallikarjun, S.S. (2020). Linear free energy relationships for transition metal chemistry: case study of CH activation with copper–oxygen complexes. *Phys. Chem. Chem. Phys.* 22(14): 7155–7159.

64 Schneider, J.E., Goetz, M.K., and Anderson, J.S. (2021). Statistical analysis of C–H activation by oxo complexes supports diverse thermodynamic control over reactivity. *Chem. Sci.* 12(11): 4173–4183.

65 Andrikopoulos, P.C., Michel, C., Chouzier, S., and Sautet, P. (2015). In silico screening of iron-oxo catalysts for CH bond cleavage. *ACS Catal.* 5(4): 2490–2499.

66 Jia, H., Nandy, A., Liu, M., and Kulik, H.J. (2022). Modeling the roles of rigidity and dopants in single-atom methane-to-methanol catalysts. *J. Mater. Chem. A* 10(11): 6193–6203.

67 Wodrich, M.D., Busch, M., and Corminboeuf, C. (2016). Accessing and predicting the kinetic profiles of homogeneous catalysts from volcano plots. *Chem. Sci.* 7(9): 5723–5735.

68 Verloop, A. (1983). The Sterimol approach: further development of the method and new applications. *Pesticide Chemistry: Human Welfare and the Environment, Proceedings of the International Congress of Pesticide Chemistry*. Pergamon. pp. 339–344. https://doi.org/10.1016/B978-0-08-029222-9.50051-2.

69 Zandkarimi, B. and Alexandrova, A.N. (2019). Dynamics of subnanometer Pt clusters can break the scaling relationships in catalysis. *J. Phys. Chem. Lett.* 10(3): 460–467.

70 Xu, W., Andersen, M., and Reuter, K. (2020). Data-driven descriptor engineering and refined scaling relations for predicting transition metal oxide reactivity. *ACS Catal.* 11(2): 734–742.

71 Andersen, M., Levchenko, S.V., Scheffler, M., and Reuter, K. (2019). Beyond scaling relations for the description of catalytic materials. *ACS Catal.* 9(4): 2752–2759.

72 Kulik, H.J. (2015). Perspective: treating electron over-delocalization with the DFT+U method. *J. Chem. Phys.* 142(24): 240901.

73 Gaggioli, C.A., Stoneburner, S.J., Cramer, C.J., and Gagliardi, L. (2019). Beyond density functional theory: the multiconfigurational approach to model heterogeneous catalysis. *ACS Catal.* 9(9): 8481–8502.

74 Ioannidis, E.I. and Kulik, H.J. (2015). Towards quantifying the role of exact exchange in predictions of transition metal complex properties. *J. Chem. Phys.* 143(3): 034104.

75 Gani, T.Z.H. and Kulik, H.J. (2017). Unifying exchange sensitivity in transition-metal spin-state ordering and catalysis through bond valence metrics. *J. Chem. Theory Comput.* 13(11): 5443–5457.

76 Zhao, Q. and Kulik, H.J. (2019). Stable surfaces that bind too tightly: can range-separated hybrids or DFT+U improve paradoxical descriptions of surface chemistry? *J. Phys. Chem. Lett.* 10(17): 5090–5098.

77 Vennelakanti, V., Nandy, A., and Kulik, H.J. (2021). The effect of Hartree–Fock exchange on scaling relations and reaction energetics for C–H activation catalysts. *Top. Catal.* 65(1–4): 296–311.

78 Lininger, C.N., Gauthier, J.A., Li, W.-L. et al. (2021). Challenges for density functional theory: calculation of CO adsorption on electrocatalytically relevant metals. *Phys. Chem. Chem. Phys.* 23(15): 9394–9406.

79 Zhao, Q., Ioannidis, E.I., and Kulik, H.J. (2016). Global and local curvature in density functional theory. *J. Chem. Phys.* 145(5): 054109.

80 Himmetoglu, B., Floris, A., de Gironcoli, S., and Cococcioni, M. (2014). Hubbard-corrected DFT energy functionals: the LDA+U description of correlated systems. *Int. J. Quantum Chem.* 114(1): 14–49.

81 Bajaj, A., Janet, J.P., and Kulik, H.J. (2017). Communication: recovering the flat-plane condition in electronic structure theory at semi-local DFT cost. *J. Chem. Phys.* 147(19): 191101.

82 Bajaj, A. and Kulik, H.J. (2022). Eliminating delocalization error to improve heterogeneous catalysis predictions with molecular DFT + U. *J. Chem. Theory Comput.* 18(2): 1142–1155.

83 Bajaj, A. and Kulik, H.J. (2021). Molecular DFT+U: a transferable, low-cost approach to eliminate delocalization error. *J. Phys. Chem. Lett.* 12(14): 3633–3640.

84 Bajaj, A., Duan, C., Nandy, A. et al. (2022). Molecular orbital projectors in non-empirical jmDFT recover exact conditions in transition-metal chemistry. *J. Chem. Phys.* 156(18): 184112.

85 Wellendorff, J., Lundgaard, K.T., Møgelhøj, A. et al. (2012). Density functionals for surface science: exchange-correlation model development with Bayesian error estimation. *Phys. Rev. B* 85(23): 235149.

86 Medford, A.J., Wellendorff, J., Vojvodic, A. et al. (2014). Assessing the reliability of calculated catalytic ammonia synthesis rates. *Science* 345(6193): 197–200.

87 Duan, C., Chen, S., Taylor, M.G. et al. (2021). Machine learning to tame divergent density functional approximations: a new path to consensus materials design principles. *Chem. Sci.* 12(39): 13021–13036.

88 Duan, C., Nandy, A., Terrones, G.G. et al. (2022). Active learning exploration of transition-metal complexes to discover method-insensitive and synthetically accessible chromophores. *JACS Au* 3(2): 391–401.

89 Duan, C., Nandy, A., Meyer, R. et al. (2022). A transferable recommender approach for selecting the best density functional approximations in chemical discovery. *Nat. Comput. Sci.* 3(1): 38–47.

90 Janet, J.P., Chan, L., and Kulik, H.J. (2018). Accelerating chemical discovery with machine learning: simulated evolution of spin crossover complexes with an artificial neural network. *J. Phys. Chem. Lett.* 9(5): 1064–1071.

91 Janet, J.P., Gani, T.Z.H., Steeves, A.H. et al. (2017). Leveraging cheminformatics strategies for inorganic discovery: application to redox potential design. *Ind. Eng. Chem. Res.* 56(17): 4898–4910.

92 Janet, J.P. and Kulik, H.J. (2017). Predicting electronic structure properties of transition metal complexes with neural networks. *Chem. Sci.* 8: 5137–5152.

93 Janet, J.P. and Kulik, H.J. (2017). Resolving transition metal chemical space: feature selection for machine learning and structure–property relationships. *J. Phys. Chem. A* 121(46): 8939–8954.

94 Janet, J.P., Ramesh, S., Duan, C., and Kulik, H.J. (2020). Accurate multiobjective design in a space of millions of transition metal complexes with neural-network-driven efficient global optimization. *ACS Cent. Sci.* 6(4): 513–524.

95 Nandy, A., Duan, C., Janet, J.P. et al. (2018). Strategies and software for machine learning accelerated discovery in transition metal chemistry. *Ind. Eng. Chem. Res.* 57(42): 13973–13986.

96 Duan, C., Liu, F., Nandy, A., and Kulik, H.J. (2020). Semi-supervised machine learning enables the robust detection of multireference character at low cost. *J. Phys. Chem. Lett.* 11: 6640–6648.

97 Duan, C., Liu, F., Nandy, A., and Kulik, H.J. (2020). Data-driven approaches can overcome the cost-accuracy tradeoff in multireference diagnostics. *J. Chem. Theory Comput.* 16: 4373–4387.

98 Duan, C., Chu, D.B.K., Nandy, A., and Kulik, H.J. (2022). Detection of multi-reference character imbalances enables a transfer learning approach for virtual high throughput screening with coupled cluster accuracy at DFT cost. *Chem. Sci.* 13(17): 4962–4971.

99 Taylor, M.G., Nandy, A., Lu, C.C., and Kulik, H.J. (2021). Deciphering cryptic behavior in bimetallic transition-metal complexes with machine learning. *J. Phys. Chem. Lett.* 12(40): 9812–9820.

100 Janet, J.P., Duan, C., Yang, T. et al. (2019). A quantitative uncertainty metric controls error in neural network-driven chemical discovery. *Chem. Sci.* 10(34): 7913–7922.

101 Taylor, M.G., Yang, T., Lin, S. et al. (2020). Seeing is believing: experimental spin states from machine learning model structure predictions. *J. Phys. Chem. A* 124(16): 3286–3299.

102 Moreau, G. and Broto, P. (1980). The autocorrelation of a topological structure: a new molecular descriptor. *Nouv. J. Chim.* 4(6): 359–360.

103 Broto, P., Moreau, G., and Vandycke, C. (1984). Molecular structures: perception, autocorrelation descriptor and SAR studies: system of atomic contributions for the calculation of the n-octanol/water partition coefficients. *Eur. J. Med. Chem.* 19(1): 71–78.

104 Gómez-Bombarelli, R., Wei, J.N., Duvenaud, D. et al. (2018). Automatic chemical design using a data-driven continuous representation of molecules. *ACS Cent. Sci.* 4(2): 268–276.

105 White, A.D. (2021). Deep learning for molecules & materials. *Living J. Comput. Mol. Sci.* 3: 1499. http://dmol.pub.

106 Thomas, N., Smidt, T., Kearnes, S. et al. (2018). Tensor field networks: rotation- and translation-equivariant neural networks for 3D point clouds. *arXiv* https://doi.org/10.48550/arXiv.1802.08219.

107 Raza, A., Sturluson, A., Simon, C.M., and Fern, X. (2020). Message passing neural networks for partial charge assignment to metal–organic frameworks. *J. Phys. Chem. C* 124(35): 19070–19082.

108 Reiser, P., Neubert, M., Eberhard, A. et al. (2022). Graph neural networks for materials science and chemistry. *Commun. Mater.* 3(1): 93.

109 Moosavi, S.M., Nandy, A., Jablonka, K.M. et al. (2020). Understanding the diversity of the metal-organic framework ecosystem. *Nat. Commun.* 11(1): 4068.

110 Martin, R.L., Smit, B., and Haranczyk, M. (2011). Addressing challenges of identifying geometrically diverse sets of crystalline porous materials. *J. Chem. Information Model.* 52(2): 308–318.

111 Willems, T.F., Rycroft, C.H., Kazi, M. et al. (2012). Algorithms and tools for high-throughput geometry-based analysis of crystalline porous materials. *Microporous Mesoporous Mater.* 149(1): 134–141.

112 Townsend, J., Micucci, C.P., Hymel, J.H. et al. (2020). Representation of molecular structures with persistent homology for machine learning applications in chemistry. *Nat. Commun.* 11(1): 3230.

113 Lee, Y., Barthel, S.D., Dłotko, P. et al. (2017). Quantifying similarity of pore-geometry in nanoporous materials. *Nat. Commun.* 8(1): 15396.

114 Krishnapriyan, A.S., Montoya, J., Haranczyk, M. et al. (2021). Machine learning with persistent homology and chemical word embeddings improves prediction accuracy and interpretability in metal-organic frameworks. *Sci. Rep.* 11(1): 8888.

115 Behler, J. (2011). Atom-centered symmetry functions for constructing high-dimensional neural network potentials. *J. Chem. Phys.* 134(7): 074106.

116 Rupp, M., Tkatchenko, A., Müller, K.-R., and von Lilienfeld, O.A. (2012). Fast and accurate modeling of molecular atomization energies with machine learning. *Phys. Rev. Lett.* 108(5): 058301.

117 Smith, J.S., Isayev, O., and Roitberg, A.E. (2017). ANI-1: an extensible neural network potential with DFT accuracy at force field computational cost. *Chem. Sci.* 8(4): 3192–3203.

118 Unke, O.T., Chmiela, S., Sauceda, H.E. et al. (2021). Machine learning force fields. *Chem. Rev.* 121(16): 10142–10186.

119 Li, Y., Li, H., Pickard, F.C. et al. (2017). Machine learning force field parameters from ab initio data. *J. Chem. Theory Comput.* 13(9): 4492–4503.

120 Duan, C., Janet, J.P., Liu, F. et al. (2019). Learning from failure: predicting electronic structure calculation outcomes with machine learning models. *J. Chem. Theory Comput.* 15(4): 2331–2345.

121 Duan, C., Liu, F., Nandy, A., and Kulik, H.J. (2021). Putting density functional theory to the test in machine-learning-accelerated materials discovery. *J. Phys. Chem. Lett.* 12(19): 4628–4637.

122 Duan, C., Nandy, A., Adamji, H. et al. (2022). Machine learning models predict calculation outcomes with the transferability necessary for computational catalysis. *J. Chem. Theory Comput.* 18(7): 4282–4292.

123 Selvaraju, R.R., Cogswell, M., Das, A. et al. (2016). Grad-CAM: visual explanations from deep networks via gradient-based localization. *arXiv* https://doi.org/10.48550/arXiv.1610.02391.

124 Cordova, M., Wodrich, M.D., Meyer, B. et al. (2020). Data-driven advancement of homogeneous nickel catalyst activity for aryl ether cleavage. *ACS Catal.* 10(13): 7021–7031.

125 Hansen, K., Biegler, F., Ramakrishnan, R. et al. (2015). Machine learning predictions of molecular properties: accurate many-body potentials and nonlocality in chemical space. *J. Phys. Chem. Lett.* 6(12): 2326–2331.

126 Meyer, B., Sawatlon, B., Heinen, S. et al. (2018). Machine learning meets volcano plots: computational discovery of cross-coupling catalysts. *Chem. Sci.* 9(35): 7069–7077.

127 Tran, K. and Ulissi, Z.W. (2018). Active learning across intermetallics to guide discovery of electrocatalysts for CO_2 reduction and H_2 evolution. *Nat. Catal.* 1(9): 696.

128 Wodrich, M.D., Fabrizio, A., Meyer, B., and Corminboeuf, C. (2020). Data-powered augmented volcano plots for homogeneous catalysis. *Chem. Sci.* 11(44): 12070–12080.

129 Kwon, D.-H., Fuller, J.T., Kilgore, U.J. et al. (2018). Computational transition-state design provides experimentally verified Cr(P,N) catalysts for control of ethylene trimerization and tetramerization. *ACS Catal.* 8(2): 1138–1142.

130 Maley Steven, M., Kwon, D.-H., Rollins, N. et al. (2020). Quantum-mechanical transition-state model combined with machine learning provides catalyst design features for selective Cr olefin oligomerization. *Chem. Sci.* 11(35): 9665–9674.

131 Lakuntza, O., Besora, M., and Maseras, F. (2018). Searching for hidden descriptors in the metal–ligand bond through statistical analysis of density functional theory (DFT) results. *Inorg. Chem.* 57(23): 14660–14670.

132 Breiman, L. (2001). Random forests. *Mach. Learn.* 45(1): 5–32.

133 Chen, S., Nielson, T., Zalit, E. et al. (2021). Automated construction and optimization combined with machine learning to generate Pt(II) methane C–H activation transition states. *Top. Catal.* 65(1-4): 312–324.

134 Santiago, C.B., Guo, J.-Y., and Sigman, M.S. (2018). Predictive and mechanistic multivariate linear regression models for reaction development. *Chem. Sci.* 9(9): 2398–2412.

135 Reid, J.P. and Sigman, M.S. (2018). Comparing quantitative prediction methods for the discovery of small-molecule chiral catalysts. *Nat. Rev. Chem.* 2(10): 290–305.

136 Xu, E.Y., Werth, J., Roos, C.B. et al. (2022). Noncovalent stabilization of radical intermediates in the enantioselective hydroamination of alkenes with sulfonamides. *J. Am. Chem. Soc.* 144(41): 18948–18958.

137 Harper, K.C., Bess, E.N., and Sigman, M.S. (2012). Multidimensional steric parameters in the analysis of asymmetric catalytic reactions. *Nat. Chem.* 4(5): 366–374.

138 Haas, B.C., Goetz, A.E., Bahamonde, A. et al. (2022). Predicting relative efficiency of amide bond formation using multivariate linear regression. *Proc. Natl. Acad. Sci.* 119(16): e2118451119.

139 Vishwakarma, G., Sonpal, A., and Hachmann, J. (2021). Metrics for benchmarking and uncertainty quantification: quality, applicability, and best practices for machine learning in chemistry. *Trends Chem.* 3(2): 146–156.

140 Gal, Y. and Ghahramani, Z. (2016). Dropout as a Bayesian approximation: representing model uncertainty in deep learning. *International Conference on Machine Learning*. pp. 1050–1059. https://proceedings.mlr.press/v48/gal16.html.

141 Deringer, V.L., Bartók, A.P., Bernstein, N. et al. (2021). Gaussian process regression for materials and molecules. *Chem. Rev.* 121(16): 10073–10141.

142 Ulissi, Z.W., Medford, A.J., Bligaard, T., and Nørskov, J.K. (2017). To address surface reaction network complexity using scaling relations machine learning and DFT calculations. *Nat. Commun.* 8: 14621.

143 Simm, G.N. and Reiher, M. (2018). Error-controlled exploration of chemical reaction networks with Gaussian processes. *J. Chem. Theory Comput.* 14(10): 5238–5248.

144 Møller, C. and Plesset, M.S. (1934). Note on an approximation treatment for many-electron systems. *Phys. Rev.* 46(7): 618–622.

145 Lee, T.J. and Taylor, P.R. (2009). A diagnostic for determining the quality of single-reference electron correlation methods. *Int. J. Quantum Chem.* 36(S23): 199–207.

146 Cheng, L., Welborn, M., Christensen, A.S., and Miller, T.F. (2019). A universal density matrix functional from molecular orbital-based machine learning: transferability across organic molecules. *J. Chem. Phys.* 150(13): 131103.

147 Welborn, M., Cheng, L., and Miller, T.F. III, (2018). Transferability in machine learning for electronic structure via the molecular orbital basis. *J. Chem. Theory Comput.* 14(9): 4772–4779.

148 Husch, T., Sun, J., Cheng, L. et al. (2021). Improved accuracy and transferability of molecular-orbital-based machine learning: organics, transition-metal complexes, non-covalent interactions, and transition states. *J. Chem. Phys.* 154(6): 064108.

149 Lee, S., Kim, B., Cho, H. et al. (2021). Computational screening of trillions of metal–organic frameworks for high-performance methane storage. *ACS Appl. Mater. Interfaces* 13(20): 23647–23654.

150 Xue, D., Balachandran, P.V., Hogden, J. et al. (2016). Accelerated search for materials with targeted properties by adaptive design. *Nat. Commun.* 7(1): 11241.

151 Herbol, H.C., Hu, W., Frazier, P. et al. (2018). Efficient search of compositional space for hybrid organic–inorganic perovskites via Bayesian optimization. *NPJ Comput. Mater.* 4(1): 51.

152 Seko, A., Hayashi, H., Nakayama, K. et al. (2017). Representation of compounds for machine-learning prediction of physical properties. *Phys. Rev. B* 95(14): 144110.

153 del Rosario, Z., Rupp, M., Kim, Y. et al. (2020). Assessing the frontier: active learning, model accuracy, and multi-objective candidate discovery and optimization. *J. Chem. Phys.* 153(2): 024112.

154 Keane, A.J. (2006). Statistical improvement criteria for use in multiobjective design optimization. *AIAA J.* 44(4): 879–891.

155 Doan, H.A., Agarwal, G., Qian, H. et al. (2020). Quantum chemistry-informed active learning to accelerate the design and discovery of sustainable energy storage materials. *Chem. Mater.* 32(15): 6338–6346.

156 Nandy, A., Duan, C., and Kulik, H.J. (2022). Audacity of huge: overcoming challenges of data scarcity and data quality for machine learning in computational materials discovery. *Curr. Opin. Chem. Eng.* 36: 100778.

157 Jia, X., Lynch, A., Huang, Y. et al. (2019). Anthropogenic biases in chemical reaction data hinder exploratory inorganic synthesis. *Nature* 573(7773): 251–255.

158 Groom, C.R., Bruno, I.J., Lightfoot, M.P., and Ward, S.C. (2016). The Cambridge structural database. *Acta Crystallogr. B Struct. Sci. Cryst.* 72(2): 171–179.

159 Allen, F.H. (2002). The Cambridge structural database: a quarter of a million crystal structures and rising. *Acta Crystallogr. B Struct. Sci.* 58(3): 380–388.

160 Chung, Y.G., Haldoupis, E., Bucior, B.J. et al. (2019). Advances, updates, and analytics for the computation-ready, experimental metal–organic framework database: CoRE MOF 2019. *J. Chem. Eng. Data* 64(12): 5985–5998.

161 Chung, Y.G., Camp, J., Haranczyk, M. et al. (2014). Computation-ready, experimental metal–organic frameworks: a tool to enable high-throughput screening of nanoporous crystals. *Chem. Mater.* 26(21): 6185–6192.

162 Rosen, A.S., Iyer, S.M., Ray, D. et al. (2021). Machine learning the quantum-chemical properties of metal–organic frameworks for accelerated materials discovery. *Matter* 4(5): 1578–1597.

163 Balcells, D. and Skjelstad, B.B. (2020). tmQM dataset—quantum geometries and properties of 86k transition metal complexes. *J. Chem. Information Model.* 60(12): 6135–6146.

164 Kim, E., Jensen, Z., van Grootel, A. et al. (2020). Inorganic materials synthesis planning with literature-trained neural networks. *J. Chem. Information Model.* 60(3): 1194–1201.

165 Jensen, Z., Kim, E., Kwon, S. et al. (2019). A machine learning approach to zeolite synthesis enabled by automatic literature data extraction. *ACS Cent. Sci.* 5(5): 892–899.

166 Kim, E., Huang, K., Saunders, A. et al. (2017). Materials synthesis insights from scientific literature via text extraction and machine learning. *Chem. Mater.* 29(21): 9436–9444.

167 Kim, E., Huang, K., Jegelka, S., and Olivetti, E. (2017). Virtual screening of inorganic materials synthesis parameters with deep learning. *NPJ Comput. Mater.* 3(1): 53.

168 Jensen, Z., Kwon, S., Schwalbe-Koda, D. et al. (2021). Discovering relationships between OSDAs and zeolites through data mining and generative neural networks. *ACS Cent. Sci.* 7(5): 858–867.

169 Nandy, A., Duan, C., and Kulik, H.J. (2021). Using machine learning and data mining to leverage community knowledge for the engineering of stable metal–organic frameworks. *J. Am. Chem. Soc.* 143(42): 17535–17547.

170 Janet, J.P., Liu, F., Nandy, A. et al. (2019). Designing in the face of uncertainty: exploiting electronic structure and machine learning models for discovery in inorganic chemistry. *Inorg. Chem.* 58(16): 10592–10606.

171 Eisenhart, R.J., Clouston, L.J., and Lu, C.C. (2015). Configuring bonds between first-row transition metals. *Acc. Chem. Res.* 48(11): 2885–2894.

172 Rudd, P.A., Liu, S., Planas, N. et al. (2013). Multiple metal-metal bonds in iron-chromium complexes. *Angew. Chem. Int. Ed.* 52(16): 4449–4452.

173 Eisenhart, R.J., Rudd, P.A., Planas, N. et al. (2015). Pushing the limits of delta bonding in metal–chromium complexes with redox changes and metal swapping. *Inorg. Chem.* 54(15): 7579–7592.

174 Yang, L., Haber, J.A., Armstrong, Z. et al. (2021). Discovery of complex oxides via automated experiments and data science. *Proc. Natl. Acad. Sci.* 118(37): e2106042118.

175 Stein, H.S., Guevarra, D., Newhouse, P.F. et al. (2019). Machine learning of optical properties of materials – predicting spectra from images and images from spectra. *Chem. Sci.* 10(1): 47–55.

176 Batra, R., Chen, C., Evans, T.G. et al. (2020). Prediction of water stability of metal–organic frameworks using machine learning. *Nat. Mach. Intell.* 2(11): 704–710.

177 Agrawal, M., Han, R., Herath, D., and Sholl, D.S. (2020). Does repeat synthesis in materials chemistry obey a power law? *Proc. Natl. Acad. Sci.* 117(2): 877–882.

178 Nandy, A., Terrones, G., Arunachalam, N. et al. (2022). MOF simplify, machine learning models with extracted stability data of three thousand metal–organic frameworks. *Sci. Data* 9(1): 74.

179 Nandy, A., Yue, S., Oh, C. et al. (2023). A database of ultrastable MOFs reassembled from stable fragments with machine learning models. *Matter.* 6(5): 1585–1603.

180 Jablonka, K.M., Ongari, D., Moosavi, S.M., and Smit, B. (2021). Using collective knowledge to assign oxidation states of metal cations in metal–organic frameworks. *Nat. Chem.* 13(8): 771–777.

181 Szczypiński, F.T., Bennett, S., and Jelfs, K.E. (2021). Can we predict materials that can be synthesised? *Chem. Sci.* 12(3): 830–840.

182 Gomez-Bombarelli, R., Aguilera-Iparraguirre, J., Hirzel, T.D. et al. (2016). Design of efficient molecular organic light-emitting diodes by a high-throughput virtual screening and experimental approach. *Nat. Mater.* 15(10): 1120–1127.

183 Jablonka, K.M., Moosavi, S.M., Asgari, M. et al. (2021). A data-driven perspective on the colours of metal–organic frameworks. *Chem. Sci.* 12(10): 3587–3598.

Index

a

ab initio electronic structure 485
ab initio molecular dynamics (AIMD) 337, 339, 496
ab initio theory 456
absorption spectra of aggregates 392–397
acid and bases, strengths of
  electronegativity 265
  electrophilicity 266–267
  hardness 265–266
  ionic product 262
  ionization constants 263–264
  pH scale 262–263
  proton affinity 264, 265
acid-base reactions 52, 252–254, 256, 261–262, 266, 267, 269–271
adiabatic resonance energy (ARE) 32–33, 420–421
adiabatic-to-diabatic (ATD) unitary transformation 384–386
aggregation-induced emission 379
alkali bond 306
all purpose reactivity indicator 63
alpha dicarbonyl 419–422
ampholytes 253–254
angular symmetry functions (ASFs) 494
anisotropy effect 164
anisotropy of the induced current density (ACID) 226, 242, 243
antibonding orbitals 212, 286, 295, 338, 340, 353
anti-symmetrical wavefunction 433
aromaticity
  and antiaromaticity 90–92
  definition of 223–224
  electronic descriptors of 230–232
  energetic descriptors of 227, 229–230
  geometric descriptors of 227
  magnetic descriptors of 232–233
  physical foundation 224–226
  stability and 225
  strengths and weaknesses 228–229
  three-dimensional rules for 237–239
  two-dimensional rules for 234–237
aromaticity cube 236
aromatic stabilization energy (ASE) 91–92, 225, 227–230, 242–243
Arrhenius theory 252–253
asynchronicity factor 516
atom-condensed Fukui functions 55–57
atomic basin 61, 76, 169, 194, 200–201, 218, 231, 290
atomic dipole moment corrected Hirshfeld charges (ADCH) 166, 175–180
atomic electronegativity 130, 175, 487
atomic orbital 1, 26, 29–30, 114, 141–142, 209, 237, 274, 275, 409, 414, 457, 460–464, 469, 470, 477, 485
atomic partial charges 78, 127, 129, 130
atoms-in-molecules (AIM)
  method 169, 217
  schemes 216–219

atoms-in-molecules (AIM) (*contd.*)
  theory 34, 76, 78, 93, 165, 169, 170, 177, 179, 194, 207, 209, 210, 214–220
attention algorithms 488

b

Bader charge analysis 337
base pair interactions 446–450
Bayesian methods 539, 540
Bent's rule 298
benzene chain 92
Berkowitz–Parr relation 61
bimetallic catalysts 487, 488
binding energy 202, 203, 337, 444, 491, 518
Bloch's theorem 456–460
Boltzmann constant 75
bond critical points (BCPs) 76, 127, 191–193, 217
bond dissociation energy (BDE) 34, 111, 114, 117–119, 132, 139, 162, 498–499, 502
bond distorted orbitals (BDOs) 30
bonding interactions 79, 81, 120, 125, 139, 265, 268, 285–288, 294, 295, 297, 448, 467
Born–Oppenheimer approximations 4, 384
boron nitride (BN)-based materials 338
boys localized molecular orbitals 8
Brønsted–Evans–Polanyi (BEP) relation 514, 515, 518, 520–522, 524, 545, 546
Brønsted–Lowry acidity and basicity 90, 91
Brønsted–Lowry theory 253–254, 276
butadiene 6, 18, 27, 28, 374, 375, 407, 409–411, 416, 417, 419–422
butadiyne 295, 410, 411, 419–422

c

Cage Critical Points (CCPs) 192, 193
canonical MOs (CMOs) 4–7, 10–13, 15–17, 126, 163, 198, 210, 381, 459

carbolongs 240
carbon-based materials 338, 343
carbon-nitrogen bond 422–425
carbonyl association mechanisms 351–352
carbonyl dissociation mechanisms 351
catalyst design
  degree of rate control and energetic span model 516–518
  with density functional theory 513–516
  scaling relationships for 518–522
catalytic CO_2 fixation 348–350
catalytic coupling
  of CO_2 with CH_4 345–348
catalytic reaction mechanisms 344
cellular methods 462
chalcogen compounds 297
charge-shift bond (CSB) 126, 436
charge transfer (CT) states 380–381
charge-transfer excitations (CTEs) 381
chemical bonding
  atomic partial charges and atomic electronegativity 130
  dative and electron-sharing bonds 120–124
  length and strength 111–120
  main-group atoms 135–142
  model 108–110
  N_2, CO, BF, LiF 131–135
  physical mechanism of 103–108
  polar bonds 124–129
  transition metal complexes 143–146
chemical concepts in solids
  applications 470–477
  basis sets 460–462
  Bloch's theorem 457–460
  interpretational tools 462–470
  three schisms of solid-state chemistry 455–457
chemical hardness and softness 51–54
chemical shifts 162, 232, 233, 505, 507
chemistry-based descriptors
  electronic and structural property descriptors 487–488

intrinsic atomic property descriptors 486–488
MA-GCNN 488, 490
SISSO method-constructed descriptors 489–491
spectral descriptors 491–493
chemistry, definition of 43
chemistry-informed molecular graph (CIMG) 507
Clar π-sextet rule 237
classical approach 104
climbing image nudged elastic band (CINEB) 337
CM5 charge 176, 177
CO_2 conversion
   homogeneous catalysis for 348–352
CO_2 electroreduction reaction (CO_2ER) 337, 339–342
CO_2 hydrogenation catalyzed 350–352
CO_2 hydrogenation reduction 342–345
coarse-grained RDMs 197
comparative molecular field analysis (CoMFA) 163
computational catalyst design
   doping of metal and nonmetal atoms 338–339
   external electric fields 339–340
   structural modification 339
computational hydrogen electrode (CHE) model 337, 341
conceptual density functional theory (CDFT) 43–64, 75, 257–261
   and acidity 270–273
   based electronic structure principles
      Drago–Wayland equation 261–262
      equalization 259
      HSAB 259–260
      MEP 260–261
      MPP 260–261
   and ITA 272–276
Condon's approximation 318, 387
configuration interaction (CI) 4, 30, 386, 409, 436
configuration state function (CSF) 30
conformational stability 79, 94

conjugated system 33, 230, 269, 270, 407–409, 412, 413, 417, 419, 420, 422, 426
conjugation effect 32, 33
continuous transformation of the origin of the current density (CTOCD) 232
cooperation and frustration 79–82, 84, 94
correlation/dispersion 434, 436, 437, 440, 445
Coulomb-decay RACs (CD-RACs) 531–532
Coulombic interactions 103, 107, 474
Coulson–Chirgwin formula 25
coupled cluster theory (CC) 4, 30, 201, 538
coupled perturbed approaches 60
Coupled-Perturbed Kohn–Sham (CPKS) approximation 60, 61
crystal orbital bond index (COBI) 467–471, 473–475, 477
C-squared Population Analysis (SCPA) 168
cyanogen 411, 419–422
cyclic carbonates 336, 348–349
cycloaddition reaction 241, 349, 350
cyclobutadiene 27, 28

d

dative bonds 117, 120–124, 126, 129, 131–132, 143, 146, 286–287, 301, 304–305
decision trees 483, 502
deficient compound 469
degree of rate control 516–518
delocalization effect 5, 32
delocalized Frenkel exciton states (FEs) 380
delocalized molecular orbitals 5
density-associated quantities (DAQs) 71, 76, 94
density-derived electrostatic and chemical (DDEC) 165, 171, 180
density functional theory (DFT) 46, 456
   aromaticity and antiaromaticity 90–92

density functional theory (DFT) (contd.)
    based EDA   435–436
    bonding and noncovalent interactions   79–81
    Brønsted–Lowry acidity and basicity   90, 91
    catalyst design with   513–516
    conceptual DFT   75
    cooperation and frustration   80–82, 84
    density-associated quantities   76
    electrophilicity and nucleophilicity   86
    homochirality and principle of chirality hierarchy   82–85
    information-theoretic approach   77–79
    method sensitivity in   525–529
    molecular isomeric and conformational stability   79
    molecular properties   91–94
    orbital-free DFT   72–75
    quantifying active site environments with   522–526
    regioselectivity and stereoselectivity   86–90
DFT exchange-correlation (XC) potential   415
DFT–SAPT   435, 436, 439
diabatic states, VB theory   34, 36
diatropic ring current   226, 232, 242
Diels–Alder (DA) reactions   16
dihydrogen   103, 105, 106, 458
dimer method   337
dinitrogen tetroxide (N_2O_4)   425–426
direct C–C coupling   346
disparity   411–412, 423
di-thiophene chain   92
Divide-and-Conquer (DC) method   14
domain natural orbitals (DNO)   198, 199
donor-acceptor interaction diagram   287
donor-acceptor interactions spectrum   286
doping of metal   338–339
d-orbital participation   143
Drago–Wayland equation   261–262
dual descriptor   54–58, 75, 162, 258, 364–366, 371, 372
Duschinsky rotation matrix   388
dynamic classifiers   533, 538
dynamic correlation   30–32, 38, 212, 435, 439

e

ECRE   33
Edmiston–Ruedenberg localized molecular orbitals   8
effective core potential   215, 461
effective oxidation states (EOS) analysis   213–216
    AIM schemes   216–219
efficient global optimization (EGO)   541
Einstein's spontaneous emission coefficient   328
electrochemical reduction reactions   487
electron density, of noninteracting system   48
electron-donating groups (EDG)   267, 269
electron-electron repulsion   46, 47, 60
electron-pair bonding   33–34, 120
electron-pair model   102, 103, 108, 109, 131, 141, 146
electron-phonon coupling   317, 379, 381, 382, 387–390
electron-sharing bonds   120–124, 129, 131, 132, 143, 146
electron-withdrawing groups (EWG)   267, 269
electronegativity   265
    equalization   45, 49, 174, 175, 259
    Sanderson   174
electronegativity equalization method (EEM)   49, 166, 174, 175, 180
electronic and structural property descriptors   487–488
electronic chemical potential   45, 48–51, 64, 266
electronic descriptors   230–232
electronic excited-state Hamiltonian   383–387
electrophilicity
    acid and bases   266–267

and nucleophilicity 79, 86, 94, 485
electrostatic energy 73
electrostatic interaction 33, 59, 79, 82, 112, 163, 164, 172, 202, 288, 293, 296, 299, 369, 424, 433, 438, 446, 465, 469, 470
electrostatic potential (ESP) 163, 288
  fitting method 172–173
  molecular 90
Eley–Rideal (ER) mechanism 344
energetic descriptors 227–230, 242
Energetic Span Model 516–518
energy decomposition analysis (EDA)
  approach 110, 433
  definitions of 437
  DFT based 435–436
  GKS 437–440
  GKS–EDA(BS) 440–441
  GKS–EDA(TD) 441–442
  multi-reference wavefunction based 436–437
  single-determinant MO based 434–435
ensemble state theory 362
equalization principles 45, 49, 259
equivariance 530
Euler–Lagrange equation 72–74
exchange-correlation effect 74
excitation energy transfer (EET) 380–382, 384, 386, 387, 389
excited states in conceptual DFT
  chemical hardness revisited 362–363
  context and justification 361–362
  local chemical potential 371–373
  polarisation in excited states 373–375
  polarisation interaction 367–370
  state-specific dual descriptors 364–366
exclusive orbitals 8
expected improvement (EI) criterion 541
external electric fields (EEF) 161, 338–340, 346, 352
external perturbation
  inductive effect 268–269

periodicity 268
resonance effect 269–270
solvent effects 267–268
steric effects 267
extra-trees regression (ETR) 492, 493

f

face-to-face transition dipole arrangement 380
featurizations 529, 530, 532–534
Fermi's Golden rule 318, 389
Fermionic quantum effect 74
first Hohenberg and Kohn theorem 72
Fisher information 77, 91, 273, 274
fitness function 539, 540
Foster theory 322–325
Fragment Energy Assembler (FEA) 5
free energy barrier 346, 349, 351
fragment particle-hole density (FPHD) 381, 384–387, 394, 397
frontier molecular orbital theory (FMOT) 5–6, 55, 258, 366
frontier molecular orbitalets (FMOLs) 15–18
frontier orbitals 17, 91–94, 260, 266, 444
frozen orbital approximation 363–366
frustrated Lewis pair (FLP) 339
  acid/base pair 339, 344, 352, 353
  catalysts 339, 344, 352
Fukui function 75, 260
  atom-condensed 55–57
  and dual descriptor 54–58
fully connected neural network (FCNN) 499
fuzzy partition 170

g

Gasteiger charge 175, 179
Gasteiger–Marsili charges 175
gauge-including magnetically induced currents (GIMIC) 232
Gaussian process (GP) models 537
gedankenexperiment 458
generalised Kohn–Sham equations 362

generalised Kohn–Sham orbitals energies 362
generalized atomic polar tensor (GAPT) 166, 176, 179
generalized gradient approximation (GGA) 415, 443, 456, 525, 526
geometric descriptors 227, 488, 496, 499, 535
geometry-dependent featurization 534
geometry-free 530
Gibbs free 271, 337–338, 516
GKS–EDA 435, 437, 440
  radical-pairing interactions 444–445
  strong chemical bonds 443–444
GKS–EDA(BS) 440–441
GKS–EDA(TD) 441–442
golden-rule based rate expressions 317–318
gradient-weighted class activation map (Grad-CAM) 533
graph-based featurizations 530
graph convolutions 488, 490, 530–531
gravitational forces 103, 107
Grignard reagents 305

h

H_2 adsorption 343, 344
H-aggregate 380, 394
halogen bonding 289–296
hard and soft acids and bases (HSAB) principle 45, 52, 58, 63, 64, 75, 257, 259–261, 266, 278, 368, 515
hardness, acid and bases 265–266
harmonic oscillator model of aromaticity (HOMA) index 227
Hartree–Fock (HF) method 3–5, 9, 10, 12, 23, 60, 110, 201, 252, 291, 414, 415, 434, 435, 438, 439, 525, 528, 529
Hartree–Fock–Roothaan (HFR)
  equation 3–5, 29, 415
  SCF procedure 415
H_2 chemisorption 344
Heitler–London–Slater–Pauling (HLSP) functions 24, 25, 408

heterolytic H_2-splitting 350
heterolytic splitting 344
hexahydro-Diels–Alder reaction (HDDA) 225
hierarchy of pure states (HOPS) 390
highest occupied molecular orbital (HOMO) 529, 534
Hilbert-space analyses 209
Hirshfeld method 170–171, 176, 177
Hirshfeld–I method 171–172
Hohenberg–Kohn theorems 46, 72, 361
HOMO-LUMO energy gap 91, 94, 259, 266
homochirality 71, 79, 82–85, 94, 95
homodesmotic reactions 223, 227, 229
homogeneous catalytic conversion 337, 348–352
HSAB principle 45, 52, 58, 63, 75, 257, 259–260, 266, 278
Huang–Rhys factor 319, 320, 388, 389
Hubbard U correction 456, 526
Huckel molecular orbital method (HMO) 2
Hückel's rule 234–236, 407
Hückel theory 374, 460, 466
Hund–Mulliken theory 3
hydrogen atom transfer (HAT) 513, 514, 521, 523, 525, 526, 528, 535
hydrogen bonding (HB) interactions 288, 489, 490
hydrogen bonds
  aromatic heterocycles vs. H_2O 445–447
hydrogen evolution reduction (HER) 340, 342
hydrogen sources 335
hydrolysis constant 263
hyperconjugation effect 32, 33
hyperconjugative interaction 410, 418, 419, 424
hypervalent compounds 141, 469

i

independent particle approximation (IPA) 60

inductive effect   61, 268–269, 279, 308, 422, 423
information conservation principle   79, 86
information theoretical approach (ITA)   17, 77–79, 82, 86, 90–94, 272–276
information-theoretic (IT) quantities   273
infrared (IR) spectroscopy   494, 496
interacting quantum atoms (IQA)   110, 200–203
interatomic interactions   101, 102, 104, 108, 112, 114, 115, 125, 131, 147, 170, 465
intramolecular multi-bond strain   412–413, 416–422, 426
intrinsic atomic property descriptors   486–488
intrinsic reaction coordinate (IRC) method   337
ionic approximation (IA)   208, 210–212
ionic bond   23, 31, 33, 79, 124, 125, 161, 168, 193, 433, 465, 466, 468, 470, 472, 473
ionic product   262
ionization constants   263–264
isodesmic reaction   227, 410, 421
isomerization stabilization energies (ISEs)   230

j

Jacobi rotation   414
Jacob's ladder   525, 527
J-aggregates   380

k

Kasha exciton model   380
Kato's cusp theorem   189
Kato's theorem   191
Kekulé structures   408, 415, 416
kernel quantity   61
kernel ridge regression (KRR)   537
kinetic energy, for chemical bond formation   107
Kistiakowsky resonance energy   409

KM-EDA   434
K-Nearest Neighbors (KNN)   484
Kohn–Sham density functional theory (KS-DFT)   4, 31, 210, 415, 434–437
 based EDA methods   436
Kohn–Sham equation   48, 60, 72–74, 110, 362–364, 440, 457
Kohn–Sham orbitals   48, 60, 73, 74, 110, 362–364, 440
 densities   364
Kohn–Sham potential   48
Konkoli–Cremer force   118

l

Landau–Zener theory   320
Langmuir–Hinshelwood (LH) mechanism   344
LCAO-MO framework   3, 209, 211
least absolute shrinkage and selection operator (LASSO)   483, 484
Lewis acids, for dative adducts   303
Lewis theory   2, 254–255
Lewis type acid-base   286
linear combination of atomic orbitals (LCAO)   1, 3, 214, 457, 458, 460, 463
 definition of   3
linear free energy relationships (LFERs)   485, 518–525, 545, 546
linear regression   272, 278, 483, 521, 524, 537
linear response function   51–62
lithium bonding   296
LMO-EDA   434, 437
local chemical potential   371–373, 376
localized molecular orbitals (LMOs)   2, 5–11
 orthogonal   7–11
Local-Orbital Basis Suite Towards Electronic-Structure Reconstruction (LOBSTER)   463
local softness and hardness   58–60
Local Space Approximation (LSA)   5
logistic regression   483

Löwdin method 168
Löwdin weight formula 25
lowest unoccupied molecular orbital
 (LUMO) 5, 6, 16, 18, 49, 55, 56,
 91–94, 265, 300, 363, 364, 374, 375
Lux-Flood, definition 256

m

machine learning (ML)
 bond dissociation energy (BDE)
  498–499, 502
 catalytical properties 499–503, 505
 chemistry-based descriptors 485–493
 databases 482
 datasets 481
 descriptors 483
 imperfect and small chemistry data
  503–505
 in improved catalysis design 529–533
 molecular design 542–545
 in nutshell 481–485
 optimization algorithms 539–542
 prediction of IR/Raman spectroscopy
  494–496
 protein circular dichroism spectra
  498, 501
 reactions and retrosynthesis 505–508
 strong and weak scaling relationships
  534–537
 surface-enhanced Raman spectroscopy
  496–498
 ultraviolet absorption spectroscopy
  497–498, 500
 uncertainty quantification in 537–539
magnetic descriptors 232–233, 242, 243
many-body perturbation theory
 (MBPT) 4
Marcus–Levich–Jortner quantum
 expression 320–322, 325, 328, 329
Marcus theory
 electron transfer 325–327
 golden Rule rate expression 318–325
 TET and other energy transfer process
  with spin exchange 329–330
 using spectra for FCWD 327–328

maximum hardness principle (MHP) 45,
 52, 75, 260–261, 515
Maxwell equation 58
Maxwell relation 54, 258
Mayer's effective atomic orbitals (eff-AOs)
 210, 213, 214, 216, 217
mechanically interlocked molecules
 (MIMs) 444
mesomeric effect 60, 61, 269
message passing 488, 530
metallabenzene 239–243
metallocorrole 349
metalloporphyrin complexes 349
metal-nitrido compounds 514
metal-oxo compounds 514
metal PNP-pincer complexes 350–352
Mingos' rule 237
minima and transition states 44
minimal basis iterative stockholder
 (MBIS) 165, 171, 177, 180
minimum electrophilicity principle
 (MEP) 90, 260–261, 271, 275,
 337, 515
minimum energy path (MEP) 90,
 260–261, 271, 275, 337, 515
minimum polarizability principle (MPP)
 260–261
mixed-metal-oxide materials 544
MMFF94 166, 176, 180
Möbius aromaticity 235, 236
model Hamiltonian 381–383, 391, 392,
 398, 495
MO/DFT methods 409, 411
modified Mulliken population analysis
 (MMPA) 167–168
molecular acidity 75, 90, 273–275, 485
molecular descriptors 485, 497
molecular dipoles 493
molecular electrostatic potential (MEP)
 59, 90, 274, 275, 485
molecular isomeric 79
molecular orbitalet 15–18
molecular orbitals (MOs)
 definition of 3
 theory 2–5

or valence bond approach 44
molecular photophysics 379
Molecular Physics 207
molecular strain 412, 413
Møller–Plesset second-order perturbation theory 538
Monkhorst–Pack method 336
Moreau–Broto autocorrelations 530
Mössbauer spectroscopy 542
Mulliken–Hush method 384
Mulliken method 166–169, 177, 180
Mulliken's population analysis 464, 466
multi-configuration self-consistent field method (MCSCF) 4, 436
multilevel attention graph convolution neural network (MA-GCNN) 488, 489
multiphilic descriptor 258
multiple linear regression (MLR) models 521, 524, 535, 537
multi-reference wavefunction based EDA 436–437
MXene materials 338

n

Naive Bayes 484
naphthalene chain 92
natural atomic orbital (NAO) 90, 209, 217, 219, 274–276
natural bond orbital (NBO) theory 9, 10, 110, 168, 537
natural language processing (NLP) 544
natural population analysis (NPA) 165, 168–169, 176–180, 291, 423
N-doped graphene 339–341, 343, 352
N-doped nanotubes 339
neural networks (NN) 484, 494–497, 502–503, 506, 532, 537
neutralization reaction 255, 263
Niels Bohr's atomic model 2
nitrobenzene 366, 422–426
nitrogen reduction reaction (NRR) process 489
non-acidic 252
nonadiabatic couplings (NACs) 35

noncovalent interaction (NCI) 76, 79–81, 94, 286–287, 304, 306, 308–309, 413, 433, 435–437, 442–445
nonlinear ML models 535, 537
non-Markovian stochastic Schrodinger equation (NMSSE) 381, 390–393, 397
nonmetal atoms 338–339
nonorthogonal atomic orbitals 23, 409
non-orthogonal localized molecular orbitals (NOLMOs) 2, 11–15
nonpolar linear CO_2 338
normalization constant 362, 408, 414
normalized DNOs 198, 199
Nose–Hoover thermostat method 337
nuclear critical points (NCPs) 191, 192
nuclear magnetic resonance (NMR) 161, 162, 207, 232, 233, 493, 505, 507
nuclear-momentum couplings (NMCs) 34, 35

o

OER electrocatalysis 525
one-electron bond 119
optical electron transfer 327
optimization algorithms 37, 539–542, 546
orbital-free DFT (OF-DFT) 72–75
orbital localization methods 211
orbital symmetry 105, 109, 140, 141, 143, 147
oriented external electric field (OEEF) 346
original multicenter bond index 469
orthogonal localized molecular orbitals 7
  boys localized molecular orbitals 8
  Edmiston–Ruedenberg localized molecular orbitals 8
  Pipek–Mezey localized molecular orbitals 8
  regional localized molecular orbitals 10–11
overlap enhanced orbitals (OEOs) 30, 31

oxidation state (OS) 91, 94, 161
  analysis 213–216
  concept of 207–208
  molecular orbital 210–212
  partial charge 208–210

p

paratropic ring currents 232
Parr–Pearson hardness 174
partial atomic charge 49, 130, 161, 207–209
partial charge
  AIM method 169
  calculation 164–165
  classification of 165–166
  computer codes for 179
  defined 161–162
  equalization of electronegativity 174–175
  ESP fitting method 172–173
  Hirshfeld method 170–171
  Hirshfeld-I method 171–172
  Löwdin method 168
  limitations of 163–164
  MMPA method 167–168
  of molecules 176–179
  Mulliken method 166–167
  NPA method 168–169
  on other ideas 175–176
  oxidation state 208–210
  practical applications of 162–163
  RESP and relevant method 173–174
  theoretical significances of 162–163
  Voronoi and VDD method 170
Pauli energy 74, 75, 79
Pauli exclusion principle 4, 74, 79, 80, 438
Pauli repulsion 112–114, 119, 132, 147, 308, 411, 412, 418, 419, 421, 424, 433, 437
Pearson's chemical hardness 45
PEOE (partial equalization of orbital electronegativity) charge 175
periodicity 173, 268, 279, 395, 471
perturbation theory approach 63, 362

pH scale 262–263
phenanthrene 92, 225, 226, 235, 236
phenanthrene chain 92
phosphoric acids 253, 272
photo-physical dynamics 389–392
π–π repulsion 416, 417
Pipek–Mezey localized molecular orbitals 8–10
Pipek–Mezey scheme 8, 211
planar tetracoordinate silicon (ptSi) 340, 352
Platt's ring perimeter model 236
Pnictogens 298–299
polar bonds 124–129, 131, 298, 302, 304
polarisation interaction 367–370
polarizability 51, 91–94, 260–261, 269, 485
polyprotic acids 253
p orbitals 112, 115, 136, 143, 234, 290, 295, 303, 304, 346, 407, 416, 417, 470
post-SCF methods 30
potential energy surfaces (PESs) 34, 44, 319, 415, 450
principle of chirality hierarchy 79, 82–85
principle of Sanderson 166
projector-augmented wave (PAW) strategy 461–465, 467
protein circular dichroism spectra 498, 501
proton affinity 264–265
proton-assisted multielectron transfer 340
pseudoatoms 169
pseudopotential 165, 289, 461, 462, 477

q

QEq 166, 175
QSAR 485
quantitative structure-property relationships (QSPRs) 485, 515, 529
quantum chemistry methods 1, 44
quantum mechanical calculations 190, 336, 434

quantum mechanics 1–3, 23, 163, 189, 190, 207, 224, 387, 433, 456, 470, 531
quantum mechanics/molecular mechanics (QM/MM) 163, 387
quantum Monte Carlo (QMC) 4
quantum theoretical approach 104
quantum theory of atoms in molecules (QTAIM) 76, 209, 218, 228
　atomic observables and properties 191
　basic of 190–194
　benzene-1,2-diol 192
　open quantum systems 194–200
　　RDMs of atoms in molecules 197–200
　　sector density operators 195–197
　open subsystems 191
　valence-shell charge concentration 193

r

radial symmetry functions (RSFs) 494
radical-pairing interactions 444–445
Raman Spectroscopy 494–497
random forest 483, 484, 502, 535
rate-determining step 344, 346, 349–351, 516, 518
Rayleigh–Schrödinger formalism 362
Rayleigh–Schrodinger perturbation theory 51, 361, 367, 373
reactive functional groups 17
reactivity descriptors
　CDFT-based 257–258
reactivity principles 515
redistributed Löwdin population analysis (RLPA) 168
reduced densities (RDs) 189
reduced density matrices (RDMs) 189, 197–200
regional localized molecular orbitals 10–11
regioselectivity and stereoselectivity 79, 86–90, 275
reorganization energy 319–321, 327, 328, 330, 388, 389

Repeating Electrostatic Potential Extracted ATomic (REPEAT) 166, 173, 179, 180
σ–σ repulsion 417
　in model molecule B_2H_4 417–418
　in model molecule B_3H_3 418–419
resonance effect 32, 269–270
resonance structures 23, 61, 109, 120, 211, 233, 234, 269, 270, 408, 409, 411, 414, 422
resonance theory 31–32, 408, 410, 419
response functions 45
　linear 51–62
　tree 45, 57, 58, 62
Restrained ElectroStatic Potential (RESP) 166, 173, 174, 179, 180
restricted Hartree–Fock (RHF) wave function 434
retrosynthesis 505–508
reverse water-gas-shift (RWGS) pathway 344
Ring Critical Points (RCPs) 76, 192, 226
Rogers' claim 409–412
Rumer's rule 23, 26–29

s

Sanderson electronegativity 174
SAPT(MC) method 436
scaling relationships 518–522, 524, 525, 534
Schrodinger equation 4, 7, 44, 47, 94, 237, 367, 381, 389, 390, 462
second Hohenberg–Kohn theorem 46, 73
π-sextet model 236
sigma hole interactions
　chalcogens 297–298
　contested interpretations 308
　donor-acceptor interactions spectrum 286–287
　group 2 atomic centers 304–306
　halogen bonding 293–296
　halogen bond, group 17 atoms 289–292
　hydrogen bonding 288–289

sigma hole interactions (contd.)
  hypovalent group 2   305
  origins   292–293
  terminology   285–286
  Pnictogens   298–299
  Tetrels   299–301
  Triels   302–304
  whole story   296
single-atom-alloy catalysts (SAACs)   491
single-determinant MO based EDA   434–435
Slater determinant   4, 24, 60, 198, 374, 408, 413, 414
Slater's ingenious augmented plane-wave (APW) method   462
small chemistry data   503–505
S_N2 reactions   74, 79, 299–301
Softness and Hardness Kernels   60–62
solvent effects   267–268
solvent system, definition   256
spectral descriptors   491–493, 503
spherical jellium model   238
spin-free quantum chemistry   25
state-specific dual descriptors   57, 364–366, 371, 372
static correlation   30, 216, 436, 437, 439
steric effects   74, 89, 261, 267, 417, 424
sterimol descriptors   524
strong and weak scaling relationships   534–537
strong Brønsted acids   276–278
strong covalent interaction (SCI)   76, 79–81
strong Lewis acids   276–278
structure–property mappings   529
sulfuric acids   253
supervised learning approaches   534
support vector machines (SVM)   484
sure-independence screening and sparsifying operator (SISSO)   493, 503–506, 525
  method-constructed descriptors   489–491
surface-enhanced Raman spectroscopy (SERS)   496–498
symmetry adapted perturbation theory (SAPT)   435
symmetry function descriptor   494
system-bath coupling   317, 318, 325

t

Tamm–Dancoff approximation   386
Taylor series expansion   45, 46, 63, 64
terminal metal-oxo moieties   513
Tetrels   299–301
theoretical descriptors   486, 487
thermodynamic equilibrium   340
thermodynamic factor   516
thermodynamic stability   223, 257, 286, 335, 348
Thomas–Fermi kinetic energy density   79
three schisms of solid-state chemistry   455–457
tight-binding theory   460
Tikhonov regularization   484
time-dependent DFT (TDDFT)   60, 367, 374, 387, 394, 395, 398, 499, 544
topological fuzzy Voronoi cells (TFVC)   209, 217
total local hardness   62
trans-ethylene chain   92
transition metal (TM) complexes   109, 121, 143–146
transition-metal doping   346
Triels   302–304
triplet-triplet annihilation (TTA)   317, 318, 325, 329, 330, 398
turnover-determining intermediate (TDI)   517, 518, 528
turnover-determining transition state (TDTS)   516, 517, 528

u

ultra-strong interactions (USI)   76, 80, 81
ultraviolet absorption spectroscopy   497–500
uncertainty quantification   537–539
Usanovich, definition   255–256

v

valence bond self-consistent field
(VBSCF)   24–26, 28–32, 34–36, 38, 439

valence bond (VB) approach   109, 125

valence-shell electron   339

van der Waals radii   294

VB-based compression approach for diabatization (VBCAD)   36

VB–EDA   436, 437

VB state correlation diagram (VBSCD)   35, 36

VB theory
aromaticity   32–33
conjugation effect   32–33
diabatic states   34–36
dynamic correlation   30–31
electron-pair bonding   33–34
hyperconjugation effect   32–33
resonance theory   31–32
Rumer's rule   26–29
VBSCF   24–26
wave function   29–30

vertical resonance energy (VRE)   419–421, 423

vibrational Hamiltonian   382, 383

vibrational spectroscopy   494

Vienna ab initio simulation package (VASP)   179, 336, 337, 462, 463

Voronoi deformation density (VDD) method   170

w

Wade–Mingos rule   237, 239

water molecule, structure of   288

wave function ψ 1, 46, 103–105, 457

wave-function-based theory   92

wave-function (WF) methods   207

weak interactions   107, 163, 293, 309

Weizsäcker kinetic energy   74

Wiberg bond orders   126

x

XEDA   442

Xiamen Valence Bond (XMVB)   24
input files   36–37
output files   38

y

Young tableaux   28, 29

z

zero bonding information   456